Available in MyMathLab® for your College Algebra Course

Achieve Your Potential

Success in math can make a difference in your life. MyMathLab is a learning experience with resources to help you Achieve Your Potential in this course and beyond. MyMathLab will help you learn the new skills required and also help you learn the bigger concepts and make connections for future courses and careers.

Visualization and Conceptual Understanding

These MyMathLab resources will help you think visually and connect the concepts.

NEW! Guided Visualizations

These engaging, interactive figures bring mathematical concepts to life, helping students visualize the concepts through directed explorations and purposeful manipulation. *Guided Visualizations* are assignable in MyMathLab and encourage active learning, critical thinking, and conceptual learning.

Video Assessment Exercises

Now assignable in MyMathLab, these videos provide a detailed walk-through of the solution process for key topics. Exercises correlated to the videos check for conceptual understanding.

Preparedness and Study Skills

MyMathLab® gives access to many learning resources that refresh knowledge of topics previously learned. An *Integrated Review* MyMathLab course, *Getting Ready* material, and *Note-taking Guide* are some of the tools available.

Integrated Review Course

This new type of MyMathLab course offers a complete college algebra course with select topics from developmental algebra. Each chapter begins with a *Skills Check* assignment to pinpoint which prerequisite topics, if any, need review. A personalized review homework assignment is automatically generated and provides extra support. Additional review materials (worksheets, videos, and more) are available at the start of each chapter.

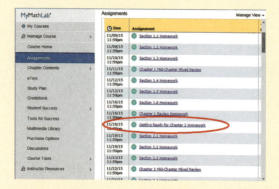

Getting Ready

Students refresh prerequisite topics through skill review quizzes and personalized homework integrated in MyMathLab. With *Getting Ready* content in MyMathLab students get just the help they need to be prepared to learn the new material.

Skills for Success Modules

Skills for Success modules help foster success in collegiate courses and prepare students for future professions. Topics such as "Time Management," "Stress Management," and "Financial Literacy" are available within the MyMathLab course.

For more information on how MyMathLab can help you Achieve Your Potential visit
http://www.pearsonhighered.com/achieve-your-potential/

5th Edition

College Algebra
IN CONTEXT

with applications for the managerial, life, and social sciences

Ronald J. Harshbarger

University of South Carolina—Beaufort

Lisa S. Yocco

Georgia Southern University

PEARSON

Boston Columbus Indianapolis New York San Francisco Amsterdam
Cape Town Dubai London Madrid Milan Munich Paris Montreal Toronto
Delhi Mexico City São Paulo Sydney Hong Kong Seoul Singapore Taipei Tokyo

Editor in Chief: Anne Kelly
Acquisitions Editor: Chelsea Kharakozova
Editorial Assistant: Ashley Gordon
Project Manager: Mary Sanger
Program Management Team Lead: Karen Wernholm
Project Management Team Lead: Christina Lepre
Media Producer: Erin Carreiro
MathXL Content Manager: Kristina Evans
Marketing Manager: Claire Kozar

Marketing Assistant: Fiona Murray
Senior Author Support/Technology Specialist: Joe Vetere
Rights and Permissions Project Manager: Gina Cheselka
Procurement Specialist: Carol Melville
Associate Director of Design: Andrea Nix
Program Design Lead: Heather Scott
Cover Image: Danilo Tee/Shutterstock
Composition, Coordination, and Interior Design: Cenveo Publisher Services

Library of Congress Cataloging-in-Publication Data

Harshbarger, Ronald J., 1938-
  College algebra in context : with applications for managerial, life, and social sciences / Ronald J. Harshbarger, University of South Carolina — Beaufort, Lisa S. Yocco, Georgia Southern University. — 5th edition.
    pages cm
  ISBN 0-13-417902-1
  Includes index.
  1. Algebra—Textbooks. 2. Social sciences—Mathematics—Textbooks.
I. Yocco, Lisa S. II. Title.
  QA152.3.H33 2017
  512.9—dc23
                          2015027344

1 2 3 4 5 6 7 8 9 10–RRD–19 18 17 16 15

www.pearsonhighered.com

ISBN 10: 0-13-417902-1
ISBN 13: 978-0-13-417902-5

Contents

CHAPTER 3   Quadratic, Piecewise-Defined, and Power Functions   159

CHAPTER 4   Additional Topics with Functions   239

CHAPTER 5   Exponential and Logarithmic Functions   308

Preface

College Algebra in Context is designed for a course in algebra that is based on data analysis, modeling, and real-life applications from the management, life, and social sciences. The text is intended to show students how to analyze, solve, and interpret problems in this course, in future courses, and in future careers. At the heart of this text is its emphasis on problem solving in meaningful contexts.

The text is application-driven and uses real data problems that motivate interest in the skills and concepts of algebra. Modeling is introduced early, in the discussion of linear functions and in the discussion of quadratic and power functions. Additional models are introduced when exponential, logarithmic, logistic, cubic, and quartic functions are discussed. Mathematical concepts are introduced informally with an emphasis on applications. Each chapter contains real data problems and extended application projects that can be solved by students working collaboratively.

The text features a constructive chapter-opening Algebra Toolbox, which reviews previously learned algebra concepts by presenting the prerequisite skills needed for successful completion of the chapter. In addition, a section on calculus preparation at the end of the text emphasizes how students can use their new knowledge in a variety of calculus courses.

Changes in the Fifth Edition

We have made a number of changes based on suggestions from users and reviewers of the fourth edition, as well as our own classroom experiences.

- Additional modeling problems that involve decision making and critical thinking have been added throughout the text.

- To keep the applications current, nearly all of the examples and exercises using "date-related" real data have been updated or replaced.

- Over 300 additional algebra skills, examples, and applied exercises have been added.

- More specific Excel steps are presented within the text, with references to Appendix B for more details.

- Excel procedures are given to solve additional types of problems, especially in investment applications in Chapter 5.

- The Excel discussion and figures have been updated to Excel 2013 throughout the text and in Appendix B. Excel 2010 procedures are retained in Appendix B.

- Throughout the text and in Appendix A, the TI graphing calculator graphs have been updated to the latest format (TI-84 Plus Silver with Color).

- Appendix B has been updated to include instructions for Excel 2013 and Excel 2010.

- The MyMathLab course has been expanded to include additional problem types, such as Video Assessment exercises and Guided Visualizations.

- A new *Note-taking Guide*, written by the authors, is available to assist students in taking thorough, organized, and understandable notes during lectures.

To improve the exposition, new content has been added, and the organization of some sections has been changed.

- Multiplying Monomials and Binomials was moved to the Chapter 1 Algebra Toolbox, and Division of Polynomials by Monomials has been added there.

- The discussion of Aligning Data was moved to Section 1.1.

- The discussion of Difference Quotients in Section 1.4 was expanded.

- A new subsection, Rounding Function Coefficients and Constants, which includes a discussion of significant digits, was added to the Chapter 2 Algebra Toolbox.

- A new subsection, Equations in Quadratic Form, was added to Section 4.4.

- A summary table of Transformations of Graphs was added to Section 5.1.

- New graphing calculator and Excel instructions and examples are given for solving investment problems in Section 5.5, Exponential Functions and Investing, and in Section 5.6, Annuities; Loan Repayment.

- A new subsection, Simplifying Complex Fractions, was added to the Chapter 6 Algebra Toolbox.

- Instructions and examples using Excel 2013 to perform matrix operations have been added in Section 7.3.

- Instructions and examples using Excel 2013 to solve systems of linear equations with inverse matrices have been added in Section 7.4.

- A new Section 7.5, Determinants and Cramer's Rule, has been added to Chapter 7.

- A new Section 8.6, The Binomial Theorem, has been added to Chapter 8.

- A new Section 8.7, Conic Sections: Circles and Parabolas, has been added to Chapter 8.

- A new Section 8.8, Conic Sections: Ellipses and Hyperbolas, has been added to Chapter 8.

Continued Features

Features of the text include the following:

- The development of algebra is motivated by the need to use algebra to find the solutions to **real data–based applications**.

  Real-life problems demonstrate the need for specific algebraic concepts and techniques. Each section begins with a motivational problem couched in a real-life setting. The problem is solved after the necessary skills have been presented in that section. The aim is to prepare students to solve problems of all types by first introducing them to various functions and then encouraging them to take advantage of available technology. Special business and finance models are included to demonstrate the application of functions to the business world.

- **Technology** has been integrated into the text.

  The text discusses the use of graphing calculators and Excel to solve problems, but there are no specific technology requirements. When a new calculator or spreadsheet skill becomes useful in a section, students can find the required keystrokes or commands in the text and in the *Graphing Calculator and Excel® Manual*, located within MyMathLab, as well as in Appendixes A and B discussed below. Technology is used to enhance and support learning when appropriate—not to supplant learning.

- The text contains two technology appendixes: a **Basic Graphing Calculator Guide** and a **Basic Guide to Excel 2010 and Excel 2013**. Footnotes throughout the text refer students to these guides for a detailed exposition when a new use of technology is introduced. Additional Excel solution procedures have been added, but, as before, they can be omitted without loss of continuity in the text.

- Each of the first seven chapters begins with an **Algebra Toolbox** section that reviews the prerequisite skills needed for successful completion of the chapter.

Topics discussed in the Toolbox are topics that are prerequisite to a college algebra course (they are often found in a Chapter R or appendix of a college algebra text). Key objectives are listed at the beginning of each Toolbox, and topics are introduced "just in time" to be used in the chapter under consideration.

■ Many problems posed in the text are **multi-part and multi-level problems**.

Many problems require thoughtful, real-world answers adapted to varying conditions, rather than numerical answers. Questions such as "When will this model no longer be valid?" "What additional limitations must be placed on your answer?" and "Interpret your answer in the context of the application" are commonplace in the text.

■ Chapter objectives are listed at the beginning of each chapter, and key objectives are given at the beginning of each section.

■ Each chapter has a **Chapter Summary**, a **Chapter Skills Check**, and a **Chapter Review**.

The Chapter Summary lists the key terms and formulas discussed in the chapter, with section references. Chapter Skills Check and Chapter Review exercises provide additional review problems.

■ The text encourages **collaborative learning**.

Each chapter ends with one or more Group Activities/Extended Applications that require students to solve multi-level problems involving real data or situations, making it desirable for students to collaborate in their solutions. These activities provide opportunities for students to work together to solve real problems that involve the use of technology and that frequently require modeling.

■ In Chapter 8, a **Preparing for Calculus** section is included that shows how algebra skills from the first four chapters are used in a calculus context.

Many students have difficulty in calculus because they have trouble applying algebra skills to calculus. This section reviews earlier topics and shows how they apply to the development and application of calculus.

■ The text encourages students to improve **communication skills and research skills**.

The Group Activities/Extended Applications require written reports and frequently require use of the Internet or a library. Some Extended Applications call for students to use literature or the Internet to find a graph or table of discrete data describing an issue. They are then required to make a scatter plot of the data, determine the function type that is the best fit for the data, create the model, discuss how well the model fits the data, and discuss how it can be used to analyze the issue.

■ **Answers to Selected Exercises** include answers to all Chapter Skills Checks, Chapter Reviews, and odd-numbered application exercises so students have feedback regarding the exercises they work.

■ **Supplements** are provided that will help students and instructors use technology to improve the learning and teaching experience. See the instructor and student resources lists.

Acknowledgments

Many individuals contributed to the development of this text. We would like to thank the following reviewers, whose comments and suggestions were invaluable in preparing it.

Jay Abramson, *Arizona State University*
* Ali Ahmad, *New Mexico State University*
Khadija Ahmed, *Monroe County Community College*
* Janet Arnold, *Indiana University—Southeast Campus*
Jamie Ashby, *Texarkana College*
Sohrab Bakhtyari, *St. Petersburg College*
Jean Bevis, *Georgia State University*
Thomas Bird, *Austin Community College*
Len Brin, *Southern Connecticut State University*
* Lisa Buckelew, *Oklahoma City Community College*
Marc Campbell, *Daytona Beach Community College*
* Rodica Cazacu, *Georgia College*
Florence Chambers, *Southern Maine Community College*
Floyd Downs, *Arizona State University*
Aniekan Ebiefung, *University of Tennessee at Chattanooga*
Marjorie Fernandez Karwowski, *Valencia Community College*
Toni W. Fountain, *Chattanooga State Technical Community College*
* Javier Gomez, *The Citadel*
John Gosselin, *University of Georgia*
David J. Graser, *Yavapai College*
Lee Graubner, *Valencia Community College East*
Linda Green, *Santa Fe Community College*
Richard Brent Griffin, *Georgia Highlands College*
Lee Hanna, *Clemson University*
Deborah Hanus, *Brookhaven College*
Steve Heath, *Southern Utah University*
Todd A. Hendricks, *Georgia Perimeter College*
Suzanne Hill, *New Mexico State University*
Sue Hitchcock, *Palm Beach Community College*

Sandee House, *Georgia Perimeter College*
Mary Hudacheck-Buswell, *Clayton State University*
David Jabon, *DePaul University*
Arlene Kleinstein, *State University of New York—Farmingdale*
Danny Lau, *Kennesaw State University*
Ann H. Lawrence, *Wake Technical Community College*
Kit Lumley, *Columbus State University*
* Phoebe Lutz, *Delta College*
Antonio Magliaro, *Southern Connecticut State University*
Beverly K. Michael, *University of Pittsburgh*
Phillip Miller, *Indiana University—Southeast Campus*
Nancy R. Moseley, *University of South Carolina Aiken*
Demetria Neal, *Gwinnett Technical College*
Malissa Peery, *University of Tennessee*
Ingrid Peterson, *University of Kansas*
Beverly Reed, *Kent State University*
Jeri Rogers, *Seminole Community College—Oviedo*
Michael Rosenthal, *Florida International University*
* Steven Rosin, *Delta College*
Sharon Sanders, *Georgia Perimeter College*
Carolyn Spillman, *Georgia Perimeter College*
Susan Staats, *University of Minnesota*
* Marvin Stick, *University of Massachusetts Lowell*
* Theresa Thomas, *Blue Ridge Community College*
Jacqueline Underwood, *Chandler-Gilbert Community College*
* Erwin Walker, *Clemson University*
Denise Widup, *University of Wisconsin—Parkside*
Sandi Wilbur, *University of Tennessee—Knoxville*
Jeffrey Winslow, *Yavapai College*
* Asmamaw Yimer, *Chicago State University*

Many thanks to Helen Medley and Paul Lorczak for an outstanding job checking the accuracy of this text. Special thanks go to the Pearson team for their assistance, encouragement, and direction throughout this project: Greg Tobin, Anne Kelly, Chelsea Kharakozova, Mary Sanger, Heather Scott, Claire Kozar, Ashley Gordon, and Fiona Murray.

Ronald J. Harshbarger
Lisa S. Yocco

* Denotes reviewers of the fifth edition.

Get the most out of
MyMathLab®

MyMathLab is the world's leading online resource for teaching and learning mathematics. MyMathLab helps students and instructors improve results, and provides engaging experiences and personalized learning for each student so learning can happen in any environment. Plus, it offers flexible and time-saving course management features to allow instructors to easily manage their classes while remaining in complete control, regardless of course format.

Personalized Support for Students

- MyMathLab comes with many learning resources—eText, animations, videos, and more—all designed to support your students as they progress through their course.

- The Adaptive Study Plan acts as a personal tutor, updating in real time based on student performance to provide personalized recommendations on what to work on next. With the new Companion Study Plan assignments, instructors can now assign the Study Plan as a prerequisite to a test or quiz, helping to guide students through concepts they need to master.

- Personalized Homework allows instructors to create homework assignments tailored to each student's specific needs, focused on just the topics they have not yet mastered.

Used by nearly 4 million students each year, the MyMathLab and MyStatLab family of products delivers consistent, measurable gains in student learning outcomes, retention, and subsequent course success.

MyMathLab®
for Harshbarger's College Algebra

MyMathLab delivers proven results in helping individual students succeed. The authors, Ronald Harshbarger and Lisa Yocco, have developed specific content in MyMathLab to give students the practice they need to become adept problem solvers in their future courses and careers. The MyMathLab features described here will help:

- Apply math in real-world scenarios
- Review math skills and forgotten concepts
- Organize the note-taking process

Apply the Math

The exercise sets are fashioned for students majoring in business, social sciences, and life sciences. The focus is on developing students' problem-solving skills to prepare them for their future courses and careers. Exercise sets incorporate modeling techniques, application problems, and real data.

Refresh Skills

Students who need help with prerequisite content are supported when and where they need it. *Getting Ready* assignments are available at the start of each chapter to refresh the algebra skills needed to learn the new material. *Getting Ready* assignments are available in both quiz and homework format.

Organize Notes

New! The *Note-taking Guide* is a workbook that provides a framework for students as they take notes for this course. Students have a guide to help them organize definitions, examples, and processes. The guide is available for download in MyMathLab and also as a print supplement.

Instructor Resources

Additional resources can be downloaded from www.mymathlab.com or www.pearsonhighered.com or hardcopy resources can be ordered from your sales representative.

Annotated Instructor's Edition

The instructor's edition includes all answers to the exercise sets. Shorter answers are presented on the same page as the exercise; longer answers are in the back of the text. Sample homework assignments, selected by the authors, are indicated by a blue underline and may be assigned in **MyMathLab®**.

TestGen®

TestGen (**www.pearsoned.com/testgen**) enables instructors to build, edit, print, and administer tests using a computerized bank of questions developed to cover all the objectives of the text.

Instructor's Solutions Manual

This manual provides complete solutions to all problems in the text (download only).

PowerPoint® Lecture Slides

Written specifically for this text, these fully editable lecture slides provide an outline for presenting definitions, figures, and key examples.

Instructor's Testing Manual

This manual contains three alternative forms of tests per chapter with answer keys that contain more applications (download only).

Student Resources

Additional resources to promote student success:

Additional Skill and Exercise Manual

This manual provides additional practice and test preparation for students (download only).

Student's Solutions Manual

This manual provides detailed, worked-out solutions to odd-numbered exercises.

Note-taking Guide

This workbook, written by Ron Harshbarger and Lisa Yocco, provides a framework for students to help them take thorough, organized, and understandable notes for the course.

To the Student

College Algebra in Context was written to help you develop the math skills you need to model problems and analyze data—tasks that are required in many jobs in the fields of management, life science, and social science. As you read this text, you may be surprised by how many ways professionals use algebra—from predicting how a population will vote in an upcoming election to projecting sales of a particular good.

There are many ways in which this text will help you succeed in your algebra course, and you will be a key partner in that success. Consider the following suggestions, which our own students have found helpful.

1. **Take careful notes in an organized notebook**. Good organization is essential in a math course so that you do not fall behind and so that you can quickly refer to a topic when you need it.
   - **Separate your notebook into three sections:** class notes and examples, homework, and a problem log consisting of problems worked in class (which will provide a sample test).
   - **Begin each set of notes with identifying information:** the date, section of the book, page number from the book, and topic.
   - **Write explanations in words**, rather than just enumerating the steps in solving a problem, so that you will understand later what was done in each step. **Use abbreviations and short phrases**, rather than complete sentences, so that you can keep up with the explanation as you write.
   - **Write step-by-step instructions** for each process.
   - **Keep handouts and tests in your notebook**, using either a spiral notebook with pockets or a loose-leaf notebook.

2. **Read the textbook**. Reading a mathematics textbook is different from reading other textbooks. Some suggestions follow.
   - **Skim the material** to get a general idea of the major topics. As you skim the material, circle any words that you do not understand. Read the Chapter Summary and look at the Exercises at the end of each section.
   - **Make note cards** for terms, symbols, and formulas. Review these note cards often to retain the information.
   - **Read for explanation and study the steps to work a problem**. It is essential that you learn *how* to work a problem and *why* the process works, rather than just memorizing sample problems.
   - **Study any illustrations and other aids** provided to help you understand a sample problem; then cover up the solution and try to work the problem on your own.
   - **Practice the process**. The more problems you do, the more confident you will become in your ability to do math and perform on tests. When doing your homework, don't give in to frustration. Put your homework aside for a while and come back to it later.
   - **Recite and review**. You should know and understand the example problems in your text well enough to be able to work similar problems on a test. Make note cards with example problems on one side and solutions on the other side.

3. **Work through the Algebra Toolbox**. The Algebra Toolbox will give you a great review of the skills needed for success in each chapter. Note the Key Objectives listed at the beginning of the Toolbox; reread them once you have completed the Toolbox Exercises.

4. **Practice with Skills Check exercises**. Skills Check exercises provide a way to practice your algebra skills before moving on to more applied problems.

5. **Work the Exercises carefully**. The examples and exercises in this book model ways in which mathematics is used in the world. Look for connections between the

examples and what you are learning in your other classes. The examples will help you work through the applied exercises in each section.

6. **Prepare for your exams**.

- **Make a study schedule**. Begin to study at least three days before the test. You should make a schedule, listing those sections of the book that you will study each day. Schedule a sample test to be taken upon completion of those sections. This sample test should be taken at least two days before the test date so that you have time to work on areas of difficulty.

- **Rework problems**. You should actively prepare for a test. Do *more* than read your notes and the textbook. Do *more* than look over your homework. Review the note cards prepared from your class notes and text. Actually *rework* problems from each section of your book. Use the Chapter Summary at the end of each chapter to be sure you know and understand the key concepts and formulas. Then get some more practice with the Chapter Skills Checks and Review Exercises. Check your answers!

- **Get help**. Do not leave questions unanswered. Remember to utilize all resources in getting the help you need. Some resources that you might consider using are your fellow classmates, tutors, MyMathLab, the *Student's Solutions Manual*, math videos, and even your teacher. Do not take the gamble that certain questions will not be on the test!

- **Make a sample test**. Write a sample test by choosing a variety of problems from each section in the book. Then write the problems for your sample test in a different order from that in which they appear in the book. (*Hint*: If you write each problem, with directions, on a separate index card and mix them up, you will have a good sample test.)

- **Review and relax the night before the test**. The night before the test is best used *reviewing* the material. This may include working one problem from each section, reworking problems that have given you difficulty, or thinking about procedures you have used.

- **Practice taking tests online**. Ask your professor if MathXL or MyMathLab is available at your school. Both provide online homework, tutorial, and assessment systems for unlimited practice exercises correlated with your textbook. (An access code is required to use these products.)

7. **Develop better math test-taking skills**.

- **Do a memory download**. As soon as you receive your test, jot down formulas or rules that you will need but are likely to forget. If you get nervous later and forget this information, you can simply refer to the memory cues that you have written down.

- **Skip the difficult questions**. Come back to these later, or try to work at least one step for partial credit.

- **Keep a schedule**. The objective is to get as many points as possible. Don't linger over one question very long.

- **Review your work**. Check for careless errors and make sure your answers make sense.

- **Use all the time given**. There are no bonus points for turning your test in early. Use extra time for checking your work.

8. **Have fun!** Look for mathematics all around you. Read the newspaper, look at data on government websites, and observe how professionals use mathematics to do their jobs and communicate information to the world.

We have enjoyed teaching this material to our students and watching their understanding grow. We wish you the very best this semester and in your future studies.

Ronald J. Harshbarger
Lisa S. Yocco

1

Functions, Graphs, and Models; Linear Functions

With digital TV becoming more affordable by the day, the demand for high-definition home entertainment is growing rapidly, with 115 million Americans having digital TV in 2010. The worldwide total was 517 million in 2010, with 40% of TV households having digital TV. More than 401 million digital TV homes are expected to be added between the end of 2010 and the end of 2015, and by 2015 more than 1 billion sets should be in use worldwide. Cell phone use is also on the rise, not just in the United States but throughout the world. By the middle of 2006, the number of subscribers to cell phone carriers had dramatically increased, and the number of total users had reached 4.6 billion by 2010. If the numbers continue to increase at a steady rate, the number of subscribers is expected to reach into the tens of billions over the next few years. These projections and others are made by collecting real-world data and creating mathematical models. The goal of this chapter and future chapters is to use real data and mathematical models to make predictions and solve meaningful problems.

section	objectives	applications
1.1 Functions and Models	Determine graphs, tables, and equations that represent functions; find domains and ranges; evaluate functions and mathematical models; align data	Body temperature, personal computers, stock market, elderly men in the workforce, public health expenditures
1.2 Graphs of Functions	Graph and evaluate functions with technology; graph mathematical models; graph data points; scale data	Aging workers, cost-benefit, voting, U.S. diabetes
1.3 Linear Functions	Identify and graph linear functions; find and interpret intercepts and slopes; find constant rates of change; model revenue, cost, and profit; find marginal revenue, marginal cost, and marginal profit; identify special linear functions	Hispanics in the United States, loan balances, revenue, cost, profit, marginal cost, marginal revenue, marginal profit
1.4 Equations of Lines	Write equations of lines; identify parallel and perpendicular lines; find average rates of change; compute the difference quotient; model approximately linear data	Service call charges, blood alcohol percent, inmate population, Tesla automobile sales, public school enrollment, average velocity

Algebra TOOLBOX

KEY OBJECTIVES

- Write sets of numbers using description or elements
- Identify sets of real numbers as being integers, rational numbers, and/or irrational numbers
- Identify the coefficients of terms and constants in algebraic expressions
- Remove parentheses and simplify polynomials
- Express inequalities as intervals and graph inequalities
- Multiply polynomials
- Divide polynomials by monomials
- Plot points on a coordinate system
- Use subscripts to represent fixed points

The Algebra Toolbox is designed to review prerequisite skills needed for success in each chapter. In this Toolbox, we discuss sets, the real numbers, the coordinate system, algebraic expressions, equations, inequalities, absolute values, and subscripts.

Sets

In this chapter we will use sets to write domains and ranges of functions, and in future chapters we will find solution sets to equations and inequalities. A **set** is a well-defined collection of objects, including but not limited to numbers. In this section, we will discuss sets of real numbers, including natural numbers, integers, and rational numbers, and later in the text we will discuss the set of complex numbers. There are two ways to define a set. One way is by listing the **elements** (or **members**) of the set (usually between braces). For example, we may say that a set A contains 2, 3, 5, and 7 by writing $A = \{2, 3, 5, 7\}$. To say that 5 is an element of the set A, we write $5 \in A$. To indicate that 6 is not an element of the set, we write $6 \notin A$. Domains of functions and solutions to equations are sometimes given in sets with the elements listed.

If all the elements of the set can be listed, the set is said to be a **finite set**. If all elements of a set cannot be listed, the set is called an **infinite set**. To indicate that a set continues with the established pattern, we use three dots. For example, $B = \{1, 2, 3, 4, 5, \ldots, 100\}$ describes the finite set of whole numbers from 1 through 100, and $N = \{1, 2, 3, 4, 5, \ldots\}$ describes the infinite set of all whole numbers beginning with 1. This set is called the **natural numbers**.

Another way to define a set is to give its description. For example, we may write $\{x \mid x \text{ is a math book}\}$ to define the set of math books. This is read as "the set of all x such that x is a math book." $N = \{x \mid x \text{ is a natural number}\}$ defines the set of natural numbers, which was also defined by $N = \{1, 2, 3, 4, 5, \ldots\}$ above.

The set that contains no elements is called the **empty set** and is denoted by $\varnothing$.

EXAMPLE 1 ▶ Sets

Write the following sets in two ways.

a. The set A containing the natural numbers less than 7.

b. The set B of natural numbers that are at least 7.

SOLUTION

a. $A = \{1, 2, 3, 4, 5, 6\}, A = \{x \mid x \in N, x < 7\}$

b. $B = \{7, 8, 9, 10, \ldots\}, B = \{x \mid x \in N, x \geq 7\}$

The relations that can exist between two sets follow.

Relations Between Sets

1. Sets X and Y are **equal** if they contain exactly the same elements.
2. Set A is called a **subset** of set B if each element of A is an element of B. This is denoted $A \subseteq B$.
3. If sets C and D have no elements in common, they are called **disjoint**.

EXAMPLE 2 ▶   Relations Between Sets

For the sets $A = \{x \mid x \leq 9, x \text{ is a natural number}\}$, $B = \{2, 4, 6\}$, $C = \{3, 5, 8, 10\}$:

a. Which of the sets A, B, and C are subsets of A?

b. Which pairs of sets are disjoint?

c. Are any of these three sets equal?

SOLUTION

a. Every element of B is contained in A. Thus, set B is a subset of A. Because every element of A is contained in A, A is a subset of A.

b. Sets B and C have no elements in common, so they are disjoint.

c. None of these sets have exactly the same elements, so none are equal.

The Real Numbers

Because most of the mathematical applications you will encounter in an applied nontechnical setting use real numbers, the emphasis in this text is the **real number system**.* Real numbers can be rational or irrational. **Rational numbers** include integers, fractions containing only integers (with no 0 in a denominator), and decimals that either terminate or repeat. Some examples of rational numbers are

$$-9, \quad \frac{1}{2}, \quad 0, \quad 12, \quad -\frac{4}{7}, \quad 6.58, \quad -7.\overline{3}$$

Irrational numbers are real numbers that are not rational. Some examples of irrational numbers are π (a number familiar to us from the study of circles), $\sqrt{2}$, $\sqrt[3]{5}$, and $\sqrt[3]{-10}$.

The types of real numbers are described in Table 1.1.

Table 1.1

Types of Real Numbers	Descriptions
Natural numbers	1, 2, 3, 4, . . .
Integers	Natural numbers, zero, and the negatives of the natural numbers: . . . , $-3, -2, -1, 0, 1, 2, 3, . . .$
Rational numbers	All numbers that can be written in the form $\frac{p}{q}$, where p and q are both integers with $q \neq 0$. Rational numbers can be written as terminating or repeating decimals.
Irrational numbers	All real numbers that are not rational numbers. Irrational numbers cannot be written as terminating or repeating decimals.

We can represent real numbers on a **real number line**. Exactly one real number is associated with each point on the line, and we say there is a one-to-one correspondence between the real numbers and the points on the line. That is, the real number line is a graph of the real numbers (see Figure 1.1).

Figure 1.1

---

* The complex number system will be discussed in the Chapter 3 Toolbox.

Notice the number π on the real number line in Figure 1.1. This special number, which can be approximated by 3.14, results when the circumference of (distance around) any circle is divided by the diameter of the circle. Another special real number is e; it is denoted by

$$e \approx 2.71828$$

We will discuss this number, which is important in financial and biological applications, later in the text.

Inequalities and Intervals on the Number Line

In this chapter, we will sometimes use inequalities and interval notation to describe domains and ranges of functions. An **inequality** is a statement that one quantity is greater (or less) than another quantity. We say that a is less than b (written $a < b$) if the point representing a is to the left of the point representing b on the real number line. We may indicate that the number a is greater than or equal to b by writing $a \geq b$. The subset of real numbers x that lie between a and b (excluding a and b) can be denoted by the **double inequality** $a < x < b$ or by the **open interval** (a, b). This is called an open interval because neither of the endpoints is included in the interval. The **closed interval** $[a, b]$ represents the set of all real numbers satisfying $a \leq x \leq b$. Intervals containing one endpoint, such as $[a, b)$ or $(a, b]$, are called **half-open intervals**. We can represent the inequality $x \geq a$ by the interval $[a, \infty)$, and we can represent the inequality $x < a$ by the interval $(-\infty, a)$. Note that ∞ and $-\infty$ are not numbers, but ∞ is used in $[a, \infty)$ to represent the fact that x increases without bound and $-\infty$ is used in $(-\infty, a)$ to indicate that x decreases without bound. Table 1.2 shows the graphs of different types of intervals.

Table 1.2

Interval Notation	Inequality Notation	Verbal Description	Number Line Graph
(a, ∞)	$x > a$	x is greater than a	
$[a, \infty)$	$x \geq a$	x is greater than or equal to a	
$(-\infty, b)$	$x < b$	x is less than b	
$(-\infty, b]$	$x \leq b$	x is less than or equal to b	
(a, b)	$a < x < b$	x is between a and b, not including either a or b	
$[a, b)$	$a \leq x < b$	x is between a and b, including a but not including b	
$(a, b]$	$a < x \leq b$	x is between a and b, not including a but including b	
$[a, b]$	$a \leq x \leq b$	x is between a and b, including both a and b	

Note that open circles may be used instead of parentheses, and solid circles may be used instead of brackets, in the number line graphs.

EXAMPLE 3 ▶   **Intervals**

Write the interval corresponding to each of the inequalities in parts (a)–(e), and then graph the inequality.

a. $-1 \le x \le 2$ **b.** $2 < x < 4$ **c.** $-2 < x \le 3$ **d.** $x \ge 3$ **e.** $x < 5$

SOLUTION

a. $[-1, 2]$

b. $(2, 4)$

c. $(-2, 3]$

d. $[3, \infty)$

e. $(-\infty, 5)$

Algebraic Expressions

In algebra we deal with a combination of real numbers and letters. Generally, the letters are symbols used to represent unknown quantities or fixed but unspecified constants. Letters representing unknown quantities are usually called **variables**, and letters representing fixed but unspecified numbers are called **literal constants**. An expression created by performing additions, subtractions, or other arithmetic operations with one or more real numbers and variables is called an **algebraic expression**. Unless otherwise specified, the variables represent real numbers for which the algebraic expression is a real number. Examples of algebraic expressions include

$$5x - 2y, \quad \frac{3x - 5}{12 + 5y}, \quad \text{and} \quad 7z + 2$$

A **term** of an algebraic expression is the product of one or more variables and a real number; the real number is called a **numerical coefficient** or simply a **coefficient**. A constant is also considered a term of an algebraic expression and is called a **constant term**. For instance, the term $5yz$ is the product of the factors 5, y, and z; this term has coefficient 5.

Polynomials

An algebraic expression containing a finite number of additions, subtractions, and multiplications of constants and nonnegative integer powers of variables is called a **polynomial**. When simplified, a polynomial cannot contain negative powers of variables, fractional powers of variables, variables in a denominator, or variables inside a radical. The expressions $5x - 2y$ and $7z^3 + 2y$ are polynomials, but $\frac{3x - 5}{12 + 5y}$ and $3x^2 - 6\sqrt{x}$ are not polynomials. If the only variable in the polynomial is x, then the polynomial is called a **polynomial in x**. The general form of a polynomial in x is

$$a_n x^n + a_{n-1} x^{n-1} + \cdots + a_1 x + a_0$$

where a_0 and each coefficient $a_n, a_{n-1}, \ldots$ are real numbers and each exponent $n, n - 1, \ldots$ is a positive integer.

For a polynomial in the single variable x, the power of x in each term is the **degree** of that term, with the degree of a constant term equal to 0. The term that has the highest power of x is called the **leading term** of the polynomial, the coefficient of this term is the **leading coefficient**, and the degree of this term is the **degree of the polynomial**. Thus, $5x^4 + 3x^2 - 6$ is a fourth-degree polynomial with leading coefficient 5. Polynomials with one term are called **monomials**, those with two terms are called **binomials**, and those with three terms are called **trinomials**. The right side of the equation $y = 4x + 3$ is a first-degree binomial, and the right side of $y = 6x^2 - 5x + 2$ is a second-degree trinomial.

EXAMPLE 4 ▶   **Polynomials**

For each polynomial, state the constant term, the leading coefficient, and the degree of the polynomial.

a. $5x^2 - 8x + 2x^4 - 3$ **b.** $5x^2 - 6x^3 + 3x^6 + 7$

SOLUTION

a. The constant term is -3; the term of highest degree is $2x^4$, so the leading coefficient is 2 and the degree of the polynomial is 4.

b. The constant term is 7; the term of highest degree is $3x^6$, so the leading coefficient is 3 and the degree of the polynomial is 6.

Operations with Polynomials

Terms that contain exactly the same variables with exactly the same exponents are called **like terms**. For example, $3x^2y$ and $7x^2y$ are like terms, but $3x^2y$ and $3xy$ are not. We add or subtract (**combine**) algebraic expressions by combining the like terms. For example, the sum of the expressions $5sx - 2y + 7z^3$ and $2y + 5sx - 4z^3$ is

$$\left(5sx - 2y + 7z^3\right) + \left(2y + 5sx - 4z^3\right) = 5sx - 2y + 7z^3 + 2y + 5sx - 4z^3$$
$$= 10sx + 3z^3$$

The simplified form of

$$3x^2y + 7xy^2 + 6x^2y - 4xy^2 - 5xy \quad \text{is} \quad 9x^2y + 3xy^2 - 5xy$$

Removing Parentheses

We often need to remove parentheses when simplifying algebraic expressions and when solving equations. Removing parentheses frequently requires use of the **distributive property**, which says that for real numbers a, b, and c, $a(b + c) = ab + ac$. Care must be taken to avoid mistakes with signs when using the distributive property. Multiplying a sum in parentheses by a negative number changes the sign of each term in the parentheses. For example,

$$-3(x - 2y) = -3(x) + (-3)(-2y) = -3x + 6y$$

and

$$-(3xy - 5x^3) = -3xy + 5x^3$$

The difference of the expressions $5sx - 2y + 7z^3$ and $2y + 5sx - 4z^3$ is

$$(5sx - 2y + 7z^3) - (2y + 5sx - 4z^3) = 5sx - 2y + 7z^3 - 2y - 5sx + 4z^3$$
$$= -4y + 11z^3$$

Note that removing parentheses preceded by a negative sign results in the sign of every term being changed.

We multiply two monomials by multiplying the coefficients and adding the exponents of the respective variables that are in both monomials. For example,

$$(3x^3y^2)(4x^2y) = 3 \cdot 4 \cdot x^3 \cdot x^2 \cdot y^2 \cdot y = 3 \cdot 4x^{3+2}y^{2+1} = 12x^5y^3$$

We can multiply more than two monomials in the same manner.

We can use the **distributive property**,

$$a(b + c) = ab + ac$$

to multiply a monomial times a polynomial. For example,

$$x(3x + y) = x \cdot 3x + x \cdot y = 3x^2 + xy$$

We can extend the property $a(b + c) = ab + ac$ to multiply a monomial times any polynomial. For example,

$$3x(2x + xy + 6) = 3x \cdot 2x + 3x \cdot xy + 3x \cdot 6 = 6x^2 + 3x^2y + 18x$$

The product of two binomials can be found by using the distributive property as follows:

$$(a + b)(c + d) = a(c + d) + b(c + d) = ac + ad + bc + bd$$

Note that this product can be remembered as the sum of the products of the First, Outer, Inner, and Last terms of the binomials, and we use the word **FOIL** to denote this method.

EXAMPLE 5 ▶   **Multiplying Polynomials**

Find the following products.

a. $(x - 4)(x - 5)$ **b.** $(2x - 3)(3x + 2)$

SOLUTION

a. $(x - 4)(x - 5) = x \cdot x + x(-5) + (-4)x + (-4)(-5)$
$$= x^2 - 5x - 4x + 20 = x^2 - 9x + 20$$

b. $(2x - 3)(3x + 2) = (2x)(3x) + (2x)2 + (-3)(3x) + (-3)2$
$$= 6x^2 + 4x - 9x - 6 = 6x^2 - 5x - 6$$

Certain products and powers involving binomials occur frequently, so the following special products should be remembered.

Special Binomial Products

1. $(x + a)(x - a) = x^2 - a^2$ (difference of two squares)
2. $(x + a)^2 = x^2 + 2ax + a^2$ (perfect square trinomial)
3. $(x - a)^2 = x^2 - 2ax + a^2$ (perfect square trinomial)

EXAMPLE 6 ▶   **Special Binomial Products**

Find the following products by using the special binomial products formulas.

a. $(5x + 1)^2$ **b.** $(2x - 5)(2x + 5)$ **c.** $(3x - 4)^2$

SOLUTION

a. $(5x + 1)^2 = (5x)^2 + 2(5x)(1) + 1^2 = 25x^2 + 10x + 1$

b. $(2x - 5)(2x + 5) = (2x)^2 - 5^2 = 4x^2 - 25$

c. $(3x - 4)^2 = (3x)^2 - 2(3x)(4) + 4^2 = 9x^2 - 24x + 16$

Dividing Polynomials by Monomials

A polynomial can be divided by a monomial if the monomial can be divided into each term of the polynomial. In this case, the monomial is called a **factor** of the polynomial. For example, if x is not equal to 0, we can simplify the algebraic expression

$$\frac{3x^2 - 4x}{x} \quad \text{as} \quad \frac{3x^2}{x} - \frac{4x}{x} = 3x - 4$$

The Coordinate System

Much of our work in algebra involves graphing. To graph in two dimensions, we use a rectangular coordinate system, or **Cartesian coordinate system**. Such a system allows us to assign a unique point in a plane to each ordered pair of real numbers. We construct the coordinate system by drawing a horizontal number line and a vertical number line so that they intersect at their origins (Figure 1.2). The point of intersection is called the **origin** of the system, the number lines are called the coordinate **axes**, and the plane is divided into four parts called **quadrants**. In Figure 1.3, we call the horizontal axis the **x-axis** and the vertical axis the **y-axis**, and we denote any point in the plane as the ordered pair (x, y).

The ordered pair (a, b) represents the point P that is $|a|$ units from the y-axis (right if a is positive, left if a is negative) and $|b|$ units from the x-axis (up if b is positive, down if b is negative). The values of a and b are called the **rectangular coordinates** of the point. Figure 1.3 shows point P with coordinates (a, b). The point is in the second quadrant, where $a < 0$ and $b > 0$.

Figure 1.2

Figure 1.3

Subscripts

We sometimes need to distinguish between two y-values and/or x-values in the same problem or on the same graph, or to designate literal constants. It is often convenient to do this by using **subscripts**. For example, if we have two fixed but unidentified points on a graph, we can represent one point as (x_1, y_1) and the other as (x_2, y_2). Subscripts also can be used to designate different equations entered in graphing utilities; for example, $y = 2x - 5$ may appear as $y_1 = 2x - 5$ when entered in the equation editor of a graphing calculator.

Toolbox EXERCISES

1. Write "the set of all natural numbers N less than 9" in two different ways.

2. Is it true that $3 \in \{1, 3, 4, 6, 8, 9, 10\}$?

3. Is A a subset of B if $A = \{2, 3, 5, 7, 8, 9, 10\}$ and $B = \{3, 5, 8, 9\}$?

4. Is it true that $\dfrac{1}{2} \in N$ if N is the set of natural numbers?

5. Is the set of integers a subset of the set of rational numbers?

6. Are sets of rational numbers and irrational numbers disjoint sets?

Identify the sets of numbers in Exercises 7–9 as containing one or more of the following: integers, rational numbers, and/or irrational numbers.

7. $\{5, 2, 5, 8, -6\}$

8. $\left\{\dfrac{1}{2}, -4.1, \dfrac{5}{3}, 1\dfrac{2}{3}\right\}$

9. $\left\{\sqrt{3}, \pi, \dfrac{\sqrt[3]{2}}{4}, \sqrt{5}\right\}$

In Exercises 10–12, express each interval or graph as an inequality.

10.

$\xleftarrow{\hspace{1cm}} \overset{\hspace{0.3cm}}{\underset{-3}{(}} \xrightarrow{\hspace{1cm}} x$

11. $[-3, 3]$

12. $(-\infty, 3]$

In Exercises 13–15, express each inequality or graph in interval notation.

13. $x \leq 7$

14. $3 < x \leq 7$

15.

$\xleftarrow{\hspace{1.5cm}} \overset{\hspace{0.3cm}}{\underset{4}{)}} \xrightarrow{\hspace{0.5cm}} x$

In Exercises 16–18, graph the inequality or interval on a real number line.

16. $(-2, \infty)$

17. $5 > x \geq 2$

18. $x < 3$

In Exercises 19–21, plot the points on a coordinate system.

19. $(-1, 3)$

20. $(4, -2)$

21. $(-4, 3)$

22. Plot the points $(-1, 2)$, $(3, -1)$, $(4, 2)$, and $(-2, -3)$ on the same coordinate system.

23. Plot the points (x_1, y_1) and (x_2, y_2) on a coordinate system if

$$x_1 = 2, y_1 = -1, x_2 = -3, y_2 = -5$$

Determine whether each expression in Exercises 24–27 is a polynomial. If it is, state the degree of the polynomial.

24. $14x^4 - 6x^3 + 9x - 7$

25. $\dfrac{5x - 8}{3x + 2}$

26. $10x - \sqrt{y}$

27. $-12x^4 + 5x^6$

For each algebraic expression in Exercises 28 and 29, give the coefficient of each term and give the constant term.

28. $-3x^2 - 4x + 8$

29. $5x^4 + 7x^3 - 3$

30. Find the sum of $z^4 - 15z^2 + 20z - 6$ and $2z^4 + 4z^3 - 12z^2 - 5$.

31. Simplify the expression

$$3x + 2y^4 - 2x^3y^4 - 119 - 5x - 3y^2 + 5y^4 + 110$$

Remove the parentheses and simplify in Exercises 32–37.

32. $4(p + d)$

33. $-2(3x - 7y)$

34. $-a(b + 8c)$

35. $4(x - y) - (3x + 2y)$

36. $4(2x - y) + 4xy - 5(y - xy) - (2x - 4y)$

37. $2x(4yz - 4) - (5xyz - 3x)$

In Exercises 38–42, find the products.

38. $(4x^2y^3)(-3a^2x^3)$

39. $2xy^3(2x^2y + 4xz - 3z^2)$

40. $(x - 7)(2x + 3)$

41. $(k - 3)^2$

42. $(4x - 7y)(4x + 7y)$

43. Simplify $\dfrac{12x - 5x^2}{x}$.

44. Simplify $\dfrac{8x^2 + 2x}{2x}$.

KEY OBJECTIVES

- Determine if a table, graph, or equation defines a function
- Find the domains and ranges of functions
- Create a scatter plot of a set of ordered pairs
- Use function notation to evaluate functions
- Apply real-world information using a mathematical model

SECTION PREVIEW **Body Temperatures**

One indication of illness in children is elevated body temperature. There are two common measures of temperature, Fahrenheit (°F) and Celsius (°C), with temperature measured in Fahrenheit degrees in the United States and in Celsius degrees in many other countries of the world. Suppose we know that a child's normal body temperature is 98.6°F and that his or her temperature is now 37°C. Does this indicate that the child is ill? To help decide this, we could find the Fahrenheit temperature that corresponds to 37°C. We could do this easily if we knew the relationship between Fahrenheit and Celsius temperature scales. In this section, we will see that the relationship between these measurements can be defined by a **function** and that functions can be defined numerically, graphically, verbally, or by an equation. We also explore how this and other functions can be applied to help solve problems that occur in real-world situations. ■

Function Definitions

There are several techniques to show how Fahrenheit degree measurements are related to Celsius degree measurements.

One way to show the relationship between Celsius and Fahrenheit degree measurements is by listing some Celsius measurements and the corresponding Fahrenheit measurements. These measurements, and any other real-world information collected in numerical form, are called **data**. These temperature measurements can be shown in a table (Table 1.3).

Figure 1.4

Table 1.3

Celsius Degrees (°C)	−20	−10	−5	0	25	50	100
Fahrenheit Degrees (°F)	−4	14	23	32	77	122	212

This relationship is also defined by the set of ordered pairs

$$\{(-20, -4), (-10, 14), (-5, 23), (0, 32), (25, 77), (50, 122), (100, 212)\}$$

We can picture the relationship between the measurements with a graph. Figure 1.4 shows a **scatter plot** of the data—that is, a graph of the ordered pairs as points. Table 1.3, the set of ordered pairs below the table, and the graph in Figure 1.4 define a **function** with a set of Celsius temperature **inputs** (called the **domain** of the function) and a set of corresponding Fahrenheit **outputs** (called the **range** of the function). A function that will give the Fahrenheit temperature measurement F that corresponds to *any* Celsius temperature measurement C between −20°C and 100°C is the equation

$$F = \frac{9}{5}C + 32$$

This equation defines F as a function of C because each input C results in exactly one output F. Its graph, shown in Figure 1.5, is a line that contains the points on the scatter plot in Figure 1.4 as well as other points. If we consider only Celsius temperatures from −20 to 100, then the domain of this function defined by the equation above is −20 ≤ C ≤ 100 and the resulting range is −4 ≤ F ≤ 212.

Figure 1.5

> ### Function
>
> A function is a rule or correspondence that assigns to each element of one set (called the domain) exactly one element of a second set (called the range).
>
>   The function may be defined by a set of ordered pairs, a table, a graph, an equation, or a verbal description.

EXAMPLE 1 ▶   Body Temperature

Suppose a child's normal temperature is 98.6°F and the only thermometer available, which is Celsius, indicates that the child's temperature is 37°C. Does this reading indicate that the child's temperature is normal?

SOLUTION

We now have a function that will give the output temperature F that corresponds to the input temperature C. Substituting 37 for C in $F = \dfrac{9}{5}C + 32$ gives $F = \dfrac{9}{5}(37) + 32 = 98.6$. This indicates that the child's temperature is normal.

Domains and Ranges

How a function is defined determines its domain and range. For instance, the domain of the function defined by Table 1.3 or by the scatter plot in Figure 1.4 is the finite set $\{-20, -10, -5, 0, 25, 50, 100\}$ with all values measured in degrees Celsius, and the range is the set $\{-4, 14, 23, 32, 77, 122, 212\}$ with all values measured in degrees Fahrenheit. This function has a finite number of inputs in its domain.

   The inputs of the function defined by $F = \dfrac{9}{5}C + 32$ and graphed in Figure 1.5 are all real numbers from -20 to 100, meaning the temperature could be a number like 10.2 or -12.84. This set of numbers is the domain and can be represented by the inequality $-20 \le C \le 100$ or the interval $[-20, 100]$. The outputs of this function are all real numbers from -4 to 212 and can be represented by the inequality $-4 \le F \le 212$ or the interval $[-4, 212]$. Functions defined by equations can also be restricted by the context in which they are used. For example, if the function $F = \dfrac{9}{5}C + 32$ is used in measuring the temperature of water, its domain is limited to real numbers from 0 to 100 and its range is limited to real numbers from 32 to 212, because water changes state with other temperatures.

   If x represents any element in the domain, then x is called the **independent variable**, and if y represents an output of the function from an input x, then y is called the **dependent variable**. The figure at left shows a general "function machine" in which the input is called x, the rule is denoted by f, and the output is symbolized by $f(x)$. The symbol $f(x)$ is read "f of x."

   If the domain of a function is not specified or restricted by the context in which the function is used, it is assumed that the domain consists of all real number inputs that result in real number outputs in the range, and that the range is a subset of the real numbers. Sometimes domains of functions are limited to values that produce defined values, such as functions with variables in the denominator or even-root functions. These functions will be discussed later.

   We can use the graph of a function to find or to verify its domain and range. We can usually see the interval(s) on which the graph exists and therefore agree on the subset of the real numbers for which the function is defined. This set is the domain of the function. We can also usually determine if the outputs of the graph form the set of all real numbers or some subset of real numbers. This set is the range of the function.

Function Machine

EXAMPLE 2 ▶   Domains and Ranges

For each of the following graphs of functions, find the domain and range.

a.

Figure 1.6

Table 1.4

x	y
−5	−7
−3	−2
−1	6
0	−2
3	5
5	−7

b.

Figure 1.7

c.

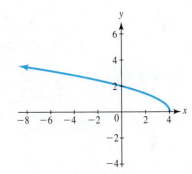

Figure 1.8

SOLUTION

a. The function defined by both Figure 1.6 and Table 1.4 contains the points whose ordered pairs are $(-5, -7)$, $(-3, -2)$, $(-1, 6)$, $(0, -2)$, $(3, 5)$, $(5, -7)$. The domain is the set of all the x-coordinates of the points, $\{-5, -3, -1, 0, 3, 5\}$. The range is the set of all y-coordinates of the points, $\{-7, -2, 6, 5\}$. Note that the y-coordinates -2 and -7 appear twice in the ordered pairs but need to be listed only once in the range.

b. To find the domain of this function, look at all the inputs on the x-axis that correspond to points on the graph. The inputs go from -3 to 3, inclusive. (Because the points at the end of the curve are closed circles, the points are included.) Thus the domain can be written as the inequality $-3 \leq x \leq 3$ or the interval $[-3, 3]$. Similarly, to find the range of the function, look at all the outputs on the y-axis that correspond to points on the graph. The outputs go from -27 to 39, inclusive, so the range can be written as $-27 \leq y \leq 39$ or $[-27, 39]$.

c. The arrow on the left side of the graph indicates that the graph continues indefinitely in that direction, so the inputs will be all real numbers less than or equal to 4. Thus the domain is $x \leq 4$ or the interval $(-\infty, 4]$. The outputs are all real numbers greater than or equal to 0, so the range is $y \geq 0$ or the interval $[0, \infty)$.

Arrow diagrams that show how each individual input results in exactly one output can also represent functions. Each arrow in Figure 1.9(a) and in Figure 1.9(c) goes from an input to exactly one output, so each of these diagrams defines a function. On the other hand, the arrow diagram in Figure 1.9(b) does not define a function because one input, 8, goes to two different outputs, 6 and 9.

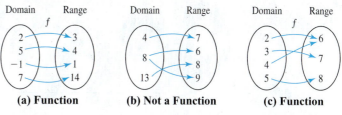

(a) Function   (b) Not a Function   (c) Function

Figure 1.9

Tests for Functions

Functions play an important role in the solution of mathematical problems. Also, to graph a relationship using graphing calculators and computer software, it is usually necessary to express the association between the variables in the form of a function. It is therefore essential for you to recognize when a relationship is a function. Recall that a function is a rule or correspondence that determines exactly one output for each input.

EXAMPLE 3 ▶ **Recognizing Functions**

For each of the following, determine whether or not the indicated relationship represents a function. Explain your reasoning. For each function that is defined, give the domain and range.

a. The PC industry celebrated its 35th anniversary in 2010. The U.S. sales S of personal computers, in millions of dollars, by year x, are shown in Table 1.5. Is S a function of x?

Table 1.5

Year, x	U.S. Sales of Personal Computers, S ($ millions)
1975	0.04
1980	0.76
1985	6.6
1990	9.5
1995	21.4
2000	46.0
2005	62.0
2010	83.8
2015	122.0

(Source: *eTForecasts*)

b. The daily profit P (in dollars) from the sale of x pounds of candy, as shown in Figure 1.10. Is P a function of x?

c. The number of tons x of coal sold determined by the profit P that is made from the sale of the product, as shown in Table 1.6. Is x a function of P?

Figure 1.10

Table 1.6

x (tons)	P ($)
0	−100,000
500	109,000
1000	480,000
1500	505,000
2000	480,000
2500	109,000
3000	−100,000

d. A is the amount in a person's checking account on a given day n. Is A a function of n?

SOLUTION

a. For each year (input) listed in Table 1.5, only one value is given for the amount of sales (output), so Table 1.5 represents S as a function of x. The set $\{1975, 1980, 1985, 1990, 1995, 2000, 2005, 2010, 2015\}$ is the domain, and the range is the set $\{0.04, 0.76, 6.6, 9.5, 21.4, 46.0, 62.0, 83.8, 122\}$ million dollars.

b. Each input x corresponds to only one daily profit P, so this scatter plot represents P as a function of x. The domain is $\{0, 50, 100, 150, 200, 250, 300\}$ pounds, and the range is $\{-200, -100, 1050, 1800, 2050\}$ dollars. For this function, x is the independent variable and P is the dependent variable.

c. The number of tons x of coal sold is not a function of the profit P that is made, because some values of P result in two values of x. For example, a profit of \$480,000 corresponds to both 1000 tons of coal and 2000 tons of coal.

d. The amount in a person's checking account on a given day changes as a consequence of deposits, withdrawals, and/or fees. Thus, there is more than one output (amount in the account) for each input (day), and this relationship is not a function.

EXAMPLE 4 ▶   Functions

a. Does the equation $y^2 = 3x - 3$ define y as a function of x?

b. Does the equation $y = -x^2 + 4x$ define y as a function of x?

c. Does the graph in Figure 1.11 give the price of Stratus Building Supply stock as a function of the day for three months in 2015?

Figure 1.11

SOLUTION

a. This indicated relationship between x and y is not a function because there can be more than one output for each input. For instance, the rule $y^2 = 3x - 3$ determines both $y = 3$ and $y = -3$ for the input $x = 4$. Note that if we solve this equation for y, we get $y = \pm\sqrt{3x - 3}$, so two values of y will result for any value of $x > 1$. In general, if y raised to an even power is contained in an equation, y cannot be solved for uniquely and thus y cannot be a function of another variable.

b. Because each value of x results in exactly one value of y, this equation defines y as a function of x.

c. The graph in Figure 1.11 gives the stock price of Stratus Building Supply for each business day for ten months. The graph shows that the price of a share of stock during each day of these months is not a function. The vertical bar above each day shows that the stock has many prices between its daily high and low. Because of the fluctuation in price during each day of the month, the price of Stratus Building Supply during these months is not a function of the day.

Vertical Line Test

Another way to determine whether an equation defines a function is to inspect its graph. If y is a function of x, no two distinct points on the graph of $y = f(x)$ can have the same first coordinate. There are two points, $(4, 3)$ and $(4, -3)$, on the graph of $y^2 = 3x - 3$ shown in Figure 1.12, so the equation does not represent y as a function of x (as we concluded in Example 4). In general, no two points on its graph can lie on the same vertical line if the relationship is a function.

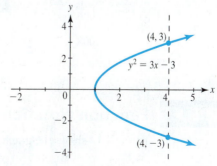

Not a function

Figure 1.12

Function

Figure 1.13

When we look at the graph of $y = -x^2 + 4x$ in Figure 1.13, we can see that any vertical line will intersect the graph in at most one point for the portion of the graph that is visible. If we know that the graph will extend indefinitely in the same pattern, we can conclude that no vertical line will intersect the graph in two points and so the equation $y = -x^2 + 4x$ represents y as a function of x.

Function Notation

We can use the function notation $y = f(x)$, read "y equals f of x," to indicate that the variable y is a function of the variable x. For specific values of x, $f(x)$ represents the resulting outputs, or y-values. In particular, the point $(a, f(a))$ lies on the graph of $y = f(x)$ for any number a in the domain of the function. We can also say that $f(a)$ is $f(x)$ evaluated at $x = a$. Thus, if

$$f(x) = 4x^2 - 2x + 3$$

then

$$f(3) = 4(3)^2 - 2(3) + 3 = 33$$
$$f(-1) = 4(-1)^2 - 2(-1) + 3 = 9$$

This means that $(3, 33)$ and $(-1, 9)$ are points on the graph of $f(x) = 4x^2 - 2x + 3$. We can find function values using an equation, values from a table, or points on a graph. For example, because the personal savings rate S, as a percent of disposable income, is a function of year x, we can write $S = f(x)$ and see that $f(2010) = 5.6$ and $f(2015) = 4.5$.

Table 1.7

Year, x	Personal Savings Rate, S	Year	Personal Savings Rate, S
1980	9.4	2009	6.1
1990	6.0	2010	5.6
2000	3.4	2011	6.0
2002	3.3	2012	7.2
2004	3.2	2013	4.9
2006	3.4	2014	5.3
2008	3.8	2015	4.5

(Source: U.S. Census Bureau)

EXAMPLE 5 ▶ **Function Notation**

Figure 1.14 shows the graph of

$$f(x) = 2x^3 + 5x^2 - 28x - 15$$

a. Use the points shown on the graph to find $f(-2)$ and $f(4)$.

b. Use the equation to find $f(-2)$ and $f(4)$.

Figure 1.14

SOLUTION

a. By observing Figure 1.14, we see that the point $(-2, 45)$ is on the graph of $f(x) = 2x^3 + 5x^2 - 28x - 15$, so $f(-2) = 45$. We also see that the point $(4, 81)$ is on the graph, so $f(4) = 81$.

b. $f(-2) = 2(-2)^3 + 5(-2)^2 - 28(-2) - 15 = 2(-8) + 5(4) - 28(-2) - 15 = 45$
$f(4) = 2(4)^3 + 5(4)^2 - 28(4) - 15 = 2(64) + 5(16) - 28(4) - 15 = 81$

Note that the values found by substitution agree with the y-coordinates of the points on the graph.

EXAMPLE 6 ▶ **Elderly Men in the Workforce**

The points on the graph in Figure 1.15 give the number of elderly men in the workforce (in millions) as a function g of the year t for selected years from 1950 through the present and projected to 2050.

a. Find and interpret $g(2020)$.

b. What is the input t if the output is $g(t) = 19.6$ million elderly men?

c. What can be said about the number of elderly men in the workforce during the period 2020–2040?

d. What is the maximum number of elderly men in the workforce during the period shown on the graph?

Figure 1.15

(Source: U.S. Department of Labor)

SOLUTION

a. The point above $t = 2020$ has coordinates $(2020, 21)$, so $g(2020) = 21$. This means that there were 21 million elderly men in the workforce in 2020.

b. The point $(2030, 19.6)$ occurs on the graph, so $g(2030) = 19.6$ and $t = 2030$ is the input when the output is $g(t) = 19.6$.

c. The number of elderly men in the workforce is projected to decrease during the period 2020–2040.

d. The maximum number of elderly men in the workforce during the period was 45.8 million, in 1950.

Mathematical Models

The process of translating real-world information into a mathematical form so that it can be applied and then interpreted in the real-world setting is called **modeling**. As we most often use the term in this text, a **mathematical model** is a functional relationship (usually in the form of an equation) that includes not only the function rule but also descriptions of all involved variables and their units of measure. For example, the function $F = \frac{9}{5}C + 32$ was used to describe the relationship between temperature scales at the beginning of the section. The **model** that describes how to convert from one temperature scale to another must include the equation $\left(F = \frac{9}{5}C + 32\right)$ and a description of the variables (F is the temperature in degrees Fahrenheit and C is the temperature in degrees Celsius). A mathematical model can sometimes provide an exact description of a real situation (such as the Celsius/Fahrenheit model), but a model frequently provides only an approximate description of a real-world situation.

Consider Table 1.8 and Figure 1.16(a), which show the number of drinks and resulting blood alcohol percent for a 90-pound woman. (One drink is equal to 1.25 oz of 80-proof liquor, 12 oz of regular beer, or 5 oz of table wine; many states have set 0.08% as the legal limit for driving under the influence.) As Figure 1.16(b) shows, the graph of the function $f(x) = 0.05x$ lies on every point of the graph in Figure 1.16(a). Thus,

Table 1.8

Number of Drinks	0	1	2	3	4	5	6	7	8	9
Blood Alcohol Percent	0	0.05	0.10	0.15	0.20	0.25	0.30	0.35	0.40	0.45

(a)

(b)

Figure 1.16

every point in Table 1.8 can be found using this function, and it is a model that gives the blood alcohol percent for a 90-pound woman as a function of the number of drinks.

The model above fits the data points exactly, but many models are approximate fits to data points.

Aligning Data

When finding a model to fit a set of data, it is often easier to use aligned inputs rather than the actual data values. **Aligned inputs** are simply input values that have been converted to smaller numbers by subtracting the same number from each input. For instance, instead of using t as the actual year in the following example, it is more convenient to use an aligned input that is the number of years after 1990.

EXAMPLE 7 ▶   **Public Health Care Expenditures**

Public health care expenditures for the period 1990–2012 can be modeled (that is, accurately approximated) by the function $E(t) = 738.1(1.065)^t$, where $E(t)$ is in billions of dollars and t is the number of years after 1990.

a. What value of t represents 2010?

b. Approximate the public health care expenditure for 2010.

c. Use the model to estimate the public health care expenditure for 2015. Can we be sure this estimate is accurate?

(Source: U.S. Department of Health and Human Services)

SOLUTION

a. Since t is the number of years after 1990, $t = 20$ represents 2010.

b. Substituting 20 for t in $E(t) = 738.1(1.065)^t$ gives a public health care expenditure of $738.1(1.065)^{20}$, or approximately 2600.8 billion dollars, for 2010.

c. Substituting 25 for t in $E(t) = 738.1(1.065)^t$ gives a public health care expenditure of $738.1(1.065)^{25}$, or 3563.3 billion dollars, for 2015. Note that this is a prediction, and many factors, including the Affordable Care Act of 2010, could make this prediction inaccurate.

SKILLS CHECK 1.1

Use the tables below in Exercises 1–6.

Table A

x	−9	−5	−7	6	12	17	20
y = f(x)	5	6	7	4	9	9	10

Table B

x	−4	−1	0	1	3	7	12
y = g(x)	5	7	3	15	8	9	10

1. Table A gives y as a function of x, with $y = f(x)$.

a. Is −5 an input or an output of this function?

b. Is $f(-5)$ an input or an output of this function?

c. State the domain and range of this function.

d. Explain why this relationship describes y as a function of x.

2. Table B gives y as a function of x, with $y = g(x)$.

   a. Is 0 an input or an output of this function?

   b. Is $g(7)$ an input or an output of this function?

   c. State the domain and range of this function.

   d. Explain why this relationship describes y as a function of x.

3. Use Table A to find $y = f(-9)$ and $y = f(17)$.

4. Use Table B to find $y = g(-4)$ and $y = g(3)$.

5. Does Table A describe x as a function of y? Why or why not?

6. Does Table B describe x as a function of y? Why or why not?

7. For each of the functions $y = f(x)$ described below, find $f(2)$.

   a.

x	−1	0	1	2	3
f(x)	5	7	2	−1	−8

   b. $y = 10 - 3x^2$

   c.

8. For each of the functions $y = f(x)$ described below, find $f(-1)$.

   a.

   b.

x	f(x)
−3	12
−1	−8
0	5
3	16

   c. $f(x) = x^2 + 3x + 8$

In Exercises 9 and 10, refer to the graph of the function $y = f(x)$ to complete the table.

9.

x	y
0	
−2	

10.

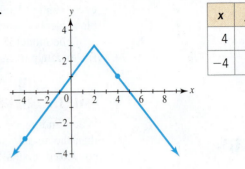

x	y
4	
−4	

11. If $R(x) = 5x + 8$, find (a) $R(-3)$, (b) $R(-1)$, and (c) $R(2)$.

12. If $C(s) = 16 - 2s^2$, find (a) $C(3)$, (b) $C(-2)$, and (c) $C(-1)$.

13. Does the table below define y as a function of x? If so, give the domain and range of f. If not, state why not.

x	−1	0	1	2	3
y	5	7	2	−1	−8

14. Does the table below define y as a function of x? If so, give the domain and range of f. If not, state why not.

x	−3	2	0	3	2
y	12	−3	5	16	4

Determine if each graph in Exercises 15–18 indicates that y is a function of x.

15.

16.

17.

18.

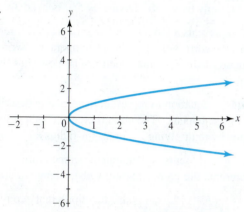

19. Determine whether the graph in the figure represents y as a function of x. Explain your reasoning.

20. Determine whether the graph below represents y as a function of x. Explain your reasoning.

21. Which of the following sets of ordered pairs defines a function?

    a. $\{(1, 6), (4, 12), (4, 8), (3, 3)\}$

    b. $\{(2, 4), (3, -2), (1, -2), (7, 7)\}$

22. Which of the following sets of ordered pairs defines a function?

    a. $\{(1, 3), (-2, 4), (3, 5), (4, 3)\}$

    b. $\{(3, 4), (-2, 5), (4, 6), (3, 6)\}$

23. Which of the following arrow diagrams defines a function?

          (a)   **(b)**

24. Which of the following arrow diagrams defines a function?

          (a)   **(b)**

In Exercises 25–30, find the domain and range for the function shown in the graph.

25.

26.

27.

28.

29.

30.

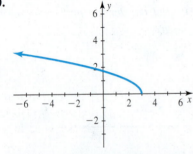

31. If a model has inputs aligned so that $x =$ the number of years after 2000, what inputs would be used to represent 2015 and 2022?

32. If a set of data from the years 1990 to 2015 is given with a model, and $x =$ the number of years after 1990, what inputs would be used for the given years?

33. Does $x^2 + y^2 = 4$ describe y as a function of x?

34. Does $x^2 + 2y = 9$ describe y as a function of x?

35. The American Lung Association distributes a Conversion Chart that gives the following steps to convert Celsius temperatures to Fahrenheit temperatures:

   a. Multiply by 9.   **b.** Divide by 5.   **c.** Add 32.

   If F represents a Fahrenheit temperature and C represents the corresponding Celsius temperature, write an equation that gives the function that describes these steps.

36. Write an equation to represent the function described by this statement: "The circumference C of a circle is found by multiplying 2π times the radius r."

37. Say F represents a Fahrenheit temperature and C represents the corresponding Celsius temperature. If $C = \dfrac{5}{9}(F - 32)$ is the function that converts Fahrenheit temperatures to Celsius temperatures for F between $32°F$ and $212°F$, write the steps similar to those used in Exercise 35 to make this conversion.

38. Write a verbal statement to represent the function $D = 3E^2 - 5$.

EXERCISES 1.1

In Exercises 39–41, determine whether the given relationship defines a function. If so, identify the independent and dependent variable and why the relationship is a function.

39. *Life Insurance*
   a. The monthly premium p for a $100,000 life insurance policy determined by age a for males aged 26–32 (as shown in the table).

   b. Ages a that get a $100,000 life insurance policy for a monthly premium of $11.81 (from the table).

Age (years)	26	27	28	29	30	31	32
Premium (dollars per month)	11.72	11.81	11.81	11.81	11.81	11.81	11.81

40. *Earnings* The mean earnings M in dollars by level of education d for females, as defined by the graph in the figure.

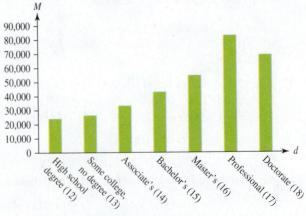

(Source: U.S. Census Bureau)

41. *Obesity* The following table represents the percent p of Americans who are obese (that is, they have body mass index (BMI) $\geq$ 30) for selected years y starting with 2000 and projected to 2030.

Year	2000	2010	2015	2020	2025	2030
% Obese	22.1	30.9	34.5	37.4	39.9	42.2

(Source: American Journal of Preventive Medicine)

42. *Temperature* The figure below shows the graph of $T = 0.43m + 76.8$, which gives the temperature T (in degrees Fahrenheit) inside a concert hall m minutes after a 40-minute power outage during a summer rock concert. State what the variables m and T represent, and explain why T is a function of m.

43. *Barcodes* A grocery scanner at Safeway connects the barcode number on a grocery item with the corresponding price.

a. Is the price a function of the barcode? Explain.

b. Is the barcode a function of the price? Explain.

44. *Piano* A child's piano has 12 keys, each of which corresponds to a note.

a. Is the note sounded a function of the key pressed? If it is, how many elements are in the domain of the function?

b. Is the key pressed a function of the note desired? If it is, how many elements are in the range of the function?

45. *Depreciation* A business property valued at $300,000 is depreciated over 30 years by the straight-line method, so its value x years after the depreciation began is

$$V = 300,000 - 10,000x$$

Explain why the value of the property is a function of the number of years.

46. *Seawater Pressure* In seawater, the pressure p is related to the depth d according to the model

$$p = \frac{18d + 496}{33}$$

where d is the depth in feet and p is in pounds per square inch. Is p a function of d? Why or why not?

47. *Weight* During the first two weeks in May, a man weighs himself daily (in pounds) and records the data in the table below.

a. Does the table define weight as a function of the day in May?

b. What is the domain of this function?

c. What is the range?

d. During what day(s) did he weigh most?

e. During what day(s) did he weigh least?

f. He claimed to be on a diet. What is the longest period of time during which his weight decreased?

May	1	2	3	4	5	6	7
Weight (lb)	178	177	178	177	176	176	175

May	8	9	10	11	12	13	14
Weight (lb)	176	175	174	173	173	172	171

48. *Test Scores*

a. Is the average score on the final exam in a course a function of the average score on a placement test for the course, if the table below defines the relationship?

b. Is the average score on a placement test a function of the average score on the final exam in a course, if the table below defines the relationship?

Average Score on Math Placement Test (%)	81	75	60	90	75
Average Score on Final Exam in Algebra Course (%)	86	70	58	95	81

49. *Car Financing* A couple wants to buy a $35,000 car and can borrow the money for the purchase at 8%, paying it off in 3, 4, or 5 years. The table below gives the monthly payment and total cost of the purchase (including the loan) for each of the payment plans.

t (years)	Monthly Payment ($)	Total Cost ($)
3	1096.78	39,484.08
4	854.46	41,014.08
5	709.68	42,580.80

(Source: Sky Financial Mortgage Tables)

Suppose that when the payment is over t years, $P(t)$ represents the monthly payment and $C(t)$ represents the total cost for the car and loan.

a. Find $P(3)$ and write a sentence that explains its meaning.

b. What is the total cost of the purchase if it is financed over 5 years? Write the answer using function notation.

c. What is t if $C(t) = 41,014.08$?

d. How much money will the couple save if they finance the car for 3 years rather than 5 years?

50. Mortgage A couple can afford $800 per month to purchase a home. As indicated in the table, if they can get an interest rate of 7.5%, the number of years t that it will take to pay off the mortgage is a function of the dollar amount A of the mortgage for the home they purchase.

Amount A ($)	t (years)
40,000	5
69,000	10
89,000	15
103,000	20
120,000	30

(Source: Comprehensive Mortgage Tables [Publication No. 492], Financial Publishing Co.)

a. If the couple wishes to finance $103,000, for how long must they make payments? Write this correspondence in function form if $t = f(A)$.

b. What is $f(120,000)$? Write a sentence that explains its meaning.

c. What is $f(3 \cdot 40,000)$?

d. What value of A makes $f(A) = 5$ true?

e. Does $f(3 \cdot 40,000) = 3 \cdot f(40,000)$? Explain your reasoning.

51. Women in the Workforce From 1930 and projected to 2015, the number of women in the workforce increased steadily. The points on the figure that follows give the number (in millions) of women in the workforce as a function f of the year for selected years.

a. Approximately how many women were in the workforce in 1960?

b. Estimate $f(1930)$ and write a sentence that explains its meaning.

c. What is the domain of this function if we consider only the indicated points?

d. How does the graph reflect the statement that the number of women in the workforce increases?

(Source: *World Almanac*)

52. Working Age The projected ratio of the working-age population (25- to 64-year-olds) to the elderly shown in the figure below defines the ratio as a function of the year shown. If this function is defined as $y = f(t)$ where t is the year, use the graph to answer the following:

a. What is the projected ratio of the working-age population to the elderly population in 2020?

b. Estimate $f(2005)$ and write a sentence that explains its meaning.

c. What is the domain of this function?

d. Is the projected ratio of the working-age population to the elderly increasing or decreasing over the domain shown in the figure?

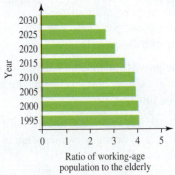

(Source: U.S. Department of Labor)

53. Age at First Marriage The table on the following page gives the U.S. median age at first marriage for selected years from 1890 to 2015.

a. If the function f is the median age at first marriage for men, find $f(1890)$ and $f(2015)$.

b. If the function g is the median age at first marriage for women, find $g(1940)$ and $g(2013)$.

c. For what value of x is $f(x) = 24.7$? Write the result in a sentence.

d. Did $f(x)$ increase or decrease from 1960 to 2015?

Median Age at First Marriage

Year	Men	Women
1890	26.1	22.0
1900	25.9	21.9
1910	25.1	21.6
1920	24.6	21.2
1930	24.3	21.3
1940	24.3	21.5
1950	22.8	20.3
1960	22.8	20.3
1970	23.2	20.8
1980	24.7	22.0
1990	26.1	23.9
2000	26.8	25.1
2004	27.1	25.8
2009	28.1	25.9
2010	28.2	26.1
2011	28.7	26.5
2012	28.6	26.6
2013	29.0	26.6
2015	29.2	27.7

(Source: U.S. Census Bureau)

54. *Population of Females under 18* The graph below gives the population of females under the age of 18, in thousands, projected from 2020 to 2060, as a function f of the year t.

a. What is $f(2030)$?

b. Interpret that result.

c. Is this function increasing or decreasing?

(*Source*: U.S. Census Bureau)

55. *Alzheimer's Disease* The figure gives the millions of U.S. citizens age 65 and older with Alzheimer's disease from 2000 and projected to 2050. If the number of millions of these U.S. citizens is the function $f(x)$, where x is the year, use the graph to answer the following.

a. How many U.S. citizens age 65 and older are expected to have Alzheimer's disease in 2020?

b. Find $f(2030)$ and explain its meaning.

c. In what year do we expect 11 million U.S. citizens age 65 and older to have Alzheimer's disease?

d. Is this function increasing or decreasing?

(*Source*: Alzheimer's Association)

56. *Farms* The points on the following figure show the number N of U.S. farms (in millions) for selected years t.

a. Is the number of U.S. farms a function of the year?

b. What is $f(1940)$ if $N = f(t)$?

c. What is t if $f(t) = 2.4$?

d. Write a sentence explaining the meaning of $f(2012) = 2.1$.

(*Source*: National Agricultural Statistics Service, U.S. Department of Agriculture)

57. Teen Birth Rate The birth rate in 2012 for U.S. girls ages 15 to 19 was the lowest since the government began tracking the statistic in 1940. The recent sharp drop is attributed to the recession. Answer the following questions using the figure below.

a. Find the output when the input is 1995, and explain its meaning.

b. For what year was the rate 40.5?

c. For what year on the graph was the birth rate at its maximum?

d. Despite the significant and steady decline in teen birth rates in recent decades, the decrease appears to have slowed recently. How would you describe the rates from 2005 to 2012?

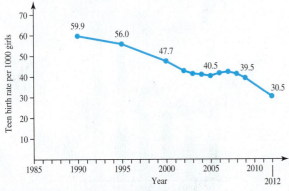

(Source: Centers for Disease Control and Prevention)

58. Social Security Funding Social Security benefits are funded by individuals who are currently employed. The following graph, based on known data and projections into the future, defines a function that gives the number of workers n supporting each retiree as a function of time t (given by the calendar year). Denote this function by $n = f(t)$.

a. Find $f(1990)$ and explain its meaning.

b. In what year will the number of workers supporting each retiree equal 2?

c. What does this function tell us about Social Security in the future?

(Source: Social Security Administration)

59. Revenue The revenue from the sale of specialty golf hats is given by the function $R(x) = 32x$ dollars, where x is the number of hats sold.

a. What is $R(200)$? Interpret this result.

b. What is the revenue from the sale of 2500 hats? Write this in function notation.

60. Cost The cost of producing specialty golf hats is given by the function $C(x) = 4000 + 12x$, where x is the number of hats produced.

a. What is $C(200)$? Interpret this result.

b. What is the cost of producing 2500 hats? Write this in function notation.

61. Utilities An electric utility company determines the monthly bill by charging 85.7 cents per kilowatt-hour (kWh) used plus a base charge of $19.35 per month. Thus, the monthly charge is given by the function

$$f(W) = 0.857W + 19.35 \text{ dollars}$$

where W is the number of kilowatt-hours.

a. Find $f(1000)$ and explain what it means.

b. What is the monthly charge if 1500 kWh are used?

62. Profit The profit from the production and sale of iPod players is given by the function $P(x) = 450x - 0.1x^2 - 2000$, where x is the number of units produced and sold.

a. What is $P(500)$? Interpret this result.

b. What is the profit from the production and sale of 4000 units? Write this in function notation.

63. Profit The daily profit from producing and selling Blue Chief bicycles is given by

$$P(x) = 32x - 0.1x^2 - 1000$$

where x is the number produced and sold and $P(x)$ is in dollars.

a. Find $P(100)$ and explain what it means.

b. Find the daily profit from producing and selling 160 bicycles.

64. Projectiles Suppose a ball thrown into the air has its height (in feet) given by the function

$$h(t) = 6 + 96t - 16t^2$$

where t is the number of seconds after the ball is thrown.

a. Find $h(1)$ and explain what it means.

b. Find the height of the ball 3 seconds after it is thrown.

c. Test other values of $h(t)$ to decide if the ball eventually falls. When does the ball stop climbing?

65. *Test Reliability* If a test that has reliability 0.7 has the number of questions increased by a factor n, the reliability R of the new test is given by

$$R(n) = \frac{0.7n}{0.3 + 0.7n}$$

a. What is the domain of the function defined by this equation?

b. If the application used requires that the size of the test be increased, what values of n make sense in the application?

66. *Body-Heat Loss* The description of body-heat loss due to convection involves a coefficient of convection K_c, which depends on wind speed s according to the equation $K_c = 4\sqrt{4s + 1}$.

a. Is K_c a function of s?

b. What is the domain of the function defined by this equation?

c. What restrictions does the physical nature of the model put on the domain?

67. *Cost–Benefit* Suppose that the cost C (in dollars) of removing $p\%$ of the particulate pollution from the smokestack of a power plant is given by

$$C(p) = \frac{237{,}000p}{100 - p}$$

a. Use the fact that percent is measured between 0 and 100 and find the domain of this function.

b. Evaluate $C(60)$ and $C(90)$.

68. *Demand* Suppose the number of units q of a product that is demanded by consumers is given as a function of p by

$$q = \frac{100}{\sqrt{2p + 1}}$$

where p is the price charged per unit.

a. What is the domain of the function defined by this equation?

b. What should the domain and range of this function be to make sense in the application?

69. *Shipping Restrictions* Some shipping restrictions say that in order for a box to be shipped, its length (longest side) plus its girth (distance around the box in the other two dimensions) must be equal to no more than 108 inches.

If the box with length plus girth equal to 108 inches has a square cross section that is x inches on each side, then the volume of the box is given by $V = x^2(108 - 4x)$ cubic inches.

a. Find $V(12)$ and $V(18)$.

b. What restrictions must be placed on x to satisfy the conditions of this model?

c. Create a table of function values to investigate the value of x that maximizes the volume. What are the dimensions of the box that has the maximum volume?

70. *Height of a Bullet* The height of a bullet shot into the air is given by $S(t) = -4.9t^2 + 98t + 2$, where t is the number of seconds after it is shot and $S(t)$ is in meters.

a. Find $S(0)$ and interpret it.

b. Find $S(9)$, $S(10)$, and $S(11)$.

c. What appears to be happening to the bullet at 10 seconds? Evaluate the function at some additional times near 10 seconds to confirm your conclusion.

1.2 Graphs of Functions

KEY OBJECTIVES

- Graph equations using the point-plotting method
- Graph equations using graphing calculators
- Graph equations using Excel spreadsheets
- Align inputs and scale outputs to model data
- Graph data points

SECTION PREVIEW **Aging Workers**

The millions of Americans who are working full time at selected ages from 20 to 65 can be modeled as a function of their age, x. A model is

$$y = -0.000362x^3 + 0.0401x^2 - 1.39x + 21.7$$

(Source: *Wall Street Journal*, June 18, 2012)

   To graph this function, we use x-values from 20 to 65 to find the corresponding y-values, which represent millions of full-time workers. These points can be used to sketch the graph of the function. If technology is used to graph the function, the **viewing window** on which the graph is shown can be determined by these values of x and y. (See Example 6.)

   In this section, we will graph functions by point plotting and with technology. We will graph application functions on windows determined by the context of the applications, and we will align data so that smaller inputs can be used in models. We will also graph data points. ■

Graphs of Functions

If an equation defines y as a function of x, we can sketch the graph of the function by plotting enough points to determine the shape of the graph and then drawing a line or curve through the points. This is called the **point-plotting method** of sketching a graph.

EXAMPLE 1 ▶ **Graphing an Equation by Plotting Points**

   a. Graph the equation $y = x^2$ by drawing a smooth curve through points determined by integer values of x between 0 and 3.

   b. Graph the equation $y = x^2$ by drawing a smooth curve through points determined by integer values of x between -3 and 3.

   SOLUTION

   a. We use the values in Table 1.9 to determine the points. The graph of the function drawn through these points is shown in Figure 1.17(a).

Table 1.9

x	0	1	2	3
$y = x^2$	0	1	4	9
Points	(0, 0)	(1, 1)	(2, 4)	(3, 9)

   b. We use Table 1.10 to find the additional points. The graph of the function drawn through these points is shown in Figure 1.17(b).

Table 1.10

x	−3	−2	−1	0	1	2	3
$y = x^2$	9	4	1	0	1	4	9
Points	(−3, 9)	(−2, 4)	(−1, 1)	(0, 0)	(1, 1)	(2, 4)	(3, 9)

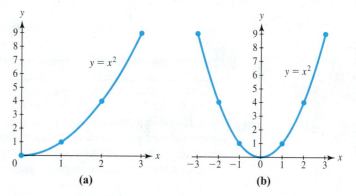

(a)   (b)

Figure 1.17

Although neither graph in Figure 1.17 shows all of the points satisfying the equation $y = x^2$, the graph in Figure 1.17(b) is a much better representation of the function (as you will learn later). When enough points are connected to determine the shape of the graph and the important parts of the graph and to suggest what the unseen parts of the graph look like, the graph is called **complete**.

Complete Graph

A graph is a complete graph if it shows the basic shape of the graph and important points on the graph (including points where the graph crosses the axes and points where the graph turns)* and suggests what the unseen portions of the graph will be.

EXAMPLE 2 ▶

Graphing a Complete Graph

Sketch the complete graph of the equation $f(x) = x^3 - 3x$, using the fact that the graph has at most two turning points.

SOLUTION

We use the values in Table 1.11 to determine some points. The graph of the function drawn through these points is shown in Figure 1.18.

Table 1.11

x	−3	−2	−1	0	1	2	3
$y = x^3 - 3x$	−18	−2	2	0	−2	2	18
Points	(−3, −18)	(−2, −2)	(−1, 2)	(0, 0)	(1, −2)	(2, 2)	(3, 18)

---

* Points where graphs turn from rising to falling or from falling to rising are called turning points.

Figure 1.18

Note that this graph has two turning points, so it is a complete graph.

The ability to draw complete graphs, with or without the aid of technology, improves with experience and requires knowledge of the shape of the basic function(s) involved in the graph. As you learn more about types of functions, you will be better able to determine when a graph is complete.

Graphing with Technology

Obtaining a sufficient number of points to sketch a graph by hand can be time consuming. Computers and graphing calculators have graphing utilities that can be used to plot many points quickly, thereby producing the graph with minimal effort. However, graphing with technology involves much more than pushing a few buttons. Unless you know what to enter into the calculator or computer (that is, the domain of the function and sometimes the range), you cannot see or use the graph that is drawn. We suggest the following steps for graphing most functions with a graphing calculator.

> ### Using a Graphing Calculator to Draw a Graph
>
> 1. Write the function with x representing the independent variable and y representing the dependent variable. Solve for y, if necessary.
>
> 2. Enter the function in the equation editor of the graphing utility. Use parentheses* as needed to ensure the mathematical correctness of the expression.
>
> 3. Activate the graph by pressing the ⟨ZOOM⟩ or ⟨GRAPH⟩ key. Most graphing utilities have several preset **viewing windows** under ⟨ZOOM⟩, including the **standard viewing window** that gives the graph of a function on a coordinate system in which the x-values range from -10 to 10 and the y-values range from -10 to 10.[†]
>
> 4. To see parts of the graph of a function other than those shown in a standard window, press ⟨WINDOW⟩ to set the x- and y-boundaries of the viewing window before pressing ⟨GRAPH⟩. Viewing window boundaries are discussed below and in the technology supplement. As you gain more knowledge of graphs of functions, determining viewing windows that give complete graphs will become less complicated.

---

* Parentheses should be placed around numerators and/or denominators of fractions, fractions that are multiplied by a variable, exponents consisting of more than one symbol, and in other places where the order of operations needs to be indicated.

† For more information on viewing windows and graphing, see Appendix A, page 666.

Although the standard viewing window is a convenient window to use, it may not show the desired graph. To see parts of the graph of a function other than those that might be shown in a standard window, we change the x- and y-boundaries of the viewing window. The values that define the viewing window can be set manually under (WINDOW) or by using the (ZOOM) key. The boundaries of a viewing window are

x_{min}: the smallest value on the x-axis (the left boundary of the window)

y_{min}: the smallest value on the y-axis (the bottom boundary of the window)

x_{max}: the largest value on the x-axis (the right boundary of the window)

y_{max}: the largest value on the y-axis (the top boundary of the window)

x_{scl}: the distance between ticks on the x-axis (helps visually find x-intercepts and other points)

y_{scl}: the distance between ticks on the y-axis (helps visually find the y-intercept and other points)

When showing viewing window boundaries on calculator graphs in this text, we write them in the form

$$[x_{min}, x_{max}] \text{ by } [y_{min}, y_{max}]$$

EXAMPLE 3 ▶ ## Graphing a Complete Graph

Sketch the graph of $y = x^3 - 3x^2 - 13$

a. using the standard window.

b. using the window $x_{min} = -10, x_{max} = 10, y_{min} = -25, y_{max} = 10$.

Which graph gives a better view of the graph of the function?

SOLUTION

a. A graph of this function in the standard window appears to be a line (Figure 1.19(a)).

b. By setting the window with $x_{min} = -10, x_{max} = 10, y_{min} = -25, y_{max} = 10$, we obtain the graph in Figure 1.19(b). This window gives a better view of the graph of this equation. As you learn more about functions, you will see that the graph in Figure 1.19(b) is a complete graph of this function.

viewing window: $[-10, 10]$ by $[-10, 10]$
This graph looks like a line.
(a)

viewing window: $[-10, 10]$ by $[-25, 10]$
This is a better graph.
(b)

Figure 1.19

Different viewing windows give different views of a graph. There are usually many different viewing windows that give complete graphs for a particular function, but some viewing windows do not show all the important parts of a graph.

EXAMPLE 4 ▶   **Cost-Benefit**

Suppose that the cost C of removing $p\%$ of the pollution from drinking water is given by the model

$$C = \frac{5350p}{100 - p} \text{ dollars}$$

a. Use the restriction on p to determine the limitations on the horizontal-axis values (which are the x-values on a calculator).

b. Graph the function on the viewing window $[0, 100]$ by $[0, 50{,}000]$. Why is it reasonable to graph this model on a viewing window with the limitation $C \geq 0$?

c. Find the point on the graph that corresponds to $p = 90$. Interpret the coordinates of this point.

SOLUTION

a. Because p represents the percent of pollution removed, it is limited to values from 0 to 100. However, $p = 100$ makes C undefined in this model, so p is restricted to $0 \leq p < 100$ for this model.

b. The graph of the function is shown in Figure 1.20(a), with x representing p and y representing C. The interval $[0, 100]$ contains all the possible values of p. The value of C is bounded below by 0 because C represents the cost of removing the pollution, which cannot be negative.

c. We can find (or estimate) the output of a function $y = f(x)$ at specific inputs with a graphing utility. We do this with ⟨TRACE⟩ or TABLE (accessed by **2ND** ⟨GRAPH⟩).* Using ⟨TRACE⟩ with the x-value 90 gives the point $(90, 48{,}150)$ (see Figure 1.20(b)). Because the values of p are represented by x-values and the values of C are represented by y, the coordinates of the point tell us that the cost of removing 90% of the pollution from the drinking water is \$48,150. Figure 1.20(c) shows the value of y for $x = 90$ and other values in a calculator table.

(a)

(b)

(c)

Figure 1.20

Spreadsheet ▶ SOLUTION   We have discussed how to create graphs with a graphing calculator, and we can also use Excel spreadsheets to generate graphs. The table in Figure 1.21 shows some input (in column A) and outputs (in column B) for the function $C = \dfrac{5350\,p}{100 - p}$ discussed in Example 4. The inputs can be entered either manually or by using the fill-down capacity of Excel. To enter the outputs, we evaluate the function using cell locations to represent variables. To evaluate $C = \dfrac{5350\,p}{100 - p}$ at $p = 0$, we type 0 in cell A2 and $= 5350*A2/(100 - A2)$ in cell B2.

---

* For more details, see Appendix A, page 667.

(This results in 0 in cell B2.) To get the remaining outputs, move the mouse to the lower right corner of cell B2 and drag the mouse to the last cell where inputs are listed.

To create the graph, we highlight the two columns containing the inputs and outputs, select the insert tab, select the scatter plot icon in the Charts group, and then select the smooth curve icon. The graph will appear. Figure 1.21 shows the table of values and the graph of the functions.*

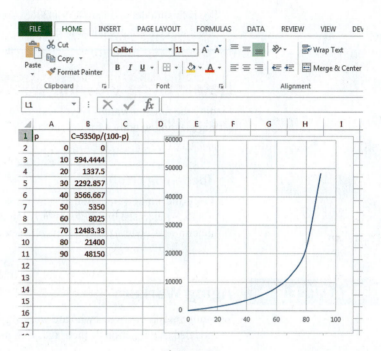

Figure 1.21

Voting

Between 1950 and 2008, the percent of the voting population who voted in presidential elections (during election years) is given by

$$f(x) = 0.0155x^2 - 1.213x + 75.26$$

where x is the number of years after 1950. That is, the input variable $x = 0$ represents 1950. (Source: Federal Election Commission)

a. What are the values of x that correspond to the years 1960 and 2008?

b. Find $f(10)$ and explain its meaning.

c. If this model is accurate for 2008, find the percent of the voting population who voted in the 2008 presidential election.

SOLUTION

a. Because 1960 is 10 years after 1950, $x = 10$ corresponds to the year 1960 in the aligned data. By similar reasoning, 2008 corresponds to $x = 58$.

---

* For detailed steps needed to graph functions, see Appendix B, pages 685-688.

b. An input value of 10 represents 10 years after 1950, which is the year 1960. To find $f(10)$, we substitute 10 for x to obtain $f(10) = 0.0155(10)^2 - 1.213(10) + 75.26 = 64.7$. Here is one possible explanation of this answer: Approximately 65% of the voting population voted in the 1960 presidential election.

c. To find the percent in 2008, we first calculate the aligned input for the function. Because 2008 is 58 years after 1950, we use $x = 58$. Thus, we find

$$f(58) = 0.0155(58)^2 - 1.213(58) + 75.26 = 57.0$$

So, if this model is accurate for 2008, approximately 57% of the voting population voted in the 2008 presidential election.

Determining Viewing Windows

Finding the function values (y-values) for selected inputs (x-values) can be useful when setting viewing windows for graphing utilities.

> **Technology Note**
>
> Once the input values for a viewing window have been selected, ⌈TRACE⌉ or TABLE (accessed by **2ND** ⌈GRAPH⌉) can be used to find enough output values to determine a y-view that gives a complete graph.*

EXAMPLE 6 ▶ Aging Workers

The millions of Americans who are working full time at selected ages from 20 to 65 can be modeled as a function of their age, x. A model is

$$y = -0.000362x^3 + 0.0401x^2 - 1.39x + 21.7$$

a. Use a viewing window with the x-values representing the ages from 20 to 65, and graph the function with a graphing calculator.

b. Use the model to estimate the number of Americans age 55 and age 64 who are working full time.

(Source: *Wall Street Journal*, June 18, 2012)

SOLUTION

a. Enter the equation in Y_1 and use the viewing window with $x_{min} = 20$ and $x_{max} = 65$ and with any values for y_{min} and y_{max}. Using TABLE with several values of x shows that we can get the desired graph using $y_{min} = 0$, $y_{max} = 10$. (See Figure 1.22 (a).) Note that this viewing window can be written in interval notation as [20, 65] by [0, 10].† Figure 1.22(b) shows the suggested viewing window, containing x-scale and y-scale values that give appropriately spaced tick marks on the graph. Figure 1.22(c) shows the graph of the function.

b. Figure 1.22(d) shows the TABLE outputs for ages 55 and 64, as well as for age 20. It shows that the model estimates that 6.32 million Americans age 55 are working and 2.09 million Americans age 64 are working. We could also find the number working at these ages with ⌈TRACE⌉ on the graph.

---

* It is occasionally necessary to make more than one attempt to find a window that gives a complete graph.
† Interval notation is discussed in the Algebra Toolbox, page 4.

(a)

(b)

(c)

(d)

Figure 1.22

Spreadsheet ▶ SOLUTION Excel can also be used to evaluate and graph the function $y = -0.000362x^3 + 0.0401x^2 - 1.39x + 21.7$, discussed in Example 6. A table of input and output values and the graph of the function are shown in Figure 1.23.

Figure 1.23

Graphing Data Points

Graphing utilities can also be used to create lists of numbers and to create graphs of the data stored in the lists. To see how lists and scatter plots are created, consider the following example.

EXAMPLE 7 ▶   U.S. Diabetes

The number of U.S. adults with diabetes, in thousands, projected to 2050 is shown in Table 1.12.

Table 1.12

Year	Thousands
2015	37,300
2020	50,000
2025	59,500
2030	68,300
2035	76,200
2040	84,100
2045	91,700
2050	100,000

(Source: Centers for Disease Control and Prevention)

a. Align the data so that $x = $ the number of years after 2015, and enter these x-values in list L1 of your graphing calculator. Enter the number of millions of U.S. adults with diabetes in L2.

b. Use a graphing command* to create the scatter plot of these data points.

SOLUTION

a. If x represents the number of years after 2015, we must subtract 2015 from each year in the original table to obtain each corresponding x-value. The number of U.S. adults with diabetes was given in thousands in the original table, and because one million is 1000 times one thousand, we must divide each number by 1000 to obtain each y-value. Figure 1.24(a) shows the aligned inputs in L1 and the outputs from the data in L2.

b. The scatter plot is shown in Figure 1.24(b), with each entry in L1 represented by an x-coordinate of a point on the graph and the corresponding entry in L2 represented by the y-coordinate of that point on the graph. The window can be set automatically or manually to include x-values from L1 and y-values from L2.

(a)   (b)

Figure 1.24

---

* Many calculators have a Stat Plot command. See Appendix A, page 666. Excel uses the points option under the Scatter icon in the Charts group, which is under the Insert tab. See Appendix B, page 690.

Spreadsheet ▶ SOLUTION   Figure 1.25 shows the aligned and scaled values from Table 1.12. To create a scatter plot of the data, we highlight the two columns containing the inputs and outputs, select the insert tab, select the scatter plot icon in the Charts group, and then select the scatter plot icon. The graph of the data will appear. (See Figure 1.25.)*

Figure 1.25

Alignment and Scaling of Data

Recall that replacing the input so that x represents the number of years after a certain year is called *alignment of the data*. Replacing the outputs by dividing or multiplying each number by a given number is called **scaling the data**.

_____

* See Appendix B, page 690, for detailed steps.

SKILLS CHECK 1.2

1. a. Complete the table of values for the function $y = x^3$, plot the values of x and y as points in a coordinate plane, and draw a smooth curve through the points.

x	−3	−2	−1	0	1	2	3
y							

b. Graph the function $y = x^3$ with a graphing calculator, using the viewing window $x_{\min} = -4$, $x_{\max} = 4$, $y_{\min} = -30$, $y_{\max} = 30$.

c. Compare the graphs in parts (a) and (b).

2. a. Complete the table of values for the function $y = 2x^2 + 1$, plot the values of x and y as points in a coordinate plane, and draw a smooth curve through the points.

x	−3	−2	−1	0	1	2	3
y							

b. Graph the function $y = 2x^2 + 1$ with a graphing calculator, using the viewing window $x_{\min} = -5$, $x_{\max} = 5$, $y_{\min} = -1$, $y_{\max} = 20$.

c. Compare the graphs in parts (a) and (b).

In Exercises 3–8, make a table of values for each function and then plot the points to graph the function by hand.

3. $f(x) = 3x - 1$

4. $f(x) = 2x - 5$

5. $f(x) = \dfrac{1}{2}x^2$

6. $f(x) = 3x^2$

7. $f(x) = \dfrac{1}{x - 2}$

8. $f(x) = \dfrac{x}{x + 3}$

In Exercises 9–14, graph the functions with a graphing calculator using a standard viewing window. State whether the graph has a turning point in this window.

9. $y = x^2 - 5$

10. $y = 4 - x^2$

11. $y = x^3 - 3x^2$

12. $y = x^3 - 3x^2 + 4$

13. $y = \dfrac{9}{x^2 + 1}$

14. $y = \dfrac{40}{x^2 + 4}$

For Exercises 15–18, graph the given function with a graphing calculator using (a) the standard viewing window and (b) the specified window. Which window gives a better view of the graph of the function?

15. $y = x + 20$ with $x_{min} = -10$, $x_{max} = 10$, $y_{min} = -10, y_{max} = 30$

16. $y = x^3 - 3x + 13$ with $x_{min} = -5$, $x_{max} = 5$, $y_{min} = -10, y_{max} = 30$

17. $y = \dfrac{0.04(x - 0.1)}{x^2 + 300}$ on $[-20, 20]$ by $[-0.002, 0.002]$

18. $y = -x^2 + 20x - 20$ on $[-10, 20]$ by $[-20, 90]$

For Exercises 19–22, find an appropriate viewing window for the function, using the given x-values. Then graph the function. Find the coordinates of any turning points.

19. $y = x^2 + 50$, for x-values between -8 and 8.

20. $y = x^2 + 60x + 30$, for x-values between -60 and 0.

21. $y = x^3 + 3x^2 - 45x$, for x-values between -10 and 10.

22. $y = (x - 28)^3$, for x-values between 25 and 31.

23. Find a complete graph of $y = 10x^2 - 90x + 300$. (A complete graph of this function shows one turning point.)

24. Find a complete graph of $y = -x^2 + 34x - 120$. (A complete graph of this function shows one turning point.)

25. Use a calculator or a spreadsheet to find $S(t)$ for the values of t given in the following table in the next column.

t	$S(t) = 5.2t - 10.5$
12	
16	
28	
43	

26. Use a calculator or a spreadsheet to find $f(q)$ for the values of q given in the following table.

q	$f(q) = 3q^2 - 5q + 8$
-8	
-5	
24	
43	

27. Enter the data below into lists and graph the scatter plot of the data, using the window $[0, 110]$ by $[0, 550]$.

x	20	30	40	50	60	70	80	90	100
y	500	320	276	80	350	270	120	225	250

28. Enter the data below into lists and graph the scatter plot of the data, using the window $[-20, 120]$ by $[0, 80]$.

x	-15	-1	25	40	71	110	116
y	12	16	10	32	43	62	74

29. Use the table below in parts (a)–(c).

x	-3	-1	1	5	7	9	11
y	-42	-18	6	54	78	102	126

a. Use a graphing utility to graph the points from the table.

b. Use a graphing utility to graph the equation $f(x) = 12x - 6$ on the same set of axes as the data in part (a).

c. Do the points fit on the graph of the equation? Do the respective values of x and y in the table satisfy the equation in part (b)?

30. Use the table below in parts (a)–(c).

x	1	3	6	8	10
y	-10	0	15	25	35

a. Use a graphing utility to graph the points from the table.

b. Use a graphing utility to graph the equation $f(x) = 5x - 15$ on the same set of axes as the data in part (a).

c. Do the points fit on the graph of the equation? Do the respective values of x and y in the table satisfy the equation in part (b)?

31. Suppose $f(x) = x^2 - 5x$ million dollars are earned, where x is the number of years after 2000.

a. What is $f(20)$?

b. The answer to part (a) gives the number of millions of dollars earned for what year?

32. Suppose $f(x) = 100x^2 - 5x$ thousand units are produced, where x is the number of years after 2000.

a. What is $f(10)$?

b. How many units are produced in 2010, according to this function?

EXERCISES 1.2

In Exercises 33–38, plot the graphs by hand.

33. *Temperature Conversion* The function that converts temperature from Fahrenheit degree measurement F to Celsius degree measurement C is $C = \dfrac{5F - 160}{9}$.
Graph this function for F from 32 to 212 degrees.

34. *Temperature Conversion* The function that converts temperature from Celsius degree measurement C to Fahrenheit degree measurement F is $F = \dfrac{9}{5}C + 32$.
Graph this function for C from 0 to 100 degrees.

35. *Height of a Ball* If a ball is thrown into the air at 64 feet per second from the top of a 100-foot-tall building, its height can be modeled by the function $S = 100 + 64t - 16t^2$, where S is in feet and t is in seconds.

a. Graph this function for t representing 0 to 6 seconds and S representing 0 to 200 feet.

b. Find the height of the ball 1 second after it is thrown and 3 seconds after it is thrown. How can these values be equal?

c. Find the maximum height the ball will reach.

36. *Depreciation* A business property valued at $600,000 is depreciated over 30 years by the straight-line method, so its value x years after the depreciation began is

$$V = 600,000 - 20,000x$$

a. Graph this function for x representing 0 to 30 years and v representing 0 to 600,000 dollars.

b. What is the value 10 years after the depreciation is started?

37. *Cost* Suppose the cost of the production and sale of x Electra dishwashers is $C(x) = 15,000 + 100x + 0.1x^2$ dollars. Graph this function on a viewing window with x between 0 and 50.

38. *Revenue* Suppose the revenue from the sale of x coffee makers is given by $R(x) = 52x - 0.1x^2$. Graph this function on a viewing window with x between 0 and 100.

39. *Women in the Workforce* Using data starting with 1950 and projected to 2050, the number of women (in millions) in the workforce is given by the function $P = 0.79x + 20.86$, where x equals the number of years after 1950.

a. Find the value of P when $x = 70$. Explain what this means.

b. What value of x represents 2030? Use the model to find the number of women in the workforce in 2030.
(Source: U.S. Census Bureau)

40. *Unemployment Rates* The unemployment rates (as percents) for the United States for selected years from 2000 to 2015 can be modeled by

$$y = -0.002x^4 + 0.051x^3 - 0.371x^2 + 1.055x + 4.234$$

where x is the number of years after 2000.

a. What are the values of x that correspond to the years 2005 and 2016?

b. Find the value of y when $x = 10$. Explain what this means.

c. What was the unemployment rate in the United States in 2015, according to the model?
(Source: Bureau of Labor Statistics, U.S. Department of Labor)

41. *Internet Access* Using data from 2005 through 2015, the percent of households with Internet access can be modeled by the function $P = 1.84t + 60.89$, with t equal to the number of years after 2005.

   a. What are the values of t that correspond to the years 2010 and 2016?

   b. $P = f(10)$ gives the value of P for what year? What is $f(10)$?

   c. What x_{min} and x_{max} should be used to set the viewing window so that t represents 2005–2020?
   (Source: Jupiter Media Metrix)

42. *Federal Tax per Capita* The amount of federal tax per capita (per person) for selected years starting with 2000 and projected to 2018 can be modeled by the function

$$S = 2.4807t^3 - 52.251t^2 + 528.68t + 5192.6$$

where t is the number of years after 2000.

   a. What are the values of t that correspond to 2012, 2016, and 2020?

   b. $S = f(18)$ gives the value of S for what year? What is $f(18)$?

   c. What x_{min} and x_{max} should be used to set a viewing window so that t represents 2000–2020?
   (Source: Internal Revenue Service)

43. *Population of Americans 85 Years of Age and Older* Using data from 1910 to 2050, the number of thousands of Americans 85 years old and older can be modeled by $P = 1.458x^2 - 124.4x + 2516$, with $x = 0$ in 1900.

   a. Use a graphing utility to graph this function on the viewing window $[30, 165]$ by $[-1000, 25,000]$.

   b. Use the graphing utility to predict the population of Americans 85 years old and older in 2030.
   (Source: U.S. Census Bureau)

44. *Life Expectancy at 65* Using data for selected years from 1950 and projected to 2050, the function $y = 0.077x + 13.827$ gives the life expectancy at age 65, with y equal to the number of additional years of expected life at age 65 and x equal to the number of years after 1950.

   a. Graph the model on the viewing window $[0, 110]$ by $[0, 25]$.

   b. What time period does the viewing window in part (a) represent?

   c. Use the graphing utility to find what the model predicts the life expectancy to be in 2038, rounded to one decimal place.
   (Source: Social Security Administration)

45. *Student Loans* Data for selected years from 2011 and projected to 2023 can be used to show that the balance of federal direct student loans y, in billions of dollars, is related to the number of years after 2010, x, by the function $y = 130.7x + 699.7$.

   a. Graph the function for x-values corresponding 2010 to 2025.

   b. Find the value of y when x is 13.

   c. What does this function predict the balance of federal direct student loans will be in 2029?
   (Source: U.S. Office of Management and Budget)

46. *Profit* The profit from the production and sale of x laser printers is given by the function $P(x) = 200x - 0.01x^2 - 5000$, where x is the number of units produced and sold. Graph this function on a viewing window with x between 0 and 1000.

47. *Profit* The profit from the production and sale of x digital cameras is given by the function $P(x) = 1500x - 8000 - 0.01x^2$, where x is the number of units produced and sold. Graph this function on a viewing window with x between 0 and 500.

48. *Tax Burden* Using data from the Internal Revenue Service, the per capita tax burden B (in hundreds of dollars) can be described by $B(x) = 2.11x^3 - 93.0x^2 + 1370x + 394$, where x is the number of years after 1990.

   a. Graph this function with a graphing utility for x-values corresponding to the years 1990 through 2020.

   b. Did the tax burden increase or decrease?
   (Source: Internal Revenue Service)

49. *Obesity* Using data from 2000 and projected to 2030, the percent of Americans who are obese can be modeled by the function $y = 0.665x + 23.4$, where x is the number of years after 2000 and y is the percent of obese Americans.

   a. What inputs correspond to the years 2000 through 2030?

   b. What outputs correspond to the inputs for 2000 and 2030?

   c. Use the answers to parts (a) and (b) and the fact that the function increases to find an appropriate viewing window, and graph this function.
   (Source: American Journal of Preventive Medicine)

50. *Cocaine Use* The percent of high school seniors during the years 1975–2010 who have ever used cocaine can be described by

$$y = 0.0152x^2 - 0.9514x + 21.5818$$

where x is the number of years after 1975.

a. What inputs correspond to the years 1975 through 2010?

b. What outputs for y could be used to estimate the percent of seniors who have ever used cocaine?

c. Based on your answers to parts (a) and (b), choose an appropriate window and graph the equation on a graphing utility.

d. Graph the function again with a new window that gives a graph nearer the center of the screen.

e. Use this function to estimate the percent in 2013.

(Source: monitoringthefuture.org)

51. *U.S. Population* The projected population of the United States for selected years from 2000 to 2060 is shown in the table below, with the population given in millions.

a. According to this table, what should the U.S. population have been in 2010?

b. Create a new table with x representing the number of years after 2000 and y representing the number of millions.

c. Use a graphing utility to graph the data from the new table as a scatter plot.

Year	Population (millions)
2000	275.3
2010	299.9
2020	324.9
2030	351.1
2040	377.4
2050	403.7
2060	432.0

(Source: U.S. Census Bureau)

52. *Smart Phones* Projections of worldwide mobile payment transaction values (in billions of dollars) for the years 2009 through 2016 are shown in the table.

a. Graph the data from the table as a scatter plot, with y equal to the payment transaction values (in billions of dollars) and x equal to the number of years after 2000.

b. Graph the function $y = 9.286x^2 - 147.619x + 605.595$ and the data points on the same axes.

Year	Transaction Values ($ billions)
2009	30
2010	60
2011	100
2012	170
2013	260
2014	360
2015	480
2016	620

53. *Social Security Benefits* The maximum monthly benefits payable to individuals who retired at full retirement age are given in the table below.

a. Use a graphing utility to graph the data, with x representing the number of years after 2000 when the retiree reached full retirement age and y representing the monthly benefits.

b. Graph the equation $y = 74.17x + 1596.99$ on the same axes as the scatter plot.

Year Attaining Full Retirement Age	Maximum Monthly Benefit ($)
2005	1939
2006	2053
2007	2116
2008	2185
2009	2323
2010	2346
2011	2366
2012	2513
2013	2533
2014	2642

(Source: Social Security Administration)

54. *Expenditures for U.S. Health Care* The cost per person per year for health care, with projections to 2018, is shown in the following table.

Year	$ per Person	Year	$ per Person
2000	4789	2010	8465
2002	5563	2012	9275
2004	6331	2014	10,289
2006	7091	2016	11,520
2008	7826	2018	12,994

(Source: U.S. Medicare and Medicaid Services)

a. Let x represent the number of years after 2000 and y represent the annual cost, and sketch the scatter plot of the data.

b. Graph the function $y = 1.149x^3 - 21.564x^2 + 473.575x + 4749.394$ on the same axes as the scatter plot.

c. Is the function a good visual fit to the data?

55. *Crude Oil* The following table gives the U.S. crude oil production, in billions of barrels, for the years from 2010 projected to 2030.

a. What does the table predict the crude oil production will be in 2026?

b. Graph the data from this table as a scatter plot, with x equal to the number of years after 2010 and y equal to the number of billions of barrels of crude oil.

c. Graph the function $y = -0.00107x^2 + 0.0369x + 1.95$ on the same axes as the scatter plot.

Year	Billions of Barrels
2010	1.94
2014	2.10
2018	2.16
2022	2.24
2026	2.26
2030	2.26

(Source: U.S. Energy Information Administration)

56. *Gross Domestic Product* The table gives the gross domestic product (GDP, the value of all goods and services) of the United States for selected years from 1970 to 2018.

a. Create a scatter plot of the data, with y representing the GDP in billions of dollars and x representing the number of years after 1970.

b. Graph the function $y = 6.20x^2 + 104x + 1110$ on the same axes as scatter plot.

c. What does the function predict the GDP will be in 2017?

Year	GDP ($ billions)
1970	1038.3
1980	2788.1
1990	5800.5
2000	9951.5
2005	12,638.4
2010	14,498.9
2015	17,936.1
2018	21,024.8

(*Source*: U.S. Bureau of Economic Analysis)

1.3 Linear Functions

KEY OBJECTIVES

- Identify linear functions
- Find the intercepts and slopes of graphs of linear functions
- Graph linear functions
- Find the rate of change of a linear function
- Identify identity and constant functions
- Apply linear revenue, cost, and profit functions
- Find marginal revenue and marginal profit from linear revenue and linear profit functions

SECTION PREVIEW **Hispanics in the United States**

Using data and projections from 1990 through 2050, the percent of Hispanics in the U.S. population can be modeled by

$$H(x) = 0.224x + 9.01$$

with x equal to the number of years after 1990. The graph of this model is shown in Figure 1.26. (Source: U.S. Census Bureau)

Because the graph of this function is a line, it is called a linear function. The rate at which the Hispanic population is increasing is constant in this model, and it is equal to the slope of the line shown in Figure 1.26. (See Example 7.) In this section, we investigate linear functions and discuss slope, constant rates of change, intercepts, revenue, cost, and profit.

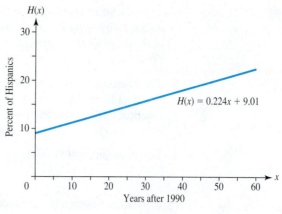

Figure 1.26

Linear Functions

A function whose graph is a line is a **linear function**.

> ### Linear function
>
> A linear function is a function that can be written in the form $f(x) = ax + b$, where a and b are constants.

If x and y are in separate terms in an equation and each appears to the first power (and not in a denominator), we can rewrite the equation relating them in the form

$$y = ax + b$$

for some constants a and b. So the original equation is a linear equation that represents a linear function. If no restrictions are stated or implied by the context of the problem situation and if its graph is not a horizontal line, both the domain and the range of a linear function consist of the set of all real numbers. Note that an equation of the form $x = d$, where d is a constant, is not a function; its graph is a vertical line.

Recall that in the form $y = ax + b$, a and b represent constants and the variables x and y can represent any variables, with x representing the independent (input) variable and y representing the dependent (output) variable. For example, the function $5q + p = 400$ can be written in the form $p = -5q + 400$, so we can say that p is a linear function of q.

EXAMPLE 1 ▶ **Linear Functions**

Determine whether each equation represents a linear function. If so, give the domain and range.

a. $0 = 2t - s + 1$ **b.** $y = 5$ **c.** $xy = 2$

SOLUTION

a. The equation

$$0 = 2t - s + 1$$

does represent a linear function because each of the variables t and s appears to the first power and each is in a separate term. We can solve this equation for s, getting

$$s = 2t + 1$$

so s is a linear function of t. Because any real number can be multiplied by 2 and increased by 1 and the result is a real number, both the domain and the range consist of the set of all real numbers. (Note that we could also solve the equation for t so that t is a linear function of s.)

b. The equation $y = 5$ is in the form $y = ax + b$, where $a = 0$ and $b = 5$, so it represents a linear function. (It is in fact a constant function, which is a special linear function.) The domain is the set of all real numbers (because $y = 5$ regardless of what x we choose), and the range is the set containing 5.

c. The equation

$$xy = 2$$

does not represent a linear function because x and y are not in separate terms and the equation cannot be written in the form $y = ax + b$.

Intercepts

Figure 1.27

The points where a graph crosses or touches the x-axis and the y-axis are called the **x-intercepts** and **y-intercepts**, respectively, of the graph. For example, Figure 1.27 shows that the graph of the linear function $2x - 3y = 12$ crosses the x-axis at $(6, 0)$, so the x-intercept is $(6, 0)$. The graph crosses the y-axis at $(0, -4)$, so the y-intercept is $(0, -4)$. In this text, we will use the widely accepted convention that the x-coordinate of the x-intercept may also be called the x-intercept, and the y-coordinate of the y-intercept may also be called the y-intercept.* The procedure for finding intercepts is a direct result of these definitions.

> ### Finding Intercepts Algebraically
>
> To find the y-intercept of a graph of $y = f(x)$, set $x = 0$ in the equation and solve for y. If the solution is b, we say the y-intercept is b and the graph intersects the y-axis at the point $(0, b)$.
>
> To find the x-intercept(s) of the graph of $y = f(x)$, set $y = 0$ in the equation and solve for x. If the solution is a, we say the x-intercept is a and the graph intersects the x-axis at the point $(a, 0)$.

EXAMPLE 2 ▶ **Finding Intercepts Algebraically**

Find the x-intercept and the y-intercept of the graph of $2x - 3y = 12$ algebraically.

SOLUTION

The y-intercept can be found by substituting 0 for x in the equation and solving for y.

$$2(0) - 3y = 12$$
$$-3y = 12$$
$$y = -4$$

Thus the y-intercept is -4, and the graph crosses the y-axis at the point $(0, -4)$.

---

* We usually call the horizontal-axis intercepts(s) the x-intercept(s) and the vertical-axis intercept(s) the y-intercept(s), but realize that symbols other than x and y can be used to represent the input and output. For example, if $p = f(q)$, the vertical intercept is called the p-intercept and the horizontal intercept is called the q-intercept.

Similarly, the x-intercept can be found by substituting 0 for y in the equation and solving for x.

$$2x - 3(0) = 12$$
$$2x = 12$$
$$x = 6$$

Thus the x-intercept is 6, and the graph crosses the x-axis at the point $(6, 0)$.

Finding Intercepts Graphically

To find the intercept(s) of a graph of $y = f(x)$, first graph the function in a window that shows all intercepts.

To find the y-intercept, ⌈TRACE⌋ to $x = 0$ and the y-intercept will be displayed. To find the x-intercept(s) of the graph of $y = f(x)$, use the zero command under the CALC menu (accessed by **2ND** ⌈TRACE⌋).*

The graph of a linear function has one y-intercept and one x-intercept unless the graph is a horizontal line. The intercepts of the graph of a linear function are often easy to calculate. If the intercepts are distinct, then plotting these two points and connecting them with a line gives the graph.

Technology Note

When graphing with a graphing utility, finding or estimating the intercepts can help you set the viewing window for the graph of a linear equation.

EXAMPLE 3 ▶ **Loan Balance**

A business property is purchased with a promise to pay off a $60,000 loan plus the $16,500 interest on this loan by making 60 monthly payments of $1275. The amount of money, y, remaining to be paid on $76,500 (the loan plus interest) is reduced by $1275 each month. Although the amount of money remaining to be paid changes every month, it can be modeled by the linear function

$$y = 76{,}500 - 1275x$$

where x is the number of monthly payments made. We recognize that only integer values of x from 0 to 60 apply to this application.

a. Find the x-intercept and the y-intercept of the graph of this linear equation.

b. Interpret the intercepts in the context of this problem situation.

c. How should x and y be limited in this model so that they make sense in the application?

d. Use the intercepts and the results of part (c) to sketch the graph of the given equation.

---

* For more details, see Appendix A, page 668.

SOLUTION

a. To find the x-intercept, set $y = 0$ and solve for x.

$$0 = 76{,}500 - 1275x$$
$$1275x = 76{,}500$$
$$x = \frac{76{,}500}{1275} = 60$$

Thus, 60 is the x-intercept.

To find the y-intercept, set $x = 0$ and solve for y.

$$y = 76{,}500 - 1275(0)$$
$$y = 76{,}500$$

Thus, 76,500 is the y-intercept.

b. The x-intercept corresponds to the number of months that must pass before the amount owed is $0. Therefore, a possible interpretation of the x-intercept is "The loan is paid off in 60 months." The y-intercept corresponds to the total (loan plus interest) that must be repaid 0 months after purchase—that is, when the purchase is made. Thus, the y-intercept tells us "A total of $76,500 must be repaid."

c. We know that the total time to pay the loan is 60 months. A value of x larger than 60 will result in a negative value of y, which makes no sense in the application, so the values of x vary from 0 to 60. The output, y, is the total amount owed at any time during the loan. The amount owed cannot be less than 0, and the value of the loan plus interest will be at its maximum, 76,500, when time is 0. Thus, the values of y vary from 0 to 76,500.

d. The graph intersects the horizontal axis at $(60, 0)$ and intersects the y-axis at $(0, 76{,}500)$, as indicated in Figure 1.28. Because we know that the graph of this equation is a line, we can simply connect the two points to obtain this first-quadrant graph.

Figure 1.28 shows the graph of the function on a viewing window determined by the context of the application.

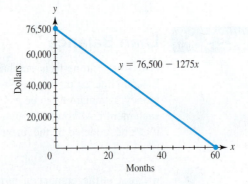

Figure 1.28

Slope of a Line

Consider two stairways: One goes from the park entrance to the shogun shrine at Nikko, Japan, and rises vertically 1 foot for every 1 foot of horizontal increase, and the second goes from the Tokyo subway to the street and rises vertically 20 centimeters for every 25 centimeters of horizontal increase. To see which set of stairs would be easier to climb, we can find the steepness, or **slope**, of each stairway. If a board is placed along the steps of each stairway, its slope is a measure of the incline of the stairway.

$$\text{shrine stairway steepness} = \frac{\text{vertical increase}}{\text{horizontal increase}} = \frac{1 \text{ foot}}{1 \text{ foot}} = 1$$

$$\text{subway stairway steepness} = \frac{\text{vertical increase}}{\text{horizontal increase}} = \frac{20 \text{ cm}}{25 \text{ cm}} = 0.8$$

The shrine stairway has a slope that is larger than that of the subway stairway, so it is steeper than the subway stairway. Thus, the subway steps would be easier to climb. In general, we define the slope of a line as follows.

Slope of a Line

The slope of a line is defined as

$$\text{slope} = \frac{\text{vertical change}}{\text{horizontal change}} = \frac{\text{rise}}{\text{run}}$$

The slope can be found by using any two points on the line (see Figure 1.29). If a nonvertical line passes through the two points, P_1 with coordinates (x_1, y_1) and P_2 with coordinates (x_2, y_2), its slope, denoted by m, is found by using

$$m = \frac{y_2 - y_1}{x_2 - x_1}$$

Figure 1.29

The slope of a vertical line is undefined because $x_2 - x_1 = 0$ and division by 0 is undefined.

The slope of any given nonvertical line is a constant. Thus, the same slope will result regardless of which two points on the line are used in its calculation.

EXAMPLE 4 ▶   **Calculating the Slope of a Line**

a. Find the slope of the line passing through the points $(-3, 2)$ and $(5, -4)$. What does the slope mean?

b. Find the slope of the line joining the x-intercept point and y-intercept point in the loan situation of Example 3.

SOLUTION

a. We choose one point as P_1 and the other point as P_2. Although it does not matter which point is chosen to be P_1, it is important to keep the correct order of the terms in the numerator and denominator of the slope formula. Letting $P_1 = (-3, 2)$ and $P_2 = (5, -4)$ and substituting in the slope formula gives

$$m = \frac{-4 - 2}{5 - (-3)} = \frac{-6}{8} = -\frac{3}{4}$$

Note that letting $P_1 = (5, -4)$ and $P_2 = (-3, 2)$ gives the same slope:

$$m = \frac{2 - (-4)}{-3 - 5} = \frac{6}{-8} = -\frac{3}{4}$$

A slope of $-\dfrac{3}{4}$ means that, from a given point on the line, by moving 3 units down and 4 units to the right or by moving 3 units up and 4 units to the left, we arrive at another point on the line.

b. Because $x = 60$ is the x-intercept in part (b) of Example 3, $(60, 0)$ is a point on the graph. The y-intercept is $y = 76{,}500$, so $(0, 76{,}500)$ is a point on the graph. Recall that x is measured in months and that y has units of dollars in the real-world setting (that is, the *context*) of Example 3. Substituting in the slope formula, we obtain

$$m = \frac{76{,}500 - 0}{0 - 60} = \frac{76{,}500}{-60} = -1275$$

This slope means that the amount owed decreases by \$1275 each month.

As Figure 1.30(a) to (d) indicates, the slope describes the direction of a line as well as the steepness.

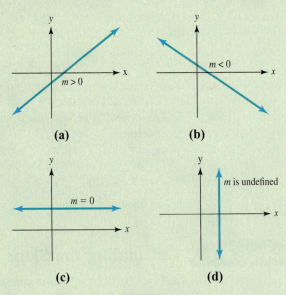

The Relation Between Orientation of a Line and Its Slope

1. The slope is *positive* if the line *rises* upward toward the right.

2. The slope is *negative* if the line *falls* downward toward the right.

(a)   (b)

(c)   (d)

Figure 1.30

3. The slope of a *horizontal line* is 0 because a horizontal line has a vertical change (rise) of 0 between any two points on the line.

4. The slope of a *vertical line* does not exist because a vertical line has a horizontal change (run) of 0 between any two points on the line.

Remembering how the orientation of a line is related to its slope can help you check that you have the correct order of the points in the slope formula. For instance, if you find that the slope of a line is positive and you see that the line falls downward from left to right when you observe its graph, you know that there is a mistake either in the slope calculation or in the graph that you are viewing.

Slope and *y*-Intercept of a Line

There is an important connection between the slope of the graph of a linear equation and its equation when written in the form $y = f(x)$. To investigate this connection, we can graph the equation

$$y = 3x + 2$$

by plotting points or by using a graphing utility. By creating a table for values of x equal to 0, 1, 2, 3, and 4 (see Table 1.13), we can see that each time x increases by 1, y increases by 3 (see Figure 1.31). Thus, the slope is

$$\frac{\text{change in } y}{\text{change in } x} = \frac{3}{1} = 3$$

From this table, we see that the *y*-intercept is 2, the same value as the constant term of the equation. Observe also that the slope of the graph of $y = 3x + 2$ is the same as the coefficient of *x*.

Table 1.13

x	y
0	2
1	5
2	8
3	11
4	14

Figure 1.31

Because we denote the slope of a line by *m*, we have the following.

Slope and *y*-Intercept of a Line

The slope of the graph of the equation $y = mx + b$ is *m* and the *y*-intercept of the graph is *b*, so the graph crosses the *y*-axis at $(0, b)$.

Thus, when the equation of a linear function is written in the form $y = mx + b$ or $f(x) = mx + b$, we can "read" the values of the slope and *y*-intercept of its graph. If a function is linear but is not written in the form $y = ax + b$, the slope and *y*-intercept can be determined by solving the system for *y* to get that form.

EXAMPLE 5 ▶   **Finding the Slope and y-Intercept of a Line**

Determine the slope and y-intercept of the linear function.

a. $y = 7x - 12$ **b.** $2x - 3y = 12$

SOLUTION

a. The equation is written in the form $y = ax + b$, so the slope is the coefficient of x and the y-intercept is the constant term; that is, $m = 7$ and $b = -12$.

b. The equation must be solved for y so that it is in the form $y = ax + b$.

$$2x - 3y = 12$$
$$2x = 12 + 3y \qquad \text{Add } 3y \text{ to both sides of the equation.}$$
$$2x - 12 = 3y \qquad \text{Subtract 12 from both sides.}$$
$$\frac{2}{3}x - 4 = y \qquad \text{Divide both sides by 3 and simplify.}$$

Thus, the slope of the line is $\frac{2}{3}$ and the y-intercept is -4.

EXAMPLE 6 ▶   **Loan Balance**

As we saw in Example 3, the amount of money y remaining to be paid on the loan of \$60,000 with \$16,500 interest is

$$y = 76{,}500 - 1275x$$

where x is the number of monthly payments that have been made.

a. What are the slope and y-intercept of the graph of this function?

b. How does the amount owed on the loan change as the number of months increases?

SOLUTION

a. Writing this equation in the form $y = mx + b$ gives $y = -1275x + 76{,}500$. The coefficient of x is -1275, so the slope is $m = -1275$. The constant term is 76,500, so the y-intercept is $b = 76{,}500$.

b. The slope of the line indicates that the amount owed decreases by \$1275 each month. This can be verified from the graph of the function in Figure 1.32 and the sample inputs and outputs in Table 1.14.

Table 1.14

Months (x)	Amount Owed (y)
1	75,225
2	73,950
3	72,675
4	71,400
5	70,125
6	68,850
7	67,575
8	66,300
9	65,025
10	63,750

Figure 1.32

Constant Rate of Change

The function $y = 76,500 - 1275x$, whose graph is shown in Figure 1.32, gives the amount owed as a function of the months remaining on the loan. The coefficient of x, -1275, indicates that for each additional month, the value of y changes by -1275 (see Table 1.14). That is, the amount owed decreases at the **constant rate** of -1275 each month. Note that the **constant rate of change** of this linear function is equal to the **slope** of its graph. This is true for all linear functions.

> ### Constant Rate of Change
>
> The rate of change of the linear function $y = mx + b$ is the constant m, the slope of the graph of the function.

Note: The rate of change in an applied context should include appropriate units of measure. The rate of change describes by how much the output changes (increases or decreases) for every input unit.

EXAMPLE 7 ▶   **Hispanics in the United States**

Using data and projections from 1990 through 2050, the percent of Hispanics in the U.S. population can be modeled by

$$H(x) = 0.224x + 9.01$$

with x equal to the number of years after 1990. The graph of this model is shown in Figure 1.33. (Source: U.S. Census Bureau)

a. What is the slope of the graph of this function?

b. What does this slope tell us about the annual rate of change in the percent of Hispanics in the United States?

SOLUTION

a. The coefficient of x in the linear function $H(x) = 0.224x + 9.01$ is 0.224, so the slope of the line that is its graph is 0.224.

b. This slope tells us that the percent of Hispanics in the U.S. population is increasing by 0.224 each year since 1990. We say that the percent has increased by 0.224 *percentage point*.

Figure 1.33

Revenue, Cost, and Profit

The **profit** that a company makes on its product is the difference between the amount received from sales (revenue) and the production and sales costs. If x units are produced and sold, we can write

$$P(x) = R(x) - C(x)$$

where

$$P(x) = \text{profit from sale of } x \text{ units}$$
$$R(x) = \text{total revenue from sale of } x \text{ units}$$
$$C(x) = \text{total cost of production and sale of } x \text{ units}$$

In general, **revenue** is found by using the equation

$$\text{revenue} = (\text{price per unit})(\text{number of units})$$

The **total cost** is composed of two parts: fixed costs and variable costs. **Fixed costs** (FC), such as depreciation, rent, and utilities, remain constant regardless of the number of units produced. **Variable costs** (VC) are those directly related to the number of units produced. Thus, the total cost, often simply called the cost, is found by using the equation

$$\text{cost} = \text{variable costs} + \text{fixed costs}$$

EXAMPLE 8 ▶ ## Cost, Revenue, and Profit

Suppose that a company manufactures 50-inch 3D plasma TVs and sells them for $1800 each. The costs incurred in the production and sale of the TVs are $400,000 plus $1000 for each TV produced and sold. Write the profit function for the production and sale of x TVs.

SOLUTION

The total revenue for x TVs is $1800x$, so the revenue function is $R(x) = 1800x$. The fixed costs are $400,000, so the total cost for x TVs is $1000x + 400,000$. Hence, $C(x) = 1000x + 400,000$. The profit function is given by $P(x) = R(x) - C(x)$, so

$$P(x) = 1800x - (1000x + 400,000)$$
$$P(x) = 800x - 400,000$$

Figures 1.34 and 1.35 show the graphs of the three linear functions: $R(x)$, $C(x)$, and $P(x)$.

Figure 1.34

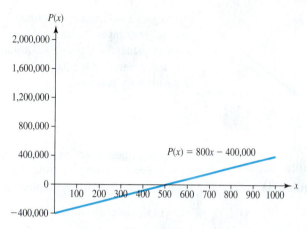

Figure 1.35

Marginal Cost, Revenue, and Profit

For total cost, total revenue, and profit functions* that are linear, the rates of change are called **marginal cost, marginal revenue**, and **marginal profit**, respectively. Suppose that the cost to produce and sell a product is $C(x) = 54.36x + 6790$ dollars, where x is the number of units produced and sold. This is a linear function, and its graph is a line with slope 54.36. Thus, the *rate of change* of this cost function, called the **marginal cost**, is $54.36 per unit produced and sold. This means that the production and sale of each additional unit will cost an additional $54.36.

EXAMPLE 9 ▶ | **Marginal Revenue and Marginal Profit**

A company produces and sells a smartphone with revenue given by $R(x) = 89.50x$ dollars and cost given by $C(x) = 54.36x + 6790$ dollars, where x is the number of smartphones produced and sold.

a. What is the marginal revenue for this smartphone, and what does it mean?

b. Find the profit function.

c. What is the marginal profit for this smartphone, and what does it mean?

SOLUTION

a. The marginal revenue for this smartphone is the rate of change of the revenue function, which is the slope of its graph. Thus, the marginal revenue is $89.50 per unit sold. This means that the sale of each additional smartphone will result in additional revenue of $89.50.

b. To find the profit function, we subtract the cost function from the revenue function.
$$P(x) = 89.50x - (54.36x + 6790) = 35.14x - 6790$$

c. The marginal profit for this smartphone is the rate of change of the profit function, which is the slope of its graph. Thus, the marginal profit is $35.14 per smartphone sold. This means that the production and sale of each additional smartphone will result in an additional profit of $35.14.

Special Linear Functions

A special linear function that has the form $y = 0x + b$, or $y = b$, where b is a real number, is called a **constant function**. The graph of the constant function $y = 3$ is shown in Figure 1.36(a). The temperature inside a sealed case containing an Egyptian

---

* In this text, we frequently use "total cost" and "total revenue" interchangeably with "cost" and "revenue" respectively.

(a) Constant Function

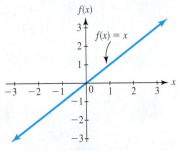

(b) Identify Function

Figure 1.36

mummy in a museum is a constant function of time because the temperature inside the case never changes. Notice that even though the range of a constant function consists of a single value, the input of a constant function is any real number or any real number that makes sense in the context of an applied problem. Another special linear function is the **identity function**

$$y = 1x + 0, \quad \text{or} \quad y = x$$

which is a linear function of the form $y = mx + b$ with slope $m = 1$ and y-intercept $b = 0$. For the general identity function $f(x) = x$, the domain and range are each the set of all real numbers. A graph of the identity function f is shown in Figure 1.36(b).

SKILLS CHECK 1.3

1. Which of the following functions are linear?

   a. $y = 3x^2 + 2$ **b.** $3x + 2y = 12$ **c.** $y = \dfrac{1}{x} + 2$

2. Is the graph in the figure below a function?

3. Find the slope of the line through $(4, 6)$ and $(28, -6)$.

4. Find the slope of the line through $(8, -10)$ and $(8, 4)$.

5. Find the slope of the line in the graph that follows.

6. Find the slope of the line in the graph below.

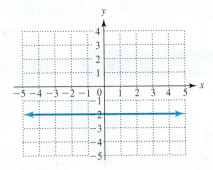

In Exercises 7–10, (a) find the x- and y-intercepts of the graph of the given equation, if they exist, and (b) graph the equation.

7. $5x - 3y = 15$ **8.** $x + 5y = 17$

9. $3y = 9 - 6x$ **10.** $y = 9x$

11. If a line is horizontal, then its slope is ____. If a line is vertical, then its slope is ____.

12. Describe the line whose slope was determined in Exercise 4.

For Exercises 13–14, determine whether the slope of the graph of the line is positive, negative, 0, or undefined.

13. a.

b.

14. a.

b.

For Exercises 15–18, (a) give the slope of the line (if it exists) and the y-intercept (if it exists) and (b) graph the line.

15. $y = 4x + 8$ **16.** $3x + 2y = 7$

17. $5y = 2$ **18.** $x = 6$

For each of the functions in Exercises 19–21, do the following:

a. *Find the slope and y-intercept (if possible) of the graph of the function.*

b. *Determine whether the graph is rising or falling.*

c. *Graph each function on a window with the given x-range and a y-range that shows a complete graph.*

19. $y = 4x + 5; [-5, 5]$

20. $y = 0.001x - 0.03; [-100, 100]$

21. $y = 50,000 - 100x; [0, 500]$

22. Rank the functions in Exercises 19–21 in order of increasing steepness.

For each of the functions in Exercises 23–26, find the rate of change.

23. $y = 4x - 3$ **24.** $y = \dfrac{1}{3}x + 2$

25. $y = 300 - 15x$ **26.** $y = 300x - 15$

27. If a linear function has the points $(-1, 3)$ and $(4, -7)$ on its graph, what is the rate of change of the function?

28. If a linear function has the points $(2, 1)$ and $(6, 3)$ on its graph, what is the rate of change of the function?

29. a. Does graph (i) or graph (ii) represent the identity function?

b. Does graph (i) or graph (ii) represent a constant function?

(i)

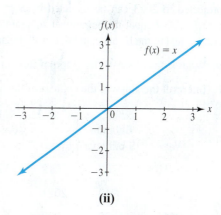

(ii)

30. What is the slope of the identity function?

31. a. What is the slope of the constant function $y = k$?

b. What is the rate of change of a constant function?

32. What is the rate of change of the identity function?

EXERCISES 1.3

33. College Enrollment The total fall enrollment in 4-year public institutions is given by $y = 0.014x + 2.290$, where x is the number of years after 1990 and y is millions of students. Is this a linear function? Why or why not? What is the independent variable?
(Source: U.S. Department of Education, National Center for Education Statistics)

34. Women in the Workforce Using data and projections from 1950 through 2050 gives the number y (in thousands) of women in the workforce as the function $y = -0.005x^2 + 1.278x + 13.332$, where x is the number of years from 1950. Is this a linear function? Why or why not?
(Source: U.S. Census Bureau, U.S. Dept. of Commerce)

35. Life Expectancy at 65 Using data for selected years from 1950 and projected to 2050, the function $y = 0.077x + 13.827$ gives the life expectancy at age 65, with y equal to the number of additional years of expected life at age 65 and x equal to the number of years after 1950.

a. Why is this a linear function?

b. Find and interpret the slope of the graph of this function.
(Source: Social Security Administration)

36. Student Loans The balance of federal direct student loans, $S(t)$, in $ billions, for selected years from 2011 and projected to 2023 can be modeled by $S = 130.7t + 699.7$, with t equal to the number of years after 2011 and S equal to the balance, in billions of dollars.

a. What is the slope of the graph of this function?

b. Interpret the slope in the context of this problem.

Year	Balance ($ billions)	Year	Balance ($ billions)
2011	702	2019	1775
2013	940	2021	2000
2015	1220	2023	2250
2017	1500		

(Source: U.S. Office of Management and Budget)

37. Marriage Rate The marriage rate per 1000 unmarried women for the years 1987–2014 is given by $132x + 1000y = 9570$ where x is the number of years after 1980.

a. Find the x-intercept of the graph of this function.

b. Find and interpret the y-intercept of the graph of this function.

c. Graph this function, using the intercepts.
(Source: National Vital Statistics Report)

38. Marijuana Use The percent p of high school seniors who ever used marijuana can be related to x, the number of years after 2000, by the equation $25p + 21x = 1215$.

a. Find the x-intercept of the graph of this function.

b. Find and interpret the p-intercept of the graph of this function.

c. Graph the function, using the intercepts. What values of x on the graph represent years 2000 and after?

39. Advertising Impact An advertising agency has found that when it promotes a new product in a city, the weekly rate of change R of the number of people who are aware of it x weeks after it is introduced is given by $R = 3500 - 70x$. Find the x-and R-intercepts and then graph the function on a viewing window that is meaningful in the application.

40. Cigarette Use The percent p of adults who have tried cigarettes can be modeled by $p = 43.3 - 0.50t$, where t is the number of years after 1960.

a. Find the p-and t-intercepts of the graph of this function.

b. Use the intercepts to graph the function.

41. Life Insurance The monthly rates for a $100,000 life insurance policy for males aged 27–32 are shown in the table.

Age (years), x	27	28	29
Premium (dollars per month), y	11.81	11.81	11.81

Age (years), x	30	31	32
Premium (dollars per month), y	11.81	11.81	11.81

a. Could the data in this table be modeled by a constant function or the identity function?

b. Write an equation whose graph contains the data points in the table.

c. What is the slope of the graph of the function found in part (b)?

d. What is the rate of change of the data in the table?

42. Asparagus Cultivation In Canada, the cultivation area (in hectares) of asparagus for the years 1999 through 2008 is shown in the table.

Year	1999	2000	2001	2002	2003
Cultivation Area (hectares)	1200	1200	1200	1200	1200

Year	2004	2005	2006	2007	2008
Cultivation Area (hectares)	1200	1200	1200	1200	1200

(Source: Sustainblog.com)

a. Sketch the data as a scatter plot, with y equal to the cultivation area and x equal to the year.

b. Could the data in the table be modeled by a constant function or the identity function?

c. Write the equation of a function that fits the data points.

d. Sketch a graph of the function you found in part (c) on the same axes as the scatter plot.

43. Disposable Income Disposable income is the amount left after taxes have been paid and is one measure of health of the economy. The total disposable income for the United States, in billions of dollars, for selected years from 2010 and projected to 2040, can be modeled by $D(x) = 0.328x + 6.32$, where x is the number of years after 2000. (*Source*: U.S. Bureau of Economic Analysis)

a. What is the slope of the graph of this function?

b. Interpret this slope as a rate of change.
(Source: Bureau of Economic Analysis)

44. Diabetes The figure below shows the projected percent of U.S. adults with diabetes for the years 2010 through 2050. What is the average rate of growth over this period of time?

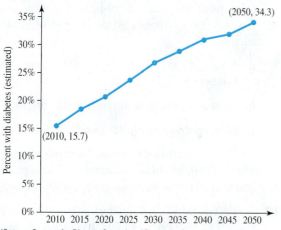

(Source: Centers for Disease Control and Prevention)

45. Women in the Workforce Using data from 1950 and projected to 2050, the number of women (in millions) in the workforce is given by the function $P = 0.79x + 20.86$, where x equals the number of years after 1950.
(Source: U.S. Census Bureau)

a. Find and interpret the P-intercept of the graph of this function.

b. What is the slope of the graph of this function?

c. Interpret the slope as a rate of change.

46. World Population The world population from 2000 and projected to 2050, in billions, is given by the function $y = 0.063x + 6.191$, where x is the number of years after 2000.
(*Source:* U.S. Census Bureau)

a. What is the slope of the graph of this function?

b. Interpret this slope as a rate of change.

47. Black Population Using data and projections from 1990 through 2050, the percent of the U.S. population that is black can be modeled by $B(x) = 0.057x + 12.3$, where x is the number of years after 1990.

a. What is the slope of the graph of this function?

b. Interpret the slope as a rate of change.
(Source: U.S. Census Bureau)

48. Seawater Pressure In seawater, the pressure p is related to the depth d according to the model $33p - 18d = 496$, where d is the depth in feet and p is in pounds per square inch.

a. What is the slope of the graph of this function?

b. Interpret the slope as a rate of change.

49. Crickets The number of times per minute n that a cricket chirps can be modeled as a function of the Fahrenheit temperature T. The data can be approximated by the function

$$n = \frac{12T}{7} - \frac{52}{7}$$

a. Is the rate of change of the number of chirps positive or negative?

b. What does this tell us about the relationship between temperature and the number of chirps?

50. Wireless Service Spending The total amount spent in the United States for wireless communication services S (in billions of dollars) can be modeled by the function

$$S = 6.205 + 11.23t$$

where t is the number of years after 1995.

a. Find the slope and the S-intercept of the graph of this equation.

b. What interpretation could be given to the y-intercept?

c. What interpretation could be given to the slope?
(Source: Cellular Telecommunications and Internet Association)

51. **Call Centers** Call centers are booming in the Philippines as multinational companies increasingly outsource these operations. The annual revenue for call centers in the Philippines from 2006 to 2010, in billions of dollars, can be modeled by $R(x) = 0.975x - 3.45$, where x is the number of years after 2000.

a. According to the model, what was the rate of change of revenue for call centers in the Philippines?

b. According to the model, what was the revenue for call centers in the Philippines in 2010?

c. Would this model be valid to estimate the revenue in 2000? Why or why not?
(Source: Business Processing Association of the Philippines)

52. **Internet Users** The percent of the U.S. population with Internet access can be modeled by

$$y = 1.36x + 68.8$$

where x is the number of years after 2000.

a. Find the slope and the y-intercept of the graph of this equation.

b. What interpretation could be given to the slope?

(Source: Jupiter Media Metrix)

53. **Depreciation** Suppose the cost of a business property is $1,920,000 and a company depreciates it with the straight-line method. Suppose V is the value of the property after x years and the line representing the value as a function of years passes through the points $(10, 1,310,000)$ and $(20, 700,000)$.

a. What is the slope of the line through these points?

b. What is the annual rate of change of the value of the property?

54. **Men in the Workforce** The number of men in the workforce (in millions) for the years from 1950 to 2050 can be approximated by the linear model determined by connecting the points $(1950, 43.8)$ and $(2050, 100.3)$.

a. Find the annual rate of change of the model whose graph is the line connecting these points.

b. What does this tell us about men in the workforce?
(Source: U.S. Census Bureau)

55. **Profit** A company charting its profits notices that the relationship between the number of units sold

x and the profit P is linear. If 300 units sold results in $4650 profit and 375 units sold results in $9000 profit, find the marginal profit, which is the rate of change of the profit.

56. **Cost** A company buys and retails baseball caps, and the total cost function is linear. The total cost for 200 caps is $2690, and the total cost for 500 caps is $3530. What is the marginal cost, which is the rate of change of the function?

57. **Marginal Cost** Suppose the monthly total cost for the manufacture of golf balls is $C(x) = 3450 + 0.56x$, where x is the number of balls produced each month.

a. What is the slope of the graph of the total cost function?

b. What is the marginal cost (rate of change of the cost function) for the product?

c. What is the cost of each additional ball that is produced in a month?

58. **Marginal Cost** Suppose the monthly total cost for the manufacture of 19-inch television sets is $C(x) = 2546 + 98x$, where x is the number of TVs produced each month.

a. What is the slope of the graph of the total cost function?

b. What is the marginal cost for the product?

c. Interpret the marginal cost for this product.

59. **Marginal Revenue** Suppose the monthly total revenue for the sale of golf balls is $R(x) = 1.60x$, where x is the number of balls sold each month.

a. What is the slope of the graph of the total revenue function?

b. What is the marginal revenue for the product?

c. Interpret the marginal revenue for this product.

60. **Marginal Revenue** Suppose the monthly total revenue from the sale of 19-inch television sets is $R(x) = 198x$, where x is the number of TVs sold each month.

a. What is the slope of the graph of the total revenue function?

b. What is the marginal revenue for the product?

c. Interpret the marginal revenue for this product.

61. **Profit** The profit for a product is given by $P(x) = 19x - 5060$, where x is the number of units produced and sold. Find the marginal profit for the product.

62. **Profit** The profit for a product is given by the function $P(x) = 939x - 12,207$, where x is the number of units produced and sold. Find the marginal profit for the product.

1.4 Equations of Lines

KEY OBJECTIVES

- Write equations of lines using the slope-intercept form and the point-slope form
- Write equations of horizontal and vertical lines
- Write the equations of lines parallel or perpendicular to given lines
- Find the average rate of change over an interval for nonlinear functions
- Find the slope of the secant line between two points on a graph
- Find the difference quotient from $(x, f(x))$ to $(x + h, f(x + h))$
- Find average rates of change for approximately linear data

SECTION PREVIEW Blood Alcohol Percent

Suppose that the blood alcohol percent for a 180-lb man is 0.11% if he has 5 drinks and that the percent increases by 0.02% for each additional drink. We can use this information to write the linear equation that models the blood alcohol percent p as a function of the number of drinks he has, because the rate of change is constant. (See Example 3.) In this section, we see how to write the equation of a linear function from information about the line, such as the slope and a point on the line or two points on the line. We also discuss **average rates of change** for data that can be modeled by nonlinear functions, slopes of secant lines, and how to create linear models that approximate data that are nearly linear. ■

Writing Equations of Lines

Creating a linear model from data involves writing a linear equation that describes the mathematical situation. If we know two points on a line or the slope of the line and one point, we can write the equation of the line.

Recall that if a linear equation has the form $y = mx + b$, then the coefficient of x is the slope of the line that is the graph of the equation, and the constant b is the y-intercept of the line. Thus, if we know the slope and the y-intercept of a line, we can write the equation of this line.

> ### Slope-Intercept Form of the Equation of a Line
>
> The slope-intercept form of the equation of a line with slope m and y-intercept b is
>
> $$y = mx + b$$
>
> In an applied context, m is the rate of change and b is the initial value (when $x = 0$).

EXAMPLE 1 ▶ Appliance Repair

An appliance repairman charges $60 for a service call plus $25 per hour for each hour spent on the repair. Assuming his service call charges can be modeled by a linear function of the number of hours spent on the repair, write the equation of the function.

SOLUTION

Let x represent the number of hours spent on the appliance repair and let y be the service call charge in dollars. The slope of the line is the amount that the charge increases for every hour of work done, so the slope is 25. Because $60 is the basic charge before any time is spent on the repair, 60 is the y-intercept of the line. Substituting these values in the slope-intercept form of the equation gives the equation of the linear function modeling this situation. When the repairman works x hours on the service call, the charge is

$$y = 25x + 60 \text{ dollars}$$

We next consider a form of an equation of a line that can be written if we know the slope and a point. If the slope of a line is m, then the slope between a fixed point (x_1, y_1) and any other point (x, y) on the line is also m. That is,

$$m = \frac{y - y_1}{x - x_1}$$

Solving for $y - y_1$ (that is, multiplying both sides of this equation by $x - x_1$) gives the **point-slope form** of the equation of a line.

Point-Slope Form of the Equation of a Line

The equation of the line with slope m that passes through a known point (x_1, y_1) is

$$y - y_1 = m(x - x_1)$$

EXAMPLE 2 ▶ **Using a Point and Slope to Write an Equation of a Line**

Write an equation for the line that passes through the point $(-1, 5)$ and has slope $\frac{3}{4}$.

SOLUTION

We are given $m = \frac{3}{4}$, $x_1 = -1$, and $y_1 = 5$. Substituting in the point-slope form, we obtain

$$y - 5 = \frac{3}{4}\big[x - (-1)\big]$$

$$y - 5 = \frac{3}{4}(x + 1)$$

$$y - 5 = \frac{3}{4}x + \frac{3}{4}$$

$$y = \frac{3}{4}x + \frac{23}{4}$$

Figure 1.37 shows the graph of this line.

Figure 1.37

EXAMPLE 3 ▶   Blood Alcohol Percent

Table 1.15 gives the number of drinks and the resulting blood alcohol percent for a 180-pound man. (One drink is equal to 1.25 oz of 80-proof liquor, 12 oz of regular beer, or 5 oz of table wine; many states have set 0.08% as the legal limit for driving under the influence.)

a. Is the rate of change of the blood alcohol percent for a 180-pound man a constant? What is it?

b. Write the equation of the function that models the blood alcohol percent as a function of the number of drinks.

Table 1.15

Number of Drinks (x)	3	4	5	6	7	8	9	10
Blood Alcohol Percent (y)	0.07	0.09	0.11	0.13	0.15	0.17	0.19	0.21

SOLUTION

a. Yes. Each additional drink increases the blood alcohol percent by 0.02, so the rate of change is 0.02 percentage point per drink.

b. The rate of change is constant, so the function is linear, with its rate of change (and slope of its graph) equal to 0.02. Using this value and any point, like $(5, 0.11)$, on the line gives us the equation we seek, with x equal to the number of drinks and y the blood alcohol percent.

$$y - 0.11 = 0.02(x - 5), \quad \text{or} \quad y = 0.02x + 0.01$$

If we know two points on a line, we can find the slope of the line and use either of the points and this slope to write the equation of the line.

EXAMPLE 4 ▶   Using Two Points to Write an Equation of a Line

Write the equation of the line that passes through the points $(-1, 5)$ and $(2, 4.)$

SOLUTION

Because we know two points on the line, we can find the slope of the line.

$$m = \frac{4 - 5}{2 - (-1)} = \frac{-1}{3} = -\frac{1}{3}$$

We can now substitute one of the points and the slope in the point-slope form to write the equation. Using the point $(-1, 5)$ gives

$$y - 5 = -\frac{1}{3}[x - (-1)], \quad \text{or} \quad y - 5 = -\frac{1}{3}x - \frac{1}{3}, \quad \text{so} \quad y = -\frac{1}{3}x + \frac{14}{3}$$

Using the point (2, 4) gives

$$y - 4 = -\frac{1}{3}(x - 2)$$

$$y - 4 = -\frac{1}{3}x + \frac{2}{3}$$

$$y = -\frac{1}{3}x + \frac{14}{3}$$

Notice that the equations are the same, regardless of which of the two given points is used in the point-slope form.

If the rate of change of the outputs with respect to the inputs is a constant and we know two points that satisfy the conditions of the application, then we can write the equation of the linear function that models the application.

EXAMPLE 5 ▶ Inmate Population

The number of people (in millions) in U.S. prisons or jails grew at a constant rate from 2001 to 2013, with 1.345 million people incarcerated in 2001 and 1.570 million incarcerated in 2013. (Source: Bureau of Justice Statistics)

a. What is the rate of growth of people incarcerated from 2001 to 2013?

b. Write the linear equation that models the number N of prisoners as a function of the year x.

c. Use the function to predict the inmate population in 2017.

SOLUTION

a. The rate of growth is constant, so the points fit on a line and a linear function can be used to find the model for the number of incarcerated people as a function of the year. The rate of change (and slope of the line) is given by

$$m = \frac{1.570 - 1.345}{2013 - 2001} = 0.01875$$

b. Substituting in the point-slope form of the equation of a line (with either point) gives the equation of the line that contains the two points and thus models the application.

$$N - 1.570 = 0.01875(x - 2013), \text{ or } N = 0.01875x - 36.17375$$

c. Evaluating the function at 2017 gives

$$N = 0.01875(2017) - 36.17375 = 1.645$$

Thus, 1.645 million people are expected to be incarcerated in 2017, according to this model.

EXAMPLE 6 ▶ **Writing Equations of Horizontal and Vertical Lines**

Write equations for the lines that pass through the point $(-1, 5)$ and have

a. slope 0.   **b.** undefined slope.

SOLUTION

a. If $m = 0$, the point-slope form gives us the equation

$$y - 5 = 0[x - (-1)]$$
$$y = 5$$

Because the output is always the same value, the graph of this linear function is a horizontal line. See Figure 1.38(a).

b. Because m is undefined, we cannot use the point-slope form to write the equation of this line. Lines with undefined slope are vertical lines. Every point on the vertical line through $(-1, 5)$ has an x-coordinate of -1. Thus, the equation of the line is $x = -1$. Note that this equation does not represent a function. See Figure 1.38(b).

(a)   (b)

Figure 1.38

As we saw in Example 6, parts (a) and (b), there are special forms when the lines are horizontal or vertical.

> **Vertical and Horizontal Lines**
>
> A vertical line has the form $x = a$, where a is a constant and a is the x-coordinate of any point on the line.
> A horizontal line has the form $y = b$, where b is a constant and b is the y-coordinate of any point on the line.

Parallel and Perpendicular Lines

Clearly horizontal lines are perpendicular to vertical lines. Two distinct nonvertical lines that have the same slope are **parallel**, and conversely. If a line has slope $m \neq 0$, any line **perpendicular** to it will have slope $-\dfrac{1}{m}$. That is, the slopes of perpendicular lines are negative reciprocals of each other if neither line is horizontal.

EXAMPLE 7 ▶   **Parallel and Perpendicular Lines**

Write the equation of the line through $(4, 5)$ and

a. parallel to the line with equation $3x + 2y = -1$.

b. perpendicular to the line with equation $3x + 2y = -1$.

SOLUTION

a. To find the slope of the line with equation $3x + 2y = -1$, we solve for y.

$$3x + 2y = -1$$
$$2y = -3x - 1$$
$$y = -\frac{3}{2}x - \frac{1}{2}$$

The line through $(4, 5)$ parallel to this line has the same slope, $-\frac{3}{2}$, and its equation is

$$y - 5 = -\frac{3}{2}(x - 4)$$

$$y = -\frac{3}{2}x + 11$$

b. The line through $(4, 5)$ perpendicular to $3x + 2y = -1$ has slope $\frac{2}{3}$, and its equation is

$$y - 5 = \frac{2}{3}(x - 4)$$

$$y = \frac{2}{3}x + \frac{7}{3}$$

In Example 7(b) we found that the equation of the line was

$$y = \frac{2}{3}x + \frac{7}{3}$$

By multiplying both sides of this equation by 3 and writing the new equation with x and y on the same side of the equation, we have a new form of the equation:

$$3y = 2x + 7, \quad \text{or} \quad 2x - 3y = -7$$

This equation is called the **general form** of the equation of the line.

General Form of the Equation of a Line

The general form of the equation of a line is $ax + by = c$, where a, b, and c are real numbers, with a and b not both equal to 0.

In summary, these are the forms we have discussed for the equation of a line.

Forms of Linear Equations

General form:	$ax + by = c$	where a, b, and c are real numbers, with a and b not both equal to 0.
Point-slope form:	$y - y_1 = m(x - x_1)$	where m is the slope of the line and (x_1, y_1) is a point on the line.
Slope-intercept form:	$y = mx + b$	where m is the slope of the line and b is the y-intercept.
Vertical line:	$x = a$	where a is a constant, and a is the x-coordinate of any point on the line. The slope is undefined.
Horizontal line:	$y = b$	where b is a constant, and b is the y-coordinate of any point on the line. The slope is 0.

Average Rate of Change

If the graph of a function is not a line, the function is nonlinear and the slope of a line joining two points on the curve may change as different points are chosen on the curve. The calculation and interpretation of the slope of a curve are topics you will study if you take a calculus course. However, there is a quantity called the **average rate of change** that can be applied to any function relating two variables.

In general, we can find the average rate of change of a function between two input values if we know how much the function outputs change between the two input values.

Average Rate of Change

The average rate of change of $f(x)$ with respect to x over the interval from $x = a$ to $x = b$ (where $a < b$) is calculated as

$$\text{average rate of change} = \frac{\text{change in } f(x) \text{ values}}{\text{corresponding change in } x \text{ values}} = \frac{f(b) - f(a)}{b - a}$$

How is the average rate of change over some interval of points related to the slope of a line connecting the points? For any function, the average rate of change between two points on its graph is the slope of the line joining the two points. Such a line is called a **secant line**.

EXAMPLE 8 ▶ Tesla Automobile Sales

Sales and projected sales for the electric-powered Tesla automobile are shown in Figure 1.39. The number of Tesla sales can be modeled by the function

$$S(t) = 3.521t^3 - 38.475t^2 + 149.054t - 170.480$$

where $S(t)$ is in thousands and t is the number of years after 2010. A graph of $S(t)$ is shown in Figure 1.39(b).

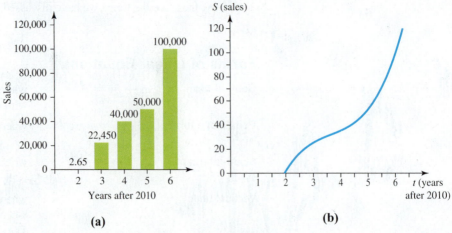

Figure 1.39

(Source: Tesla Motors)

a. Find the average rate of change of $S(t)$ between $(2, S(2))$ and $(6, S(6))$.

b. Interpret your answer to part (a).

c. What is the relationship between the slope of the secant line joining the points $(2, 1.896)$ and $(6, 99.28)$ and the answer to part (a)?

SOLUTION

a. Using the model to find the average rate of change between the two points gives

$$\frac{S(6) - S(2)}{6 - 2} = \frac{(99.28 - 1.896)}{4} = \frac{97.384}{4} = 24.346$$

b. One possible interpretation is that, on average between 2012 and 2016, sales of Tesla automobiles increased by 24,346 units per year.

c. The slope of this secant line, shown in Figure 1.40, is numerically the same as the average rate of change of sales between 2012 and 2016 found in part (a).

Figure 1.40

Difference Quotient

If we have points $(x, f(x))$ and $(x + h, f(x + h))$ on the graph of $y = f(x)$, we can find the slope of the secant line (and average rate of change) of the function from x to $x + h$, called the **difference quotient**, as follows.

> ### Difference Quotient
>
> For $h \neq 0$, the slope of the secant line joining the points $(x, f(x))$ and $(x + h, f(x + h))$, and the average rate of change of the function $f(x)$ from x to $x + h$, is
>
> $$\frac{f(x + h) - f(x)}{x + h - x} = \frac{f(x + h) - f(x)}{h}$$

EXAMPLE 9 ▶ **Difference Quotient**

For the function $f(x) = x^2 + 1$, whose graph is shown in Figure 1.41, find

a. $f(x + h)$.

b. $f(x + h) - f(x)$.

c. the average rate of change with the difference quotient $\dfrac{f(x + h) - f(x)}{h}$, for $h \neq 0$.

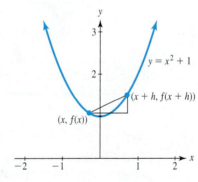

Figure 1.41

SOLUTION

a. $f(x + h)$ is found by substituting $x + h$ for x in $f(x) = x^2 + 1$:

$$f(x + h) = (x + h)^2 + 1 = (x^2 + 2xh + h^2) + 1$$

b. $f(x + h) - f(x) = (x^2 + 2xh + h^2) + 1 - (x^2 + 1)$

$$= x^2 + 2xh + h^2 + 1 - x^2 - 1 = 2xh + h^2$$

c. The slope of the secant line and the average rate of change of the function is

$$\frac{f(x + h) - f(x)}{h} = \frac{2xh + h^2}{h} = 2x + h, \text{ for } h \neq 0.$$

EXAMPLE 10 ▶ **Average Velocity**

Suppose a ball is shot straight upward at 64 feet per second. The equation that gives the height of the ball t seconds after it is shot is $f(t) = 64t - 16t^2$.

a. Find the average velocity (the average rate of change of the height per second) of this ball over the time interval from $t = 1$ to $t = 2$.

b. Find the difference quotient that can be used to find the average velocity of this ball over the time interval from $t = 1$ to $t = 1 + h$, for $h \neq 0$.

SOLUTION

a. The average velocity over the time interval from $t = 1$ to $t = 2$ is the change in the distance traveled divided by the difference in time.

$$V_{\text{ave}} = \frac{f(2) - f(1)}{2 - 1} = \frac{(128 - 64) - (64 - 16)}{1} = 16 \text{ feet per second}$$

b. To find the average velocity over the time interval from $t = 1$ to $t = 1 + h$, $h \neq 0$, we use the difference quotient.

$$V_{\text{ave}} = \frac{f(1 + h) - f(1)}{1 + h - 1} = \frac{[(64(1 + h) - 16(1 + h)^2)] - [(64 - 16)]}{h}$$

$$= \frac{64 + 64h - 16(1 + 2h + h^2) - 48}{h}$$

$$= \frac{64 + 64h - 16 - 32h - 16h^2 - 48}{h}$$

$$= \frac{32h - 16h^2}{h}$$

$$= \frac{h(32 - 16h)}{h}$$

$$= 32 - 16h, \text{ for } h \neq 0$$

Approximately Linear Data

Real-life data are rarely perfectly linear, but some sets of real data points lie sufficiently close to a line that two points can be used to create a linear function that models the data. Consider the following example.

EXAMPLE 11 ▶ **Public School Enrollment**

Table 1.16 and Figure 1.42 show the real and projected enrollment (in thousands) at U.S. public schools for selected years from 1980 through 2024. (Source: U.S. National Center for Education Statistics)

a. Create a scatter plot of the data. Does a line fit the data points exactly?

b. Find the average rate of change of the high school enrollment between 1980 and 2024.

c. Write the equation of the line determined by this rate of change and one of the two points.

Table 1.16

Year, x	Enrollment, y (thousands)
1980	41,651
2000	46,857
2010	49,373
2012	49,522
2022	51,485
2023	51,604
2024	52,113

(Source: U.S. National Center for Education Statistics)

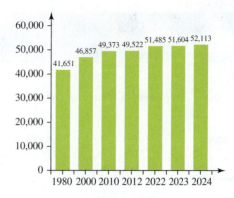

Figure 1.42

SOLUTION

a. A scatter plot is shown in Figure 1.43. The graph shows that the data do not fit a linear function exactly, but that a linear function could be used as an approximate model for the data.

Figure 1.43

b. To find the average rate of change between 1980 and 2024, we use the points $(1980, 41{,}651)$ and $(2024, 52{,}113)$.

$$\frac{52{,}113 - 41{,}651}{2024 - 1980} = 237.8$$

This means that the public school enrollments have grown at an average rate of 237.8 thousand per year (because the outputs are in thousands).

c. Because the enrollment growth is approximately linear, the average rate of change can be used as the slope of the graph of a linear function describing the enrollments from 1980 through 2024. The graph connects the points $(1980, 41{,}651)$ and $(2024, 52{,}113)$, so we can use either of these two points to write the equation of the line. Using the point $(1980, 41{,}651)$ and $m = 237.8$ gives the equation

$$y - 41{,}651 = 237.8(x - 1980)$$
$$y - 41{,}651 = 237.8x - 470{,}844$$
$$y = 237.8x - 429{,}193$$

This equation is a model of the enrollment, in thousands, as a function of the year.

The equation in Example 11 approximates a model for the data, but it is not the best possible model for the data. In Section 2.2, we will see how technology can be used to find the linear function that is the best fit for a set of data of this type.

SKILLS CHECK 1.4

For Exercises 1–18, write the equation of the line with the given conditions.

1. Slope 4 and y-intercept $\dfrac{1}{2}$

2. Slope 5 and y-intercept $\dfrac{1}{3}$

3. Slope $\dfrac{1}{3}$ and y-intercept 3

4. Slope $-\dfrac{1}{2}$ and y-intercept -8

5. Through the point $(4, -6)$ with slope $-\dfrac{3}{4}$

6. Through the point $(-4, 3)$ with slope $-\dfrac{1}{2}$

7. Vertical line, through the point $(9, -10)$

8. Horizontal line, through the point $(9, -10)$

9. Passing through $(-2, 1)$ and $(4, 7)$

10. Passing through $(-1, 3)$ and $(2, 6)$

11. Passing through $(5, 2)$ and $(-3, 2)$

12. Passing through $(9, 2)$ and $(9, 5)$

13. x-intercept -5 and y-intercept 4

14. x-intercept 4 and y-intercept -5

15. Passing through $(4, -6)$ and parallel to the line with equation $3x + y = 4$

16. Passing through $(5, -3)$ and parallel to the line with equation $2x + y = -3$

17. Passing through $(-3, 7)$ and perpendicular to the line with equation $2x + 3y = 7$

18. Passing through $(-4, 5)$ and perpendicular to the line with equation $3x + 2y = -8$

For Exercises 19–22, write the equation of the line whose graph is shown.

19.

20.

21.

22.

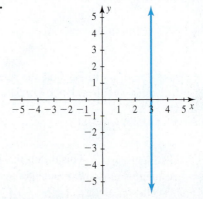

23. Write the equation of a function if its rate of change is -15 and its initial value is 12.

24. Write the equation of a function if its rate of change is -8 and its initial value is -7.

25. If a function has rate of change $\dfrac{2}{3}$ and $y = 9$ when $x = 3$, write its equation.

26. If a function has rate of change $-\dfrac{1}{5}$ and $y = 12$ when $x = -2$, write its equation.

27. For the function $y = x^2$, compute the average rate of change between $x = -1$ and $x = 2$.

28. For the function $y = x^3$, compute the average rate of change between $x = -1$ and $x = 2$.

29. For the function shown in the figure, find the average rate of change from $(-2, 7)$ to $(1, -2)$.

30. For the function shown in the figure, find the average rate of change from $(-1, 2)$ to $(2, -4)$.

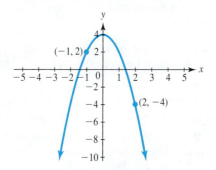

31. If one point on a line is $(1, -10)$ and the line's slope is -3, find the y-intercept.

32. If one point on a line is $(-2, 8)$ and the line's slope is $-\dfrac{3}{2}$, find the y-intercept.

For the functions given in Exercises 33–36, find

$$\frac{f(x + h) - f(x)}{h}, \text{ for } h \neq 0$$

33. $f(x) = 45 - 15x$ **34.** $f(x) = 32x + 12$

35. $f(x) = 2x^2 + 4$ **36.** $f(x) = 3x^2 + 1$

In Exercises 37–38, use the tables on the next page, which give a set of input values x and the corresponding outputs y that satisfy a function.

37. a. Would you say that a linear function could be used to model the data? Explain.

b. Write the equation of the linear function that fits the data.

x	10	20	30	40	50
y	585	615	645	675	705

38. a. Verify that the values satisfy a linear function.

b. Write the equation of the linear function that fits the data.

x	1	7	13	19
y	−0.5	8.5	17.5	26.5

EXERCISES 1.4

39. *Utility Charges* Palmetto Electric determines its monthly bills for residential customers by charging a base price of $12.00 plus an energy charge of 10.34 cents for each kilowatt-hour (kWh) used. Write an equation for the monthly charge y (in dollars) as a function of x, the number of kWh used.

40. *Phone Bills* For interstate calls, AT&T charges 10 cents per minute plus a base charge of $2.99 each month. Write an equation for the monthly charge y as a function of the number of minutes of use.

41. *Depreciation* A business uses straight-line depreciation to determine the value y of a piece of machinery over a 10-year period. Suppose the original value (when $t = 0$) is equal to $36,000 and the value is reduced by $3600 each year. Write the linear equation that models the value y of this machinery at the end of year t.

42. *Sleep* Each day a young person should sleep 8 hours plus $\dfrac{1}{4}$ hour for each year the person is under 18 years of age.

a. Based on this information, how much sleep does a 10-year-old need?

b. Based on this information, how much sleep does a 14-year-old need?

c. Use the answers from parts (a) and (b) to write a linear equation relating hours of sleep y to age x, for $6 \leq x \leq 18$.

d. Use your equation from part (c) to verify that an 18-year-old needs 8 hours of sleep.

43. **Heart Rate** The desired heart rate for weight loss during exercise is a linear function of a person's age, with the desired rate for a 60-year-old person equal to 104 beats per minute and with the desired rate increasing by 6.5 beats per minute for each decrease of 10 years in age. (Source: paramountfitness.com)

    a. Write the equation of the function, with t equal to age and H equal to desired heart rate.

    b. What does the function give as the desired rate for a 40-year-old person?

44. **Heart Rate** The desired heart rate for cardiac health during exercise is a linear function of a person's age, with the desired rate for a 50-year-old person equal to 136 beats per minute and with the desired rate increasing by 8 beats per minute for each decrease of 10 years in age. (*Source*: Paramountfitness.com)

    a. Write the equation of the function, with t equal to age and h equal to desired heart rate.

    b. What does the function give as the desired rate for a 75-year-old person?

45. **Depreciation** A business uses straight-line depreciation to determine the value y of an automobile over a 5-year period. Suppose the original value (when $t = 0$) is equal to $26,000 and the salvage value (when $t = 5$) is equal to $1000.

    a. By how much has the automobile depreciated over the 5 years?

    b. By how much is the value of the automobile reduced at the end of each of the 5 years?

    c. Write the linear equation that models the value s of this automobile at the end of year t.

46. **Retirement** For Pennsylvania state employees for whom the average of the three best yearly salaries is $75,000, the retirement plan gives an annual pension of 2.5% of 75,000, multiplied by the number of years of service. Write the linear function that models the pension P in terms of the number of years of service y. (Source: Pennsylvania State Retirement Fund)

47. **Patrol Cars** The Beaufort County Sheriff's office assigns a patrol car to each of its deputy sheriffs, who keeps the car 24 hours a day. As the population of the county grew, the number of deputies and the number of patrol cars also grew. The data in the following table could describe how many patrol cars were necessary as the number of deputies increased. Write the function that models the relationship between the number of deputies and the number of patrol cars.

Number of Deputies	10	50	75	125	180	200	250
Number of Patrol Cars in Use	10	50	75	125	180	200	250

48. **Search Market Share** Data from comScore show that Google's share of the U.S. search market has remained flat, as shown in the following table.

Date	April 2010	May 2010	June 2010	July 2010	August 2010	Sept 2010
Market Share (%)	66	66	66	66	66	66

Date	Oct 2010	Nov 2010	Dec 2010	Jan 2011	Feb 2011	March 2011
Market Share (%)	66	66	66	66	66	66

    a. Write the linear model that gives Google's market share as a function of the number of months after April 2010.

    b. What is the rate of change of Google's market share for this time period?

49. **Profit** A company charting its profits notices that the relationship between the number of units sold, x, and the profit, P, is linear. If 300 units sold results in $4650 profit and 375 units sold results in $9000 profit, write the equation that models its profit.

50. **Cost** A company buys and retails baseball caps. The total cost function is linear, the total cost for 200 caps is $2680, and the total cost for 500 caps is $3580. Write the equation that models this cost function.

51. **Depreciation** Suppose the cost of a business property is $1,920,000 and a company depreciates it with the straight-line method. If V is the value of the property after x years and the line representing the value as a function of years passes through the points $(10, 1,310,000)$ and $(20, 700,000)$, write the equation that gives the annual value of the property.

52. **Depreciation** Suppose the cost of a business property is $860,000 and a company depreciates it with the straight-line method. Suppose y is the value of the property after t years.

    a. What is the value at the beginning of the depreciation (when $t = 0$)?

    b. If the property is completely depreciated ($y = 0$) in 25 years, write the equation of the line representing the value as a function of years.

53. **Social Security** Monthly social security benefits differ with the age a recipient decides to start receiving benefits. Recipients whose monthly benefits are $2000, if benefits start being paid at a full retirement age of 66, can have their benefits increase by $160 per month for each year after 66 they delay starting their benefits. Note that the monthly benefit does not increase if the starting age is after 70.

a. Write the linear function that models the monthly benefit y as a function of age x at which benefits are started, for the starting ages from 66 to 70.

b. What monthly benefit would these recipients receive if they started receiving benefits at age 70?
(Source: Social Security Administration)

54. *Consumer Price Index* Goods and services that cost $100 in 2012 cost $108.6 in 2016, so we say that the U.S. consumer price index (CPI) is 100 in 2012 and 108.6 in 2016.

a. Write the equation of the linear function y that represents the CPI as a function of the year x.

b. Use the equation from part (a) to predict the CPI in 2020.
(Source: Social Security Administration)

55. *Drinking and Driving* The following table gives the number of drinks and the resulting blood alcohol percent for a 100-lb woman legally considered to be driving under the influence (DUI). (One drink is equal to 1.25 oz of 80-proof liquor, 12 oz of regular beer, or 5 oz of table wine; many states have set 0.08% as the legal limit for driving under the influence.)

a. The average rate of change of the blood alcohol percent with respect to the number of drinks is a constant. What is it?

b. Use the rate of change and one point determined by a number of drinks and the resulting blood alcohol percent to write the equation of a linear model for this data.

Number of Drinks	2	3	4	5	6	7
Blood Alcohol Percent	0.09	0.135	0.18	0.225	0.27	0.315

(Source: Pennsylvania Liquor Control Board)

56. *Blood Alcohol Percent* The table below gives the number of drinks and the resulting blood alcohol percent for a 220-lb man.

a. The rate of change in blood alcohol percent per drink for a 220-lb man is a constant. What is it?

b. Write the equation of the function that models the blood alcohol percent as a function of the number of drinks.

Number of Drinks	0	1	2	3	4
Blood Alcohol Percent	0	0.017	0.034	0.051	0.068

Number of Drinks	5	6	7	8	9
Blood Alcohol Percent	0.085	0.102	0.119	0.136	0.153

(Source: Pennsylvania Liquor Control Board)

57. *Men in the Workforce* The number of men in the workforce (in millions) for selected decades from 1950 to 2050 is shown in the figure. The decade is defined by the year at the beginning of the decade, and $g(t)$ is defined by the average number of men (in millions) in the workforce during the decade (indicated by the point on the graph within the decade). The data can be approximated by the linear model determined by the line connecting $(1950, 43.8)$ and $(2050, 100.3)$.

a. Write the equation of the line connecting these two points to find a linear model for the data.

b. Does this line appear to be a reasonable fit to the data points?

c. How does the slope of this line compare with the average rate of change in the function during this period?

(Source: U.S. Bureau of Labor Statistics)

58. *World Population* The figure shows the real and projected total population w, in billions, of the world for selected years t from 1950 through 2050.

a. Write the equation of the line connecting the points $(1950, 2.556)$ and $(2050, 9.346)$.

b. Does this line appear to be a reasonable fit to the data points?

c. Interpret the slope of this line as a rate of change of the world population.

d. Use the equation from part (a) to predict what the world population will be in 2025.
(Source: U.S. Census Bureau)

59. *Obesity* In 2000, 22.1% of Americans were obese, and the percent is projected to be 42.2% in 2030. Assuming that the growth in obesity is linear:

   a. Find the linear function that models this growth, with *x* representing the year and *p* representing the percent.

   b. Use the function to predict what percent of Americans will be obese in 2020.

(Source: *American Journal of Preventive Medicine*)

60. *Pay TV* The number of U.S. households with pay TV from 2010 and projected to 2017 is shown in the graph.

   a. Would the line connecting the data points for 2010 and 2011 be increasing or decreasing? What is its slope?

   b. Write the equation of the line connecting the data points for 2012 and 2017, with *x* equal to the year and *y* equal to the projected number of millions of U.S. households with pay TV.

   c. Does this indicate that the number of U.S. households with pay TV is increasing or decreasing over the period 2012–2017?

   d. What does this equation predict the number of U.S. households with pay TV will be in 2020?

Years	Pay-TV Households (millions)
2010	99.0
2011	100.9
2012	100.8
2013	99.3
2014	98.0
2015	97.2
2016	96.0
2017	94.6

(Source: Data from gigaom.com)

61. *Women in the Workforce* The number of women in the workforce, based on data and projections from 1950 to 2050, can be modeled by a linear function. The number was 18.4 million in 1950 and is projected to be 81.6 million in 2030. Let *x* represent the number of years after 1950.

   a. What is the slope of the line through $(0, 18.4)$ and $(80, 81.6)$?

   b. What is the average rate of change in the number of women in the workforce during this time period?

   c. Use the slope from part (a) and the number of millions of women in the workforce in 1950 to write the equation of the line.

(Source: U.S. Department of Labor)

62. *Teen Birth Rate* The birth rate in 2013 for U.S. girls ages 15 to 19 was the lowest since the government began tracking the statistic in 1940. The highest birth rate for U.S. teens was 96.3 per 1000 girls in 1957. In 2013, the birth rate was lowest at 26.3. Let *x* equal the number of years after 1950.

   a. What is the slope of the line joining the points $(7, 96.3)$ and $(63, 26.3)$?

   b. What is the average rate of change in the birth rate over this period?

   c. What does this average rate of change tell about the birth rate?

   d. Use the slope from part (a) and the birth rate from 2013 to write the equation of the line. Let *x* represent the number of years after 1950.

(Source: Centers for Disease Control and Prevention)

63. *Prison Population* The graph showing the total number of prisoners in state and federal prisons for the years 1960 through 2012 is shown in the figure. There were 212,953 prisoners in 1960 and 1,570,397 in 2012.

   a. What is the average rate of growth of the prison population from 1960 to 2012? Give the rate to the nearest integer.

   b. What is the slope, to the nearest integer, of the line connecting the points satisfying the conditions above?

   c. Use the slope from (b) and the 2012 point to write the equation of the secant line joining these two points on the curve.

   d. Does the secant line from part (c) give a good model of the data shown in the graph?

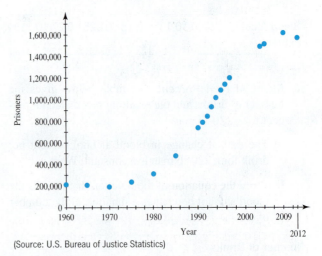

(Source: U.S. Bureau of Justice Statistics)

64. *Investment* The graph of the future value of an investment of $1000 for *x* years earning interest at a rate of 8% compounded continuously is shown in the following figure. The $1000 investment is worth approximately $1083 after 1 year and about $1492 after 5 years.

a. What is the average rate of change in the future value over the 4-year period?

b. Interpret this average rate of change.

c. What is the slope of the line connecting the points satisfying the conditions above?

d. Write the equation of the secant line joining the two given points.

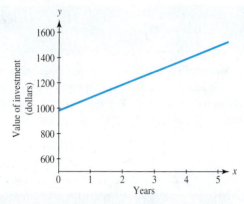

65. *Working Age* The scatter plot below projects the ratio of the working-age population to the elderly.

a. Do the data appear to fit a linear function?

b. The data points shown in the scatter plot from 2010 to 2030 are projections made from a mathematical model. Do those projections appear to be made with a linear model? Explain.

c. If the ratio is 3.9 in 2010 and projected to be 2.2 in 2030, what is the average annual rate of change of the data over this period of time?

d. Write the equation of the line joining these two points.

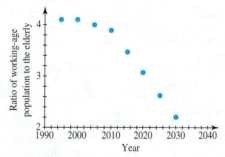

(Source: *Newsweek*)

66. *Women in the Workforce* The number of women in the workforce (in millions) for selected years from 1890 to 2010 is shown in the following figure.

a. Would the data in the scatter plot be modeled well by a linear function? Why or why not?

b. The number of women in the workforce was 10.519 million in 1930 and 16.443 million in 1950. What is the average rate of change in the number of women in the workforce during this period?

c. If the number of women in the workforce was 16.443 million in 1950 and 75.500 million in 2010, what is the average rate of change in the number of women in the workforce during this period?

d. Is it reasonable that these two average rates of change are different? How can you tell this from the graph?

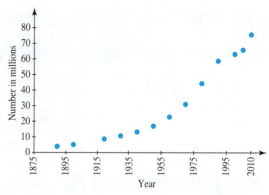

(Source: *Newsweek*)

67. *U.S. Population* The number of White non-Hispanic individuals in the U.S. civilian non-institutional population 16 years and older was 153.1 million in 2000 and is projected to be 169.4 million in 2050.
(Source: U.S. Census Bureau)

a. Find the average annual rate of change in population during the period 2000–2050, with the appropriate units.

b. Use the slope from part (a) and the population in 2000 to write the equation of the line associated with 2000 and 2050.

c. What does this model project the population to be in 2020?

68. *Social Agency* A social agency provides emergency food and shelter to two groups of clients. The first group has x clients who need an average of $300 for emergencies, and the second group has y clients who need an average of $200 for emergencies. The agency has $100,000 to spend for these two groups.

a. Write the equation that gives the number of clients who can be served in each group.

b. Find the y-intercept and the slope of the graph of this equation. Interpret each value.

c. If 10 clients are added from the first group, what happens to the number served in the second group?

69. *Average Velocity* A ball is shot straight upward, with its height, in feet, after t seconds given by the function $f(t) = -16t^2 + 128t$. Find the average velocity of the ball from $t = 2$ to $t = 4$.

70. *Average Velocity* A ball is shot straight upward, with its height, in feet, after t seconds given by the function $f(t) = 32t - 16t^2$. Find the average velocity of the ball from $t = 2$ to $t = 3$.

71. *Velocity* A ball is shot straight upward, with its height, in feet, after t seconds given by the function $f(t) = -16t^2 + 128t$. Find the difference quotient

to find the average velocity of the ball from $t = 2$ to $t = 2 + h$ seconds, with $h \neq 0$.

72. *Velocity* A ball is shot straight upward, with its height, in feet, after t seconds given by the function $f(t) = 112t - 16t^2$. Find the difference quotient to find the average velocity of the ball from $t = 3$ to $t = 3 + h$ seconds, with $h \neq 0$.

chapter 1 SUMMARY

In this chapter, we studied the basic concept of a function and how to recognize functions with tables, graphs, and equations. We studied linear functions in particular—graphing, finding slope, model formation, and equation writing. We used graphing utilities to graph functions and to evaluate functions. We solved business and economics problems involving linear functions.

Key Concepts and Formulas

1.1 Functions and Models

Scatter plot of data	A scatter plot is a graph of pairs of values that represent real-world information collected in numerical form.
Function definition	A function is a rule or correspondence that determines exactly one output for each input. The function may be defined by a set of ordered pairs, a table, a graph, or an equation.
Domain and range	The set of possible inputs for a function is called its domain, and the set of possible outputs is called its range. In general, if the domain of a function is not specified, we assume that it includes all real numbers except • values that result in a denominator of 0 • values that result in an even root of a negative number
Independent variable and dependent variable	If x represents any element in the domain, then x is called the independent variable; if y represents an output of the function from an input x, then y is called the dependent variable.
Tests for functions	We can test for a function graphically, numerically, and analytically.
Vertical line test	A set of points in a coordinate plane is the graph of a function if and only if no vertical line intersects the graph in more than one point.
Function notation	We denote that y is a function of x using function notation when we write $y = f(x)$.
Modeling	The process of translating real-world information into a mathematical form so that it can be applied and then interpreted in the real-world setting is called modeling.
Mathematical model	A mathematical model is a functional relationship (usually in the form of an equation) that includes not only the function rule but also descriptions of all involved variables and their units of measure.

1.2 Graphs of Functions

Point-plotting method	The point-plotting method of sketching a graph involves sketching the graph of a function by plotting enough points to determine the shape of the graph and then drawing a smooth curve through the points.

Complete graph	A graph is a complete graph if it shows the basic shape of the graph and important points on the graph (including points where the graph crosses the axes and points where the graph turns) and suggests what the unseen portions of the graph will be.
Using a calculator to draw a graph	After writing the function with x representing the independent variable and y representing the dependent variable, enter the function in the equation editor of the graphing utility. Activate the graph with (ZOOM) or (GRAPH). (Set the x- and y-boundaries of the viewing window before pressing (GRAPH).)
Viewing windows	The values that define the viewing window can be set manually or by using the (ZOOM) keys. The boundaries of a viewing window are x_{min}: the smallest value on the x-axis (the left boundary of the window) y_{min}: the smallest value on the y-axis (the bottom boundary of the window) x_{max}: the largest value on the x-axis (the right boundary of the window) y_{max}: the largest value on the y-axis (the top boundary of the window)
Spreadsheets	Spreadsheets like Excel can be used to create accurate graphs, sometimes better looking than those created with graphing calculators.
Aligning data	Using aligned inputs (input values that have been shifted horizontally to smaller numbers), rather than the actual data values, results in smaller coefficients in models and less involved computations.
Evaluating functions with a calculator	We can find (or estimate) the output of a function $y = f(x)$ at specific inputs with a graphing calculator by using (TRACE) and moving the cursor to (or close to) the value of the independent variable.
Graphing data points	Graphing utilities can be used to create lists of numbers and to create graphs of the data stored in the lists.
Scaling data	Replacing the outputs of a set of data by dividing or multiplying each number by a given number results in models with less involved computations.

1.3 Linear Functions

Linear functions	A linear function is a function of the form $f(x) = ax + b$, where a and b are constants. The graph of a linear function is a line.
Intercepts	A point (or the x-coordinate of the point) where a graph crosses or touches the horizontal axis is called an x-intercept, and a point (or the y-coordinate of the point) where a graph crosses or touches the vertical axis is called a y-intercept. To find the x-intercept(s) of the graph of an equation, set $y = 0$ in the equation and solve for x. To find the y-intercept(s), set $x = 0$ and solve for y.
Slope of a line	The slope of a line is a measure of its steepness and direction. The slope is defined as $$\text{slope} = \frac{\text{vertical change}}{\text{horizontal change}} = \frac{\text{rise}}{\text{run}}$$ If (x_1, y_1) and (x_2, y_2) are two points on a line, then the slope of the line is $$m = \frac{y_2 - y_1}{x_2 - x_1}$$
Slope and y-intercept of a line	The slope of the graph of the equation $y = mx + b$ is m, and the y-intercept of the graph is b. (This is called slope-intercept form.)

Constant rate of change	If a model is linear, the rate of change of the outputs with respect to the inputs will be constant, and the rate of change equals the slope of the line that is the graph of a linear function. Thus, we can determine whether a linear model fits a set of real data by determining whether the rate of change remains constant for the data.
Revenue, cost, and profit	If a company sells x units of a product for p dollars per unit, then the total revenue for this product can be modeled by the linear function $R(x) = px$. $$\text{Profit} = \text{Revenue} - \text{Cost}, \quad \text{or} \quad P(x) = R(x) - C(x)$$
Marginal cost, marginal revenue, and marginal profit	For total cost, revenue, and profit functions that are linear, the rates of change are called marginal cost, marginal revenue, and marginal profit, respectively.
Special linear functions	
• **Constant function**	A special linear function that has the form $y = c$, where c is a real number, is called a constant function.
• **Identity function**	The identity function $y = x$ is a linear function of the form $y = mx + b$ with slope $m = 1$ and y-intercept $b = 0$.

1.4 Equations of Lines

Writing equations of lines	
• **Slope-intercept form**	The slope-intercept form of the equation of a line with slope m and y-intercept b is $$y = mx + b$$
• **Point-slope form**	The point-slope form of the equation of the line with slope m and passing through a known point (x_1, y_1) is $$y - y_1 = m(x - x_1)$$
• **Horizontal line**	The equation of a horizontal line is $y = b$, where b is a constant.
• **Vertical line**	The equation of a vertical line is $x = a$, where a is a constant.
• **General form**	The general form of the equation of a line is $ax + by = c$, where a, b, and c are constants.
Parallel and perpendicular lines	Two distinct nonvertical lines are parallel if they have the same slope. Two nonvertical and nonhorizontal lines are perpendicular if their slopes are negative reciprocals.
Average rate of change	The average rate of change of a quantity over an interval describes how a change in the output of a function describing that quantity $f(x)$ is related to a change in the input x over that interval. The average rate of change of $f(x)$ with respect to x over the interval from $x = a$ to $x = b$ (where $a < b$) is calculated as $$\text{average rate of change} = \frac{\text{change in } f(x) \text{ values}}{\text{corresponding change in } x \text{ values}} = \frac{f(b) - f(a)}{b - a}$$
Secant line	When a function is not linear, the average rate of change between two points is the slope of the line joining two points on the curve, which is called a secant line.
Difference quotient	The average rate of change of a function from a point $(x, f(x))$ to the point $(x + h, f(x + h))$ is the difference quotient $\dfrac{f(x + h) - f(x)}{h}$, for $h \neq 0$.
Approximately linear data	Some sets of real data lie sufficiently close to a line that two points can be used to create a linear function that models the data.

chapter 1   SKILLS CHECK

Use the values in the table below in Exercises 1–4.

x	−3	−1	1	3	5	7	9	11	13
y	9	6	3	0	−3	−6	−9	−12	−15

1. Explain why the relationship shown by the table describes y as a function of x.

2. State the domain and range of the function.

3. If the function defined by the table is denoted by f, so that $y = f(x)$, what is $f(3)$?

4. Do the outputs in this table indicate that a linear function fits the data? If so, write the equation of the line.

5. If $C(s) = 16 - 2s^2$, find

   a. $C(3)$ b. $C(-2)$ c. $C(-1)$

6. For each of the functions $y = f(x)$ described below, find $f(-3)$.

   a.

   b.

7. Graph the function $f(x) = -2x^3 + 5x$.

8. Graph the function $y = 3x^2$.

9. Graph $y = -10x^2 + 400x + 10$ on a standard window and on a window with $x_{\min} = 0$, $x_{\max} = 40$, $y_{\min} = 0$, $y_{\max} = 5000$. Which window gives a better view of the graph of the function?

10. Use a graphing utility to graph the points (x, y) from the table.

x	10	15	20	25	30	35	40
y	−8	−6	−3	0	5	8	10

11. Find the domain of each function.

    a. $y = \sqrt{2x - 8}$ b. $f(x) = \dfrac{x + 2}{x - 6}$

12. One line passes through the points $(-12, 16)$ and $(-1, 38)$, and a second line has equation $2y + x = 23$. Are the lines parallel, perpendicular, or neither?

13. A line passes through $(-1, 4)$ and $(5, -3)$. Find the slope of a line parallel to this line and the slope of a line perpendicular to this line.

14. Find the slope of the line through $(-4, 6)$ and $(8, -16)$.

15. Given the equation $2x - 3y = 12$, (a) find the x- and y-intercepts of the graph and (b) graph the equation.

16. What is the slope of the graph of the function given in Exercise 15?

17. Find the slope and y-intercept of the graph of $y = -6x + 3$.

18. Find the rate of change of the function whose equation is given in Exercise 17.

19. Write the equation of the line that has slope $\dfrac{1}{3}$ and y-intercept 3.

20. Write the equation of a line that has slope $\dfrac{-3}{4}$ and passes through $(4, -6)$.

21. Write the equation of the line that passes through $(-1, 3)$ and $(2, 6)$.

22. For the function $y = x^2$, compute the average rate of change between $x = 0$ and $x = 3$.

For the functions given in Exercises 23 and 24, find

   a. $f(x + h)$
   b. $f(x + h) - f(x)$
   c. $\dfrac{f(x + h) - f(x)}{h}$, for $h \neq 0$.

23. $f(x) = 5 - 4x$ 24. $f(x) = 10x - 50$

chapter 1   REVIEW

25. **Voters** It wasn't until Harry Truman garnered 77% of the black vote in 1948 that a majority of blacks reported that they considered themselves Democrats. The table gives the percent p of black voters who supported Democratic candidates for president for the years 1960–2012.

    a. Is the percent p a function of the year y?

    b. Let $p = f(y)$ denote that p is a function of y. Find $f(1992)$ and explain what it means.

    c. What is y if $f(y) = 94$? What does this mean?

Year	Democrat (%)	GOP (%)
1960	68	32
1964	94	6
1968	85	15
1972	87	13
1976	85	15
1980	86	12
1984	89	9
1992	82	11
1996	84	12
2000	93	7
2004	74	26
2008	95	5
2012	93	6

(Source: Joint Center for Political and Economic Studies)

26. **Voters**
    a. What is the domain of the function defined by the table in Exercise 25?

    b. Is this function defined for 1982? Why is this value not included in the table?

27. **Voters** Graph the function defined by the table in Exercise 25 on the window [1950, 2015] by [60, 100].

28. **Voters**
    a. Find the slope of the line joining points (1972, 87) and (2012, 93).

    b. Use the answer from part (a) to find the average annual rate of change of the percent of black voters who voted for Democratic candidates for president between the years 1972 and 2012, inclusive.

    c. Is the average annual rate of change from 1972 to 1984 equal to the average annual rate of change from 1972 to 2012?

When money is borrowed to purchase an automobile, the amount borrowed A determines the monthly payment P. In particular, if a dealership offers a 5-year loan at 2.9% interest, then the amount borrowed for the car determines the payment according to the following table. Use the table to define the function $P = f(A)$ in Exercises 29–31.

Amount Borrowed ($)	Monthly Payment ($)
10,000	179.25
15,000	268.87
20,000	358.49
25,000	448.11
30,000	537.73

(Source: Sky Financial)

29. **Car Loans**
    a. Explain why the monthly payment P is a linear function of the amount borrowed A.

    b. Find $f(25,000)$ and interpret its meaning.

    c. If $f(A) = 358.49$, what is A?

30. **Car Loans**
    a. What are the domain and range of the function $P = f(A)$ defined by the table?

    b. Is this function $P = f(A)$ defined for a $12,000 loan?

31. **Car Loans** The equation of the line that fits the data points in the table is $f(A) = 0.017924A + 0.01$.

    a. Use the equation to find $f(28,000)$ and explain what it means.

    b. Can the function f be used to find the monthly payment for any dollar amount A of a loan if the interest rate and length of loan are unchanged?

32. **Life Expectancy** The figure gives the number of years the average woman is estimated to live beyond age 65 during selected years between 1950 and 2030. Let x represent the year and let y represent the expected number of years a woman will live past age 65. Write $y = f(x)$ and answer the following questions:

    a. What is $f(1960)$ and what does it mean?

    b. What is the life expectancy for the average woman in 2010?

c. In what year was the average woman expected to live 19 years past age 65?

Women
(Source: National Center for Health Statistics)

33. **Life Expectancy** The figure gives the number of years the average man is estimated to live beyond age 65 during selected years between 1950 and 2030. Let x represent the year and let y represent the expected number of years a man will live past age 65. Write $y = g(x)$ and answer the following questions:

a. What is $g(2020)$ and what does it mean?

b. What was the life expectancy for the average man in 1950?

c. Write a function expression that indicates that the average man in 1990 has a life expectancy of 80 years.

Men
(Source: National Center for Health Statistics)

34. **Marijuana Use** The function that models the percent of high school seniors using marijuana once or more in the prior 12 months is $y = 0.0123x^3 - 0.497x^2 + 5.95x + 14.9$, where x equals the number of years after 1990. Is this a linear function? Why or why not? (*Source:* Monitoringthefuture.com)

35. **Heart Rate** A Paramount elliptical machine provides the following table, which gives the desired heart rate during exercise for cardiac health based on the age of the person exercising.

a. Use the points corresponding to age 20 and age 90 to write the equation of the line connecting them. Let t represent age and h represent heart rate.

b. Does every data point in the table fit the graph of this equation?

c. Can the data in the table be expressed exactly as this linear function?

Age	Heart Rate (beats per minute)	Age	Heart Rate (beats per minute)
20	160	60	128
30	152	70	120
40	144	80	112
50	136	90	104

(Source: Paramount fitness.com)

36. **Crude Oil** The U.S. crude oil production, in billions of barrels, was 1.94 in the year 2010 and is projected to be 2.26 in 2030.

a. Find the projected average rate of change in production from 2010 through 2030.

b. Use the rate of change from part (a) and the point corresponding to 2010 to write the equation of the line that joins the points (2010, 1.94) and (2030, 2.26), with x equal to the year 2010 and y equal to the number of billions of barrels of crude oil. (Source: U.S. Energy Information Administration)

37. **Fuel** The table below shows data for the number of gallons of gas purchased each day x of a certain week by the 150 taxis owned by the Inner City Transportation Taxi Company. Write the equation that models the data.

Days from First Day	0	1	2	3	4	5	6
Gas Used (gal)	4500	4500	4500	4500	4500	4500	4500

38. **Work Hours** The average weekly hours worked by employees at PriceLo Company are given by the data in the table below.

Month and Year	Average Weekly Hours	Month and Year	Average Weekly Hours
June 2016	33.8	Oct. 2016	33.8
July 2016	33.8	Nov. 2016	33.8
Aug. 2016	33.8	Dec. 2016	33.8
Sept. 2016	33.8	Jan. 2017	33.8
		Feb. 2017	33.8

a. Write the equation of a function that describes the average weekly hours using an input x equal to the number of months past May 2016.

b. What type of function is this?

39. *Revenue* A company has revenue given by $R(x) = 564x$ dollars and total costs given by $C(x) = 40,000 + 64x$ dollars, where x is the number of units produced and sold.

    a. What is the revenue when 120 units are produced?

    b. What is the cost when 120 units are produced?

    c. What is the marginal cost and what is the marginal revenue for this product?

    d. What is the slope of the graph of $C(x) = 40,000 + 64x$?

    e. Graph $R(x)$ and $C(x)$ on the same set of axes.

40. *Profit* A company has revenue given by $R(x) = 564x$ dollars and total cost given by $C(x) = 40,000 + 64x$ dollars, where x is the number of units produced and sold. The profit can be found by forming the function $P(x) = R(x) - C(x)$.

    a. Write the profit function.

    b. Find the profit when 120 units are produced and sold.

    c. How many units give break-even?

    d. What is the marginal profit for this product?

    e. How is the marginal profit related to the marginal revenue and the marginal cost?

41. *Depreciation* A business property can be depreciated for tax purposes by using the formula $y + 3000x = 300,000$, where y is the value of the property x years after it was purchased.

    a. Find the y-intercept of the graph of this function. Interpret this value.

    b. Find the x-intercept. Interpret this value.

42. *Marginal Profit* A company has determined that its profit for a product can be described by a linear function. The profit from the production and sale of 150 units is $455, and the profit from 250 units is $895.

    a. What is the average rate of change of the profit for this product when between 150 and 250 units are sold?

    b. What is the slope of the graph of this profit function?

    c. Write the equation of the profit function for this product.

    d. What is the marginal profit for this product?

    e. How many units give break-even for this product?

Group Activities
▶ EXTENDED APPLICATIONS

1. Body Mass Index

Obesity is a risk factor for the development of medical problems, including high blood pressure, high cholesterol, heart disease, and diabetes. Of course, how much a person can safely weigh depends on his or her height. One way of comparing weights that account for height is the *body mass index (BMI)*. The table on the next page gives the BMI for a variety of heights and weights of people. Roche Pharmaceuticals state that a BMI of 30 or greater can create an increased risk of developing medical problems associated with obesity.

Describe how to assist a group of people in using the information. Some things you would want to include in your description follow.

1. How a person uses the table to determine his or her BMI.

2. How a person determines the weight that will put him or her at medical risk.

3. How a person whose weight or height is not in the table can determine whether his or her BMI is 30. To answer this question, develop and explain a formula to find the weight that would give a BMI of 30 for a person of a given height:

    a. Pick the points from the table that correspond to a BMI of 30 and create a table of these heights and weights. Change the heights to inches to simplify the data.

    b. Create a scatter plot of the data.

    c. Using the two points with the smallest and largest heights, write a linear equation that models the data.

    d. Graph the linear equation from part (c) with the scatter plot and discuss the fit.

    e. Explain how to use the model to test for obesity.

Body Mass Index for Specified Height (ft/in.) and Weight (lb)

Height/Weight	120	130	140	150	160	170	180	190	200	210	220	230	240	250
5'0"	23	25	27	29	31	33	35	37	39	41	43	45	47	49
5'1"	23	25	27	28	30	32	34	36	38	40	42	44	45	47
5'2"	22	24	26	27	29	31	33	35	37	38	40	42	44	46
5'3"	21	23	25	27	28	30	32	34	36	37	39	41	43	44
5'4"	21	22	24	26	28	29	31	33	34	36	38	40	41	43
5'5"	20	22	23	25	27	28	30	32	33	35	37	38	40	42
5'6"	19	21	23	24	26	27	29	31	32	34	36	37	39	40
5'7"	19	20	23	24	25	27	28	30	31	33	35	36	38	39
5'8"	18	20	21	23	24	26	27	29	30	32	34	35	37	38
5'9"	18	19	21	22	24	25	27	28	30	31	33	34	36	37
5'10"	17	19	20	22	23	24	25	27	28	29	31	33	35	36
5'11"	17	18	20	21	22	24	25	27	28	29	31	32	34	35
6'0"	16	18	19	20	22	23	24	26	27	29	30	31	33	34
6'1"	16	17	19	20	21	22	24	25	26	28	29	30	32	33

(Source: Roche Pharmaceuticals)

2. Total Revenue, Total Cost, and Profit

The total revenue is the amount a company receives from the sales of its products. It can be found by multiplying the selling price per unit times the number of units sold. That is, the revenue is

$$R(x) = p \cdot x$$

where p is the price per unit and x is the number of units sold.

The total cost comprises two costs: the fixed costs and the variable costs. Fixed costs (FC), such as depreciation, rent, and utilities, remain constant regardless of the number of units produced. Variable costs (VC) are those directly related to the number of units produced. The variable cost is the cost per unit (c) times the number of units produced ($VC = c \cdot x$), and the fixed cost is constant ($FC = k$), so the total cost is found by using the equation

$$C(x) = c \cdot x + k$$

where c is the cost per unit and x is the number of units produced.

Also, as discussed in Section 1.3, the profit a company makes on x units of a product is the difference between the revenue and the cost from the production and sale of x units.

$$P(x) = R(x) - C(x)$$

Suppose a company manufactures MP3 players and sells them to retailers for $98 each. It has fixed costs of $262,500 related to the production of the MP3 players, and the cost per unit for production is $23.

1. What is the total revenue function?
2. What is the marginal revenue for this product?
3. What is the total cost function?
4. What is the marginal cost for this product? Is the marginal cost equal to the variable cost or the fixed cost?
5. What is the profit function for this product? What is the marginal profit?
6. What are the cost, revenue, and profit if 0 units are produced?
7. Graph the total cost and total revenue functions on the same axes and estimate where the graphs intersect.
8. Graph the profit function and estimate where the graph intersects the x-axis.
9. What do the points of intersection in Question 7 and Question 8 give?

2

Linear Models, Equations, and Inequalities

Data from the Social Security Administration can be used to show how the monthly Social Security benefits differ with the age at which a recipient decides to start receiving benefits. These data can be used to create linear equations that give the benefit as a function of the age of the recipient. We can also find when the manufacturing sector of China reaches and surpasses that of the United States, by finding linear models and solving them simultaneously. Market equilibrium occurs when the number of units of a product demanded equals the number of units supplied. In this chapter, we solve these and other problems by creating linear models, solving linear equations and inequalities, and solving systems of equations.

section	objectives	applications
2.1 Algebraic and Graphical Solution of Linear Equations	Find solutions algebraically; relate solutions, zeros, and x-intercepts; solve linear equations; find functional forms of equations in two variables; solve literal equations; apply direct variation	Prison sentence length, credit card debt, stock market, simple interest, blood alcohol percent
2.2 Fitting Lines to Data Points: Modeling Linear Functions	Fit lines to data points with linear regression; model with linear functions; apply linear models; evaluate goodness of fit	Earnings and gender, retirement, population of females, U.S. population
2.3 Systems of Linear Equations in Two Variables	Solve systems of linear equations in two variables with graphing, substitution, and elimination; find break-even and market equilibrium; find nonunique solutions if they exist	China's manufacturing, break-even, market equilibrium, investments, medication
2.4 Solutions of Linear Inequalities	Solve linear inequalities with analytical and graphical methods; solve double inequalities	Profit, body temperature, apparent temperature, course grades, expected prison sentences

Algebra TOOLBOX

KEY OBJECTIVES

- Use properties of equations to solve linear equations
- Determine if a linear equation is a conditional equation, an identity, or a contradiction
- Use properties of inequalities to solve linear inequalities

Properties of Equations

In this chapter, we will solve equations. To solve an equation means to find the value(s) of the variable(s) that makes the equation a true statement. For example, the equation $3x = 6$ is true when $x = 2$, so 2 is a solution of this equation. Equations of this type are sometimes called **conditional equations** because they are true only for certain values of the variable. Equations that are true for all values of the variables for which both sides are defined are called **identities**. For example, the equation $7x - 4x = 5x - 2x$ is an identity. Equations that are not true for any value of the variable are called **contradictions**. For example, the equation $2(x + 3) = 2x - 1$ is a contradiction.

We frequently can solve an equation for a given variable by rewriting the equation in an equivalent form whose solution is easy to find. Two equations are **equivalent** if and only if they have the same solutions. The following operations give equivalent equations:

Properties of Equations

1. **Addition Property** Adding the same number to both sides of an equation gives an equivalent equation. For example, $x - 5 = 2$ is equivalent to $x - 5 + 5 = 2 + 5$, or to $x = 7$.

2. **Subtraction Property** Subtracting the same number from both sides of an equation gives an equivalent equation. For example, $z + 12 = -9$ is equivalent to $z + 12 - 12 = -9 - 12$, or to $z = -21$.

3. **Multiplication Property** Multiplying both sides of an equation by the same nonzero number gives an equivalent equation. For example, $\dfrac{y}{6} = 5$ is equivalent to $6\left(\dfrac{y}{6}\right) = 6(5)$, or to $y = 30$.

4. **Division Property** Dividing both sides of an equation by the same nonzero number gives an equivalent equation. For example, $17x = -34$ is equivalent to $\dfrac{17x}{17} = \dfrac{-34}{17}$, or to $x = -2$.

5. **Substitution Property** The equation formed by substituting one expression for an equal expression is equivalent to the original equation. For example, if $y = 3x$, then $x + y = 8$ is equivalent to $x + 3x = 8$, so $4x = 8$ and $x = 2$.

| EXAMPLE 1 ▶ | **Properties of Equations** |

State the property (or properties) of equations that can be used to solve each of the following equations, and then use the property (or properties) to solve the equation.

a. $3x = 6$ **b.** $\dfrac{x}{5} = 12$ **c.** $3x - 5 = 17$ **d.** $4x - 5 = 7 + 2x$

SOLUTION

a. Division Property. Dividing both sides of the equation by 3 gives the solution to the equation.

$$3x = 6$$

$$\frac{3x}{3} = \frac{6}{3}$$

$$x = 2$$

Thus, $x = 2$ is the solution to the original equation.

b. Multiplication Property. Multiplying both sides of the equation by 5 gives the solution to the equation.

$$\frac{x}{5} = 12$$

$$5\left(\frac{x}{5}\right) = 5(12)$$

$$x = 60$$

Thus, $x = 60$ is the solution to the original equation.

c. Addition Property and Division Property. Adding 5 to both sides of the equation and dividing both sides by 3 gives an equivalent equation.

$$3x - 5 = 17$$

$$3x - 5 + 5 = 17 + 5$$

$$3x = 22$$

$$\frac{3x}{3} = \frac{22}{3}$$

$$x = \frac{22}{3}$$

Thus, $x = \dfrac{22}{3}$ is the solution to the original equation.

d. Addition Property, Subtraction Property, and Division Property. Adding 5 to both sides of the equation and subtracting $2x$ from both sides of the equation gives an equivalent equation.

$$4x - 5 = 7 + 2x$$

$$4x - 5 + 5 = 7 + 2x + 5$$

$$4x = 12 + 2x$$

$$4x - 2x = 12 + 2x - 2x$$

$$2x = 12$$

Dividing both sides by 2 gives the solution to the original equation.

$$\frac{2x}{2} = \frac{12}{2}$$

$$x = 6$$

Thus, $x = 6$ is the solution to the original equation.

EXAMPLE 2 ▶   Equations

Determine whether each equation is a conditional equation, an identity, or a contradiction.

a. $3(x - 5) = 2x - 7$ **b.** $5x - 6(x + 1) = -x - 9$

c. $-3(2x - 4) = 2x + 12 - 8x$

SOLUTION

a. To solve $3(x - 5) = 2x - 7$, we first use the Distributive Property, then the Addition Property.

$$3(x - 5) = 2x - 7$$
$$3x - 15 = 2x - 7$$
$$3x - 15 - 2x = 2x - 7 - 2x$$
$$x - 15 + 15 = -7 + 15$$
$$x = 8$$

Because this equation is true for only the value $x = 8$, it is a conditional equation.

b. To solve $5x - 6(x + 1) = -x - 9$, we first use the Distributive Property, then the Addition Property.

$$5x - 6(x + 1) = -x - 9$$
$$5x - 6x - 6 = -x - 9$$
$$-x - 6 = -x - 9$$
$$-x - 6 + x = -x - 9 + x$$
$$-6 = -9$$

Because $-6 = -9$ is false, the original equation is a contradiction.

c. To solve $-3(2x - 4) = 2x + 12 - 8x$, we first use the Distributive Property, then the Addition Property.

$$-3(2x - 4) = 2x + 12 - 8x$$
$$-6x + 12 = 12 - 6x$$
$$-6x + 12 + 6x = 12 - 6x + 6x$$
$$12 = 12$$

Because $12 = 12$ is a true statement, the equation is true for all real numbers and thus is an identity.

Properties of Inequalities

We will solve linear inequalities in this chapter. As with equations, we can find solutions to inequalities by finding equivalent inequalities from which the solutions can be easily seen. We use the following properties to reduce an inequality to a simple equivalent inequality.

Properties of Inequalities	Examples	
Substitution Property The inequality formed by substituting one expression for an equal expression is equivalent to the original inequality.	$7x - 6x \geq 8$ $x \geq 8$	$3(x - 1) < 8$ $3x - 3 < 8$
Addition and Subtraction Properties The inequality formed by adding the same quantity to (or subtracting the same quantity from) both sides of an inequality is equivalent to the original inequality.	$x - 6 < 12$ $x - 6 + 6 < 12 + 6$ $x < 18$	$3x > 13 + 2x$ $3x - 2x > 13 + 2x - 2x$ $x > 13$
Multiplication Property I The inequality formed by multiplying (or dividing) both sides of an inequality by the same *positive* quantity is equivalent to the original inequality.	$\frac{1}{3}x \leq 6$ $3\left(\frac{1}{3}x\right) \leq 3(6)$ $x \leq 18$	$3x > 6$ $\frac{3x}{3} > \frac{6}{3}$ $x > 2$
Multiplication Property II The inequality formed by multiplying (or dividing) both sides of an inequality by the same *negative* number and reversing the inequality symbol is equivalent to the original inequality.	$-x > 7$ $-1(-x) < -1(7)$ $x < -7$	$-4x \geq 12$ $\frac{-4x}{-4} \leq \frac{12}{-4}$ $x \leq -3$

Of course, these properties can be used in combination to solve an inequality. This means that the steps used to solve a linear inequality are the same as those used to solve a linear equation, except that the inequality symbol is reversed if both sides are multiplied (or divided) by a negative number.

EXAMPLE 3 ▶ ## Properties of Inequalities

State the property (or properties) of inequalities that can be used to solve each of the following inequalities, and then use the property (or properties) to solve the inequality.

a. $\dfrac{x}{2} \geq 13$ **b.** $-3x < 6$ **c.** $5 - \dfrac{x}{3} \geq -4$ **d.** $-2(x + 3) < 4$

SOLUTION

a. Multiplication Property I. Multiplying both sides by 2 gives the solution to the inequality.

$$\frac{x}{2} \geq 13$$

$$2\left(\frac{x}{2}\right) \geq 2(13)$$

$$x \geq 26$$

The solution to the inequality is $x \geq 26$.

b. Multiplication Property II. Dividing both sides by -3 and reversing the inequality symbol gives the solution to the inequality.

$$-3x < 6$$
$$\frac{-3x}{-3} > \frac{6}{-3}$$
$$x > -2$$

The solution to the inequality is $x > -2$.

c. Subtraction Property, Multiplication Property II. First we subtract 5 from both sides of the inequality.

$$5 - \frac{x}{3} \geq -4$$
$$5 - \frac{x}{3} - 5 \geq -4 - 5$$
$$-\frac{x}{3} \geq -9$$

Multiplying both sides by -3 and reversing the inequality symbol gives the solution to the inequality.

$$-\frac{x}{3} \geq -9$$
$$-3\left(-\frac{x}{3}\right) \leq -3(-9)$$
$$x \leq 27$$

The solution to the inequality is $x \leq 27$.

d. Substitution Property, Addition Property, Multiplication Property II. First we distribute -2 on the left side of the inequality.

$$-2(x + 3) < 4$$
$$-2x - 6 < 4$$

Adding 6 to both sides and then dividing both sides by -2 and reversing the inequality symbol gives the solution to the inequality.

$$-2x - 6 + 6 < 4 + 6$$
$$-2x < 10$$
$$\frac{-2x}{-2} > \frac{10}{-2}$$
$$x > -5$$

The solution to the inequality is $x > -5$.

Rounding Function Coefficients and Constants

In this chapter we will create linear functions that approximately fit given data points. When we use technology to create the models, we usually have to round the coefficients and constants to report (write) the model. Unless we are instructed otherwise, it will be common practice to report the functions with three decimal places. Recall the following:

Tenths means 0.1; hundredths means 0.01, thousandths means 0.001, ten-thousandths means 0.0001, and so on.

In some cases, however, we will be asked to report the functions with three or four digits of accuracy. These **significant digits** carry meaning that contributes to the precision of the number. In general, significant digits are those from the first nonzero digit on the left to the last digit *after* the decimal point, if there is a decimal point. If there is no decimal point, any zeros to the right of the nonzero digits are not significant. Note that

1. All nonzero digits are significant.

2. All zeroes between significant digits are significant.

3. All zeroes that are both to the right of the decimal point and to the right of all nonzero significant digits are themselves significant.

4. Leading zeroes are not significant.

Thus 0.0324, 56,300 and 30.0 all have three significant digits.

EXAMPLE 4 ▶ **Significant Digits**

Determine the number of significant digits in each of the following numbers.

a. 33,465.2 **b.** 0.004432 **c.** 5600 **d.** 2.0056 **e.** 22.370

SOLUTION

a. For the number 33,465.2, there are five nonzero digits to the left of the decimal point and one nonzero digit to the right of the decimal point, so there are six significant digits.

b. For the number 0.004432, there are three leading zeros, which are not significant, and four digits after those zeros, so there are four significant digits.

c. Since there is no decimal point, the zeros after the first two digits in the number 5600 are not significant, so there are two significant digits.

d. For the number 2.0056, there are three nonzero digits, and the zeros between nonzero digits are significant, so there are five significant digits.

e. For the number 22.370, the zero to the right of the decimal is significant, so there are five significant digits.

EXAMPLE 5 ▶ **Rounding Function Coefficients and Constants**

Rewrite the function $f(x) = 0.003457x^2 - 240.3669x + 360.71$ with the coefficients (and constant) rounded to:

a. three decimal places

b. three significant digits

c. four significant digits

SOLUTION

a. Rounded to three decimal places, the function $f(x) = 0.003457x^2 - 240.3669x + 360.71$ becomes $f(x) = 0.003x^2 - 240.367x + 360.710$.

b. The function $f(x) = 0.003457x^2 - 240.3669x + 360.71$ written with three significant digits is $f(x) = 0.00346x^2 - 240.0x + 361$.

c. The function $f(x) = 0.003457x^2 - 240.3669x + 360.71$ written with four significant digits is $f(x) = 0.003457x^2 - 240.4x + 360.7$.

Toolbox EXERCISES

In Exercises 1–8, state the property (or properties) of equations that can be used to solve each of the following equations; then use the property (or properties) to solve the equation.

1. $3x = 6$

2. $x - 7 = 11$

3. $x + 3 = 8$

4. $x - 5 = -2$

5. $\dfrac{x}{3} = 6$

6. $-5x = 10$

7. $2x + 8 = -12$

8. $\dfrac{x}{4} - 3 = 5$

Solve the equations in Exercises 9–16.

9. $4x - 3 = 6 + x$

10. $3x - 2 = 4 - 7x$

11. $\dfrac{3x}{4} = 12$

12. $\dfrac{5x}{2} = -10$

13. $3(x - 5) = -2x - 5$

14. $-2(3x - 1) = 4x - 8$

15. $2x - 7 = -4\left(4x - \dfrac{1}{2}\right)$

16. $-2(2x - 6) = 3\left(3x - \dfrac{1}{3}\right)$

In Exercises 17–20, use the Substitution Property of Equations to solve the equation.

17. Solve for x if $y = 2x$ and $x + y = 12$

18. Solve for x if $y = 4x$ and $x + y = 25$

19. Solve for x if $y = 3x$ and $2x + 4y = 42$

20. Solve for x if $y = 6x$ and $3x + 2y = 75$

In Exercises 21–24, determine whether the equation is a conditional equation, an identity, or a contradiction.

21. $3x - 5x = 2x + 7$

22. $3(x + 1) = 3x - 7$

23. $9x - 2(x - 5) = 3x + 10 + 4x$

24. $\dfrac{x}{2} - 5 = \dfrac{x}{4} + 2$

In Exercises 25–32, solve the inequalities.

25. $5x + 1 > -5$

26. $1 - 3x \geq 7$

27. $\dfrac{x}{4} > -3$

28. $\dfrac{x}{6} > -2$

29. $\dfrac{x}{4} - 2 > 5x$

30. $\dfrac{x}{2} + 3 > 6x$

31. $-3(x - 5) < -4$

32. $-\dfrac{1}{2}(x + 4) < 6$

In Exercises 33–38, determine the number of significant digits in each of the following numbers.

33. 52,865.3

34. 0.003389

35. 4700

36. 4.0092

37. 58.610

38. 0.000041

39. Rewrite the function

$$f(x) = 0.007543x^2 - 632.5778x + 480.65$$

with the coefficients (and constant) rounded to:

a. three decimal places

b. three significant digits

c. four significant digits

Algebraic and Graphical Solution of Linear Equations

KEY OBJECTIVES

- Solve linear equations algebraically

- Solve real-world application problems

- Compare solutions of equations with zeros and x-intercepts of graphs of functions

- Solve linear equations graphically using the x-intercept and intersection methods

- Solve literal equations for a specified variable

- Solve direct variation problems

SECTION PREVIEW   **Prison Sentences**

The average sentence length and the average time served in state prison for various crimes are shown in Table 2.1. These data can be used to create a linear function that is an approximate model for the data:

$$y = 0.55x - 2.886$$

Table 2.1

Average Sentence (months)	Average Time Served (months)
62	30
85	45
180	95
116	66
92	46
61	33
56	26

(Source: U.S. Department of Justice)

This function describes the mean time y served in prison for a crime as a function of the mean sentence length x, where x and y are each measured in months. Assuming that this model remains valid, it predicts that the mean time served on a 5-year (60 months) sentence is $0.55(60) - 2.886 \approx 30$ months, or approximately $2\frac{1}{2}$ years.

To find the sentence that would give an expected time served of 10 years (120 months), we use algebraic or graphical methods to solve the equation

$$120 = 0.55x - 2.866$$

for x. (See Example 6.) In this section, we use additional algebraic methods and graphical methods to solve linear equations in one variable. ∎

Algebraic Solution of Linear Equations

We can use algebraic, graphical, or a combination of algebraic and graphical methods to solve linear equations. Sometimes it is more convenient to use algebraic solution methods rather than graphical solution methods, especially if an exact solution is desired. The steps used to solve a linear equation in one variable algebraically follow.

Steps for Solving a Linear Equation in One Variable

1. If a linear equation contains fractions, multiply both sides of the equation by a number that will remove all denominators from the equation. If there are two or more fractions, use the least common denominator (LCD) of the fractions.

2. Perform any multiplications or divisions to remove any parentheses or other symbols of grouping.

3. Perform any additions or subtractions to get all terms containing the variable on one side and all other terms on the other side of the equation. Combine like terms.

4. Divide both sides of the equation by the coefficient of the variable.

5. Check the solution by substitution in the original equation. If a real-world solution is desired, check the algebraic solution for reasonableness in the real-world situation.

EXAMPLE 1 ▶ **Algebraic Solutions**

a. Solve for x: $\dfrac{2x - 3}{4} = \dfrac{x}{3} + 1$

b. Solve for y: $\dfrac{y}{5} - \dfrac{1}{2}\left(\dfrac{y}{2}\right) = \dfrac{7}{20}$

c. Solve for x if $y = 0.72$: $y = 1.312x - 2.56$

SOLUTION

a. $\dfrac{2x - 3}{4} = \dfrac{x}{3} + 1$

$12\left(\dfrac{2x - 3}{4}\right) = 12\left(\dfrac{x}{3} + 1\right)$ Multiply both sides by the LCD, 12.

$3(2x - 3) = 12\left(\dfrac{x}{3} + 1\right)$ Simplify the fraction $\dfrac{12}{4}$ on the left.

$6x - 9 = 4x + 12$ Remove parentheses using the distributive property.

$2x = 21$ Subtract $4x$ from both sides and add 9 to both sides.

$x = \dfrac{21}{2}$ Divide both sides by 2.

Check the result: $\dfrac{2\left(\dfrac{21}{2}\right) - 3}{4} \stackrel{?}{=} \dfrac{\dfrac{21}{2}}{3} + 1 \Rightarrow \dfrac{9}{2} = \dfrac{9}{2}$

b. $\dfrac{y}{5} - \dfrac{1}{2}\left(\dfrac{y}{2}\right) = \dfrac{7}{20}$

$\dfrac{y}{5} - \dfrac{y}{4} = \dfrac{7}{20}$ Remove parentheses first to find the LCD.

$20\left(\dfrac{y}{5} - \dfrac{y}{4}\right) = 20\left(\dfrac{7}{20}\right)$ Multiply both sides by the LCD, 20.

$4y - 5y = 7$ Remove parentheses using the distributive property.

$-y = 7$ Combine like terms.

$y = -7$ Multiply both sides by -1.

Check the result: $\dfrac{-7}{5} - \dfrac{1}{2}\left(\dfrac{-7}{2}\right) \stackrel{?}{=} \dfrac{7}{20} \Rightarrow \dfrac{-7}{5} + \dfrac{7}{4} \stackrel{?}{=} \dfrac{7}{20} \Rightarrow \dfrac{7}{20} = \dfrac{7}{20}$

c.
$$y = 1.312x - 2.56$$
$$0.72 = 1.312x - 2.56 \qquad \text{Substitute 0.72 for } y.$$
$$3.28 = 1.312x \qquad \text{Add 2.56 to both sides.}$$
$$x = 2.5 \qquad \text{Divide both sides by 1.312.}$$

Check the result: $\qquad 0.72 \overset{2}{=} 1.312(2.5) - 2.56 \Rightarrow 0.72 = 0.72$

As the next example illustrates, we solve an application problem that is set in a real-world context by using the same solution methods. However, you must remember to include units of measure with your answer and check that your answer makes sense in the problem situation.

EXAMPLE 2 ▶ Credit Card Debt

It is hard for some people to pay off credit card debt in a reasonable period of time because of high interest rates. The interest paid on a $10,000 debt over 3 years is approximated by

$$y = 175.393x - 116.287 \text{ dollars}$$

when the interest rate is x%. What is the interest rate if the interest is $1637.60? (Source: Consumer Federation of America)

SOLUTION

To answer this question, we solve the linear equation

$$1637.60 = 175.393x - 116.287$$
$$1753.887 = 175.393x$$
$$x = 9.9998$$

Thus, if the interest rate is approximately 10%, the interest is $1637.60. (Note that if you check the approximate answer, you are checking only for the reasonableness of the estimate.)

EXAMPLE 3 ▶ Stock Market

For a period of time, a man is very successful speculating on an Internet stock, with its value growing to $100,000. However, the stock drops rapidly until its value has been reduced by 40%. What percent increase will have to occur before the latest value returns to $100,000?

SOLUTION

The value of the stock after the 40% loss is $100,000 - 0.40(100,000) = 60,000$. To find the percent p of increase that is necessary to return the value to 100,000, we solve

$$60,000 + 60,000p = 100,000$$
$$60,000p = 40,000$$
$$p = \frac{40,000}{60,000} = \frac{2}{3} = 66\frac{2}{3}\%$$

Thus, the stock value must increase by $66\frac{2}{3}\%$ to return to a value of $100,000.

Solutions, Zeros, and *x*-Intercepts

Because *x*-intercepts are *x* values that make the output of the function equal to 0, these intercepts are also called **zeros** of the function.

> ### Zero of a Function
>
> Any number a for which $f(a) = 0$ is called a **zero** of the function $f(x)$. If a is real, a is an *x*-intercept of the graph of the function.

The zeros of the function are values that make the function equal to 0, so they are also solutions to the equation $f(x) = 0$.

> The following three concepts are numerically the same:
> The *x*-intercepts of the graph of $y = f(x)$
> The real zeros of the function f
> The real solutions to the equation $f(x) = 0$

The following example illustrates the relationships that exist among *x*-intercepts of the graph of a function, zeros of the function, and solutions to associated equations.

EXAMPLE 4 ▶ **Relationships Among x-Intercepts, Zeros, and Solutions**

For the function $f(x) = 13x - 39$, find

a. $f(3)$

b. The zero of $f(x) = 13x - 39$

c. The *x*-intercept of the graph of $y = 13x - 39$

d. The solution to the equation $13x - 39 = 0$

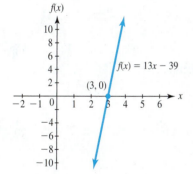

Figure 2.1

SOLUTION

a. Evaluate the output $f(3)$ by substituting the input 3 for x in the function:

$$f(3) = 13(3) - 39 = 0$$

b. Because $f(x) = 0$ when $x = 3$, we say that 3 is a zero of the function.

c. The *x*-intercept of a graph occurs at the value of x where $y = 0$, so the *x*-intercept is $x = 3$. The only *x*-intercept is $x = 3$ because the graph of $f(x) = 13x - 39$ is a line that crosses the *x*-axis at only one point (Figure 2.1).

d. The solution to the equation $13x - 39 = 0$ is $x = 3$ because $13x - 39 = 0$ gives $13x = 39$, or $x = 3$.

Figure 2.2

We can use ⬚TRACE⬚ or TABLE (accessed by ⬚2ND⬚ ⬚GRAPH⬚) on a graphing utility to check the reasonableness of a solution. The graph of $y = 13x - 39$ given in Figure 2.1 confirms that $x = 3$ is the *x*-intercept. We can also confirm that $f(3) = 0$ by using Excel. Figure 2.2 shows that when we enter 3 in cell A2 and the formula "$= 13 * A2 - 39$" in cell B2, the function $f(x) = 13x - 39$ is evaluated at $x = 3$, with a result of 0.*

---

* See Appendix B, page 686.

Graphical Solution of Linear Equations

We can also view the graph or use ⟨TRACE⟩ on a graphing utility to obtain a quick estimate of an answer. ⟨TRACE⟩ may or may not provide the exact solution to an equation, but
it will provide an estimate of the solution. If your calculator or computer has a graphical
or numerical solver, the solver can be used to obtain the solution to the equation
$f(x) = 0$. Recall that an x-intercept of the graph of $y = f(x)$, a real zero of $f(x)$, and
the real solution to the equation $f(x) = 0$ are all different names for the same input
value. If the graph of the linear function $y = f(x)$ is not a horizontal line, we can find
the one solution to the linear equation $f(x) = 0$ as described below. We call this solution method the **x-intercept method**.

> ### Solving a Linear Equation Using the x-Intercept Method with Graphing Utilities
>
> 1. Rewrite the equation to be solved with 0 (and nothing else) on one side of
>   the equation.
>
> 2. Enter the nonzero side of the equation found in the previous step in the
>   equation editor of your graphing utility and graph the line in an appropriate
>   viewing window. Be certain that you can see the line cross the horizontal
>   axis on your graph.
>
> 3. Find the x-intercept by using the zero command after pressing **2ND** ⟨TRACE⟩.
>   The x-intercept is the value of x that makes the equation equal to zero, so
>   it is the solution to the equation. The value of x displayed by this method
>   is sometimes a decimal approximation of the exact solution rather than the
>   exact solution. *Note:* Using the Frac command after pressing **MATH** will
>   often convert a decimal solution of a linear equation (approximated on the
>   display) to the exact solution.*

EXAMPLE 5 ▶   **Graphical Solution**

Solve $\dfrac{2x - 3}{4} = \dfrac{x}{3} + 1$ for x using the x-intercept method.

SOLUTION

To solve the equation using the x-intercept method, first rewrite the equation with 0 on
one side.

$$0 = \frac{x}{3} + 1 - \frac{2x - 3}{4}$$

Next, enter the right side of the equation,

$$\frac{x}{3} + 1 - \frac{2x - 3}{4}$$

in the equation editor of your graphing utility as

$$y_1 = (x/3) + 1 - (2x - 3)/4$$

and graph this function. Use parentheses as needed to preserve the order of operations.
You should obtain a graph similar to the one seen in Figure 2.3, which was obtained
using the viewing window $[-10, 15]$ by $[-4, 4]$. However, any graph in which you
can see the x-intercept will do. Using "zero" gives the x-intercept of the graph, 10.5.

Figure 2.3

---

* See Appendix A, page 670.

(See Figure 2.3.) This is the value of x that makes $y = 0$, and thus the original equation true, so it is the solution to this linear equation.

You can determine if $x = 10.5 = \dfrac{21}{2}$, found graphically, is the exact answer by substituting this value into the original equation. Both sides of the original equation are equal when $x = 10.5$, so it is the exact solution.

___

We can also use the intersection method described below to solve linear equations:

> ## Solving a Linear Equation Using the Intersection Method
> 1. Enter the left side of the equation as y_1 and the right side as y_2. Graph both of these equations on a window that shows their point of intersection.
> 2. Find the point of intersection with the intersect command after pressing **2ND** **TRACE**.* This is the point where $y_1 = y_2$. The x-value of this point is the value of x that makes the two sides of the equation equal, so it is the solution to the original equation.

Figure 2.4

The equation in Example 5,

$$\frac{2x - 3}{4} = \frac{x}{3} + 1$$

can be solved using the intersection method. Enter the left side as $y_1 = (2x - 3)/4$ and the right side as $y_2 = (x/3) + 1$. Graphing these two functions gives the graphs in Figure 2.4. The point of intersection is found to be $(10.5, 4.5)$. So the solution is $x = 10.5$ (as we found in Example 5).

EXAMPLE 6 ▶ Criminal Sentences

The function $y = 0.55x - 2.886$ describes the mean time y served in prison for a crime as a function of the mean sentence length x, where x and y are each measured in months. To find the sentence for a crime that would give an expected time served of 10 years, write an equation and solve it by using (a) the x-intercept method and (b) the intersection method.
(Source: U.S. Department of Justice)

SOLUTION

a. Note that 10 years is 120 months. We solve the linear equation $120 = 0.55x - 2.886$ by using the x-intercept method as follows:

$0 = 0.55x - 2.886 - 120$ Rewrite the equation in a form with 0 on one side.

$0 = 0.55x - 122.886$ Combine the constant terms.

Enter $y_1 = 0.55x - 122.886$ and graph this equation (Figure 2.5). Using "zero" gives an x-intercept of approximately 223 (Figure 2.5). Thus, if a prisoner receives a sentence of 223 months, we would expect him or her to serve 120 months, or 10 years.

Figure 2.5

___

* See Appendix A, page 670.

b. To use the intersection method, enter $y_1 = 0.55x - 2.886$ and $y_2 = 120$, graph these equations, and find the intersection of the lines with "intersect". Figure 2.6 shows the point of intersection, and again we see that x is approximately 223.

Figure 2.6

Spreadsheet ▶ SOLUTION   We can use the **Goal Seek** feature of Excel to find the x-intercept of the graph of a function—that is, the zeros of the function. To solve

$$120 = 0.55x - 2.886$$

we enter the nonzero side of $0 = 0.55x - 122.886$ in cell B2 as shown in Figure 2.7 and put any number in A2. We then select DATA, What-if-Analysis, and Goal Seek. In the Goal Seek menu, we select the target cell B2, set it to 0, and set the changing cell to A2. Figure 2.8 shows the solution in cell A2.*

Figure 2.7

Figure 2.8

Literal Equations; Solving an Equation for a Specified Linear Variable

An equation that contains two or more letters that represent constants or variables is called a **literal equation**. Formulas are examples of literal equations. If one of two or more of the variables in an equation is present only to the first power, we can solve for that variable by treating the other variables as constants and using the same steps that we used to solve a linear equation in one variable. This is often useful because it is necessary to get equations in the proper form to enter them in a graphing utility.

* For detailed steps, See Appendix B, page 692.

EXAMPLE 7 ▶   Simple Interest

The formula for the future value of an investment of P dollars at simple interest rate r for t years is $A = P(1 + rt)$. Solve the formula for r, the interest rate.

SOLUTION

$$A = P(1 + rt)$$

$$A = P + Prt \qquad \text{Multiply to remove parentheses.}$$

$$A - P = Prt \qquad \text{Get the term containing } r \text{ by itself on one side of the equation.}$$

$$r = \frac{A - P}{Pt} \qquad \text{Divide both sides by } Pt.$$

EXAMPLE 8 ▶   Solving an Equation for a Specified Variable

Solve the equation $2(2x - b) = \dfrac{5cx}{3}$ for x.

SOLUTION

We solve the equation for x by treating the other variables as constants:

$$2(2x - b) = \frac{5cx}{3}$$

$$6(2x - b) = 5cx \qquad \text{Clear the equation of fractions by multiplying by the LCD, 3.}$$
$$12x - 6b = 5cx \qquad \text{Multiply to remove parentheses.}$$

$$12x - 5cx = 6b \qquad \text{Get all terms containing } x \text{ on one side and all other terms on the other side.}$$

$$x(12 - 5c) = 6b \qquad \text{Factor } x \text{ from the expression. The remaining factor is the coefficient of } x.$$

$$\frac{x(12 - 5c)}{(12 - 5c)} = \frac{6b}{12 - 5c} \qquad \text{Divide both sides by the coefficient of } x.$$

$$x = \frac{6b}{12 - 5c}$$

EXAMPLE 9 ▶   Writing an Equation in Functional Form and Graphing Equations

Solve each of the following equations for y so that y is expressed as a function of x. Then graph the equation on a graphing utility with a standard viewing window.

a. $2x - 3y = 12$ **b.** $x^2 + 4y = 4$

SOLUTION

a. Because y is to the first power in the equation, we solve the equation for y using linear equation solution methods:

$$2x - 3y = 12$$

$$-3y = -2x + 12 \qquad \text{Isolate the term involving } y \text{ by subtracting } 2x \text{ from both sides.}$$

$$y = \frac{2x}{3} - 4 \qquad \text{Divide both sides by } -3, \text{ the coefficient of } y.$$

Figure 2.9

The graph of this equation is shown in Figure 2.9.

Figure 2.10

b. The variable y is to the first power in the equation

$$x^2 + 4y = 4$$

so we solve for y by using linear solution methods. (Note that to solve this equation for x would be more difficult; this method will be discussed in Chapter 3.)

$$4y = -x^2 + 4 \qquad \text{Isolate the term containing } y \text{ by subtracting } x^2 \text{ from both sides.}$$

$$y = \frac{-x^2}{4} + 1 \qquad \text{Divide both sides by 4, the coefficient of } y.$$

The graph of this equation is shown in Figure 2.10.

Direct Variation

Often in mathematics we need to express relationships between quantities. One relationship that is frequently used in applied mathematics occurs when two quantities are proportional. Two variables x and y are proportional to each other (or vary directly) if their quotient is a constant. That is, y is **directly proportional** to x, or y **varies directly** with x, if y and x are related by

$$\frac{y}{x} = k \quad \text{or, equivalently,} \quad y = kx$$

where k is called the **constant of proportionality** or the **constant of variation**.

As we saw earlier, the revenue from the sale of x units at p dollars per unit is $R(x) = px$, so we can say that the revenue is directly proportional to the number of units, with the constant of proportionality equal to p.

We know that blood alcohol percent is dependent on the number of drinks consumed and that excess consumption can lead to impairment of the drinker's judgment and motor skills (and sometimes to a DUI arrest). The blood alcohol percent p of a 130-pound man is directly proportional to the number of drinks x, where a drink is defined as 1.25 oz of 80-proof liquor, 12 oz of regular beer, or 5 oz of table wine.

EXAMPLE 10 ▶ **Blood Alcohol Percent**

The blood alcohol percent of a 130-pound man is directly proportional to the number of drinks consumed, and 3 drinks give a blood alcohol percent of 0.087. Find the constant of proportionality and the blood alcohol percent resulting from 5 drinks.

SOLUTION

The equation representing blood alcohol percent is $0.087 = k(3)$. Solving for k in $0.087 = k(3)$ gives $k = 0.029$, so the constant of proportionality is 0.029. Five drinks would result in a blood alcohol percent of $5(0.029) = 0.145$.

SKILLS CHECK 2.1

In Exercises 1–12, solve the equations.

1. $5x - 14 = 23 + 7x$ **2.** $3x - 2 = 7x - 24$

3. $3(x - 7) = 19 - x$ **4.** $5(y - 6) = 18 - 2y$

5. $x - \dfrac{5}{6} = 3x + \dfrac{1}{4}$ **6.** $3x - \dfrac{1}{3} = 5x + \dfrac{3}{4}$

7. $\dfrac{5(x - 3)}{6} - x = 1 - \dfrac{x}{9}$

8. $\dfrac{4(y-2)}{5} - y = 6 - \dfrac{y}{3}$

9. $5.92t = 1.78t - 4.14$

10. $0.023x + 0.8 = 0.36x - 5.266$

11. $\dfrac{3}{4} + \dfrac{1}{5}x - \dfrac{1}{3} = \dfrac{4}{5}x$

12. $\dfrac{2}{3}x - \dfrac{6}{5} = \dfrac{1}{2} + \dfrac{5}{6}x$

13. $3(x-1) + 5 = 4(x-3) - 2(2x-3)$

14. $5x - (x-2) + 7 = -(2x-9) - 8(3x+6)$

15. $2[3(x-10) - 3(2x+1)] = $
$$3[-(x-6) - 5(2x-4)]$$

16. $\dfrac{3x}{5} - \dfrac{x-2}{6} = \dfrac{2x+3}{4} - 1$

17. $\dfrac{6-x}{4} - 2 - \dfrac{6-2x}{3} = -\left[\dfrac{5x-2}{3} - \dfrac{x}{5}\right]$

18. $12 - \dfrac{3-2x}{5} - \dfrac{4-x}{3} = -5\left[\dfrac{5x}{2} - \dfrac{2(3x+2)}{5}\right]$

For Exercises 19–22, find (a) the solution to the equation $f(x) = 0$, (b) the x-intercept of the graph of $y = f(x)$, and (c) the zero of $f(x)$.

19. $f(x) = 32 + 1.6x$　　**20.** $f(x) = 15x - 60$

21. $f(x) = \dfrac{3}{2}x - 6$　　**22.** $f(x) = \dfrac{x-5}{4}$

In Exercises 23 and 24, you are given a table showing input and output values for a given function $y_1 = f(x)$. Using this table, find (if possible) (a) the x-intercept of the graph of $y = f(x)$, (b) the y-intercept of the graph of $y = f(x)$, and (c) the solution to the equation $f(x) = 0$.

23.　　　　　　　**24.**

In Exercises 25 and 26, you are given the graph of a certain function $y = f(x)$ and the zero of that function. Using this graph, find (a) the x-intercept of the graph of $y = f(x)$ and (b) the solution to the equation $f(x) = 0$.

25.

26.

In Exercises 27 and 28, use the Y= screen to write the equation being solved if $Y_1 = Y_2$. Then use the table to solve the equation.

27.

X	Y1	Y2
-5	-15	-21
-4	-13	-18
-3	-11	-15
-2	-9	-12
-1	-7	-9
0	-5	-6
1	-3	-3
2	-1	0
3	1	3
4	3	6
5	5	9

Plot1 Plot2 Plot3
Y1＝2X−5
Y2＝3(X−2)
X=-5

28.

X	Y1	Y2
-5	30	-22
-4	20	-19
-3	10	-16
-2	0	-13
-1	-10	-10
0	-20	-7
1	-30	-4
2	-40	-1
3	-50	2
4	-60	5
5	-70	8

Plot1 Plot2 Plot3
Y1＝-5(2X+4)
Y2＝3X−7
X=-5

In Exercises 29–32, you are given the equation of a function. For each function, (a) find the zero of the function, (b) find the x-intercept of the graph of the function, and (c) solve the equation $f(x) = 0$.

29. $f(x) = 4x - 100$

30. $f(x) = 6x - 120$

31. $f(x) = 330 + 40x$

32. $f(x) = 250 + 45x$

In Exercises 33–40, solve the equations using graphical methods.

33. $14x - 24 = 27 - 3x$　**34.** $3x - 8 = 15x + 4$

35. $3(s-8) = 5(s-4) + 6$

36. $5(2x+1) + 5 = 5(x-2)$

37. $\dfrac{3t}{4} - 2 = \dfrac{5t-1}{3} + 2$

38. $4 - \dfrac{x}{6} = \dfrac{3(x-2)}{4}$　**39.** $\dfrac{t}{3} - \dfrac{1}{2} = \dfrac{t+4}{9}$

40. $\dfrac{x-5}{4} + x = \dfrac{x}{2} + \dfrac{1}{3}$

41. Solve $A = P(1 + rt)$ for r.

42. Solve $V = \dfrac{1}{3}\pi r^2 h$ for h.

43. Solve $5F - 9C = 160$ for F.

44. Solve $4(a - 2x) = 5x + \dfrac{c}{3}$ for x.

45. Solve $\dfrac{P}{2} + A = 5m - 2n$ for n.

46. Solve $y - y_1 = m(x - x_1)$ for x.

In Exercises 47–50, solve the equations for y and graph them with a standard window on a graphing utility.

47. $5x - 3y = 5$

48. $3x + 2y = 6$

49. $x^2 + 2y = 6$

50. $4x^2 + 2y = 8$

EXERCISES 2.1

51. Depreciation An \$828,000 building is depreciated for tax purposes by its owner using the straight-line depreciation method. The value of the building y, after x months of use, is given by $y = 828{,}000 - 2300x$ dollars. After how many years will the value of the building be \$690,000?

52. Temperature Conversion The equation $5F - 9C = 160$ gives the relationship between Fahrenheit and Celsius temperature measurements. What Fahrenheit measure is equivalent to a Celsius measurement of $20°$?

53. Investments The future value of a simple interest investment is given by $S = P(1 + rt)$. What principal P must be invested for $t = 5$ years at the simple interest rate $r = 10\%$ so that the future value grows to \$9000?

54. Temperature–Humidity Index The temperature–humidity index I is $I = t - 0.55(1 - h)(t - 58)$, where t is the air temperature in degrees Fahrenheit and h is the relative humidity expressed as a decimal. If the air temperature is $88°$, find the humidity h (as a percent) that gives an index value of 84.7.

55. Earnings and Minorities The median annual salary (in thousands of dollars) of minorities M is related to the median salary (in thousands of dollars) of whites W by $M = 0.819W - 3.214$. What is the median annual salary for whites when the median annual salary for minorities is \$41,520?
(Source: *Statistical Abstract of the United States*)

56. Game Show Question The following question was worth \$32,000 on the game show *Who Wants to Be a Millionaire?*: "At what temperature are the Fahrenheit and Celsius temperature scales the same?" Answer this question. Recall that Fahrenheit and Celsius temperatures are related by $5F - 9C = 160$.

57. Federal Income Tax Per Capita Using data from 2010 and projected to 2018, the federal income tax per capita is given by the function $y = f(x) = 486.48x + 3486.84$, where x is the number of years after 2010 and y is in dollars. If this model remains valid, in what year will the federal income tax per capita be \$8351.64?

58. Reading Score The average reading score on the National Assessment of Progress tests is given by $y = 0.155x + 255.37$, where x is the number of years after 1970. In what year would the average reading score be 259.4 if this model is accurate?

59. Banks The number of banks in the United States for selected years from 2007 to 2014 is given by $y = -286.8x + 8972$, where x is the number of years after 2006. If this model is accurate after 2014, in what year would the number of banks be 3236?
(Source: Federal Deposit Insurance Corporation)

60. Profit The profit from the production and sale of specialty golf hats is given by the function $P(x) = 20x - 4000$, where x is the number of hats produced and sold.

a. Producing and selling how many units will give a profit of $8000?

b. How many units must be produced and sold to avoid a loss?

61. *Marriage Rate* The marriage rate per 1000 unmarried women for the years 1987–2014 can be modeled by the function $132x + 1000y = 9570$, where x is the number of years after 1980. During what year does the equation indicate the marriage rate will be 4.29?
(Source: *National Vital Statistics Report 2010*)

62. *World Forest Area* When x is the number of years after 1990, the world forest area (natural forest or planted stands) as a percent of land area is given by $f(x) = -0.065x + 31.39$. In what year will the percent be 29.44 if this model is accurate?
(Source: The World Bank)

63. *U.S. Population* The U.S. population can be modeled for the years 1960–2060 by the function $p = 2.6x + 177$, where p is in millions of people and x is in years from 1960. During what year does the model estimate the population to be 320,000,000?
(Source: www.census.gov/statab)

64. *Disposable Income* Disposable income is the amount left after taxes have been paid and is one measure of health of the economy. The total disposable income for the United States, in billions of dollars, for selected years from 2010 and projected to 2040 can be modeled by $D(x) = 0.328x + 6.32$, where x is the number of years after 2000. In what year does this model project the total disposable income for the United States to be $12.88 billion?
(*Source*: U.S. Bureau of Economic Analysis)

65. *Heart Rate* The desired heart rate for weight loss during exercise is given by the linear function $H = -0.65t + 143$, with t equal to the person's age and H equal to the heart rate in beats per minute.

a. If a person's desired heart rate for weight loss is 130, what is the person's age?

b. If a person's desired heart rate for weight loss is 104, what is the person's age?
(Source: paramountfitness.com)

66. *Population of Females Under 18* Using projected data from 2020 to 2060, the population of females under 18 in the United States is given by $y = 154.03x + 34,319$, with x equal to the number of years after 2000 and y equal to the projected population, in thousands. In what year will the population reach 38,169,750, according to this model?
(Source: U.S. Census Bureau)

67. *Hispanic Population* Using data and projections from 1980 through 2050, the number (in millions) of Hispanics in the U.S. civilian noninstitutional population is given by $y = 0.876x + 6.084$, where x is the number of years after 1980. During what year was the number of Hispanics 14.8 million, if this model is accurate?
(Source: U.S. Census Bureau)

68. *Internet Users* The percent of the U.S. population with Internet access can be modeled by $y = 1.36x + 68.8$, with x equal to the number of years after 2000. When does this model indicate that the U.S. population with Internet access will reach 96%?
(Source: Jupiter Media Metrix)

69. *Life Expectancy at 65* Using data for selected years from 1950 and projected to 2050, the equation $y = 0.077x + 14$ gives the number of additional years of life expected at age 65, with x equal to the number of years after 1950. In what year does this model predict that the life expectancy at age 65 will reach 20 additional years?
(Source: Social Security Administration)

70. *Cell Phone Subscribers* The number of cell phone subscribers (in billions) between 2014 and 2020 can be modeled by $y = 0.234x + 6.856$, where x is the number of years after 2010. In what year does this model indicate that there were 9.43 billion subscribers?
(Source: Semiannual CTIA Wireless Survey)

71. *Grades* To earn an A in a course, a student must get an average score of at least 90 on five tests. If her first four test scores are 92, 86, 79, and 96, what score does she need on the last test to obtain a 90 average?

72. *Grades* To earn an A in a course, a student must get an average score of at least 90 on three tests and a final exam. If his final exam score is higher than his lowest score, then the lowest score is removed and the final exam score counts double. If his first three test scores are 86, 79, and 96, what is the lowest score he can get on the last test and still obtain a 90 average?

73. *Tobacco Judgment* As part of the largest-to-date damage award in a jury trial in history, the penalty handed down against the tobacco industry included a $74 billion judgment against Philip Morris. If this amount was 94% of this company's 1999 revenue, how much was Philip Morris's 1999 revenue?
(Source: *Newsweek*)

74. *Tobacco Judgment* As part of the largest-to-date damage award in a jury trial in history, the penalty handed down against the tobacco industry included a $36 billion judgment against R. J.

Reynolds. If this amount was 479% of this company's 1999 revenue, how much was R. J. Reynolds' 1999 revenue?
(Source: *Newsweek*)

75. **Sales Commission** A salesman earns $50,000 in commission in 1 year and then has his commission reduced by 20% the next year. What percent increase in commission over the second year will give him $50,000 in the third year?

76. **Salaries** A man earning $100,000 per year has his salary reduced by 5% because of a reduction in his company's market. A year later, he receives $104,500 in salary. What percent raise from the reduced salary does this represent?

77. **Sales Tax** The total cost of a new automobile, including a 6% sales tax on the price of the automobile, is $29,998. How much of the total cost of this new automobile is sales tax?

78. **Wildlife Management** In wildlife management, the capture–mark–recapture technique is used to estimate the population of certain types of fish or animals. To estimate the population, we enter information in the equation

$$\frac{\text{total in population}}{\text{total number marked}} = \frac{\text{total number in second capture}}{\text{number found marked in second capture}}$$

Suppose that 50 sharks are caught along a certain shoreline, marked, and then released. If a second capture of 50 sharks yields 20 sharks that have been marked, what is the resulting population estimate?

79. **Investment** The formula for the future value A of a simple interest investment is $A = P + Prt$, where P is the original investment, r is the annual interest rate, and t is the time in years. Solve this formula for t.

80. **Investment** The formula for the future value A of a simple interest investment is $A = P + Prt$, where P is the original investment, r is the annual interest rate, and t is the time in years. Solve this formula for P.

81. **Investment** If P dollars are invested for t years at simple interest rate r, the future value of the investment is $A = P + Prt$. If $2000 invested for 6 years gives a future value of $3200, what is the simple interest rate of this investment?

82. **Investment** If an investment at 7% simple interest has a future value of $5888 in 12 years, what is the original investment?

83. **Investment** The simple interest earned in 9 years is directly proportional to the interest rate r. If the interest is $920 when r is 12%, what is the amount of interest earned in 9 years at 8%?

84. **Investment** The interest earned at 9% simple interest is directly proportional to the number of years the money is invested. If the interest is $4903.65 in 5 years, in how many years will the interest earned at 9% be $7845.84?

85. **Circles** Does the circumference of a circle vary directly with the radius of the circle? If so, what is the constant of variation?

86. **Calories** The amount of heat produced in the human body by burning protein is directly proportional to the amount of protein burned. If burning 1 gram of protein produces 32 calories of heat, how much protein should be burned to produce 180 calories?

87. **Body Mass Index** Body mass index (BMI) is a measure that helps determine obesity, with a BMI of 30 or greater indicating that the person is obese. A BMI table for heights in inches and weights in pounds is shown on page 83. BMI was originally defined in the metric system of measure, and the BMI is directly proportional to the weight of a person of a given height.

   a. If the BMI of a person who is 1.5 meters tall is 20 when the person weighs 45 kilograms, what is the constant of variation?

   b. If a woman of this height has a BMI of 32, what does she weigh?
   (Source: Roche Pharmaceuticals)

88. **Land Cost** The cost of land in Savannah is directly proportional to the size of the land. If a 2500-square-foot piece of land costs $172,800, what is the cost of a piece of land that is 5500 square feet?

Fitting Lines to Data Points: Modeling Linear Functions

KEY OBJECTIVES

- Find exact linear models for data
- Determine whether a set of data can be modeled exactly or approximately
- Create scatter plots for sets of data
- Find approximate linear models for data
- Visually determine whether a linear model is a "good" fit for data
- Solve problems using linear models

SECTION PREVIEW Earnings and Gender

Table 2.2 shows the median weekly earnings of men and women in 2013. Figure 2.11 shows the scatter plot of the data, and it appears that a line would approximately fit along these data points. We can determine the relationship between the two sets of earnings by creating a linear equation that gives male weekly earnings as a function of female weekly earnings. (See Example 3.) In this section, we will create linear models from data points, use graphing utilities to create linear models, and use the model to answer questions about the data.

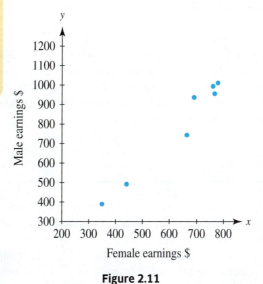

Figure 2.11

Table 2.2

Age	Women	Men
16–19	350	390
20–24	442	492
25–34	665	744
35–44	767	956
45–54	761	994
55–64	779	1011
65 and older	691	937

(Source: U.S. Bureau of Labor Statistics)

Exact and Approximate Linear Models

We have seen that if data points fit exactly on a line, we can use two of the points to model the linear function (write the equation of the line). We can determine that the data points fit exactly on a line by determining that the changes in output values are equal for equal changes in the input values. In this case, we say that the inputs are **uniform** and the **first differences** are constant.

> - If the first differences of data outputs are constant for uniform inputs, the rate of change is constant and a linear function can be found that fits the data exactly.
> - If the first differences are "nearly constant," a linear function can be found that is an approximate fit for the data.

If the first differences of data outputs are constant for uniform inputs, we can use two of the points to write the linear equation that models the data. If the first differences of data outputs are constant for inputs differing by 1, this constant difference is the rate of change of the function, which is the slope of the line fitting the points exactly.

EXAMPLE 1 ▶   **Retirement**

Table 2.3 gives the annual retirement payment to a 62-year-old retiree with 21 or more years of service at Clarion State University as a function of the number of years of service and the first differences of the outputs.

Table 2.3

Year	21	22	23	24	25
Retirement Payment	40,950	42,900	44,850	46,800	48,750
First Differences		1950	1950	1950	1950

Because the first differences of the outputs of this function are constant (equal to 1950) for each unit change of the input (years), the rate of change is the constant 1950. Using this rate of change and a point gives the equation of the line that contains all the points. Representing the annual retirement payment by y and the years of service by x and using the point $(21, 40{,}950)$, we obtain the equation

$$y - 40{,}950 = 1950(x - 21)$$
$$y = 1950x$$

Note that the values in Table 2.3 represent points satisfying a **discrete function** (a function with a finite number of inputs), with each input representing the number of years of service. Although only points with integer inputs represent the annual retirement payments, we can model the application with the **continuous function** $y = 1950x$, whose graph is a line that passes through the 5 data points. Informally, a continuous function can be defined as a function whose graph can be drawn over its domain without lifting the pen from the paper.

In the context of this application, we must give a discrete interpretation to the model. This is because the only inputs of the function that make sense in this case are nonnegative integers representing the number of years of service. The graph of the discrete function defined in Table 2.3 is the scatter plot shown in Figure 2.12(a), and the continuous function that fits these data points, $y = 1950x$, is shown in Figure 2.12(b).

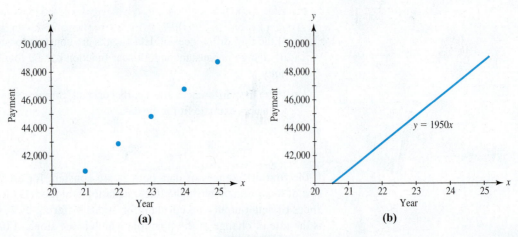

Figure 2.12

When a scatter plot of data can be approximately fitted by a line, we attempt to find the graph that visually gives the best fit for the data and find its equation. For now, we informally define the "best-fit" line as the one that appears to come closest to all the data points.

EXAMPLE 2 ▶   Population of Females Under 18

Table 2.4 gives the projected population, in millions, of females under the age of 18 projected for the years 2020 through 2060.

Table 2.4

Year	Population (millions)	Year	Population (millions)
2020	37.2	2045	41.0
2025	38.2	2050	42.0
2030	39.2	2055	42.9
2035	39.8	2060	43.6
2040	40.4		

(Source: U.S. Census Bureau)

a. Draw a scatter plot of the data with the x-value of each point representing the number of years after 2000 and the y-value representing the population of females under the age of 18, in millions.

b. Graph the equation $y = 0.1x + 36$ and the data points on the same axes. Is this a good fit for the data?

c. Use a piece of (uncooked!) spaghetti or a mechanical pencil "lead" to find a line through two of the data points that is a good visual fit for the points, and use the two points to write the equation of this line.

SOLUTION

a. By using x as the number of years after 2000, we have aligned the data with $x = 0$ representing 2000, with $x = 20$ representing 2020, and so forth. In the lists of a graphing calculator (or columns of an Excel spreadsheet), we enter the aligned input data representing the years in Table 2.4 and the output data representing the population of females under the age of 18 (in millions). Table 2.5 shows the lists containing the data. The window for this scatter plot can be set manually or with a command on the graphing utility such as ZoomStat, which is used to automatically set the window and display the graph. The graph of these points is shown in Figure 2.13.

Table 2.5

L1	L2
20	37.2
25	38.2
30	39.2
35	39.8
40	40.4
45	41.0
50	42.0
55	42.9
60	43.6

Figure 2.13

b. The graph of the function $y = 0.1x + 36$ and the scatter plot of the data are shown in Figure 2.14. The line does not appear to be the best possible fit to the data points.

Figure 2.14

c. Placing a piece of spaghetti or a pencil lead along the points shows a line that passes through the points $(25, 38.2)$ and $(60, 43.6)$ and appears to be a good fit (see Figure 2.15). The equation of this line is found as follows.

$$m = \frac{43.6 - 38.2}{60 - 25} \approx 0.154 \quad \text{so} \quad y - 43.6 = 0.154(x - 60)$$

$$\text{or} \quad y = 0.154x + 34.36$$

Comparing Figure 2.14 and Figure 2.16 shows that this function is a much better fit than $y = 0.1x + 36$ from part (b).

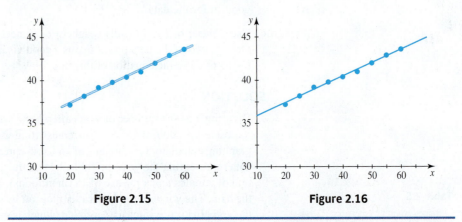

Figure 2.15   **Figure 2.16**

Is either of the lines in Figure 2.14 and 2.16 the best-fitting of all possible lines that could be drawn on the scatter plot? How do we determine the best-fit line? We now discuss the answers to these questions.

Fitting Lines to Data Points; Linear Regression

The points determined by the data in Table 2.4 on page 107 do not all lie on a line, but we can determine the equation of the line that is the "best fit" for these points by using a procedure called **linear regression**. This procedure defines the *best-fit line* as the line for which the sum of the squares of the vertical distances from the data points to the line is a minimum. For this reason, the linear regression procedure is also called the **least squares method**.

The vertical distance between a data point and the corresponding point on a line is simply the amount by which the line misses going through the point—that is, the difference in the y-values of the data point and the point on the line. If we call this difference in outputs d_i (where i takes on the values from 1 to n for n data points), the least squares method* requires that

$$d_1^2 + d_2^2 + d_3^2 + \cdots + d_n^2$$

be as small as possible. The line for which this sum of squared differences is as small as possible is called the *linear regression line* or *least squares line* and is the one that we consider to be the **best-fit line** for the data.

For an illustration of the linear regression process, consider again the two lines in Figure 2.14 and 2.16 and the data points in Table 2.4. Figures 2.17 and 2.18 indicate the vertical distances for each of these lines. To the right of each figure is the calculation of the sum of squared differences for the line.

Figure 2.17

$$d_1^2 + d_2^2 + d_3^2 + \cdots + d_9^2$$
$$= (37.2 - 38)^2 + (38.2 - 38.5)^2 + (39.2 - 39)^2$$
$$+ (39.8 - 39.5)^2 + (40.4 - 40.0)^2 + (41.0 - 40.5)^2$$
$$+ (42.0 - 41.0)^2 + (42.9 - 41.5)^2 + (43.6 - 42)^2$$
$$= 6.79$$

Figure 2.18

$$d_1^2 + d_2^2 + d_3^2 + \cdots + d_9^2$$
$$= (37.2 - 37.44)^2 + (38.2 - 38.21)^2 + (39.2 - 38.98)^2$$
$$+ (39.8 - 39.75)^2 + (40.4 - 40.52)^2 + (41.0 - 41.29)^2$$
$$+ (42.0 - 42.06)^2 + (42.9 - 42.83)^2 + (43.6 - 43.6)^2$$
$$= 0.2156$$

We see numerically from the sum of squared differences calculations, as well as visually from Figures 2.14 and 2.15, that the line

$$y_2 = 0.154x + 34.36$$

is a better model for the data than the line y_1. However, is this line the best-fit line for these data? Remember that the best-fit line must have the smallest sum of squared differences of *all* lines that can be drawn through the data! We will see that the line which is the best fit for these data points, with coefficients rounded to three decimal places, is

$$y = 0.155x + 34.278$$

The sum of squared differences for the regression line is approximately 0.198, which is slightly smaller than the sum 0.2156, which was found for $y_2 = 0.154x + 34.3$.

Development of the formulas that give the best-fit line for a set of data is beyond the scope of this text, but graphing calculators, computer programs, and spreadsheets have built-in formulas or programs that give the equation of the best-fit line. That is,

---

* The sum of the squared differences is often called SSE, the *sum of squared errors*. The development of the equations that lead to the minimum SSE is a calculus topic.

we can use technology to find the best linear model for the data. The calculation of the sum of squared differences for the dental employees data was given only to illustrate what the best-fit line means. You will not be asked to calculate—nor do we use from this point on—the value of the sum of squared differences in this text.

We illustrate the use of technology to find the regression line by returning to the data of Example 2. To find the equation that gives the population of females under 18 as a function of the years after 2000, we use the following steps.

Modeling Data

1. Enter the data into lists of a graphing utility. Figure 2.19(a) shows the data from Example 2 entered in the lists of a TI-84 Plus C.

(a)
Figure 2.19

2. Create a scatter plot of the data to see whether a linear model is reasonable. The data should appear to follow a linear pattern with no distinct curvature. Figures 2.19(b) and (c) show the scatter plot of the data from Example 2 and the window that was used.

(b)   **(c)**
Figure 2.19

3. To use a graphing utility to obtain the linear equation that is the best fit for the data, press **STAT**, move to CALC (accessed by **2ND** **TRACE**), and select 4:LinReg. Then move to Store RegEQ:, press **VARS**, move to Y-VARS, and press **ENTER** four times. This will show the equation of the best-fitting line on the screen and on the Y= screen.* Figure 2.19(d) shows the equation for Example 2, which can be approximated by $y = 0.155x + 34.278$, and Figure 2.19(e) shows the equation on the Y= screen.

(d)   **(e)**
Figure 2.19

---
* See Appendix A, page 673, for details.

4. Graph the linear function (unrounded) and the data points on the same graph to see how well the function fits the data. The graph of the data and the best-fitting line is shown in Figure 2.19(f).

(f)
Figure 2.19

5. Report the function and/or numerical results in a way that makes sense in the context of the problem, with the appropriate units and with the variables identified. Unless otherwise indicated, report functions with coefficients rounded to three decimal places. Sometimes you will be asked to report a model with three or four significant digits. For example, the model above could be reported with three significant digits as $y = 0.155x + 34.3$. See the Toolbox, page 90, for a discussion of significant digits.

Figure 2.19(f) shows that

$$y = 0.155x + 34.278$$

is a good model of the population of females under 18, where y is in millions and x is the number of years after 2000.

The screens for the graphing utility that you use may vary slightly from those given in this text. Also, the regression line you obtain is dependent on your particular technology and may have some decimal places slightly different from those shown here.

It is important to be careful when rounding coefficients in equations and when rounding during calculations with the model. We will use the following guidelines in this text.

Technology Note

After a model for a data set has been found, it can be rounded for reporting purposes. However, do not use a rounded model in calculations, and do not round answers during the calculation process unless instructed to do so. When the model is used to find numerical answers, the answers should be rounded in a way that agrees with the context of the problem.

EXAMPLE 3 ▶   **Earnings and Gender**

Table 2.6 shows the median weekly earnings of women and men in 2013, in dollars.

a. Let x represent earnings for females, let y represent earnings for males, and create a scatter plot of the data.

b. Create a linear model that expresses male weekly earnings as a function of female weekly earnings.

c. Graph the linear function and the data points on the same graph, and discuss how well the function models the data.

Table 2.6

Age	Women	Men
16 – 19	350	390
20 – 24	442	492
25 – 34	665	744
35 – 44	767	956
45 – 54	761	994
55 – 64	779	1011
65 and older	691	937

(Source: U.S. Bureau of Labor Statistics)

SOLUTION

a. Enter the data from Table 2.6 in the lists of a graphing utility. Figure 2.20(a) shows a list of the data points. The scatter plot of the data points from Table 2.6 is shown in Figure 2.20(b).

(a)

(b)

Figure 2.20

Figure 2.21

b. The points in the scatter plot in Figure 2.20(b) exhibit a nearly linear pattern, so a linear function could be used to model these data. The equation of the line that is the best fit for the data points can be found with a graphing calculator (Figure 2.21). The equation, rounded to three decimal places, is

$$y = 1.460x - 140.325$$

Remember that a *model* gives not only the equation but also a description of the variables and their units of measure. The rounded model

$$y = 1.460x - 140.325$$

where x and y are in dollars, expresses male weekly earnings y as a function of female weekly earnings x.

c. Using the unrounded function in the equation editor (Figure 2.22(a)) and graphing it along with the data points,* we observe that the line follows the general trend

---

* Even though we write rounded models found from data given in this text, all graphs and calculations use the unrounded model found by the graphing utility.

indicated by the data (Figure 2.22(b)). Note that not all points lie on the graph of the equation, even though this is the line that is the best fit for the data.

(a)   (b)

Figure 2.22

EXAMPLE 4 ▶ U.S. Population

The total U.S. population for selected years beginning in 1960 and projected to 2050 is shown in Table 2.7, with the population given in millions.

a. Align the data to represent the number of years after 1960, and draw a scatter plot of the data.

b. Create the linear equation that is the best fit for these data, where y is in millions and x is the number of years after 1960.

c. Graph the equation of the linear model on the same graph with the scatter plot and discuss how well the model fits the data.

d. Align the data to represent the years after 1950 and create the linear equation that is the best fit for the data, where y is in millions.

e. How do the x-values for a given year differ?

f. Use both unrounded models to estimate the population in 2000 and in 2010. Are the estimates equal?

Table 2.7

Year	Population (millions)	Year	Population (millions)
1960	180.671	1995	263.044
1965	194.303	1998	270.561
1970	205.052	2000	281.422
1975	215.973	2003	294.043
1980	227.726	2025	358.030
1985	238.466	2050	408.695
1990	249.948		

(Source: U.S. Census Bureau)

SOLUTION

a. The aligned data have $x = 0$ representing 1960, $x = 5$ representing 1965, and so forth. Figure 2.23(a) shows the first eleven entries using the aligned data. The scatter plot of these data is shown in Figure 2.23(b).

Figure 2.23

b. The equation of the best-fit line is found by using linear regression with a graphing calculator. With the decimals rounded to three places, the linear model for the U.S. population is

$$y = 2.607x + 177.195 \text{ million}$$

where x is the number of years after 1960.

c. Using the unrounded function in the equation editor (Figure 2.24(a)) and graphing it along with the scatter plot shows that the graph of the best-fit line is very close to the data points (Figure 2.24(b)). However, the points do not all fit the line because the U.S. population did not increase by exactly the same amount each year.

Figure 2.24

d. If we align the data to represent the years after 1950, then $x = 10$ corresponds to 1960, $x = 15$ corresponds to 1965, and so forth. Figure 2.25(a) shows the first eleven entries using the aligned data, and Figure 2.25(b) shows the scatter plot. The equation that best fits the data, found using linear regression with a calculator, is

$$y = 2.607x + 151.128$$

where x is the number of years from 1950 and y is in millions. Figure 2.25(c) shows the regression equation.

Figure 2.25

e. The x-values for a given year are 10 more with this model than with the first model.

f. Both models estimate the population to be 281.464 million in 2000 and to be 307.531 million in 2010. They are equal estimates. In fact, if you substitute $x - 10$ for x in the first model, you will get the second model.

Spreadsheet ▶ SOLUTION   Using Excel to find the best linear function that fits data involves several steps.

1. Enter the data in two columns and find the scatter plot of the data, which we have done in Chapter 1.

2. Right click on one point of the scatter plot and select Add Trendline.

3. In the menu at the right, select Linear and check Display Equation on the chart at the bottom, and the best-fitting line and its equation will appear. Figure 2.26 shows the best-fitting line for the data in Example 4, with x equal to the number of years after 1960, and its equation.*

Figure 2.26

Applying Models

Because 2000 is a year between two given values in the table of Example 4, using the model to find the population in 2000 is called **interpolation**. When a model is evaluated for a *prediction* using input(s) outside the given data points, the process is called **extrapolation**.

A question arises whenever data that involve time are used. If we label the x-coordinate of a point on the input axis as 1999, what time during 1999 do we mean? Does 1999 refer to the beginning of the year, the middle, the end, or some other time? Because most data from which the functions in applications are derived represent end-of-year totals, we adopt the following convention when modeling, unless otherwise specified:

A point on the input axis indicating a time refers to the end of the time period.

For instance, the point representing the year 1999 means "at the end of 1999." Notice that this instant in time also represents the beginning of the year 2000. If, for example, data are aligned with x equal to the number of years after 1990, then any x-value greater than 9 and less than or equal to 10 represents some time in the year 2000. If a point represents anything other than the end of the period, this information will be clearly

---

* See Appendix B, page 691 for details.

indicated. Also, if a and b are points in time, we use the phrases "from a to b" and "between a and b" to indicate the same time interval.

Goodness of Fit

Consider again Example 4, where we modeled the U.S. population for selected years. Looking at Table 2.8 which is a partial table of the data, we see that for uniform inputs the first differences of the outputs appear to be relatively close to the same constant, especially compared to the size of the population. (If these differences were closer to a constant, the fit would be better.)

Table 2.8

Uniform Inputs (year)	Outputs Population (millions)	First Differences in Output (difference in population)
1960	180.671	
1965	194.303	**13.632**
1970	205.052	**10.749**
1975	215.973	**10.921**
1980	227.726	**11.753**
1985	238.466	**10.740**
1990	249.948	**11.482**
1995	263.044	**13.096**
2000	281.422	**18.378**

So how "good" is the fit of the linear model $y = 2.6067295436133x + 151.1276669$, with x equal to the number of years after 1960, to the data in Example 4? Based on observation of the graph of the line and the data points on the same set of axes (see Figure 2.26), it is reasonable to say that the regression line provides a very good fit, but not an exact fit, to the data.

The goodness of fit of a line to a set of data points can be observed from the graph of the line and the data points on the same set of axes, and it can be measured if your graphing utility computes the **correlation coefficient**.* The correlation coefficient is a number r, $-1 \leq r \leq 1$, that measures the strength of the linear relationship that exists between the two variables. The closer $|r|$ is to 1, the more closely the data points fit the linear regression line. (There is no linear relationship between the two variables if $r = 0$.) Positive values of r indicate that the output variable increases as the input variable increases, and negative values of r indicate that the output variable decreases as the input variable increases. But the *strength* of the relationship is indicated by how close $|r|$ is to 1. For the data of Example 4, computing the correlation coefficient gives $r = .997$ (Figure 2.27), which means that the linear relationship is strong and that the linear model is an excellent fit for the data. When a calculator feature or computer program is used to fit a linear model to data, the resulting equation can be considered the best linear fit for the data. As we will see later in this text, other mathematical models may be better fits to some sets of data, especially if the first differences of the outputs are not close to being constant.

Figure 2.27

---

* See Appendix A, page 673.

SKILLS CHECK 2.2

Report models to three decimal places unless otherwise specified. Use unrounded models to graph and calculate unless otherwise specified.

Discuss whether the data shown in the scatter plots in the figures for Exercises 1 and 2 should be modeled by a linear function.

1.

2.

Discuss whether the data shown in the scatter plots in the figures for Exercises 3 and 4 should be modeled by a linear function exactly or approximately.

3.

4.

Create a scatter plot for each of the sets of data in Exercises 5 and 6.

5.

x	1	3	5	7	9
y	4	7	10	13	16

6.

x	1	2	3	5	7	9	12
y	1	3	6	1	9	2	6

7. Can the scatter plot in Exercise 5 be fit exactly or only approximately by a linear function? How do you know?

8. Can the scatter plot in Exercise 6 be fit exactly or only approximately by a linear function? How do you know?

9. Find the linear function that is the best fit for the data in Exercise 5.

10. Find the linear function that is the best fit for the data in Exercise 6.

Use the data in the following table for Exercises 11–14.

x	5	8	11	14	17	20
y	7	14	20	28	36	43

11. Construct a scatter plot of the data in the table.

12. Determine whether the points plotted in Exercise 11 appear to lie near some line.

13. Create a linear model for the data in the table.

14. Use the function $y = f(x)$ created in Exercise 13 to evaluate $f(3)$ and $f(5)$.

Use the data in the table for Exercises 15–18.

x	2	5	8	9	10	12	16
y	5	10	14	16	18	21	27

15. Construct a scatter plot of the data in the table.

16. Determine if the points plotted in Exercise 15 appear to lie near some line.

17. Create a linear model for the data in the table.

18. Use the rounded function $y = f(x)$ that was found in Exercise 17 to evaluate $f(3)$ and $f(5)$.

19. Determine which of the equations, $y = -2x + 8$ or $y = -1.5x + 8$, is the better fit for the data points $(0, 8), (1, 6), (2, 5), (3, 3)$.

20. Determine which of the equations, $y = 2.3x + 4$ or $y = 2.1x + 6$, is the better fit for the data points $(20, 50), (30, 73), (40, 96), (50, 119), (60, 142)$.

21. Without graphing, determine which of the following data sets are exactly linear, approximately linear, or nonlinear.

a.

x	y
1	5
2	8
3	11
4	14
5	17

b.

x	y
1	2
2	5
3	10
4	5
5	2

c.

x	y
1	6
2	8
3	12
4	14
5	18

22. Why can't first differences be used to tell if the following data are linear? $(1, 3), (4, 5), (5, 7), (7, 9)$

EXERCISES 2.2

Report models to three decimal places unless otherwise specified. Use unrounded models to graph and calculate unless otherwise specified.

23. **Women in the Workforce** The number of women in the workforce for selected years from 1930 through 2015 is shown in the following figures.

   a. Do the points on the graph in the figure define the number of working women as a discrete or continuous function of the year?

   b. Does the graph of $y = W(x)$ shown in the figure define the number of working women as a discrete or continuous function of the year?

   c. Would the data in the scatter plot be better modeled by a linear function than by the nonlinear function $y = W(x)$? Why or why not?

(Source: U.S. Census Bureau)

24. *The Cloud* World IT cloud services revenues from 2009 and projected to 2015 are shown in the table.

   a. Create a scatter plot for the data with y equal to the billions of dollars of revenue and x equal to the number of years after 2000.

   b. Use the scatter plot to determine whether the data can be fit exactly by a linear function.

Year	Revenue ($ billions)	Year	Revenue ($ billions)
2009	18	2013	45
2010	21	2014	58
2011	27	2015	73
2012	35		

(Source: International Data Corporation)

25. *Social Security Benefits* The figure shows how monthly Social Security benefits differ with the age a recipient decides to start receiving benefits. The example shown in this table is for recipients whose monthly benefit is $1000 if benefits start being paid at a full retirement age of 66. Note that benefits do not start being paid before age 62, and that the monthly benefit does not increase if the starting age is after 70.

   a. Can the data shown be modeled exactly by a linear function? Explain.

   b. Can the data shown from age 66 to age 70 be modeled exactly by a linear function? Explain.

   c. Write the linear function $S = f(x)$ that models the data shown from age 66 to age 70, where x is the age.

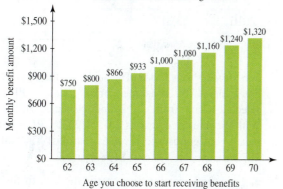

Monthly benefit amounts differ based on the age you decide to start receiving benefits
This example assumes a benefit of $1,000 at a full retirement age of 66

(Source: Social Security Administration)

26. *Future Value of an Investment* If $1000 is invested at 6% simple interest, the initial value and the future

value S at the end of each of 5 years are shown in the table that follows.

   a. Can a linear function model exactly the points from the table? Explain.

   b. If so, find a linear function $S = f(t)$ that models the points.

   c. Use the model to find the future value of this investment at the end of the 7th year. Is this an interpolation or an extrapolation from the data?

   d. Should this model be interpreted discretely or continuously?

Year (t)	0	1	2	3	4	5
Future Value (S)	1000	1060	1120	1180	1240	1300

27. *Drug Doses* The table below shows the usual dosage for a certain prescription drug that combats bacterial infections for a person's weight.

Weight (lb)	Usual Dosage (mg)
88	40
99	45
110	50
121	55
132	60
143	65
154	70
165	75
176	80
187	85
198	90

   a. Can the data be modeled exactly by a linear function? Explain.

   b. If so, how many of the data points are needed to find the linear function that models the data exactly?

   c. Find the linear function $D = f(w)$ that models the data exactly.

   d. If the model is interpreted continuously, what dosage would be recommended for a 209-pound patient?

28. *Cardiac Health* A Paramount elliptical machine provides the table on the next page, which gives the desired heart rate during exercise for cardiac health based on the age of the person exercising on the elliptical machine.

a. Can the desired heart rate be expressed exactly as a linear function of age, based on this data?

b. Write a linear function that models this data.

Age	Heart Rate (beats per minute)	Age	Heart Rate (beats per minute)
20	160	60	128
30	152	70	120
40	144	80	112
50	136	90	104

(Source: Paramount fitness.com)

29. *Disposable Income* Disposable income is the amount left after taxes have been paid and is one measure of health of the economy. The table gives the total disposable income for the United States, in billions of dollars, for selected years from 2010 and projected to 2040.

a. Can this data be fitted exactly by a linear model?

b. Find a linear function that models the data, where x is the number of years after 2000 and y is the total disposable income in billions of dollars.

c. What does the model predict the total disposable income will be in 2023?

Year	Income ($ billions)
2010	10.017
2015	11.120
2020	12.655
2025	14.259
2030	15.948
2035	17.752
2040	19.785

30. *Oil Imports* Crude oil and petroleum products are imported continuously by the United States. The table shows the billions of dollars of net expenditures for the imports of these products for several years.

a. Write a linear function that models this data, with x equal to the number of years after 2015 and y equal to the billions of dollars of expenditures. Report the model with two decimal places.

b. Is the model an exact fit for the data?

c. What does the model predict the billions of dollars of net expenditures will be for the imports in 2023?

d. When does the model predict the billions of dollars of net expenditures for the imports will be 208.2 $billion?

Year	No Expenditures ($ billions)
2018	191.0
2020	198.9
2022	214.5
2024	228.4

(Source: Energy Information Administration)

31. *Annual Wage* The table gives the average annual wage of U.S. workers for selected years from 2011 projected through 2021.

a. Graph the data, with y equal to the average annual wage in dollars and x equal to the number of years after 2010.

b. Find a linear function that models the data, with y equal to the average annual wage in dollars and x equal to the number of years after 2010. Report the model with four significant digits.

c. Find the slope of the reported model, and interpret it in this context.

Year	Average Annual Wage	Year	Average Annual Wage
2011	$43,009	2017	$55,989
2012	44,644	2018	58,698
2013	46,496	2019	61,178
2014	48,595	2020	63,676
2015	50,893	2021	66,161
2016	53,317		

(Source: Social Security Administration)

32. *Consumer Price Index* The table on the next page gives the U.S. consumer price index (CPI) for selected years from 2012 and projected to 2050. With the reference year as 2012, a 2025 CPI of 138.4 means that goods and services that cost $100.00 in 2012 are expected to cost $138.4 in 2025.

a. Find the linear function that is the best fit for the data, with x as the number of years after 2010 and y as the CPI.

b. Graph the model and the data on the same axes.

c. Find the slope of the reported model, and interpret it in this context.

Year	CPI	Year	CPI
2012	100.0	2035	182.4
2016	108.6	2040	209.4
2020	120.6	2045	240.5
2025	138.4	2050	276.0
2030	158.9		

(Source: Social Security Administration)

33. *Diabetes* As the following table shows, projections indicate that the percent of U.S. adults with diabetes could dramatically increase.

a. Find a linear model that fits the data in the table, with x equal to the number of years after 2010.

b. Use the model to predict the percent of U.S. adults with diabetes in 2027.

c. Graphically find what year this model predicts the percent be 31.1%.

Year	Percent
2010	15.7
2015	18.9
2020	21.1
2025	24.2
2030	27.2
2035	29.0
2040	31.4
2045	32.1
2050	34.3

(Source: Centers for Disease Control and Prevention)

34. *Carbon Dioxide Emissions* The table gives the millions of metric tons of carbon dioxide emissions from biomass energy combustion in the United States for selected years from 2010 projected to 2032.

a. Find a linear function that gives the millions of metric tons of carbon dioxide emissions, y, as function of x, the number of years after 2010.

b. Graph the model on the same axes with the data. Is it a good fit for the data?

c. What does the (unrounded) model predict the millions of metric tons of carbon dioxide emissions will be in 2027?

d. When will the millions of metric tons of carbon dioxide emissions reach 776.6, according to the model?

Year	CO_2 Emissions (millions of metric tons)
2010	338.5
2012	364.5
2014	396.1
2016	425.8
2018	453.1
2020	498.4
2022	556.2
2024	590.9
2026	629.7
2028	663.1
2030	701.1
2032	743.7

(Source: U.S. Department of Energy)

35. *U.S. Population Over Age 16* The table gives the projections of the population, in millions, of U.S. residents over age 16 for selected years 2015 to 2060.

a. Graph the data, with y equal to the population in millions and x equal to the number of years after 2010.

b. Find a linear function that models the data, with y equal to the population in millions and x equal to the number of years after 2010.

c. Graph the model and the data on the same axes and comment on the fit of the model to the data.

d. What does the model predict that the population will be in 2042?

e. In what year does the model predict that the population will be 336.827 million?

Year	Population (millions)	Year	Population (millions)
2015	255.161	2040	306.634
2020	266.024	2045	315.152
2025	276.558	2050	323.314
2030	286.967	2055	331.770
2035	297.259	2060	340.868

36. *U.S. Population* The following table gives projections of the U.S. population from 2000 to 2100.

   a. Find a linear function that models the data, with x equal to the number of years after 2000 and $f(x)$ equal to the population in millions.

   b. Find $f(65)$ and state what it means.

   c. What does this model predict the population to be in 2080? How does this compare with the value for 2080 in the table?

Year	Population (millions)	Year	Population (millions)
2000	275.3	2060	432.0
2010	299.9	2070	463.6
2020	324.9	2080	497.8
2030	351.1	2090	533.6
2040	377.4	2100	571.0
2050	403.7		

(Source: www.census.gov/population/projections)

37. *Gross Domestic Product* The following table gives the gross domestic product (the value of all goods and services, in billions of dollars) of the United States for selected years from 2000 to 2018.

   a. Create a scatter plot of the data, with y representing the GDP in billions of dollars and x representing the number of years after 2000.

   b. Find the linear function that best fits the data, with x equal to the number of years after 2000. Report the model with three significant digits.

   c. Graph the model with the scatter plot to see if the line is an excellent fit for the data.

Year	GDP ($ billions)	Year	GDP ($ billions)
2000	9951.5	2012	15,684.8
2005	12,638.4	2013	16,202.7
2007	14,061.8	2014	17,011.4
2008	14,369.1	2015	17,936.1
2009	14,119.0	2016	18,934.2
2010	14,498.9	2017	19,980.0
2011	15,075.7	2018	21,024.8

(Source: U.S. Bureau of Economic Analysis)

38. *Non-White Population* The graph gives the number of millions of individuals in the U.S. civilian non-institutional population 16 years and older who are non-White or Hispanic for the years 2000 and projected to 2050.

   a. Create a scatter plot of the data with x equal to the number of years after 2000 y equal to the population 16 years and older who are non-White or Hispanic, in millions.

   b. Find the best linear function modeling y, the size of this population, as a function of x, the number of years after 2000.

   c. Use the model to predict that population in 2035.

   d. Graph the unrounded model on the same axes as the scatter plot of the data. Is the model a "good" fit?

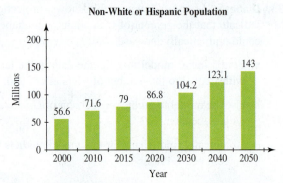

Non-White or Hispanic Population

(Source: U.S. Census Bureau)

39. *Life Expectancy at 65* The graph gives the number of additional years of life expectancy at age 65 for selected years 1950 projected to 2050.

   a. Graph the data, with y equal to the number of additional years of life expectancy and x equal to the number of years after 1950.

   b. Find a linear function that models the data, with y equal to the life expectancy and x equal to the number of years after 1950.

   c. Graph the model and the data on the same axes, and comment on the fit of the model to the data.

   d. What does the model predict that the additional life expectancy will be in 2022?

   e. Graphically find the year that the unrounded model predicts the additional life expectancy will be 21 years.

Life Expectancy

(Source: Social Security Administration)

40. *Civilian Labor Force* The table gives the size of the U.S. workforce, in millions, for selected years from 1950 and projected to 2050.

    a. Graph the data, with y equal to the number in the U.S. workforce, in millions, and x equal to the number of years after 1950.

    b. Find a linear function that models the data, with y equal to the number in millions and x equal to the number of years after 1950.

    c. Graph the data and the model on the same set of axes. Does the model fit the data exactly?

    d. What does the model predict the size of the workforce to be in 2045?

    e. Graphically determine the year in which the workforce is projected to be 182.5 million according to the model.

Year	Total Workforce (millions)
1950	62.2
1960	69.6
1970	82.7
1980	107.0
1990	125.8
2000	140.9
2010	157.7
2015	162.8
2020	164.7
2030	170.1
2040	180.5
2050	191.8

    (Source: U.S. Bureau of Labor Statistics)

41. *Obesity* People who are severely obese (BMI ≥ 40) are at most risk for serious health problems, which are the most expensive to treat. The percent of Americans who are severely obese from 2000 and projected to 2030 are shown in the table at the top of the next column.

    a. Find the linear function that models severe obesity, with x equal to the number of years after 2000 and y equal to the percent of severely obese Americans.

    b. Graph the model and the data set on the same axes, and comment on the fit.

    c. At what rate does the model indicate that severe obesity is growing during this period?

Year	2000	2010	2015	2020	2025	2030
% Severely Obese	2.2	4.9	6.4	7.9	9.5	11.1

(Source: American Journal of Preventive Medicine, vol. 42, (June 2012): 563– 570. ajpmonline.org)

42. *Male Civilian Labor Force* The table gives the male civilian labor force, in millions, for selected years from 1950 projected to 2050.

    a. Graph the data, with y equal to the number of men in the workforce, in millions, and x equal to the number of years after 1950.

    b. Find a linear function that models the data, with y equal to the number of men in the workforce, in millions, and x equal to the number of years after 1950.

    c. Graph the model and the data on the same axes, and comment on the fit of the model to the data.

    d. What does the model predict that the number of men in the workforce will be in 2051?

    e. In what year does the model predict that the population will be 99.5 million?

Year	Workforce (millions)
1950	43.8
1960	46.4
1970	51.2
1980	61.5
1990	69.0
2000	75.2
2010	82.2
2020	85.4
2030	88.5
2040	94.0
2050	100.3

    (Source: U.S. Bureau of Labor Statistics)

43. *Population of the World* The table on the following page gives the total population of the world from 1950 and projected to 2050.

    a. Create a scatter plot of the data, with y equal to the world population in billions and x equal to the number of years after 1950.

b. Find a function that models the data, with y equal to the world population in billions and x equal to the number of years after 1950. Report the model with three significant digits.

c. Graph the data points and the model on the same axes. Is the model a good fit?

d. What does the model estimate that the world population was in 2007?

e. In what year does the model estimate the world population will be 10 billion?

Year	Total World Population (billions)
1950	2.556
1960	3.039
1970	3.707
1980	4.454
1990	5.279
2000	6.083
2010	6.849
2020	7.585
2030	8.247
2040	8.850
2050	9.346

(Source: U.S. Census Bureau, International Database.)

44. *Income Tax per Capita* The table shows the U.S. personal income tax per capita from 2010 and projected to 2018.

a. Find the model that gives y, the federal income tax per capita (in dollars), as a function of x, the number of years after 2010. Report the model with four significant digits.

b. What is the predicted income tax per capita in 2020 if this model still applies?

c. When does the model predict that the income tax per capita will be $10,784?

Year	Tax ($)
2010	3530
2011	4085
2012	4376
2013	4803
2014	5369
2015	5980
2016	6465
2017	6939
2018	7348

45. *National Health Care* The table shows the national expenditures for health care, in billions, in the United States for selected years from 1995 with projections to 2020.

a. Find a linear function that models the data, with x equal to the number of years after 1990 and y equal to the expenditures for health care in billions of dollars.

b. What does this model predict the expenditure will be in 2018?

c. During what year does the model predict that the expenditures for health care will be $4659 billion?

Year	Expenditures ($ billions)
1995	1017
2000	1353
2005	1981
2010	2954
2015	3308
2020	4487

(Source: U.S. Centers for Medicare and Medicaid Services)

Systems of Linear Equations in Two Variables

SECTION PREVIEW China's Manufacturing

Figure 2.28 shows that the size of the manufacturing sector of China exceeded that of the United States in this decade. If we find the linear functions that model these graphs, with x representing the number of years past 2000 and y representing the sizes of the manufacturing sector in trillions of 2005 dollars, the point of intersection of the graphs of these functions will represent the *simultaneous* solution of the two equations because both equations will be satisfied by the coordinates of the point. (See Example 5.)

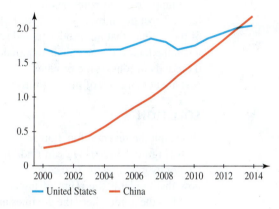

Figure 2.28

(Source: IHS Global Insight)

In this section, we solve systems of linear equations in two variables graphically, by substitution, and by the elimination method. ∎

System of Equations

A **system of linear equations** is a collection of linear equations containing the same set of variables. A system of equations can have exactly one solution, no solution, or infinitely many solutions. A *solution to a system of equations* in two variables is an ordered pair that satisfies both equations in the system. We will solve systems of two equations in two variables by graphing, by substitution, and by the elimination method.

Graphical Solution of Systems

In Section 2.1, we used the intersection method to solve a linear equation by first graphing functions representing the expressions on each side of the equation and then finding the intersection of these graphs. For example, to solve

$$3000x - 7200 = 5800x - 8600$$

we can graph

$$y_1 = 3000x - 7200 \quad \text{and} \quad y_2 = 5800x - 8600$$

and find the point of intersection to be $(0.5, -5700)$. The x-coordinate of the point of intersection of the lines is the value of x that satisfies the original equation, $3000x - 7200 = 5800x - 8600$. Thus, the solution to this equation is $x = 0.5$ (Figure 2.29). In this example, we were actually using a graphical method to solve a **system of two equations in two variables** denoted by

$$\begin{cases} y = 3000x - 7200 \\ y = 5800x - 8600 \end{cases}$$

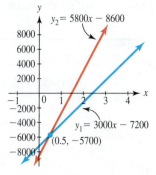

Figure 2.29

The coordinates of the point of intersection of the two graphs give the *x*- and *y*-values that satisfy both equations **simultaneously**, and these values are called the **solution** to the system. The following example uses the graphical method to solve a system of equations in two variables.*

EXAMPLE 1 ▶ Break-Even

A company is said to **break even** from the production and sale of a product if the total revenue equals the total cost—that is, if $R(x) = C(x)$. Because profit $P(x) = R(x) - C(x)$, we can also say that the company breaks even if the profit for the product is zero.

Suppose a company has its total revenue for a product given by $R = 5585x$ and its total cost given by $C = 61,740 + 440x$, where *x* is the number of thousands of tons of the product that are produced and sold per year. The company is said to break even when the total revenue equals the total cost—that is, when $R = C$. Find the number of thousands of tons of the product that gives break-even and how much the revenue and cost are at that level of production.

SOLUTION

We graph the revenue function as $y_1 = 5585x$ and the cost function as $y_2 = 61,740 + 440x$ (Figure 2.30(a)). We can find break-even with the **intersection method** on a calculator by graphing the two equations, $y_1 = 5585x$ and $y_2 = 61,740 + 440x$, on a window that contains the point of intersection and then finding the point of intersection, which is the point where the *y*-values are equal.* This point, which gives break-even, is $(12, 67,020)$ (Figure 2.30(b)).

(a)   (b)

Figure 2.30

Thus, the company will break even on this product if 12 thousand tons of the product are sold, when both the cost and revenue equal $67,020.

It is frequently necessary to solve each equation for a variable so that the equation can be graphed with a graphing utility. It is also necessary to find a viewing window that contains the point of intersection. Consider the following example.

EXAMPLE 2 ▶ Solving a System of Linear Equations

Solve the system

$$\begin{cases} 3x - 4y = 21 \\ 2x + 5y = -9 \end{cases}$$

---
* See Appendix A, page 671, for details.

SOLUTION

To solve this system with a graphing utility, we first solve both equations for y:

$$3x - 4y = 21 \qquad\qquad 2x + 5y = -9$$
$$-4y = 21 - 3x \qquad\qquad 5y = -9 - 2x$$
$$y = \frac{21 - 3x}{-4} = \frac{3x - 21}{4} \qquad y = \frac{-9 - 2x}{5}$$

Graphing these equations with a window that contains the point of intersection (Figure 2.31 (a)) and finding the point of intersection (Figure 2.31(b)) gives $x = 3$, $y = -3$, so the solution is $(3, -3)$.*

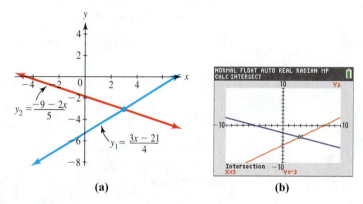

(a)   **(b)**

Figure 2.31

Solution by Substitution

Graphing with a graphing utility is not always the easiest method to use to solve a system of equations because the equations must be solved for y to be entered in the utility and an appropriate window must be found. A second solution method for a system of linear equations is the **substitution method**, where one equation is solved for a variable and that variable is replaced by the equivalent expression in the other equation.

The substitution method is illustrated by the following example, which finds market equilibrium. The quantity of a product that is demanded by consumers is called the **demand** for the product, and the quantity that is supplied is called the **supply**. In a free economy, both demand and supply are related to the price, and the price where the number of units demanded equals the number of units supplied is called the **equilibrium price**.

EXAMPLE 3 ▶ | **Market Equilibrium**

Suppose that the daily demand for a product is given by $p = 200 - 2q$, where q is the number of units demanded and p is the price per unit in dollars, and that the daily supply is given by $p = 60 + 5q$, where q is the number of units supplied and p is the price in dollars. If a price results in more units being supplied than demanded, we say there is a *surplus*, and if the price results in fewer units being supplied than demanded,

---

* For a discussion of solving systems of equations using Excel, see Appendix B, page 696.

we say there is a *shortfall*. **Market equilibrium** occurs when the supply quantity equals the demand quantity (and when the prices are equal)—that is, when q and p both satisfy the system

$$\begin{cases} p = 200 - 2q \\ p = 60 + 5q \end{cases}$$

a. If the price is $140, how many units are supplied and how many are demanded?

b. Does this price give a surplus or a shortfall of the product?

c. What price gives market equilibrium?

SOLUTION

a. If the price is $140, the number of units demanded satisfies $140 = 200 - 2q$, or $q = 30$, and the number of units supplied satisfies $140 = 60 + 5q$, or $q = 16$.

b. At this price, the quantity supplied is less than the quantity demanded, so a shortfall occurs.

c. Because market equilibrium occurs where q and p both satisfy the system

$$\begin{cases} p = 200 - 2q \\ p = 60 + 5q \end{cases}$$

we seek the solution to this system.

We can solve this system by substitution. Substituting $60 + 5q$ for p in the first equation gives

$$60 + 5q = 200 - 2q$$

Solving this equation gives

$$60 + 5q = 200 - 2q$$
$$7q = 140$$
$$q = 20$$

Thus, market equilibrium occurs when the number of units is 20, and the equilibrium price is

$$p = 200 - 2(20) = 60 + 5(20) = 160 \text{ dollars per unit}$$

The substitution in Example 3 was not difficult because both equations were solved for p. In general, we use the following steps to solve systems of two equations in two variables by substitution.

Solution of Systems of Equations by Substitution

1. Solve one of the equations for one of the variables in terms of the other variable.

2. Substitute the expression from step 1 into the other equation to give an equation in one variable.

3. Solve the linear equation for the variable.

4. Substitute this solution into the equation from step 1 or into one of the original equations and solve this equation for the second variable.

5. Check the solution in both original equations or check graphically.

EXAMPLE 4 ▶ Solution by Substitution

Solve the system $\begin{cases} 3x + 4y = 10 \\ 4x - 2y = 6 \end{cases}$ by substitution.

SOLUTION

To solve this system by substitution, we can solve either equation for either variable and substitute the resulting expression into the other equation. Solving the second equation for y gives

$$4x - 2y = 6$$
$$-2y = -4x + 6$$
$$y = 2x - 3$$

Substituting this expression for y in the first equation gives

$$3x + 4(2x - 3) = 10$$

Solving this equation gives

$$3x + 8x - 12 = 10$$
$$11x = 22$$
$$x = 2$$

Substituting $x = 2$ into $y = 2x - 3$ gives $y = 2(2) - 3 = 1$, so the solution to the system is $x = 2, y = 1$, or $(2, 1)$.

Checking shows that this solution satisfies both original equations.

EXAMPLE 5 ▶ China's Manufacturing

Figure 2.32 shows the size of the manufacturing sector of China and that of the United States in this century. Suppose these graphs for China and the United States can be modeled by the functions

$$y = 0.158x - 0.0949 \quad \text{and} \quad y = 0.037x + 1.477$$

respectively, with x representing the years after 2000 and y representing the sizes of the manufacturing sector in trillions of 2005 dollars. Find the year during which China reached the United States on its way to becoming the world's largest manufacturer.

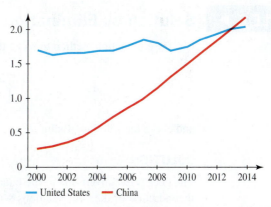

Figure 2.32

(Source: IHS Global Insight)

SOLUTION

China's manufacturing sector reaches that of the United States when the y-values of the two models are equal, so we solve the following system by substitution:

$$\begin{cases} y = 0.158x - 0.0949 \\ y = 0.037x + 1.477 \end{cases}$$

$$0.158x - 0.0949 = 0.037x + 1.477$$
$$0.121x = 1.5719$$
$$x = 12.99 \approx 13$$

Thus, the size of China's manufacturing sector is estimated to have reached that of the United States in 2013. We can check the solution graphically by graphing the two functions and using intersect on a graphing calculator. (See Figure 2.33.)

Figure 2.33

Solution by Elimination

A second algebraic method, called the **elimination method**, is frequently an easier method to use to solve a system of linear equations. The elimination method is based on rewriting one or both of the equations in an equivalent form that allows us to eliminate one of the variables by adding or subtracting the equations.

> ### Solving a System of Two Equations in Two Variables by Elimination
>
> 1. If necessary, multiply one or both equations by a nonzero number that will make the coefficients of one of the variables in the equations equal, except perhaps for sign.
> 2. Add or subtract the equations to eliminate one of the variables.
> 3. Solve for the variable in the resulting equation.
> 4. Substitute the solution from step 3 into one of the original equations and solve for the second variable.
> 5. Check the solutions in the remaining original equation, or check graphically.

EXAMPLE 6 ▶   **Solution by Elimination**

Use the elimination method to solve the system

$$\begin{cases} 3x + 4y = 10 \\ 4x - 2y = 6 \end{cases}$$

and check the solution graphically.

SOLUTION

The goal is to convert one of the equations into an equivalent equation of a form such that addition of the two equations will eliminate one of the variables. Notice that the coefficient of y in the second equation, -2, is a factor of the coefficient of y in the first equation, 4. If we multiply both sides of the second equation by 2 and add the two equations, this will eliminate the y-variable.

$$\begin{cases} 3x + 4y = 10 & (1) \\ 4x - 2y = 6 & (2) \end{cases}$$

$$\begin{cases} 3x + 4y = 10 & (1) \\ 8x - 4y = 12 & (3) \end{cases}$$ Multiply 2 times Equation (2), getting equivalent Equation (3).

$$11x \quad = 22$$ Add Equations (1) and (3) to eliminate y.

$$x \quad = 2$$ Solve the new equation for x.

Substituting $x = 2$ in the first equation gives $3(2) + 4y = 10$, or $y = 1$. Thus, the solution to the system is $x = 2$, $y = 1$, or $(2, 1)$.

To check graphically, we solve both equations for y, getting $y_1 = \dfrac{10 - 3x}{4}$ and $y_2 = \dfrac{4x - 6}{2}$, then graph the equations and find the point of intersection to be $(2, 1)$ (Figure 2.34).

Figure 2.34

Modeling Systems of Linear Equations

Solving some real problems requires us to create two or more equations whose simultaneous solution is the solution to the problem. Consider the following examples.

EXAMPLE 7 ▶   Investments

An investor has $300,000 to invest, part at 12% and the remainder in a less risky investment at 7%. If her investment goal is to have an annual income of $27,000, how much should she put in each investment?

SOLUTION

If we denote the amount invested at 12% as x and the amount invested at 7% as y, the sum of the investments is $x + y$, so we have the equation

$$x + y = 300,000$$

The annual income from the 12% investment is $0.12x$, and the annual income from the 7% investment is $0.07y$. Thus, the desired annual income from the two investments is

$$0.12x + 0.07y = 27,000$$

We can write the given information as a system of equations:

$$\begin{cases} x + y = 300,000 \\ 0.12x + 0.07y = 27,000 \end{cases}$$

To solve this system, we multiply the first equation by -0.12 and add the two equations. This results in an equation with one variable:

$$\begin{cases} -0.12x - 0.12y = -36,000 \\ 0.12x + 0.07y = 27,000 \end{cases}$$

$$-0.05y = -9000$$

$$y = 180,000$$

Substituting 180,000 for y in the first original equation and solving for x gives $x = 120{,}000$. Thus, \$120,000 should be invested at 12%, and \$180,000 should be invested at 7%.

To check this solution, we see that the total investment is \$120,000 + \$180,000, which equals \$300,000. The interest earned at 12% is $\$120{,}000(0.12) = \$14{,}400$, and the interest earned at 7% is $\$180{,}000(0.07) = \$12{,}600$. The total interest is \$14,400 + \$12,600, which equals \$27,000. This agrees with the given information.

EXAMPLE 8 ▶   ## Medication

A nurse has two solutions that contain different concentrations of a certain medication. One is a 12% concentration, and the other is an 8% concentration. How many cubic centimeters (cc) of each should she mix together to obtain 20 cc of a 9% solution?

SOLUTION

We begin by denoting the amount of the first solution by x and the amount of the second solution by y. The total amount of solution is the sum of x and y, so

$$x + y = 20$$

The total medication in the combined solution is 9% of 20 cc, or $0.09(20) = 1.8$ cc, and the mixture is obtained by adding $0.12x$ and $0.08y$, so

$$0.12x + 0.08y = 1.8$$

We can use substitution to solve the system

$$\begin{cases} x + y = 20 \\ 0.12x + 0.08y = 1.8 \end{cases}$$

Substituting $20 - x$ for y in $0.12x + 0.08y = 1.8$ gives $0.12x + 0.08(20 - x) = 1.8$, and solving this equation gives

$$0.12x + 0.08(20 - x) = 1.8$$
$$0.12x + 1.6 - 0.08x = 1.8$$
$$0.04x = 0.2$$
$$x = 5$$

Thus, combining 5 cc of the first solution with $20 - 5 = 15$ cc of the second solution gives 20 cc of the 9% solution.

Dependent and Inconsistent Systems

The system of linear equations discussed in Example 2 has a unique solution, shown as the point of intersection of the graphs. It is possible that two equations in a system of linear equations in two variables describe the same line. When this happens, the equations are equivalent, and the values that satisfy one equation are also solutions to the other equation and to the system. Such a system is a **dependent system**. If a system contains two equations whose graphs are parallel lines, they have no point in common, and thus the system has no solution. Such a system of equations is **inconsistent**. Figure 2.35(a)–(c) represents these three situations: systems that have a unique solution, many solutions (dependent system), and no solution (inconsistent system), respectively. Note that the slopes of the lines are equal in Figure 2.35(b) and in Figure 2.35(c).

Figure 2.35

EXAMPLE 9 ▶ ## Systems with Nonunique Solutions

Use the elimination method to solve each of the following systems, if possible. Verify the solution graphically.

a. $\begin{cases} 2x - 3y = 4 \\ 6x - 9y = 12 \end{cases}$ **b.** $\begin{cases} 2x - 3y = 4 \\ 6x - 9y = 36 \end{cases}$

SOLUTION

a. To solve $\begin{cases} 2x - 3y = 4 \\ 6x - 9y = 12 \end{cases}$, we multiply the first equation by -3 and add the equations, getting

$$\begin{cases} -6x + 9y = -12 \\ 6x - 9y = 12 \\ \hspace{1.5em} 0 = 0 \end{cases}$$

This indicates that the graphs of the equations intersect when $0 = 0$, *which is always true*. Thus, any values of x and y that satisfy one of these equations also satisfy the other, and there are *infinitely many* solutions. Figure 2.36(a) shows that the graphs of the equations lie on the same line. Notice that the second equation is a multiple of the first, so the equations are equivalent. This system is *dependent*.

The infinitely many solutions all satisfy both of the two equations. That is, they are values of x and y that satisfy

$$2x - 3y = 4, \quad \text{or} \quad y = \frac{2}{3}x - \frac{4}{3}$$

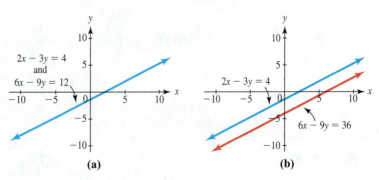

Figure 2.36

b. To solve $\begin{cases} 2x - 3y = 4 \\ 6x - 9y = 36 \end{cases}$, we multiply the first equation by -3 and add the equations, getting

$$\begin{cases} -6x + 9y = -12 \\ 6x - 9y = 36 \end{cases}$$
$$0 = 24$$

This indicates that the equations intersect when $0 = 24$, *which is never true*. Thus, no values of x and y satisfy both of the equations. Figure 2.36(b) shows that the graphs of the equations are parallel. This system is *inconsistent*.

EXAMPLE 10 ▶   Investment

Members of an investment club have set a goal of earning 15% on the money they invest in stocks. They are considering buying two stocks, for which the cost per share and the projected growth per share (both in dollars) are summarized in Table 2.9.

Table 2.9

	Utility	Technology
Cost/share	$30	$45
Growth/share	$4.50	$6.75

a. If they have $180,000 to invest, how many shares of each stock should they buy to meet their goal?

b. If they buy 1800 shares of the utility stock, how many shares of the technology stock should they buy to meet their goal?

SOLUTION

a. The money available to invest in stocks is $180,000, so if x is the number of utility shares and y is the number of technology shares purchased, we have

$$30x + 45y = 180,000$$

A 15% return on their investment would be $0.15(180,000) = 27,000$ dollars, so we have

$$4.50x + 6.75y = 27,000$$

To find x and y, we solve the system

$$\begin{cases} 30x + 45y = 180,000 \\ 4.50x + 6.75y = 27,000 \end{cases}$$

Multiplying 4.5 times both sides of the first equation and -30 times both sides of the second equation gives

$$\begin{cases} 135x + 202.5y = 810,000 \\ -135x - 202.5y = -810,000 \end{cases}$$

Adding the equations gives $0 = 0$, so the system is dependent, with many solutions.

The number of shares of each stock that can be purchased satisfies both of the two original equations. In particular, it satisfies $30x + 45y = 180{,}000$, so

$$y = \frac{180{,}000 - 30x}{45}, \quad \text{or} \quad y = \frac{12{,}000 - 2x}{3}$$

with x between 0 and 6000 shares and y between 0 and 4000 shares (because neither x nor y can be negative).

b. Substituting 1800 for x in the equation gives $y = 2800$, so if they buy 1800 shares of the utility stock, they should buy 2800 shares of the technology stock to meet their goal.

SKILLS CHECK 2.3

Determine if each ordered pair is a solution of the system of equations given.

1. $\begin{cases} 2x + 3y = -1 \\ x - 4y = -6 \end{cases}$

    a. $(2, 1)$ **b.** $(-2, 1)$

2. $\begin{cases} 4x - 2y = 7 \\ -2x + 2y = -4 \end{cases}$

    a. $\left(\dfrac{3}{2}, -\dfrac{1}{2} \right)$ **b.** $\left(\dfrac{1}{2}, -\dfrac{3}{2} \right)$

3. What are the coordinates of the point of intersection of $y = 3x - 2$ and $y = 3 - 2x$?

4. Give the coordinates of the point of intersection of $3x + 2y = 5$ and $5x - 3y = 21$.

In Exercises 5–8, solve the systems of equations graphically.

5. $\begin{cases} y = 3x - 12 \\ y = 4x + 2 \end{cases}$ **6.** $\begin{cases} 2x - 4y = 6 \\ 3x + 5y = 20 \end{cases}$

7. $\begin{cases} 4x - 3y = -4 \\ 2x - 5y = -4 \end{cases}$ **8.** $\begin{cases} 5x - 6y = 22 \\ 4x - 4y = 16 \end{cases}$

9. Does the system $\begin{cases} 2x + 5y = 6 \\ x + 2.5y = 3 \end{cases}$ have a unique solution, no solution, or many solutions? What does this mean graphically?

10. Does the system $\begin{cases} 6x + 4y = 3 \\ 3x + 2y = 3 \end{cases}$ have a unique solution, no solution, or many solutions? What does this mean graphically?

In Exercises 11–14, solve the systems of equations by substitution.

11. $\begin{cases} x = 5y + 12 \\ 3x + 4y = -2 \end{cases}$ **12.** $\begin{cases} 2x - 3y = 2 \\ y = 5x - 18 \end{cases}$

13. $\begin{cases} 2x - 3y = 5 \\ 5x + 4y = 1 \end{cases}$ **14.** $\begin{cases} 4x - 5y = -17 \\ 3x + 2y = -7 \end{cases}$

In Exercises 15–24, solve the systems of equations by elimination, if a solution exists.

15. $\begin{cases} x + 3y = 5 \\ 2x + 4y = 8 \end{cases}$ **16.** $\begin{cases} 4x - 3y = -13 \\ 5x + 6y = 13 \end{cases}$

17. $\begin{cases} 5x = 8 - 3y \\ 2x + 4y = 8 \end{cases}$ **18.** $\begin{cases} 3y = 5 - 3x \\ 2x + 4y = 8 \end{cases}$

19. $\begin{cases} 0.3x + 0.4y = 2.4 \\ 5x - 3y = 11 \end{cases}$ **20.** $\begin{cases} 8x - 4y = 0 \\ 0.5x + 0.3y = 2.2 \end{cases}$

21. $\begin{cases} 3x + 6y = 12 \\ 4y - 8 = -2x \end{cases}$ **22.** $\begin{cases} 6y - 12 = 4x \\ 10x - 15y = -30 \end{cases}$

23. $\begin{cases} 6x - 9y = 12 \\ 3x - 4.5y = -6 \end{cases}$ **24.** $\begin{cases} 4x - 8y = 5 \\ 6x - 12y = 10 \end{cases}$

In Exercises 25–34, solve the systems of equations by any convenient method, if a solution exists.

25. $\begin{cases} y = 3x - 2 \\ y = 5x - 6 \end{cases}$ **26.** $\begin{cases} y = 8x - 6 \\ y = 14x - 12 \end{cases}$

27. $\begin{cases} 4x + 6y = 4 \\ x = 4y + 8 \end{cases}$ **28.** $\begin{cases} y = 4x - 5 \\ 3x - 4y = 7 \end{cases}$

29. $\begin{cases} 2x - 5y = 16 \\ 6x - 8y = 34 \end{cases}$ **30.** $\begin{cases} 4x - y = 4 \\ 6x + 3y = 15 \end{cases}$

31. $\begin{cases} 3x = 7y - 1 \\ 4x = 11 - 3y \end{cases}$ **32.** $\begin{cases} 5x = 12 + 3y \\ -5y = 8 - 3x \end{cases}$

33. $\begin{cases} 4x - 3y = 9 \\ 8x - 6y = 16 \end{cases}$ **34.** $\begin{cases} 5x - 4y = 8 \\ -15x + 12y = -12 \end{cases}$

EXERCISES 2.3

35. Break-Even A manufacturer of kitchen sinks has total revenue given by $R = 76.50x$ and has total cost given by $C = 2970 + 27x$, where x is the number of sinks produced and sold. Use graphical methods to find the number of units that gives break-even for the product.

36. Break-Even A jewelry maker has total revenue for her bracelets given by $R = 89.75x$ and incurs a total cost of $C = 23.50x + 1192.50$, where x is the number of bracelets produced and sold. Use graphical methods to find the number of units that gives break-even for the product.

37. Break-Even A manufacturer of reading lamps has total revenue given by $R = 15.80x$ and total cost given by $C = 8593.20 + 3.20x$, where x is the number of units produced and sold. Use a nongraphical method to find the number of units that gives break-even for this product.

38. Break-Even A manufacturer of automobile air conditioners has total revenue given by $R = 136.50x$ and total cost given by $C = 9661.60 + 43.60x$, where x is the number of units produced and sold. Use a nongraphical method to find the number of units that gives break-even for this product.

39. Market Equilibrium The demand for a brand of clock radio is given by $p + 2q = 320$, and the supply for these radios is given by $p - 8q = 20$, where p is the price and q is the number of clock radios. Solve the system containing these two equations to find (a) the price at which the quantity demanded equals the quantity supplied and (b) the equilibrium quantity.

40. Supply and Demand A certain product has supply and demand functions given by $p = 5q + 20$ and $p = 128 - 4q$, respectively.

   a. If the price p is $60, how many units q are supplied and how many are demanded?

   b. What price gives market equilibrium, and how many units are demanded and supplied at this price?

41. Concerta and Ritalin Concerta and Ritalin are two different brands of a drug used to treat ADHD.

   a. Use the fact that market share for Concerta was $y = 2.4\%$ for August 25, 2000 $(x = 0)$, and was 10% eleven weeks later to write a linear function representing its market share as a function of time.

   b. Use the fact that market share for Ritalin was $y = 7.7\%$ for August 25, 2000 $(x = 0)$, and was 6.9% eleven weeks later to write a linear function representing its market share as a function of time.

   c. Find the number of weeks past the release date of Concerta (August 25) that the weekly market share of Concerta reached that of Ritalin.
(Source: *Newsweek*, December 4, 2000)

42. Market Equilibrium Wholesalers' willingness to sell laser printers is given by the supply function $p = 50.50 + 0.80q$, and retailers' willingness to buy the printers is given by $p = 400 - 0.70q$, where p is the price per printer in dollars and q is the number of printers. What price will give market equilibrium for the printers?

43. Population Demographics The number of millions of individuals in the U.S. civilian noninstitutional population 16 years and older who are White, non-Hispanic can be modeled by $y = 0.293x + 157.8454$, and the remainder of the population can be modeled by $1.73x + 54.094$, where x is the number of years after 2000. If the growth continues according to these models, in what year will the populations be equal?
(Source: U.S. Census Bureau)

44. College Enrollment Suppose the percent of males who enrolled in college within 12 months of high school graduation is given by $y = -0.126x + 55.72$ and the percent of females who enrolled in college within 12 months of high school graduation is given by $y = 0.73x + 39.7$, where x is the number of years after 1960. Use graphical methods to find the year these models indicate that the percent of females equaled the percent of males.
(Source: *Statistical Abstract of the United States*)

45. U.S. Population Using data and projections from 1980 through 2050, the percent of Hispanics in the U.S. civilian noninstitutional population is given by $y = 0.224x + 9.0$ and the percent of blacks is given by $y = 0.057x + 12.3$, where x is the number of years after 1990. During what year did the percent of Hispanics equal the percent of blacks in the United States?

46. Age at First Marriage and First Birth The graph shows the "Great Crossover" where the median age of women at first marriage and the median age at giving first birth are equal. After the year at this crossover point, many women are having children before getting married.

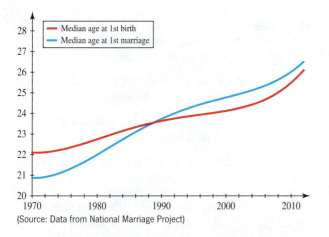

(Source: Data from National Marriage Project)

   a. Use the points $(1970, 22.1)$ and $(2010, 25.5)$ to find the equation of a line that models the median age at giving first birth as a function of the year.

   b. Use the points $(1970, 20.9)$ and $(2010, 26.1)$ to find the equation of a line that models the median age for women at first marriage as a function of the year.

   c. Use the models from parts (a) and (b) to approximate the year in which the median age at giving first birth is equal to the median age at first marriage. Does this approximation agree with the crossover point in the accompanying graph

47. Revenue The sum of the 2016 revenue and twice the 2013 revenue for Mama Joan's International, Inc., is $2144.9 million. The difference between the 2016 and 2013 revenues is $135.5 million. If Mama Joan's revenue between 2013 and 2016 is an increasing linear function, find the 2013 and 2016 revenues.

48. Stock Prices The sum of the high and low prices of a share of stock in Johns, Inc., in 2012 is $83.50, and the difference between these two prices in 2012 is $21.88. Find the high and low prices.

49. Pricing A concert promoter needs to make $84,000 from the sale of 2400 tickets. The promoter charges $30 for some tickets and $45 for the others.

   a. If there are x of the $30 tickets sold and y of the $45 tickets sold, write an equation that states that the total number of tickets sold is 2400.

   b. How much money is received from the sale of x tickets for $30 each?

   c. How much money is received from the sale of y tickets for $45 each?

   d. Write an equation that states that the total amount received from the sale is $84,000.

   e. Solve the equations simultaneously to find how many tickets of each type must be sold to yield the $84,000.

50. Rental Income A woman has $500,000 invested in two rental properties. One yields an annual return of 10% of her investment, and the other returns 12% per year on her investment. Her total annual return from the two investments is $53,000. Let x represent the amount of the 10% investment and y represent the amount of the 12% investment.

   a. Write an equation that states that the sum of the investments is $500,000.

   b. What is the annual return on the 10% investment?

   c. What is the annual return on the 12% investment?

   d. Write an equation that states that the sum of the annual returns is $53,000.

   e. Solve these two equations simultaneously to find how much is invested in each property.

51. Investment One safe investment pays 8% per year, and a more risky investment pays 12% per year.

   a. How much must be invested in each account if an investor of $100,000 would like a return of $9000 per year?

   b. Why might the investor use two accounts rather than put all the money in the 12% investment?

52. Investment A woman invests $52,000 in two different mutual funds, one that averages 10% per year and another that averages 14% per year. If her average annual return on the two mutual funds is $5720, how much did she invest in each fund?

53. Investment Jake has $250,000 to invest. He chooses one money market fund that pays 6.6% and a mutual fund that has more risk but has averaged 8.6% per year. If his goal is to average 7% per year with minimal risk, how much should he invest in each fund?

54. *Investment* Sue chooses one money market fund that pays 6.2% and a mutual fund that has more risk but has averaged 9.2% per year. If she has $300,000 to invest and her goal is to average 7.6% per year with minimal risk, how much should she invest in each fund?

55. *Medication* A pharmacist wants to mix two solutions to obtain 100 cc of a solution that has an 8% concentration of a certain medicine. If one solution has a 10% concentration of the medicine and the second has a 5% concentration, how much of each of these solutions should she mix?

56. *Medication* A pharmacist wants to mix two solutions to obtain 200 cc of a solution that has a 12% concentration of a certain medicine. If one solution has a 16% concentration of the medicine and the second has a 6% concentration, how much of each solution should she mix?

57. *Nutrition* A glass of skim milk supplies 0.1 mg of iron and 8.5 g of protein. A quarter pound of lean meat provides 3.4 mg of iron and 22 g of protein. A person on a special diet is to have 7.1 mg of iron and 69.5 g of protein. How many glasses of skim milk and how many quarter-pound servings of meat will provide this?

58. *Nutrition* Each ounce of substance A supplies 6% of a nutrient a patient needs, and each ounce of substance B supplies 10% of the required nutrient. If the total number of ounces given to the patient was 14 and 100% of the nutrient was supplied, how many ounces of each substance was given?

59. *Alcohol Use* According to the National Household Survey on Drug Abuse by the U.S. Department of Health and Human Services, the pattern of higher rates of current alcohol use, binge alcohol use, and heavy alcohol use among full-time college students than among others aged 18 to 22 has remained consistent since 2002. (See the figure.)

(Source: National Survey on Drug Abuse, U.S. Department of Health and Human Services)

Using data from 2002 to 2008, the equations representing the percents of young adults aged 18 to 22 who used alcohol are

Enrolled in college: $y = -0.282x + 19.553$

Not enrolled: $y = -0.086x + 13.643$

where x represents the number of years after 2000. Solve this system, if possible, to determine when the percent for those enrolled in college will equal that for those not enrolled. What will the percent be?

60. *Medication* A nurse has two solutions that contain different concentrations of a certain medication. One is a 30% concentration, and the other is a 15% concentration. How many cubic centimeters (cc) of each should she mix to obtain 45 cc of a 20% solution?

61. *Supply and Demand* The table below gives the quantity of graphing calculators demanded and the quantity supplied for selected prices.

   a. Find the linear equation that gives the price as a function of the quantity demanded.

   b. Find the linear equation that gives the price as a function of the quantity supplied.

   c. Use these equations to find the market equilibrium price.

Price ($)	Quantity Demanded (thousands)	Quantity Supplied (thousands)
50	210	0
60	190	40
70	170	80
80	150	120
100	110	200

62. *Market Analysis* The supply function and the demand function for a product are linear and are determined by the table that follows. Create the supply and demand functions and find the price that gives market equilibrium.

Supply Price	Function Quantity	Demand Price	Function Quantity
200	400	400	400
400	800	200	800
600	1200	0	1200

63. *Social Security* Persons scheduled to receive a $1000 monthly Social Security benefit at a full retirement age of 66 can begin drawing benefits at age 62 and receive $750 per month. At what age will these per-

sons who started taking benefits at age 66 have the same in total benefits paid as those who started taking benefits at age 62?
(Source: Social Security Administration)

64. *Medication* Suppose combining x cubic centimeters (cc) of a 20% concentration of a medication and y cc of a 5% concentration of the medication gives $(x + y)$ cc of a 15.5% concentration. If 7 cc of the 20% concentration are added, by how much must the amount of 5% concentration be increased to keep the same concentration?

65. *Social Agency* A social agency provides emergency food and shelter to two groups of clients. The first group has x clients who need an average of $300 for emergencies, and the second group has y clients who need an average of $200 for emergencies. The agency has $100,000 to spend for these two groups.

   a. Write an equation that describes the maximum number of clients who can be served with the $100,000.

   b. If the first group has twice as many clients as the second group, how many clients are in each group if all the money is spent?

66. *Market Equilibrium* A retail chain will buy 800 televisions if the price is $350 each and 1200 if the price is $300. A wholesaler will supply 700 of these televisions at $280 each and 1400 at $385 each. Assuming that the supply and demand functions are linear, find the market equilibrium point and explain what it means.

67. *Market Equilibrium* A retail chain will buy 900 cordless phones if the price is $10 each and 400 if the price is $60. A wholesaler will supply 700 phones at $30 each and 1400 at $50 each. Assuming that the supply and demand functions are linear, find the market equilibrium point and explain what it means.

2.4 Solutions of Linear Inequalities

KEY OBJECTIVES

- Solve linear inequalities algebraically

- Solve linear inequalities graphically with the intersection and x-intercept methods

- Solve double inequalities algebraically and graphically

SECTION PREVIEW   Profit

For an electronic reading device, the respective weekly revenue and weekly cost are given by

$$R(x) = 400x \quad \text{and} \quad C(x) = 200x + 16,000$$

where x is the number of units produced and sold. For what levels of production will a profit result?

Profit will occur when revenue is greater than cost. So we find the level of production and sale x that gives a profit by solving the **linear inequality**

$$R(x) > C(x), \quad \text{or} \quad 400x > 200x + 16,000$$

(See Example 2.) In this section, we will solve linear inequalities of this type algebraically and graphically. ■

Algebraic Solution of Linear Inequalities

An **inequality** is a statement that one quantity or expression is greater than, less than, greater than or equal to, or less than or equal to another.

> ### Linear Inequality
>
> A linear inequality (or first-degree inequality) in the variable x is an inequality that can be written in the form $ax + b > 0$, where $a \neq 0$. (The inequality symbol can be $>$, $\geq$, $<$, or $\leq$.)

The inequality $4x + 3 < 7x - 6$ is a linear inequality (or first-degree inequality) because the highest power of the variable (x) is 1. The values of x that satisfy the inequality form the solution set for the inequality. For example, 5 is in the solution set of this inequality because substituting 5 into the inequality gives

$$4 \cdot 5 + 3 < 7 \cdot 5 - 6, \quad \text{or} \quad 23 < 29$$

which is a true statement. On the other hand, 2 is not in the solution set because

$$4 \cdot 2 + 3 \not< 7 \cdot 2 - 6$$

Solving an inequality means finding its solution set. The solution to an inequality can be written as an inequality or in interval notation. The solution can also be represented by a graph on a real number line.

Two inequalities are *equivalent* if they have the same solution set.

We use the properties of inequalities discussed in the Algebra Toolbox to solve an inequality. In general, the steps used to solve a linear inequality are the same as those used to solve a linear equation, except that the inequality symbol is reversed if both sides are multiplied (or divided) by a negative number.

Steps for Solving a Linear Inequality Algebraically

1. If a linear inequality contains fractions with constant denominators, multiply both sides of the inequality by a positive number that will remove all denominators in the inequality. If there are two or more fractions, use the least common denominator (LCD) of the fractions.

2. Remove any parentheses by multiplication.

3. Perform any additions or subtractions to get all terms containing the variable on one side and all other terms on the other side of the inequality. Combine like terms.

4. Divide both sides of the inequality by the coefficient of the variable. *Reverse the inequality symbol if this number is negative.*

5. Check the solution by substitution or with a graphing utility. If a real-world solution is desired, check the algebraic solution for reasonableness in the real-world situation.

EXAMPLE 1 ▶   Solution of a Linear Inequality

Solve the inequality $3x - \dfrac{1}{3} \le -4 + x$.

SOLUTION

To solve the inequality

$$3x - \frac{1}{3} \le -4 + x$$

first multiply both sides by 3:

$$3\left(3x - \frac{1}{3}\right) \le 3(-4 + x)$$

Removing parentheses gives

$$9x - 1 \le -12 + 3x$$

Performing additions and subtractions to both sides to get the variables on one side and the constants on the other side gives

$$6x \leq -11$$

Dividing both sides by the coefficient of the variable gives

$$x \leq -\frac{11}{6}$$

The solution set contains all real numbers less than or equal to $-\frac{11}{6}$. The graph of the solution set $\left(-\infty, -\frac{11}{6}\right]$ is shown in Figure 2.37.

Figure 2.37

EXAMPLE 2 ▶ Profit

For an electronic reading device, the weekly revenue and weekly cost (in dollars) are given by

$$R(x) = 400x \quad \text{and} \quad C(x) = 200x + 16{,}000$$

respectively, where x is the number of units produced and sold. For what levels of production will a profit result?

SOLUTION

Profit will occur when revenue is greater than cost. So we find the level of production and sale that gives a profit by solving the linear inequality $R(x) > C(x)$, or $400x > 200x + 16{,}000$.

$$400x > 200x + 16{,}000$$
$$200x > 16{,}000 \qquad \text{Subtract } 200x \text{ from both sides.}$$
$$x > 80 \qquad \text{Divide both sides by 200.}$$

Thus, a profit occurs if more than 80 units are produced and sold.

EXAMPLE 3 ▶ Body Temperature

A child's health is at risk when his or her body temperature is 103°F or higher. What Celsius temperature reading would indicate that a child's health was at risk?

SOLUTION

A child's health is at risk if $F \geq 103$, and $F = \frac{9}{5}C + 32$, where F is the temperature in degrees Fahrenheit and C is the temperature in degrees Celsius. Substituting $\frac{9}{5}C + 32$ for F, we have

$$\frac{9}{5}C + 32 \geq 103$$

Now we solve the inequality for C:

$$\frac{9}{5}C + 32 \geq 103$$

$$9C + 160 \geq 515 \qquad \text{Multiply both sides by 5 to clear fractions.}$$

$$9C \geq 355 \qquad \text{Subtract 160 from both sides.}$$

$$C \geq 39.\overline{4} \qquad \text{Divide both sides by 9.}$$

Thus, a child's health is at risk if his or her Celsius temperature is approximately 39.4° or higher.

Graphical Solution of Linear Inequalities

In Section 2.1, we used graphical methods to solve linear equations. In a similar manner, graphical methods can be used to solve linear inequalities. We will illustrate both the intersection of graphs method and the x-intercept method.*

Intersection Method

To solve an inequality by the intersection method, we use the following steps.

> ### Steps for Solving a Linear Inequality with the Intersection Method
>
> 1. Set the left side of the inequality equal to y_1, set the right side equal to y_2, and graph the equations using your graphing utility.
>
> 2. Choose a viewing window that contains the point of intersection and find the point of intersection, with x-coordinate a. This is the value of x where $y_1 = y_2$.
>
> 3. The values of x that satisfy the inequality represented by $y_1 < y_2$ are those values for which the graph of y_1 is below the graph of y_2. The values of x that satisfy the inequality represented by $y_1 > y_2$ are those values of x for which the graph of y_1 is above the graph of y_2.

To solve the inequality

$$5x + 2 < 2x + 6$$

by using the intersection method, let

$$y_1 = 5x + 2 \quad \text{and} \quad y_2 = 2x + 6$$

Graphing the equations (Figure 2.38) shows that their point of intersection occurs at $x = \frac{4}{3}$. Some graphing utilities show this answer in the form $x = 1.3333333$. The exact x-value $\left(x = \frac{4}{3} \right)$ can also be found by solving the equation $5x + 2 = 2x + 6$ algebraically. Figure 2.38 shows that the graph of y_1 is below the graph of y_2 when x is

---

* See Appendix A, page 672.

less than $\frac{4}{3}$. Thus, the solution to the inequality $5x + 2 < 2x + 6$ is $x < \frac{4}{3}$, which can be written in interval notation as $\left(-\infty, \frac{4}{3}\right)$.

Figure 2.38

EXAMPLE 4 ▶   **Intersection Method of Solution**

Solve $\frac{5x + 2}{5} \geq \frac{4x - 7}{8}$ using the intersection of graphs method.

SOLUTION

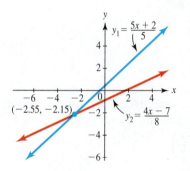

Figure 2.39

Enter the left side of the inequality as $y_1 = (5x + 2)/5$, enter the right side of the inequality as $y_2 = (4x - 7)/8$, graph these lines, and find their point of intersection. As seen in Figure 2.39, the two lines intersect at the point where $x = -2.55$ and $y = -2.15$.

The solution to the inequality is the x-interval for which the graph of y_1 is above the graph of y_2, or the x-value for which the graph of y_1 intersects the graph of y_2. Figure 2.39 indicates that this is the interval to the right of and including the input value of the point of intersection of the two lines. Thus, the solution is $x \geq -2.55$, or $[-2.55, \infty)$.

---

x-Intercept Method

To use the x-intercept method to solve a linear inequality, we use the following steps.

> ### Solving Linear Inequalities with the x-Intercept Method
>
> 1. Rewrite the inequality with all nonzero terms on one side of the inequality and combine like terms, getting $f(x) > 0, f(x) < 0, f(x) \leq 0$, or $f(x) \geq 0$.
>
> 2. Graph the nonzero side of this inequality. (Any window in which the x-intercept can be clearly seen is appropriate.)
>
> 3. Find the x-intercept of the graph to find the solution to the equation $f(x) = 0$. (The exact solution can be found algebraically.)
>
> 4. Use the graph to determine where the inequality is satisfied.

To use the *x*-intercept method to solve the inequality $5x + 2 < 2x + 6$, we rewrite the inequality with all nonzero terms on one side of the inequality and combine like terms:

$$5x + 2 < 2x + 6$$

$$3x - 4 < 0 \qquad \text{Subtract } 2x \text{ and 6 from both sides of the inequality.}$$

Graphing the nonzero side of this inequality as the linear function $f(x) = 3x - 4$ gives the graph in Figure 2.40. Finding the *x*-intercept of the graph (Figure 2.40) gives the solution to the equation $3x - 4 = 0$. The *x*-intercept (and zero of the function) is $x = 1.3333 \cdots = \dfrac{4}{3}$.

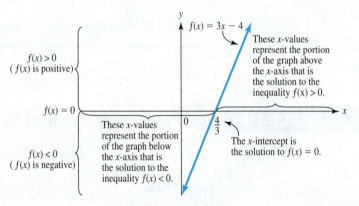

Figure 2.40

We now want to find where $f(x) = 3x - 4$ is less than 0. Notice that the portion of the graph *below* the *x*-axis gives $3x - 4 < 0$. Thus, the solution to $3x - 4 < 0$, and thus to $5x + 2 < 2x + 6$, is $x < \dfrac{4}{3}$, or $\left(-\infty, \dfrac{4}{3}\right)$.

EXAMPLE 5 ▶ **Apparent Temperature**

During a recent summer, Dallas, Texas, endured 29 consecutive days when the temperature was at least 100°F. On many of these days, the combination of heat and humidity made it feel even hotter than it was. When the temperature is 100°F, the apparent temperature *A* (or heat index) depends on the humidity *h* (expressed as a decimal) according to

$$A = 90.2 + 41.3h$$

For what humidity levels is the apparent temperature at least 110°F? (Source: W. Bosch and C. Cobb, "Temperature-Humidity Indices," *UMAP Journal*)

SOLUTION

If the apparent temperature is at least 110°F, the inequality to be solved is

$$A \geq 110, \quad \text{or} \quad 90.2 + 41.3h \geq 110$$

Rewriting this inequality with 0 on the right side gives $41.3h - 19.8 \geq 0$. Entering $y_1 = 41.3x - 19.8$ and graphing with a graphing utility gives the graph in Figure 2.41. The x-intercept of the graph is (approximately) 0.479.

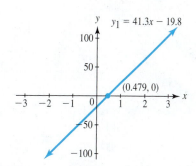

Figure 2.41

The x-interval where the graph is on or above the x-axis is the solution that we seek, so the solution to the inequality is $[0.479, \infty)$. However, humidity is limited to 100%, so the solution is $0.479 \leq h \leq 1.00$, and we say that the apparent temperature is at least 110° when the humidity is between 47.9% and 100%, inclusive.

Double Inequalities

The inequality $0.479 \leq h \leq 1.00$ in Example 5 is a **double inequality**. A double inequality represents two inequalities connected by the word *and* or *or*. The inequality $0.479 \leq h \leq 1.00$ is a compact way of saying $0.479 \leq h$ and $h \leq 1.00$. Double inequalities can be solved algebraically or graphically, as illustrated in the following example. Note that any arithmetic operation is performed to *all three* parts of a double inequality.

EXAMPLE 6 ▶   **Course Grades**

A student has taken four tests and has earned grades of 90%, 88%, 93%, and 85%. If all five tests count the same, what grade must the student earn on the final test so that his course average is a B (that is, so his average is at least 80% and less than 90%)?

ALGEBRAIC SOLUTION

To receive a B, the final test score, represented by x, must satisfy

$$80 \leq \frac{90 + 88 + 93 + 85 + x}{5} < 90$$

Solving this inequality gives

$$80 \leq \frac{356 + x}{5} < 90$$

$$400 \leq 356 + x < 450 \qquad \text{Multiply all three parts by 5.}$$

$$44 \leq x < 94 \qquad \text{Subtract 356 from all three parts.}$$

Thus, he will receive a grade of B if his final test score is at least 44 but less than 94.

GRAPHICAL SOLUTION

To solve this inequality graphically, we assign the left side of the inequality to y_1, the middle to y_2, and the right side to y_3, and we graph these equations to obtain the graph in Figure 2.42.

$$y_1 = 80$$
$$y_2 = (356 + x)/5$$
$$y_3 = 90$$

| Figure 2.42 | Figure 2.43 |

We seek the values of x where the graph of y_2 is above or on the graph of y_1 and below the graph of y_3. The left endpoint of this x-interval occurs at the point of intersection of y_2 and y_1, and the right endpoint of the interval occurs at the intersection of y_2 and y_3. These two points can be found using the intersection method. Figure 2.43 shows the points of intersection of these graphs.

The x-values of the points of intersection are 44 and 94, so the solution to $80 \leq \dfrac{356 + x}{5} < 90$ is $44 \leq x < 94$, which agrees with our algebraic solution.

EXAMPLE 7 ▶ **Expected Prison Sentences**

The mean (expected) time y served in prison for a serious crime can be approximated by a function of the mean sentence length x, with $y = 0.55x - 2.886$, where x and y are measured in months. According to this model, to how many months should a judge sentence a convicted criminal so that the criminal will serve between 37 and 78 months? (Source: National Center for Policy Analysis)

SOLUTION

We seek values of x that give y-values between 37 and 78, so we solve the inequality $37 \leq 0.55x - 2.886 \leq 78$ for x:

$$37 \leq 0.55x - 2.886 \leq 78$$

$$37 + 2.886 \leq 0.55x \leq 78 + 2.886$$

$$39.886 \leq 0.55x \leq 80.886$$

$$72.52 \leq x \leq 147.07$$

Thus, the judge could impose a sentence of 73 to 147 months if she wants the criminal to actually serve between 37 and 78 months.

SKILLS CHECK 2.4

In Exercises 1–12, solve the inequalities both algebraically and graphically. Draw a number line graph of each solution and give interval notation.

1. $6x - 1 \leq 11 + 2x$

2. $2x + 6 < 4x + 5$

3. $4(3x - 2) \leq 5x - 9$

4. $5(2x - 3) > 4x + 6$

5. $4x + 1 < -\dfrac{3}{5}x + 5$

6. $4x - \dfrac{1}{2} \leq -2 + \dfrac{x}{3}$

7. $\dfrac{x - 5}{2} < \dfrac{18}{5}$

8. $\dfrac{x - 3}{4} < \dfrac{16}{3}$

9. $\dfrac{3(x - 6)}{2} \geq \dfrac{2x}{5} - 12$

10. $\dfrac{2(x - 4)}{3} \geq \dfrac{3x}{5} - 8$

11. $2.2x - 2.6 \geq 6 - 0.8x$

12. $3.5x - 6.2 \leq 8 - 0.5x$

In Exercises 13 and 14, solve graphically by the intersection method. Give the solution in interval notation.

13. $7x + 3 < 2x - 7$

14. $3x + 4 \leq 6x - 5$

In Exercises 15 and 16, solve graphically by the x-intercept method. Give the solution in interval notation.

15. $5(2x + 4) \geq 6(x - 2)$

16. $-3(x - 4) < 2(3x - 1)$

17. The graphs of two linear functions f and g are shown in the following figure. (Domains are all real numbers.)

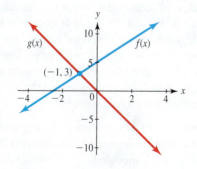

a. Solve the equation $f(x) = g(x)$.

b. Solve the inequality $f(x) < g(x)$.

18. The graphs of three linear functions f, g, and h are shown in the following figure.

a. Solve the equation $f(x) = g(x)$.

b. Solve the inequality $h(x) \leq g(x)$.

c. Solve the inequality $f(x) \leq g(x) \leq h(x)$.

In Exercises 19–28, solve the double inequalities.

19. $17 \leq 3x - 5 < 31$

20. $120 < 20x - 40 \leq 160$

21. $2x + 1 \geq 6$ and $2x + 1 \leq 21$

22. $16x - 8 > 12$ and $16x - 8 < 32$

23. $3x + 1 < -7$ and $2x - 5 > 6$

24. $6x - 2 \leq -5$ or $3x + 4 > 9$

25. $\dfrac{3}{4}x - 2 \geq 6 - 2x$ or $\dfrac{2}{3}x - 1 \geq 2x - 2$

26. $\dfrac{1}{2}x - 3 < 5x$ or $\dfrac{2}{5}x - 5 > 6x$

27. $37.002 \leq 0.554x - 2.886 \leq 77.998$

28. $70 \leq \dfrac{60 + 88 + 73 + 65 + x}{5} < 80$

EXERCISES 2.4

29. **Depreciation** Suppose a business purchases equipment for $12,000 and depreciates it over 5 years with the straight-line method until it reaches its salvage value of $2000 (see the figure below). Assuming that the depreciation can be for any part of a year, do the following:

    a. Write an equation that represents the depreciated value V as a function of the years t.

    b. Write an inequality that indicates that the depreciated value V of the equipment is less than $8000.

    c. Write an inequality that describes the time t during which the depreciated value is at least half of the original value.

Depreciated Value

30. **Blood Alcohol Percent** The blood alcohol percent p of a 220-pound male is a function of the number of 12-oz beers consumed, and the percent at which a person is considered legally intoxicated (and guilty of DUI if driving) is 0.1% or higher (see the following figure).

    a. Use an inequality to indicate the percent of alcohol in the blood when a person is considered legally intoxicated.

Number of 12-oz beers

(Source: Pennsylvania Liquor Control Board)

b. If x is the number of beers consumed by a 220-pound male, write an inequality that gives the number of beers that will cause him to be legally intoxicated.

31. **Freezing** The equation $F = \frac{9}{5}C + 32$ gives the relationship between temperatures measured in degrees Celsius and degrees Fahrenheit. We know that a temperature at or below 32°F is "freezing." Use an inequality to represent the corresponding "freezing" Celsius temperature.

32. **Boiling** The equation $C = \frac{5}{9}(F - 32)$ gives the relationship between temperatures measured in degrees Celsius and degrees Fahrenheit. We know that a temperature at or above 100°C is "boiling." Use an inequality to represent the corresponding "boiling" Fahrenheit temperature.

33. **Job Selection** Deb Cook is given the choice of two positions, one paying $3100 per month and the other paying $2000 per month plus a 5% commission on all sales made during the month. What amount must she sell in a month for the second position to be more profitable?

34. **Stock Market** Susan Mason purchased 1000 shares of stock for $22 per share, and 3 months later the value had dropped by 20%. What is the minimum percent increase required for her to make a profit?

35. **Grades** If Stan Cook has a course average score between 80 and 89, he will earn a grade of B in his algebra course. Suppose that he has four exam scores of 78, 69, 92, and 81 and that his teacher said the final exam score has twice the weight of each of the other exams. What range of scores on the final exam will result in Stan earning a grade of B?

36. **Grades** If John Deal has a course average score between 70 and 79, he will earn a grade of C in his algebra course. Suppose that he has three exam scores of 78, 62, and 82 and that his teacher said the final exam score has twice the weight of the other exams. What range of scores on the final exam will result in John earning a grade of C?

37. **Cigarette Use** The percent of 12th graders who have ever used cigarettes for each of the years 2000 through 2014 can be modeled by $p = -1.873t + 60.643$, where p is equal to the percent and t is equal to number of the years after 2000. When will the percent be less than 27%, according to this model?

38. *SAT Scores* The College Board now reports SAT scores with a new scale, which has the new scale score y defined as a function of the old scale score x by the equation $y = 0.97x + 128.3829$. Suppose a college requires a new scale score greater than or equal to 1000 to admit a student. To determine what old score values would be equivalent to the new scores that would result in admission to this college, do the following:

a. Write an inequality to represent the problem, and solve it algebraically.

b. Solve the inequality from part (a) graphically to verify your result.

39. *Social Security* A person scheduled to receive a $2000 monthly Social Security benefit at a full retirement age of 66 can delay drawing benefits until age 70 and receive $2640 per month. At what age will the person who started taking benefits at age 70 have more in total benefits paid than if he or she had started taking benefits at age 66?

40. *Internet Access* The percent of the U.S. population with Internet access can be modeled by $y = 1.36x + 68.8$, with x equal to the number of years after 2000. When does this model indicate that the U.S. population with Internet access exceeds 89%?

41. *HID Headlights* The new high-intensity discharge (HID) headlights containing xenon gas have an expected life of 1500 hours. Because a complete system costs $1000, it is hoped that these lights will last for the life of the car. Suppose that the actual life of the lights could be 10% longer or shorter than the advertised expected life. Write an inequality that gives the range of life of these new lights.
(Source: *Automobile*, July 2000)

42. *Prison Sentences* The mean time y spent in prison for a crime can be found from the mean sentence length x, using the equation $y = 0.554x - 2.886$, where x and y are measured in months. To how many months should a judge sentence a convicted criminal if she wants the criminal to actually serve between 4 and 6 years?
(Source: Index of Leading Cultural Indicators)

43. *Annual Wage* The average annual wage of U.S. workers for selected years from 2011 projected through 2021 can be modeled by $y = 2375x + 39,630$, with y equal to the average annual wage in dollars and x equal to the number of years after 2010. When does the model project that the average annual wage of U.S. workers will be at least $68,130?
(Source: Social Security Administration)

44. *Doctorates* The number of new doctorates in mathematics employed in academic positions per year can be modeled by $y = 28.5x + 50.5$, where x is the number of years after 2000.

a. If the model is accurate, algebraically determine the year in which the number of these doctorates employed was 250.

b. Use a graph to verify your answer to part (a).

c. Use your graph to find when the number of these doctorates employed was below 250 per year.
(Source: www.ams.org)

45. *Home Appraisal* A home purchased in 1996 for $190,000 was appraised at $270,000 in 2000. Assuming the rate of increase in the value of the home is constant, do the following:

a. Write an equation for the value of the home as a function of the number of years, x, after 1996.

b. Assuming that the equation in part (a) remained accurate, write an inequality that gives the range of years (until the end of 2010) when the value of the home was greater than $400,000.

c. Does it seem reasonable that this model remained accurate until the end of 2010?

46. *Car Sales Profit* A car dealer purchases 12 new cars for $32,500 each and sells 11 of them at a profit of 5.5%. For how much must he sell the remaining car to average a profit of at least 6% on the 12 cars?

47. *Electrical Components Profit* A company's daily profit from the production and sale of electrical components can be described by the equation $P(x) = 6.45x - 2000$ dollars, where x is the number of units produced and sold. What level of production and sales will give a daily profit of more than $10,900?

48. *Profit* The yearly profit from the production and sale of Plumber's Helpers is $P(x) = -40,255 + 9.80x$ dollars, where x is the number of Plumber's Helpers produced and sold. What level of production and sales gives a yearly profit of more than $84,355?

49. *Break-Even* A large hardware store's monthly profit from the sale of PVC pipe can be described by the equation $P(x) = 6.45x - 9675$ dollars, where x is the number of feet of PVC pipe sold. What level of monthly sales is necessary to avoid a loss?

50. *Break-Even* The yearly profit from the production and sale of Plumber's Helpers is $P(x) = -40,255 + 9.80x$ dollars, where x is the number of Plumber's Helpers produced and sold. What level of production and sales will result in a loss?

51. *Break-Even* A company produces a logic board for computers. The annual fixed cost for the board is $345,000, and the variable cost is $125 per board. If the logic board sells for $489, write an inequality that gives the number of logic boards that will give a profit for the product.

52. *Temperature* The temperature T (in degrees Fahrenheit) inside a concert hall m minutes after a 40-minute power outage during a summer rock concert is given by $T = 0.43m + 76.8$. Write and solve an inequality that describes when the temperature in the hall is not more than 85°F.

53. *Hispanic Population* Using data and projections from 1990 through 2050, the percent of Hispanics in the U.S. population is given by $H(x) = 0.224x + 9.0$, where x is the number of years after 1990. Find the years when the Hispanic population is projected to be at least 14.6% of the U.S. population.
(Source: U.S. Census Bureau)

54. *Reading Tests* The average reading score of 17-year-olds on the National Assessment of Progress tests is given by $y = 0.155x + 244.37$ points, where x is the number of years after 1970. Assuming that this model was valid, write and solve an inequality that describes when the average 17-year-old's reading score on this test was between but not including 245 and 248. (Your answer should be interpreted discretely.)
(Source: U.S. Department of Education)

55. *National Health Care* The national expenditures for health care in the United States for selected years from 1995 with projections to 2020 are given by $y = 138.2x + 97.87$, with x equal to the number of years after 1990 and y equal to the expenditures for health care in billions of dollars. When does the model project that the expenditures for health care will exceed $4658.5 billion?
(Source: U.S. Centers for Medicare and Medicaid Services)

56. *Black Population* Using data and projections from 1990 through 2050, the percent of the U.S. population that is black can be modeled by $B(x) = 0.057x + 12.3$, where x is the number of years after 1990. When does this model call for the percent of blacks to be at least 13.44%?
(Source: U.S. Census Bureau)

---

chapter 2 SUMMARY

In this chapter, we studied the solution of linear equations and systems of linear equations. We used graphing utilities to solve linear equations. We solved business and economics problems involving linear functions, solved application problems, solved linear inequalities, and used graphing utilities to model linear functions.

Key Concepts and Formulas

2.1 Algebraic and Graphical Solution of Linear Equations

Algebraic solution of linear equations	If a linear equation contains fractions, multiply both sides of the equation by a number that will remove all denominators in the equation. Next remove any parentheses or other symbols of grouping, and then perform any additions or subtractions to get all terms containing the variable on one side and all other terms on the other side of the equation. Combine like terms. Divide both sides of the equation by the coefficient of the variable. Check the solution by substitution in the original equation.
Solving real-world application problems	To solve an application problem that is set in a real-world context, use the same solution methods. However, remember to include units of measure with your answer and check that your answer makes sense in the problem situation.
Zero of a function	Any number a for which $f(a) = 0$ is called a zero of the function $f(x)$.
Solutions, zeros, and x-intercepts	If a is an x-intercept of the graph of a function f, then a is a real zero of the function f, and a is a real solution to the equation $f(x) = 0$.

Graphical solution of linear equations	
• *x*-intercept method	Rewrite the equation with 0 on one side, enter the nonzero side into the equation editor of a graphing calculator, and find the *x*-intercept of the graph. This is the solution to the equation.
• **Intersection method**	Enter the left side of the equation into y_1, enter the right side of the equation into y_2, and find the point of intersection. The *x*-coordinate of the point of intersection is the solution to the equation.
Literal equations; solving an equation for a specified linear variable	To solve an equation for one variable if two or more variables are in the equation and if that variable is to the first power in the equation, solve for that variable by treating the other variables as constants and using the same steps as are used to solve a linear equation in one variable.
Direct variation	Two variables *x* and *y* are directly proportional to each other if their quotient is a constant.

2.2 Fitting Lines to Data Points: Modeling Linear Functions

Fitting lines to data points	When real-world information is collected as numerical information called *data*, technology can be used to determine the pattern exhibited by the data (provided that a recognizable pattern exists). Such patterns can often be described by mathematical functions.
Constant first differences	If the first differences of data outputs are constant (for equally spaced inputs), a linear model can be found that fits the data exactly. If the first differences are "nearly constant," a linear model can be found that is an approximate fit for the data.
Linear regression	We can determine the equation of the line that is the best fit for a set of points by using a procedure called linear regression (or the least squares method), which defines the best-fit line as the line for which the sum of the squares of the vertical distances from the data points to the line is a minimum.
Modeling data	We can model a set of data by entering the data into a graphing utility, obtaining a scatter plot, and using the graphing utility to obtain the linear equation that is the best fit for the data. The equation and/or numerical results should be reported in a way that makes sense in the context of the problem, with the appropriate units and with the variables identified.
Discrete versus continuous	We use the term *discrete* to describe data or a function that is presented in the form of a table or a scatter plot. We use the term *continuous* to describe a function or graph when the inputs can be any real number or any real number between two specified values.
Applying models	Using a model to find an output for an input between two given data points is called *interpolation*. When a model is used to predict an output for an input outside the given data points, the process is called *extrapolation*.
Goodness of fit	The goodness of fit of a linear model can be observed from a graph of the model and the data points and/or measured with the correlation coefficient.

2.3 Systems of Linear Equations in Two Variables

System of equations	A system of linear equations is a set of equations in two or more variables. A solution of the system must satisfy every equation in the system.

Solving a system of linear equations in two variables	
• **Graphing**	Graph the equations and find their point of intersection.
• **Substitution**	Solve one of the equations for one variable and substitute that expression into the other equation, thus giving an equation in one variable.
• **Elimination**	Rewrite one or both equations in a form that allows us to eliminate one of the variables by adding or subtracting the equations.
Break-even analysis	A company is said to break even from the production and sale of a product if the total revenue equals the total cost—that is, if the profit for that product is zero.
Market equilibrium	*Market equilibrium* is said to occur when the quantity of a commodity demanded is equal to the quantity supplied. The price at this point is called the *equilibrium price*, and the quantity at this point is called the *equilibrium quantity*.
Dependent and inconsistent systems	
• **Unique solution**	Graphs are intersecting lines.
• **No solution**	Graphs are parallel lines; the system is *inconsistent*.
• **Many solutions**	Graphs are the same line; the system is *dependent*.
Modeling systems of equations	Solution of real problems sometimes requires us to create two or more equations whose simultaneous solution is the solution to the problem.

2.4 Solutions of Linear Inequalities

Linear inequality	A linear inequality (or first-degree inequality) is an inequality that can be written in the form $ax + b > 0$, where $a \neq 0$. (The inequality symbol can be $>$, $\geq$, $<$, or $\leq$.)
Algebraically solving linear inequalities	The steps used to solve a linear inequality are the same as those used to solve a linear equation, except that the inequality symbol is reversed if both sides are multiplied (or divided) by a negative number.
Graphical solution of linear inequalities	
• **Intersection method**	Set the left side of the inequality equal to y_1 and set the right side equal to y_2, graph the equations using a graphing utility, and find the x-coordinate of the point of intersection. The values of x that satisfy the inequality represented by $y_1 < y_2$ are those values for which the graph of y_1 is below the graph of y_2.
• **x-intercept method**	To use the x-intercept method to solve an inequality, rewrite the inequality with all nonzero terms on one side and zero on the other side of the inequality and combine like terms. Graph the nonzero side of this inequality and find the x-intercept of the graph. If the inequality to be solved is $f(x) > 0$, the solution will be the interval of x-values representing the portion of the graph above the x-axis. If the inequality to be solved is $f(x) < 0$, the solution will be the interval of x-values representing the portion of the graph below the x-axis.
Double inequalities	A double inequality represents two inequalities connected by the word *and* or *or*. Double inequalities can be solved algebraically or graphically. Any operation performed on a double inequality must be performed on *all three* parts.

chapter 2 SKILLS CHECK

In Exercises 1–6, solve the equation for x algebraically and graphically.

1. $3x + 22 = 8x - 12$

2. $2(x - 7) = 5(x + 3) - x$

3. $\dfrac{3(x - 2)}{5} - x = 8 - \dfrac{x}{3}$

4. $\dfrac{6x + 5}{2} = \dfrac{5(2 - x)}{3}$

5. $\dfrac{3x}{4} - \dfrac{1}{3} = 1 - \dfrac{2}{3}\left(x - \dfrac{1}{6}\right)$

6. $3.259x - 198.8546 = -3.8(8.625x + 4.917)$

7. For the function $f(x) = 7x - 105$, (a) find the zero of the function, (b) find the x-intercept of the graph of the function, and (c) solve the equation $f(x) = 0$.

8. Solve $P(a - y) = 1 + \dfrac{m}{3}$ for y.

9. Solve $4x - 3y = 6$ for y and graph it on a graphing utility with a standard window.

Use the table of data below in Exercises 10–13.

x	1	3	6	8	10
y	-9	-1	5	12	18

10. Create a scatter plot of the data.

11. Find the linear function that is the best fit for the data in the table.

12. Use a graphing utility to graph the function found in Exercise 11 on the same set of axes as the scatter plot in Exercise 10, with $x_{min} = 0$, $x_{max} = 12$, $y_{min} = -10$, and $y_{max} = 20$.

13. Do the data points in the table fit exactly on the graph of the function from Exercise 12?

Solve the systems of linear equations in Exercises 14–19, if possible.

14. $\begin{cases} 3x + 2y = 0 \\ 2x - y = 7 \end{cases}$

15. $\begin{cases} 3x + 2y = -3 \\ 2x - 3y = 3 \end{cases}$

16. $\begin{cases} -4x + 2y = -14 \\ 2x - y = 7 \end{cases}$

17. $\begin{cases} -6x + 4y = 10 \\ 3x - 2y = 5 \end{cases}$

18. $\begin{cases} 2x + 3y = 9 \\ -x - y = -2 \end{cases}$

19. $\begin{cases} 2x + y = -3 \\ 4x - 2y = 10 \end{cases}$

In Exercises 20–22, solve the inequalities both algebraically and graphically.

20. $3x + 8 < 4 - 2x$

21. $3x - \dfrac{1}{2} \le \dfrac{x}{5} + 2$

22. $18 \le 2x + 6 < 42$

chapter 2 REVIEW

When money is borrowed to purchase an automobile, the amount borrowed A determines the monthly payment P. In particular, if a dealership offers a 5-year loan at 2.9% interest, then the amount borrowed for the car determines the payment according to the following table. Use the table to define the function $P = f(A)$ in Exercises 23–24.

Amount Borrowed ($)	Monthly Payment ($)
10,000	179.25
15,000	268.87
20,000	358.49
25,000	448.11
30,000	537.73

(Source: Sky Financial)

23. *Car Loans*
   a. Are the first differences of the outputs in the table constant?

   b. Is there a line on which these data points fit exactly?

   c. Write the equation $P = f(A)$ of the line that fits the data points in the table, with coefficients rounded to three decimal places.

24. *Car Loans*
   a. Use the rounded linear model found in part (c) of Exercise 23 to find $P = f(28,000)$ and explain what it means.

   b. Can the function f be used to find the monthly payment for any dollar amount A of a loan if the interest rate and length of loan are unchanged?

c. Determine the amount of a loan that will keep the payment less than or equal to $500, using the unrounded model.

25. *Crude Oil* Using data for the years from 2010 and projected to 2030, the U.S. Crude Oil production is given by $y = 0.0154x + 1.85$ billions of barrels, with x equal to the number of years after 2010. In what year does the model predict the crude oil production will be 2.158 billion barrels?
(Source: U.S. Energy Information Administration)

26. *Fuel* The table below shows data for the number of gallons of gas purchased each day of a certain week by the 250 taxis owned by the Inner City Transportation Taxi Company. Write the equation that models the data.

Days from First Day	0	1	2	3	4	5	6
Gas Used (gal)	4500	4500	4500	4500	4500	4500	4500

27. *Marginal Profit* A company has determined that its profit for a product can be described by a linear function. The profit from the production and sale of 150 units is $455, and the profit from 250 units is $895.

a. Write the equation of the profit function for this product.

b. How many units must be produced and sold to make a profit on this product?

28. *Job Selection* A job candidate is given the choice of two positions, one paying $2100 per month and one paying $1000 per month plus a 5% commission on all sales made during the month.

a. How much (in dollars) must the employee sell in a month for the second position to pay as much as the first?

b. To be sure that the second position will pay more than the first, how much (in dollars) must the employee sell each month?

29. *Marketing* A car dealer purchased 12 automobiles for $24,000 each. If she sells 8 of them at an average profit of 12%, for how much must she sell the remaining 4 to obtain an average profit of 10% on all 12?

30. *Investment* A retired couple has $420,000 to invest. They chose one relatively safe investment fund that has an annual yield of 6% and another riskier investment that has a 10% annual yield. How much should they invest in each fund to earn $30,000 per year?

31. *Life-Income Plan* Pomona College offers a life-income plan whereby a donor can transfer cash or

securities to the college, the college pays the donor a fixed income for life, and the remaining balance passes to Pomona College when the contract ends at the death of the donor. With the Pomona Plan, the donor can get monthly income based on the amount of the donation and the age of the donor. Suppose the annual rate of returns on the donation for selected years are as given in the table.

a. Write equation of the line connecting the points $(70, 7.0)$ and $(90, 13.0)$ from the table. Write the equation in the form $y = f(x)$, where y is the annual percent of return and x is the age of the donor.

b. Graph the function and the data points on the same axes.

c. Do the points from the table fit the graph exactly?

d. What does this function give as the annual rate of return on a donation at age 78?

e. At what age could a donation be made to get a return with an annual rate of 12.4%?

Age at Donation	Annual Return Percent
70	7.0
75	8.5
80	10.0
85	11.5
90	13.0

(Source: pomonaplan.pomona.edu)

32. *Profit* A company has revenue given by $R(x) = 500x$ dollars and total costs given by $C(x) = 48,000 + 100x$ dollars, where x is the number of units produced and sold. How many units will give a profit?

33. *Profit* A company has revenue given by $R(x) = 564x$ dollars and total cost given by $C(x) = 40,000 + 64x$ dollars, where x is the number of units produced and sold. The profit can be found by forming the function $P(x) = R(x) - C(x)$.

a. Write the profit function.

b. For what values of x is $P(x) > 0$?

c. For how many units is there a profit?

34. *Depreciation* A business property can be depreciated for tax purposes by using the formula $y + 15,000x = 300,000$, where y is the value of the property x years after it was purchased.

a. For what x-values is the property value below $150,000?

b. After how many years is the property value below $150,000?

35. *Cigarette Use* The percent of 12th-graders who have ever used cigarettes for each of the years 2000 through 2014 is shown in the table.

a. Find a linear function that models the data, with p equal to the percent and t equal to the number of years after 2000.

b. What does the model predict that the percent will be in 2020? Is this interpolation or extrapolation?

Year	Percent	Year	Percent
2000	62.5	2008	44.7
2001	61.0	2009	43.6
2002	57.2	2010	42.2
2003	53.7	2011	40.0
2004	52.8	2012	39.5
2005	50.0	2013	38.1
2006	47.1	2014	34.4
2007	46.2		

(Source: monitoringthefuture.org)

36. *Life Expectancy*

a. Find a linear function $y = f(x)$ that models the data shown in the figure that follows, with x equal to the number of years after 1950 and y equal to the number of years the average 65-year-old woman is estimated to live beyond age 65.

b. Graph the data and the model on the same set of axes.

c. Use the model to estimate $f(99)$ and explain what it means.

d. Determine the time period (in years) for which the average 65-year-old woman can expect to live more than 84 years.

(Source: National Center for Health Statistics)

37. *Life Expectancy*

a. Find a linear function $y = g(x)$ that models the data shown in the figure, with x equal to the number of years after 1950 and y equal to the number of years the average 65-year-old man is estimated to live beyond age 65.

b. Graph the data and the model on the same set of axes.

c. Use the model to estimate $g(130)$ and explain what it means.

d. In what year would the average 65-year-old man expect to live to age 90?

e. Determine the time period for which the average man could expect to live less than 81 years.

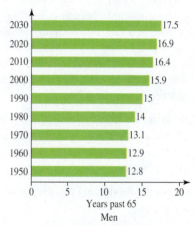

(Source: National Center for Health Statistics)

38. *Social Security* The figure on the next page shows how the monthly Social Security benefits differ with the age at which a recipient decides to start receiving benefits. The example shown in this table is for recipients whose monthly benefit is $1000 if benefits start being paid at a full retirement age of 66. Note that benefits do not start being paid before age 62 and that the monthly benefit does not increase if the starting age is after 70.

a. What is the linear function that best models the monthly benefit y as a function of age x at which benefits are started, using only the data points from 66 to 70?

b. Graph the model and the data points on the same axes.

c. Is this model an exact model or an approximate model?

Monthly benefit amounts differ based on the age you decide to start receiving benefits
This example assumes a benefit of $1,000 at a full retirement age of 66

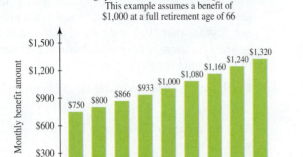

(Source: Social Security Administration)

39. *Social Security* The figure in Exercise 38 shows how the monthly Social Security benefits differ for recipients whose monthly benefit is $1000 if benefits start being paid at a full retirement age of 66 based on the age at which a recipient starts receiving the benefits. Note that benefits do not start being paid before age 62 and that the monthly benefit does not increase if the starting age is after 70.

a. What is the linear function that best models the monthly benefit y as a function of age x at which benefits are started, for the data points from 62 to 70?

b. Graph the model and the data points on the same axes.

c. Is this model an exact model or an approximate model?

(Source: Social Security Administration)

40. *Earnings per Share* The table below gives the earnings ($ thousands) per share (EPS) for ACS stock for the years 2010–2015.

a. Graph the data points to determine if a linear equation is a reasonable model for the data.

b. If it is reasonable, find the linear model that is the best fit for the data. Let x equal the number of years after 2005 and y equal the EPS, in thousands of dollars.

c. Graph the data points and the function on the same axes and discuss the goodness of fit.

Year	2010	2011	2012	2013	2014	2015
EPS ($ thousands)	0.85	1.05	1.29	1.46	1.68	1.90

41. *Marriage Rate* According to data from the *National Vital Statistics Report*, the marriage rate (the number of marriages per 1000 unmarried women) can be

described by $132x + 1000y = 9570$, where x is the number of years after 1980. For what years does this model indicate that the rate will be above 6.25 per 1000? Will be below 3.63 per 1000?

42. *Heart Rate* The table gives the desired heart rate during exercise for weight loss based on the age of the person exercising on an elliptical machine. (*Source:* Paramount fitness.com)

a. Can the desired heart rate be expressed exactly as a linear function of age, based on this data?

Age	Heart Rate (beats per minute)	Age	Heart Rate (beats per minute)
20	130	60	104
30	123	70	97
40	117	80	91
50	110	90	84

b. Write a linear function that models the data above.

c. Can the desired heart rate be expressed exactly as a linear function of age, based on the data in the following table? If yes, what is the function?

Age	Heart Rate	Age	Heart Rate
20	130	60	104
30	123.5	70	97.5
40	117	80	91
50	110.5	90	84.5

43. *Prison Sentences* The mean time y in prison for a crime can be found as a function of the mean sentence length x, using $y = 0.554x - 2.886$, where x and y are in months. If a judge sentences a convicted criminal to serve between 3 and 5 years, how many months would we expect the criminal to serve?
(Source: Index of Leading Cultural Indicators)

44. *Investment* A retired couple has $240,000 to invest. They chose one relatively safe investment fund that has an annual yield of 8% and another riskier investment that has a 12% annual yield. How much should they invest in each fund to earn $23,200 per year?

45. *Break-Even* A computer manufacturer has a new product with daily total revenue given by $R = 565x$ and daily total cost given by $C = 6000 + 325x$. How many units per day must be produced and sold to give break-even for the product?

46. *Medication* Medication A is given six times per day, and medication B is given twice per day. For a certain patient, the total intake of the two medications is limited to 25.2 mg per day. If the ratio of the dosage of medication A to the dosage of medication B is 2 to 3, how many milligrams are in each dosage?

47. **Market Equilibrium** The demand for a certain brand of women's shoes is given by $3q + p = 340$, and the supply of these shoes is given by $p - 4q = -220$, where p is the price in dollars and q is the number of pairs at price p. Solve the system containing these two equations to find the equilibrium price and the equilibrium quantity.

48. **Market Analysis** Suppose that, for a certain product, the supply and demand functions are $p = \dfrac{q}{10} + 8$ and $10p + q = 1500$, respectively, where p is in dollars and q is in units. Find the equilibrium price and quantity.

49. **Pricing** A concert promoter needs to make $120,000 from the sale of 2600 tickets. The promoter charges $40 for some tickets and $60 for the others.

   a. If there are x of the $40 tickets and y of the $60 tickets, write an equation that states that the total number of the tickets sold is 2600.

   b. How much money is made from the sale of x tickets for $40 each?

   c. How much money is made from the sale of y tickets for $60 each?

   d. Write an equation that states that the total amount made from the sale is $120,000.

   e. Solve the equations simultaneously to find how many tickets of each type must be sold to yield the $120,000.

50. **Rental Income** A woman has $500,000 invested in two rental properties. One yields an annual return of 12% of her investment, and the other returns 15% per year on her investment. Her total annual return from the two investments is $64,500. If x represents the 12% investment and y represents the 15% investment, answer the following:

   a. Write an equation that states that the sum of the investments is $500,000.

   b. What is the annual return on the 12% investment?

   c. What is the annual return on the 15% investment?

   d. Write an equation that states that the sum of the annual returns is $64,500.

   e. Solve these two equations simultaneously to find how much is invested in each property.

Group Activities
▶ EXTENDED APPLICATIONS

1. Taxes

The table below gives the income tax due, $f(x)$, on each given taxable income, x.

1. What are the domain and the range of the function in the table?
2. Create a scatter plot of the data.
3. Do the points appear to lie on a line?
4. Do the inputs change by the same amount? Do the outputs change by the same amount?
5. Is the rate of change in tax per $1 of income constant? What is the rate of change?
6. Will a linear function fit the data points exactly?
7. Write a linear function $y = g(x)$ that fits the data points.
8. Verify that the linear model fits the data points by evaluating the linear function at $x = 63,900$ and $x = 64,100$ and comparing the resulting y-values with the income tax due for these taxable incomes.
9. Is the model a discrete or continuous function?

10. Can the model be used to find the tax due on any taxable income between $63,700 and $64,300?
11. What is the tax due on a taxable income of $64,150, according to this model?

U.S. Federal Taxes

Taxable Income ($)	Income Tax Due ($)
63,700	8779
63,800	8804
63,900	8829
64,000	8854
64,100	8879
64,200	8904
64,300	8929

(Source: U.S. Internal Revenue Service)

2. Research

Linear functions can be used to model many types of real data. Graphs displaying linear growth are frequently found in periodicals such as *Newsweek* and *Time*, in newspapers such as *USA Today* and the *Wall Street Journal*, and on numerous websites on the Internet. Tables of data can also be found in these sources, especially on federal and state government websites, such as www.census.gov. (This website is the source of the Florida resident population data used in the Chapter Review, for example.)

Your mission is to find a company sales record, a stock price, a biological growth pattern, or a sociological trend over a period of years (using at least four points) that is linear or "nearly linear," and then to create a linear function that is a model for the data.

A linear model will be a good fit for the data if

- the data are presented as a graph that is linear or "nearly linear."
- the data are presented in a table and the plot of the data points lies near some line.
- the data are presented in a table and the first differences of the outputs are nearly constant for equally spaced inputs.

After you have created the model, you should test the goodness of fit of the model to the data and discuss uses that you could make of the model.

Your completed project should include

a. A complete citation of the source of the data you are using.
b. An original copy or photocopy of the data being used.
c. A scatter plot of the data.
d. The equation that you have created.
e. A graph containing the scatter plot and the modeled equation.
f. A statement about how the model could be used to make estimations or predictions about the trend you are observing.

Some helpful hints:

1. If you decide to use a relation determined by a graph that you have found, read the graph very carefully to determine the data points or contact the source of the data to get the data from which the graph was drawn.
2. Align the independent variable by letting x represent the number of years after some convenient year and then enter the data into a graphing utility and create a scatter plot.
3. Use your graphing utility to create the equation of the function that is the best fit for the data. Graph this equation and the data points on the same axes to see if the equation is reasonable.

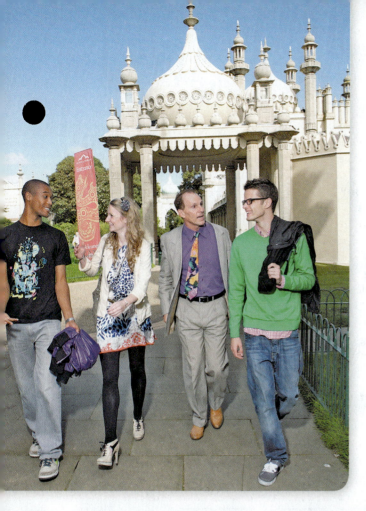

3

Quadratic, Piecewise-Defined, and Power Functions

Revenue and profit from the sale of products frequently cannot be modeled by linear functions because they increase at rates that are not constant. In this chapter, we use nonlinear functions, including quadratic and power functions, to model numerous applications in business, economics, and the life and social sciences.

sections	objectives	applications
3.1 Quadratic Functions; Parabolas	Graph quadratic functions; find vertices of parabolas; identify increasing and decreasing functions	Maximum revenue from sales, foreign-born population, height of a ball, minimizing cost
3.2 Solving Quadratic Equations	Solve equations by factoring; solve equations graphically; combine graphs and factoring methods; solve equations with the square root method and by completing the square; solve equations using the quadratic formula; find complex solutions; find discriminants	National health care, height of a ball, profit, hospital expenditures
3.3 Piecewise-Defined Functions and Power Functions	Graph and apply power, root, reciprocal, piecewise-defined, and absolute value functions; apply direct variation as a power; apply inverse variation	Residential power costs, postage, wind chill factor, service calls, wingspan, production, illumination
3.4 Quadratic and Power Models	Model with quadratic functions; compare linear and quadratic models; model with power functions; compare power and quadratic models	Mobile Internet advertising, the cloud, diabetes, cohabiting households, purchasing power

Algebra TOOLBOX

KEY OBJECTIVES

- Simplify expressions involving integer and rational exponents
- Find the absolute values of numbers
- Simplify expressions involving radicals
- Convert rational exponents to radicals and vice versa
- Special binomial products
- Factor polynomial expressions completely
- Identify numbers as real, imaginary, or pure imaginary
- Determine values that make complex numbers equal

In this Toolbox, we discuss absolute value, integer and rational exponents, and radicals. We also discuss the multiplication of monomials and binomials, factoring, and complex numbers.

Integer Exponents

In this chapter and future ones, we will discuss functions and equations containing integer powers of variables. For example, we will discuss the function $y = x^{-1} = \dfrac{1}{x}$.

> If a is a real number and n is a positive integer, then a^n represents a as a factor n times in a product
>
> $$a^n = \underbrace{a \cdot a \cdot a \ldots \cdot a}_{n \text{ times}}$$
>
> In a^n, a is called the base and n is called the exponent.

In particular, $a^2 = a \cdot a$ and $a^1 = a$. Note that, for positive integers m and n,

$$a^m \cdot a^n = \underbrace{a \cdot a \cdot a \ldots \cdot a}_{m \text{ times}} \cdot \underbrace{a \cdot a \ldots \cdot a}_{n \text{ times}} = \underbrace{a \cdot a \cdot a \ldots \cdot a}_{m+n \text{ times}} = a^{m+n}$$

and that, for $m \geq n$,

$$\frac{a^m}{a^n} = \frac{\overbrace{a \cdot a \cdot a \ldots \cdot a}^{m \text{ times}}}{\underbrace{a \cdot a \ldots \cdot a}_{n \text{ times}}} = \overbrace{a \cdot a \cdot a \ldots \cdot a}^{m-n \text{ times}} = a^{m-n} \ \text{ if } a \neq 0$$

These two important properties of exponents can be extended to all integers.

> ### Properties of Exponents
>
> For real numbers a and b and integers m and n,
>
> 1. $a^m \cdot a^n = a^{m+n}$ (Product Property)
> 2. $\dfrac{a^m}{a^n} = a^{m-n}, a \neq 0$ (Quotient Property)

We define an expression raised to zero and to a negative power as follows.

> ### Zero and Negative Exponents
>
> For $a \neq 0, b \neq 0$,
>
> 1. $a^0 = 1$ 2. $a^{-1} = \dfrac{1}{a}$
>
> 3. $a^{-n} = \dfrac{1}{a^n}$ 4. $\left(\dfrac{a}{b}\right)^{-n} = \left(\dfrac{b}{a}\right)^n$

EXAMPLE 1 ▶ **Zero and Negative Exponents**

Simplify the following expressions by removing all zero and negative exponents, for nonzero a, b, and c.

a. $(4c)^0$ **b.** $4c^0$ **c.** $(5b)^{-1}$ **d.** $5b^{-1}$ **e.** $\left(\dfrac{a}{b}\right)^{-3}$ **f.** $6a^{-3}$

SOLUTION

a. $(4c)^0 = 1$ **b.** $4c^0 = 4(1) = 4$ **c.** $(5b)^{-1} = \dfrac{1}{(5b)} = \dfrac{1}{5b}$

d. $5b^{-1} = 5 \cdot \dfrac{1}{b} = \dfrac{5}{b}$ **e.** $\left(\dfrac{a}{b}\right)^{-3} = \left(\dfrac{b}{a}\right)^3 = \dfrac{b^3}{a^3}$ **f.** $6a^{-3} = 6 \cdot \dfrac{1}{a^3} = \dfrac{6}{a^3}$

Absolute Value

The distance the number a is from 0 on a number line is the **absolute value** of a, denoted by $|a|$. The absolute value of any nonzero number is positive, and the absolute value of 0 is 0. For example, the distance from 5 to 0 is 5, so $|5| = 5$, and the distance from -8 to 0 is 8, so $|-8| = 8$. Note that if a is a nonnegative number, then $|a| = a$, but if a is negative, then $|a|$ is the positive number $-a$. Formally, we say

$$|a| = \begin{cases} a & \text{if } a \geq 0 \\ -a & \text{if } a < 0 \end{cases}$$

For example, $|5| = 5$ and $|-5| = -(-5) = 5$.

Rational Exponents and Radicals

In this chapter we will study functions involving a variable raised to a rational power (called **power functions**), and we will solve equations involving rational exponents and radicals. This may involve converting expressions involving radicals to expressions involving rational exponents, or vice versa.

Exponential expressions are defined for rational numbers in terms of radicals. Note that, for $a \geq 0$ and $b \geq 0$,

$$\sqrt{a} = b \text{ only if } a = b^2$$

Thus,

$$(\sqrt{a})^2 = b^2 = a, \text{ so } (\sqrt{a})^2 = a$$

We define $a^{1/2} = \sqrt{a}$, so $(a^{1/2})^2 = a$ for $a \geq 0$.

The following definitions show the connection between rational exponents and radicals.

> ### Rational Exponents
> 1. If a is a real number, variable, or algebraic expression and n is a positive integer $n \geq 2$, then
>
> $$a^{1/n} = \sqrt[n]{a}$$
>
> provided that $\sqrt[n]{a}$ exists.
>
> 2. If a is a real number and if m and n are integers containing no common factor with $n \geq 2$, then
>
> $$a^{m/n} = \sqrt[n]{a^m} = (\sqrt[n]{a})^m$$
>
> provided that $\sqrt[n]{a}$ exists.

EXAMPLE 2 ▶ **Rational Exponents**

Write the following expressions with exponents rather than radicals.

a. $\sqrt[3]{x^2}$ **b.** $\sqrt[4]{x^3}$ **c.** $\sqrt{(3xy)^5}$ **d.** $3\sqrt{(xy)^5}$

SOLUTION

a. $\sqrt[3]{x^2} = x^{2/3}$ **b.** $\sqrt[4]{x^3} = x^{3/4}$ **c.** $\sqrt{(3xy)^5} = (3xy)^{5/2}$ **d.** $3(xy)^{5/2}$

EXAMPLE 3 ▶ **Radical Notation**

Write the following in radical form.

a. $y^{1/2}$ **b.** $(3x)^{3/7}$ **c.** $12x^{3/5}$

SOLUTION

a. $y^{1/2} = \sqrt{y}$ **b.** $(3x)^{3/7} = \sqrt[7]{(3x)^3} = \sqrt[7]{27x^3}$ **c.** $12x^{3/5} = 12\sqrt[5]{x^3}$

Special Binomial Products and Factoring

We introduced the following special products in the Chapter 1 Toolbox. We show them here again because, as we shall see, they are very important in factoring.

> **Special Binomial Products**
>
> 1. $(x + a)(x - a) = x^2 - a^2$ (difference of two squares)
> 2. $(x + a)^2 = x^2 + 2ax + a^2$ (perfect square trinomial)
> 3. $(x - a)^2 = x^2 - 2ax + a^2$ (perfect square trinomial)

EXAMPLE 4 ▶ **Special Products**

Find the following products by using the special binomial products formulas.

a. $(3x - 1)^2$ **b.** $(4x - 5y)(4x + 5y)$ **c.** $(4x + 3)^2$

SOLUTION

a. $(3x - 1)^2 = (3x)^2 - 2(3x)(1) + (1)^2 = 9x^2 - 6x + 1$

b. $(4x - 5y)(4x + 5y) = (4x)^2 - (5y)^2 = 16x^2 - 25y^2$

c. $(4x + 3)^2 = (4x)^2 + 2(4x)(3) + (3)^2 = 16x^2 + 24x + 9$

Factoring is the process of writing a number or an algebraic expression as the product of two or more numbers or expressions. For example, the distributive property justifies factoring of monomials from polynomials, as in

$$5x^2 - 10x = 5x(x - 2)$$

Factoring out the **greatest common factor** (gcf) from a polynomial is the first step in factoring.

EXAMPLE 5 ▶ Greatest Common Factor

Factor out the greatest common factor.

a. $4x^2y^3 - 18xy^4$ **b.** $3x(a - b) - 2y(a - b)$

SOLUTION

a. The gcf of 4 and 18 is 2. The gcf of x^2 and x is the lower power of x, which is x. The gcf of y^3 and y^4 is the lower power of y, which is y^3. Thus, the gcf of $4x^2y^3$ and $18xy^4$ is $2xy^3$. Factoring out the gcf gives

$$4x^2y^3 - 18xy^4 = 2xy^3(2x - 9y)$$

b. The gcf of $3x(a - b)$ and $2y(a - b)$ is the binomial $a - b$. Factoring out $a - b$ from each term gives

$$3x(a - b) - 2y(a - b) = (a - b)(3x - 2y)$$

By recognizing that a polynomial has the form of one of the special products given on the previous page, we can factor that polynomial.

EXAMPLE 6 ▶ Factoring Special Products

Use knowledge of binomial products to factor the following algebraic expressions.

a. $9x^2 - 25$ **b.** $4x^2 - 12x + 9$

SOLUTION

a. Both terms are squares, so the polynomial can be recognized as the **difference of two squares**. It will then factor as the product of the sum and the difference of the square roots of the terms (see Special Binomial Products, Formula 1).

$$9x^2 - 25 = (3x + 5)(3x - 5)$$

b. Recognizing that the second-degree term and the constant term are squares leads us to investigate whether $12x$ is twice the product of the square roots of these two terms (see Special Binomial Products, Formula 3). The answer is yes, so the polynomial is a **perfect square**, and it can be factored as follows:

$$4x^2 - 12x + 9 = (2x - 3)^2$$

This can be verified by expanding $(2x - 3)^2$.

The first step in factoring is to look for common factors. All applicable factoring techniques should be applied to factor a polynomial completely.

EXAMPLE 7 ▶   **Factoring Completely**

Factor the following polynomials completely.

a. $3x^2 - 33x + 72$ **b.** $6x^2 - x - 1$

SOLUTION

a. The number 3 can be factored from all three terms, giving

$$3x^2 - 33x + 72 = 3(x^2 - 11x + 24)$$

If the trinomial can be factored into the product of two binomials, the first term of each binomial must be x, and we seek two numbers whose product is 24 and whose sum is -11. Because -3 and -8 satisfy these requirements, we get

$$3(x^2 - 11x + 24) = 3(x - 3)(x - 8)$$

b. The four possible factorizations of $6x^2 - x - 1$ that give $6x^2$ as the product of the first terms and -1 as the product of the last terms follow:

$$(6x - 1)(x + 1) \qquad (6x + 1)(x - 1)$$
$$(2x - 1)(3x + 1) \qquad (2x + 1)(3x - 1)$$

The factorization that gives a product with middle term $-x$ is the correct factorization.

$$(2x - 1)(3x + 1) = 6x^2 - x - 1$$

Some polynomials, such as $6x^2 + 9x - 8x - 12$, can be factored by **grouping**. To do this, we factor out common factors from pairs of terms and then factor out a common binomial expression if it exists. For example,

$$6x^2 + 9x - 8x - 12 = 3x(2x + 3) - 4(2x + 3) = (2x + 3)(3x - 4)$$

When a second-degree trinomial can be factored but there are many possible factors to test to find the correct one, an alternative method of factoring can be used. The steps used to factor a trinomial using factoring by grouping techniques follow.

Factoring a Trinomial into the Product of Two Binomials Using Grouping

Steps	Example
To factor a quadratic trinomial in the variable x:	Factor $5x - 6 + 6x^2$:
1. Arrange the trinomial with the powers of x in descending order.	1. $6x^2 + 5x - 6$
2. Form the product of the second-degree term and the constant term (first and third terms).	2. $6x^2(-6) = -36x^2$
3. Determine whether there are two factors of the product in step 2 that will sum to the middle (first-degree) term. (If there are no such factors, the trinomial will not factor into two binomials.)	3. $-36x^2 = (-4x)(9x)$ and $-4x + 9x = 5x$
4. Rewrite the middle term from step 1 as a sum of the two factors from step 3.	4. $6x^2 + 5x - 6 = 6x^2 - 4x + 9x - 6$
5. Factor the four-term polynomial from step 4 by grouping.	5. $6x^2 - 4x + 9x - 6 = 2x(3x - 2) + 3(3x - 2) = (3x - 2)(2x + 3)$

EXAMPLE 8 ▶ Factor $10x^2 + 23x - 5$, using grouping.

SOLUTION

To use this method to factor $10x^2 + 23x - 5$, we

1. Note that the trinomial has powers of x in descending order.

   $10x^2 + 23x - 5$

2. Multiply the second-degree and constant terms.

   $10x^2(-5) = -50x^2$

3. Factor $-50x^2$ so the sum of the factors is $23x$.

   $-50x^2 = 25x(-2x),\ 25x + (-2x) = 23x$

4. Rewrite the middle term of the expression as a sum of the two factors from step 3.

   $10x^2 + 23x - 5 = 10x^2 + 25x - 2x - 5$

5. Factor by grouping.

$$10x^2 + 25x - 2x - 5$$
$$= 5x(2x + 5) - (2x + 5)$$
$$= (2x + 5)(5x - 1)$$

Complex Numbers

In this chapter, some equations do not have real number solutions but do have solutions that are complex numbers.

The numbers discussed up to this point are real numbers (either rational numbers such as 2, -3, $\frac{5}{8}$, and $-\frac{2}{3}$ or irrational numbers such as $\sqrt{3}$, $\sqrt[3]{6}$, and π). But some equations do not have real solutions. For example, if

$$x^2 + 1 = 0, \quad \text{then} \quad x^2 = -1$$

but there is no real number that, when squared, will equal -1. However, we can denote one solution to this equation as the **imaginary unit i**, defined by

$$i = \sqrt{-1}$$

Note that $i^2 = \sqrt{-1} \cdot \sqrt{-1} = -1$.

The set of **complex numbers** is formed by adding real numbers and multiples of i.

> **Complex Number**
>
> The number $a + bi$, in which a and b are real numbers, is said to be a **complex number in standard form**. The a is the real part of the number, and bi is the imaginary part. If $b = 0$, the number $a + bi = a$ is a real number, and if $b \neq 0$, the number $a + bi$ is an **imaginary number**. If $a = 0$, bi is a **pure imaginary number**.

The complex number system includes the real numbers as well as the imaginary numbers (Figure 3.1). Examples of imaginary numbers are $3 + 2i$, $5 - 4i$, and $\sqrt{3} - \frac{1}{2}i$; examples of pure imaginary numbers are $-i$, $2i$, $12i$, $-4i$, $i\sqrt{3}$, and πi.

Figure 3.1

Figure 3.2

The complex number system is an extension of the real numbers that includes imaginary numbers. The term *imaginary number* seems to imply that the numbers do not exist, but in fact they do have important theoretical and technical applications. Complex numbers are used in the design of electrical circuits and airplanes, and they were used in the development of quantum physics. For example, one way of accounting for the amount as well as the phase of the current or voltage in an alternating current electrical circuit involves complex numbers. Special sets of complex numbers can be graphed on the **complex coordinate system** to create pictures called **fractal images**. (Figure 3.2 shows a fractal image called the Mandelbrot set, which can be generated with the complex number i.)

EXAMPLE 9 ▶ **Simplifying Complex Numbers**

Simplify the following numbers by writing them in the form a, bi, or $a + bi$.

a. $6 + \sqrt{-8}$ **b.** $\dfrac{4 - \sqrt{-6}}{2}$

SOLUTION

a. $6 + \sqrt{-8} = 6 + i\sqrt{8} = 6 + 2i\sqrt{2}$
because $\sqrt{8} = \sqrt{4 \cdot 2} = \sqrt{4} \cdot \sqrt{2} = 2\sqrt{2}$

b. $\dfrac{4 - \sqrt{-6}}{2} = \dfrac{4 - i\sqrt{6}}{2} = \dfrac{4}{2} - \dfrac{i\sqrt{6}}{2} = 2 - \dfrac{i\sqrt{6}}{2}$

EXAMPLE 10 ▶ **Complex Numbers**

Identify each number as one or more of the following: real, imaginary, pure imaginary.

a. $6 + 4i$ **b.** $3i - \sqrt{4}$ **c.** $3 - 2i^2$ **d.** $4 - \sqrt{-16}$ **e.** $\sqrt{-3}$

SOLUTION

a. Imaginary because it contains i.

b. Imaginary because it contains i.

c. Real because $3 - 2i^2 = 3 - 2(-1) = 3 + 2 = 5.$

d. Imaginary because it contains $\sqrt{-16} = 4i$, which gives $4 - 4i$.

e. Imaginary because it contains $\sqrt{-3} = i\sqrt{3}$, and pure imaginary because the real part is 0.

Toolbox EXERCISES

In Exercises 1–6, use the rules of exponents to simplify the following expressions and remove all zero and negative exponents. Assume that all variables are nonzero.

1. $\left(\dfrac{2}{3}\right)^{-2}$

2. $\left(\dfrac{3}{2}\right)^{-3}$

3. $10^{-2} \cdot 10^{0}$

4. $8^{-2} \cdot 8^{0}$

5. $(2^{-1})^{3}$

6. $(4^{-2})^{2}$

7. $(2x^{-2}y)^{-4}$

8. $\left(\dfrac{x^2}{y^3}\right)^{5}$

9. $(-32x^5)^{-2}$

10. $(4x^3y^{-1}z)^{0}$

11. $\left(\dfrac{x^{-2}y}{z}\right)^{-3}$

12. $\left(\dfrac{a^{-2}b^{-1}c^{-4}}{a^4b^{-3}c^0}\right)^{-3}$

Find the absolute values in Exercises 13 and 14.

13. $|-6|$

14. $|7 - 11|$

15. Write each of the following expressions in simplified exponential form.

    a. $\sqrt{x^3}$ **b.** $\sqrt[4]{x^3}$ **c.** $\sqrt[5]{x^3}$

    d. $\sqrt[6]{27y^9}$ **e.** $27\sqrt[6]{y^9}$

16. Write each of the following in radical form.

    a. $a^{3/4}$ **b.** $-15x^{5/8}$

    c. $(-15x)^{5/8}$

In Exercises 17–22, find the products.

17. $(6x - 5)^2$

18. $(7s - 2t)(7s + 2t)$

19. $(8w + 2)^2$

20. $(6x - y)^2$

21. $(2y + 5z)^2$

22. $(3x^2 + 5y)(3x^2 - 5y)$

In Exercises 23–42, factor each of the polynomials completely.

23. $3x^2 - 12x$

24. $12x^5 - 24x^3$

25. $9x^2 - 25m^2$

26. $x^2 - 8x + 15$

27. $x^2 - 2x - 35$

28. $3x^2 - 5x - 2$

29. $8x^2 - 22x + 5$

30. $6n^2 + 18 + 39n$

31. $9ab - 12ab^2 + 18b^2$

32. $8x^2y - 160x + 48x^2$

33. $12y^3z + 4y^2x^2 - 8y^2z^3$

34. $x^2(x^2 + 4)^2 + 2(x^2 + 4)^2$

35. $x^2 + 8x + 16$

36. $x^2 - 4x + 4$

37. $4x^2 + 20x + 25$

38. $x^4 - 6x^2y + 9y^2$

39. $5x^5 - 80x$

40. $3y^4 + 9y^2 - 12y^2 - 36$

41. $18p^2 + 12p - 3p - 2$

42. $5x^2 - 10xy - 3x + 6y$

In Exercises 43 and 44, identify each number as one of the following: real, imaginary, pure imaginary.

43. a. $2 - i\sqrt{2}$ **b.** $5i$ **c.** $4 + 0i$

    d. $2 - 5i^2$

44. a. $3 + i\sqrt{5}$ **b.** $3 + 0i$ **c.** $8i$

    d. $2i^2 - i$

In Exercises 45–47, find values for a and b that make the statement true.

45. $a + bi = 4$

46. $a + 3i = 15 - bi$

47. $a + bi = 2 + 4i$

Quadratic Functions; Parabolas

SECTION PREVIEW   Revenue

When products are sold with variable discounts or with prices affected by supply and demand, revenue functions for these products may be nonlinear. Suppose the monthly revenue from the sale of Carlson 42-inch 3D televisions is given by the function

$$R(x) = -0.1x^2 + 600x \text{ dollars}$$

where x is the number of TVs sold. In this case, the revenue of this product is represented by a **second-degree function**, or **quadratic function**. A quadratic function is a function that can be written in the form

$$f(x) = ax^2 + bx + c$$

where a, b, and c are real numbers with $a \neq 0$.

The graph of the quadratic function $f(x) = ax^2 + bx + c$ is a **parabola** with a turning point called the **vertex**. Figure 3.3 shows the graph of the function $R(x) = -0.1x^2 + 600x$, which is a parabola that opens downward, and the vertex occurs where the function has its maximum value. Notice also that the graph of this revenue function is symmetric about a vertical line through the vertex. This vertical line is called the **axis of symmetry**.

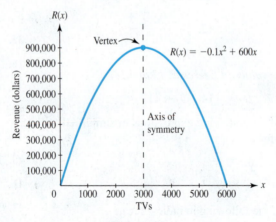

Figure 3.3

Table 3.1

x	y
−4	16
−3	9
−2	4
−1	1
0	0
1	1
2	4
3	9
4	16

Knowing where the maximum value of $R(x)$ occurs can show the company how many units must be sold to obtain the largest revenue. Knowing the maximum output of $R(x)$ helps the company plan its sales campaign. (See Example 2.) In this section, we graph and apply quadratic functions. ∎

Parabolas

The graph of every quadratic function has the distinctive shape known as a parabola. The graph of a quadratic function is determined by the location of the vertex and whether the parabola opens upward or downward.

Consider the basic quadratic function $y = x^2$. Each output y is obtained by squaring an input x (Table 3.1). The graph of $y = x^2$, shown in Figure 3.4, is a parabola that opens upward with the vertex (turning point) at the origin, (0, 0).

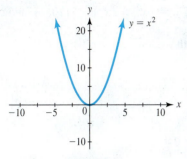

Figure 3.4

Observe that for $x > 0$, the graph of $y = x^2$ rises as it moves from left to right (that is, as the x-values increase), so the function $y = x^2$ is **increasing** for $x > 0$. For values of $x < 0$, the graph falls as it moves from left to right (as the x-values increase), so the function $y = x^2$ is **decreasing** for $x < 0$.

Increasing and Decreasing Functions

A function f is **increasing** on an interval if, for any x_1 and x_2 in the interval, when $x_2 > x_1$, it is true that $f(x_2) > f(x_1)$.

A function f is **decreasing** on an interval if, for any x_1 and x_2 in the interval, when $x_2 > x_1$, it is true that $f(x_2) < f(x_1)$.

The quadratic function $y = -x^2$ has the form $y = ax^2$ with $a < 0$, and its graph is a parabola that opens downward with vertex at $(0, 0)$ (Figure 3.5(a)). The function $y = \frac{1}{2}x^2$ has the form $y = ax^2$ with $a > 0$, and its graph is a parabola opening upward (Figure 3.5(b)).

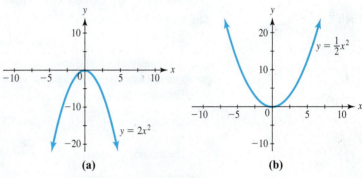

Figure 3.5

In general, the graph of a quadratic function of the form $y = ax^2$ is a parabola that opens upward (is **concave up**) if a is positive and opens downward (is **concave down**) if a is negative.* The vertex, which is the point where the parabola turns, is a **minimum point** if a is positive and is a **maximum point** if a is negative. The vertical line through the vertex is called the **axis of symmetry** because this line divides the graph into two halves that are reflections of each other (Figure 3.6(a)).

We can find the x-coordinate of the vertex of the graph of $y = ax^2 + bx + c$ by using the fact that the axis of symmetry of a parabola passes through the vertex. As

Figure 3.6

---

* A parabola that is concave up will appear to "hold water," and a parabola that is concave down will appear to "shed water."

Figure 3.6(b) shows, the y-intercept of the graph of $y = ax^2 + bx + c$ is $(0, c)$, and there is another point on the graph with y-coordinate c.

The x-coordinates of the points on this graph with y-coordinate c satisfy

$$c = ax^2 + bx + c$$

Solving this equation gives

$$0 = ax^2 + bx$$
$$0 = x(ax + b)$$
$$x = 0 \quad \text{or} \quad x = \frac{-b}{a}$$

The x-coordinate of the vertex is on the axis of symmetry, which is halfway from $x = 0$ to $x = \dfrac{-b}{a}$, so it is at $x = \dfrac{-b}{2a}$. The y-coordinate of the vertex can be found by evaluating the function at the x-coordinate of the vertex.

Graph of a Quadratic Function

The graph of the function

$$f(x) = ax^2 + bx + c$$

is a parabola that opens upward, and the vertex is a minimum, if $a > 0$. The parabola opens downward, and the vertex is a maximum, if $a < 0$.

The larger the value of $|a|$, the more narrow the parabola will be. Its vertex is at the point $\left(\dfrac{-b}{2a}, f\left(\dfrac{-b}{2a} \right) \right)$ (Figure 3.7).

The axis of symmetry of the parabola has equation $x = \dfrac{-b}{2a}$.

Figure 3.7

Observe that the graph of $y = x^2$ (Figure 3.4) is narrower than the graph of $y = \dfrac{1}{2}x^2$ (Figure 3.5(b)), and that 1 (the coefficient of x^2 in $y = x^2$) is larger than $\dfrac{1}{2}$ (the coefficient of x^2 in $y = \dfrac{1}{2}x^2$).

If we know the location of the vertex and the direction in which the parabola opens, we can make a good sketch of the graph by plotting just a few more points.

EXAMPLE 1 ▶   **Graphing a Quadratic Function**

Find the vertex and graph the quadratic function $f(x) = -2x^2 - 4x + 6$.

SOLUTION

Note that a, the coefficient of x^2, is -2, so the parabola opens downward. The x-coordinate of the vertex is $\dfrac{-b}{2a} = \dfrac{-(-4)}{2(-2)} = -1$, and the y-coordinate of the vertex is $f(-1) = -2(-1)^2 - 4(-1) + 6 = 8$. Thus, the vertex is $(-1, 8)$. The x-intercepts can be found by setting $f(x) = 0$ and solving for x:

$$-2x^2 - 4x + 6 = 0$$
$$-2(x^2 + 2x - 3) = 0$$
$$-2(x + 3)(x - 1) = 0$$
$$x = -3 \quad \text{or} \quad x = 1$$

The y-intercept is easily found by computing $f(0)$:

$$f(0) = -2(0)^2 - 4(0) + 6 = 6$$

The axis of symmetry is the vertical line $x = -1$. The graph is shown in Figure 3.8.

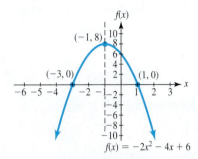

Figure 3.8

$f(x) = -2x^2 - 4x + 6$

Spreadsheet ▶ SOLUTION   To graph the function $f(x) = -2x^2 - 4x + 6$ using Excel, first create a table containing values for x and $f(x)$. Then highlight the two columns containing these values, select insert, scatter plot, and the smooth curve option. Figure 3.9 shows the table of values and the graph.*

Figure 3.9

EXAMPLE 2 ▶   ## Maximizing Revenue

Suppose the monthly revenue from the sale of Carlson 42-inch 3D televisions is given by the function

$$R(x) = -0.1x^2 + 600x \text{ dollars}$$

where x is the number of televisions sold.

---

* For detailed steps, see Appendix B, page 687.

a. Find the vertex and the axis of symmetry of the graph of this function.

b. Determine if the vertex represents a maximum or minimum point.

c. Interpret the vertex in the context of the application.

d. Graph the function.

e. For what x-values is the function increasing? decreasing? What does this mean in the context of the application?

SOLUTION

a. The function is a quadratic function with $a = -0.1$, $b = 600$, and $c = 0$. The x-coordinate of the vertex is $\dfrac{-b}{2a} = \dfrac{-600}{2(-0.1)} = 3000$, and the axis of symmetry is the line $x = 3000$. The y-coordinate of the vertex is

$$R(3000) = -0.1(3000)^2 + 600(3000) = 900{,}000$$

so the vertex is (3000, 900,000).

b. Because $a < 0$, the parabola opens downward, so the vertex is a maximum point.

c. The x-coordinate of the vertex gives the number of televisions that must be sold to maximize revenue, so selling 3000 sets will result in the maximum revenue. The y-coordinate of the vertex gives the maximum revenue, $900,000.

d. Table 3.2 lists some values that satisfy $R(x) = -0.1x^2 + 600x$. The graph is shown in Figure 3.10.

Table 3.2

x	R(x)
0	0
1000	500,000
2000	800,000
3000	900,000
4000	800,000
5000	500,000
6000	0

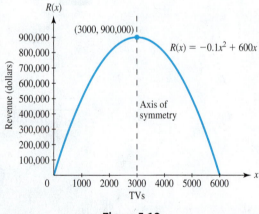

Figure 3.10

e. The function is increasing on the interval $(-\infty, 3000)$ and decreasing on $(3000, \infty)$. However, in the context of the application, negative inputs and outputs do not make sense, so we may say the revenue increases on (0, 3000) and decreases on (3000, 6000).

Even when using a graphing utility to graph a quadratic function, it is important to recognize that the graph is a parabola and to locate the vertex. Determining which way the parabola opens and the x-coordinate of the vertex is very useful in setting the viewing window so that a complete graph (that includes the vertex and the intercepts) is shown.

EXAMPLE 3 ▶   **Foreign-Born Population**

Using data from 1900 and projected to 2020, the percent of the U.S. population that was foreign born can be modeled by the equation

$$y = 0.0027x^2 - 0.3692x + 18.9149$$

where x is the number of years after 1900.

a. During what year does the model indicate that the percent of foreign-born population was a minimum?

b. What is the minimum percent?
(Source: U.S. Census Bureau)

SOLUTION

a. This equation is in the form $f(x) = ax^2 + bx + c$, so $a = 0.0027$. Because $a > 0$, the parabola opens upward, the vertex is a minimum, and the x-coordinate of the vertex is where the minimum percent occurs.

$$x = \frac{-b}{2a} = \frac{-(-0.3692)}{2(0.0027)} = 68.4$$

So the percent of the U.S. population that was foreign born was a minimum in the 69th year after 1900, or in 1969.

b. The minimum for the model is found by evaluating the function at $x = 68.4$. This value, 6.3, can be found with ⌈TRACE⌉ or TABLE (accessed by **2ND** ⌈GRAPH⌉) or with direct evaluation (Figure 3.11). So the minimum percent is 6.3%.

Figure 3.11

Suppose an object is shot or thrown into the air and then falls. If air resistance is ignored, the height in feet of the object after t seconds can be modeled by

$$S(t) = -16t^2 + v_0 t + h_0$$

where -16 ft/sec^2 is the acceleration due to gravity, v_0 ft/sec is the initial velocity (at $t = 0$ sec), and h_0 is the initial height in feet (at $t = 0$).

EXAMPLE 4 ▶ **Height of a Ball**

A ball is thrown upward at 64 feet per second from the top of an 80-foot-high building.

a. Write the quadratic function that models the height (in feet) of the ball as a function of the time t (in seconds).

b. Find the t-coordinate and S-coordinate of the vertex of the graph of this quadratic function.

c. Graph the model.

d. Explain the meaning of the coordinates of the vertex for this model.

SOLUTION

a. The model has the form $S(t) = -16t^2 + v_0 t + h_0$, where $v_0 = 64$ and $h_0 = 80$. Thus, the model is

$$S(t) = -16t^2 + 64t + 80 \text{ (feet)}$$

b. The height S is a function of the time t, and the t-coordinate of the vertex is

$$t = \frac{-b}{2a} = \frac{-64}{2(-16)} = 2$$

The S-coordinate of the vertex is the value of S at $t = 2$, so $S = -16(2)^2 + 64(2) + 80 = 144$ is the S-coordinate of the vertex. The vertex is $(2, 144)$.

c. The function is quadratic and the coefficient in the second-degree term is negative, so the graph is a parabola that opens down with vertex $(2, 144)$. To graph the function, we choose a window that includes the vertex $(2, 144)$ near the center top of the screen. Using the window $[0, 6]$ by $[-20, 150]$ gives the graph shown in Figure 3.12(a). Using CALC (accessed by **2ND** (TRACE)) maximum* verifies that the vertex is $(2, 144)$ (see Figure 3.12(b)).

(a)   (b)

Figure 3.12

d. The graph is a parabola that opens down, so the vertex is the highest point on the graph and the function has its maximum there. The t-coordinate of the vertex, 2, is the time (in seconds) at which the ball reaches its maximum height, and the S-coordinate, 144, is the maximum height (in feet) that the ball reaches.

Vertex Form of a Quadratic Function

When a quadratic function is written in the form $f(x) = ax^2 + bx + c$, we can calculate the coordinates of the vertex. But if a quadratic function is written in the form

$$y = a(x - h)^2 + k$$

the vertex of the parabola is at (h, k) (Figure 3.13(a)). For example, the graph of $y = (x - 2)^2 + 3$ is a parabola opening upward with vertex $(2, 3)$ (Figure 3.13(b)).

(a)   (b)   (c)

Figure 3.13

* For more details, see Appendix A, page 669.

Graph of a Quadratic Function

In general, the graph of the function

$$y = a(x - h)^2 + k$$

is a parabola with its vertex at the point (h, k).

The parabola opens upward if $a > 0$, and the vertex is a minimum.

The parabola opens downward if $a < 0$, and the vertex is a maximum.

The axis of symmetry of the parabola has equation $x = h$.

The a is the same as the leading coefficient in $y = ax^2 + bx + c$, so the larger the value of $|a|$, the narrower the parabola will be.

Note that the graph of $y = 2(x - 2)^2 + 3$ in Figure 3.13(c) is narrower than the graph of $y = (x - 2)^2 + 3$ in Figure 3.13(b).

EXAMPLE 5 ▶ Minimizing Cost

The cost for producing x Champions golf hats is given by the function

$$C(x) = 0.2(x - 40)^2 + 200 \text{ dollars}$$

a. Find the vertex of this function.

b. Is the vertex a maximum or minimum? Interpret the vertex in the context of the application.

c. Graph the function using a window that includes the vertex.

d. Describe what happens to the function between $x = 0$ and the x-coordinate of the vertex. What does this mean in the context of the application?

SOLUTION

a. This function is in the form $y = a(x - h)^2 + k$ with $h = 40$ and $k = 200$. Thus, the vertex of $C(x) = 0.2(x - 40)^2 + 200$ is $(40, 200)$.

b. Because $a = 0.2$, which is positive, the vertex is a minimum. This means that the cost of producing golf hats is at a minimum of \$200 when 40 hats are produced.

c. We know that the vertex of the graph of this function is $(40, 200)$ and that it is a minimum, so we can choose a window with the $x = 40$ near the center of the screen and $y = 200$ near the bottom of the screen. The graph using the window $[0, 100]$ by $[-50, 1000]$ is shown in Figure 3.14.

d. For x-values between 0 and 40, the graph decreases. Thus, the cost of producing golf hats is decreasing until 40 hats are produced; after 40 hats are produced, the cost begins to increase.

Figure 3.14

We can use the vertex form $y = a(x - h)^2 + k$ to write the equation of a quadratic function if we know the vertex and a point on its graph.

EXAMPLE 6 ▶ Profit

Right Sports Management had its monthly maximum profit, \$450,000, when it produced and sold 5500 Waist Trimmers. Its fixed cost is \$155,000. If the profit can be modeled by a quadratic function of x, the number of Waist Trimmers produced and sold each month, find this quadratic function $P(x)$.

SOLUTION

When 0 units are produced, the cost is $155,000 and the revenue is $0. Thus, the profit is −$155,000 when 0 units are produced, and the y-intercept of the graph of the function is $(0, -155,000)$. The vertex of the graph of the quadratic function is $(5500, 450,000)$. Using these points gives

$$P(x) = a(x - 5500)^2 + 450,000$$

and

$$-155,000 = a(0 - 5500)^2 + 450,000$$

which gives

$$a = -0.02$$

Thus, the quadratic function that models the profit is $P(x) = -0.02(x - 5500)^2 + 450,000$, or $P(x) = -0.02x^2 + 220x - 155,000$, where $P(x)$ is in dollars and x is the number of units produced and sold.

EXAMPLE 7 ▶   **Equation of a Quadratic Function**

If the points in the table lie on a parabola, write the equation whose graph is the parabola.

x	−1	0	1	2
y	13	−2	−7	−2

SOLUTION

The x-values are a uniform distance apart. Because the points $(0, -2)$ and $(2, -2)$ both have a y-coordinate of -2, the symmetry of a parabola indicates that the vertex will be halfway between $x = 0$ and $x = 2$. Thus, the vertex of this parabola is at $(1, -7)$. (See Figure 3.15.)

The equation of the function is

$$y = a(x - 1)^2 - 7$$

The point $(2, -2)$ or any other point in the table besides $(1, -7)$ can be used to find a.

$$-2 = a(2 - 1)^2 - 7, \quad \text{or} \quad a = 5$$

Thus, the equation is

$$y = 5(x - 1)^2 - 7, \quad \text{or} \quad y = 5x^2 - 10x - 2$$

Figure 3.15

EXAMPLE 8 ▶   **Vertex Form of a Quadratic Function**

Write the vertex form of the equation of the quadratic function from the general form $y = 2x^2 - 8x + 5$ by finding the vertex and using the value of a.

SOLUTION

The vertex is at $x = \dfrac{-b}{2a} = \dfrac{-(-8)}{2(2)} = 2$, and $y = 2(2^2) - 8(2) + 5 = -3$. We know a is 2, because a is the same in both forms, so the equation has the form $y = a(x - 2)^2 - 3$. Thus,

$$y = 2(x - 2)^2 - 3$$

is the vertex form of the equation.

SKILLS CHECK 3.1

In Exercises 1–6, (a) determine whether the function is quadratic. If it is, (b) determine whether the graph is concave up or concave down. (c) Determine whether the vertex of the graph is a maximum point or a minimum point.

1. $y = 2x^2 - 8x + 6$ **2.** $y = 4x - 3$

3. $y = 2x^3 - 3x^2$ **4.** $f(x) = x^2 + 4x + 4$

5. $g(x) = -5x^2 - 6x + 8$

6. $h(x) = -2x^2 - 4x + 6$

In Exercises 7–14, (a) graph each quadratic function on $[-10, 10]$ by $[-10, 10]$. (b) Does this window give a complete graph?

7. $y = 2x^2 - 8x + 6$ **8.** $f(x) = x^2 + 4x + 4$

9. $g(x) = -5x^2 - 6x + 8$

10. $h(x) = -2x^2 - 4x + 6$

11. $y = x^2 + 8x + 19$

12. $y = x^2 - 4x + 5$

13. $y = 0.01x^2 - 8x$

14. $y = 0.1x^2 + 8x + 2$

15. Write the equation of the quadratic function whose graph is shown.

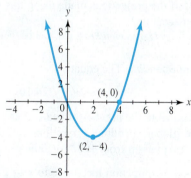

16. Write the equation of the quadratic function whose graph is shown.

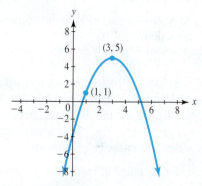

17. The two graphs shown have equations of the form $y = a(x - 2)^2 + 1$. Is the value of a larger for y_1 or y_2?

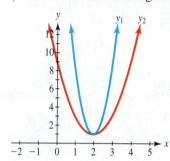

18. The two graphs shown have equations of the form $y = -a(x - 3)^2 + 5$. Is the value of $|a|$ larger for y_1 or y_2?

19. If the points in the table lie on a parabola, write the equation whose graph is the parabola.

x	−1	1	3	5
y	−7	13	−7	−67

20. If the points in the table lie on a parabola, write the equation whose graph is the parabola.

x	−6	−5	−4	−3	−2	−1
y	33	12	−3	−12	−15	−12

In Exercises 21–30, (a) give the coordinates of the vertex of the graph of each function. (b) Graph each function on a window that includes the vertex.

21. $y = (x - 1)^2 + 3$ **22.** $y = (x + 10)^2 - 6$

23. $y = (x + 8)^2 + 8$ **24.** $y = (x - 12)^2 + 1$

25. $f(x) = 2(x - 4)^2 - 6$

26. $f(x) = -0.5(x - 2)^2 + 1$

27. $y = 12x - 3x^2$ **28.** $y = 3x + 18x^2$

29. $y = 3x^2 + 18x - 3$ **30.** $y = 5x^2 + 75x + 8$

For Exercises 31–34, (a) find the x-coordinate of the vertex of the graph. (b) Set the viewing window so that the x-coordinate of the vertex is near the center of the window and the vertex is visible, and then graph the given equation. (c) State the coordinates of the vertex.

31. $y = 2x^2 - 40x + 10$

32. $y = -3x^2 - 66x + 12$

33. $y = -0.2x^2 - 32x + 2$

34. $y = 0.3x^2 + 12x - 8$

In Exercises 35–40, sketch complete graphs of the functions.

35. $y = x^2 + 24x + 144$

36. $y = x^2 - 36x + 324$

37. $y = -x^2 - 100x + 1600$

38. $y = -x^2 - 80x - 2000$

39. $y = 2x^2 + 10x - 600$

40. $y = 2x^2 - 75x - 450$

Use the graph of each function in Exercises 41–46 to estimate the x-intercepts.

41. $y = 2x^2 - 8x + 6$ **42.** $f(x) = x^2 + 4x + 4$

43. $y = x^2 - x - 110$ **44.** $y = x^2 + 9x - 36$

45. $g(x) = -5x^2 - 6x + 8$

46. $h(x) = -2x^2 - 4x + 6$

EXERCISES 3.1

47. **Profit** The daily profit for a product is given by $P = 32x - 0.01x^2 - 1000$, where x is the number of units produced and sold.

   a. Graph this function for x between 0 and 3200.

   b. Describe what happens to the profit for this product when the number of units produced is between 1 and 1600.

   c. What happens to the profit after 1600 units are produced?

48. **Profit** The daily profit for a product is given by $P = 420x - 0.1x^2 - 4100$ dollars, where x is the number of units produced and sold.

   a. Graph this function for x between 0 and 4200.

   b. Is the graph of the function concave up or down?

49. **Worldwide Mobile Phone Shipments** For selected years from 2009 and projected to 2016, the number of worldwide mobile phone shipments, in millions, can be modeled by $N(t) = -7.88t^2 + 335.33t - 1202.14$, where t is the number of years after 2000. Because of new technology and market saturation, mobile phone shipments may decrease in the future.

   a. Graph the function for the years 2005 through 2020.

   b. Graph the function for the years 2005 through 2030 and find the year in which worldwide mobile shipments are a maximum.

   c. If the model is valid for years after 2020, what will happen to worldwide mobile shipments?

   (Source: portioresearch.com)

50. **World Population** A low-projection scenario of world population for the years 1995–2150 by the United Nations is given by the function $y = -0.36x^2 + 38.52x + 5822.86$, where x is the number of years after 1990 and the world population is measured in millions of people.

   a. Graph this function for $x = 0$ to $x = 120$.

   b. What would the world population have been in 2010 if the projections made using this model had been accurate?

   (Source: *World Population Prospects,* United Nations)

51. **Tourism Spending** The equation

$$y = 1.69x^2 - 0.92x + 324.10,$$

   with x equal to the number of years after 1998, models the global spending (in billions of dollars) on travel and tourism from 1998 to 2009.

   a. Graph this function for $x = 0$ to $x = 20$.

   b. If $x = 0$ in 1998, find the spending projected by this model for 2015.

   c. Is the value in part (b) an interpolation or an extrapolation?

   (Source: *Statistical Abstract of the United States*)

52. **Flight of a Ball** If a ball is thrown upward at 96 feet per second from the top of a building that is 100 feet high, the height of the ball can be modeled by $S(t) = 100 + 96t - 16t^2$ feet, where t is the number of seconds after the ball is thrown.

   a. Describe the graph of the model.

b. Find the *t*-coordinate and *S*-coordinate of the vertex of the graph of this quadratic function.

c. Explain the meaning of the coordinates of the vertex for this model.

53. *Flight of a Ball* If a ball is thrown upward at 39.2 meters per second from the top of a building that is 30 meters high, the height of the ball can be modeled by $S(t) = 30 + 39.2t - 9.8t^2$ meters, where *t* is the number of seconds after the ball is thrown.

a. Find the *t*-coordinate and *S*-coordinate of the vertex of the graph of this quadratic function.

b. Explain the meaning of the coordinates of the vertex for this function.

c. Over what time interval is the function increasing? What does this mean in relation to the ball?

54. *Photosynthesis* The rate of photosynthesis *R* for a certain plant depends on the intensity of light *x*, in lumens, according to $R(x) = 270x - 90x^2$.

a. Sketch the graph of this function on a meaningful window.

b. Determine the intensity *x* that gives the maximum rate of photosynthesis.

55. *Crude Oil* The U.S. Crude Oil production, in billions of barrels, for the years from 2010 projected to 2030, can be modeled by $y = -0.001x^2 + 0.037x + 1.949$, with *x* equal to the number of years after 2010 and *y* equal to the number of billions of barrels of crude oil.

a. Find and interpret the vertex of the graph of this model.

b. What does the model predict the crude oil production will be in 2033?

c. Graph the function for the years 2010 to 2030. (Source: U.S. Energy Information Administration)

56. *Workers and Output* The weekly output of graphing calculators is $Q(x) = 200x + 6x^2$. Graph this function for values of *x* and *Q* that make sense in this application, if *x* is the number of weeks, $x \leq 10$.

57. *Profit* The profit for a certain brand of MP3 player can be described by the function $P(x) = 40x - 3000 - 0.01x^2$ dollars, where *x* is the number of MP3 players produced and sold.

a. To maximize profit, how many MP3 players must be produced and sold?

b. What is the maximum possible profit?

58. *Profit* The profit for Easy-Cut lawnmowers can be described by the function $P(x) = 840x - 75.6 - $

$0.4x^2$ dollars, where *x* is the number of mowers produced and sold.

a. To maximize profit, how many mowers must be produced and sold?

b. What is the maximum possible profit?

59. *Revenue* The annual total revenue for Pilot V5 pens is given by $R(x) = 1500x - 0.02x^2$ dollars, where *x* is the number of pens sold.

a. To maximize the annual revenue, how many pens must be sold?

b. What is the maximum possible annual revenue?

60. *Revenue* The monthly total revenue for satellite radios is given by $R(x) = 300x - 0.01x^2$ dollars, where *x* is the number of radios sold.

a. To maximize the monthly revenue, how many radios must be sold?

b. What is the maximum possible monthly revenue?

61. *Area* If 200 feet of fence are used to enclose a rectangular pen, the resulting area of the pen is $A = x(100 - x)$, where *x* is the width of the pen.

(100 − x)

x

a. Is *A* a quadratic function of *x*?

b. What is the maximum possible area of the pen?

62. *Area* If 25,000 feet of fence are used to enclose a rectangular field, the resulting area of the field is $A = (12,500 - x)x$, where *x* is the width of the pen. What is the maximum possible area of the pen?

63. *Obesity* Obesity (BMI $\geq$ 30) increases the risk of diabetes, heart disease, and many other ailments. The percent of Americans who are obese, from 2000 projected to 2030, can be modeled by $y = -0.010x^2 + 0.971x + 22.1$, where *x* equals the number of years after 2000 and *y* is the percent of obese Americans. (Source: *American Journal of Preventive Medicine*)

a. Graph the function for the years 2000 through 2030.

b. Find and interpret the vertex of the graph.

c. If the model remains valid, describe what happens to the percent of obese Americans after 2030.

64. *U.S. Visitors* The number of international visitors to the United States, in millions, can be modeled by $y = 0.592x^2 - 3.277x + 48.493$, where *x* is the number of years after 2000.

a. Does the model estimate that a maximum or minimum number of visitors will occur during this period? How can you tell without graphing the equation?

b. Find the input and output at the vertex.

c. Interpret the results of part (b).
(Source: *Statistical Abstract of the United States*)

65. **Poverty** The number of millions of people in the United States who lived below the poverty level for can be modeled by $y = 0.114x^2 - 2.322x + 45.445$, where x is the number of years after 1990.

   a. Is the vertex of the graph of this function a maximum or minimum?

   b. Find the coordinates of the vertex and interpret them in the context of the problem.

   c. Use the vertex to set the window and graph the model.
(Source: U.S. Census Bureau, U.S. Department of Commerce)

66. **Abortions** The number of abortions in the United States per 1000 women aged 15–44 can be modeled by the function $y = -0.026x^2 + 0.951x + 18.161$, where x is the number of years after 1970 and y is measured in thousands.

   a. In what year does this model indicate the number was a maximum?

   b. What was the maximum number of abortions in this year if this model is accurate?

   c. For what values of x can we be sure that this model no longer applies?
(Source: Alan Guttmacher Institute)

67. **Wind and Pollution** The amount of particulate pollution p in the air depends on the wind speed s, among other things, with the relationship between p and s approximated by $p = 25 - 0.01s^2$, where p is in ounces per cubic yard and s is in miles per hour.

   a. Sketch the graph of this model with s on the horizontal axis and with nonnegative values of s and p.

   b. Is the function increasing or decreasing on this domain?

   c. What is the p-intercept of the graph?

   d. What does the p-intercept mean in the context of this application?

68. **Drug Sensitivity** The sensitivity S to a drug is related to the dosage size x by $S = 1000x - x^2$.

   a. Sketch the graph of this model using a domain and range with nonnegative x and S.

   b. Is the function increasing or decreasing for x between 0 and 500?

c. What is the positive x-intercept of the graph?

d. Why is this x-intercept important in the context of this application?

69. **Falling Object** A tennis ball is thrown downward into a swimming pool from the top of a tall hotel. The height of the ball from the pool is given by $D(t) = -16t^2 - 4t + 210$ feet, where t is the time, in seconds, after the ball is thrown. Graphically find the t-intercepts for this function. Interpret the value(s) that make sense in this problem context.

70. **Break-Even** The profit for a product is given by $P = 1600 - 100x + x^2$, where x is the number of units produced and sold. Graphically find the x-intercepts of the graph of this function to find how many units will give break-even (that is, return a profit of zero).

71. **Flight of a Ball** A softball is hit with upward velocity 32 feet per second when $t = 0$, from a height of 3 feet.

   a. Find the function that models the height of the ball as a function of time.

   b. Find the maximum height of the ball.

72. **Flight of a Ball** A baseball is hit with upward velocity 48 feet per second when $t = 0$, from a height of 4 feet.

   a. Find the function that models the height of the ball as a function of time.

   b. Find the maximum height of the ball and in how many seconds the ball will reach that height.

73. **Apartment Rental** The owner of an apartment building can rent all 100 apartments if he charges $1200 per apartment per month, but the number of apartments rented is reduced by 2 for every $40 increase in the monthly rent.

   a. Construct a table that gives the revenue if the rent charged is $1240, $1280, and $1320.

   b. Does $R(x) = (1200 + 40x)(100 - 2x)$ model the revenue from these apartments if x represents the number of $40 increases?

   c. What monthly rent gives the maximum revenue for the apartments?

74. **Rink Rental** The owner of a skating rink rents the rink for parties at $720 if 60 or fewer skaters attend, so the cost is $12 per person if 60 attend. For each 6 skaters above 60, she reduces the price per skater by $0.50.

   a. Construct a table that gives the revenue if the number attending is 66, 72, and 78.

   b. Does the function $R(x) = (60 + 6x)(12 - 0.5x)$ model the revenue from the party if x represents the number of increases of 6 people each?

c. How many people should attend for the rink's revenue to be a maximum?

75. World Population A low-projection scenario for world population for 1995–2150 by the United Nations is given by the function $y = -0.36x^2 + 38.52x + 5822.86$, where x is the number of years after 1990 and the world population is measured in millions of people.

a. Find the input and output at the vertex of the graph of this model.

b. Interpret the values from part (a).

c. For what years after 1995 does this model predict that the population will increase?

(Source: *World Population Prospects*, United Nations)

3.2 Solving Quadratic Equations

KEY OBJECTIVES

- Solve quadratic equations using factoring
- Solve quadratic equations graphically using the x-intercept method and the intersection method
- Solve quadratic equations by combining graphical and factoring methods
- Solve quadratic equations using the square root method
- Solve quadratic equations by completing the square
- Solve quadratic equations using the quadratic formula
- Solve quadratic equations having complex solutions

SECTION PREVIEW National Health Care

The national health care expenditures in the United States for selected years from 1980 and projected to 2020 can be modeled by

$$C(x) = 2.224x^2 - 29.841x + 353.300$$

with x equal to the number of years after 1970 and y equal to the national health care expenditures in billions of dollars. (Source: Centers for Medicare and Medicaid Services)

Health care expenditures have a large impact on consumer spending, so controlling them is important. To find when these expenditures reach $5227 billion, we solve the equation

$$2.224x^2 - 29.841x + 353.300 = 5227 \quad \text{or} \quad 2.224x^2 - 29.841x - 4873.700 = 0$$

The values of x that satisfy this last equation are also the zeros of the function

$$y = 2.224x^2 - 29.841x - 4873.700$$

and the x-values of intersection of the functions $y = 2.224x^2 - 29.841x + 353.300$ and $y = 5227$. (See Example 9.) In this section, we learn how to solve quadratic equations by factoring methods, graphical methods, the square root method, completing the square, and the quadratic formula. ■

Factoring Methods

An equation that can be written in the form $ax^2 + bx + c = 0$, with $a \neq 0$, is called a **quadratic equation**. Solutions to some quadratic equations can be found exactly by factoring; other quadratic equations require different types of solution methods to find or to approximate solutions.

Solution by factoring is based on the following property of real numbers.

Zero Product Property

If the product of two real numbers is 0, then at least one of them must be 0. That is, for real numbers a and b, if the product $ab = 0$, then either $a = 0$ or $b = 0$ or both a and b are equal to 0.

To use this property to solve a quadratic equation by factoring, we must first make sure that the equation is written in a form with zero on one side. If the resulting nonzero expression is factorable, we factor it and use the zero product property to convert the equation into two linear equations that are easily solved.* Before applying this technique to real-world applications, we consider the following example.

EXAMPLE 1 ▶ ## Solving a Quadratic Equation by Factoring

Solve the equation $3x^2 + 7x = 6$.

SOLUTION

We first subtract 6 from both sides of the equation to rewrite the equation with 0 on one side:

$$3x^2 + 7x - 6 = 0$$

To begin factoring the trinomial $3x^2 + 7x - 6$, we seek factors of $3x^2$ (that is, $3x$ and x) as the first terms of two binomials and factors of -6 as the last terms of the binomials. The factorization whose inner and outer products combine to $7x$ is $(3x - 2)(x + 3)$, so we have

$$(3x - 2)(x + 3) = 0$$

Using the zero product property gives

$$3x - 2 = 0, \quad \text{or} \quad x + 3 = 0$$

Solving these linear equations gives the two solutions to the original equation:

$$x = \frac{2}{3} \quad \text{or} \quad x = -3$$

EXAMPLE 2 ▶ ## Height of a Ball

The height above ground of a ball thrown upward at 64 feet per second from the top of an 80-foot-high building is modeled by $S(t) = 80 + 64t - 16t^2$ feet, where t is the number of seconds after the ball is thrown. How long will the ball be in the air?

SOLUTION

The ball will be in the air from $t = 0$ (with the height $S = 80$) until it reaches the ground ($S = 0$). Thus, we can find the time in the air by solving

$$0 = -16t^2 + 64t + 80$$

Because 16 is a factor of each of the terms, we can get a simpler but equivalent equation by dividing both sides of the equation by -16:

$$0 = t^2 - 4t - 5$$

This equation can be solved easily by factoring the right side:

$$0 = (t - 5)(t + 1)$$

$$0 = t - 5 \quad \text{or} \quad 0 = t + 1$$

$$t = 5 \quad \text{or} \quad t = -1$$

---

* For a review of factoring methods, see page 163 in the Algebra Toolbox.

Figure 3.16

The time in the air starts at $t = 0$, so $t = -1$ has no meaning in this application. S also equals 0 at $t = 5$, which means that the ball is on the ground 5 seconds after it was thrown; that is, the ball is in the air 5 seconds. Figure 3.16 shows a graph of the function, which confirms that the height of the ball is 0 at $t = 5$.

Graphical Methods

In cases where factoring $f(x)$ to solve $f(x) = 0$ is difficult or impossible, graphing $y = f(x)$ can be helpful in finding the solution. Recall that if a is a real number, the following three statements are equivalent:

- a is a real solution to the equation $f(x) = 0$.
- a is a real zero of the function f.
- a is an x-intercept of the graph of $y = f(x)$.

It is important to remember that the above three statements are equivalent because connecting these concepts allows us to use different methods for solving equations. Sometimes graphical methods are the easiest way to find or approximate solutions to real data problems. If the x-intercepts of the graph of $y = f(x)$ are easily found, then graphical methods may be helpful in finding the solutions. Note that if the graph of $y = f(x)$ does not cross or touch the x-axis, there are no real solutions to the equation $f(x) = 0$.

We can also find solutions or decimal approximations of solutions to quadratic equations by using the intersection method with a graphing utility.

EXAMPLE 3 ▶ Profit

Consider the daily profit from the production and sale of x units of a product, given by

$$P(x) = -0.01x^2 + 20x - 500 \text{ dollars}$$

a. Use a graph to find the levels of production and sales that give a daily profit of $1400.

b. Is it possible for the profit to be greater than $1400?

SOLUTION

a. To find the level of production and sales, x, that gives a daily profit of 1400 dollars, we solve

$$1400 = -0.01x^2 + 20x - 500$$

To solve this equation by the intersection method, we graph

$$y_1 = -0.01x^2 + 20x - 500 \quad \text{and} \quad y_2 = 1400$$

To find the appropriate window for this graph, we note that the graph of the function $y_1 = -0.01x^2 + 20x - 500$ is a parabola with the vertex at

$$x = \frac{-b}{2a} = \frac{-20}{2(-0.01)} = 1000 \quad \text{and} \quad y = P(1000) = 9500$$

We use a viewing window containing this point, with $x = 1000$ near the center, to graph the function (Figure 3.17). Using the intersection method, we see that $(100, 1400)$ and $(1900, 1400)$ are the points of intersection of the graphs of the two functions. Thus, $x = 100$ and $x = 1900$ are solutions to the equation $1400 = -0.01x^2 + 20x - 500$, and the profit is $1400 when $x = 100$ units or $x = 1900$ units of the product are produced and sold.

Figure 3.17

b. We can see from the graph in Figure 3.17 that the profit is more than \$1400 for many values of x. Because the graph of this profit function is a parabola that opens down, the maximum profit occurs at the vertex of the graph. As we found in part (a), the vertex occurs at $x = 1000$, and the maximum possible profit is $P(1000) = \$9500$, which is more than \$1400.

Combining Graphs and Factoring

If the factors of a quadratic function are not easily found, its graph may be helpful in finding the factors. Note once more the important correspondences that exist among solutions, zeros, and x-intercepts: the x-intercepts of the graph of $y = P(x)$ are the real solutions of the equation $0 = P(x)$ and the real zeros of $P(x)$. In addition, the factors of $P(x)$ are related to the zeros of $P(x)$. The relationship among the factors, solutions, and zeros is true for any polynomial function f and can be generalized by the following theorem.

> ### Factor Theorem
>
> The polynomial function f has a factor $(x - a)$ if and only if $f(a) = 0$. Thus, $(x - a)$ is a factor of $f(x)$ if and only if $x = a$ is a solution to $f(x) = 0$.

This means that we can verify our factorization of f and the real solutions to $0 = f(x)$ by graphing $y = f(x)$ and observing where the graph crosses the x-axis. We can also sometimes use our observation of the graph to assist us in the factorization of a quadratic function:

> *If one solution can be found exactly from the graph, it can be used to find one of the factors of the function. The second factor can then be found easily, leading to the second solution.*

EXAMPLE 4 ▶   **Graphing and Factoring Methods Combined**

Solve $0 = 3x^2 - x - 10$ by using the following steps.

a. Graphically find one of the x-intercepts of $y = 3x^2 - x - 10$.

b. Algebraically verify that the zero found in part (a) is an exact solution to $0 = 3x^2 - x - 10$.

c. Use the method of factoring to find the other solution to $0 = 3x^2 - x - 10$.

SOLUTION

a. The vertex of the graph of $y = 3x^2 - x - 10$ is at $x = \dfrac{1}{6}$ and $y \approx -10.08$, and because the parabola opens up, we set a viewing window that includes this point near the bottom center of the window. Graphing the function and using the x-intercept method, we find that the graph crosses the x-axis at $x = 2$ (Figure 3.18). This means that 2 is a zero of $f(x) = 3x^2 - x - 10$.

(a)

(b)

Figure 3.18

b. Because $x = 2$ was obtained graphically, it may be an approximation of the exact solution. To verify that it is exact, we show that $x = 2$ makes the equation $0 = 3x^2 - x - 10$ a true statement:

$$3(2)^2 - 2 - 10 = 12 - 2 - 10 = 0$$

c. Because $x = 2$ is a solution of $0 = 3x^2 - x - 10$, one factor of $3x^2 - x - 10$ is $(x - 2)$. Thus, we use $(x - 2)$ as one factor and seek a second binomial factor so that the product of the two binomials is $3x^2 - x - 10$:

$$3x^2 - x - 10 = 0$$
$$(x - 2)(\qquad) = 0$$
$$(x - 2)(3x + 5) = 0$$

The remaining factor is $3x + 5$, which we set equal to 0 to obtain the other solution:

$$3x = -5$$
$$x = -\frac{5}{3}$$

Thus, the two solutions to $0 = 3x^2 - x - 10$ are $x = 2$ and $x = -\dfrac{5}{3}$.

Graphical and Numerical Methods

When approximate solutions to quadratic equations are sufficient, graphical and/or numerical solution methods can be used. These methods of solving are illustrated in Example 5.

EXAMPLE 5 ▶   **National Hospital Expenditures**

In 2015, hospital spending is projected to increase 5.6% as a consequence of the continued effects of the Affordable Care Act insurance expansion, combined with the effect of faster economic growth. For 2016 through 2022, continued population aging and the impacts of improved economic conditions are expected to result in projected average annual growth of 6.4%. The function

$$E(x) = 2.171x^2 - 5.051x + 628.152$$

gives the national expenditures for hospitals, in billions of dollars, for the years 2006 and projected to 2022, where x equals the number of years after 2000.

a. Graph the function for the years 2006 to 2022.

b. Find the year after 2000 during which the national hospital expenditures are expected to reach $1,400,000,000,000, using a graphical method.

c. Verify the solution numerically.
(Source: Centers for Medicare & Medicaid Services)

SOLUTION

a. The graph of the function for the years 2006 to 2022 is shown in Figure 3.19(a). The inputs that represent these years are $x = 6$ to $x = 22$.

b. If expenditures are $1,400,000,000,000, the output for the function will be $E(x) = 1400$, so we solve the equation

$$2.171x^2 - 5.051x + 628.152 = 1400$$

We solve this equation with the intersection method. Figure 3.19(b) shows the graphs of $Y_1 = 2.171x^2 - 5.051x + 628.152$ and $Y_2 = 1400$ and the point where the graphs intersect.

The intersection point indicates that the hospital expenditures will reach 1400 billion dollars, or $1,400,000,000,000, when x is more than 20, so the expenditures reach $ 1400 billion in 2021.

Figure 3.20

(a)

(b)

Figure 3.19

c. Figure 3.20 shows a table of inputs near 21 and the corresponding outputs. The table shows that the expenditures are close to $1400 billion in 2020 but do not actually reach $1400 billion until during 2021.

Spreadsheet ▸ SOLUTION   To find the intersection of the graphs of $y = 2.171x^2 - 5.051x + 628.152$ and $y = 40$ with Excel, we enter their outputs in columns B and C, respectively, enter B − C in column D, and then use the Goal Seek command to find the x-value that makes the function in column D equal to 0, as we did with linear equations in Chapter 2. The x-value that makes the difference of these two functions equal to 0 is approximately 20.05 (see Figure 3.21).* This value agrees with the value found with a calculator in Figure 3.19(b). The graph of the two functions in Figure 3.21 confirms the solution.

---

* For details, see Appendix B, page 694.

Figure 3.21

The Square Root Method

We have solved quadratic equations by factoring and by graphical methods. Another method can be used to solve quadratic equations that are written in a particular form. In general, we can find the solutions of quadratic equations of the form $x^2 = C$, where C is a constant, by taking the square root of both sides. For example, to solve $x^2 = 25$, we take the square root of both sides of the equation, getting $x = \pm 5$. Note that there are two solutions because $5^2 = 25$ and $(-5)^2 = 25$.

> ### Square Root Method
>
> The solutions of the quadratic equation $x^2 = C$ are $x = \pm\sqrt{C}$. Note that, when we take the square root of both sides, we use a $\pm$ symbol because there are both a positive and a negative value that, when squared, give C.

Note that this method can also be used to solve equations of the form $(ax + b)^2 = C$.

EXAMPLE 6 ▶ ### Square Root Solution Method

Solve the following equations by using the square root method.

a. $2x^2 - 16 = 0$

b. $(x - 6)^2 = 18$

SOLUTION

a. This equation can be written in the form $x^2 = C$, so the square root method can be used. We want to rewrite the equation so that the x^2-term is isolated and its coefficient is 1 and then take the square root of both sides.

$$2x^2 - 16 = 0$$
$$2x^2 = 16$$
$$x^2 = 8$$
$$x = \pm\sqrt{8} = \pm 2\sqrt{2}$$

The exact solutions to $2x^2 - 16 = 0$ are $x = 2\sqrt{2}$ and $x = -2\sqrt{2}$.

b. The left side of this equation is a square, so we can take the square root of both sides to find x.

$$(x - 6)^2 = 18$$
$$x - 6 = \pm\sqrt{18}$$
$$x = 6 \pm \sqrt{18}$$
$$= 6 \pm \sqrt{9}\sqrt{2}$$
$$= 6 \pm 3\sqrt{2}$$

Completing the Square

When we convert one side of a quadratic equation to a perfect binomial square and use the square root method to solve the equation, the method used is called **completing the square**.

EXAMPLE 7 ▶   Completing the Square

Solve the equation $x^2 - 12x + 7 = 0$ by completing the square.

SOLUTION

Because the left side of this equation is not a perfect square, we rewrite it in a form where we can more easily get a perfect square. We subtract 7 from both sides, getting

$$x^2 - 12x = -7$$

We complete the square on the left side of this equation by adding the appropriate number to both sides of the equation to make the left side a perfect square trinomial. Because $(x + a)^2 = x^2 + 2ax + a^2$, the constant term of a perfect square trinomial will be the *square of half the coefficient of x*. Using this rule with the equation $x^2 - 12x = -7$, we would need to take half the coefficient of x and add the square of this number to get a perfect square trinomial. Half of -12 is -6, so adding $(-6)^2 = 36$ to $x^2 - 12x$ gives the perfect square $x^2 - 12x + 36$. We also have to add 36 to the other side of the equation (to preserve the equality), giving

$$x^2 - 12x + 36 = -7 + 36$$

Factoring this perfect square trinomial gives

$$(x - 6)^2 = 29$$

We can now solve the equation with the square root method.

$$(x - 6)^2 = 29$$

$$x - 6 = \pm \sqrt{29}$$

$$x = 6 \pm \sqrt{29}$$

The Quadratic Formula

We can generalize the method of completing the square to derive a general formula that gives the solution to any quadratic equation. We use this method to find the general solution to $ax^2 + bx + c = 0, a \neq 0$.

$$ax^2 + bx + c = 0 \qquad \text{Standard form}$$

$$ax^2 + bx = -c \qquad \text{Subtract } c \text{ from both sides.}$$

$$x^2 + \frac{b}{a}x = -\frac{c}{a} \qquad \text{Divide both sides by } a.$$

We would like to make the left side of the last equation a perfect square trinomial. Half the coefficient of x is $\dfrac{b}{2a}$, and squaring this gives $\dfrac{b^2}{4a^2}$. Hence, adding $\dfrac{b^2}{4a^2}$ to both sides of the equation gives a perfect square trinomial on the left side, and we can continue with the solution.

$$x^2 + \frac{b}{a}x + \frac{b^2}{4a^2} = \frac{b^2}{4a^2} - \frac{c}{a} \qquad \text{Add } \frac{b^2}{4a^2} \text{ to both sides of the equation.}$$

$$\left(x + \frac{b}{2a}\right)^2 = \frac{b^2 - 4ac}{4a^2} \qquad \text{Factor the left side and combine the fractions on the right side.}$$

$$x + \frac{b}{2a} = \pm \sqrt{\frac{b^2 - 4ac}{4a^2}} \qquad \text{Take the square root of both sides.}$$

$$x = -\frac{b}{2a} \pm \frac{\sqrt{b^2 - 4ac}}{2|a|} \qquad \text{Subtract } \frac{b}{2a} \text{ from both sides and simplify.}$$

$$x = -\frac{b}{2a} \pm \frac{\sqrt{b^2 - 4ac}}{2a} \qquad \pm 2|a| = \pm 2a$$

$$x = \frac{-b \pm \sqrt{b^2 - 4ac}}{2a} \qquad \text{Combine the fractions.}$$

The formula we have developed is called the **quadratic formula**.

Quadratic Formula

The solutions of the quadratic equation $ax^2 + bx + c = 0$ are given by the formula

$$x = \frac{-b \pm \sqrt{b^2 - 4ac}}{2a}$$

Note that a is the coefficient of x^2, b is the coefficient of x, and c is the constant term.

Because of the $\pm$ sign, the solutions can be written as

$$x = \frac{-b + \sqrt{b^2 - 4ac}}{2a} \quad \text{and} \quad x = \frac{-b - \sqrt{b^2 - 4ac}}{2a}$$

We can use the quadratic formula to solve all quadratic equations exactly, but it is especially useful for finding exact solutions to those equations for which factorization is difficult or impossible. For example, the solutions to $2.171x^2 - 5.051x + 628.152 = 1400$ were found approximately by graphical methods in Example 5. If we need to find the exact solutions, we could use the quadratic formula.

EXAMPLE 8 ▶ Solving Using the Quadratic Formula

Solve $6 - 3x^2 + 4x = 0$ using the quadratic formula.

SOLUTION

The equation $6 - 3x^2 + 4x = 0$ can be rewritten as $-3x^2 + 4x + 6 = 0$, so $a = -3$, $b = 4$, and $c = 6$. The two solutions to this equation are

$$x = \frac{-4 \pm \sqrt{4^2 - 4(-3)(6)}}{2(-3)} = \frac{-4 \pm \sqrt{88}}{-6} = \frac{-4 \pm 2\sqrt{22}}{-6} = \frac{2 \pm \sqrt{22}}{3}$$

Thus, the exact solutions are the irrational numbers $x = \dfrac{2 + \sqrt{22}}{3}$ and $x = \dfrac{2 - \sqrt{22}}{3}$. Three-place decimal approximations for these solutions are $x \approx 2.230$ and $x \approx -0.897$.

Decimal approximations of irrational solutions found with the quadratic formula will often suffice as answers to an applied problem. The quadratic formula is especially useful when the coefficients of a quadratic equation are decimal values that make factorization impractical. This occurs in many applied problems, such as the one discussed at the beginning of this section. If the graph of the quadratic function $y = f(x)$ does not intersect the x-axis at "nice" values of x, the solutions may be irrational numbers, and using the quadratic formula allows us to find these solutions exactly.

EXAMPLE 9 ▶ National Health Care

The national health care expenditures in the United States for selected years from 1980 and projected to 2020 can be modeled by

$$C(x) = 2.224x^2 - 29.841x + 353.300$$

with x equal to the number of years after 1970 and y equal to the national health care expenditures in billions of dollars. Find the year after 1970 that the model predicts the expenditures will reach $5227 billion.
(Source: Centers for Medicare & Medicaid Services)

SOLUTION

To answer this question, we solve (using algebraic, numerical, and graphical methods) the equation

$$2.224x^2 - 29.841x + 353.300 = 5227$$

We choose to use the quadratic formula (even though an approximate answer is all that is needed). We first write $2.224x^2 - 29.841x + 353.300 = 5227$ with 0 on one side:

$$2.224x^2 - 29.841x - 4873.700 = 0$$

This gives $a = 2.224$, $b = -29.841$, and $c = -4873.700$. Substituting in the quadratic formula gives

$$x = \frac{-(-29.841) \pm \sqrt{(-29.841)^2 - 4(2.224)(-4873.700)}}{2(2.224)}$$

$$= \frac{29.841 \pm 210.3495198}{4.448}$$

$$x \approx 54.0 \quad \text{or} \quad x \approx -40.6$$

Thus, the model predicts the national health care expenditures will reach \$5227 billion in the 54th year after 1970—that is, in 2024. (Note that a negative value for x will give a year before 1970.) Figure 3.22(a) shows the outputs of

$$Y_1 = 2.224x^2 - 29.841x + 353.300$$

at integer values near the solution above, and Figure 3.22(b) shows the graph of the intersection of $Y_1 = 2.224x^2 - 29.841x + 353.300$ and $Y_2 = 5227$, which confirms the conclusions we found numerically and graphically.

(a)

(b)

Figure 3.22

The Discriminant

We can also determine the type of solutions a quadratic equation has by looking at the expression $b^2 - 4ac$, which is called the **discriminant** of the quadratic equation $ax^2 + bx + c = 0$. The discriminant is the expression inside the radical in the quadratic formula $x = \dfrac{-b \pm \sqrt{b^2 - 4ac}}{2a}$, so it determines if the quantity inside the radical is positive, zero, or negative.

- If $b^2 - 4ac > 0$, there are two different real solutions.
- If $b^2 - 4ac = 0$, there is one real solution.
- If $b^2 - 4ac < 0$, there is no real solution.

For example, the equation $3x^2 + 4x + 2 = 0$ has no real solution because $4^2 - 4(3)(2) = -8 < 0$, and the equation $x^2 + 4x + 2 = 0$ has two different real solutions because $4^2 - 4(1)(2) = 8 > 0$.

Some quadratic equations have a solution that occurs twice. For example, the equation $x^2 - 6x + 9 = 0$ can be solved by factoring:

$$x^2 - 6x + 9 = 0$$

$$(x - 3)(x - 3) = 0$$

$$x = 3, x = 3$$

Because $x - 3$ is a factor twice, the solution $x = 3$ is said to have **multiplicity** 2. Note that the discriminant of this equation is $(-6)^2 - 4(1)(9) = 36 - 36 = 0$. When the discriminant equals 0, there will be only one solution to the quadratic equation, and it will have multiplicity 2.

Aids for Solving Quadratic Equations

The x-intercepts of the graph of a quadratic function $y = f(x)$ can be used to determine how to solve the quadratic equation $f(x) = 0$. Table 3.3 summarizes these ideas with suggested methods for solving and the graphical representations of the solutions.

Table 3.3   Connections Between Graphs of Quadratic Functions and Solution Methods

Graph			
Type of x-Intercepts	Graph crosses x-axis twice.	Graph touches but does not cross x-axis.	Graph does not cross x-axis.
Type of Solutions	Equation has two real solutions.	Equation has one real solution of multiplicity 2.	Equation has no real solutions.
Suggested Solution Methods*	Use factoring, graphing, or the quadratic formula.	Use factoring, graphing, or the square root method.	Use the quadratic formula to verify that solutions are not real.

Equations with Complex Solutions

Recall that the solutions of the quadratic equation $x^2 = C$ are $x = \pm\sqrt{C}$, so the solutions of the equation $x^2 = -a$ for $a > 0$ are

$$x = \pm\sqrt{-a} = \pm\sqrt{-1}\sqrt{a} = \pm i\sqrt{a}$$

EXAMPLE 10 ▶   **Solution Using the Square Root Method**

Solve the equations.

a. $x^2 = -9$ **b.** $3x^2 + 24 = 0$

SOLUTION

a. Taking the square root of both sides of the equation gives the solution of $x^2 = -9$:

$$x = \pm\sqrt{-9} = \pm\sqrt{-1}\sqrt{9} = \pm 3i$$

b. We solve $3x^2 + 24 = 0$ using the square root method, as follows:

$$3x^2 = -24$$
$$x^2 = -8$$
$$x = \pm\sqrt{-8} = \pm\sqrt{-1}\sqrt{4 \cdot 2} = \pm 2i\sqrt{2}$$

_____

* Solution methods other than the suggested methods may also be successful.

We used the square root method to solve the equations in Example 10 because neither equation contained a first-degree term (that is, a term containing x to the first power). We can also find complex solutions by using the quadratic formula.* Recall that the solutions of the quadratic equation $ax^2 + bx + c = 0$ are given by the formula

$$x = \frac{-b \pm \sqrt{b^2 - 4ac}}{2a}$$

EXAMPLE 11 ▶ **Complex Solutions of Quadratic Equations**

Solve the equations.

a. $x^2 - 3x + 5 = 0$ **b.** $3x^2 + 4x = -3$

SOLUTION

a. Using the quadratic formula, with $a = 1$, $b = -3$, and $c = 5$, gives

$$x = \frac{-(-3) \pm \sqrt{(-3)^2 - 4(1)(5)}}{2(1)} = \frac{3 \pm \sqrt{-11}}{2} = \frac{3 \pm i\sqrt{11}}{2}$$

Note that the solutions can also be written in the form $\dfrac{3}{2} \pm \dfrac{\sqrt{11}}{2}i$.

Thus, the solutions are the complex numbers $\dfrac{3}{2} + \dfrac{\sqrt{11}}{2}i$ and $\dfrac{3}{2} - \dfrac{\sqrt{11}}{2}i$.

b. Writing $3x^2 + 4x = -3$ in the form $3x^2 + 4x + 3 = 0$ gives $a = 3$, $b = 4$, and $c = 3$, so the solutions are

$$x = \frac{-4 \pm \sqrt{(4)^2 - 4(3)(3)}}{2(3)} = \frac{-4 \pm \sqrt{-20}}{6} = \frac{-4 \pm \sqrt{-1}\sqrt{4}\sqrt{5}}{6}$$

$$= \frac{-4 \pm 2i\sqrt{5}}{6} = \frac{2(-2 \pm i\sqrt{5})}{2 \cdot 3} = \frac{-2 \pm i\sqrt{5}}{3}$$

Thus,

$$x = -\frac{2}{3} + \frac{\sqrt{5}}{3}i \quad \text{and} \quad x = -\frac{2}{3} - \frac{\sqrt{5}}{3}i$$

---

* The solutions could also be found by completing the square. Recall that the quadratic formula was developed by completing the square on the quadratic equation $ax^2 + bx + c = 0$.

SKILLS CHECK 3.2

In Exercises 1–10, use factoring to solve the equations.

1. $x^2 - 3x - 10 = 0$ **2.** $x^2 - 9x + 18 = 0$ **7.** $0 = 2t^2 - 11t + 12$ **8.** $6x^2 - 13x + 6 = 0$

3. $x^2 - 11x + 24 = 0$ **4.** $x^2 + 3x - 10 = 0$ **9.** $6x^2 + 10x = 4$ **10.** $10x^2 + 11x = 6$

5. $2x^2 + 2x - 12 = 0$ **6.** $2s^2 + s - 6 = 0$

In Exercises 11–16, find the x-intercepts algebraically.

11. $f(x) = 3x^2 - 5x - 2$

12. $f(x) = 5x^2 + 7x + 2$

13. $f(x) = 4x^2 - 9$

14. $f(x) = 4x^2 + 20x + 25$

15. $f(x) = 3x^2 + 4x - 4$

16. $f(x) = 9x^2 - 1$

Use a graphing utility to find or to approximate the x-intercepts of the graph of each function in Exercises 17–22.

17. $y = x^2 + 7x + 10$ **18.** $y = x^2 + 4x - 32$

19. $y = 3x^2 - 8x + 4$ **20.** $y = 2x^2 + 8x - 10$

21. $y = 2x^2 + 7x - 4$ **22.** $y = 5x^2 - 17x + 6$

Use a graphing utility as an aid in factoring to solve the equations in Exercises 23–28.

23. $2w^2 - 5w - 3 = 0$ **24.** $3x^2 - 4x - 4 = 0$

25. $x^2 - 40x + 256 = 0$ **26.** $x^2 - 32x + 112 = 0$

27. $2s^2 - 70s = 1500$ **28.** $3s^2 - 130s = -1000$

In Exercises 29–34, use the square root method to solve the quadratic equations.

29. $4x^2 - 9 = 0$ **30.** $x^2 - 20 = 0$

31. $x^2 - 32 = 0$ **32.** $5x^2 - 25 = 0$

33. $(x - 5)^2 = 9$ **34.** $(2x + 1)^2 = 20$

In Exercises 35–38, complete the square to solve the quadratic equations.

35. $x^2 - 4x - 9 = 0$ **36.** $x^2 - 6x + 1 = 0$

37. $x^2 - 3x + 2 = 0$ **38.** $2x^2 - 9x + 8 = 0$

In Exercises 39–42, use the quadratic formula to solve the equations.

39. $x^2 - 5x + 2 = 0$ **40.** $3x^2 - 6x - 12 = 0$

41. $5x + 3x^2 = 8$ **42.** $3x^2 - 30x - 180 = 0$

In Exercises 43–48, use a graphing utility to find or approximate solutions of the equations.

43. $2x^2 + 2x - 12 = 0$ **44.** $2x^2 + x - 6 = 0$

45. $0 = 6x^2 + 5x - 6$ **46.** $10x^2 = 22x - 4$

47. $4x + 2 = 6x^2 + 3x$ **48.** $(x - 3)(x + 2) = -4$

In Exercises 49–54, find the exact solutions to $f(x) = 0$ in the complex numbers and confirm that the solutions are not real by showing that the graph of $y = f(x)$ does not cross the x-axis.

49. $x^2 + 25 = 0$ **50.** $2x^2 + 40 = 0$

51. $(x - 1)^2 = -4$ **52.** $(2x + 1)^2 + 7 = 0$

53. $x^2 + 4x + 8 = 0$ **54.** $x^2 - 5x + 7 = 0$

In Exercises 55–58, you are given the graphs of several functions of the form $f(x) = ax^2 + bx + c$ for different values of a, b, and c. For each function,

a. *Determine whether the discriminant is positive, negative, or zero.*

b. *Determine whether there are 0, 1, or 2 real solutions to $f(x) = 0$.*

c. *Solve the equation $f(x) = 0$, if possible.*

55.

56.

57.

58.

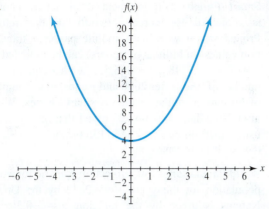

For each function in Exercise 59–62,

a. Calculate the discriminant.

b. Determine whether there are 0, 1 or 2 real solutions to $f(x) = 0$.

59. $f(x) = 3x^2 - 5x - 2$ **60.** $f(x) = 2x^2 - x + 1$

61. $f(x) = 4x^2 + 4x + 1$ **62.** $f(x) = 4x^2 + 25$

EXERCISES 3.2

In Exercises 63–74, solve analytically and then check graphically.

63. *Flight of a Ball* If a ball is thrown upward at 96 feet per second from the top of a building that is 100 feet high, the height of the ball can be modeled by $S(t) = 100 + 96t - 16t^2$ feet, where t is the number of seconds after the ball is thrown. How long after the ball is thrown is the height 228 feet?

64. *Falling Object* A tennis ball is thrown into a swimming pool from the top of a tall hotel. The height of the ball above the pool is modeled by $D(t) = -16t^2 - 4t + 200$ feet, where t is the time, in seconds, after the ball is thrown. How long after the ball is thrown is it 44 feet above the pool?

65. *Break-Even* The profit for an electronic reader is given by $P(x) = -12x^2 + 1320x - 21,600$, where x is the number of readers produced and sold. How many readers give break-even (that is, give zero profit) for this product?

66. *Break-Even* The profit for Coffee Exchange coffee beans is given by $P(x) = -15x^2 + 180x - 405$ thousand dollars, where x is the number of tons of coffee beans produced and sold. How many tons give break-even (that is, give zero profit) for this product?

67. *Break-Even* The total revenue function for French door refrigerators is given by $R = 550x$ dollars, and the total cost function for this same product is given by $C = 10,000 + 30x + x^2$, where C is measured in dollars. For both functions, the input x is the number of refrigerators produced and sold.

   a. Form the profit function for the refrigerators from the two given functions.

   b. What is the profit when 18 refrigerators are produced and sold?

   c. What is the profit when 32 refrigerators are produced and sold?

   d. How many refrigerators must be sold to break even on this product?

68. *Break-Even* The total revenue function for a home theater system is given by $R = 266x$, and the total cost function for the system is $C = 2000 + 46x + 2x^2$, where R and C are each measured in dollars and x is the number of units produced and sold.

   a. Form the profit function for this product from the two given functions.

   b. What is the profit when 55 systems are produced and sold?

   c. How many systems must be sold to break even on this product?

69. *Wind and Pollution* The amount of particulate pollution p from a power plant in the air above the plant depends on the wind speed s, among other things, with the relationship between p and s approximated by $p = 25 - 0.01s^2$, with s in miles per hour.

   a. Find the value(s) of s that will make $p = 0$.

   b. What does $p = 0$ mean in this application?

   c. What solution to $0 = 25 - 0.01s^2$ makes sense in the context of this application?

70. *Velocity of Blood* Because of friction from the walls of an artery, the velocity of a blood corpuscle in an artery is greatest at the center of the artery and decreases as the distance r from the center increases. The velocity of the blood in the artery can be modeled by the function

$$v = k(R^2 - r^2)$$

where R is the radius of the artery and k is a constant that is determined by the pressure, viscosity of the blood, and the length of the artery. In the case where $k = 2$ and $R = 0.1$ centimeter, the velocity is $v = 2(0.01 - r^2)$ centimeters per second (cm/sec).

a. What distance r would give a velocity of 0.02 cm/sec?

b. What distance r would give a velocity of 0.015 cm/sec?

c. What distance r would give a velocity of 0 cm/sec? Where is the blood corpuscle?

71. *Drug Sensitivity* The sensitivity S to a drug is related to the dosage size x by $S = 100x - x^2$, where x is the dosage size in milliliters.

a. What dosage(s) would give zero sensitivity?

b. Explain what your answer in part (a) might mean.

72. *Body-Heat Loss* The model for body-heat loss depends on the coefficient of convection K, which depends on wind speed s according to the equation $K^2 = 16s + 4$, where s is in miles per hour. Find the positive coefficient of convection when the wind speed is

a. 20 mph.

b. 60 mph.

c. What is the change in K for a change in speed from 20 mph to 60 mph?

73. *Market Equilibrium* Suppose that the demand for artificial Christmas trees is given by the function

$$p = 109.70 - 0.10q$$

and that the supply of these trees is given by

$$p = 0.01q^2 + 5.91$$

where p is the price of a tree in dollars and q is the quantity of trees that are demanded/supplied in hundreds. Find the price that gives the market equilibrium price and the number of trees that will be sold/bought at this price.

74. *Market Equilibrium* The demand for diamond-studded watches is given by $p = 7000 - 2x$ dollars, and the supply of watches is given by $p = 0.01x^2 + 2x + 1000$ dollars, where x is the number of watches demanded and supplied when the price per watch is p dollars. Find the equilibrium quantity and the equilibrium price.

75. *Smart Phones* Bill payment using smart phones is projected to increase rapidly in the future. Projections of worldwide mobile payment transaction values (in billions of dollars) can be modeled by $y = 9.3x^2 - 148.8x + 605.595$, with x equal to the number of years after 2000 and y equal to the number of billions of dollars paid via smart phones. When after 2010 does this model project that bill payment using smart phones will be \$1135 billion? (Source: Portioresearch.com)

76. *World Population* A low-projection scenario for world population for the years 1995–2150 by the United Nations is given by the function $y = -0.36x^2 + 38.52x + 5822.86$, where x is the number of years after 1990 and the world population is measured in millions of people. In what year will the world population reach 6581 million? (Source: World Population Prospects, United Nations)

77. *U.S. Energy Consumption* Energy consumption in the United States in quadrillion BTUs can be modeled by $C(x) = -0.013x^2 + 1.281x + 67.147$, where x is the number of years after 1970.

a. One solution to the equation $87.567 = -0.013x^2 + 1.281x + 67.147$ is $x = 20$. What does this mean?

b. Graphically verify that $x = 20$ is a solution to $87.567 = -0.013x^2 + 1.281x + 67.147$.

c. To find when after 2020 U.S. energy consumption will be 87.567 quadrillion BTUs according to the model, do we need to find the second solution to this equation? Why or why not? (Source: U.S. Energy Information Administration)

78. *Tourism Spending* The global spending on travel and tourism (in billions of dollars) can be described by the equation $y = 1.69x^2 - 0.92x + 324.10$, where x equals the number of years after 1990. Graphically find the year in which spending is projected to reach \$1817.5 billion. (Source: *Statistical Abstract of the United States*)

79. *China's Labor Pool* Using UN data from 1975 and projections to 2050, the number of millions of people age 15 to 59 in China can be modeled by $y = -0.224x^2 + 22x + 370.7$, with x equal to the number of years after 1970 and y equal to the number of millions of people in this labor pool. Use technology with the model to find when this population is expected to be equal to 903.1 million. (Source: United Nations)

80. *Alzheimer's Disease* Partially because of Americans living longer, the number with Alzheimer's disease and other dementia is projected to grow each year. The millions of U.S. citizens with Alzheimer's from

2000 and projected to 2050 can be modeled by $y = 0.002107x^2 + 0.09779x + 4.207$, with x equal to the number of years after 2000 and y equal to the number of millions of Americans with Alzheimer's disease. In what year does the model predict that 9.037 million Americans will have Alzheimer's disease or other dementia?

Year	2000	2010	2020	2030	2040	2050
Projected (millions affected)	4.0	5.9	6.8	8.7	11.8	14.3

(Source: National Academy on an Aging Society)

81. **Federal Funds** The amount of federal funds, in billions, spent on child nutrition programs can be modeled by $N = 0.645x^2 - 0.165x + 10.298$, where x is the number of years after 2000.

   a. What was the increase in federal funds spent on child nutrition programs between 2000 and 2010?

   b. Use graphical methods to find in what year after 2000 the model estimates that the amount of funds spent will be $134.408 billion.

   c. Use the model to estimate the funds that will be spent in 2020. Is this interpolation or extrapolation? What assumptions are you making when you make this estimation?
   (Source: *Statistical Abstract of the United States*)

82. **International Travel to the United States** The number of millions of visitors to the United States can be modeled by $V(x) = 0.592x^2 - 3.277x + 48.493$, where x is the number of years after 2000.

   a. Use the model to find the increase in the number of visitors to the United States from 2000 to 2008.

   b. When would the number of visitors exceed 84,000,000, according to the model?
   (Source: Office of Travel and Tourism Industries, U.S. Department of Commerce)

83. **Gold Prices** The price of an ounce of gold in U.S. dollars for the years 1997–2011 can be modeled by the function $G(x) = 11.532x^2 - 259.978x + 1666.555$, where x is the number of years after 1990.

   a. Graph this function for values of x representing 1997–2011.

   b. According to the model, what will the price of gold be in 2020?

   c. Use graphical or numerical methods to estimate when the price of gold will reach $2702.80 per ounce.
   (Source: goldprice.org)

84. **Non-White Population** The number of millions of individuals in the U.S civilian non-institutional population 16 years and older who are non-White or Hispanic for the years 2000 and projected to 2050 can be modeled by $y = 0.007x^2 + 1.24x + 43.71$, with x equal to the number of years after 1990 and y equal to this population, in millions. Use technology to find the year after 1990 when this population is projected to be 104.51 million.
(Source: U.S. Census Bureau)

85. **Retail Sales** November and December retail sales, excluding autos, for the years 2001–2010 can be modeled by the function $S(x) = -1.751x^2 + 38.167x + 388.997$ billion dollars, where x is the number of years after 2000.

   a. Graph the function for values of x representing 2001–2010.

   b. During what years does the model estimate the sales to be $550 billion?

   c. The recession in 2008 caused retail sales to drop. Does the model agree with the facts; that is, does it indicate that a maximum occurred in 2008?
   (Source: U.S. Census Bureau)

86. **Cell Phones** Using the CTIA Wireless Survey for 1985–2009, the number of U.S. cell phone subscribers (in millions) can be modeled by

   $$y = 0.632x^2 - 2.651x + 1.209$$

   where x is the number of years after 1985.

   a. Graphically find when the number of U.S. subscribers was 301,617,000.

   b. When does the model estimate that the number of U.S. subscribers would reach 359,515,000?

   c. What does the answer to (b) tell about this model?

87. **World Population** One projection of the world population by the United Nations (a low-projection scenario) is given in the table below. The data can be modeled by $y = -0.36x^2 + 38.52x + 5822.86$ million people, where x is the number of years after 1990. In what year after 1990 does this model predict the world population will first reach 6,702,000,000?

Year	Projected Population (millions)
1995	5666
2000	6028
2025	7275
2050	7343
2075	6402
2100	5153
2125	4074
2150	3236

(Source: *World Population Prospects,* United Nations)

Piecewise-Defined Functions and Power Functions

SECTION PREVIEW   Residential Power Costs

The data in Table 3.4 give the rates that Georgia Power Company charges its residential customers for electricity during the months of October through May, excluding fuel adjustment costs and taxes. The monthly charges can be modeled by a **piecewise-defined function**, as we will see in Example 3. Piecewise-defined functions are used in applications in the life, social, and physical sciences, as well as in business, when there is not a single function that accurately represents the application.

Power functions are another type of function that is a basic building block for many mathematical applications. In this section, we evaluate, graph, and apply piecewise-defined and power functions.

Table 3.4

Monthly Kilowatt-hours (kWh)	Monthly Charge
0 to 650	$7.50 plus $0.045991 per kWh
More than 650, up to 1000	$37.40 plus $0.03946 per kWh above 650
More than 1000	$51.21 plus $0.038737 per kWh above 1000

Piecewise-Defined Functions

It is possible that a set of data cannot be modeled with a single equation. We can use a piecewise-defined function when there is not a single function that accurately represents the situation. A piecewise-defined function is so named because it is defined with different pieces for different parts of its domain rather than one equation.

EXAMPLE 1 ▶   Postage

The postage paid for a first-class letter "jumps" by 21 cents for each ounce after the first ounce, but does not increase until the weight increases by 1 ounce. Table 3.5 gives the 2015 postage for letters up to 3.5 ounces. Write a piecewise-defined function that models the price of postage and graph the function.

Table 3.5

Weight x (oz)	Postage y (cents)
$0 < x \le 1$	49
$1 < x \le 2$	70
$2 < x \le 3$	91
$3 < x \le 3.5$	112

SOLUTION

The function that models the price P of postage for x ounces, where x is between 0 and 3.5, is

$$P(x) = \begin{cases} 49 & \text{if } 0 < x \le 1 \\ 70 & \text{if } 1 < x \le 2 \\ 91 & \text{if } 2 < x \le 3 \\ 112 & \text{if } 3 < x \le 3.5 \end{cases}$$

The graph of this function (Figure 3.23) shows that each output is a constant with discontinuous "steps" at $x = 1, 2$, and 3. This is a special piecewise-defined function, called a **step function**. Other piecewise-defined functions can be defined by polynomial functions over limited domains.

$$P(x) = \begin{cases} 49 & \text{if } 0 < x \le 1 \\ 70 & \text{if } 1 < x \le 2 \\ 91 & \text{if } 2 < x \le 3 \\ 112 & \text{if } 3 < x \le 3.5 \end{cases}$$

Figure 3.23

EXAMPLE 2 ▶ Piecewise-Defined Function

Graph the function

$$f(x) = \begin{cases} 5x + 2 & \text{if } 0 \le x < 3 \\ x^3 & \text{if } 3 \le x \le 5 \end{cases}$$

Table 3.6(a)

x	0	1	2	2.99
f(x)	2	7	12	16.95

Table 3.6(b)

x	3	4	5
f(x)	27	64	125

SOLUTION

To graph this function, we can construct a table of values for each of the pieces. Table 3.6(a) gives outputs of the function for some sample inputs x on the interval $[0, 3)$; on this interval, $f(x)$ is defined by $f(x) = 5x + 2$. Table 3.6(b) gives outputs for some sample inputs on the interval $[3, 5]$; on this interval, $f(x)$ is defined by $f(x) = x^3$.

Plotting the points from Table 3.6(a) and connecting them with a smooth curve gives the graph of $y = f(x)$ on the x-interval $[0, 3)$ (Figure 3.24). The open circle on this piece of the graph indicates that this piece of the function is not defined for $x = 3$. Plotting the points from Table 3.6(b) and connecting them with a smooth curve gives the graph of $y = f(x)$ on the x-interval $[3, 5]$ (Figure 3.24). The closed circles on this piece of the graph indicate that this piece of the function is defined for $x = 3$ and $x = 5$. The graph of $y = f(x)$, shown in Figure 3.24, consists of these two pieces.

$$f(x) = \begin{cases} 5x + 2 & \text{if } 0 \le x < 3 \\ x^3 & \text{if } 3 \le x \le 5 \end{cases}$$

Figure 3.24

EXAMPLE 3 ▶ **Residential Power Costs**

Excluding fuel adjustment costs and taxes, the rates Georgia Power Company charges its residential customers for electricity during the months of October through May are shown in Table 3.7.

Table 3.7

Monthly Kilowatt-hours (kWh)	Monthly Charge
0 to 650	$7.50 plus $0.045991 per kWh
More than 650, up to 1000	$37.40 plus $0.03946 per kWh above 650
More than 1000	$51.21 plus $0.038737 per kWh above 1000

a. Write the piecewise-defined function C that gives the monthly charge for residential electricity, with input x equal to the monthly number of kilowatt-hours.

b. Find $C(950)$ and explain what it means.

c. Find the charge for using 1560 kWh in a month.

d. Use a graph of the function to determine how many kilowatt-hours would be used if the monthly charge were $122.87.

SOLUTION

a. The monthly charge is given by the function

$$C(x) = \begin{cases} 7.50 + 0.045991x & \text{if } 0 \le x \le 650 \\ 37.40 + 0.03946(x - 650) & \text{if } 650 < x \le 1000 \\ 51.21 + 0.038737(x - 1000) & \text{if } x > 1000 \end{cases}$$

where $C(x)$ is the charge in dollars for x kWh of electricity.

b. To evaluate $C(950)$, we must determine which "piece" defines the function when $x = 950$. Because 950 is between 650 and 1000, we use the "middle piece" of the function:

$$C(950) = 37.40 + 0.03946(950 - 650) = 49.238$$

Companies regularly round charges *up* to the next cent if any part of a cent is due. This means that if 950 kWh are used in a month, the bill is $49.24.

c. To find the charge for 1560 kWh, we evaluate $C(1560)$, using the "bottom piece" of the function because $1560 > 1000$. Evaluating gives

$$C(1560) = 51.21 + 0.038737(1560 - 1000) = 72.903$$

so the charge for the month is $72.91.

d. The graph of the function is shown in Figure 3.25(a). We graph the function $y = 122.87$ and find the point of intersection of the two graphs. See Figure 3.25(b). Note that the intersection occurs with the third piece of the piecewise-defined function, where $x > 1000$. The intersection point shows that when 2850 kWh are used, the monthly cost is $122.87.

(a)

(b)

Figure 3.25

EXAMPLE 4 ▶ Wind Chill Factor

Wind chill factors are used to measure the effect of the combination of temperature and wind speed on human comfort, by providing equivalent air temperatures with no wind blowing. One formula that gives the wind chill factor for a 30°F temperature and a wind with velocity V in miles per hour is

$$W = \begin{cases} 30 & \text{if } 0 \le V < 4 \\ 1.259V - 18.611\sqrt{V} + 62.255 & \text{if } 4 \le V \le 55.9 \\ -6.5 & \text{if } V > 55.9 \end{cases}$$

a. Find the wind chill factor for the 30°F temperature if the wind is 40 mph.

b. Find the wind chill factor for the 30°F temperature if the wind is 65 mph.

c. Graph this function for $0 \le V \le 70$.

d. What are the domain and range of the function graphed in (c)?
(Source: The National Weather Service)

SOLUTION

a. Because $V = 40$ is in the interval $4 \le V \le 55.9$, the wind chill factor is

$$1.259(40) - 18.611\sqrt{40} + 62.255 \approx -5.091 \approx -5$$

This means that, if the wind is 40 mph, a temperature of 30°F would actually feel like -5°F.

b. Because $V = 65$ is in the interval $V > 55.9$, the wind chill factor is -6.5. This means that, if the wind is 65 mph, a temperature of 30°F would actually feel like -6.5°F.

c. The graph is shown in Figure 3.26.

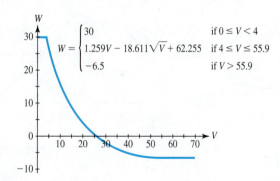

Figure 3.26

d. The domain is specified to be $0 \le V \le 70$, and the range, which is the set of outputs for W, is $-6.5 \le W \le 30$. That is, for wind speeds between 0 and 70 mph, the wind chill factor is no lower than -6.5°F and no higher than 30°F.

EXAMPLE 5 ▶ Service Calls

The cost of weekend service calls by Airtech Services is shown by the graph in Figure 3.27. Write a piecewise-defined function that represents the cost for service during the first five hours, as defined by the graph.

Figure 3.27

SOLUTION

The graph indicates that there are three pieces to the function, and each piece is a constant function. For t-values between 0 and 1, $C = 330$; for t-values between 1 and 2, $C = 550$; and for t-values between 2 and 5, $C = 770$. The open and closed circles on the graph tell us where the endpoints of each interval of the domain are defined. Thus, the cost function, with C in dollars and t in hours, is

$$C = \begin{cases} 330 & \text{if } 0 < t \le 1 \\ 550 & \text{if } 1 < t \le 2 \\ 770 & \text{if } 2 < t \le 5 \end{cases}$$

Absolute Value Function

We can construct a new function by combining two functions into a special piecewise-defined function. For example, we can write a function with the form

$$f(x) = \begin{cases} x & \text{if } x \ge 0 \\ -x & \text{if } x < 0 \end{cases}$$

This function is called the **absolute value function**, which is denoted by $f(x) = |x|$ and is derived from the definition of the absolute value of a number. Recall that the definition of the absolute value of a number is

$$|x| = \begin{cases} x & \text{if } x \ge 0 \\ -x & \text{if } x < 0 \end{cases}$$

To graph $f(x) = |x|$, we graph the portion of the line $y_1 = x$ for $x \ge 0$ (Figure 3.28(a)) and the portion of the line $y_2 = -x$ for $x < 0$ (Figure 3.28(b)). When these pieces are joined on the same graph (Figure 3.28(c)), we have the graph of $y = |x|$.

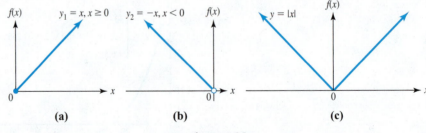

Figure 3.28

Solving Absolute Value Equations

We now consider the solution of equations that contain absolute value symbols, called **absolute value equations**. Consider the equation $|x| = 5$. We know that $|5| = 5$ and $|-5| = 5$. Therefore, the solution to $|x| = 5$ is $x = 5$ or $x = -5$. Also, if $|x| = 0$, then x must be 0. Finally, because the absolute value of a number is never negative, we cannot solve $|x| = a$ if a is negative. We generalize this as follows.

Absolute Value Equation

If $|x| = a$ and $a > 0$, then $x = a$ or $x = -a$.
There is no solution to $|x| = a$ if $a < 0$; $|x| = 0$ has solution $x = 0$.

If one side of an equation is a function contained in an absolute value and the other side is a nonnegative constant, we can solve the equation by using the method above.

EXAMPLE 6 ▶ **Absolute Value Equations**

Solve the following equations.

a. $|x - 3| = 9$ **b.** $|2x - 4| = 8$ **c.** $|x^2 - 5x| = 6$

SOLUTION

a. If $|x - 3| = 9$, then $x - 3 = 9$ or $x - 3 = -9$. Thus, the solution is $x = 12$ or $x = -6$.

b. If $|2x - 4| = 8$, then

$$2x - 4 = 8 \quad \text{or} \quad 2x - 4 = -8$$
$$2x = 12 \qquad\qquad 2x = -4$$
$$x = 6 \qquad\qquad x = -2$$

Thus, the solution is $x = 6$ or $x = -2$.

c. If $|x^2 - 5x| = 6$, then

$$x^2 - 5x = 6 \quad \text{or} \quad x^2 - 5x = -6$$
$$x^2 - 5x - 6 = 0 \qquad\qquad x^2 - 5x + 6 = 0$$
$$(x - 6)(x + 1) = 0 \qquad\qquad (x - 3)(x - 2) = 0$$
$$x = 6 \quad \text{or} \quad x = -1 \qquad\qquad x = 3 \quad \text{or} \quad x = 2$$

Thus, four values of x satisfy the equation. We can check these solutions by graphing or by substitution. Using the intersection method, we graph $y_1 = |x^2 - 5x|$ and $y_2 = 6$ and find the points of intersection. The solutions to the equation, $x = 6, -1,$ 3, and 2 found above, are also the x-values of the points of intersection (Figure 3.29).

Figure 3.29

Power Functions

We define power functions as follows.

Power Functions

A **power function** is a function of the form $y = ax^b$, where a and b are real numbers, $b \neq 0$.

We know that the area of a rectangle is $A = lw$ square units, where l and w are the length and the width, respectively, of the rectangle. Because $l = w$ when the rectangle is a square, the area of a square can be found with the formula $A = x^2$ if each side is x units long.

This is the function $A = x^2$ for $x \geq 0$. A related function is $y = x^2$, defined on the set of all real numbers; this is a power function with power 2. This function is also a basic quadratic function, called the **squaring function**, and its complete graph is a parabola (see Figure 3.30(a)).

If we fill a larger cube with cubes that are 1 unit on each edge and have a volume of 1 cubic unit, the number of these small cubes that fit gives the volume of the larger cube. It is easily seen that if the length of an edge of a cube is x units, its volume is x^3 cubic units for any nonnegative value of x. The related function $y = x^3$ defined for all real numbers x is called the **cubing function**. A complete graph of this function is shown in Figure 3.30(b).

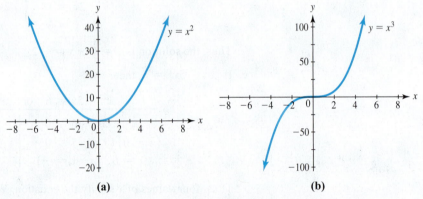

(a)   (b)

Figure 3.30

The functions $y = x$, $y = x^2$, and $y = x^3$ are examples of **power functions** with positive integer powers. Additional examples of power functions include functions with noninteger powers of x. For example, the graphs of the functions $y = x^{2/3}$ and $y = x^{3/2}$ are shown in Figures 3.31(a) and 3.31(b).

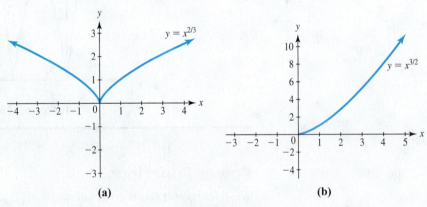

(a)   (b)

Figure 3.31

Note that $y = x^{2/3}$ is a power function with power less than 1, and its graph increases in the first quadrant but not as fast as $y = x$ does for values of x greater than 1. We say that the graph is *concave down* in the first quadrant. Also, $y = x^{3/2}$ is a power function with power greater than 1, and its graph increases in the first quadrant faster than $y = x$ does for values of x greater than 1. We say that the graph is *concave up* in the first quadrant.

In general, if $a > 0$ and $x > 0$, the graph of $y = ax^b$ is concave up if $b > 1$ and concave down if $0 < b < 1$. Additionally, if a function is increasing and its graph is concave up on an interval, it is increasing at an increasing rate, and if it is increasing and its graph is concave down on an interval, it is increasing at a decreasing rate. Figures 3.32(a) and 3.32(b) show the first-quadrant portion of typical graphs of $y = ax^b$ for $a > 0$ and different values of b.

              (a)   (b)

Figure 3.32

EXAMPLE 7 ▶ Wingspan of Birds

At 17 feet, the wingspan of a prehistoric bony-toothed bird may exceed that of any other bird that ever existed. The animal weighed about 64 pounds and soared the Chilean skies 5–10 million years ago. The wingspan of a bird can be estimated using the model

$$L = 2.43W^{0.3326}$$

where L is the wingspan in feet and W is the weight in pounds.
(Source: Discovery News)

a. Graph this function in the context of the application.

b. Use the function to compute the wingspan of a mute swan that weighs 40 pounds.

c. Use the function to approximate the weight of an albatross whose wingspan is 11 feet 4 inches.

d. Is the graph of this function concave up or concave down? What does this mean?

SOLUTION

a. It is appropriate to graph this model only in the first quadrant because both the weight and the wingspan must be nonnegative. The graph is shown in Figure 3.33(a).

b. To compute the wingspan of a mute swan that weighs 40 pounds, we evaluate the function with an input of 40, giving an output of $L = 2.43 (40)^{0.3326} = 8.288$. Thus, the wingspan of the mute swan is approximately 8.3 feet.

c. If the wingspan of an albatross is 11 feet 4 inches, we graphically solve the equation $11.333 = 2.43W^{0.3326}$. Using the intersection method (Figure 3.33(b)) gives the weight of the albatross to be approximately 102.48 pounds.

d. The graph is concave down. This tells us that as the weight of the bird increases, the wingspan increases, but at a decreasing rate.

Figure 3.33

Note that the power in $L = 2.43W^{0.3326}$ is 0.3326, which can be written in the form $\dfrac{3326}{10,000}$ and is therefore a rational number.

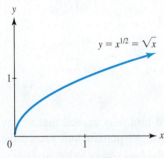

(a) Square root function

Root Functions

Functions with rational powers can also be written with radicals. For example, $y = x^{1/3}$ can be written in the form $y = \sqrt[3]{x}$, and $y = x^{1/2}$ can be written in the form $y = \sqrt{x}$. Functions like $y = \sqrt[3]{x}$ and $y = \sqrt{x}$ are special power functions called **root functions**.

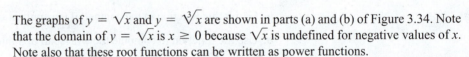

Root Functions

A root function is a function of the form $y = ax^{1/n}$, or $y = a\sqrt[n]{x}$, where n is an integer, $n \geq 2$.

(b) Cube root function

Figure 3.34

The graphs of $y = \sqrt{x}$ and $y = \sqrt[3]{x}$ are shown in parts (a) and (b) of Figure 3.34. Note that the domain of $y = \sqrt{x}$ is $x \geq 0$ because $\sqrt{x}$ is undefined for negative values of x. Note also that these root functions can be written as power functions.

All root functions of the form $f(x) = \sqrt[n]{x}$, with n an even positive integer, have the same shape as shown in Figure 3.34(a); the domain and the range are both $[0, \infty)$. Root functions with n odd have the same shape as shown in Figure 3.34(b); the domain and range are both $(-\infty, \infty)$.

Reciprocal Function

One special power function is the function

$$y = x^{-1} = \frac{1}{x}$$

This function is usually called the **reciprocal function**. Its graph, shown in Figure 3.35, is called a rectangular hyperbola. Note that x cannot equal 0, so 0 is not in the domain of this function. As the graph of $y = \dfrac{1}{x}$ shows, as x gets close to 0, $|y|$ gets large and the graph approaches but does not touch the y-axis. We say that the y-axis is a **vertical asymptote** of this graph.

Figure 3.35

Note also that the values of $y = \dfrac{1}{x}$ get very small as $|x|$ gets large, and the graph approaches but does not touch the x-axis. We say that the x-axis is a **horizontal asymptote** for this graph.

Variation

In Section 2.1, we discussed direct variation. If a quantity y varies directly as a power of x, we say that y is directly proportional to the nth power of x.

Direct Variation as the *n*th Power

A quantity y varies directly as the nth power $(n > 0)$ of x if there is a constant k such that

$$y = kx^n$$

The number k is called the constant of variation or the constant of proportionality.

EXAMPLE 8 ▶ | **Production**

In the production of an item, the number of units of one raw material required varies as the cube of the number of units of a second raw material that is required. Suppose 500 units of the first and 5 units of the second raw material are required to produce 100 units of the item. How many units of the first raw material are required if the number of units produced requires 10 units of the second raw material?

SOLUTION

If x is the number of units of the second raw material required and y is the number of units of the first raw material required, then y varies as the cube of x, or

$$y = kx^3$$

Because $y = 500$ when $x = 5$, we have

$$500 = k \cdot 5^3, \quad \text{or} \quad k = 4$$

Then $y = 4x^3$ and $y = 4(10^3) = 4000$. Thus, 4000 units of the first raw material are required if the number of units produced requires 10 units of the second raw material.

When two variables x and y are **inversely proportional** (or vary inversely), an increase in one variable results in a decrease in the other.

Inverse Variation

A quantity y is inversely proportional to x (or y varies inversely as x) if there exists a nonzero number k such that

$$y = \frac{k}{x}, \quad \text{or} \quad y = kx^{-1}$$

Also, y is inversely proportional to the nth power of x if there exists a nonzero number k such that

$$y = \frac{k}{x^n}, \quad \text{or} \quad y = kx^{-n}$$

We can also say that y varies inversely as the nth power of x.

EXAMPLE 9 ▶ **Illumination**

The illumination produced by a light varies inversely as the square of the distance from the source of the light. If the illumination 30 feet from a light source is 60 candela, what is the illumination 20 feet from the source?

SOLUTION

If L represents the illumination and d represents the distance, the relation is

$$L = \frac{k}{d^2}$$

Substituting for L and d and solving for k gives

$$60 = \frac{k}{30^2} \Rightarrow k = 54{,}000$$

Thus, the relation is $L = \dfrac{54{,}000}{d^2}$, and when $d = 20$ feet,

$$L = \frac{54{,}000}{20^2} = 135 \text{ candela}$$

So the illumination 20 feet from the source of the light is 135 candela.

SKILLS CHECK 3.3

In Exercises 1–10, sketch the graph of each function using a window that gives a complete graph.

1. $y = x^3$

2. $y = x^4$

3. $y = x^{1/3}$

4. $y = x^{4/3}$

5. $y = \sqrt{x} + 2$

6. $y = \sqrt[3]{x} - 2$

7. $y = \dfrac{1}{x} - 3$

8. $y = 4 - \dfrac{1}{x}$

9. $y = \begin{cases} -1 & \text{if } x < 0 \\ 1 & \text{if } x \geq 0 \end{cases}$

10. $y = \begin{cases} 2 & \text{if } x \geq 2 \\ 6 & \text{if } x < 2 \end{cases}$

11. a. Graph the function $f(x) = \begin{cases} 5 & \text{if } 0 \leq x < 2 \\ 10 & \text{if } 2 \leq x < 4 \\ 15 & \text{if } 4 \leq x < 6 \\ 20 & \text{if } 6 \leq x < 8 \end{cases}$

b. What type of function is this?

12. a. Graph the function

$$f(x) = \begin{cases} 100 & \text{if } 0 \leq x < 20 \\ 200 & \text{if } 20 \leq x < 40 \\ 300 & \text{if } 40 \leq x < 60 \\ 400 & \text{if } 60 \leq x < 80 \end{cases}$$

b. What type of function is this?

13. a. Graph $f(x) = \begin{cases} 4x - 3 & \text{if } x \leq 3 \\ x^2 & \text{if } x > 3 \end{cases}$

b. Find $f(2)$ and $f(4)$.

c. State the domain of the function.

14. a. Graph $f(x) = \begin{cases} 3 - x & \text{if } x \leq 2 \\ x^2 & \text{if } x > 2 \end{cases}$

b. Find $f(2)$ and $f(3)$.

c. State the domain of the function.

15. One type of step function is the **greatest integer function**, denoted int(x) or $[\![x]\!]$, where $[\![x]\!] =$ the greatest integer that is less than or equal to x.

a. If $f(x) = [\![x]\!]$, find $f(0.2)$, $f(3.8)$, $f(-2.6)$, and $f(5)$.

b. Graph $f(x) = [\![x]\!]$ for domain $[-5, 5]$.

16. To graph $f(x) = [\![x]\!]$ on a graphing calculator, use the function int(x), found in the Catalog. Graph this function for domain $[-5, 5]$ with your calculator in Dot mode.

17. a. Graph $f(x) = |x|$.

b. Find $f(-2)$ and $f(5)$.

c. State the domain of the function.

18. a. Graph $f(x) = |x - 4|$.

   b. Find $f(-2)$ and $f(5)$.

   c. State the domain of the function.

For each of the functions in Exercises 19–22, find the value of (a) $f(-1)$ and (b) $f(3)$, if possible.

19. $y = \begin{cases} 5 & \text{if } x \leq 1 \\ 6 & \text{if } x > 1 \end{cases}$

20. $y = \begin{cases} -2 & \text{if } x < -1 \\ 4 & \text{if } x \geq -1 \end{cases}$

21. $y = \begin{cases} x^2 - 1 & \text{if } x \leq 0 \\ x^3 + 2 & \text{if } x > 0 \end{cases}$

22. $y = \begin{cases} 3x + 1 & \text{if } x < 3 \\ x^2 & \text{if } x \geq 3 \end{cases}$

23. Determine whether the function $y = 4x^3$ is increasing or decreasing for

   a. $x < 0$.   **b.** $x > 0$.

24. Determine whether the function $y = -3x^4$ is increasing or decreasing for

   a. $x < 0$.   **b.** $x > 0$.

For each of the functions in Exercises 25–28, determine if the function is concave up or concave down in the first quadrant.

25. $y = x^{1/2}$ **26.** $y = x^{3/2}$

27. $y = x^{1.4}$ **28.** $y = x^{0.6}$

29. Graph $f(x) = \begin{cases} x & \text{if } x \geq 0 \\ -x & \text{if } x < 0 \end{cases}$

30. Graph $f(x) = \begin{cases} x - 4 & \text{if } x \geq 4 \\ 4 - x & \text{if } x < 4 \end{cases}$

31. Graph $f(x) = \begin{cases} x & \text{if } x < 0 \\ -x & \text{if } x \geq 0 \end{cases}$

32. Compare the graph in Exercise 29 with the graph in Exercise 17(a).

33. Compare the graph in Exercise 30 with the graph in Exercise 18(a).

In Exercises 34–38, solve the equations and check graphically.

34. $|2x - 5| = 3$ **35.** $\left| x - \dfrac{1}{2} \right| = 3$

36. $|x| = x^2 + 4x$ **37.** $|3x - 1| = 4x$

38. $|x - 5| = x^2 - 5x$

39. Suppose that S varies directly as the 2/3 power of T, and that $S = 64$ when $T = 64$. Find S when $T = 8$.

40. Suppose that y varies directly as the square root of x, and that $y = 16$ when $x = 4$. Find x when $y = 24$.

41. If y varies inversely as the 4th power of x and $y = 5$ when $x = -1$, what is y when $x = 0.5$?

42. If S varies inversely as the square root of T and $S = 4$ when $T = 4$, what is S when $T = 16$?

EXERCISES 3.3

43. *Electric Charges* For the nonextreme weather months, Palmetto Electric charges $7.10 plus 6.747 cents per kilowatt-hour (kWh) for customers using up to 1200 kWh, and charges $88.06 plus 5.788 cents per kWh above 1200 for customers using more than 1200 kWh.

   a. Write the function that gives the monthly charge in dollars as a function of the kilowatt-hours used.

   b. What is the monthly charge if 960 kWh are used?

   c. What is the monthly charge if 1580 kWh are used?

44. *Postal Rates* The table that follows gives the 2014 postal rates as a function of the weight for media mail.

Write a step function that gives the postage P as a function of the weight in pounds x for $0 < x \leq 5$.

Weight (lb)	Postal Rate ($)
$0 < x \leq 1$	2.69
$1 < x \leq 2$	3.17
$2 < x \leq 3$	3.65
$3 < x \leq 4$	4.13
$4 < x \leq 5$	4.61

(Source: USPS)

45. *First-Class Postage* The first-class postage charged for a letter in a large envelope is a function of its weight. The U.S. Postal Service uses the following table to describe the rates for 2015.

Weight Increment x (oz)	First-Class Postage P(x)
First ounce or fraction of an ounce	98¢
Each additional ounce or fraction	21¢

(Source: pe.usps.gov/text)

a. Convert this table to a piecewise-defined function that represents first-class postage for letters weighing up to 4 ounces, using x as the weight in ounces and P as the postage in cents.

b. Find $P(1.2)$ and explain what it means.

c. Give the domain of P as it is defined above.

d. Find $P(2)$ and $P(2.01)$.

e. Find the postage for 2-ounce and 2.01-ounce letters in large envelopes.

46. *Income Tax* The U.S. federal income tax owed by a married couple filing jointly can be found from the following table.

Filing Status: Married Filing Jointly

If Taxable Income Is between	Tax Due Is	of the Amount over
$0–$16,750	10%	$0
$16,750–$68,000	$1675 + 15%	$16,750
$68,000–$137,300	$9362.50 + 25%	$68,000
$137,300–$209,250	$26,687.50 + 28%	$137,300
$209,250–$373,650	$46,833.50 + 33%	$209,250
$373,650–up	$101,085.50 + 35%	$373,650

(Source: Internal Revenue Service, Form 1040 Instructions)

a. Write the piecewise-defined function T with input x that models the federal tax dollars owed as a function of x, the taxable income dollars earned, with $0 < x \leq 137,300$.

b. Use the function to find $T(42,000)$.

c. Find the tax owed on a taxable income of $65,000.

d. A friend tells Jack Waddell not to earn anything over $68,000 because it would raise his tax rate to 25% on all of his taxable income. Test this statement by finding the tax on $68,000 and $68,000 + $1. What do you conclude?

47. *Banks* The number of banks in the United States for the years 1935 through 2009 is given by

$$f(x) = \begin{cases} 84.3x + 12,365 & \text{if } x < 80 \\ -376.1x + 48,681 & \text{if } x \geq 80 \end{cases}$$

where x is the number of years after 1900.

a. What does this model give as the number of banks in 1970? 1990? 2015?

b. Graph the function for $50 \leq x \leq 112$.
(Source: Federal Deposit Insurance Corporation)

48. *Wind Chill* The formula that gives the wind chill factor for a 60°F temperature and a wind with velocity V in miles per hour is

$$W = \begin{cases} 60 & \text{if } 0 \leq V < 4 \\ 0.644V - 9.518\sqrt{V} + 76.495 & \text{if } 4 \leq V \leq 55.9 \\ 41 & \text{if } V > 55.9 \end{cases}$$

a. Find the wind chill factor for the 60° temperature if the wind is 20 mph.

b. Find the wind chill factor for the 60° temperature if the wind is 65 mph.

c. Graph the function for $0 \leq V \leq 80$.

d. What are the domain and range of the function graphed in part (c)?

49. *Female Physicians* Representation of females in medicine continues to show steady increases. The number of female physicians can be modeled by $F(x) = 0.623x^{1.552}$, where x is the number of years after 1960 and $F(x)$ is the number of female physicians in thousands.

a. What type of function is this?

b. What is $F(35)$? What does this mean?

c. How many female physicians will there be in 2020, according to the model?
(Source: American Medical Association)

50. *Taxi Miles* The Inner City Taxi Company estimated, on the basis of collected data, that the number of taxi miles driven each day can be modeled by the function $Q = 489L^{0.6}$, when they employ L drivers per day.

a. Graph this function for $0 \leq L \leq 35$.

b. How many taxi miles are driven each day if there are 32 drivers employed?

c. Does this model indicate that the number of taxi miles increases or decreases as the number of drivers increases? Is this reasonable?

d. Is the graph of this function concave up or concave down? What does this mean?

51. *College Enrollment* The total percent of individuals aged 16 to 24 enrolled in college as of October of each year who completed high school during the preceding 12 months is $f(t) = 61.925t^{0.041}$, where t is the number of years after 1999.

   a. Use this model to estimate the total percent in 2000 and 2008.

   b. Graph this function for the years 2000 to 2015.

   c. According to the model, will the percent enrolled in college ever reach 100%?

(Source: National Center for Education Statistics)

52. *U.S. Population* The U.S. population can be modeled by the function $y = 165.6x^{1.345}$, where y is in thousands and x is the number of years after 1800.

   a. What was the population in 1960, according to this model?

   b. Is the graph of this function concave up or concave down? What does this mean?

   c. Use numerical or graphical methods to find when the model estimates the population to be 93,330,000.

(Source: U.S. Census Bureau)

53. *Personal Expenditures* Personal consumption expenditures for durable goods in the United States, in billions of dollars, can be modeled by the function $P(x) = 306.472x^{0.464}$, where x is the number of years after 2000.

   a. Is this function increasing or decreasing for the years 2000–2020?

   b. Is this function concave up or concave down during this period of time? What does this mean?

   c. Use numerical or graphical methods to find when the model predicts that personal consumption expenditures will reach $1532.35 billion.

(Source: Bureau of Economic Analysis, U.S. Department of Commerce)

54. *Production Output* The monthly output of a product (in units) is given by $P = 1200x^{5/2}$, where x is the capital investment in thousands of dollars.

   a. Graph this function for x from 0 to 10 and P from 0 to 200,000.

   b. Is the graph concave up or concave down? What does this mean?

55. *Diabetes* The projected percent of the U.S. adult population with diabetes (diagnosed and undiagnosed) can be modeled by $y = 4.97x^{0.495}$, where x is the number of years after 2000.

   a. Does this model indicate that the percent of the U.S. adult population with diabetes is projected to increase or decrease?

   b. What percent is projected for 2022?

   c. In what year does this model project the percent to be 17?

(Source: Centers for Disease Control and Prevention)

56. *Harvesting* A farmer's main cash crop is tomatoes, and the tomato harvest begins in the month of May. The number of bushels of tomatoes harvested on the xth day of May is given by the equation $B(x) = 6(x + 1)^{3/2}$. How many bushels did the farmer harvest on May 8?

57. *Purchasing Power* The purchasing power of a 1983 dollar is given by the function $y = 34.394x^{-1.1088}$, where x is the number of years after 1960.

   a. Graph the function for $10 \le x \le 60$.

   b. What is the projected purchasing power of a 1983 dollar in 2020?

(Source: Bureau of Labor Statistics)

58. *Trust in the Government* The percent of people who say they trust the government in Washington always or most of the time is given by $y = 154.131x^{-0.492}$, with x equal to the number of years after 1960.

   a. Graph the function for the years 1960 through 2025.

   b. Does this model indicate that trust in the government is increasing or decreasing?

   c. Use the graph to estimate the percent of people who trusted the government in 2020.

(Source: Pew Research Center)

59. *Body Weight* The weight of a body varies inversely as the square of its distance from the center of Earth. If the radius of Earth is 4000 miles, how much would a 180-pound man weigh 1000 miles above the surface of Earth?

60. *Concentration of Body Substances* The concentration C of a substance in the body depends on the quantity of substance Q and the volume V through which it is distributed. For a static substance, the concentration is given by

$$C = \frac{Q}{V}$$

   a. For $Q = 1000$ milliliters (mL) of a substance, graph the concentration as a function of the volume on the interval from $V = 1000$ mL to $V = 5000$ mL.

b. For a fixed quantity of a substance, does the concentration of the substance in the body increase or decrease as the volume through which it is distributed increases?

61. *Investing* If money is invested for 3 years with interest compounded annually, the future value of the investment varies directly as the cube of $1 + r$, where r is the annual interest rate. If the future value of the investment is $6298.56 when the interest rate is 8%, what rate gives a future value of $5955.08?

62. *Investing* If money is invested for 4 years with interest compounded annually, the future value of the investment varies directly as the fourth power of $1 + r$, where r is the annual interest rate. If the future value of the investment is $17,569.20 when the interest rate is 10%, what rate gives a future value of $24,883.20?

63. *Investing* The present value that will give a future value S in 3 years with interest compounded annually varies inversely as the cube of $1 + r$, where r is the annual interest rate. If the present value of $8396.19 gives a future value of $10,000, what would be the present value of $16,500?

64. *Investing* The present value that will give a future value S in 4 years with interest compounded annually varies inversely as the fourth power of $1 + r$, where r is the annual interest rate. If the present value of $1525.79 gives a future value of $2000, what would be the present value of $8000?

3.4   Quadratic and Power Models

KEY OBJECTIVES

- Find the exact quadratic function that fits three points on a parabola
- Model data approximately using quadratic functions
- Model data using power functions
- Use first and second differences and visual comparison to determine whether a linear or quadratic function is the better fit to a set of data
- Determine whether a quadratic or power function gives the better fit to a given set of data

SECTION PREVIEW   **Mobile Internet Advertising**

Table 3.8 shows the dollar value or projected dollar value of the U.S. mobile Internet advertising market for the years from 2006 to 2012. The scatter plot of the data, in Figure 3.36, shows that a curve would fit the data better than a line, so a linear function is not a good model for the data. (See Example 3.) Graphing calculators and Excel permit us to model nonlinear data with other types of functions by using steps similar to those that we used to model data with linear functions. In this section, we model data with quadratic and power functions.

Table 3.8

Year	Market ($ billions)
2006	0.045
2007	0.114
2008	0.273
2009	0.409
2010	0.545
2011	0.791
2012	1.023

(Source: *Wall Street Journal,* August 16, 2010)

Figure 3.36

Modeling a Quadratic Function from Three Points on Its Graph

If we know three (or more) points that fit exactly on a parabola, we can find the quadratic function whose graph is the parabola.

EXAMPLE 1 ▶ **Writing the Equation of a Quadratic Function**

A parabola passes through the points $(0, 5)$, $(4, 13)$, and $(-2, 25)$. Write the equation of the quadratic function whose graph is this parabola.

SOLUTION

The general form of a quadratic function is $y = ax^2 + bx + c$. Because $(0, 5)$ is on the graph of the parabola, we can substitute 0 for x and 5 for y in $y = ax^2 + bx + c$, which gives $5 = a(0)^2 + b(0) + c$, so $c = 5$. Substituting the values of x and y for the other two points on the parabola and $c = 5$ in $y = ax^2 + bx + c$ gives

$$\begin{cases} 13 = a \cdot 4^2 + b \cdot 4 + 5 \\ 25 = a \cdot (-2)^2 + b \cdot (-2) + 5 \end{cases} \quad \text{or} \quad \begin{cases} 8 = 16a + 4b \\ 20 = 4a - 2b \end{cases}$$

Multiplying both sides of the bottom equation by 2 and adding the equations gives

$$\begin{cases} 8 \;\; = 16a + 4b \\ 40 = 8a - 4b \end{cases}$$
$$\overline{ 48 = 24a \qquad \Rightarrow a = 2}$$

Substituting $a = 2$ in $8 = 16a + 4b$ gives $8 = 32 + 4b$, so $b = -6$. Thus, $a = 2$, $b = -6$, and $c = 5$, and the equation of this parabola is

$$y = 2x^2 - 6x + 5$$

EXAMPLE 2 ▶ **Equation of a Quadratic Function**

Find the equation of the quadratic function whose graph is a parabola containing the points $(-1, 9)$, $(2, 6)$, and $(3, 13)$.

SOLUTION

Using the three points $(-1, 9)$, $(2, 6)$, and $(3, 13)$, we substitute the values for x and y in the general equation $y = ax^2 + bx + c$, getting three equations.

$$\begin{cases} 9 = a(-1)^2 + b(-1) + c \\ 6 = a(2)^2 + b(2) + c \\ 13 = a(3)^2 + b(3) + c \end{cases} \quad \text{or} \quad \begin{cases} 9 = a - b + c \\ 6 = 4a + 2b + c \\ 13 = 9a + 3b + c \end{cases}$$

We use these three equations to solve for a, b, and c, using techniques similar to those used to solve two equations in two variables.*

Subtracting the third equation, $13 = 9a + 3b + c$, from each of the first and second equations gives a system of two equations in two variables.

$$\begin{cases} -4 = -8a - 4b \\ -7 = -5a - b \end{cases}$$

_____

* We will discuss solution of systems of three equations in three variables further in Chapter 7.

Multiplying the second equation by -4 and adding the equations gives

$$\begin{cases} -4 = -8a - 4b \\ 28 = 20a + 4b \end{cases}$$
$$\overline{24 = 12a \quad \Rightarrow a = 2}$$

Substituting $a = 2$ in $-4 = -8a - 4b$ gives

$$-4 = -16 - 4b, \text{ or } b = -3$$

and substituting $a = 2$ and $b = -3$ in the original first equation, $9 = a - b + c$, gives

$$9 = 2 - (-3) + c, \quad \text{or} \quad c = 4$$

Thus, $a = 2$, $b = -3$, and $c = 4$, and the quadratic function whose graph contains the points is

$$y = 2x^2 - 3x + 4$$

Modeling with Quadratic Functions

If the graph of a set of data has a pattern that approximates the shape of a parabola or part of a parabola, a quadratic function may be appropriate to model the data. In Example 3, we will use technology to model the dollar value of the U.S. mobile advertising market as a function of the number of years after 2000.

EXAMPLE 3 ▶   Mobile Internet Advertising

Table 3.9 shows the dollar value or projected dollar value of the U.S. mobile Internet advertising market for the years from 2006 to 2012.

Table 3.9

Year	Advertising Market ($ billions)	Year	Advertising Market ($ billions)
2006	0.045	2010	0.545
2007	0.114	2011	0.791
2008	0.273	2012	1.023
2009	0.409		

(Source: *Wall Street Journal*, August 16, 2010)

a. Create a scatter plot of the data, with x equal to the number of years after 2000.

b. Find the quadratic function that is the best fit for the data, with x equal to the number of years after 2000.

c. Graph the aligned data and the model on the same axes. Does this model seem like a reasonable fit?

d. Use the model to find the year in which the mobile Internet advertising market is projected to reach $3 billion.

SOLUTION

a. The scatter plot of the data is shown in Figure 3.37(a). The shape looks as though it could be part of a parabola, so it is reasonable to find a quadratic function using the data points.

b. Enter the aligned inputs 6 through 12 in one list of a graphing calculator and the corresponding outputs in a second list. Using quadratic regression (select **STAT**, CALC, 5: QuadReg) gives a quadratic function that models the data. The function, with the coefficients rounded to three decimal places, is $y = 0.0149x^2 - 0.105x + 0.137$, with $x = 0$ for 2000.

c. The scatter plot of the aligned data and the graph of the function are shown in Figure 3.37(b). The model is a good fit for the data.

d. Intersecting the (unrounded) model with $y = 3$ gives $x = 17.84$, so the mobile Internet advertising market should reach \$3 billion during 2018 (Figure 3.37(c)).

(a)   (b)   (c)

Figure 3.37

EXAMPLE 4 ▶   The Cloud

North American cloud computing revenues from 2008 and projected to 2020 are shown in Table 3.10.

a. Create a scatter plot for the data with y equal to the number of billions of euros of revenue and x equal to the number of years after 2000, and determine the function types that can best be used to model the data.

b. Find the best-fitting quadratic model for the data.

Table 3.10

Year	Revenue (billions of euros)	Year	Revenue (billions of euros)
2008	7.800	2015	29.503
2009	9.553	2016	34.158
2010	11.665	2017	39.330
2011	14.568	2018	45.150
2012	17.831	2019	51.785
2013	21.393	2020	59.453
2014	25.270		

c. Graph the aligned data and the function on the same axes. Does the model seem like a reasonable fit to the data?

d. Use the model to predict the revenue in 2019. Is this estimate close to the data projection?

e. In what year does the model predict the revenue will be 65.98 billion euros?

SOLUTION

a. The scatter plot, with x equal to the number of years after 2000 and y equal to the revenue in billions of euros, is shown in Figure 3.38(a). The shape appears as though it could be part of a parabola, so it is reasonable to find a quadratic function using the aligned data points.

b. Using quadratic regression on a graphing calculator gives the quadratic function that is the best fit for the data. The function, with the coefficients rounded to three decimal places, is

$$y = 0.231x^2 - 2.246x + 11.131$$

Note that we will use the unrounded model in graphing and performing calculations.

c. The scatter plot of the aligned data and the graph of the function are shown in Figure 3.38(b). The function appears to be an excellent fit to the data.

d. Evaluating the function at $x = 19$ gives $y = 51.966$, so the model predicts that the revenue will be 51.966 billion euros in 2019.

e. Finding the intersection of the model and the equation $y = 65.98$ gives $x = 21$ (see Figure 3.38(c)), so the model predicts that the revenue will be 65.98 billion euros in 2021.

(a)　　　　　　　　(b)　　　　　　　　(c)

Figure 3.38

Spreadsheet ▸ SOLUTION   We can use graphing utilities, software programs, and spreadsheets to find the quadratic function that is the best fit for data. Figure 3.39 shows a partial Excel spreadsheet for the aligned data of Example 4. Selecting all the cells containing the data, getting the scatter plot of the data, selecting Add Trendline, and picking Polynomial with order 2 gives the equation of the quadratic function that is the best fit for the data. Observe that this model is the same as the one found in Example 4. Figure 3.39 shown this equation and its graph, along with the scatter plot of the data.*

_____

* See Appendix B, page 691, for details.

Figure 3.39

Comparison of Linear and Quadratic Models

Recall that when the changes in inputs are constant and the (first) differences of the outputs are constant or nearly constant, a linear model will give a good fit for the data. In a similar manner, we can compare the differences of the first differences, which are called the **second differences**. If the second differences are constant for equally spaced inputs, the data can be modeled exactly by a quadratic function.

Consider the data in Table 3.11, which gives the measured height y of a toy rocket x seconds after it has been shot into the air from the ground.

Figure 3.40 gives the first differences and second differences for the equally spaced inputs for the rocket height data in Table 3.11. Excel is especially useful for finding first and second differences.

Table 3.11   Height of a Rocket

Time x (seconds)	Height (meters)
1	68.6
2	117.6
3	147
4	156.8
5	147
6	117.6
7	68.6

	A	B	C	D
1	Time	Height	First Differences	Second Differences
2	1	68.6		
3	2	117.6	49	
4	3	147	29.4	-19.6
5	4	156.8	9.8	-19.6
6	5	147	-9.8	-19.6
7	6	117.6	-29.4	-19.6
8	7	68.6	-49	-19.6

Figure 3.40

The first differences are not constant, but each of the second differences is -19.6, which indicates that the data can be fit exactly by a quadratic function. We can find that the quadratic model for the height of the toy rocket is $y = 78.4x - 9.8x^2$ meters, where x is the time in seconds.

Modeling with Power Functions

We can model some experimental data by observing patterns. For example, by observing how the area of each of the (square) faces of a cube is found and that a cube has six faces, we can deduce that the surface area of a cube that is x units on each edge is

$$S = 6x^2 \text{ square units}$$

We can also measure and record the surface areas for cubes of different sizes to investigate the relationship between the edge length and the surface area for cubes. Table 3.12 contains selected measures of edges and the resulting surface areas.

Table 3.12

Edge Length x (units)	Surface Area of Cube (square units)
1	6
2	24
3	54
4	96
5	150

We can enter the lengths from the table as the independent (x) variable and the corresponding surface areas as the dependent (y) variable in a graphing utility, and then have the utility create the **power function** that is the best model for the data (Figure 3.41). This model also has the equation $y = 6x^2$ square units.

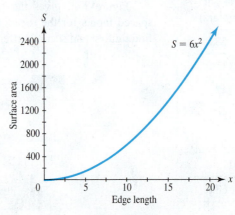

Figure 3.41

EXAMPLE 5 ▶   **Diabetes**

Figure 3.42 and Table 3.13 show that the percent of the U.S. adult population with diabetes (diagnosed and undiagnosed) is projected to grow rapidly in the future.

a. Find the power model that fits the data, with x equal to the number of years after 2000.

b. Use the model to predict the percent of U.S. adults with diabetes in 2015.

c. In what year does this model predict the percent to be 29.6?

Table 3.13

Year	Percent
2010	15.7
2015	18.9
2020	21.1
2025	24.2
2030	27.2
2035	29.0
2040	31.4
2045	32.1
2050	34.3

(Source: Centers for Disease Control and Prevention)

Figure 3.42

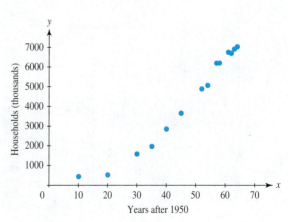

Figure 3.43

SOLUTION

a. The power model that fits the data is $f(x) = 4.947x^{0.495}$, with x equal to the number of years after 2000.

b. Evaluating the unrounded model at 15 gives the percent of U.S. adults with diabetes in 2015 to be

$$f(15) = 18.9\%$$

c. By intersecting the graphs of the unrounded model and $y = 29.6$ (Figure 3.43), we find the percent to be 29.6 when $x = 36.99$, or during 2037.

EXAMPLE 6 ▶   Cohabiting Households

The data in Table 3.14 give the number of cohabiting (without marriage) households (in thousands) for selected years from 1960 to 2013. The scatter plot of the data, with x representing the number of years after 1950 and y representing thousands of households, is shown in Figure 3.44.

a. Find the power function that models the data.

b. Graph the data and the model on the same axes.

Table 3.14   Cohabiting Households

Year	Cohabiting Households (thousands)	Year	Cohabiting Households (thousands)
1960	439	2004	5080
1970	523	2007	6209
1980	1589	2008	6214
1985	1983	2010	6768
1990	2856	2011	6715
1995	3668	2012	6928
2002	4898	2013	7046

(Source: U.S. Census Bureau)

Figure 3.44

SOLUTION

a. Using power regression in a graphing utility gives the power function that models the data. The power function that is the best fit is

$$y = 5.105x^{1.731}$$

where y is in thousands and x is the number of years after 1950.

b. The graphs of the data and the power function that models it are shown in Figure 3.45.

Figure 3.45

Spreadsheet
▶ SOLUTION   Power models can also be found with Excel. The Excel spreadsheet in Figure 3.46 shows the number of cohabiting households
(in thousands) for 1960–2013, listed as years after 1950. Selecting the cells containing the data, getting the scatter plot of the data, selecting Add Trendline, and picking
Power gives the equation of the power function that is the best fit for the data, along
with the scatter plot and the graph of the best-fitting power function. The equation
and graph of the function are shown in Figure 3.46. This model agrees with the one
found in Example 6.

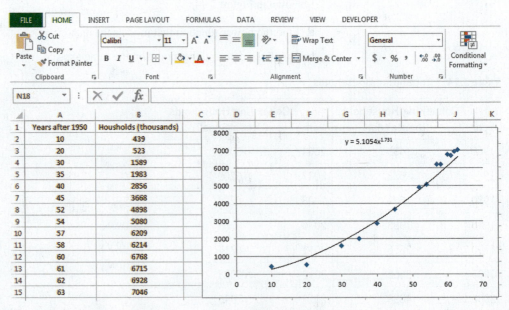

Figure 3.46

Comparison of Power and Quadratic Models

We found a power function that is a good fit for the data in Figure 3.46, but a linear function or a quadratic function may also be a good fit for the data. A quadratic function may fit data points even if there is no obvious turning point in the graph of the data points. If the data points appear to rise (or fall) more rapidly than a line, then a quadratic model or a power model may fit the data well. In some cases, it may be necessary to find both models to determine which is the better fit for the data.

EXAMPLE 7 ▶   **Purchasing Power**

Inflation causes a decrease in the value of money used to purchase goods and services. Table 3.15 gives the purchasing power of a 1983 dollar based on consumer prices for 1968–2010. Note that a 2006 dollar will purchase half as much as a 1983 dollar; that is, a given good or service costs twice as much in 2006.

a. Find a quadratic function that models the data if x is the number of years after 1960 and y is the purchasing power.

b. Find a power function that models the data if x is the number of years after 1960 and y is the purchasing power.

c. Graph each model with the aligned data points and discuss which model is the better fit.

d. Use the power model to predict the purchasing power of a 1983 dollar in 2055.

e. Find when the purchasing power is $0.40, according to the power model.

Table 3.15

Year	Purchasing Power	Year	Purchasing Power	Year	Purchasing Power
1968	2.873	1983	1.003	1998	0.613
1969	2.726	1984	0.961	1999	0.600
1970	2.574	1985	0.928	2000	0.581
1971	2.466	1986	0.913	2001	0.565
1972	2.391	1987	0.88	2002	0.556
1973	2.251	1988	0.846	2003	0.543
1974	2.029	1989	0.807	2004	0.529
1975	1.859	1990	0.766	2005	0.512
1976	1.757	1991	0.734	2006	0.500
1977	1.649	1992	0.713	2007	0.482
1978	1.532	1993	0.692	2008	0.464
1979	1.38	1994	0.675	2009	0.463
1980	1.215	1995	0.656	2010	0.456
1981	1.098	1996	0.637		
1982	1.035	1997	0.623		

(Source: Bureau of Labor Statistics)

SOLUTION

a. The quadratic function that is the best fit is $y = 0.00211x^2 - 0.173x + 4.030$.

b. The power function that is the best fit is $y = 34.394x^{-1.109}$.

c. The graphs are shown in Figure 3.47. The graph of the power model in part (a) appears to be a better fit than the graph of the quadratic model in part (b). The quadratic model begins to rise after 2001, while the actual purchasing power continues to fall.

d. The value of y is 0.221 when x is 95, so the purchasing power of a 1983 dollar is predicted to be 22 cents in 2055.

e. The value of x that gives 0.40 for y is 55.55, so the purchasing power of a 1983 dollar is predicted to be 40 cents in 2016.

Figure 3.47

SKILLS CHECK 3.4

In Exercises 1–6, write the equation of the quadratic function whose graph is a parabola containing the given points.

1. $(0, 1)$, $(3, 10)$, and $(-2, 15)$

2. $(0, -3)$, $(4, 37)$, and $(-3, 30)$

3. $(6, 30)$, $(0, -3)$, and $(-3, 7.5)$

4. $(6, -22)$, $(-3, 23)$, and $(0, 2)$

5. $(0, 6)$, $(2, \frac{22}{3})$, and $(-9, \frac{99}{2})$

6. $(0, 7)$, $(2, 8.5)$, and $(-3, 12.25)$

7. A ball is thrown upward from the top of a 48-foot-high building. The ball is 64 feet above ground level after 1 second, and it reaches ground level in 3 seconds. The height above ground is a quadratic function of the time after the ball is thrown. Write the equation of this function.

8. A ball is dropped from the top of a 256-foot-high building. The ball is 192 feet above ground after 2 seconds, and it reaches ground level in 4 seconds. The height above ground is a quadratic function of the time after the ball is thrown. Write the equation of this function.

9. Find the equation of the quadratic function whose graph is a parabola containing the points $(-1, 6)$, $(2, 3)$, and $(3, 10)$.

10. Find the equation of the quadratic function whose graph is a parabola containing the points $(-2, -4)$, $(3, 1)$, and $(2, 4)$.

11. Find the quadratic function that models the data in the table below.

x	−2	−1	0	1	2	3	4
y	16	5	0	1	8	21	40

x	5	6	7	8	9	10
y	65	96	133	176	225	280

12. The following table has the inputs, x, and the outputs for three functions, f, g, and h. Use second differences to determine which function is exactly quadratic, which is approximately quadratic, and which is not quadratic.

x	f(x)	g(x)	h(x)
0	0	2	0
2	399	0.8	110
4	1601	1.2	300
6	3600	3.2	195
8	6402	6.8	230
10	9998	12	290

13. As you can verify, the following data points give constant second differences, but the points do not fit on the graph of a quadratic function. How can this be?

x	1	2	4	8	16	32	64
y	1	6	15	28	45	66	91

14. a. Make a scatter plot of the data in the table below.

 b. Does it appear that a quadratic model or a power model is the better fit for the data?

x	y
1	4
2	9
3	11
4	21
5	32
6	45

15. Find the quadratic function that is the best fit for $f(x)$ defined by the table in Exercise 12.

16. Find the quadratic function that is the best fit for $g(x)$ defined by the table in Exercise 12.

17. a. Find a power function that models the data in the table in Exercise 14.

 b. Find a linear function that models the data.

 c. Visually determine which function is the better fit for the data.

18. a. Make a scatter plot of the data in the table below.

 b. Does it appear that a linear model or a power model is the better fit for the data?

x	y
3	4.7
5	8.6
7	13
9	17

19. a. Find a power function that models the data in the table in Exercise 18.

 b. Find a linear function that models the data.

 c. Visually determine if each model is a good fit.

20. Find the quadratic function that models the data in the tables that follow.

x	−2	−1	0	1	2	3
y	15	5	2	1	3	10

x	4	5	6	7	8
y	20	35	55	75	176

21. Find the power function that models the data in the table below.

x	1	2	3	4	5	6	7	8
y	3	4.5	5.8	7	8	9	10	10.5

22. Is a power function or a quadratic function the better model for the data below?

x	0.5	1	2	3	4	5	6
y	1	7	17	32	49	70	90

EXERCISES 3.4

Calculate numerical results with the unrounded models, unless otherwise instructed. Report models to 3 decimal places unless otherwise stated.

23. **Hepatitis C** The table shows projections of world-wide sales (in billions of dollars) of hepatitis C treatments from 2012 to 2016.

    a. What is the quadratic function that best models these sales as a function of the number of years after 2010? Let x represent the number of years after 2010 and y represent the number of billions of dollars in sales of treatments.

    b. Graph the model and the data on the same axes, with x equal to the number of years after 2010.

    c. What are the projected sales in 2018, if the model still applies?

Year	Sales ($ billions)
2012	4.5
2013	4.6
2014	6.3
2015	8.1
2016	9.7

(Source: Data from "Evaluate Pharma," in the *Wall Street Journal*, November 15, 2012)

24. **National Health Care** The table shows national expenditures for health care in the United States for selected years, with projections to 2015.

    a. Use a scatter plot with x as the number of years after 1950 and y as the total expenditures for health care (in billions) to identify what type (or types) of function would make a good model for these data.

    b. Find a power model and a quadratic model for the data.

    c. Which model from (b) more accurately estimates the 2010 expenditures for national health care?

    d. Use the better model from (c) to estimate the 2020 expenditures for national health care.

Year	National Expenditures for Health Care (in billions)
1960	$28
1970	75
1980	255
1990	717
1995	1020
2000	1359
2005	2016
2010	2879
2015	4032

(Source: U.S. Centers for Medicare and Medicaid Services)

25. **Smart Phones** Bill payment using smart phones is projected to increase rapidly. Projections of world-wide mobile payment transaction values (in billions of dollars) for the years 2009 through 2016 are shown in the table below.

    a. Find a quadratic function that models y, the payment transaction values, as a function of x, the number of years after 2000.

    b. Graph the model and the data points on the same axes. Comment on the fit.

    c. What does the model project the world-wide mobile payment transaction values to be in 2018?

Year	Transaction Values ($ billions)
2009	30
2010	60
2011	100
2012	170
2013	260
2014	360
2015	480
2016	620

(Source: Portioresearch.com)

26. *Obesity* Obesity $(\text{BMI} \geq 30)$ increases the risk of diabetes, heart disease, and many other ailments. The percent of Americans who are obese from 2000 projected to 2030 is shown in the table below.

a. Graph the data points to determine whether a quadratic function is a reasonable model for the data.

b. If it is reasonable, find the quadratic function that models obesity, with x equal to the number of years after 2000 and y equal to the percent of obese Americans. Report the model with three significant digits.

c. Use the unrounded model to predict the percent with obesity in 2023.

Year	2000	2010	2015	2020	2025	2030
% Obese	22.1	30.9	34.5	37.4	39.9	42.2

(Source: *American Journal of Preventive Medicine*)

27. *Wind Chill* The table gives the wind chill temperature when the outside temperature is 20°F.

Wind (mph)	Wind Chill (°F)	Wind (mph)	Wind Chill (°F)
5	13	35	0
10	9	40	−1
15	6	45	−2
20	4	50	−3
25	3	55	−3
30	1	60	−4

(Source: National Weather Service)

a. Use x as the wind speed and create a quadratic model for the data.

b. At what wind speed does the model estimate that the wind chill temperature will be −3°F?

c. Do you think the model found in part (a) is valid for $x > 60$? Explain.

28. *Foreign-Born Population* The following table gives the percent of the U.S. population that is foreign born.

a. Create a scatter plot for the data, with x equal to the number of years after 1900 and y equal to the percent.

b. Does it appear that the data could be modeled with a quadratic function?

c. Find the best-fitting quadratic function for the data. Report your answer to 4 decimal places.

d. Use the function to estimate the percent in 2028.

Year	Foreign Born (%)	Year	Foreign Born (%)
1910	14.7	1980	6.2
1920	13.2	1990	8.0
1930	11.6	2000	10.4
1940	8.8	2005	11.7
1950	6.9	2010	12.4
1960	5.4	2016	11.6
1970	4.7	2018	11.8

(Source: U.S. Census Bureau)

29. *Aging Workers* The table shows the millions of Americans who are working full time at selected ages from 20 to 62.

a. Find a quadratic model that gives the number (in millions) y of Americans working full time as a function of their age, x. Report the model with three significant digits.

b. Graph the model and the data points on the same axes.

c. Use technology to find the age at which the model indicates that the number of workers is greatest.

Age	Millions Working Full Time
20	3.62
27	6.30
32	6.33
37	6.96
42	7.07
47	7.52
52	7.10
57	5.58
62	3.54

(Source: Wall Street Journal, June 18, 2012)

30. *U.S. Population* The table gives the U.S. population, in millions, for selected years, with projections to 2050.

a. Create a scatter plot for the data in the table, with x equal to the number of years after 1960.

b. Use the scatter plot to determine the type of function that can be used to model the data, and create a function that best fits the data, with x equal to the number of years after 1960.

Year	U.S. Population (millions)	Year	U.S. Population (millions)
1960	180.671	1995	263.044
1965	194.303	1998	270.561
1970	205.052	2000	281.422
1975	215.973	2003	294.043
1980	227.726	2025	358.030
1985	238.466	2050	408.695
1990	249.948		

(Source: U.S. Census Bureau)

31. *Alzheimer's Disease* Partially because of Americans living longer, the number with Alzheimer's disease and other dementia is projected to grow each year. The table below gives the millions of U.S. citizens with Alzheimer's from 2000 and projected to 2050.

Year	2000	2010	2020	2030	2040	2050
Number	4.0	5.9	6.8	8.7	11.8	14.3

(Source: National Academy on an Aging Society)

a. Create a scatter plot of the data, with x equal to the number of years after 2000 and y equal to the number of millions of Americans with Alzheimer's disease or other dementia.

b. Find a quadratic function that models this data. Report the model with four significant digits.

c. Graph the data and the model on the same axes, and comment on the fit.

d. How many Americans does the model predict will have Alzheimer's disease or other dementia in 2056?

e. In what year after 2000 does the model predict that 10.2 million Americans will have Alzheimer's or other dementia?

32. *Volume* The measured volume of a pyramid with each edge of the base equal to x units and with its altitude (height) equal to x units is given in the table below.

a. Determine whether the second differences of the outputs are constant.

b. If the answer is yes, find the quadratic model that is the best fit for the data. Otherwise, find the power function that is the best fit.

Edge Length x (units)	Volume of Pyramid (cubic units)
1	1/3
2	8/3
3	9
4	64/3
5	125/3
6	72

33. *Worldwide Cell Phone Subscribers* The following table gives the number of worldwide cell phone subscribers, in billions, for selected years from 2010 and projected to 2020.

a. Find a quadratic function that models the number of subscribers, $Q(t)$, in billions, as a function of t, the number of years after 2010.

b. What does the model predict the number of subscribers will be in 2021?

c. What does the model predict the number of subscribers will be in 2022? What do the results of parts (b) and (c) tell us?

d. Graphically find when the model reaches its maximum.

e. Is this model a good predictor of future subscribership after 2021?

Year	Subscribers (billions)	Year	Subscribers (billions)
2010	5.40	2016	8.35
2011	6.05	2017	8.60
2012	6.65	2018	8.79
2013	7.19	2019	8.95
2014	7.65	2020	9.07
2015	8.03		

(Source: Key Global Telecom Indicators)

34. *World Population* One projection of the world population by the United Nations for selected years (a low-projection scenario) is given in the table below.

Year	Projected Population (millions)	Year	Projected Population (millions)
1995	5666	2075	6402
2000	6028	2100	5153
2025	7275	2125	4074
2050	7343	2150	3236

(Source: *World Population Prospects,* United Nations)

a. Find a quadratic function that fits the data, using the number of years after 1990 as the input.

b. Find the positive *x*-intercept of this graph, to the nearest year.

c. When can we be certain that this model no longer applies?

35. *China's Labor Pool* The table shows United Nation data from 1975 and projections to 2050 of the number of millions of people age 15 to 59 in China.

a. What is the quadratic function that best models this population as a function of the number of years after 1970? Let *x* represent the number of years after 1970 and *y* represent the number of millions of people in this labor pool.

b. Use technology with the model to find the maximum size of this population before it begins to shrink.

Year	Labor Pool (millions)	Year	Labor Pool (millions)
1975	490	2015	920
1980	560	2020	920
1985	650	2025	905
1990	730	2030	875
1995	760	2035	830
2000	800	2040	820
2005	875	2045	800
2010	910	2050	670

(Source: United Nations)

36. *Personal Income* Total personal income in the United States (in billions of dollars) for selected years from 1960 to 2009 is given in the following table.

Year	Personal Income ($ billions)	Year	Personal Income ($ billions)
1960	411.3	2003	9378.1
1970	838.6	2004	9937.2
1980	2301.5	2005	10,485.9
1990	4846.7	2006	11,268.1
1995	6200.9	2007	11,912.3
2000	8559.7	2008	12,391.1
2001	8883.3	2009	12,174.9
2002	9060.1		

(Source: Bureau of Economic Analysis, U.S. Department of Commerce)

a. These data can be modeled by a power function. Write the equation of this function, with *x* as the number of years after 1950.

b. What does this model predict for total U.S. personal income in 2012?

c. Find the quadratic function that is the best fit for the data, with *x* as the number of years after 1950.

d. Which model is the better fit for the data?

37. *Consumer Price Index* The table below gives the U.S. consumer price index (CPI) for selected years from 2012 and projected to 2050. With the reference year 2012, a 2020 CPI = 120.56 means goods and services that cost $100.00 in 2012 are expected to cost $120.56 in 2020.

a. Find the quadratic function that is the best fit for the data, with *x* as the number of years after 2010 and *y* as the CPI in dollars.

b. Graph the model and the data on the same axes and comment on the fit of the model to the data.

c. Use the model to predict the CPI in 2022.

d. According to the model, in what year will the CPI reach 216?

Year	CPI	Year	CPI	Year	CPI
2012	100.00	2025	138.41	2040	209.44
2016	108.58	2030	158.90	2045	240.45
2020	120.56	2035	182.43	2050	276.05

(Source: Social Security Administration)

38. *International Visitors* The number of international visitors to the United States for selected years 1986–2010 is given in the table below.

Year	U.S. Visitors (millions)	Year	U.S. Visitors (millions)
1986	26	2000	50.9
1987	29.5	2001	44.9
1988	34.1	2002	41.9
1989	36.6	2003	41.2
1990	39.5	2004	46.1
1991	43	2005	49.4
1992	47.3	2006	51.0
1994	45.5	2007	56.0
1995	44	2008	57.9
1996	46.3	2010	55
1997	48.9		

(Source: Office of Travel and Tourism, U.S. Department of Commerce)

a. Using an input equal to the number of years after 1980, graph the aligned data points and both of the given equations to determine which is the better model for the aligned data.

   i. $y = 17\sqrt[3]{x}$

   ii. $y = -0.031x^2 + 2x + 21$

b. If you had to pick one of these models to predict the number of international visitors in the year 2020, which model would be the more reasonable choice?

39. **Mortgages** The balance owed y on a \$50,000 mortgage after x monthly payments is shown in the table that follows. Graph the data points with each of the equations below to determine which is the better model for the data, if x is the number of months that payments have been made.

   a. $y = 338,111.278x^{-0.676}$

   b. $y = 4700\sqrt{110 - x}$

Monthly Payments	Balance Owed ($)
12	47,243
24	44,136
48	36,693
72	27,241
96	15,239
108	8074

40. **Poverty** The table shows the number of millions of people in the United States who lived below the poverty level for selected years.

   a. Find a quadratic model that approximately fits the data, using x as the number of years after 1990.

   b. Use the model to predict the number of people living below the poverty level in 2013.

Year	People Living Below the Poverty Level (millions)	Year	People Living Below the Poverty Level (millions)
1993	39.3	2002	34.6
1994	38.1	2003	35.9
1995	36.4	2004	37.0
1996	36.5	2005	37.0
1997	35.6	2006	36.5
1998	34.5	2007	37.2
1999	32.3	2008	39.8
2000	31.1	2009	43.6

(Source: U.S. Bureau of the Census, U.S. Department of Commerce)

41. **Energy Use** The following table shows the energy use per dollar of GDP, as a percent, for years from 2000 and projected to 2035. (GDP, gross domestic product, is the monetary value of all the finished goods and services produced within a country's borders in a specific time period, though GDP is usually calculated on an annual basis. It includes all of private and public consumption, government outlays, investments, and exports less imports that occur within a defined territory.) These data are indexed to 1980, which means that energy use for any year is viewed as a percent of the use per dollar of GDP in 1980.

   a. Find a quadratic function that models these data, with x equal to the number of years after 2000 and y equal to the energy use per dollar of GDP, as a percent of the energy use per dollar of GDP in 1980. Report the model with three significant digits.

   b. Graph the model and the data on the same axes.

   c. Use technology to find when this model predicts that energy use per dollar of GDP will reach a minimum.

Energy Use per Dollar of GDP Indexed to 1980

Year	Percent	Year	Percent	Year	Percent
2000	67	2015	51	2030	37
2005	60	2020	45	2035	34
2010	56	2025	41		

(Source: U.S. Department of Energy)

42. **Average Annual Wage** The following table shows the U.S. average annual wage in thousands of dollars for selected years from 2014 and projected to 2050.

   a. Find a power function that models this data, with x equal to the number of years after 2000 and y equal to the average annual wage in thousands of dollars. Report the model with four significant digits.

   b. What does the model predict that the average annual wage will be in 2043?

   c. Use technology to find when the model predicts that the average annual wage will be \$159.6 thousand.

Year	Average Annual Wage ($ thousands)	Year	Average Annual Wage ($ thousands)
2014	48.6	2030	93.2
2016	53.3	2035	113.2
2018	58.7	2040	137.6
2020	63.7	2045	167.1
2025	76.8	2050	202.5

(Source: Social Security Administration)

43. *Insurance Rates* The following table gives the monthly insurance rates for a $100,000 life insurance policy for smokers 35–50 years of age.

a. Create a scatter plot for the data.

b. Does it appear that a quadratic function can be used to model the data? If so, find the best-fitting quadratic model.

c. Find the power model that is the best fit for the data.

d. Compare the two models by graphing each model on the same axes with the data points. Which model appears to be the better fit?

Age (yr)	Monthly Insurance Rate ($)	Age (yr)	Monthly Insurance Rate ($)
35	17.32	43	23.71
36	17.67	44	25.11
37	18.02	45	26.60
38	18.46	46	28.00
39	19.07	47	29.40
40	19.95	48	30.80
41	21.00	49	32.55
42	22.22	50	34.47

(Source: American General Life Insurance Company)

44. *Banks* The table gives the number of banks in the United States for selected years from 1980 to 2014.

a. Create a scatter plot of the data, with *x* equal to the number of years after 1980 and *y* equal to the number of banks.

b. Find a quadratic function that models the data.

c. Use the model to estimate the number of banks in 2025.

d. Use technology to find the year after 2000 in which the model estimates that the number of banks in the U.S. will equal 5473?

Year	Number of Banks	Year	Number of Banks
1980	18,763	2010	7821
1990	15,158	2011	7523
2000	9905	2012	7255
2007	8605	2013	6950
2008	8841	2014	6669
2009	8185		

(Source: Federal Deposit Insurance Corporation)

45. *U.S. Gross Domestic Product* The table gives the U.S. gross domestic product (GDP) (in trillions of dollars) for selected years from 2005 and projected to 2070.

a. Find the best-fitting quadratic model for the data, with *x* equal to number of years after 2000.

b. Find the power model that is the best fit for the data, with *x* equal to number of years after 2000.

c. Compare the two models by graphing each model on the same axes with the data points. Which model appears to be the better fit?

Year	GDP ($ trillions)	Year	GDP ($ trillions)
2005	12.145	2040	79.680
2010	16.174	2045	103.444
2015	21.270	2050	133.925
2020	27.683	2055	173.175
2025	35.919	2060	224.044
2030	46.765	2065	290.042
2035	61.100	2070	375.219

(Source: U.S. Bureau of Economic Analysis)

46. *Auto Noise* The noise level of a Volvo S60 increases as the speed of the car increases. The table gives the noise, in decibels (db), at different speeds.

Speed (mph)	Noise Level (db)
10	42
30	57
50	64
70	66
100	71

(Source: *Road & Track*, 2011)

a. Fit a power function model to the data.

b. Graph the data points and model on the same axes.

c. Use the result from part (a) to estimate the noise level at 80 mph.

47. *Crude Oil* The following table gives the U.S. crude oil production, in billions of barrels, for the years from 2010 projected to 2030.

a. Graph a scatter plot of the data, with *x* equal to the number of years after 2000 and *y* equal to the number of billions of barrels of crude oil.

b. Find the power function that models the data, with *x* equal to the number of years after 2000 and *y* equal to the number of billions of barrels of crude oil.

c. Find the quadratic function that models the data, with x equal to the number of years after 2000 and y equal to the number of billions of barrels of crude oil.

d. Use the power model to predict the number of billions of barrels of crude oil in 2040.

e. Use the quadratic model to predict the number of billions of barrels of crude oil in 2040.

f. If the crude oil production begins to decrease after 2030, which model is the better fit for the data?

Year	Billions of Barrels
2010	1.94
2014	2.10
2018	2.16
2022	2.24
2026	2.26
2030	2.26

(Source: U.S. Energy Information Administration)

48. *Box-Office Revenues* The data in the table in the next column give the box-office revenues, in billions of dollars, for movies released in selected years between 1980 and 2014.

Year	Revenue ($ billions)	Year	Revenue ($ billions)
1980	2.8	2007	9.6
1985	3.8	2008	9.6
1990	5.0	2009	10.6
1995	5.3	2010	10.6
2000	7.5	2011	10.2
2004	9.3	2012	10.8
2005	8.8	2013	10.9
2006	9.2	2014	10.4

(Source: Motion Picture Association of America)

a. Find the power model that best fits the revenues as a function of the number of years after 1970. Round to four decimal places.

b. What does the unrounded model estimate as the revenue in 2023?

c. Discuss the reliability of this estimation.

In this chapter, we discussed in depth quadratic functions, including realistic applications that involve the vertex and x-intercepts of parabolas and solving quadratic equations. We then studied piecewise-defined functions, power functions, and other nonlinear functions. Real-world data are provided throughout the chapter, and we fit power and quadratic functions to some of these data.

Key Concepts and Formulas

3.1 Quadratic Functions; Parabolas

Quadratic function	Also called a *second-degree polynomial function*, this function can be written in the form $f(x) = ax^2 + bx + c$, where $a \neq 0$.
Parabola	A parabola is the graph of a quadratic function.
Vertex	The turning point on a parabola is the vertex.
Maximum point	If a quadratic function has $a < 0$, the vertex of the parabola is the maximum point, and the parabola opens downward.

Minimum point	If a quadratic function has $a > 0$, the vertex of the parabola is the minimum point, and the parabola opens upward.
Axis of symmetry	The axis of symmetry is the vertical line through the vertex of the parabola.
Forms of quadratic functions	$y = a(x - h)^2 + k$ $y = ax^2 + bx + c$ Vertex at (h, k) x-coordinate of vertex at $x = -\dfrac{b}{2a}$ Axis of symmetry is the line $x = h$ Axis of symmetry is the line $x = -\dfrac{b}{2a}$ Parabola opens up if a is positive and down if a is negative. Parabola opens up if a is positive and down if a is negative.

3.2 Solving Quadratic Equations

Solving quadratic equations	An equation that can be written in the form $ax^2 + bx + c = 0$, $a \neq 0$, is called a *quadratic equation*.
Zero product property	For real numbers a and b, if the product $ab = 0$, then at least one of the numbers a or b must equal 0.
Solving by factoring	To solve a quadratic equation by factoring, write the equation in a form with 0 on one side. Then factor the nonzero side of the equation, if possible, and use the zero product property to convert the equation into two linear equations that are easily solved.
Solving and checking graphically	Solutions or decimal approximations of solutions to quadratic equations can be found by using $\boxed{\text{TRACE}}$, zero, or intersect with a graphing utility.
Factor Theorem	The factorization of a polynomial function $f(x)$ and the real solutions to $f(x) = 0$ can be verified by graphing $y = f(x)$ and observing where the graph crosses the x-axis. If $x = a$ is a solution to $f(x) = 0$, then $(x - a)$ is a factor of f.
Solving using the square root method	When a quadratic equation has the simplified form $x^2 = C$, the solutions are $x = \pm \sqrt{C}$. This method can also be used to solve equations of the form $(ax + b)^2 = C$.
Completing the square	A quadratic equation can be solved by converting one side to a perfect binomial square and taking the square root of both sides.
Solving using the quadratic formula	The solutions of the quadratic equation $ax^2 + bx + c = 0$, with $a \neq 0$, are given by the formula $$x = \frac{-b \pm \sqrt{b^2 - 4ac}}{2a}$$

Discriminant	The expression $b^2 - 4ac$ of the quandratic equation $y = ax^2 + bx + c$ determines the nature of the solutions of $ax^2 + bx + c = 0$.
Solutions, zeros, *x*-intercepts, and factors	If a is a real number, the following three statements are equivalent: • a is a real solution to the equation $f(x) = 0$. • a is a real zero of the function $f(x)$. • a is an *x*-intercept of the graph of $y = f(x)$.

3.3 Piecewise-Defined Functions and Power Functions

Piecewise-defined functions	This is a function that is defined in pieces for different intervals of the domain. Piecewise-defined functions can be graphed by graphing their pieces on the same axes. A familiar example of a piecewise-defined function is the *absolute value function*.
Absolute value function	$$\|x\| = \begin{cases} x & \text{if } x \geq 0 \\ -x & \text{if } x < 0 \end{cases}$$
Absolute value equation	The solution of the absolute value equation $\|x\| = a$ is $x = a$ or $x = -a$ when $a \geq 0$. There is no solution to $\|x\| = a$ if $a < 0$.
Power functions	A power function is a function of the form $y = ax^b$, where a and b are real numbers, $b \neq 0$.
Cubing function	A special power function, $y = x^3$, is called the cubing function.
Root functions	A root function is a function of the form $y = ax^{1/n}$, or $y = a\sqrt[n]{x}$, where n is an integer, $n \geq 2$.
Reciprocal function	This function is formed by the quotient of one and the identity function, $f(x) = \dfrac{1}{x}$, and can be written $f(x) = x^{-1}$.
Direct variation as the *n*th power	If a quantity y varies directly as the *n*th power of x, then y is directly proportional to the *n*th power of x and $$y = kx^n$$ The number k is called the constant of variation or the constant of proportionality.
Inverse variation	A quantity y is inversely proportional to the *n*th power of x if there exists a nonzero number k such that $y = \dfrac{k}{x^n}$.

3.4 Quadratic and Power Models

Modeling a quadratic function from three points on its graph	If we know three (or more) points that fit exactly on a parabola, we can find the quadratic function whose graph is the parabola.
Quadratic modeling	The use of graphing utilities permits us to fit quadratic functions to data by using technology.
Second differences	If the second differences of data are constant for equally spaced inputs, a quadratic function is an exact fit for the data.
Power modeling	The use of graphing utilities permits us to fit power functions to nonlinear data by using technology.

chapter 3   SKILLS CHECK

In Exercises 1–8, (a) give the coordinates of the vertex of the graph of each quadratic function and (b) graph each function in a window that includes the vertex and all intercepts.

1. $y = (x - 5)^2 + 3$ **2.** $y = (x + 7)^2 - 2$

3. $y = 3x^2 - 6x - 24$ **4.** $y = 2x^2 + 8x - 10$

5. $y = -x^2 + 30x - 145$

6. $y = -2x^2 + 120x - 2200$

7. $y = x^2 - 0.1x - 59.998$

8. $y = x^2 + 0.4x - 99.96$

In Exercises 9 and 10, use factoring to solve the equations.

9. $x^2 - 5x + 4 = 0$ **10.** $6x^2 + x - 2 = 0$

Use a graphing utility as an aid in factoring to solve the equations in Exercises 11 and 12.

11. $5x^2 - x - 4 = 0$ **12.** $3x^2 + 4x - 4 = 0$

In Exercises 13 and 14, use the quadratic formula to solve the equations.

13. $x^2 - 4x + 3 = 0$ **14.** $4x^2 + 4x - 3 = 0$

15. a. Use graphical and algebraic methods to find the x-intercepts of the graph of $f(x) = 3x^2 - 6x - 24$.

b. Find the solutions to $f(x) = 0$ if $f(x) = 3x^2 - 6x - 24$.

16. a. Use graphical and algebraic methods to find the x-intercepts of the graph of $f(x) = 2x^2 + 8x - 10$.

b. Find the solutions to $f(x) = 0$ if $f(x) = 2x^2 + 8x - 10$.

In Exercises 17 and 18, use the square root method to solve the equations.

17. $5x^2 - 20 = 0$ **18.** $(x - 4)^2 = 25$

In Exercises 19–22, find the exact solutions to the equations in the complex numbers.

19. $z^2 - 4z + 6 = 0$ **20.** $w^2 - 4w + 5 = 0$

21. $4x^2 - 5x + 3 = 0$ **22.** $4x^2 + 2x + 1 = 0$

In Exercises 23–30, graph each function.

23. $f(x) = \begin{cases} 3x - 2 & \text{if } x < -1 \\ 4 - x^2 & \text{if } x \geq -1 \end{cases}$

24. $f(x) = \begin{cases} 4 - x & \text{if } x \leq 3 \\ x^2 - 5 & \text{if } x > 3 \end{cases}$

25. $f(x) = 2x^3$ **26.** $f(x) = x^{3/2}$

27. $f(x) = \sqrt{x - 4}$ **28.** $f(x) = \dfrac{1}{x} - 2$

29. $y = x^{4/5}$ **30.** $y = \sqrt[3]{x + 2}$

31. Determine whether the function $y = -3x^2$ is increasing or decreasing

    a. For $x < 0$.   **b.** For $x > 0$.

32. For each of the functions, determine whether the function is concave up or concave down.

    a. $y = x^{5/4}$ **b.** $y = x^{4/5}$ for $x > 0$

33. Solve $|3x - 6| = 24$.

34. Solve $|2x + 3| = 13$.

35. Find the equation of a quadratic function whose graph is a parabola passing through the points $(0, -2)$, $(-2, 12)$, and $(3, 7)$.

36. Find the equation of a quadratic function whose graph is a parabola passing through the points $(-2, -9)$, $(2, 7)$, and $(4, -9)$.

37. Find a power function that models the data below.

x	1	2	3	4	5	6
y	4	9	11	21	32	45

38. Find a quadratic function that models the data below.

x	1	3	4	6	8
y	2	8	15	37	63

39. Suppose that q varies directly as the 3/2 power of p and that $q = 16$ when $p = 4$. Find q when $p = 16$.

40. If $f(x) = \begin{cases} 3x - 2 & \text{if } -8 \leq x < 0 \\ x^2 - 4 & \text{if } 0 \leq x < 3, \\ -5 & \text{if } x \geq 3 \end{cases}$ find

$f(-8), f(0),$ and $f(4)$.

41. Maximizing Profit The monthly profit from producing and selling x units of a product is given by the function $P(x) = -0.01x^2 + 62x - 12{,}000$.

a. Producing and selling how many units will result in the maximum profit for this product?

b. What is the maximum possible profit for the product?

42. Profit The revenue from sales of x units of a product is given by $R(x) = 200x - 0.01x^2$, and the cost of producing and selling the product can be described by $C(x) = 38x + 0.01x^2 + 16{,}000$.

a. Producing and selling how many units will give maximum profit?

b. What is the maximum possible profit from producing and selling the product?

43. Height of a Ball If a ball is thrown into the air at 64 feet per second from a height of 192 feet, its height (in feet) is given by $S(t) = 192 + 64t - 16t^2$, where t is in seconds.

a. In how many seconds will the ball reach its maximum height?

b. What is the maximum possible height for the ball?

44. Height of a Ball If a ball is thrown into the air at 29.4 meters per second from a height of 60 meters, its height (in meters) is given by $S(t) = 60 + 29.4t - 9.8t^2$, where t is in seconds.

a. In how many seconds will the ball reach its maximum height?

b. What is the maximum possible height for the ball?

45. Visas The number of skilled workers' visas issued in the United States can be modeled by $y = -1.48x^2 + 38.901x - 118.429$, where x is the number of years after 1990 and y is the number of visas in thousands.

a. Use numerical methods to determine the year when the number of visas issued was a maximum.

b. What was the maximum number of visas issued?

c. During what year after 2000 does the model indicate that the number of visas issued would be 100 thousand?

(Source: Department of Homeland Security)

46. Break-Even The profit for a product is given by $P = -3600 + 150x - x^2$, where x is the number of

units produced and sold. How many units will give break-even (that is, return a profit of 0)?

47. Falling Ball If a ball is dropped from the top of a 400-foot-high building, its height S in feet is given by $S(t) = 400 - 16t^2$, where t is in seconds. In how many seconds will it hit the ground?

48. Profit The profit from producing and selling x units of a product is given by the function $P(x) = -0.3x^2 + 1230x - 120{,}000$ dollars. Producing and selling how many units will result in a profit of $324,000 for this product?

49. Millionaire's Tax Rate The effective tax rate for a head of household earning the equivalent of $1 million in non-investment income in 2010 dollars can be modeled by the function

$$T(x) = \begin{cases} 0.08x^2 - 2.64x + 22.35 & \text{if } 15 \le x \le 45 \\ -0.525x + 89.82 & \text{if } 45 < x \le 110 \end{cases}$$

where x is the number of years after 1900.

a. Graph the function for $15 \le x \le 110$.

b. According to the model, what was the tax rate for a millionaire head of household in 1990?

c. In 2010, with President Bush's tax cuts in effect, the tax rate was 32.4%. Does the model agree with this rate?

(Source: The Tax Foundation)

50. Diabetes As the following table shows, projections indicate that the percent of U.S. adults with diabetes could dramatically increase.

Year	Percent	Year	Percent
2010	15.7	2035	29.0
2015	18.9	2040	31.4
2020	21.1	2045	32.1
2025	24.2	2050	34.3
2030	27.2		

(Source: Centers for Disease Control and Prevention)

a. Find a quadratic model that fits the data in the table, with $x = 0$ in 2000.

b. Use the model to predict the percent of U.S. adults with diabetes in 2022.

c. In what year does this model predict the percent of U.S. adults with diabetes will be 30.2%?

51. *ATV Deaths* The number of ATV-related deaths for people of all ages from 1999 to 2009 can be modeled by the function

$$f(x) = \begin{cases} 3.607x^2 - 16.607x + 254.000 & \text{if } 9 \le x \le 15 \\ -43.25x^2 + 1361.65x - 9881.75 & \text{if } 15 < x \le 19 \end{cases}$$

where *x* is the number of years after 1990.

a. Graph the function for $9 \le x \le 19$.

b. According to the model, how many deaths were there in 2003?

c. According to the model, how many deaths were there in 2009?

d. Use the model to predict when, after 2000, the number of deaths would be 200.
(Source: U.S. Consumer Product Safety Commission)

52. *Home Range* The home range of an animal is the region to which the animal confines its movements. The area, in hectares, of the home range of a meat-eating mammal can be modeled by the function $H(x) = 0.11x^{1.36}$, where *x* is the mass of the animal in grams. What would the home range be for a bobcat weighing 1.6 kg?
(Source: *The American Naturalist*)

53. *Internet Usage* Around 40% of the world's population had an internet connection in 2014. In 1995, it was less than 1%. The table shows the percent of worldwide Internet users from 2000 to 2014.

a. Find the quadratic function that models the data, with *x* equal to the number of years after 2000.

b. Graph the data and the function that models the data.

c. Does this model appear to be a good fit for the data?

d. Use the result of part (a) to estimate when Internet usage would be 55%.

Year	Percent of Worldwide Internet Users	Year	Percent of Worldwide Internet Users
2000	6.7	2008	23.1
2001	8.1	2009	25.6
2002	10.6	2010	29.4
2003	12.2	2011	32.5
2004	14.1	2012	35.5
2005	15.8	2013	37.9
2006	17.6	2014	40.4
2007	20.6		

(Source: *Internet Live Stats*)

54. *Personal Income* The income received by people from all sources minus their personal contributions for Social Security insurance is called *personal income*. The table lists the personal income, in billions of dollars, of people living in the United States for the indicated years.

Year	Personal Income ($ billions)	Year	Personal Income ($ billions)
1990	4847	2008	12,239
2000	8559	2009	12,026
2004	9937	2010	12,429
2005	10,486	2011	13,202
2006	11,268	2012	13,888
2007	11,894	2013	14,167

(Source: Bureau of Economic Analysis)

a. Using an input equal to the number of years after 1980, find a quadratic function that models these data.

b. Use your unrounded model to estimate when after 1980 the personal income was $9500 billion.

c. When does the model estimate that the personal income will be double its 2009 value?

55. *Non-White Population* The table gives the number of millions of individuals in the U.S. civilian non-institutional population 16 years and older who are non-White or Hispanic for the years 2000 and projected to 2050.

a. Find a power function $y = f(x)$ modeling this population as a function of the number of years after 1990.

b. Graph the model on the same axes with the scatter plot of the data to determine whether the model is a good fit to the data.

c. What does the model predict that this population will be in 2033?

Year	Population (billions)
2000	56.6
2010	71.6
2015	79.0
2020	86.8
2030	104.2
2040	123.1
2050	143.0

(Source: U.S. Census Bureau)

56. *Insurance Premiums* The following table gives the monthly premiums required for a $250,000 term life insurance policy on a 35-year-old female nonsmoker for different guaranteed term periods.

a. Find a quadratic function that models the monthly premium as a function of the length of term for a 35-year-old female nonsmoking policyholder. Report your answer to five decimal places.

b. Assuming that the domain of the function contains integer values between 10 years and 30 years, what term in years could a 35-year-old nonsmoking female purchase for $130 a month?

Term Period (years)	Monthly Premium for 35-Year-Old Female ($)
10	103
15	125
20	145
25	183
30	205

(Source: Quotesmith.com)

57. *National Health Expenditures* Health spending is projected to grow at an average rate of 5.8% from 2012 through 2022, 1.0 percentage point faster than the expected average annual growth in the gross domestic product (GDP). The table gives the national health care expenditures, in billions of dollars, from 2006 and projected to 2022.

Year	Expenditures ($ billions)	Year	Expenditures ($ billions)
2006	2,163	2015	3,273
2007	2,298	2016	3,458
2008	2,407	2017	3,660
2009	2,501	2018	3,889
2010	2,600	2019	4,142
2011	2,701	2020	4,416
2012	2,807	2021	4,702
2013	2,915	2022	5,009
2014	3,093		

(Source: Centers for Medicare & Medicaid Services)

a. Create a scatter plot of the data, with x equal to the number of years after 2000 and y equal to the national health care expenditures in billions of dollars.

b. Choose an appropriate function to model the data, and write the equation of that model.

c. Graph your model and the scatter plot on the same axes.

d. Use the rounded model to find the average rate of change of health care expenditures from 2006 to 2012; from 2012 to 2022. Round to one decimal place.

e. Use the rounded model to find the percent increase in health care expenditures from 2012 to 2013; from 2021 to 2022. Round to one decimal place. Is the statement that health spending is projected to grow at an average rate of 5.8 percent from 2012 through 2022 reasonable?

58. *Hawaii Population* The data in the table below give the population of Hawaii from 1990 through 2012.

Year	Population (thousands)	Year	Population (thousands)
1990	1108	1997	1187
1991	1131	2000	1212
1992	1150	2006	1285
1993	1160	2009	1295
1994	1173	2010	1341
1995	1179	2012	1390
1996	1183		

(Source: *Statistical Abstract of the United States*)

a. Align the data with $x =$ the number of years after 1980 and find a power function to fit the aligned data.

b. Is the graph of this function concave up or concave down?

c. If the pattern indicated by the model remains valid, estimate in what year Hawaii's population will rise to 1.5 million people.

59. *Pet Industry* The table on the following page shows U.S. pet industry expenditures for selected years from 1994 to 2014.

a. Find a linear model for the data, with x equal to the number of years after 1990, and discuss the fit.

b. Find a quadratic model for the data, with x equal to the number of years after 1990.

c. Which model appears to be the best fit?

d. Use both models to predict the amount of money Americans will spend on their pets in 2020.

Year	Expenditure ($ billions)	Year	Expenditure ($ billions)
1994	17	2007	41.2
1996	21	2008	43.2
1998	23	2009	45.5
2001	28.5	2010	48.35
2002	29.5	2011	50.96
2003	32.4	2012	53.33
2004	34.4	2013	55.72
2005	36.3	2014	58.51
2006	38.5		

(Source: statista.com)

60. *Prescription Drug Expenditures* In 2012, prescription drug spending is estimated to have accounted for $260.8 billion of national health spending, a decline of 0.8%, compared to 2.9% growth in 2011. This decline was due to increased adoption of generic drugs as a number of popular brand-name drugs lost patent protection, increases in cost-sharing requirements, and lower spending on new medicines. The table gives expenditures for prescription drugs, in billions of dollars, for the years 2006 and projected to 2022.

a. Create a scatter plot of the data, with x equal to the number of years after 2000 and y equal to the prescription drug expenditures in billions of dollars.

b. For the data from 2006 to 2013, find a quadratic function to model the data with x equal to the number of years after 2000. Round to two decimal places.

c. For the data from 2014 to 2022, find a quadratic function to model the data with x equal to the number of years after 2000. Round to two decimal places.

d. Join the two functions from parts (b) and (c) to create a piecewise function that models the data from 2006 to 2022.

e. Use the model to estimate the prescription drug expenditures for 2010 and 2022.

Year	Expenditures ($ billions)	Year	Expenditures ($ billions)
2006	224.1	2015	294.9
2007	235.9	2016	311.6
2008	242.6	2017	330.7
2009	254.6	2018	350.6
2010	255.7	2019	372.7
2011	263.0	2020	397.9
2012	260.8	2021	425.5
2013	262.3	2022	455.0
2014	275.9		

(Source: Centers for Medicare & Medicaid Services)

Group Activities
▶ EXTENDED APPLICATIONS

1. Modeling Annual Earnings by Gender

Graphs displaying linear and nonlinear growth are frequently found in periodicals such as *Newsweek* and *Time*, in newspapers such as *USA Today* and the *Wall Street Journal*, and on websites on the Internet. Tables of data can also be found in these sources, especially on federal and state government websites such as www.census.gov.

Your mission is to find a company sales, stock price, biological growth, or sociological trend over a period of years that is nonlinear, to determine which type of nonlinear function is the best fit for the data, and to find a nonlinear function that is a model for the data.

A quadratic model will be a good fit for the data under the following conditions:

- If the data are presented as a graph that resembles a parabola or part of a parabola.
- If the data are presented in a table and a plot of the data points shows that they lie near some parabola or part of a parabola.
- If the data are presented in a table and the second differences of the outputs are nearly constant for equally spaced inputs.

A power model may be a good fit if the shape of the graph or of the plot of the data points resembles part of a parabola but a parabola is not a good fit for the data.

After creating the model, you should test the goodness of fit of the model to the data and discuss uses that you could make of the model.

Your completed project should include

a. A complete citation of the source of the data that you are using.
b. An original copy or photocopy of the data being used.
c. A scatter plot of the data.
d. The equation that you have created.
e. A graph containing the scatter plot and the modeled equation.
f. A statement about how the model could be used to make estimations or predictions about the trend you are observing.

Some helpful hints:

1. If you decide to use a relation determined by a graph that you have found, read the graph very carefully to determine the data points or contact the source of the data to get the data from which the graph was drawn.
2. Align the independent variable by letting x represent the number of years after some convenient year and then enter the data into a graphing utility and create a scatter plot.
3. Use your graphing utility to create the equation of the function that is the best fit for the data. Graph this equation and the data points on the same axes to see if the equation is reasonable.

2. Gender and Earnings

Median Annual Earnings by Gender

Year	Males Number (millions)	Males Earnings ($)	Females Number (millions)	Females Earnings ($)	Female-to-Male Earnings Ratio
1948	47.370	20,239	22.725	8,523	
1950	47.585	21,709	24.651	8,050	
1960	55.172	28,064	36.526	8,680	
1970	65.008	35,709	51.647	11,976	
1980	78.661	33,726	80.826	13,243	
1990	88.220	35,062	92.245	17,399	
2000	98.504	38,340	101.704	21,729	
2010	105.191	34,408	107.220	22,196	
2012	107.534	34,397	109.383	21,833	
2013	108.706	35,228	109.956	22,063	

(Source: U.S. Census Bureau)

The table gives the numbers of male and female employees (in millions) in the United States and the median earnings (in 2013 dollars) for persons of age 15 and over for selected years from 1940 through 2013. To investigate equity of employment of females:

1. Find the linear function $y = f(x)$ that models the number of employed males as a function of the number of years after 1940.
2. Find the quadratic function $y = g(x)$ that models the number of employed females as a function of the number of years after 1940.
3. Graph the unrounded functions $y = f(x)$ and $y = g(x)$ and graphically find when the number of female employees reaches the number of male employees.
4. When do these models predict that the number of male employees will again exceed number of the female employees?

5. Calculate the female-to-male earnings ratio for each of the given years, and enter the ratios in the table. Round the ratios to three decimal places.
6. Use each of the given years and the corresponding female-to-male earnings ratio to create a quadratic function that models the female-to-male earnings ratio as a function of the number of years after 1940. Report the model with three significant digits.
7. Use the unrounded function in Exercise 6 to find the year after 1940 when the female median earnings are expected to equal male median earnings.
8. One possible reason why it might take a long time for median female earnings to reach median male earnings is gender discrimination. What other reasons may play a role?

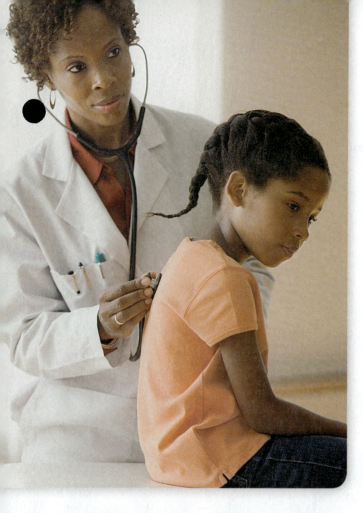

4

Additional Topics with Functions

A profit function can be formed by algebraically combining cost and revenue functions. An average cost function can be constructed by finding the quotient of two functions. Other new functions can be created using function composition and inverse functions. In this chapter, we study additional topics with functions to create functions and solve applied problems.

sections	objectives	applications
4.1 Transformations of Graphs and Symmetry	Find equations and graphs of vertically and horizontally shifted functions; stretch, compress, and reflect graphs; identify symmetry with respect to the *x*-axis, *y*-axis, and origin	Diabetes, profit, stimulus and response, pollution, velocity of blood
4.2 Combining Functions; Composite Functions	Add, subtract, multiply, and divide functions; find composition of functions	Profit, unconventional vehicle sales, average cost, canning orange juice
4.3 One-to-One and Inverse Functions	Identify inverse functions; find the inverse of a function; graph inverse functions; find inverse functions on limited domains	Loans, temperature measurement, investments, velocity of blood
4.4 Additional Equations and Inequalities	Solve radical equations, equations with rational powers, equations in quadratic form, and absolute value inequalities; solve quadratic inequalities analytically and graphically; solve power inequalities	Profit, height of a model rocket, obesity, investments

Algebra TOOLBOX

KEY OBJECTIVES

- Identify functions
- Find domains and ranges
- Determine if a function is increasing or decreasing over given intervals
- Determine over what intervals a function is increasing or decreasing

In this Toolbox, we create a library of functions studied so far, which sets the stage for the study of properties and operations with functions.

Linear Functions

$y = mx + b$

- m is the slope of the line, b is the y-intercept:

$$m = \frac{y_2 - y_1}{x_2 - x_1}$$

- The function is increasing if $m > 0$.

- The function is decreasing if $m < 0$.

Special Linear Functions

IDENTITY FUNCTION	CONSTANT FUNCTION
$f(x) = x$ Domain: $(-\infty, \infty)$ Range: $(-\infty, \infty)$	$f(x) = c$ Domain: $(-\infty, \infty)$ Range: $\{c\}$, where c is a constant
• Increases on its domain	• Graph is a horizontal line.

Quadratic Functions

$f(x) = ax^2 + bx + c$ with $a > 0$

Axis of symmetry $x = \frac{-b}{2a}$

Vertex $\left(\frac{-b}{2a}, f\left(\frac{-b}{2a}\right)\right)$

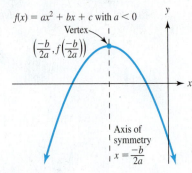

$f(x) = ax^2 + bx + c$ with $a < 0$

Vertex $\left(\frac{-b}{2a}, f\left(\frac{-b}{2a}\right)\right)$

Axis of symmetry $x = \frac{-b}{2a}$

- Vertex is at the point $\left(\dfrac{-b}{2a}, f\left(\dfrac{-b}{2a}\right)\right)$.

Piecewise-Defined Functions

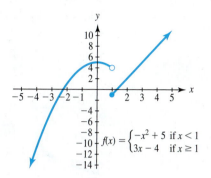

$$f(x) = \begin{cases} -x^2 + 5 & \text{if } x < 1 \\ 3x - 4 & \text{if } x \geq 1 \end{cases}$$

Special Piecewise-Defined Functions

ABSOLUTE VALUE FUNCTION

$f(x) = |x|$

Domain: $(-\infty, \infty)$
Range: $[0, \infty)$

- Decreases on $(-\infty, 0)$, increases on $(0, \infty)$

GREATEST INTEGER FUNCTION

$f(x) = [\![x]\!]$

Domain: $(-\infty, \infty)$
Range: $\{x \mid x \text{ is an integer}\} = \{\ldots, -3, -2, -1, 0, 1, 2, 3, \ldots\}$

- $f(x) = [\![x]\!]$ is a special **step function**.

Power Functions

POWER FUNCTION, n EVEN

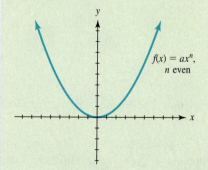

$f(x) = ax^n$, n even

- Decreases on $(-\infty, 0)$, increases on $(0, \infty)$ for $a > 0$
- Increases on $(-\infty, 0)$, decreases on $(0, \infty)$ for $a < 0$

POWER FUNCTION, n ODD

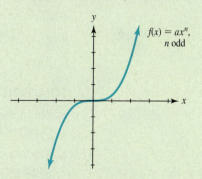

$f(x) = ax^n$, n odd

- Increases on $(-\infty, \infty)$ for $a > 0$
- Decreases on $(-\infty, \infty)$ for $a < 0$

Special Power Functions

ROOT FUNCTION, *n* EVEN

Domain: $[0, \infty)$
Range: $[0, \infty)$

• Increases on its domain

ROOT FUNCTION, *n* ODD

Domain: $(-\infty, \infty)$
Range: $(-\infty, \infty)$

• Increases on its domain

RECIPROCAL FUNCTION

Domain: $(-\infty, 0) \cup (0, \infty)$
Range: $(-\infty, 0) \cup (0, \infty)$

• Decreases on $(-\infty, 0)$ and $(0, \infty)$

EXAMPLE 1 ▶ **Special Functions**

Identify the type of each of the following functions from its equation. Then find the domain and range, and graph the function.

a. $f(x) = x$

b. $f(x) = \sqrt{x}$

c. $f(x) = 3$

SOLUTION

a. This is the identity function, a special linear function. The domain and range are both the set of all real numbers, $(-\infty, \infty)$. See Figure 4.1(a).

b. This is a special power function, $f(x) = x^{1/2}$, called the square root function. Only nonnegative values can be used for x, so the domain is $[0, \infty)$. The outputs will also be nonnegative values, so the range is $[0, \infty)$. See Figure 4.1(b).

c. This is a constant function, a special linear function. The domain is all real numbers, $(-\infty, \infty)$, and the range consists of the single value 3. See Figure 4.1(c).

(a)

(b)

(c)

Figure 4.1

EXAMPLE 2 ▶   **Special Functions**

Identify the type of each of the following functions from its equation. Then find the domain and range, and graph the function.

a. $f(x) = x^3$

b. $f(x) = \sqrt[3]{x}$

c. $f(x) = \dfrac{5}{x^2}$

SOLUTION

a. This is a special power function, with power 3. It is called the cubing function, and both the domain and the range are the set of all real numbers, $(-\infty, \infty)$. See Figure 4.2(a).

b. This is a special power function, with power $1/3$. It is called the cube root function. Any real number input results in a real number output, so both the domain and the range are the set of all real numbers, $(-\infty, \infty)$. See Figure 4.2(b).

c. This function can be written as $f(x) = 5x^{-2}$, so it is a special power function, with power -2. If 0 is substituted for x, the result is undefined, so only nonzero real numbers can be used as inputs and only positive real numbers result as outputs. Thus, the domain is $(-\infty, 0) \cup (0, \infty)$, and the range is $(0, \infty)$. We will see in Chapter 6 that $f(x) = \dfrac{5}{x^2}$ is also called a **rational function**. See Figure 4.2(c).

(a)

(b)

(c)

Figure 4.2

EXAMPLE 3 ▶ **Functions**

Identify the type of each of the following functions from its equation. Then find the domain and range, and graph the function.

a. $g(x) = 2x^2 - 4x + 3$ **b.** $f(x) = 3x - 18$ **c.** $f(x) = \begin{cases} 3x - 4 \text{ if } x < 2 \\ 8 - 2x \text{ if } x \geq 2 \end{cases}$

SOLUTION

a. This is a quadratic function. Any real number can be used as an input, and the resulting outputs are all real numbers greater than or equal to 1. Thus, the domain is $(-\infty, \infty)$, and the range is $[1, \infty)$. Its graph is a parabola that opens up. See Figure 4.3(a).

b. This is a linear function. Its graph is a line that rises as x increases, because its slope is positive. The domain and range are both $(-\infty, \infty)$. See Figure 4.3(b).

c. This is a piecewise-defined function, where each piece is a linear function. The domain is $(-\infty, \infty)$, and the range is $(-\infty, 4]$. See Figure 4.3(c).

(a)

(b)

(c)

Figure 4.3

EXAMPLE 4 ▶ **Functions**

Determine whether each of the graphs represents a function.

a.   **b.**

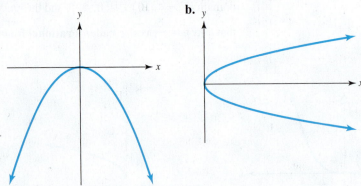

SOLUTION

a. The vertical line test confirms that this graph represents a function. It is a parabola opening downward, so it is the graph of a quadratic function.

b. By the vertical line test, we see that this is not the graph of a function. This graph has the shape of a parabola and is called a **horizontal parabola**.

Toolbox **EXERCISES**

1. The domain of the reciprocal function is _____, and its range is _____.

2. The domain of the constant function $g(x) = k$ is _____, and its range is ____.

3. The reciprocal function decreases on _____.

4. The absolute value function increases on the interval _____ and decreases on _____.

5. The range of the squaring function is _____.

6. The domain of the squaring function is _____.

In Exercises 7–12, determine whether the function is increasing or decreasing on the given interval.

7. $g(x) = \sqrt[5]{x};\ (-\infty, \infty)$ 8. $h(x) = \sqrt[4]{x};\ [0, \infty)$

9. $f(x) = 5 - 0.8x;\ (-\infty, \infty)$

10. $f(x) = \dfrac{-1}{x};\ (-\infty, 0)$

11. $g(x) = -2x^{2/3};\ (0, \infty)$

12. $h(x) = 3x^{1/6};\ [0, \infty)$

Identify the type of each of the following functions from its equation. Then graph the function.

13. $f(x) = x^3$

14. $f(x) = \sqrt{x}$

15. $f(x) = 8\sqrt[6]{x}$

16. $f(x) = \dfrac{-4}{x^3}$

Which of the following graphs represents a function?

17.

18.

19.

20.

21.

22.

KEY OBJECTIVES

- Find equations and graphs of functions whose graphs have been vertically shifted, horizontally shifted, stretched, compressed, and reflected

- Determine if a graph is symmetric about the y-axis

- Determine if a graph is symmetric about the origin

- Determine if a graph is symmetric about the x-axis

- Determine if a function is even, odd, or neither

SECTION PREVIEW   Diabetes

The projected percent of the U.S. adult population with diabetes (diagnosed and undiagnosed) for 2010 through 2050 can be modeled by the power function $f(x) = 4.947x^{0.495}$, with x equal to the number of years after 2000. To find the model that gives this projection with x equal to the number of years after 2010, we can rewrite the equation in a "shifted" form. (See Example 5.)

In this section, we discuss shifting, stretching, compressing, and reflecting the graph of a function. This is useful in obtaining graphs of additional functions and also in finding windows in which to graph them. We also discuss symmetry of graphs. ■

Shifts of Graphs of Functions

Consider the quadratic function in the form $y = a(x - h)^2 + k$. Substituting h for x in this equation gives $y = k$, so the graph of this function contains the point (h, k). Because the graph is symmetric about the vertical line through the vertex, there is a second point on the graph with y-coordinate k, *unless* the point (h, k) is the vertex of the parabola (Figure 4.4). To see whether the second point exists or if (h, k) is the vertex, we solve

$$k = a(x - h)^2 + k$$

This gives

$$0 = a(x - h)^2$$
$$0 = (x - h)^2$$
$$0 = x - h$$
$$x = h$$

Figure 4.4

Thus, only one value of x corresponds to $y = k$, so the point (h, k) is the vertex of any parabola with an equation of the form $y = a(x - h)^2 + k$.* We can say that the vertex of $y = a(x - h)^2 + k$ has been **shifted** from $(0, 0)$ to the point (h, k), and this entire graph is in fact the graph of $y = ax^2$ shifted h units horizontally and k units vertically. We will see that graphs of other functions can be shifted similarly.

In Section 3.3, we studied absolute value functions. In the following example, we investigate shifts of the absolute value function.

EXAMPLE 1 ▶   **Shifts of Functions**

Graph the functions $f(x) = |x|$ and $g(x) = |x| + 3$ on the same axes. What relationship do you notice between the two graphs?

SOLUTION

Table 4.1 shows inputs for x and outputs for both $f(x) = |x|$ and $g(x) = |x| + 3$. Observe that for a given input, each output value of $g(x) = |x| + 3$ is 3 more than the corresponding output value for $f(x) = |x|$. Thus, the y-coordinate of each point on the graph of $g(x) = |x| + 3$ is 3 more than the y-coordinate of the point on the graph of $f(x) = |x|$ with the same x-coordinate. We say that the graph of $g(x) = |x| + 3$ is the graph of $f(x) = |x|$ shifted up 3 units (Figure 4.5).

---

* This confirms use of the vertex form of a quadratic function in Section 3.1.

Table 4.1

x	$f(x) = \lvert x \rvert$	$g(x) = \lvert x \rvert + 3$
−3	3	6
−2	2	5
−1	1	4
0	0	3
1	1	4
2	2	5
3	3	6
4	4	7

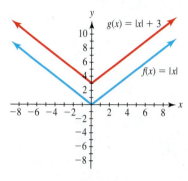

Figure 4.5

In Example 1, the equation of the shifted graph is $g(x) = \lvert x \rvert + 3 = f(x) + 3$. In general, we have the following.

> ### Vertical Shifts of Graphs
>
> If k is a positive real number:
>
> The graph of $g(x) = f(x) + k$ can be obtained by shifting the graph of $f(x)$ upward k units.
>
> The graph of $g(x) = f(x) - k$ can be obtained by shifting the graph of $f(x)$ downward k units.

EXAMPLE 2 ▶   ## Shifts of Functions

Graph the functions $f(x) = \lvert x \rvert$ and $g(x) = \lvert x - 5 \rvert$ on the same axes. What relationship do you notice between the two graphs?

SOLUTION

Table 4.2 shows inputs for x and outputs for both $f(x) = \lvert x \rvert$ and $g(x) = \lvert x - 5 \rvert$. Observe that the output values for both functions are equal *if* the input value for $g(x) = \lvert x - 5 \rvert$ is 5 more than the corresponding input value for $f(x) = \lvert x \rvert$. Thus, the x-coordinate of each point on the graph of $g(x) = \lvert x - 5 \rvert$ is 5 more than the x-coordinate of the point on the graph of $f(x) = \lvert x \rvert$ with the same y-coordinate. We say that the graph of $g(x) = \lvert x - 5 \rvert$ is the graph of $f(x) = \lvert x \rvert$ shifted 5 units to the right (Figure 4.6).

Table 4.2

x	$f(x) = \lvert x \rvert$	x	$g(x) = \lvert x - 5 \rvert$
−3	3	2	3
−2	2	3	2
−1	1	4	1
0	0	5	0
1	1	6	1
2	2	7	2
3	3	8	3
4	4	9	4

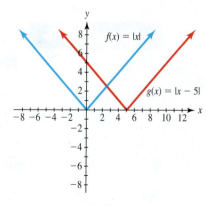

Figure 4.6

In Example 2, the equation of the shifted graph is $g(x) = |x - 5| = f(x - 5)$. In general, we have the following.

> ### Horizontal Shifts of Graphs
>
> If h is a positive real number:
>
> > The graph of $g(x) = f(x - h)$ can be obtained by shifting the graph of $f(x)$ to the right h units.
> > The graph of $g(x) = f(x + h)$ can be obtained by shifting the graph of $f(x)$ to the left h units.

EXAMPLE 3 ▶ Shifts of Functions

Graph the functions $f(x) = \sqrt{x}$ and $g(x) = \sqrt{x + 3} - 4$ on the same axes. What relationship do you notice between the two graphs?

SOLUTION

The graphs of $f(x) = \sqrt{x}$ and $g(x) = \sqrt{x + 3} - 4$ are shown in Figure 4.7. We see that the graph of $g(x) = \sqrt{x + 3} - 4$ can be obtained by shifting the graph of $f(x) = \sqrt{x}$ to the left 3 units and down 4 units. To verify this, consider a few points from each graph in Table 4.3.

Table 4.3

$f(x) = \sqrt{x}$		$g(x) = \sqrt{x + 3} - 4$	
x	$f(x) = \sqrt{x}$	x	$g(x) = \sqrt{x + 3} - 4$
0	0	-3	-4
1	1	-2	-3
2	$\sqrt{2}$	-1	$\sqrt{2} - 4$
3	$\sqrt{3}$	0	$\sqrt{3} - 4$
4	2	1	-2

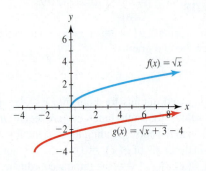

Figure 4.7

EXAMPLE 4 ▶ Profit

Right Sports Management sells elliptical machines, and the monthly profit from these machines can be modeled by

$$P(x) = -(x - 55)^2 + 4500$$

where x is the number of elliptical machines produced and sold.

a. Use the graph of $y = -x^2$ and the appropriate transformation to set the window and graph the profit function.

b. How many elliptical machines must be produced and sold to yield the maximum monthly profit? What is the maximum monthly profit?

(a)   (b)

Figure 4.8

SOLUTION

a. Shifting the graph of $y = -x^2$ (Figure 4.8(a)) 55 units to the right and up 4500 units gives the graph of $P(x) = -(x - 55)^2 + 4500$ (Figure 4.8(b)). Note that we graph the function only in the first quadrant, because only nonnegative values make sense for this application.

b. The maximum value of the function occurs at the vertex, and the vertex of the graph is $(55, 4500)$. Thus, maximum monthly profit of $4500 is made when 55 elliptical machines are produced and sold.

EXAMPLE 5 ▶   **Diabetes**

The projected percent of the U.S. adult population with diabetes (diagnosed and undiagnosed) for 2010 through 2050 can be modeled by the power function $f(x) = 4.947x^{0.495}$, with x equal to the number of years after 2000. Find the power function that models this projection with x equal to the number of years after 2010.

SOLUTION

Because y represents the percent in both models, there is no vertical shift from one function to the other. To find the power model that will give the percent where x is the number of years after 2010 rather than the number of years after 2000, we rewrite the equation in a new form where $x = 0$ represents the same input as $x = 10$ does in the original model. This can be accomplished by replacing x in the original model with $x + 10$, giving the new model

$$f(x) = 4.947(x + 10)^{0.495}$$

where x is the number of years after 2010.

To see that these two models are equivalent, observe the inputs and outputs for selected years in Table 4.4 and the graphs in Figure 4.9.

Table 4.4

Original Model		New Model	
Years from 2000	Percent	Years from 2010	Percent
10	15.46	0	15.46
20	21.79	10	21.79
30	26.64	20	26.64
50	34.30	40	34.30

Figure 4.9

Stretching and Compressing Graphs

EXAMPLE 6 ▶ **Stimulus-Response**

One of the early results in psychology relates the magnitude of a stimulus x to the magnitude of the response y with the model $y = kx^2$, where k is an experimental constant. Compare the graphs of $y = kx^2$ for $k = 1, 2$, and $\frac{1}{2}$.

SOLUTION

Table 4.5 shows the values of y for $y = f(x) = x^2, y = 2f(x) = 2x^2$, and $y = \frac{1}{2}f(x) = \frac{1}{2}x^2$ for selected values of x. Observe that for these x-values, the y-values for $y = 2x^2$ are 2 times the y-values for $y = x^2$, and the points on the graph of $y = 2x^2$ have y-values that are 2 times the y-values on the graph of $y = x^2$ for equal x-values. We say that the graph of $y = 2x^2$ is a **vertical stretch** of $y = x^2$ by a factor of 2. (Compare Figure 4.10(a) and Figure 4.10 (b).)

Observe also that for these x-values, the y-values for $y = \frac{1}{2}x^2$ are $\frac{1}{2}$ of the y-values for $y = x^2$, and the points on the graph of $y = \frac{1}{2}x^2$ have y-values that are $\frac{1}{2}$ of the y-values on the graph of $y = x^2$ for equal x-values. We say that the graph of $y = \frac{1}{2}x^2$ is a **vertical compression** of $y = x^2$ using a factor of $\frac{1}{2}$ (Figure 4.10(c)).

Table 4.5

x	0	1	2
$f(x)$	0	1	4

x	0	1	2
$2f(x)$	0	2	8

x	0	1	2
$\frac{1}{2}f(x)$	0	$\frac{1}{2}$	2

Figure 4.10

Reflections of Graphs

In Figure 4.11(a), we see that the graph of $y = -x^2$ is a parabola that *opens down*. It can be obtained by reflecting the graph of $f(x) = x^2$, which *opens up*, across the x-axis. We can compare the y-coordinates of the graphs of these two functions by looking at Table 4.6 and Figure 4.11(b). Notice that for a given value of x, the y-coordinates of $y = x^2$ and $y = -x^2$ are negatives of each other.

Table 4.6

x	$y = x^2$	$y = -x^2$
-2	4	-4
-1	1	-1
0	0	0
1	1	-1
2	4	-4

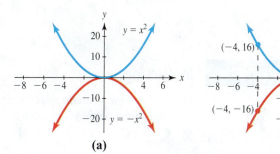

(a)   **(b)**

Figure 4.11

We can also see that the graph of $y = (-x)^3$ is a reflection of the graph of $y = x^3$ across the y-axis by looking at Table 4.7 and Figure 4.12. Notice that for a given value of y, the x-coordinates of $y = x^3$ and $y = (-x)^3$ are negatives of each other.

Table 4.7

x	$y = x^3$	x	$y = (-x)^3$
2	8	-2	8
1	1	-1	1
0	0	0	0
-1	-1	1	-1
-2	-8	2	-8

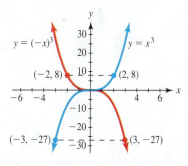

Figure 4.12

In general, we have the following:

A summary of the transformations of a graph follows.

Graph Transformations

For a given function $y = f(x)$,

Vertical Shift	$y = f(x) + k$	Graph is shifted k units up if $k > 0$ and k units down if $k < 0$.
Horizontal Shift	$y = f(x - h)$	Graph is shifted h units right if $h > 0$ and h units left if $h < 0$.
Stretch/Compress	$y = af(x)$	Graph is vertically stretched using a factor of $|a|$ if $|a| > 1$. Graph is compressed using a factor of $|a|$ if $|a| < 1$.
Reflection	$y = -f(x)$	Graph is reflected across the x-axis.
	$y = f(-x)$	Graph is reflected across the y-axis.

EXAMPLE 7 ▶   **Pollution**

Suppose that for a certain city, the cost C of obtaining drinking water that contains $p\%$ impurities (by volume) is given by

$$C = \frac{120{,}000}{p} - 1200$$

a. Determine the domain of this function and graph the function without concern for the context of the problem.

b. Use knowledge of the context of the problem to graph the function in a window that applies to the application.

c. What is the cost of obtaining drinking water that contains 5% impurities?

SOLUTION

a. All values of p except $p = 0$ result in real values for the function. Thus, the domain of C is all real numbers except 0. The graph of this function is a transformation of the graph of $C = \dfrac{1}{p}$, stretched by a factor of 120,000, then shifted downward 1200 units. The viewing window should have its center near $p = 0$ (horizontally) and $C = -1200$ (vertically). We increase the vertical view to allow for the large stretching factor, using the viewing window $[-100, 100]$ by $[-20{,}000, 20{,}000]$. The graph, shown in Figure 4.13(a), has the same shape as the graph of $y = \dfrac{1}{x}$.

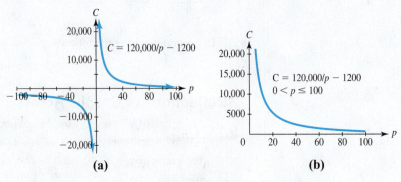

(a)   (b)

Figure 4.13

b. Because p represents the percent of impurities, the viewing window is set for values of p from 0 to 100. (Recall from part (a) that p cannot be 0.) The cost of reducing the impurities cannot be negative, so the vertical view is set from 0 to 20,000. The graph is shown in Figure 4.13(b).

c. To determine the cost of obtaining drinking water that contains 5% impurities, we substitute 5 for p in $C = \dfrac{120{,}000}{p} - 1200$, giving

$$C = \frac{120{,}000}{5} - 1200 = 22{,}800$$

Thus, the cost is $22,800 to obtain drinking water that contains 5% impurities.

EXAMPLE 8 ▶   **Velocity of Blood**

Because of friction from the walls of an artery, the velocity of blood is greatest at the center of the artery and decreases as the distance r from the center increases. The velocity of the blood in an artery can be modeled by the function

$$v = k\left(R^2 - r^2\right)$$

where R is the radius of the artery and k is a constant that is determined by the pressure, the viscosity of the blood, and the length of the artery. In the case where $k = 2$ and $R = 0.1$ centimeter, the velocity is

$$v = 2\left(0.01 - r^2\right) \text{ centimeter per second}$$

a. Graph this function and the functions $v = r^2$ and $v = -2r^2$ in the viewing window $[-0.1, 0.1]$ by $[-0.05, 0.05]$.

b. How is the function $v = 2\left(0.01 - r^2\right)$ related to the second-degree power function $v = r^2$?

SOLUTION

a. The graph of the function $v = 2\left(0.01 - r^2\right)$ is shown in Figure 4.14(a), the graph of $v = r^2$ is shown in Figure 4.14(b), and the graph of $v = -2r^2$ is shown in Figure 4.14(c).

Figure 4.14

b. The function $v = 2\left(0.01 - r^2\right)$ can also be written in the form $v = -2r^2 + 0.02$, so it should be no surprise that the graph is a reflected, stretched, and shifted form of the graph of $v = r^2$. If the graph of $v = r^2$ is reflected about the r-axis, has each r^2-value doubled, and then is shifted upward 0.02 unit, the graph of $v = -2r^2 + 0.02 = 2\left(0.01 - r^2\right)$ results.

Figure 4.15

Symmetry; Even and Odd Functions

In Chapter 3, we observed that any parabola that is a graph of a quadratic function is symmetric about a vertical line called the axis of symmetry. This means that the two halves of the parabola are reflections of each other. In particular, the graph of $y = x^2$ in Figure 4.15 is symmetric (that is, is a reflection of itself) about the y-axis. For example, the point $(3, 9)$ is on the graph, and its mirror image across the y-axis is the point $(-3, 9)$. This suggests an algebraic way to determine whether the graph of an equation is symmetric with respect to the y-axis—that is, replace x with $-x$ and simplify. If the resulting equation is equivalent to the original equation, then the graph is symmetric with respect to the y-axis.

Symmetry with Respect to the y-axis

The graph of $y = f(x)$ is symmetric with respect to the y-axis if, for every point (x, y) on the graph, the point $(-x, y)$ is also on the graph—that is,

$$f(-x) = f(x)$$

for all x in the domain of f. Such a function is called an **even function**.

All power functions with even integer exponents will have graphs that are symmetric about the y-axis. This is why all functions satisfying this condition are called even functions, although functions other than power functions exist that are even functions.

A graph of $y = x^3$ is shown in Figure 4.16(a). Notice in Figure 4.16(b) that if we draw a line through the origin and any point on the graph, it will intersect the graph at a second point, and the distances from the two points to the origin will be equal.

Figure 4.16

For example, the point $(3, 27)$ is on the graph of $y = x^3$, and the line through $(3, 27)$ and the origin intersects the graph at the point $(-3, -27)$, which is the same distance from the origin as $(3, 27)$. In general, the graph of a function is called symmetric with respect to the origin if for every point (x, y) on the graph, the point $(-x, -y)$ is also on the graph.

To determine algebraically whether the graph of an equation is symmetric about the origin, replace x with $-x$, replace y with $-y$, and simplify. If the resulting equation is equivalent to the original equation, then the graph is symmetric with respect to the origin.

Symmetry with Respect to the Origin

The graph of $y = f(x)$ is symmetric with respect to the origin if, for every point (x, y) on the graph, the point $(-x, -y)$ is also on the graph—that is,

$$f(-x) = -f(x)$$

for all x in the domain of f. Such a function is called an **odd function**.

The graphs of some equations can be symmetric with respect to the x-axis (see Figure 4.17). Such graphs do not represent y as a function of x (because they do not pass the vertical line test).

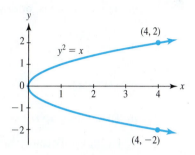

Figure 4.17

For example, we can graph the equation $x^2 + y^2 = 9$ by hand or with technology after solving it for y. Solving gives

$$y^2 = 9 - x^2$$
$$y = \pm\sqrt{9 - x^2}$$

Entering values for x in the equation gives points on the **circle**, and entering the two functions in a calculator, as in Figure 4.18(a), gives the graph. Note that the calculator graph will not look like a circle unless the calculator window is square. Figure 4.18(b) gives the graph with the standard window, Figure 4.18(c) gives the graph if ⟨ZOOM⟩ Square is used after ⟨ZOOM⟩ Standard, and Figure 4.18(d) gives the graph with ⟨ZOOM⟩ Decimal.* In general, if the equation contains one and only one even power of y, its graph will be symmetric about the x-axis.

(a)

(b)

(c)

(d)

Figure 4.18

To determine algebraically whether the graph of an equation is symmetric about the x-axis, replace y with $-y$ and simplify. If the resulting equation is equivalent to the original equation, then the graph is symmetric with respect to the x-axis.

Symmetry with Respect to the *x*-axis

The graph of an equation is symmetric with respect to the x-axis if, for every point (x, y) on the graph, the point $(x, -y)$ is also on the graph.

---

* See Appendix A, page 666 for details.

EXAMPLE 9 ▶ ## Symmetry

Determine algebraically whether the graph of each equation is symmetric with respect to the x-axis, y-axis, or origin. Confirm your conclusion graphically.

a. $y = \dfrac{2x^2}{x^2 + 1}$ **b.** $y = x^3 - 3x$ **c.** $x^2 + y^2 = 16$

SOLUTION

a. To test the graph of $y = \dfrac{2x^2}{x^2 + 1}$ for symmetry with respect to the x-axis, we replace y with $-y$. Because the result,

$$-y = \frac{2x^2}{x^2 + 1}$$

is not equivalent to the original equation, the graph is not symmetric with respect to the x-axis. This equation gives y as a function of x.

To test for y-axis symmetry, we replace x with $-x$:

$$y = \frac{2(-x)^2}{(-x)^2 + 1} = \frac{2x^2}{x^2 + 1}$$

which is equivalent to the original equation. Thus, the graph is symmetric with respect to the y-axis, and y is an even function of x.

To test for symmetry with respect to the origin, we replace x with $-x$ and y with $-y$:

$$-y = \frac{2(-x)^2}{(-x)^2 + 1} = \frac{2x^2}{x^2 + 1}$$

which is not equivalent to the original equation. Thus, the graph is not symmetric with respect to the origin.

Figure 4.19 confirms the fact that the graph of $y = \dfrac{2x^2}{x^2 + 1}$ is symmetric with respect to the y-axis.

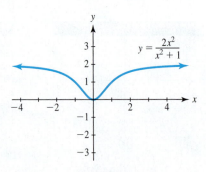

Figure 4.19

b. To test the graph of $y = x^3 - 3x$ for symmetry with respect to the x-axis, we replace y with $-y$. Because the result,

$$-y = x^3 - 3x$$

is not equivalent to the original equation, the graph is not symmetric with respect to the x-axis.

To test for y-axis symmetry, we replace x with $-x$:

$$y = (-x)^3 - 3(-x) = -x^3 + 3x$$

which is not equivalent to the original equation. Thus, the graph is not symmetric with respect to the y-axis.

To test for symmetry with respect to the origin, we replace x with $-x$ and y with $-y$:

$$-y = (-x)^3 - 3(-x) = -x^3 + 3x, \text{ or } y = x^3 - 3x$$

which is equivalent to the original equation. Thus, the graph is symmetric with respect to the origin, and y is an odd function of x.

Figure 4.20 confirms this.

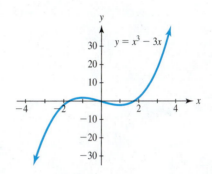

Figure 4.20

c. To test the graph of $x^2 + y^2 = 16$ for symmetry with respect to the x-axis, we replace y with $-y$:

$$x^2 + (-y)^2 = 16 \quad \text{or} \quad x^2 + y^2 = 16$$

This is equivalent to the original equation, so the graph is symmetric with respect to the x-axis. Thus, y is not a function of x in this equation.

To test for y-axis symmetry, we replace x with $-x$:

$$(-x)^2 + y^2 = 16 \quad \text{or} \quad x^2 + y^2 = 16$$

This is equivalent to the original equation, so the graph is symmetric with respect to the y-axis.

To test for symmetry with respect to the origin, we replace x with $-x$ and y with $-y$:

$$(-x)^2 + (-y)^2 = 16, \quad \text{or} \quad x^2 + y^2 = 16$$

which is equivalent to the original equation. Thus, the graph is symmetric with respect to the origin.

Solving for y gives the two functions shown in Figure 4.21(a). Choosing a window $[-7, 7]$ by $[-5, 5]$ and then selecting $\boxed{\text{ZOOM}}$ 5: ZSquare gives the graph in Figure 4.21(b), which confirms the fact that $y = x^2 + y^2$ is symmetric with respect to the x-axis, the y-axis, and the origin.

(a)

(b)

Figure 4.21

SKILLS CHECK 4.1

In Exercises 1–16, (a) sketch the graph of each pair of functions using a standard window, and (b) describe the transformations used to obtain the graph of the second function from the first function.

1. $y = x^3, y = x^3 + 5$

2. $y = x^2, y = x^2 + 3$

3. $y = \sqrt{x}, y = \sqrt{x-4}$

4. $y = x^2, y = (x+2)^2$

5. $y = \sqrt[3]{x}, y = \sqrt[3]{x+2} - 1$

6. $y = x^3, y = (x-5)^3 - 3$

7. $y = |x|, y = |x-2| + 1$

8. $y = |x|, y = |x+3| - 4$

9. $y = x^2, y = -x^2 + 5$

10. $y = \sqrt{x}, y = -\sqrt{x-2}$

11. $y = \dfrac{1}{x}, y = \dfrac{1}{x} - 3$

12. $y = \dfrac{1}{x}, y = \dfrac{2}{x-1}$

13. $f(x) = x^2, g(x) = \dfrac{1}{3}x^2$

14. $f(x) = x^3, g(x) = 0.4x^3$

15. $f(x) = |x|, g(x) = 3|x|$

16. $f(x) = \sqrt{x}, g(x) = 4\sqrt{x}$

17. How is the graph of $y = (x-2)^2 + 3$ transformed from the graph of $y = x^2$?

18. How is the graph of $y = (x+4)^3 - 2$ transformed from the graph of $y = x^3$?

19. Suppose the graph of $y = x^{3/2}$ is shifted to the left 4 units. What is the equation that gives the new graph?

20. Suppose the graph of $y = x^{3/2}$ is shifted down 5 units and to the right 4 units. What is the equation that gives the new graph?

21. Suppose the graph of $y = x^{3/2}$ is stretched by a factor of 3 and then shifted up 5 units. What is the equation that gives the new graph?

22. Suppose the graph of $y = x^{2/3}$ is compressed by a factor of $\dfrac{1}{5}$ and then shifted right 6 units. What is the equation that gives the new graph?

23. Suppose the graph of $f(x) = \sqrt{x}$ is shifted up 3 units and to the left 6 units. What is the equation of the new graph? Verify the result graphically.

24. Suppose the graph of $f(x) = x^2$ is shifted up 2 units and to the right 5 units. What is the equation of the new graph? Verify the result graphically.

25. Suppose the graph of $f(x) = \dfrac{1}{x}$ is stretched vertically by a factor of 4 and then shifted down 3 units. What is the equation of the new graph? Verify the result graphically.

26. Suppose the graph of $f(x) = x^3$ is compressed vertically using a factor of $\dfrac{1}{3}$ and then shifted to the left 5 units. What is the equation of the new graph? Verify the result graphically.

27. Suppose the graph of $f(x) = \sqrt{x}$ is reflected across the x-axis and shifted up 2 units. What is the equation of the new graph? Verify the result graphically.

28. Suppose the graph of $f(x) = \sqrt[3]{x}$ is reflected across the x-axis and shifted to the right 1 unit. What is the equation of the new graph? Verify the result graphically.

In Exercises 29–32, write the equation of the function g(x) that is transformed from the given function f(x) and whose graph is shown.

29. $f(x) = x^2$

30. $f(x) = x^2$

31. $f(x) = |x|$

32. $f(x) = x^2$

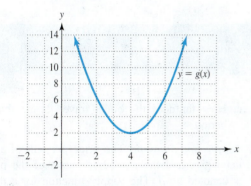

In Exercises 33–36, determine visually whether each of the graphs is symmetric about the x-axis, the y-axis, or neither. If the graph represents a function, determine whether the function is even, odd, or neither.

33.

34.

35.

36.

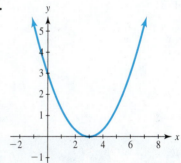

In Exercises 37 and 38, determine whether the graph of the equation in Y_1 is symmetric about the y-axis, based on the ordered pairs shown in each pair of tables.

37.

X	Y₁			
0	0			
1	2			
2	8			
3	18			
4	32			
5	50			

X=

X	Y₁			
0	0			
-1	2			
-2	8			
-3	18			
-4	32			
-5	50			

X=

38.

X	Y₁			
0	0			
1	.5			
2	2			
3	4.5			
4	8			
5	12.5			

X=

X	Y₁			
0	0			
-1	.5			
-2	2			
-3	4.5			
-4	8			
-5	12.5			

X=0

In Exercises 39–46, determine algebraically whether the graph of the given equation is symmetric with respect to the x-axis, the y-axis, and/or the origin. Confirm graphically.

39. $y = 2x^2 - 3$

40. $y = -x^2 + 4$

41. $y = x^3 - x$

42. $y = -x^3 + 5x$

43. $y = \dfrac{6}{x}$　　　　　　**44.** $x = 3y^2$

45. $x^2 + y^2 = 25$　　　　**46.** $x^2 - y^2 = 25$

In Exercises 47–52, determine whether the function is even, odd, or neither.

47. $f(x) = |x| - 5$

48. $f(x) = |x - 2|$

49. $g(x) = \sqrt{x^2 + 3}$

50. $f(x) = \dfrac{1}{2}x^3 - x$

51. $g(x) = \dfrac{5}{x}$

52. $g(x) = 4x + x^2$

Determine whether each of the complete graphs in Exercises 53 and 54 represents a function that is even, odd, or neither.

53.

54.

55. Graph $(y + 1)^2 = x + 4$. Does this graph have any type of symmetry?

56. About what line is the function $y = 3 \pm \sqrt{x - 2}$ symmetric?

EXERCISES 4.1

57. Marijuana Use The number of millions of people age 12 and older in the United States who ever used marijuana is described by the function $M(x) = -0.062(x - 4.8)^2 + 25.4$ for $3 \le x \le 8$, where x is the number of years after 2000.

　a. The graph of this function is a shifted graph of which basic function?

　b. Find and interpret $M(3)$.

　c. Sketch a graph of $y = M(x)$ for $3 \le x \le 8$.
　(Source: 2008 National Survey on Drug Use and Health, U.S. Department of Health and Human Services)

58. Ballistics Ballistic experts are able to identify the weapon that fired a certain bullet by studying the markings on the bullet. If a test is conducted by firing a bullet into a water tank, the distance that a bullet will travel is given by $s = 27 - (3 - 10t)^3$ inches for $0 \le t \le 0.3$, t in seconds.

　a. The graph of this function is a shifted graph of which basic function?

　b. Graph this function for $0 \le t \le 0.3$.

　c. How far does the bullet travel during this time period?

59. Supply and Demand The price per unit of a product is \$$p$, and the number of units of the product is denoted by q. The supply function for a product is given by $p = \dfrac{180 + q}{6}$, and the demand for the product is given by $p = \dfrac{30{,}000}{q} - 20$.

　a. Is the supply function a linear function or a shifted reciprocal function?

　b. Is the demand function a shifted linear function or a shifted reciprocal function? Describe the transformations needed to obtain the specific function from the basic function.

60. Supply and Demand The supply function for a commodity is given by $p = 58 + \dfrac{q}{2}$, and the demand function for this commodity is given by $p = \dfrac{2555}{q + 5}$.

　a. Is the supply function a linear function or a shifted reciprocal function?

b. Is the demand function a shifted linear function or a shifted reciprocal function? Describe the transformations needed to obtain the specific function from the basic function.

61. *Average Cost* The monthly average cost of producing x sofas is $\overline{C}(x) = \dfrac{30,000}{x} + 150$ dollars.

a. Will the average cost function decrease or increase as the number of units produced increases?

b. What transformations of the graph of the reciprocal function give the graph of this function?

62. *Pollution* The daily cost C (in dollars) of removing pollution from the smokestack of a coal-fired electric power plant is related to the percent of pollution p being removed according to the equation

$$C = \frac{10,500}{100 - p}$$

a. Describe the transformations needed to obtain this function from the function $C = \dfrac{1}{p}$.

b. Graph the function for $0 \le p < 100$.

c. What is the daily cost of removing 80% of the pollution?

63. *Population Growth* Suppose the population of a certain microorganism at time t (in minutes) is given by

$$P = -1000\left(\frac{1}{t + 10} - 1\right)$$

a. Describe the transformations needed to obtain this function from the function $f(t) = \dfrac{1}{t}$.

b. Graph this function for values of t representing 0 to 100 minutes.

64. *Mortgages* The balance owed y on a \$50,000 mortgage after x monthly payments is shown in the table that follows. The function that models the data is

$$y = 4700\sqrt{110 - x}$$

a. Is this a shifted root function?

b. What is the domain of the function in the context of this application?

c. Describe the transformations needed to obtain the graph from the graph of $y = \sqrt{x}$.

Monthly Payments	Balance Owed ($)
12	47,243
24	44,136
48	36,693
72	27,241
96	15,239
108	8074

65. *Personal Expenditures* Personal consumption expenditures for durable goods in the United States, in billions of dollars, can be modeled by the function $P(x) = 306.472x^{0.464}$, where x is the number of years after 1990. Rewrite the function with x equal to the number of years after 1995. (*Hint*: Shift the x-value by the difference in years.)
(Source: Bureau of Economic Analysis, U.S. Department of Commerce)

66. *Unconventional Vehicle Sales* The number of E85 flex fuel vehicles, in millions, projected to be sold in the United States can be modeled by the function $F(x) = 0.084x^{0.675}$, where x is the number of years after 2000. Convert the function so that x equals the number of years after 1990.
(Source: www.eia.gov)

67. *Tobacco Sales* Sales of fine-cut cigarettes (in millions) in Canada can be described by $y = -13.898x^2 + 255.467x + 5425.618$, where x is the number of years after 1980. Convert the function so that x represents the number of years after 1990.

68. *Aircraft Accidents* The number of U.S. aircraft accidents can be modeled by $y = -17.560x^2 + 126.488x + 7986.786$, with $x = 0$ in 2000. Rewrite the model with x equal to the number of years after 1990.

69. *U.S. Cell Phone Subscribers* The number of U.S. cell phone subscriberships, in millions, can be modeled by the function $S(t) = 0.00061t^{3.9}$, where t is the number of years after 1980.
(Source: Semiannual CTIA Wireless Survey)

a. Use the model to estimate the number of subscribers in 2016.

b. Write a new model $C(t)$ with t equal to the number of years after 1985.

c. With the model in part (b), what value of t should be used to determine the number of subscribers in 2016? Does the value of $C(t)$ agree with the answer to part (a)?

70. **Cost-Benefit** Suppose for a certain city the cost C of obtaining drinking water that contains $p\%$ impurities (by volume) is given by

$$C = \frac{120,000}{p} - 1200$$

a. What is the cost of drinking water that is 100% impure?

b. What is the cost of drinking water that is 50% impure?

c. What transformations of the graph of the reciprocal function give the graph of this function?

71. **U.S. Poverty Threshold** The table below shows the yearly income poverty thresholds for a single person for selected years from 1990 to 2013.

Year	Income ($)	Year	Income ($)
1990	6652	2004	9827
1995	7763	2005	10,160
1998	8316	2008	10,991
2000	8794	2010	11,137
2002	9182	2013	11,888

(Source: U.S. Census Bureau)

The poverty threshold income for a single person can be modeled by the function $P(x) = 2029\sqrt{x}$, for $x \geq 0$, where x is the number of years after 1980.

a. Assuming that the model is accurate after 2013, find the poverty threshold income in 2017.

b. Use a transformation to rewrite the model from (a) as $S(x)$, with x representing the number of years after 1985.

c. Using the model in part (b), what value of x should be used to determine the poverty threshold income in 2017? Does the value of $S(x)$ agree with the answer to part (a)?

72. **Pollution** The daily cost C (in dollars) of removing pollution from the smokestack of a coal-fired electric power plant is related to the percent of pollution p being removed according to the equation $100C - Cp = 10,500$.

a. Solve this equation for C to write the daily cost as a function of p, the percent of pollution removed.

b. What is the daily cost of removing 50% of the pollution?

c. Why would this company probably resist removing 99% of the pollution?

4.2　Combining Functions; Composite Functions

KEY OBJECTIVES

- Find sums, differences, products, and quotients of two functions
- Form average cost functions
- Find the composition of two functions

SECTION PREVIEW　Profit

If the daily total cost to produce x units of a product is

$$C(x) = 360 + 40x + 0.1x^2 \text{ thousand dollars}$$

and the daily revenue from the sale of x units of this product is

$$R(x) = 60x \text{ thousand dollars}$$

we can model the profit function as $P(x) = R(x) - C(x)$, giving

$$P(x) = 60x - (360 + 40x + 0.1x^2)$$

or

$$P(x) = -0.1x^2 + 20x - 360$$

In a manner similar to the one used to form this function, we can construct new functions by performing algebraic operations with two or more functions. For example, we can build an average cost function by finding the quotient of two functions. (See Example 4.) In addition to using arithmetic operations with functions, we can create new functions using function composition. ∎

Operations with Functions

New functions that are the sum, difference, product, and quotient of two functions are defined as follows:

Operation	Formula	Example with $f(x) = \sqrt{x}$ and $g(x) = x^3$
Sum	$(f + g)(x) = f(x) + g(x)$	$(f + g)(x) = \sqrt{x} + x^3$
Difference	$(f - g)(x) = f(x) - g(x)$	$(f - g)(x) = \sqrt{x} - x^3$
Product	$(f \cdot g)(x) = f(x) \cdot g(x)$	$(f \cdot g)(x) = \sqrt{x} \cdot x^3 = x^3\sqrt{x}$
Quotient	$\left(\dfrac{f}{g}\right)(x) = \dfrac{f(x)}{g(x)} \quad (g(x) \neq 0)$	$\left(\dfrac{f}{g}\right)(x) = \dfrac{\sqrt{x}}{x^3} \quad (x \neq 0)$

The domain of the sum, difference, and product of f and g consists of all real numbers of the input variable for which f and g are defined. The domain of the quotient function consists of all real numbers for which f and g are defined and $g \neq 0$.

EXAMPLE 1 ▶ Operations with Functions

If $f(x) = x^3$ and $g(x) = x - 1$, find the following functions and give their domains.

a. $(f + g)(x)$ **b.** $(f - g)(x)$ **c.** $(f \cdot g)(x)$ **d.** $\left(\dfrac{f}{g}\right)(x)$

SOLUTION

a. $(f + g)(x) = f(x) + g(x) = x^3 + x - 1$; all real numbers

b. $(f - g)(x) = f(x) - g(x) = x^3 - (x - 1) = x^3 - x + 1$; all real numbers

c. $(f \cdot g)(x) = f(x) \cdot g(x) = x^3(x - 1) = x^4 - x^3$; all real numbers

d. $\left(\dfrac{f}{g}\right)(x) = \dfrac{f(x)}{g(x)} = \dfrac{x^3}{x - 1}$; all real numbers except 1

EXAMPLE 2 ▶ Revenue, Cost, and Profit

The demand for a certain electronic component is given by $p(x) = 1000 - 2x$. Producing and selling x components involves a monthly fixed cost of \$1999 and a production cost of \$4 for each component.

a. Write the equations that model total revenue and total cost as functions of the components produced and sold in a month.

b. Write the equation that models the profit as a function of the components produced and sold during a month.

c. Find the maximum possible monthly profit.

SOLUTION

a. The revenue for the components is given by the product of $p(x) = 1000 - 2x$ and x, the number of units sold.

$$R(x) = p(x) \cdot x = (1000 - 2x)x = 1000x - 2x^2 \text{ dollars}$$

The monthly total cost is the sum of the variable cost, $4x$, and the fixed cost, 1999.

$$C(x) = 4x + 1999 \text{ dollars}$$

b. The monthly profit for the production and sale of the electronic components is the difference between the revenue and cost functions.

$$P(x) = R(x) - C(x) = (1000x - 2x^2) - (4x + 1999) = -2x^2 + 996x - 1999$$

c. The maximum monthly profit occurs where $x = \dfrac{-996}{2(-2)} = 249$. The maximum profit is $P(249) = 122{,}003$ dollars.

EXAMPLE 3 ▶　Unconventional Vehicle Sales

In December 2010, the U.S. Energy Information Administration (EIA) presented results projecting rapid growth in sales of unconventional vehicles through 2035. The EIA expects unconventional vehicles—vehicles using diesel or alternative fuels and/or hybrid electric systems—to account for over 40% of U.S. light-duty vehicles sold in 2035. Figure 4.22 shows the number of light-duty cars and trucks in each category sold, in millions, projected to 2035.

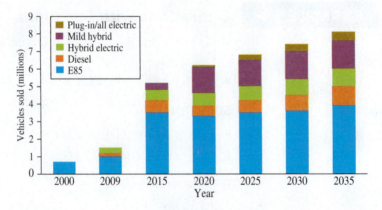

Figure 4.22

The functions that model the number of each type of vehicle sold, in millions, are

E85 flex fuel:	$F(x) = -0.0027x^2 + 0.193x + 0.439$
Diesel:	$D(x) = 0.028x + 0.077$
Hybrid electric:	$H(x) = 0.025x + 0.154$
Mild hybrid:	$M(x) = -0.006x^2 + 0.350x - 3.380$
Plug-in electric and all electric:	$E(x) = -0.001x^2 + 0.081x - 1.115$

where x is the number of years after 2000.

a. Add functions F and D to obtain a function that gives the total number of flex fuel and diesel vehicles sold.

b. Add functions H, M, and E to obtain a function that gives the total number of hybrid and electric vehicles sold.

c. Graph the functions from parts (a) and (b) together on the same axes for the years 2000–2035.

d. Graph the functions from parts (a) and (b) together on the same axes through 2050 $(x = 50)$. Will the number of hybrid and electric vehicles exceed the number of flex fuel and diesel vehicles before 2050, according to the models? Based on this graph, do the models appear to be good predictors of unconventional vehicle sales after 2035?

e. Add the functions from parts (a) and (b) to obtain a function that gives the total number of unconventional vehicles sold. According to the EIA projection, there will be 8.1 million unconventional vehicles sold in 2035. Does the model agree with this projection?

SOLUTION

a. Adding functions F and D gives

$$F(x) + D(x) = -0.0027x^2 + 0.193x + 0.439 + 0.028x + 0.077$$
$$= -0.0027x^2 + 0.221x + 0.516$$

b. Adding functions H, M, and E gives

$$H(x) + M(x) + E(x) = 0.025x + 0.154 - 0.006x^2 + 0.350x - 3.380$$
$$-0.001x^2 + 0.081x - 1.115$$
$$= -0.007x^2 + 0.456x - 4.341$$

c. The graphs of the functions from parts (a) and (b) through $x = 35$ are shown in Figure 4.23.

Figure 4.23   **Figure 4.24**

d. The graphs of the functions from parts (a) and (b) through $x = 50$ are shown in Figure 4.24. The graphs do not intersect in this time interval, so the number of hybrid and electric vehicles will not exceed the number of flex fuel and diesel vehicles before 2050. Based on these graphs, the models do not appear to be good predictors after 2035.

e. Adding the two functions from parts (a) and (b) gives

$$U(x) = -0.0027x^2 + 0.221x + 0.516 - 0.007x^2 + 0.456x - 4.341$$
$$= -0.0097x^2 + 0.677x - 3.825$$

Substituting $x = 35$ into function U gives

$$U(35) = -0.0097(35)^2 + 0.677(35) - 3.825 = 7.99$$

Thus, the model predicts $7.99 \approx 8.0$ million, which is close to the EIA projection.

Average Cost

A company's **average cost** per unit, when x units are produced, is the quotient of the function $C(x)$ (the total production cost) and the identity function $I(x) = x$ (the number of units produced). That is, the **average cost** function is

$$\overline{C}(x) = \frac{C(x)}{x}$$

EXAMPLE 4 ▶   **Average Cost**

Sunny's Greenhouse produces roses, and their total cost for the production of x hundred roses is

$$C(x) = 50x + 500$$

a. Form the average cost function.

b. For which input values is $\overline{C}$ defined? Give a real-world explanation of this answer.

c. Graph $\overline{C}$ for 0 to 5000 roses (50 units). What can you say about the average cost?

SOLUTION

a. The average cost function is the quotient of the cost function $C(x) = 50x + 500$ and the identity function $I(x) = x$.

$$\overline{C}(x) = \frac{50x + 500}{x}$$

b. The average cost function is defined for all real numbers such that $x > 0$ because producing negative units is not possible and the function is undefined for $x = 0$. (This is reasonable, because if nothing is produced, it does not make sense to discuss average cost per unit of product.)

c. The graph is shown in Figure 4.25. It can be seen from the graph that as the number of roses produced increases, the average cost decreases.

Figure 4.25

Composition of Functions

We have seen that we can convert a temperature of $x°$ Celsius to a Fahrenheit temperature by using the formula $F(x) = \dfrac{9}{5}x + 32$. There are also temperature scales in which $0°$ is absolute zero, which is set at $-273.15°C$. One of these, called the Kelvin temperature scale (K), uses the same degree size as Celsius until it reaches absolute zero, which is 0 K. To convert a Celsius temperature to a Kelvin temperature, we can add 273.15 to the Celsius temperature, and to convert a Kelvin temperature to a Celsius temperature, we can use the formula

$$C(x) = x - 273.15$$

To convert from x kelvins to a Fahrenheit temperature, we find $F(C(x))$ by substituting $C(x) = x - 273.15$ for x in $F(x) = \dfrac{9}{5}x + 32$, getting

$$F(C(x)) = F(x - 273.15) = \frac{9}{5}(x - 273.15) + 32 = \frac{9}{5}x - 459.67$$

That is, we can convert from the Kelvin scale to the Fahrenheit scale by using

$$F(K) = \frac{9}{5}K - 459.67$$

The process we have used to get a new function from these two functions is called **composition of functions**. The function $F(K(x))$ is called a **composite function**.

Composite Function

The composite function f of g is denoted by $f \circ g$ and defined by

$$(f \circ g)(x) = f(g(x))$$

The domain of $f \circ g$ is the subset of the domain of g for which $f \circ g$ is defined. The composite function $g \circ f$ is defined by $(g \circ f)(x) = g(f(x))$. The domain of $g \circ f$ is the subset of the domain of f for which $g \circ f$ is defined.

When computing the composite function $f \circ g$, keep in mind that the output of g becomes the input for f and that the rule for f is applied to this new input.

A composite function machine can be thought of as a machine within a machine. Figure 4.26 shows a composite "function machine" in which the input is denoted by x, the inside function is g, the outside function is f, the composite function rule is denoted by $f \circ g$, and the composite function output is symbolized by $f(g(x))$. Note that for most functions f and g, $f \circ g \neq g \circ f$.

Figure 4.26

EXAMPLE 5 ▶ Function Composition and Orange Juice

There are two machines in a room; one of the machines squeezes oranges to make orange juice, and the other is a canning machine that puts the orange juice into cans. Thinking of these two processes as functions, we name the squeezing function g and the canning function f.

a. Describe the composite function $(f \circ g)(\text{orange}) = f(g(\text{orange}))$.

b. Describe the composite function $(g \circ f)(\text{orange}) = g(f(\text{orange}))$.

c. Which function, $f(g(\text{orange})), g(f(\text{orange}))$, neither, or both, makes sense in context?

SOLUTION

a. The process by which the output $f(g(\text{orange}))$ is obtained is easiest to understand by thinking "from the inside out." Think of the input as an orange. The orange first goes into the *g* machine, so it is squeezed. The output of *g*, liquid orange juice, is then put into the *f* machine, which puts it into a can. The result is a can containing orange juice (Figure 4.27).

Figure 4.27

b. Again thinking from the inside out, the output $g(f(\text{orange}))$ is obtained by first putting an orange into the *f* machine, which puts it into a can. The output of *f*, the canned orange, is then put into the squeezing machine, *g*. The result is a compacted can, containing an orange (Figure 4.28).

Figure 4.28

c. As can be seen from the results of parts (a) and (b), the order in which the functions appear in the composite function symbol does make a difference. The process that makes sense is the one in part (a), $f(g(\text{orange}))$.

EXAMPLE 6 ▶ **Finding Composite Function Outputs**

Find the following composite function outputs, using $f(x) = 2x - 5$, $g(x) = 6 - x^2$, and $h(x) = \dfrac{1}{x}$. Give the domain of each new function formed.

a. $(h \circ f)(x) = h(f(x))$ **b.** $(f \circ g)(x) = f(g(x))$ **c.** $(g \circ f)(x) = g(f(x))$

SOLUTION

a. First, the output of *f* becomes the input for *h*.

$$(h \circ f)(x) = h(f(x)) = h(2x - 5) = \frac{1}{2x - 5}$$

Next, the rule for h is applied to this input.

The domain of this function is all $x \neq \dfrac{5}{2}$ because $\dfrac{1}{2x - 5}$ is undefined if $x = \dfrac{5}{2}$.

b. $(f \circ g)(x) = f(g(x)) = f(6 - x^2) = 2(6 - x^2) - 5 = 12 - 2x^2 - 5$

$$= -2x^2 + 7$$

The domain of this function is the set of all real numbers.

c. $(g \circ f)(x) = g(f(x)) = g(2x - 5) = 6 - (2x - 5)^2 = 6 - (4x^2 - 20x + 25)$

$$= 6 - 4x^2 + 20x - 25 = -4x^2 + 20x - 19$$

The domain of this function is the set of all real numbers.

> **Technology Note**
>
> If you have two functions input as Y_1 and Y_2, you can enter the sum, difference, product, quotient, or composition of the two functions in Y_3. These combinations of functions can then be graphed or evaluated for input values of x.

EXAMPLE 7 ▶   **Combinations with Functions**

If $f(x) = \sqrt{x - 5}$ and $g(x) = 2x^2 - 4$,

a. Graph $(f + g)(x)$ using the window $[-1, 10]$ by $[0, 200]$. What is the domain of $f + g$?

b. Graph $\left(\dfrac{f}{g}\right)(x)$ using the window $[5, 15]$ by $[0, 0.016]$. What is the domain of $\dfrac{f}{g}$?

c. Compute $(f \circ g)(-3)$ and $(g \circ f)(9)$.

SOLUTION

a. Functions f and g are entered as Y_1 and Y_2 in a graphing calculator and the sum of f and g is entered in Y_3, as shown in Figure 4.29(a). The graph of $(f + g)(x)$ is shown in Figure 4.29(b).

(a)   (b)

Figure 4.29

The domain of $(f + g)(x)$ is $x \geq 5$ because values of x less than 5 make $f(x) + g(x)$ undefined. This can also be seen in the graph in Figure 4.29(b).

b. The quotient of f and g is entered in Y_3, as shown in Figure 4.30(a). The graph of $\left(\dfrac{f}{g}\right)(x)$ is shown in Figure 4.30(b). The domain of $\left(\dfrac{f}{g}\right)(x)$ is $x \geq 5$ because values of x less than 5 make $\dfrac{f(x)}{g(x)}$ undefined. This can also be seen in the graph in Figure 4.30(b).

(a)　　　　　　(b)

Figure 4.30

c. To compute $(f \circ g)(-3)$, we enter $Y_1(Y_2(-3))$, obtaining 3 (Figure 4.31(a)). To compute $(g \circ f)(9)$, we enter $Y_2(Y_1(9))$, obtaining 4 (Figure 4.31(b)).

(a)　　　　　　(b)

Figure 4.31

SKILLS CHECK 4.2

In Exercises 1–8, find the following: (a) $(f + g)(x)$, (b) $(f - g)(x)$, (c) $(f \cdot g)(x)$, (d) $\left(\dfrac{f}{g}\right)(x)$, and (e) the domain of $\dfrac{f}{g}$.

1. $f(x) = 3x - 5; g(x) = 4 - x$

2. $f(x) = 2x - 3; g(x) = 5 - x$

3. $f(x) = x^2 - 2x; g(x) = 1 + x$

4. $f(x) = 2x^2 - x; g(x) = 2x + 1$

5. $f(x) = \dfrac{1}{x}; g(x) = \dfrac{x + 1}{5}$

6. $f(x) = \dfrac{x - 2}{3}; g(x) = \dfrac{1}{x}$

7. $f(x) = \sqrt{x}; g(x) = 1 - x^2$

8. $f(x) = x^3; g(x) = \sqrt{x + 3}$

9. If $f(x) = x^2 - 5x$ and $g(x) = 6 - x^3$, evaluate

    a. $(f + g)(2)$　　　　**b.** $(g - f)(-1)$

    c. $(f \cdot g)(-2)$　　　**d.** $\left(\dfrac{g}{f}\right)(3)$

10. If $f(x) = 4 - x^2$ and $g(x) = x^3 + x$, evaluate

    a. $(f + g)(1)$　　　　**b.** $(f - g)(-2)$

    c. $(f \cdot g)(-3)$　　　**d.** $\left(\dfrac{g}{f}\right)(2)$

In Exercises 11–20, find (a) $(f \circ g)(x)$ and (b) $(g \circ f)(x)$.

11. $f(x) = 2x - 6; g(x) = 3x - 1$

12. $f(x) = 3x - 2; g(x) = 2x - 2$

13. $f(x) = x^2; g(x) = \dfrac{1}{x}$ **14.** $f(x) = x^3; g(x) = \dfrac{2}{x}$

15. $f(x) = \sqrt{x - 1}; g(x) = 2x - 7$

16. $f(x) = \sqrt{3 - x}; g(x) = x - 5$

17. $f(x) = |x - 3|; g(x) = 4x$

18. $f(x) = |4 - x|; g(x) = 2x + 1$

19. $f(x) = \dfrac{3x + 1}{2}; g(x) = \dfrac{2x - 1}{3}$

20. $f(x) = \sqrt[3]{x + 1}; g(x) = x^3 + 1$

In Exercises 21 and 22, use f(x) and g(x) to evaluate each expression.

21. $f(x) = 2x^2; g(x) = \dfrac{x - 5}{3}$

   a. $(f \circ g)(2)$ **b.** $(g \circ f)(-2)$

22. $f(x) = (x - 1)^2; g(x) = 3x - 1$

   a. $(f \circ g)(2)$ **b.** $(g \circ f)(-2)$

In Exercises 23 and 24, use the following graphs of f and g to evaluate the functions.

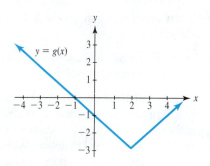

23. a. $(f + g)(2)$ **b.** $(f \circ g)(-1)$

   c. $\left(\dfrac{f}{g}\right)(4)$ **d.** $(f \circ g)(1)$

   e. $(g \circ f)(-2)$

24. a. $(g - f)(-2)$ **b.** $(f \circ g)(3)$

   c. $\left(\dfrac{f}{g}\right)(0)$ **d.** $(f \circ g)(-2)$

   e. $(g \circ f)(2)$

EXERCISES 4.2

25. Profit Suppose that the total weekly cost for the production and sale of x bicycles is $C(x) = 23x + 3420$ dollars and that the total revenue is given by $R(x) = 89x$ dollars, where x is the number of bicycles.

   a. Write the equation of the function that models the weekly profit from the production and sale of x bicycles.

   b. What is the profit on the production and sale of 150 bicycles?

   c. Write the function that gives the average profit per bicycle.

   d. What is the average profit per bicycle if 150 are produced and sold?

26. Profit Suppose that the total weekly cost for the production and sale of televisions is $C(x) = 189x + 5460$ and that the total revenue is given by

$R(x) = 988x$, where x is the number of televisions and $C(x)$ and $R(x)$ are in dollars.

   a. Write the equation of the function that models the weekly profit from the production and sale of x televisions.

   b. What is the profit on the production and sale of 80 television sets in a given week?

   c. Write the function that gives the average profit per television.

   d. What is the average profit per television if 80 are produced and sold?

27. Revenue and Cost The total revenue function for LED TVs is given by $R = 1050x$ dollars, and the total cost function for the TVs is $C = 10,000 + 30x + x^2$ dollars, where x is the number of TVs that are produced and sold.

   a. Which function is quadratic, and which is linear?

b. Form the profit function for the TVs from these two functions.

c. Is the profit function a linear function, a quadratic function, or neither of these?

28. *Revenue and Cost* The total monthly revenue function for Easy-Ride golf carts is given by $R = 26{,}600x$ dollars, and the total monthly cost function for the carts is $C = 200{,}000 + 4600x + 2x^2$ dollars, where x is the number of golf carts that are produced and sold.

a. Which function is quadratic, and which is linear?

b. Form the profit function for the golf carts from these two functions.

c. Is the profit function a linear function, a quadratic function, or neither of these?

29. *Revenue and Cost* The total weekly revenue function for a certain digital camera is given by $R = 550x$ dollars, and the total weekly cost function for the cameras is $C = 10{,}000 + 30x + x^2$ dollars, where x is the number of cameras that are produced and sold.

a. Find the profit function.

b. Find the number of cameras that gives maximum profit.

c. Find the maximum possible profit.

30. *Revenue and Cost* The total monthly revenue function for camcorders is given by $R = 6600x$ dollars, and the total monthly cost function for the camcorders is $C = 2000 + 4800x + 2x^2$ dollars, where x is the number of camcorders that are produced and sold.

a. Find the profit function.

b. Find the number of camcorders that gives maximum profit.

c. Find the maximum possible profit.

31. *Average Cost* If the monthly total cost of producing 27-inch television sets is given by $C(x) = 50{,}000 + 105x$, where x is the number of sets produced per month, then the average cost per set is given by

$$\overline{C}(x) = \frac{50{,}000 + 105x}{x}$$

a. Explain how $C(x)$ and another function can be combined to obtain the average cost function.

b. What is the average cost per set if 3000 sets are produced?

32. *Cost-Benefit* Suppose that for a certain city the cost C of obtaining drinking water that contains $p\%$ impurities (by volume) is given by

$$C = \frac{120{,}000}{p} - 1200$$

a. This function can be considered as the difference of what two functions?

b. What is the cost of drinking water that is 80% impure?

33. *Printers* The weekly total cost function for producing a laser printer is $C(x) = 3000 + 72x$, where x is the number of printers produced per week.

a. Form the weekly average cost function for this product.

b. Find the average cost for the production of 100 printers.

34. *Electronic Components* The monthly cost of producing x electronic components is $C(x) = 2.15x + 2350$.

a. Find the monthly average cost function.

b. Find the average cost for the production of 100 components.

35. *Football Tickets* At a certain school, the number of student tickets sold for a home football game can be modeled by $S(p) = 62p + 8500$, where p is the winning percent of the home team. The number of nonstudent tickets sold for these home games is given by $N(p) = 0.5p^2 + 16p + 4400$.

a. Write an equation for the total number of tickets sold for a home football game at this school as a function of the winning percent p.

b. What is the domain for the function in part (a) in this context?

c. Assuming that the football stadium is filled to capacity when the team wins 90% of its home games, what is the capacity of the school's stadium?

36. *T-Shirt Sales* Let $T(c)$ be the number of T-shirts that are sold when the shirts have c colors and $P(c)$ be the price, in dollars, of a T-shirt that has c colors. Write a sentence explaining the meaning of the function $(T \cdot P)(c)$.

37. *Harvesting* A farmer's main cash crop is tomatoes, and the tomato harvest begins in the month of May. The number of bushels of tomatoes harvested on the xth day of May is given by the equation $B(x) = 6(x + 1)^{3/2}$. The market price in dollars of 1 bushel of tomatoes on the xth day of May is given by the formula $P(x) = 8.5 - 0.12x$.

a. How many bushels did the farmer harvest on May 8?

b. What was the market price of tomatoes on May 8?

c. How much was the farmer's tomato harvest worth on May 8?

d. Write a model for the worth W of the tomato harvest on the xth day of May.

38. Total Cost If the fixed cost for producing a product is $3000 and the variable cost for the production is $3.30x^2$ dollars, where x is the number of units produced, form the total cost function $C(x)$.

39. Profit A manufacturer of satellite systems has monthly fixed costs of $32,000 and variable costs of $432 per system, and it sells the systems for $592 per unit.

a. Write the function that models the profit P from the production and sale of x units of the system.

b. What is the profit if 600 satellite systems are produced and sold in 1 month?

c. At what rate does the profit grow as the number of units increases?

40. Profit A manufacturer of computers has monthly fixed costs of $87,500 and variable costs of $87 per computer, and it sells the computers for $295 per unit.

a. Write the function that models the profit P from the production and sale of x computers.

b. What is the profit if 700 computers are produced and sold in 1 month?

c. What is the y-intercept of the graph of the profit function? What does it mean?

41. Educational Attainment The table below shows the percent of male and female individuals aged 16 to 24 enrolled in college as of October of each year who completed high school during the preceding 12 months.

Year	Male (%)	Female (%)
2000	59.9	66.2
2001	60.1	63.5
2002	62.1	68.4
2003	61.2	66.5
2004	61.4	71.5
2005	66.5	70.4
2006	65.8	66.1
2007	66.1	68.3
2008	65.9	71.6
2009	66.0	73.8
2010	62.8	74.0
2011	64.7	72.2
2012	61.3	71.3

(Source: National Center for Education Statistics)

a. Will adding the percents for males and females create a new function that gives the total percent of all 16–24-year-old high school graduates enrolled in college? Why or why not?

b. A function that approximately models the total percent of all 16–24-year-old high school graduates enrolled in college is $f(t) = 61.925t^{0.041}$, where t is the number of years after 1999. Use this model to estimate the total percent in 2002 and 2012.

c. Add the percent of males and the percent of females for the 2002 data and the 2012 data in the table and divide by 2. Do these values conflict with your answers in part (b)?

42. Population of Children The following table gives the estimated population (in millions) of U.S. boys age 5 and under and U.S. girls age 5 and under in selected years.

	1995	2000	2005	2010
Boys	10.02	9.71	9.79	10.24
Girls	9.57	9.27	9.43	9.77

(Source: U.S. Department of Commerce)

A function that models the population (in millions) of U.S. boys age 5 and under t years after 1990 is $B(t) = 0.0076t^2 - 0.1752t + 10.705$, and a function that models the population (in millions) of U.S. girls age 5 and under t years after 1990 is $G(t) = 0.0064t^2 - 0.1448t + 10.12$.

a. Find the equation of a function that models the estimated U.S. population (in millions) of children age 5 and under t years after 1990.

b. Use the result of part (a) to estimate the U.S. population of children age 5 and under in 2003.

43. Function Composition Think of each of the following processes as a function designated by the indicated letter: f, placing in a styrofoam container; g, grinding. Describe each of the functions in parts (a) to (e), then answer part (f).

a. $f(\text{meat})$ **b.** $g(\text{meat})$ **c.** $(g \circ g)(\text{meat})$

d. $f(g(\text{meat}))$ **e.** $g(f(\text{meat}))$

f. Which gives a sensible operation: the function in part (d) or the function in part (e)?

44. Function Composition Think of each of the following processes as a function designated by the indicated letter: f, putting on a sock; g, taking off a sock. Describe each of the functions in parts (a) to (c).

a. $f(\text{left foot})$ **b.** $f(f(\text{left foot}))$

c. $(g \circ f)(\text{right foot})$

45. **Shoe Sizes** A woman's shoe that is size x in Japan is size $s(x)$ in the United States, where $s(x) = x - 17$. A woman's shoe that is size x in the United States is size $p(x)$ in Britain, where $p(x) = x - 1.5$. Find a function that will convert Japanese shoe size to British shoe size.
(Source: Kuru International Exchange Association)

46. **Shoe Sizes** A man's shoe that is size x in Britain is size $d(x)$ in the United States, where $d(x) = x + 0.5$. A man's shoe that is size x in the United States is size $t(x)$ in Continental size, where $t(x) = x + 34.5$. Find a function that will convert British shoe size to Continental shoe size.
(Source: Kuru International Exchange Association)

47. **Exchange Rates** On a certain date, each Japanese yen was worth 0.34954 Russian ruble and each Chilean peso was worth 0.171718 Japanese yen. Find the value of 100 Chilean pesos in Russian rubles at that time. Round the answer to two decimal places.
(Source: x-rates.com)

48. **Exchange Rates** On a certain date, each euro was worth 1.3773 U.S. dollars and each Mexican peso was worth 0.06047 euro. Find the value of 100 Mexican pesos in U.S. dollars at that time.

49. **Facebook** If $f(x)$ represents the number of Facebook users x years after 2000 and $g(x)$ represents the number of MySpace users x years after 2000, what function gives the ratio of MySpace users to Facebook users x years after 2000?

50. **Home Computers** If $f(x)$ represents the percent of American homes with computers and $g(x)$ represents the number of American homes, with x equal to the number of years after 1990, then what function represents the number of American homes with computers, with x equal to the number of years after 1990?

51. **Education** If the function $f(x)$ gives the number of female PhDs produced by American universities x years after 2005 and the function $g(x)$ gives the number of male PhDs produced by American universities x years after 2005, what function gives the total number of PhDs produced by American universities x years after 2005?

52. **Wind Chill** If the air temperature is 25°F, the wind chill temperature C is given by $C = 59.914 - 2.35s - 20.14\sqrt{s}$, where s is the wind speed in miles per hour.

    a. State two functions whose difference gives this function.

    b. Graph this function for $3 \le s \le 12$.

    c. Is the function increasing or decreasing on this domain?

53. **Discount Prices** Half-Price Books has a sale with an additional 20% off the regular $(1/2)$ price of their books. What percent of the retail price is charged during this sale?
(Source: Half-Price Books, Cleveland, Ohio)

54. **Average Cost** The monthly average cost of producing 42-inch plasma TVs is given by $\overline{C}(x) = \dfrac{100,000}{x} + 150$ dollars, where x is the number of sets produced.

    a. Graph this function for $x > 0$.

    b. Will the average cost function decrease or increase as the number of sets produced increases?

    c. What transformations of the graph of the reciprocal function give the graph of this function?

4.3   One-to-One and Inverse Functions

KEY OBJECTIVES

- Determine whether two functions are inverses
- Determine whether a function is one-to-one
- Find the inverse of a function
- Graph inverse functions
- Find inverse functions on limited domains

SECTION PREVIEW   Investment

If $1000 is invested for three years at interest rate x, compounded annually, the future value y of the investment is given by

$$y = f(x) = 1000(1 + x)^3$$

If we want a specific future value to be available in three years, and we want to know what interest rate is necessary to get this amount, we can find the inverse of the above function. (See Example 6.)

In this section, we determine whether two functions are inverses of each other and find the inverse of a function if it has one. We also solve applied problems by using inverse functions. ■

Inverse Functions

Suppose a property costing $153,000 is purchased with the promise of paying $1275 per month for 10 years. If all payments are made on time, the amount remaining to be paid when x monthly payments have been made is

$$f(x) = 153,000 - 1275x$$

and the function that can be used to find directly how many monthly payments have been made for any given amount x remaining to be paid is

$$g(x) = \frac{153,000 - x}{1275}$$

To see that one of these functions "undoes" what the other one does, see Table 4.8, which gives selected inputs and the resulting outputs for both functions. Observe that if we input the number 80 into $f(x)$, the output after it is operated on by f is 51,000, and if 51,000 is then operated on by g, the output is 80, which means that $g(f(80)) = 80$. In addition, if 51,000 is operated on by g, the output is 80, and if 80 is operated on by f, the output is 51,000, so $f(g(51,000)) = 51,000$.

Table 4.8

x	0	40	80	120
f(x)	153,000	102,000	51,000	0

x	153,000	102,000	51,000	0
g(x)	0	40	80	120

In fact, $g(f(x)) = x$ and $f(g(x)) = x$ for any input x. This means that both of these composite functions act as identity functions because their outputs are the same as their inputs. This happens because the second function performs the inverse operations of the first function. Because of this, we say that $g(x)$ and $f(x)$ are **inverse functions**.

Inverse Functions

Functions f and g for which $f(g(x)) = x$ for all x in the domain of g and $g(f(x)) = x$ for all x in the domain of f are called inverse functions. In this case, we denote g by f^{-1}, read as "f inverse."

EXAMPLE 1 ▶ **Temperature Conversion**

To convert from x degrees Celsius to y degrees Fahrenheit, we can use the function $y = f(x) = \frac{9}{5}x + 32$. To convert x degrees Fahrenheit to y degrees Celsius, we can use the function $y = g(x) = \frac{5x - 160}{9}$. Show that f and g are inverse functions.

SOLUTION

To show that these functions are inverse functions, we show that $f(g(x)) = x$ and $g(f(x)) = x$.

$$f(g(x)) = f\left(\frac{5x - 160}{9}\right) = \frac{9}{5}\left(\frac{5x - 160}{9}\right) + 32 = \frac{5x - 160}{5} + 32$$
$$= x - 32 + 32 = x$$

$$g(f(x)) = g\left(\frac{9}{5}x + 32\right) = \frac{5\left(\frac{9}{5}x + 32\right) - 160}{9} = \frac{9x + 160 - 160}{9} = \frac{9x}{9} = x$$

Thus the functions are inverse functions.

Consider a function that doubles each input. Its inverse takes half of each input. For example, the discrete function f with equation $f(x) = 2x$ and domain $\{1, 4, 5, 7\}$ has the range $\{2, 8, 10, 14\}$. The inverse of this function has the equation $f^{-1}(x) = \frac{x}{2}$.

The domain of the inverse function is the set of outputs of the original function, $\{2, 8, 10, 14\}$. The outputs of the inverse function form the set $\{1, 4, 5, 7\}$, which is the domain of the original function. Figure 4.32 illustrates the relationship between the domains and ranges of the function f and its inverse f^{-1}. In this case and in every case, the domain of the inverse function is the range of the original function, and the range of the inverse function is the domain of the original function.

Figure 4.32

It is very important to note that the "$^{-1}$" used in f^{-1} and g^{-1} is *not* an exponent, but rather a symbol used to denote the inverse of the function. The expression f^{-1} *always* refers to the inverse function of f and *never* to the reciprocal $\frac{1}{f}$ of f; that is, $f^{-1}(x) \neq \frac{1}{f(x)}$.

One-to-One Functions

In general, we can show that a function has an inverse if it is a **one-to-one function**. We define a one-to-one function as follows.

One-to-One Function

A function f is a one-to-one function if each output of the function corresponds to exactly one input in the domain of the function. This means there is a one-to-one correspondence between the elements of the domain and the elements of the range.

This statement means that for a one-to-one function f, $f(a) \neq f(b)$ if $a \neq b$.

EXAMPLE 2 ▶ **One-to-One Functions**

Determine if each of the functions is a one-to-one function.

a. $f(x) = 3x^4$ **b.** $f(x) = x^3 - 1$

SOLUTION

a. Clearly $a = -2$ and $b = 2$ are different inputs, but they both give the same output for $f(x) = 3x^4$.

$$f(-2) = 3(-2)^4 = 48 \quad \text{and} \quad f(2) = 3(2)^4 = 48$$

Thus, $y = 3x^4$ is not a one-to-one function.

b. Suppose that $a \neq b$. Then $a^3 \neq b^3$ and $a^3 - 1 \neq b^3 - 1$, so if $f(x) = x^3 - 1$, $f(a) \neq f(b)$. This satisfies the condition $f(a) \neq f(b)$ if $a \neq b$, so the function $f(x) = x^3 - 1$ is one-to-one.

Recall that no vertical line can intersect the graph of a function in more than one point. The definition of a one-to-one function means that if a function is one-to-one, a horizontal line will intersect its graph in at most one point.

> **Horizontal Line Test**
>
> A function is one-to-one if no horizontal line can intersect the graph of the function in more than one point.

EXAMPLE 3 ▶ **Horizontal Line Test**

Determine if each of the functions is one-to-one by using the horizontal line test.

a. $y = 3x^4$ **b.** $y = x^3 - 1$

SOLUTION

a. We can see that $y = 3x^4$ is not a one-to-one function because we can observe that the graph of this function does not pass the horizontal line test (Figure 4.33(a)). Note that when $x = 2$, $y = 48$, and when $x = -2$, $y = 48$.

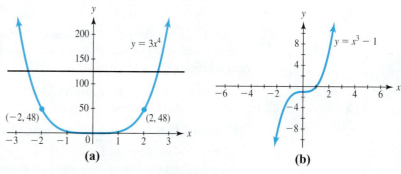

Figure 4.33

b. The graph in Figure 4.33(b) and the horizontal line test indicate that the function $y = x^3 - 1$ is one-to-one. That is, no horizontal line will intersect this graph in more than one point.

We summarize the information about inverse functions as follows.

Inverse Functions

The functions f and g are inverse functions if, whenever the pair (a, b) satisfies $y = f(x)$, the pair (b, a) satisfies $y = g(x)$. Note that when this happens,

$$f(g(x)) = x \quad \text{and} \quad g(f(x)) = x$$

for all x in the domain of g and f, respectively. The domain of the function f is the range of its inverse g, and the domain of g is the range of f.

EXAMPLE 4 ▶   Inverse Functions

a. Determine whether $f(x) = x^3 - 1$ has an inverse function.

b. Verify that $g(x) = \sqrt[3]{x} + 1$ is the inverse function of $f(x) = x^3 - 1$.

c. Find the domain and range of each function.

SOLUTION

a. Because each output of the function $f(x) = x^3 - 1$ corresponds to exactly one input, the function is one-to-one. Thus, it has an inverse function.

b. To verify that the functions $f(x) = x^3 - 1$ and $g(x) = \sqrt[3]{x} + 1$ are inverse functions, we show that the compositions $f(g(x))$ and $g(f(x))$ are each equal to the identity function, x.

$$f(g(x)) = f(\sqrt[3]{x} + 1) = (\sqrt[3]{x} + 1)^3 - 1 = (x + 1) - 1 = x \quad \text{and}$$

$$g(f(x)) = g(x^3 - 1) = \sqrt[3]{(x^3 - 1) + 1} = \sqrt[3]{x^3} = x$$

Thus, we have verified that these functions are inverses.

c. The cube of any real number decreased by 1 is a real number, so the domain of f is the set of all real numbers. The cube root of any real number plus 1 is also a real number, so the domain of g is the set of all real numbers. The ranges of f and g are the set of real numbers.

By using the definition of inverse functions, we can find the equation for the inverse function of f by interchanging x and y in the equation $y = f(x)$ and solving the new equation for y.

Finding the Inverse of a Function

To find the inverse of the function f that is defined by the equation $y = f(x)$:

1. Rewrite the equation replacing $f(x)$ with y.

2. Interchange x and y in the equation defining the function.

3. Solve the new equation for y. If this equation cannot be solved uniquely for y, the original function has no inverse function.

4. Replace y with $f^{-1}(x)$.

EXAMPLE 5 ▶ Finding an Inverse Function

a. Find the inverse function of $f(x) = \dfrac{2x-1}{3}$.

b. Graph $f(x) = \dfrac{2x-1}{3}$ and its inverse function on the same axes.

SOLUTION

a. Using the steps for finding the inverse of a function, we have

$$y = \frac{2x-1}{3} \qquad \text{Replace } f(x) \text{ with } y.$$

$$x = \frac{2y-1}{3} \qquad \text{Interchange } x \text{ and } y.$$

$$3x = 2y - 1 \qquad \text{Solve for } y.$$

$$3x + 1 = 2y$$

$$\frac{3x+1}{2} = y$$

$$f^{-1}(x) = \frac{3x+1}{2} \qquad \text{Replace } y \text{ with } f^{-1}(x).$$

b. The graphs of $f(x) = \dfrac{2x-1}{3}$ and its inverse $f^{-1}(x) = \dfrac{3x+1}{2}$ are shown in Figure 4.34.

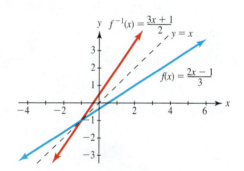

Figure 4.34

Notice that the graphs of $y = f(x)$ and $y = f^{-1}(x)$ in Figure 4.34 appear to be reflections of each other about the line $y = x$. This occurs because if you were to choose any point (a, b) on the graph of the function f and interchange the x- and y-coordinates, the new point (b, a) would be on the graph of the inverse function f^{-1}. This should make sense because the inverse function is formed by interchanging x and y in the equation defining the original function. In fact, this relationship occurs for every function and its inverse.

Graphs of Inverse Functions

The graphs of a function and its inverse are symmetric with respect to the line $y = x$.

We again illustrate the symmetry of graphs of inverse functions in Figure 4.35, which shows the inverse functions $f(x) = x^3 - 2$ and $f^{-1}(x) = \sqrt[3]{x} + 2$. Note that the point $(0, -2)$ is on the graph of $f(x) = x^3 - 2$ and the point $(-2, 0)$ is on the graph of $f^{-1}(x) = \sqrt[3]{x} + 2$.

Figure 4.35

EXAMPLE 6 ▶   **Investment**

If $1000 is invested for 3 years at interest rate x, compounded annually, the future value y of the investment is given by

$$y = f(x) = 1000(1 + x)^3$$

a. To find the interest rate x that results in future value y, find the inverse function of

$$f(x) = 1000(1 + x)^3$$

b. Use the inverse function to find the interest rate that gives a future value of $1331 in 3 years.

SOLUTION

a. To find the inverse of $f(x) = 1000(1 + x)^3$, we proceed as follows.

1. Replace $f(x)$ with y. $y = 1000(1 + x)^3$

2. Interchange x and y. $x = 1000(1 + y)^3$

3. Solve this new equation for y.

$$\frac{x}{1000} = (1 + y)^3$$

$$\sqrt[3]{\frac{x}{1000}} = \sqrt[3]{(1 + y)^3}$$

$$\frac{\sqrt[3]{x}}{\sqrt[3]{1000}} = 1 + y$$

$$y = \frac{\sqrt[3]{x}}{10} - 1$$

4. Replace y with $f^{-1}(x)$.

$$f^{-1}(x) = \frac{\sqrt[3]{x}}{10} - 1$$

b. To find the interest rate that gives a future value of $1331 in 3 years, we evaluate

$$y = f^{-1}(x) = \frac{\sqrt[3]{x}}{10} - 1$$

at $x = 1331$, getting $y = f^{-1}(1331) = \dfrac{\sqrt[3]{1331}}{10} - 1 = \dfrac{11}{10} - 1 = 0.10 = 10\%$.

Inverse Functions on Limited Domains

As we have stated, a function cannot have an inverse function if it is not a one-to-one function. However, if there is a limited domain over which such a function is a one-to-one function, then it has an inverse function for this domain. Consider the function $f(x) = x^2$. The horizontal line test on the graph of this function (Figure 4.36(a)) indicates that this function is not a one-to-one function and that the function f does not have an inverse function. To see why, consider the attempt to find the inverse function:

$$y = x^2$$
$$x = y^2 \qquad \text{Interchange } x \text{ and } y.$$
$$\pm\sqrt{x} = y \qquad \text{Solve for } y.$$

This equation is not the inverse function because it is not a function (one value of x gives two values for y). Because we cannot solve $x = y^2$ uniquely for y, $f(x) = x^2$ does not have an inverse function.

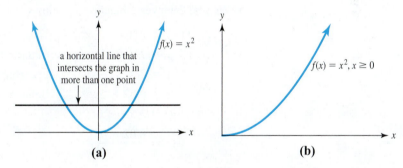

(a)   (b)

Figure 4.36

However, if we limit the domain of the function to $x \geq 0$, no horizontal line intersects the graph of $f(x) = x^2$ more than once, so the function is one-to-one on this limited domain (Figure 4.36(b)), and the function has an inverse. If we restrict the domain of the original function by requiring that $x \geq 0$, then the range of the inverse function is restricted to $y \geq 0$, and the equation $y = \sqrt{x}$, or $f^{-1}(x) = \sqrt{x}$, defines the inverse function. Figure 4.37 shows the graph of $f(x) = x^2$ for $x \geq 0$ and its inverse $f^{-1}(x) = \sqrt{x}$. Note that the graphs are symmetric about the line $y = x$.

Figure 4.37

EXAMPLE 7 ▶ **Velocity of Blood**

Because of friction from the walls of an artery, the velocity of blood is greatest at the center of the artery and decreases as the distance x from the center increases. The (nonnegative) velocity in centimeters per second of a blood corpuscle in an artery can be modeled by the function

$$v = k(R^2 - x^2), k > 0, 0 \le x \le R$$

where R is the radius of the artery and k is a constant that is determined by the pressure, the viscosity of the blood, and the length of the artery. In the case where $k = 2$ and $R = 0.1$ centimeter, the velocity can be written as a function of the distance x from the center as

$$v(x) = 2(0.01 - x^2) \text{ centimeter/second}$$

a. Assuming the velocity is nonnegative, does the inverse of this function exist on a domain limited by the context of the application?

b. What is the inverse function?

c. What does the inverse function mean in the context of this application?

SOLUTION

a. Because x represents distance, it is nonnegative, and because $R = 0.1$ and x equals the distance of blood in an artery from the center of the artery, $0 \le x \le 0.1$. Hence, the function is one-to-one for $0 \le x \le 0.1$ and we can find its inverse.

b. We find the inverse function as follows.

$$f(x) = 2(0.01 - x^2)$$
$$y = 2(0.01 - x^2) \qquad \text{Replace } f(x) \text{ with } y.$$
$$x = 2(0.01 - y^2) \qquad \text{Interchange } y \text{ and } x.$$
$$x = 0.02 - 2y^2 \qquad \text{Solve for } y.$$
$$2y^2 = 0.02 - x$$
$$y^2 = \frac{0.02 - x}{2}$$
$$y = \sqrt{\frac{0.02 - x}{2}} \qquad \text{Only nonnegative values of } y \text{ are possible.}$$
$$f^{-1}(x) = \sqrt{\frac{0.02 - x}{2}} \qquad \text{Replace } y \text{ with } f^{-1}(x).$$

c. The inverse function gives the distance from the center of the artery as a function of the velocity of a blood corpuscle.

SKILLS CHECK 4.3

In Exercises 1 and 2, determine if the function f defined by the arrow diagram has an inverse. If it does, create an arrow diagram that defines the inverse. If it does not, explain why not.

1.

2.

In Exercises 3 and 4, determine whether the function f defined by the set of ordered pairs has an inverse. If it does, find the inverse.

3. $\{(5,2),(4,1),(3,7),(6,2)\}$

4. $\{(2,8),(3,9),(4,10),(5,11)\}$

5. If $f(x)=3x$ and $g(x)=\dfrac{x}{3}$,

　　a. What are $f(g(x))$ and $g(f(x))$?

　　b. Are $f(x)$ and $g(x)$ inverse functions?

6. If $f(x)=4x-1$ and $g(x)=\dfrac{x+1}{4}$,

　　a. What are $f(g(x))$ and $g(f(x))$?

　　b. Are $f(x)$ and $g(x)$ inverse functions?

7. If $f(x)=x^3+1$ and $g(x)=\sqrt[3]{x-1}$, are $f(x)$ and $g(x)$ inverse functions?

8. If $f(x)=(x-2)^3$ and $g(x)=\sqrt[3]{x}+2$, are $f(x)$ and $g(x)$ inverse functions?

9. For the function f defined by $f(x)=3x-4$, complete the tables below for f and f^{-1}.

x	f(x)
−1	−7
0	
1	
2	
3	

x	f⁻¹(x)
−7	−1
−4	
−1	
2	
5	

10. For the function g defined by $g(x)=2x^3-1$, complete the tables below for g and g^{-1}.

x	g(x)
−2	−17
−1	
0	
1	
2	

x	g⁻¹(x)
−17	−2
−3	
−1	
1	
15	

In Exercises 11–14, determine whether the function is one-to-one.

11. $\{(1,5),(2,6),(3,7),(4,5)\}$

12. $\{(2,-4),(5,-8),(8,-12),(11,-16)\}$

13. $f(x)=(x-3)^3$　　**14.** $f(x)=\dfrac{1}{x}$

In Exercises 15 and 16, determine whether each graph is the graph of a one-to-one function.

15.

16.

In Exercises 17 and 18, determine whether the function is one-to-one.

17. $y=-2x^4$　　　　　**18.** $y=\sqrt{x+3}$

19. a. Write the inverse of $f(x)=3x-4$.

　　b. Do the values for f^{-1} in the table of Exercise 9 fit the equation for f^{-1}?

20. a. Write the inverse of $g(x)=2x^3-1$.

　　b. Do the values for g^{-1} in the table of Exercise 10 fit the equation for g^{-1}?

21. If function h has an inverse and $h^{-1}(-2)=3$, find $h(3)$.

22. Find the inverse of $f(x)=\dfrac{1}{x}$.

23. Find the inverse of $g(x)=4x+1$.

24. Find the inverse of $f(x)=4x^2$ for $x\ge 0$.

25. Find the inverse of $g(x)=x^2-3$ for $x\ge 0$.

26. Graph $g(x)=\sqrt{x}$ and its inverse $g^{-1}(x)$ for $x\ge 0$ on the same axes.

27. Graph $g(x)=\sqrt[3]{x}$ and its inverse $g^{-1}(x)$ on the same axes.

28. $f(x)=(x-2)^2$ and $g(x)=\sqrt{x}+2$ are inverse functions for what values of x?

29. Is the function $f(x)=2x^3+1$ a one-to-one function? Does it have an inverse?

30. Sketch the graph of $y = f^{-1}(x)$ on the axes with the graph of $y = f(x)$, shown below.

31. Sketch the graph of $y = f^{-1}(x)$ on the axes with the graph of $y = f(x)$, shown below.

EXERCISES 4.3

32. *Shoe Sizes* If x is the size of a man's shoe in Britain, then $d(x) = x + 0.5$ is its size in the United States.

 a. Find the inverse of the function.

 b. Use the inverse function to find the British size of a shoe if it is U.S. size $8\frac{1}{2}$.

(Source: Kuru International Exchange Association)

33. *Shoe Sizes* If x is the size of a man's shoe in the United States, then $t(x) = x + 34.5$ is its Continental size.

 a. Find a function that will convert Continental shoe size to U.S. shoe size.

 b. Use the inverse function to find the U.S. size if the Continental size of a shoe is 43.

(Source: Kuru International Exchange Association)

34. *Investments* If x dollars are invested at 10% for 6 years, the future value of the investment is given by $S(x) = x + 0.6x$.

 a. Find the inverse of this function.

 b. What do the outputs of the inverse function represent?

 c. Use this function to find the amount of money that must be invested for 6 years at 10% to have a future value of $24,000.

35. *Currency Conversion* Suppose the function that converts from Canadian dollars to U.S. dollars is $f(x) = 1.0136x$, where x is the number of Canadian dollars and $f(x)$ is the number of U.S. dollars.

 a. Find the inverse function for f and interpret its meaning.

 b. Use f and f^{-1} to determine the money you will have if you take 500 U.S. dollars to Canada, convert them to Canadian dollars, don't spend any, and then convert them back to U.S. dollars. (Assume that there is no fee for conversion and the conversion rate remains the same.)

(Source: Expedia.com)

36. *Apparent Temperature* If the outside temperature is 90°F, the apparent temperature is given by $A(x) = 82.35 + 29.3x$, where x is the humidity written as a decimal. Find the inverse of this function and use it to find the percent humidity that will give an apparent temperature of 97° if the temperature is 90°F.

(Source: "Temperature-Humidity Indices," *The UMAP Journal*, Fall 1989)

37. *Antidepressants* The function that models the percent of children ages 0–19 taking antidepressants from 2004 to 2009 is $f(x) = -0.085x + 2.97$, where x is the number of years after 2000.

 a. Find the inverse of this function. What do the outputs of the inverse function represent?

 b. Use the inverse function to find when the percentage is 2.3%.

(Source: Medco Health Solutions)

38. *Body-Heat Loss* The model for body-heat loss depends on the coefficient of convection $K = f(x)$, which depends on wind speed x according to the equation $f(x) = 4\sqrt{4x + 1}$.

 a. What are the domain and range of this function without regard to the context of this application?

 b. Find the inverse of this function.

 c. What are the domain and range of the inverse function?

d. In the context of the application, what are the domain and range of the inverse function?

39. *Algorithmic Relationship* For many species of fish, the weight W is a function of the length x, given by $W = kx^3$, where k is a constant depending on the species. Suppose $k = 0.002$, W is in pounds, and x is in inches, so the weight is $W(x) = 0.002x^3$.

a. Find the inverse function of this function.

b. What does the inverse function give?

c. Use the inverse function to find the length of a fish that weighs 2 pounds.

d. In the context of the application, what are the domain and range of the inverse function?

40. *Decoding Messages* If we assign numbers to the letters of the alphabet as follows and assign 27 to a blank space, we can convert a message to a numerical sequence. We can "encode" a message by adding 3 to each number that represents a letter in a message.

A	B	C	D	E	F	G	H	I	J	K	L	M
1	2	3	4	5	6	7	8	9	10	11	12	13

N	O	P	Q	R	S	T	U	V	W	X	Y	Z
14	15	16	17	18	19	20	21	22	23	24	25	26

Thus, the message "Go for it" can be encoded by using the numbers to represent the letters and further encoded by using the function $C(x) = x + 3$. The coded message would be 10 18 30 9 18 21 30 12 23. Find the inverse of the function and use it to decode 23 11 8 30 21 8 4 15 30 23 11 12 17 10.

41. *Decoding Messages* Use the numerical representation from Exercise 40 and the inverse of the encoding function $C(x) = 3x + 2$ to decode 41 5 35 17 83 41 77 83 14 5 77.

42. *Social Security Numbers and Income Taxes* Consider the function that assigns each person who pays federal income tax his or her Social Security number. Is this a one-to-one function? Explain.

43. *Checkbook Balance* Consider the function with the check number in your checkbook as input and the dollar amount of the check as output. Is this a one-to-one function? Explain.

44. *Volume of a Cube* The volume of a cube is $f(x) = x^3$ cubic inches, where x is the length of the edge of the cube in inches.

a. Is this function one-to-one?

b. Find the inverse of this function.

c. What are the domain and range of this inverse function in the context of the application?

d. How could this inverse function be used?

45. *Volume of a Sphere* The volume of a sphere is $f(x) = \frac{4}{3}\pi x^3$ cubic inches, where x is the radius of the sphere in inches.

a. Is this function one-to-one?

b. Find the inverse of this function.

c. What are the domain and range of the inverse function for this application?

d. How could this inverse function be used?

e. What is the radius of a sphere if its volume is 65,450 cubic inches?

46. *Surface Area* The surface area of a cube is $f(x) = 6x^2 \text{ cm}^2$, where x is the length of the edge of the cube in centimeters.

a. For what values of x does this model make sense? Is the model a one-to-one function for these values of x?

b. What is the inverse of this function on this interval?

c. How could the inverse function be used?

47. *Internet Access* Using data from 2009 through 2015, the percent of households with Internet access can be modeled by the function $y = 53.265x^{0.170}$, with x equal to the number of years after 2005.

a. Find the formula for the inverse of the function f.

b. If the inverse is f^{-1}, find $f^{-1}(f(18))$.

48. *Supply* The supply function for a product is $p(x) = \frac{1}{4}x^2 + 20$, where x is the number of thousands of units a manufacturer will supply if the price is $p(x)$ dollars.

a. Is this function a one-to-one function?

b. What is the domain of this function in the context of the application?

c. Is the function one-to-one for the domain in part (b)?

d. Find the inverse of this function and use it to find how many units the manufacturer is willing to supply if the price is $101.

49. *Illumination* The intensity of illumination of a light is a function of the distance from the light. For a given light, the intensity is given by $I(x) = \dfrac{300,000}{x^2}$ candlepower, where x is the distance in feet from the light.

a. Is this function a one-to-one function if it is not limited by the context of the application?

b. What is the domain of this function in the context of the application?

c. Is the function one-to-one for the domain in part (b)?

d. Find the inverse of this function on the domain from part (b) and use it to find the distance at which the intensity of the light is 75,000 candlepower.

50. *First-Class Postage* The postage charged for first-class mail is a function of its weight. The U.S. Postal Service uses the following table to describe the rates for 2015.

Weight Increment, *x*	Postal Rate
First ounce or fraction of an ounce	$0.49
Each additional ounce or fraction	$0.21

(Source: pe.usps.gov/text)

a. Convert this table to a piecewise-defined function $P(x)$ that represents postage for letters weighing more than 0 and no more than 3 ounces, using x as the weight in ounces and $P(x)$ as the postage in cents.

b. Does P have an inverse function? Why or why not?

51. *Currency Conversion* Suppose the function that converts United Kingdom (U.K.) pounds to U.S. dollars is $f(x) = 1.6249x$, where x is the number of pounds and $f(x)$ is the number of U.S. dollars.

a. Find the inverse function for f and interpret its meaning.

b. Use f and f^{-1} to determine the money you will have if you take 1000 U.S. dollars to the United Kingdom, convert them to pounds, don't spend any, and then convert them back to U.S. dollars. (Assume that there is no fee for conversion and the conversion rate remains the same.)

(Source: International Monetary Fund)

52. *Path of a Ball* If a ball is thrown into the air at a velocity of 96 feet per second from a building that is 256 feet high, the height of the ball after x seconds is $f(x) = 256 + 96x - 16x^2$ feet.

a. For how many seconds will the ball be in the air?

b. Is this function a one-to-one function over this time interval?

c. Give an interval over which the function is one-to-one.

d. Find the inverse of the function over the interval $0 \le x \le 3$. What does it give?

4.4 Additional Equations and Inequalities

KEY OBJECTIVES

- Solve radical equations
- Solve equations with rational powers
- Solve equations in quadratic form
- Solve quadratic inequalities
- Solve power inequalities
- Solve inequalities involving absolute value

SECTION PREVIEW Profit

The daily profit from the production and sale of x units of a product is given by

$$P(x) = -0.01x^2 + 20.25x - 500$$

Because a profit occurs when $P(x)$ is positive, there is a profit for those values of x that make $P(x) > 0$. Thus, the values of x that give a profit are the solutions to

$$-0.01x^2 + 20.25x - 500 > 0$$

In this section, we solve quadratic inequalities by using both analytical and graphical methods. We also solve equations and inequalities involving radicals, rational powers, and absolute values. ∎

Radical Equations; Equations Involving Rational Powers

An equation containing a radical can frequently be converted to an equation that does not contain a radical by raising both sides of the equation to an appropriate power. For example, an equation containing a square root radical can usually be converted by squaring both sides of the equation. It is possible, however, that raising both sides of an equation to a power may produce an *extraneous* solution (a value that does not satisfy the original equation). For this reason, we must check solutions when using this technique. To solve an equation containing a radical, we use the following steps.

Solving Radical Equations

1. Isolate a single radical on one side of the equation.

2. Square both sides of the equation. (Or raise both sides to a power that is equal to the index of the radical.)

3. If a radical remains, repeat steps 1 and 2.

4. Solve the resulting equation.

5. All solutions must be checked in the original equation, and only those that satisfy the original equation are actual solutions.

EXAMPLE 1 ▶ **Radical Equation**

Solve $\sqrt{x + 5} + 1 = x$.

SOLUTION

To eliminate the radical, we isolate it on one side and square both sides:

$$\sqrt{x + 5} = x - 1$$
$$x + 5 = x^2 - 2x + 1$$
$$0 = x^2 - 3x - 4$$
$$0 = (x - 4)(x + 1)$$
$$x = 4 \quad \text{or} \quad x = -1$$

Checking these values shows that 4 is a solution and -1 is not.

$x = 4$	$x = -1$
$\sqrt{4 + 5} + 1 \overset{?}{=} 4$	$\sqrt{-1 + 5} + 1 \overset{?}{=} -1$
$\sqrt{9} + 1 \overset{?}{=} 4$	$\sqrt{4} + 1 \overset{?}{=} -1$
$3 + 1 \overset{?}{=} 4$	$2 + 1 \overset{?}{=} -1$
$4 = 4$	$3 \neq -1$

We can also check graphically by the intersection method. Graphing $y = \sqrt{x + 5} + 1$ and $y = x$, we find the point of intersection to be (4, 4) (Figure 4.38). Note that there is *not* a point of intersection at $x = -1$.

Thus, the solution to the equation $\sqrt{x + 5} + 1 = x$ is $x = 4$.

Figure 4.38

EXAMPLE 2 ▶   Radical Equation

Solve $\sqrt{4x - 8} - 1 = \sqrt{2x - 5}$.

SOLUTION

We begin by confirming that one of the radicals is isolated. Then we square both sides of the equation:

$$\left(\sqrt{4x - 8} - 1 \right)^2 = \left(\sqrt{2x - 5} \right)^2$$

$$4x - 8 - 2\sqrt{4x - 8} + 1 = 2x - 5$$

$$4x - 7 - 2\sqrt{4x - 8} = 2x - 5$$

Now we must isolate the remaining radical and square both sides of the equation again:

$$2x - 2 = 2\sqrt{4x - 8}$$

$$(2x - 2)^2 = \left(2\sqrt{4x - 8} \right)^2$$

$$4x^2 - 8x + 4 = 4(4x - 8)$$

$$4x^2 - 8x + 4 = 16x - 32$$

$$4x^2 - 24x + 36 = 0$$

$$4(x^2 - 6x + 9) = 0$$

$$4(x - 3)^2 = 0$$

$$x = 3$$

We can check by substituting or graphing. Graphing $y = \sqrt{4x - 8} - 1$ and $y = \sqrt{2x - 5}$, we find the point of intersection at $x = 3$ (Figure 4.39). Thus, $x = 3$ is the solution.

Figure 4.39

Equations Containing Rational Powers

Some equations containing rational powers can be solved by writing the equation as a radical equation.

EXAMPLE 3 ▶ **Equation with Rational Powers**

Solve the equation $(x - 3)^{2/3} - 4 = 0$.

SOLUTION

Rewriting the equation as a radical equation and with the radical isolated gives

$$(x - 3)^{2/3} = 4$$
$$\sqrt[3]{(x - 3)^2} = 4$$

Cubing both sides of the equation and solving gives

$$(x - 3)^2 = 64$$
$$\sqrt{(x - 3)^2} = \pm \sqrt{64}$$
$$x - 3 = \pm 8$$
$$x = 3 \pm 8$$
$$x = 11 \quad \text{or} \quad x = -5$$

Entering each solution in the original equation shows that each checks there.

$x = 11$	$x = -5$
$(11 - 3)^{2/3} - 4 \overset{?}{=} 0$	$(-5 - 3)^{2/3} - 4 \overset{?}{=} 0$
$8^{2/3} - 4 \overset{?}{=} 0$	$(-8)^{2/3} - 4 \overset{?}{=} 0$
$(\sqrt[3]{8})^2 - 4 \overset{?}{=} 0$	$(\sqrt[3]{-8})^2 - 4 \overset{?}{=} 0$
$2^2 - 4 \overset{?}{=} 0$	$(-2)^2 - 4 \overset{?}{=} 0$
$0 = 0$	$0 = 0$

Equations in Quadratic Form

Some equations that are not quadratic can be written as quadratic equations using substitution. These equations are said to be in **quadratic form**. For example, the equation $x^4 - 5x^2 - 36 = 0$ can be written as $(x^2)^2 - 5(x^2) - 36 = 0$ and then can be more easily solved if we substitute a simpler variable for x^2. If we let $u = x^2$, then the equation $x^4 - 5x^2 - 36 = 0$ can be written in the quadratic form $u^2 - 5u - 36 = 0$. In general, an equation in quadratic form will have the form $au^2 + bu + c = 0$, with u equal to some expression. We can then solve this quadratic equation for u, substitute the expression for u, and complete the solution, as Example 4 shows.

EXAMPLE 4 ▶   **Solving an Equation in Quadratic Form**

Solve the equation $x^4 - 5x^2 - 36 = 0$.

SOLUTION

The equation contains an expression to a power, x^2, that expression to a power, $(x^2)^2$, and a constant. If we let $u = x^2$, we can solve the equation as follows.

$$x^4 - 5x^2 - 36 = 0$$

$$(x^2)^2 - 5(x^2) - 36 = 0 \qquad \text{Write the equation in quadratic form.}$$

$$u^2 - 5u - 36 = 0 \qquad \text{Substitute } u \text{ for } x^2.$$

$$(u - 9)(u + 4) = 0 \qquad \text{Factor the quadratic equation.}$$

$$u - 9 = 0 \quad u + 4 = 0 \qquad \text{Use the zero-product property.}$$

$$u = 9 \quad u = -4 \qquad \text{Solve for } u.$$

Now we have solved our new equation for u, but we must go back and solve the original equation for x. To do this, we substitute x^2 for u in the last two lines of the solution above:

$$u = 9 \qquad\qquad u = -4$$

$$x^2 = 9 \qquad\qquad x^2 = -4$$

$$x = \pm\sqrt{9} \qquad x = \pm\sqrt{-4}$$

$$x = 3, -3 \qquad x = 2i, -2i$$

We can check the solutions graphically. Figure 4.40 shows the graph of $y = x^4 - 5x^2 - 36$. Note that there are only two x-intercepts, 3 and -3. This is so because the solutions $2i$ and $-2i$ are imaginary and therefore are not real zeros of the function.

Figure 4.40

EXAMPLE 5 ▶ **Solving an Equation in Quadratic Form**

Solve the equation $x^{2/3} - x^{1/3} - 6 = 0$.

SOLUTION

Note that $x^{2/3} = (x^{1/3})^2$. The equation contains the expression $x^{1/3}$, that expression to the second power, $(x^{1/3})^2$, and a constant. If we let $u = x^{1/3}$, we can rewrite the equation as

$$x^{2/3} - x^{1/3} - 6 = 0$$
$$(x^{1/3})^2 - (x^{1/3}) - 6 = 0 \qquad \text{Write the equation in quadratic form.}$$
$$u^2 - u - 6 = 0 \qquad \text{Substitute } u \text{ for } x^{1/3}.$$
$$(u - 3)(u + 2) = 0 \qquad \text{Factor the quadratic equation.}$$
$$u = 3 \quad u = -2 \qquad \text{Solve for } u.$$
$$x^{1/3} = 3 \quad x^{1/3} = -2 \qquad \text{Substitute } x^{1/3} \text{ for } u.$$
$$x = 3^3 \quad x = (-2)^3 \qquad \text{Cube both sides to eliminate the rational exponent.}$$
$$x = 27 \quad x = -8$$

Thus the solutions are 27 and −8, which can be verified graphically (Figure 4.41).

Figure 4.41

Quadratic Inequalities

A **quadratic inequality** is an inequality that can be written in the form

$$ax^2 + bx + c > 0$$

where a, b, and c are real numbers and $a \neq 0$ (or with $>$ replaced by $<, \geq$, or $\leq$).

Algebraic Solution of Quadratic Inequalities

To solve a quadratic inequality $f(x) > 0$ or $f(x) < 0$ algebraically, we first need to find the zeros of $f(x)$. The zeros can be found by factoring or the quadratic formula. If the quadratic function has two real zeros, these two zeros divide the real number line into three intervals. Within each interval, the value of the quadratic function is either always positive or always negative, so we can use one value in each interval to test the function on the interval. We can use the following steps, which summarize these ideas, to solve quadratic inequalities.

Solving a Quadratic Inequality Algebraically

1. Write an equivalent inequality with 0 on one side and with the function $f(x)$ on the other side.

2. Solve $f(x) = 0$.

3. Create a sign diagram that uses the solutions from step 2 to divide the number line into intervals. Pick a test value in each interval and determine whether $f(x)$ is positive or negative in that interval to create a sign diagram.*

4. Identify the intervals that satisfy the inequality in step 1. The values of x that define these intervals are solutions to the original inequality.

EXAMPLE 6 ▶   ## Solving Quadratic Inequalities

Solve the inequality $x^2 - 3x > 3 - 5x$.

SOLUTION

Rewriting the inequality with 0 on the right side of the inequality gives $f(x) > 0$ with $f(x) = x^2 + 2x - 3$:

$$x^2 + 2x - 3 > 0$$

Writing the equation $f(x) = 0$ and solving for x gives

$$x^2 + 2x - 3 = 0$$
$$(x + 3)(x - 1) = 0$$
$$x = -3 \quad \text{or} \quad x = 1$$

The two values of x, -3 and 1, divide the number line into three intervals: the numbers less than -3, the numbers between -3 and 1, and the numbers greater than 1. We need only find the signs of $(x + 3)$ and $(x - 1)$ in each interval and then find the sign of their product to find the solution to the original inequality. Testing a value in each interval determines the sign of each factor in each interval. (See the sign diagram in Figure 4.42.)

Figure 4.42

The function $f(x) = (x + 3)(x - 1)$ is positive on the intervals $(-\infty, -3)$ and $(1, \infty)$, so the solution to $x^2 + 2x - 3 > 0$ and thus the original inequality is

$$x < -3 \quad \text{or} \quad x > 1$$

---

* The numerical feature of your graphing utility can be used to test the x-values.

Graphical Solution of Quadratic Inequalities

Recall that we solved a linear inequality $f(x) > 0$ (or $f(x) < 0$) graphically by graphing the related equation $f(x) = 0$ and observing where the graph is above (or below) the x-axis—that is, where $f(x)$ is positive (or negative).

Similarly, we can solve a quadratic inequality $f(x) > 0$ (or $f(x) < 0$) graphically by graphing the related equation $f(x) = 0$ and observing the x-values of the intervals where $f(x)$ is positive (or negative).

For example, we can graphically solve the inequality

$$x^2 - 3x - 4 \leq 0$$

by graphing

$$y = x^2 - 3x - 4$$

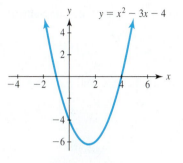

Figure 4.43

and observing the part of the graph that is below or on the x-axis. Using the x-intercept method, we find that $x = -1$ and $x = 4$ give $y = 0$, and we see that the graph is below or on the x-axis for values of x satisfying $-1 \leq x \leq 4$ (Figure 4.43). Thus, the solution to the inequality is $-1 \leq x \leq 4$.

Table 4.9 shows the possible graphs of quadratic functions and the solutions to the related inequalities. Note that if the graph of $y = f(x)$ lies entirely above the x-axis, the solution to $f(x) > 0$ is the set of all real numbers, and there is no solution to $f(x) < 0$. (The graph of $y = f(x)$ can also touch the x-axis in one point.)

Table 4.9   Quadratic Inequalities

Orientation of Graph of $y = f(x)$	Inequality to Be Solved	Part of Graph That Satisfies the Inequality	Solution to Inequality
	$f(x) > 0$	where the graph is above the x-axis	$x < a$ or $x > b$
	$f(x) < 0$	where the graph is below the x-axis	$a < x < b$
	$f(x) > 0$	where the graph is above the x-axis	$a < x < b$
	$f(x) < 0$	where the graph is below the x-axis	$x < a$ or $x > b$
	$f(x) > 0$	the entire graph	all real numbers
	$f(x) < 0$	none of the graph	no solution
	$f(x) > 0$	none of the graph	no solution
	$f(x) < 0$	the entire graph	all real numbers

EXAMPLE 7 ▶   # Height of a Model Rocket

A model rocket is projected straight upward from ground level according to the equation

$$h = -16t^2 + 192t, \; t \geq 0$$

where h is the height in feet and t is the time in seconds. During what time interval will the height of the rocket exceed 320 feet?

ALGEBRAIC SOLUTION

To find the time interval when the rocket is higher than 320 feet, we find the values of t for which

$$-16t^2 + 192t > 320$$

Getting 0 on the right side of the inequality, we have

$$-16t^2 + 192t - 320 > 0$$

First we determine when $-16t^2 + 192t - 320$ is equal to 0:

$$-16t^2 + 192t - 320 = 0$$
$$-16(t^2 - 12t + 20) = 0$$
$$-16(t - 2)(t - 10) = 0$$
$$t = 2 \quad \text{or} \quad t = 10$$

The values of $t = 2$ and $t = 10$ divide the number line into three intervals: the numbers less than 2, the numbers between 2 and 10, and the numbers greater than 10 (Figure 4.44). To solve the inequality, we find the sign of the product of -16, $(t - 2)$, and $(t - 10)$ in each interval. We do this by testing a value in each interval.

Figure 4.44

This shows that the product $-16(t - 2)(t - 10)$ is positive only for $2 < t < 10$, so the solution to $-16t^2 + 192t - 320 > 0$ is $2 < t < 10$. This solution is also a solution to the original inequality, so the rocket exceeds 320 feet between 2 and 10 seconds after launch.

Note that if the inequality problem is applied, we must check that the solution makes sense in the context of the problem.

GRAPHICAL SOLUTION

To find the interval when the height exceeds 320 feet, we can also solve $-16t^2 + 192t > 320$ using a graphing utility.* To use the intercept method, we rewrite the inequality with 0 on the right side:

$$-16t^2 + 192t - 320 > 0$$

_____

* In a graphing calculator, we would use the independent variable x instead of t.

We enter $y = -16t^2 + 192t - 320$ and graph the function (Figure 4.45). The t-intercepts of the graph are $t = 2$ and $t = 10$.

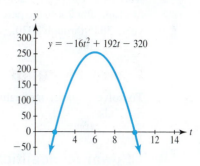

Figure 4.45

The solution to $-16t^2 + 192t - 320 > 0$ is the interval where the graph is above the t-axis, which is

$$2 < t < 10$$

Thus, the height of the rocket exceeds 320 feet between 2 and 10 seconds.

EXAMPLE 8 ▶   **Obesity**

Obesity (BMI ≥ 30) increases the risk of diabetes, heart disease, and many other ailments. The percent of Americans who are obese, from 2000 projected to 2030, can be modeled by $y = -0.0102x^2 + 0.971x + 22.1$, with x equal to the number of years after 2000 and y equal to the percent of obese Americans. For what years from 2000 to 2030 does this model estimate the percent to be at least 40%?

SOLUTION

To find the years from 2000 to 2030 when the percent of Americans who are obese is at least 40%, we solve the inequality

$$-0.0102x^2 + 0.971x + 22.1 \geq 40$$

We begin solving this inequality by graphing $y = -0.0102x^2 + 0.971x + 22.1$. We set the viewing window from $x = 0$ (2000) to $x = 30$ (2030). The graph of this function is shown in Figure 4.46(a).

Figure 4.46

We seek the point(s) where the graph of $y = 40$ intersects the graph of $y = -0.0102x^2 + 0.971x + 22.1$. Graphing the line $y = 40$ and intersecting it with the graph of the quadratic function in the given window gives the point of intersection (25, 40), as shown in Figure 4.46(b). This graph shows that the percent of Americans who are obese is at or above 40% for $25 \leq x \leq 30$. Because x represents the number of years after 2000, the percent of obese Americans is projected to be above 40% for the years 2025 through 2030.

Power Inequalities

To solve a power inequality, we will use a combination of analytical and graphical methods.

> ### Power Inequalities
>
> To solve a power inequality, first solve the related equation. Then use graphical methods to find the values of the variable that satisfy the inequality.

EXAMPLE 9 ▶   **Investment**

The future value of $3000 invested for 3 years at rate r, compounded annually, is given by $S = 3000(1 + r)^3$. What interest rate will give a future value of at least $3630?

SOLUTION

To solve this problem, we solve the inequality

$$3000(1 + r)^3 \geq 3630$$

We begin by solving the related equation using the root method.

$$3000(1 + r)^3 = 3630$$

$$(1 + r)^3 = 1.21 \qquad \text{Divide both sides by 3000.}$$

$$1 + r = \sqrt[3]{1.21} \qquad \text{Take the cube root} \left(\frac{1}{3} \text{ power}\right) \text{ of both sides.}$$

$$1 + r \approx 1.0656$$

$$r \approx 0.0656$$

This tells us that the investment will have a future value of $3630 in 3 years if the interest rate is approximately 6.56%, and it helps us set the window to solve the inequality graphically. Graphing $y_1 = 3000(1 + r)^3$ and $y_2 = 3630$ on the same axes shows that $3000(1 + r)^3 \geq 3630$ if $r \geq 0.0656$ (Figure 4.47). Thus, the investment will result in at least $3630 if the interest rate is 6.56% or higher. Note that in the context of this problem, only values of r from 0 (0%) to 1 (100%) make sense.

Figure 4.47

Inequalities Involving Absolute Values

Recall that if u is an algebraic expression, then for $a \geq 0$, $|u| = a$ means that $u = a$ or $u = -a$. We can represent inequalities involving absolute values as follows.

> For $a \geq 0$:
>
> $|u| < a$ means that $-a < u < a$.
>
> $|u| \leq a$ means that $-a \leq u \leq a$.
>
> $|u| > a$ means that $u < -a$ or $u > a$.
>
> $|u| \geq a$ means that $u \leq -a$ or $u \geq a$.

We can solve inequalities involving absolute values algebraically or graphically.

EXAMPLE 10 ▶

Solving Absolute Value Inequalities

Solve the following inequalities and verify the solutions graphically.

a. $|2x - 3| \leq 5$ **b.** $|3x + 4| - 5 > 0$

SOLUTION

a. This inequality is equivalent to $-5 \leq 2x - 3 \leq 5$. If we add 3 to all three parts of this inequality and then divide all parts by 2, we get the solution.

$$-5 \leq 2x - 3 \leq 5$$
$$-2 \leq 2x \leq 8$$
$$-1 \leq x \leq 4$$

Figure 4.48(a) shows that the graph of $y = |2x - 3|$ is on or below the graph of $y = 5$ for $-1 \leq x \leq 4$.

b. This inequality is equivalent to $|3x + 4| > 5$, which is equivalent to $3x + 4 < -5$ or $3x + 4 > 5$. We solve each of the inequalities as follows:

$$3x + 4 < -5 \quad \text{or} \quad 3x + 4 > 5$$
$$3x < -9 \qquad\qquad 3x > 1$$
$$x < -3 \qquad\qquad x > \frac{1}{3}$$

Figure 4.48(b) shows that the graph of $y = |3x + 4| - 5$ is above the x-axis for $x < -3$ or $x > \frac{1}{3}$.

(a)

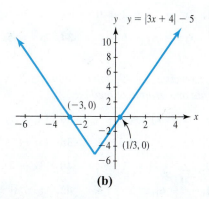

(b)

Figure 4.48

SKILLS CHECK 4.4

In Exercises 1–20, solve the equations algebraically and check graphically or by substitution.

1. $\sqrt{2x^2 - 1} - x = 0$ 2. $\sqrt{3x^2 + 4} - 2x = 0$

3. $\sqrt[3]{x - 1} = -2$ 4. $\sqrt[3]{4 - x} = 3$

5. $\sqrt{3x - 2} + 2 = x$ 6. $\sqrt{x - 2} + 2 = x$

7. $\sqrt[3]{4x + 5} = \sqrt[3]{x^2 - 7}$

8. $\sqrt{5x - 6} = \sqrt{x^2 - 2x}$

9. $\sqrt{x - 1} = \sqrt{x - 5}$

10. $\sqrt{x - 10} = -\sqrt{x - 20}$

11. $(x + 4)^{2/3} = 9$ 12. $(x - 5)^{3/2} = 64$

13. $4x^{5/2} - 8 = 0$ 14. $6x^{5/3} - 18 = 0$

15. $x^4 - 5x^2 + 4 = 0$

16. $x^4 - 11x^2 + 18 = 0$

17. $x^{-2} - x^{-1} - 30 = 0$

18 $2x^{-2} + 3x^{-1} - 2 = 0$

19. $2x^{2/3} + 5x^{1/3} - 12 = 0$

20. $x^{2/5} - x^{1/5} - 6 = 0$

In Exercises 21–30, use algebraic methods to solve the inequalities.

21. $x^2 + 4x < 0$ 22. $x^2 - 25x < 0$

23. $9 - x^2 \geq 0$ 24. $x > x^2$

25. $-x^2 + 9x - 20 > 0$ 26. $2x^2 - 8x < 0$

27. $2x^2 - 8x \geq 24$ 28. $t^2 + 17t \leq 8t - 14$

29. $x^2 - 6x < 7$ 30. $4x^2 - 4x + 1 > 0$

In Exercises 31–34, use graphical methods to solve the inequalities.

31. $2x^2 - 7x + 2 \geq 0$ 32. $w^2 - 5w + 4 > 0$

33. $5x^2 \geq 2x + 6$ 34. $2x^2 \leq 5x + 6$

In Exercises 35–42, solve the inequalities by using algebraic and graphical methods.

35. $(x + 1)^3 < 4$ 36. $(x - 2)^3 \geq -2$

37. $(x - 3)^5 < 32$ 38. $(x + 5)^4 > 16$

39. $|2x - 1| < 3$ 40. $|3x + 1| \leq 5$

41. $|x - 6| \geq 2$ 42. $|x + 8| > 7$

In Exercises 43–46, you are given the graphs of several functions of the form $f(x) = ax^2 + bx + c$ for different values of a, b, and c. For each function, (a) solve $f(x) \geq 0$, and (b) solve $f(x) < 0$.

43.

44.

45.

46.

In Exercises 47 and 48, you are given the graphs of two functions $f(x)$ and $g(x)$. Solve $f(x) \leq g(x)$.

47.

48.

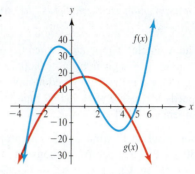

EXERCISES 4.4

Use algebraic and/or graphical methods to solve Exercises 49–58.

49. Profit The monthly profit from producing and selling x units of a product is given by

$$P(x) = -0.3x^2 + 1230x - 120,000$$

Producing and selling how many units will result in a profit for this product?

50. Profit The monthly profit from producing and selling x units of a product is given by

$$P(x) = -0.01x^2 + 62x - 12,000$$

Producing and selling how many units will result in a profit for this product?

51. Profit The revenue from sales of x units of a product is given by $R(x) = 200x - 0.01x^2$, and the cost of producing and selling the product is $C(x) = 38x + 0.01x^2 + 16,000$. Producing and selling how many units will result in a profit?

52. Profit The revenue from sales of x units of a product is given by $R(x) = 600x - 0.01x^2$, and the cost of producing and selling the product is $C(x) = 77x + 0.02x^2 + 52,000$. Producing and selling how many units will result in a profit?

53. Projectiles Two projectiles are fired into the air over a lake, with the height of the first projectile given by $y = 100 + 130t - 16t^2$ and the height of the second projectile given by $y = -16t^2 + 180t$, where y is in feet and t is in seconds. Over what time interval, before the lower one hits the lake, is the second projectile above the first?

54. Projectile A rocket shot into the air has height $s = 128t - 16t^2$ feet, where t is the number of seconds after the rocket is shot. During what time after the rocket is shot is it at least 240 feet high?

55. Tobacco Sales Sales of fine-cut cigarettes (in millions) in Canada are given by $y = -13.898x^2 + 255.467x + 5425.618$, where x is the number of years after 1980. During what years does this model indicate that the sales would be at least 6 billion?
(Source: Health Canada)

56. World Population The low long-range world population numbers and projections for the years 1995–2150 are given by the equation $y = -0.00036x^2 + 0.0385x + 5.823$, where x is the number of years after 1990 and y is in billions. During what years does this model estimate the population to be above 6 billion?
(Source: U.N. Department of Economic and Social Affairs)

57. Foreign-Born Population Suppose the percent of the U.S. population that is foreign born is given by $y = 0.0027x^2 - 0.3692x + 18.914$, where x is the number of years after 1900. During what years after 2000 does this model indicate that the percent is at least 13.49?
(Source: U.S. Census Bureau)

58. Gross Domestic Product The U.S. gross domestic product (GDP) (in trillions of dollars) for selected years from 2005 and projected to 2070 can be modeled by $y = 0.116x^2 - 3.792x + 45.330$, where x is the number of years after 2000. During what years between 2005 and 2070 was the gross domestic product no more than $23.03 trillion?
(Source: Microsoft Encarta 98)

Use graphical and/or numerical methods to solve Exercises 59–64.

59. **Wind Chill** The wind chill temperature when the outside temperature is 20°F is given by $y = 0.0052x^2 - 0.62x + 15.0$, where x is the wind speed in mph. For what wind speeds between 0 mph and 90 mph is the wind chill temperature −3°F or below?

60. **China's Labor Pool** Data from 1975 and projections to 2050 of the number of millions of people age 15 to 59 in China can be modeled by $y = -0.224x^2 + 22.005x + 370.705$, where x is the number of years after 1970 and y represents the number of millions of people in this labor pool. During what years from 1970 to 2050 is number of people age 15 to 59 at least 903.38 million? (Source: United Nations)

61. **Drug Use** The percent of high school students who have ever used marijuana is given by $y = -0.1967x^2 + 4.063x + 27.7455$, where x is the number of years after 1990. Use the model to estimate the years when the percent who ever used marijuana is greater than 43.1%.

62. **Airplane Crashes** The number of all airplane crashes (in thousands) is given by the equation $y = 0.0057x^2 - 0.197x + 3.613$, where x is the number of years after 1980. During what years from 1980 through 2015 does the model indicate the number of crashes as being below 3146?

63. **High School Smokers** The percent of high school students who smoked cigarettes on 1 or more of the 30 days preceding the Youth Risk Behavior Survey is given by $y = -0.061x^2 + 0.275x + 33.698$, where x is the number of years after 1990. During what years from 1990 on is the percent greater than 20%?

64. **Car Design** Sports cars are designed so that the driver's seat is comfortable for people of height 5 feet 8 inches, plus or minus 8 inches.

   a. Write an absolute value inequality that gives the height x of a person who will be uncomfortable.

   b. Solve this inequality for x to identify the heights of people who are uncomfortable.

65. **Retail Sales** November and December retail sales, excluding autos, for the years 2001–2010 can be modeled by the function $S(x) = -1.751x^2 + 38.167x + 388.997$ billion dollars, where x is the number of years after 2000. During what years does the model estimate retail sales to be above $500 billion?

66. **Voltage** Required voltage for an electric oven is 220 volts, but it will function normally if the voltage varies from 220 by 10 volts.

   a. Write an absolute value inequality that gives the voltage x for which the oven will work normally.

   b. Solve this inequality for x.

67. **Purchasing Power** Inflation causes a decrease in the value of money used to purchase goods and services. The purchasing power of a 1983 dollar based on consumer prices for 1968–2010 can be modeled by the function $y = 34.394x^{-1.109}$, where x is the number of years after 1960. For what years through 2012 is the purchasing power of a 1983 dollar less than $1.00, according to the model?

chapter 4	**SUMMARY**

In this chapter, we presented the building blocks for function construction. We can construct new functions by horizontal or vertical transformation, stretching or compressing, or reflection. We can also construct new functions by combining functions algebraically or with function composition. Inverse functions can be used to "undo" the operations in a function. We finished the chapter by solving additional equations and inequalities.

Key Concepts and Formulas

4.1 Transformations of Graphs and Symmetry

Vertical shifts of graphs	If k is a positive real number, • The graph of $g(x) = f(x) + k$ can be obtained by shifting the graph of $f(x)$ upward k units. • The graph of $g(x) = f(x) - k$ can be obtained by shifting the graph of $f(x)$ downward k units.
Horizontal shifts of graphs	If h is a positive real number, • The graph of $g(x) = f(x - h)$ can be obtained by shifting the graph of $f(x)$ to the right h units. • The graph of $g(x) = f(x + h)$ can be obtained by shifting the graph of $f(x)$ to the left h units.
Stretching and compressing graphs	The graph of $y = af(x)$ is obtained by vertically stretching the graph of $f(x)$ using a factor of $|a|$ if $|a| > 1$ and by vertically compressing the graph of $f(x)$ using a factor of $|a|$ if $|a| < 1$.
Reflections of graphs across the coordinate axes	The graph of $y = -f(x)$ can be obtained by reflecting the graph of $y = f(x)$ across the x-axis. The graph of $y = f(-x)$ can be obtained by reflecting the graph of $y = f(x)$ across the y-axis.
Symmetry with respect to the y-axis	The graph of $y = f(x)$ is symmetric with respect to the y-axis if $f(-x) = f(x)$. Such a function is called an even function.
Symmetry with respect to the origin	The graph of $y = f(x)$ is symmetric with respect to the origin if $f(-x) = -f(x)$. Such a function is called an odd function.
Symmetry with respect to the x-axis	The graph of an equation is symmetric with respect to the x-axis if, for every point (x, y) on the graph, the point $(x, -y)$ is also on the graph.

4.2 Combining Functions; Composite Functions

Operations with functions	*Sum:* $(f + g)(x) = f(x) + g(x)$ *Difference:* $(f - g)(x) = f(x) - g(x)$ *Product:* $(f \cdot g)(x) = f(x) \cdot g(x)$ *Quotient:* $\left(\dfrac{f}{g}\right)(x) = \dfrac{f(x)}{g(x)}, g(x) \neq 0$

	The domain of the sum, difference, and product of f and g consists of all real numbers of the input variable for which f and g are defined. The domain of the quotient function consists of all real numbers for which f and g are defined and $g \neq 0$.
Average cost	The quotient of the cost function $C(x)$ and the identity function $I(x) = x$ gives the average cost function $\overline{C}(x) = \dfrac{C(x)}{x}$.
Composite functions	The notation for the composite function "f of g" is $(f \circ g)(x) = f(g(x))$. The domain of $f \circ g$ is the subset of the domain of g for which $f \circ g$ is defined.

4.3 One-to-One and Inverse Functions

Inverse functions	If $f(g(x)) = x$ and $g(f(x)) = x$ for all x in the domain of g and f, then f and g are inverse functions. In addition, the functions f and g are inverse functions if, whenever the pair (a, b) satisfies $y = f(x)$, the pair (b, a) satisfies $y = g(x)$. We denote g by f^{-1}, read as "f inverse."
One-to-one functions	For a function f to have an inverse, f must be one-to-one. A function f is one-to-one if each output of f corresponds to exactly one input of f.
Horizontal line test	If no horizontal line can intersect the graph of a function in more than one point, then the function is one-to-one.
Finding the inverse functions	To find the inverse of the function f that is defined by the equation $y = f(x)$: **1.** Rewrite the equation with y replacing $f(x)$. **2.** Interchange x and y in the equation defining the function. **3.** Solve the new equation for y. If this equation cannot be solved uniquely for y, the function has no inverse. **4.** Replace y by $f^{-1}(x)$.
Graphs of inverse functions	The graphs of a function and its inverse are symmetric with respect to the line $y = x$.
Inverse functions on limited domains	If a function is one-to-one on a limited domain, then it has an inverse function on that domain.

4.4 Additional Equations and Inequalities

Radical equations	An equation containing radicals can frequently be converted to an equation that does not contain radicals by raising both sides of the equation to a power that is equal to the index of the radical. The solutions must be checked when using this technique.
Equations with rational powers	Some equations with rational powers can be solved by writing the equation as a radical equation.
Equations in quadratic form	An equation of the form $au^2 + bu + c = 0$, where u is an algebraic expression, can be solved by substitution and a quadratic method.
Quadratic inequality	A quadratic inequality is an inequality that can be written in the form $ax^2 + bx + c > 0$, where a, b, c are real numbers and $a \neq 0$ (or with $>$ replaced by $<$, $\geq$, or $\leq$).
Solving a quadratic inequality algebraically	To solve a quadratic inequality $f(x) < 0$ or $f(x) > 0$ algebraically, solve $f(x) = 0$ and use the solutions to divide the number line into intervals. Pick a test value in each interval to determine whether $f(x)$ is positive or negative in that interval and identify the intervals that satisfy the original inequality.

Solving a quadratic inequality graphically	To solve a quadratic inequality $f(x) < 0$ (or $f(x) > 0$) graphically, graph the related equation $y = f(x)$ and observe the x-values of the intervals where the graph of $f(x)$ is below (or above) the x-axis.
Power inequalities	To solve a power inequality, first solve the related equation. Then use graphical methods to find the values of the variable that satisfy the inequality.
Absolute value inequalities	If $a > 0$, the solution to $|u| < a$ is $-a < u < a$. If $a > 0$, the solution to $|u| > a$ is $u < -a$ or $u > a$.

chapter 4   SKILLS CHECK

1. How is the graph of $g(x) = (x - 8)^2 + 7$ transformed from the graph of $f(x) = x^2$?

2. How is the graph of $g(x) = -2(x + 1)^3$ transformed from the graph of $f(x) = x^3$?

3. **a.** Graph the functions $f(x) = \sqrt{x}$ and $g(x) = \sqrt{x + 2} - 3$.

   b. How are the graphs related?

4. What is the domain of the function $g(x) = \sqrt{x + 2} - 3$?

5. Suppose the graph of $f(x) = x^{1/3}$ is shifted up 4 units and to the right 6 units. What is the equation that gives the new graph?

6. Suppose the graph of $f(x) = x^{1/3}$ is vertically stretched by a factor of 3 and then shifted down 5 units. What is the equation that gives the new graph?

For Exercises 7–9, match each graph with the correct equation.

   a. $y = |x| + 2$ **b.** $y = x^3$

   c. $y = \dfrac{3}{x - 1} + 1$ **d.** $y = \dfrac{-3}{x + 1} + 1$

   e. $y = 2x^3$ **f.** $y = |x + 2|$

7.

8.

9.

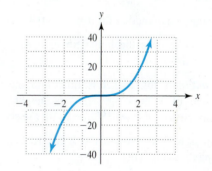

In Exercises 10 and 11, determine algebraically whether the graph of the function is symmetric with respect to the x-axis, y-axis, and/or origin. Confirm graphically.

10. $f(x) = x^3 - 4x$ 11. $f(x) = -x^2 + 5$

12. Determine whether the function $f(x) = -\dfrac{2}{x}$ is even, odd, or neither.

For Exercises 13–20, use the functions $f(x) = 3x^2 - 5x$, $g(x) = 6x - 4$, $h(x) = 5 - x^3$ to find the following.

13. $(f + g)(x)$ 14. $(h - g)(x)$

15. $(g \cdot f)(x)$ 16. $\left(\dfrac{h}{g}\right)(x)$

17. $(f - g)(-2)$ **18.** $(f \circ g)(x)$

19. $(g \circ f)(x)$ **20.** $(g \circ h)(-3)$

21. For the functions $f(x) = 2x - 5$ and $g(x) = \dfrac{x + 5}{2}$,

     a. Find $f(g(x))$ and $g(f(x))$.

     b. What is the relationship between $f(x)$ and $g(x)$?

22. Find the inverse of $f(x) = 3x - 2$.

23. Find the inverse of $g(x) = \sqrt[3]{x - 1}$.

24. Graph $f(x) = (x + 1)^2$ and its inverse $f^{-1}(x)$ on the domain $[-1, 10]$.

25. Is the function $f(x) = \dfrac{x}{x - 1}$ a one-to-one function?

26. Does the function $f(x) = x^2 - 3x$ have an inverse?

27. Solve $\sqrt{4x^2 + 1} = 2x + 2$.

28. Solve $\sqrt{3x^2 - 8} + x = 0$.

29. Solve $4x - 5x^{1/2} + 1 = 0$ —

30. Solve $(x - 6)^2 - 3(x - 6) - 10 = 0$

31. Solve the inequality $x^2 - 7x \le 18$.

32. Solve the inequality $2x^2 + 5x \ge 3$.

33. Solve $|2x - 4| \le 8$.

34. Solve $|4x - 3| \ge 15$.

35. Solve $(x - 4)^3 < 4096$.

36. Solve $(x + 2)^2 \ge 512$.

chapter 4   REVIEW

37. *Ballistics* Ballistic experts are able to identify the weapon that fired a certain bullet by studying the markings on the bullet. If a test is conducted by firing a bullet into a bale of paper, the distance that the bullet will travel is given by $s = 64 - (4 - 10t)^3$ inches, for $0 \le t \le 0.4$, t in seconds.

     a. Graph this function for $0 \le t \le 0.4$.

     b. How far does the bullet travel during this time period?

38. *Projectile* A toy rocket is fired into the air from the top of a building, with its height given by $S = -16(t - 4)^2 + 380$ feet, with t in seconds.

     a. In how many seconds will the rocket reach its maximum height?

     b. What is the maximum possible height?

     c. Describe the transformations needed to obtain the graph of $S(t)$ from the graph of $y = t^2$.

39. *Airline Traffic* The number of millions of passengers traveling on U.S. airlines can be modeled by $f(x) = -7.232x^2 + 95.117x + 441.138$, where x is the number of years after 2000.

     a. During what year does the model indicate that the number of passengers was at a maximum?

     b. Determine the years during which the number of passengers was above 700 million.

40. *Pet Industry* U.S. pet industry expenditures can be modeled by $y = 0.037x^2 + 1.018x + 12.721$ million

dollars, where x equals the number of years after 1990. During what years from 1990 through 2010 did Americans spend more than $40 million on their pets?

41. *Degrees Earned* The number of postsecondary degrees earned, in thousands (including associate's, bachelor's, master's, first professional, and doctoral degrees), can be modeled by $y = 3.980x^2 - 17.597x + 2180.899$, where x is the number of years after 1990. If this model is accurate, during what years from 1990 on would the number of postsecondary degrees earned be less than 3500 thousand?
(Source: National Center for Education Statistics, *Digest of Education Statistics*)

42. *Average Cost* Suppose the total cost function for a product is determined to be $C(x) = 30x + 3150$, where x is the number of units produced and sold.

     a. Write the average cost function, which is the quotient of two functions.

     b. Graph this function for $0 < x \le 20$.

43. *Average Cost* The monthly average cost of producing x sofas is $\overline{C}(x) = \dfrac{50{,}000}{x} + 120$ dollars.

     a. Graph this function for $x > 0$.

     b. Will the average cost function decrease or increase as the number of units produced increases?

     c. What transformations of the graph of the reciprocal function give the graph of this function?

44. *Supply and Demand* The price per unit of a product is p, and the number of units of the product is denoted by q. Suppose the supply function for a product is given by $p = \dfrac{180 + q}{6}$, and the demand for the product is given by $p = \dfrac{300}{q} - 20$.

a. Which of these functions is a shifted reciprocal function?

b. Graph the supply and demand functions on the same axes.

45. *Supply and Demand* The supply function for a commodity is given by $p = 58 + \dfrac{q}{2}$, and the demand function for this commodity is given by $p = \dfrac{2555}{q + 5}$.

a. Which of these functions is a shifted reciprocal function?

b. Graph the supply and demand functions on the same axes.

46. *Marijuana Use* Suppose the percent M of high school seniors who have tried marijuana can be considered as a function of the time x according to the model

$$M(x) = -0.2(x - 10.3)^2 + 48.968$$

where x is the number of years after 1990.

a. Describe the transformations of the graph of $f(x) = x^2$ needed to obtain the graph of M.

b. Graph the model on a window representing the years 1990 through 2015.

c. When can we be certain that this model is no longer valid?

47. *Southwest Airlines* The number of employees, in thousands, of Southwest Airlines can be modeled by the function $E(t) = 0.017t^2 + 2.164t + 8.061$, where t is the number of years after 1990. Southwest Airlines' revenue during the same time period is given by the model $R(E) = 0.165E - 0.226$ billion dollars when there are E thousand employees of the company.

a. Find and interpret the meaning of the function $R(E(t))$.

b. Use the result of part (a) to find $R(E(3))$. Interpret this result.

c. How many employees worked for Southwest Airlines in 1997?

d. What was the 1997 revenue for this airline?

(Source: Hoover's Online Capsules)

48. *Prison Sentences* The mean time in prison y for certain crimes can be found as a function of the mean sentence length x, using $f(x) = 0.554x - 2.886$, where x and y are measured in months.

a. Find the inverse of this function.

b. Interpret the inverse function from part (a).

(Source: Index of Leading Cultural Indicators)

49. *Internet Access* Using data from 2009 through 2015, the percent of U.S. households with Internet access can be modeled by function $y = 53.265x^{0.170}$, with x equal to the number of years after 2005.

a. Find the formula for the inverse of the function f.

b. If the inverse is f^{-1}, find $f^{-1}(f(15))$.

50. *Purchasing Power* The purchasing power of a 1983 dollar based on consumer prices for 1968–2010 can be modeled by the function $P(x) = 34.394x^{-1.109}$, where x is the number of years after 1960.

a. Another function that might be used to model purchasing power is $f(x) = \dfrac{50}{1.6x} - 0.2$. What is the basic function that can be transformed to obtain f?

b. The purchasing power of a 1983 dollar in 2010 was $0.456. Which model gives the better estimate of the actual value?

51. *Profit* The monthly profit from producing and selling x units of a product is given by

$$P(x) = -0.01x^2 + 62x - 12{,}000$$

Producing and selling how many units will result in profit for this product ($P(x) > 0$)?

52. *The Cloud* North American cloud computing revenues from 2008 and projected to 2020 can be modeled by $y = 0.231x^2 - 2.242x + 11.101$, where x equals the number of years after 2000 and y equals the revenue in billions of euros. For what years from 2008 to 2025 does this model estimate the North American cloud computing revenues to be at least 65.89 billion?

Group Activities
▶ EXTENDED APPLICATIONS

1. Cost, Revenue, and Profit

The following table gives the weekly revenue and cost, respectively, for selected numbers of units of production and sale of a product by the Quest Manufacturing Company.

Number of Units	Revenue ($)	Number of Units	Cost ($)
100	6800	100	32,900
300	20,400	300	39,300
500	34,000	500	46,500
900	61,200	900	63,300
1400	95,200	1400	88,800
1800	122,400	1800	112,800
2500	170,000	2500	162,500

Provide the information requested and answer the questions.

1. Use technology to determine the functions that model revenue and cost functions for this product, using x as the number of units produced and sold.
2. **a.** Combine the revenue and cost functions with the correct operation to create the profit function for this product.
   b. Use the profit function to complete the following table:

x (units)	Profit, P(x) ($)
0	
100	
600	
1600	
2000	
2500	

3. Find the number of units of this product that must be produced and sold to break even.
4. Find the maximum possible profit and the number of units that gives the maximum profit.
5. **a.** Use operations with functions to create the average cost function for the product.
   b. Complete the following table.

x (units)	Average Cost, C(x) ($/unit)
1	
100	
300	
1400	
2000	
2500	

   c. Graph this function using the viewing window $[0, 2500]$ by $[0, 400]$.
6. Graph the average cost function using the viewing window $[0, 4000]$ by $[0, 100]$. Determine the number of units that should be produced to minimize the average cost and the minimum average cost.
7. Compare the number of units that produced the minimum average cost with the number of units that produced the maximum profit. Are they the same number of units? Discuss which of these values is more important to the manufacturer and why.

2. Cell Phone Revenue

The table in the right column gives the number of U.S. cellular telephone subscribers and the average monthly bill for the years 1998–2008. To investigate the average monthly revenue for cell phone companies over the time period from 1998 to 2008, answer the following.

1. Find a quadratic function $S = f(t)$ that models the number of subscribers, in millions, as a function of the number of years after 1995. Write the model with three decimal places.
2. Graph the data and the model to visually determine if the model is a good fit for the subscriber data.
3. Find a quadratic function $D = g(t)$ that models the average monthly cell phone bill, in dollars, as a function of the number of years after 1995. Write the model with three decimal places.
4. Graph the data and the model to visually determine if the model is a good fit for the billing data.
5. Using models from steps 3 and 4 with coefficients rounded to one decimal place, take the product of these two functions to find a function that models the average monthly revenue for the cell phone companies.
6. Graph the data points representing the years and average revenue on the same axes as the function found in step 5. Is it a good fit? Explain.
7. Create a new column of revenue data by multiplying the number of subscribers in each year times the average monthly bill for that year.

8. Find a quadratic function that is a good fit for the data representing the years after 1995 and average revenue. Is your model a better fit than the function found in step 5?
9. Find a 4th degree (quartic on a graphing calculator) function that models the data representing the years after 1995 and average revenue. Does this model or the model found in step 5 fit the data better?

Year	Number of Subscribers (millions)	Average Monthly Bill ($)
1998	69.209	39.43
1999	86.047	41.24
2000	109.478	45.27
2001	128.375	47.37
2002	140.766	48.40
2003	158.722	49.91
2004	182.140	50.64
2005	207.896	49.98
2006	233.041	50.56
2007	255.396	49.94
2008	270.334	49.79

(Source: CTIA–The Wireless Association)

5

Exponential and Logarithmic Functions

The intensities of earthquakes like the one in Japan in 2011 are measured with the Richter scale, which uses logarithmic functions. Logarithmic functions are also used in measuring loudness (in decibels) and stellar magnitude and in calculating pH values. Exponential functions are used in many real-world settings, such as the growth of investments, the growth of populations, sales decay, and radioactive decay.

sections	objectives	applications
5.1 Exponential Functions	Graph exponential functions; apply exponential growth and decay functions; compare transformations of graphs; graph exponential functions with base e	Growth of paramecia, inflation, personal income, sales decay, investments with continuous compounding, carbon-14 dating
5.2 Logarithmic Functions; Properties of Logarithms	Graph and evaluate logarithmic functions; convert equations to exponential and logarithmic forms; evaluate common and natural logarithms; simplify using logarithmic properties	Diabetes, pH scale, Richter scale, investments, decibel scale
5.3 Exponential and Logarithmic Equations	Solve exponential equations; solve logarithmic equations; convert logarithms using change of base; solve exponential and logarithmic inequalities	Carbon-14 dating, doubling time for investments, growth of investments, global warming, sales decay
5.4 Exponential and Logarithmic Models	Model with exponential functions; use constant percent change to determine if data fit an exponential model; find logarithmic models; compare different models of data	Diabetes, insurance premiums, e-commerce sales, sales decay, inflation, women in the labor force
5.5 Exponential Functions and Investing	Find the future value when interest is compounded periodically and when it is compounded continuously; find the present value of an investment; use graphing utilities to model investment data	Future value of an investment, present value of an investment, mutual fund growth
5.6 Annuities; Loan Repayment	Find the future value of an annuity; find the present value of an annuity; find the payments needed to amortize debt	Future value of an ordinary annuity, present value of an ordinary annuity, home mortgage
5.7 Logistic and Gompertz Functions	Graph and apply logistic growth functions; graph and apply logistic decay functions; graph and apply Gompertz functions	Women in the workforce, expected life span, uninsured Americans, deer populations, company growth

Algebra TOOLBOX

KEY OBJECTIVES

- Use properties of exponents with integers
- Use properties of exponents with real numbers
- Simplify exponential expressions
- Write numbers in scientific notation
- Convert numbers in scientific notation to standard notation

Additional Properties of Exponents

In this Toolbox, we discuss properties of integer and real exponents and exponential expressions. These properties are useful in the discussion of exponential functions and of logarithmic functions, which are related to exponential functions. Integer exponents and rational exponents were discussed in Chapter 3, as well as the Product Property and Quotient Property.

For real numbers a and b and integers m and n,

1. $a^m \cdot a^n = a^{m+n}$ (Product Property)

2. $\dfrac{a^m}{a^n} = a^{m-n}, a \neq 0$ (Quotient Property)

Additional properties of exponents, which can be developed using the properties above, follow:

For real numbers a and b and integers m and n,

3. $(ab)^m = a^m b^m$ (Power of a Product Property)

4. $\left(\dfrac{a}{b}\right)^m = \dfrac{a^m}{b^m}$, for $b \neq 0$ (Power of a Quotient Property)

5. $(a^m)^n = a^{mn}$ (Power of a Power Property)

6. $a^{-m} = \dfrac{1}{a^m}$, for $a \neq 0$

For example, we can prove Property 3 for positive integer m as follows:

$$(ab)^m = \underbrace{(ab)(ab) \cdots (ab)}_{m \text{ times}} = \underbrace{(a \cdot a \cdots a)}_{m \text{ times}} \underbrace{(b \cdot b \cdots b)}_{m \text{ times}} = a^m b^m$$

EXAMPLE 1 ▶ Using the Properties of Exponents

Use properties of exponents to simplify each of the following. Assume that denominators are nonzero.

a. $\dfrac{5^6}{5^4}$ b. $\dfrac{y^2}{y^5}$ c. $(3xy)^3$ d. $\left(\dfrac{y}{z}\right)^4$ e. $3^{15-2m} \cdot 3^{2m}$

SOLUTION

a. $\dfrac{5^6}{5^4} = 5^{6-4} = 5^2 = 25$ b. $\dfrac{y^2}{y^5} = y^{2-5} = y^{-3} = \dfrac{1}{y^3}$ c. $(3xy)^3 = 3^3 x^3 y^3 = 27x^3 y^3$

d. $\left(\dfrac{y}{z}\right)^4 = \dfrac{y^4}{z^4}$ e. $3^{15-2m} \cdot 3^{2m} = 3^{(15-2m)+2m} = 3^{15-2m+2m} = 3^{15}$

We can also simplify and evaluate expressions involving powers of powers.

EXAMPLE 2 ▶ **Powers of Powers**

a. Simplify $(x^4)^5$. **b.** Evaluate $(2^3)^5$. **c.** Simplify $(x^2y^3)^5$. **d.** Evaluate 2^{3^2}.

SOLUTION

a. $(x^4)^5 = x^{4 \cdot 5} = x^{20}$

b. We can evaluate $(2^3)^5$ in two ways:

$$(2^3)^5 = 2^{15} = 32{,}768 \quad \text{or} \quad (2^3)^5 = (8)^5 = 32{,}768$$

c. $(x^2y^3)^5 = (x^2)^5(y^3)^5 = x^{10}y^{15}$ **d.** $2^{3^2} = 2^9 = 512$

EXAMPLE 3 ▶ **Applying Exponent Properties**

Compute the following products and quotients and write the answer with positive exponents.

a. $(-2x^{-2}y)(5x^{-2}y^{-3})$ **b.** $\dfrac{8xy^{-2}}{2x^4y^{-6}}$ **c.** $\dfrac{\dfrac{2x^{-1}y}{3a}}{\dfrac{6xy^{-2}}{5a}}$ **d.** $(4^0x^3y^{-2})^{-3}$

SOLUTION

a. $(-2x^{-2}y)(5x^{-2}y^{-3}) = -10x^{-2+(-2)}y^{1+(-3)} = -10x^{-4}y^{-2}$

$$= -10 \cdot \frac{1}{x^4} \cdot \frac{1}{y^2} = \frac{-10}{x^4y^2}$$

b. $\dfrac{8xy^{-2}}{2x^4y^{-6}} = 4x^{1-4}y^{-2-(-6)} = 4x^{-3}y^4 = 4 \cdot \dfrac{1}{x^3} \cdot y^4 = \dfrac{4y^4}{x^3}$

c. To perform this division, we invert and multiply.

$$\frac{\dfrac{2x^{-1}y}{3a}}{\dfrac{6xy^{-2}}{5a}} = \frac{2x^{-1}y}{3a} \cdot \frac{5a}{6xy^{-2}} = \frac{10ax^{-1}y}{18axy^{-2}} = \frac{5x^{-1-1}y^{1-(-2)}}{9} = \frac{5x^{-2}y^3}{9} = \frac{5y^3}{9x^2}$$

d. $(4^0x^3y^{-2})^{-3} = 4^0x^{-9}y^6 = 1 \cdot \dfrac{1}{x^9} \cdot y^6 = \dfrac{y^6}{x^9}$

Real Number Exponents

All the above properties also hold for rational exponents so long as no negative numbers in even powered roots result, and they also apply for all real numbers that give expressions that are real numbers.

EXAMPLE 4 ▶ **Operations with Real Exponents**

Perform the indicated operations.

a. $(x^{1/3})(x^{3/4})$ **b.** $\dfrac{y^{1/2}}{y^{2/5}}$ **c.** $(c^{1/3})^{3/4}$

SOLUTION

a. $(x^{1/3})(x^{3/4}) = x^{1/3+3/4} = x^{13/12}$ **b.** $\dfrac{y^{1/2}}{y^{2/5}} = y^{1/2-2/5} = y^{1/10}$

c. $(c^{1/3})^{3/4} = c^{1/3 \cdot 3/4} = c^{1/4}$

Scientific Notation

Evaluating exponential and logarithmic functions in this chapter may result in outputs that are very large or very close to 0. **Scientific notation** is a convenient way to write very large (positive or negative) numbers or numbers close to 0. Numbers in scientific notation have the form

$$N \times 10^p \text{ (where } 1 \leq N < 10 \text{ and } p \text{ is an integer)}$$

For instance, the scientific notation form of 2,654,000 is 2.654×10^6. When evaluating functions or solving equations with a calculator, we may see an expression that looks like 5.122E-8 appear on the screen. This calculator expression represents 5.122×10^{-8}, which is scientific notation for $0.00000005122 \approx 0$. Figure 5.1 shows two numbers expressed in standard notation and in scientific notation on a calculator display.

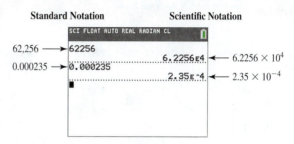

Standard Notation   Scientific Notation

$62,256 \rightarrow$ 62256
 6.2256E4 $\leftarrow 6.2256 \times 10^4$
$0.000235 \rightarrow$ 0.000235
 2.35E-4 $\leftarrow 2.35 \times 10^{-4}$

Figure 5.1

A number written in scientific notation can be converted to standard notation by multiplying (when the exponent on 10 is positive) or by dividing (when the exponent on 10 is negative). For example, we convert 7.983×10^5 to standard notation by multiplying 7.983 by $10^5 = 100,000$:

$$7.983 \times 10^5 = 7.983 \cdot 100,000 = 798,300$$

We convert 4.563×10^{-7} to standard notation by dividing 4.563 by $10^7 = 10,000,000$:

$$4.563 \times 10^{-7} = \frac{4.563}{10,000,000} = 0.0000004563$$

Multiplying two numbers in scientific notation involves adding the powers of 10, and dividing them involves subtracting the powers of 10.

EXAMPLE 5 ▶ Scientific Notation

Compute the following and write the answers in scientific notation.

a. $(7.983 \times 10^5)(4.563 \times 10^{-7})$ **b.** $(7.983 \times 10^5)/(4.563 \times 10^{-7})$

SOLUTION

a. $(7.983 \times 10^5)(4.563 \times 10^{-7}) = 36.426429 \times 10^{5+(-7)}$

$$= 36.426429 \times 10^{-2} = (3.6426429 \times 10^1) \times 10^{-2} = 3.6426429 \times 10^{-1}$$

The calculation using technology is shown in Figure 5.2(a).

b. $(7.983 \times 10^5)/(4.563 \times 10^{-7}) = 1.749506903 \times 10^{5-(-7)}$

$$= 1.749506903 \times 10^{12}. \text{ Figure 5.2(b) shows the calculation using technology.}$$

(a)   (b)

Figure 5.2

Toolbox EXERCISES

In Exercises 1 and 2, use the properties of exponents to simplify.

1. **a.** $x^4 \cdot x^3$ **b.** $\dfrac{x^{12}}{x^7}$ **c.** $(4ay)^4$ **d.** $\left(\dfrac{3}{z}\right)^4$
   e. $2^3 \cdot 2^2$ **f.** $(x^4)^2$

2. **a.** $y^5 \cdot y$ **b.** $\dfrac{w^{10}}{w^4}$ **c.** $(6bx)^3$ **d.** $\left(\dfrac{5z}{2}\right)^3$
   e. $3^2 \cdot 3^3$ **f.** $(2y^3)^4$

In Exercises 3–18, use the properties of exponents to simplify the expressions and remove all zero and negative exponents. Assume that all variables are nonzero.

3. 10^{5^0} **4.** 4^{2^2}

5. $x^{-4} \cdot x^{-3}$ **6.** $y^{-5} \cdot y^{-3}$

7. $(c^{-6})^3$ **8.** $(x^{-2})^4$

9. $\dfrac{a^{-4}}{a^{-5}}$ **10.** $\dfrac{b^{-6}}{b^{-8}}$

11. $(x^{-1/2})(x^{2/3})$ **12.** $(y^{-1/3})(y^{2/5})$

13. $(3a^{-3}b^2)(2a^2b^{-4})$ **14.** $(4a^{-2}b^3)(-2a^4b^{-5})$

15. $\left(\dfrac{2x^{-3}}{x^2}\right)^{-2}$ **16.** $\left(\dfrac{3y^{-4}}{2y^2}\right)^{-3}$

17. $\dfrac{28a^4b^{-3}}{-4a^6b^{-2}}$ **18.** $\dfrac{36x^5y^{-2}}{-6x^6y^{-4}}$

In Exercises 19–22, write the numbers in scientific notation.

19. 46,000,000   **20.** 862,000,000,000

21. 0.000094   **22.** 0.00000278

In Exercises 23–26, write the numbers in standard form.

23. 4.372×10^5

24. 7.91×10^6

25. 5.6294×10^{-4}

26. 6.3478×10^{-3}

In Exercises 27 and 28, multiply or divide, as indicated, and write the result in scientific notation.

27. $(6.250 \times 10^7)(5.933 \times 10^{-2})$

28. $\dfrac{2.961 \times 10^{-2}}{4.583 \times 10^{-4}}$

In Exercises 29–38, use the properties of exponents to simplify the expressions. Assume that all variables are nonzero.

29. $x^{1/2} \cdot x^{5/6}$ **30.** $y^{2/5} \cdot y^{1/4}$

31. $(c^{2/3})^{5/2}$ **32.** $(x^{3/2})^{3/4}$

33. $\dfrac{x^{3/4}}{x^{1/2}}$ **34.** $\dfrac{y^{3/8}}{y^{1/4}}$

35. $\left(\dfrac{2b^{1/3}}{6c^{2/3}}\right)^3$

36. $\left(\dfrac{9x^{2/3}}{16y^{5/6}}\right)^{3/2}$

37. $(x^{4/9}y^{2/3}z^{-1/3})^{3/2}$

38. $\left(\dfrac{-8x^9y^{1/2}}{27z^{3/2}}\right)^{4/3}$

Exponential Functions

KEY OBJECTIVES

- Graph and apply exponential functions
- Find horizontal asymptotes
- Graph and apply exponential growth functions
- Graph and apply exponential decay functions
- Compare transformations of graphs of exponential functions

SECTION PREVIEW **Paramecia**

The primitive single-cell animal called a paramecium reproduces by splitting into two pieces (a process called binary fission), so the population doubles each time there is a split. If we assume that the population begins with 1 paramecium and doubles each hour, then there will be 2 paramecia after 1 hour, 4 paramecia after 2 hours, 8 after 3 hours, and so on. If we let y represent the number of paramecia in the population after x hours have passed, the points (x, y) that satisfy this function for the first 9 hours of growth are described by the data in Table 5.1 and the scatter plot in Figure 5.3(a). We can show that each of these points and the data points for $x = 10, 11$, and so forth lie on the graph of the function $y = 2^x$, with $x \geq 0$ (Figure 5.3(b)).

Functions like $y = 2^x$, which have a constant base raised to a variable power, are called exponential functions. We discuss exponential functions and their applications in this section. ∎

Table 5.1

x (hours)	0	1	2	3	4	5	6	7	8	9
y (paramecia)	1	2	4	8	16	32	64	128	256	512

(a)

(b)

Figure 5.3

Exponential Functions

Recall that a linear function has a constant rate of change; that is, the outputs change by the same amount for each unit increase in the inputs. An exponential function has outputs that are *multiplied* by a fixed number for each unit increase in the inputs.

Note that in Example 1, the number y of paramecia is multiplied by 2 for each hour increase in time x, so the number is an exponential function of time.

EXAMPLE 1 ▶ **Paramecia**

As discussed in the Section Preview, the primitive single-cell animal called a paramecium reproduces by splitting into two pieces, so the population doubles each time there is a split. The number y of paramecia in the population after x hours have passed fits the graph of $y = 2^x$.

a. In the context of the paramecium application, what inputs can be used?

b. Is the function $y = 2^x$ discrete or continuous?

c. Graph the function $y = 2^x$, without regard to restricting inputs to those that make sense for the paramecium application.

d. What are the domain and range of this function?

SOLUTION

a. Not every point on the graph of $y = 2^x$ in Figure 5.3(b) describes a number of paramecia. For example, the point $(7.1, 2^{7.1})$ is on this graph, but $2^{7.1} \approx 137.187$ does not represent a number of paramecia because it is not a whole number. Since only whole number inputs will give whole number outputs, only whole number inputs can be used for this application.

b. Every point describing a number of paramecia is on the graph of $y = 2^x$, so the function $y = 2^x$ is a continuous model for this discrete paramecium growth function.

c. Some values satisfying the function $y = 2^x$ are shown in Table 5.2, and the graph of the function is shown in Figure 5.4.

d. The function that models the growth of paramecia was restricted because of the physical setting, but if the domain of the function $y = 2^x$ is not restricted, it is the set of real numbers. Note that the range of this function is the set of all positive real numbers because there is no power of 2 that results in a value of 0 or a negative number.

Table 5.2

x	y
−3	$2^{-3} = 0.125$
−1.5	$2^{-1.5} \approx 0.35$
−0.5	$2^{-0.5} \approx 0.71$
0	$2^0 = 1$
0.5	$2^{0.5} \approx 1.41$
1.5	$2^{1.5} \approx 2.83$
3	$2^3 = 8$
5	$2^5 = 32$

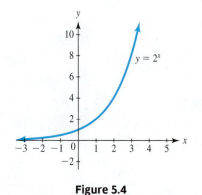

Figure 5.4

The graph in Figure 5.4 approaches but never touches the x-axis as x becomes more negative (approaching $-\infty$), so the x-axis is a **horizontal asymptote** for the graph. As previously mentioned, the function $y = 2^x$ is an example of a special class of functions called **exponential functions**. In general, we define an exponential function as follows.

Exponential Function

If b is a positive real number, $b \neq 1$, then the function $f(x) = b^x$ is an exponential function. The constant b is called the *base* of the function, and the variable x is the *exponent*.

If we graph another exponential function $y = b^x$ that has a base b greater than 1, the graph will have the same basic shape as the graph of $y = 2^x$. (See the graphs of $y = 1.5^x$ and $y = 12^x$ in Figure 5.5.)

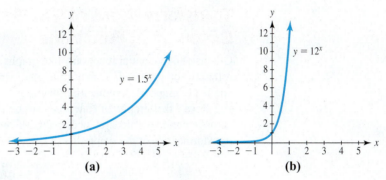

Figure 5.5

The exponential function $y = b^x$ with base b satisfying $0 < b < 1$ is a *decreasing* function. See the graph of $y = \left(\frac{1}{3}\right)^x$ in Figure 5.6(a). We can also write the function $y = \left(\frac{1}{3}\right)^x$ in the form $y = 3^{-x}$ because $3^{-x} = \frac{1}{3^x} = \left(\frac{1}{3}\right)^x$. (See Figure 5.6(b).) In general, a function of the form $y = b^{-x}$ with $b > 1$ can also be written as $y = c^x$ with $0 < c < 1$ and $c = \frac{1}{b}$, and their graphs are identical.

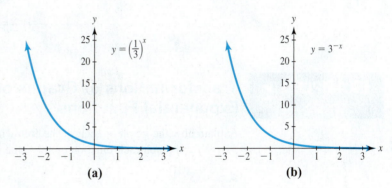

Figure 5.6

Graphs of Exponential Functions

Equation:	$y = b^x$
x-intercept:	none
y-intercept:	$(0, 1)$
Domain:	all real numbers
Range:	all real numbers $y > 0$
Horizontal asymptote:	*x*-axis
Shape:	an increasing function if $b > 1$, $y = b^x$ a decreasing function if $y = b^x$ with $0 < b < 1$ or if $y = b^{-x}$ with $b > 1$

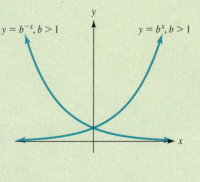

Transformations of Graphs of Exponential Functions

Graphs of exponential functions, like graphs of other types of functions, can be shifted, reflected, or stretched. Transformations, which were introduced in Chapter 4, can be applied to graphs of exponential functions.

Recall that graphs of functions can be shifted vertically or horizontally, reflected across either axis, and stretched or shrunk vertically. The following is a summary of the transformations of the graph of $f(x) = b^x$.

Vertical shift	$g(x) = b^x + k$	Graph is shifted k units up if k is positive, and $|k|$ units down if k is negative.
Horizontal shift	$g(x) = b^{x-h}$	Graph is shifted $|h|$ units right if h is positive, and $|h|$ units left if h is negative.
Reflection across the x-axis	$g(x) = -b^x$	Graph is reflected across the x-axis.
Reflection across the y-axis	$g(x) = b^{-x}$	Graph is reflected across the y-axis.
Vertical stretch/compress	$g(x) = a \cdot b^x$	Graph is vertically stretched if $|a| > 1$; graph is vertically compressed if $|a| < 1$.
Horizontal stretch/compress	$g(x) = b^{a \cdot x}$	Graph is horizontally stretched if $|a| < 1$; graph is horizontally compressed if $|a| > 1$.

EXAMPLE 2 ▶ **Transformations of Graphs of Exponential Functions**

Explain how the graph of each of the following functions compares with the graph of $y = 3^x$, and graph each function on the same axes as $y = 3^x$.

a. $y = 3^{x-4}$

b. $y = 2 + 3^{x-4}$

c. $y = 5(3^x)$

SOLUTION

a. The graph of $y = 3^{x-4}$ has the same shape as the graph of $y = 3^x$, but it is shifted 4 units to the right. The graph of $y = 3^x$ and the graph of $y = 3^{x-4}$ are shown in Figure 5.7(a).

b. The graph of $y = 2 + 3^{x-4}$ has the same shape as the graph of $y = 3^x$, but it is shifted 4 units to the right and 2 units up. The graph of $y = 2 + 3^{x-4}$ is shown in Figure 5.7(b). The horizontal asymptote is $y = 2$.

c. As we saw in Chapter 4, multiplication of a function by a constant greater than 1 stretches the graph of the function by a factor equal to that constant. Each of the y-values of $y = 5(3^x)$ is 5 times the corresponding y-value of $y = 3^x$. The graph of $y = 5(3^x)$ is shown in Figure 5.7(c).

Figure 5.7

The graphs of exponential functions can also be horizontally stretched and compressed. In general, the graph of $f(x) = b^{ax}$ is the graph of $f(x) = b^x$ stretched horizontally if $|a| < 1$ and horizontally compressed if $|a| > 1$.

EXAMPLE 3 ▶   ## Horizontal Stretching and Compressing

Use a table of values from $x = -3$ to $x = 3$ to determine how the graphs of $y = 3^{2x}$ and $y = 3^{(1/2)x}$ are related to the graph of $y = 3^x$. Then graph each function on the same axes as $y = 3^x$.

SOLUTION

Table 5.3 shows the outputs of the functions $y_2 = 3^{2x}$ and $y_3 = 3^{(1/2)x}$ compared to those of $y_1 = 3^x$. It appears that the outputs of $y_2 = 3^{2x}$ are the squares of the outputs of $y_1 = 3^x$, and the outputs of $y_3 = 3^{(1/2)x}$ are the square roots of the outputs of $y_1 = 3^x$.

Table 5.3

x	$y_1 = 3^x$	$y_2 = 3^{2x}$	$y_3 = 3^{(1/2)x}$
-3	$\dfrac{1}{27}$	$\dfrac{1}{729}$	$\sqrt{\dfrac{1}{27}}$
-2	$\dfrac{1}{9}$	$\dfrac{1}{81}$	$\sqrt{\dfrac{1}{9}}$
-1	$\dfrac{1}{3}$	$\dfrac{1}{9}$	$\sqrt{\dfrac{1}{3}}$
0	1	1	1
1	3	9	$\sqrt{3}$
2	9	81	3
3	27	729	$\sqrt{27}$

The graphs of each of these functions confirm our findings. The graph of $y = 3^{2x}$ is compressed horizontally from the graph of $y = 3^x$ (Figure 5.8(a)), and the graph of $y = 3^{(1/2)x}$ is stretched horizontally from the graph of $y = 3^x$ (Figure 5.8(b)).

(a)   (b)

Figure 5.8

Exponential Growth Models

Functions of the form $y = b^x$ with $b > 1$ and, more generally, functions of the form $y = a(b^{kx})$ with $b > 1$, $a > 0$, and $k > 0$ are increasing and can be used to model growth.

Exponential Growth Models

Equation:	$y = a(b^{kx}), b > 1, a > 0, k > 0$
x-intercept:	none
y-intercept:	$(0, a)$
Domain:	all real numbers
Range:	all real numbers $y > 0$
Horizontal asymptote:	*x*-axis (the line $y = 0$)
Shape:	increasing on domain and concave up

As we will see in Section 5.5, the future value S of an investment of P dollars invested for t years at interest rate r, compounded annually, is given by the exponential growth function

$$S = P(1 + r)^t$$

EXAMPLE 4 ▶   **Inflation**

Suppose that inflation is predicted to average 4% per year for each year from 2012 to 2025. This means that an item that costs $10,000 one year will cost $10,000(1.04) the next year and $10,000(1.04)(1.04) = $10,000(1.04^2)$ the following year.

a. Write the function that gives the cost of a $10,000 item t years after 2012.

b. Graph the growth model found in part (a) for $t = 0$ to $t = 13$.

c. If an item costs $10,000 in 2012, use the model to predict its cost in 2025.

SOLUTION

a. The function that gives the cost of a $10,000 item t years after 2012 is

$$y = 10{,}000\,(1.04^t)$$

b. The graph of $y = 10{,}000\,(1.04^t)$ is shown in Figure 5.9.

c. The year 2025 is 13 years from 2012, so the predicted cost of this item in 2025 is

$$y = 10{,}000\,(1.04^{13}) = 16{,}650.74 \text{ dollars}$$

Figure 5.9

EXAMPLE 5 ▶	**Personal Income**

Total personal income in the United States (in billions of dollars) for selected years from 1960 and projected to 2018 can be modeled by

$$y = 492.4\,(1.07^x)$$

with x equal to the number of years after 1960.

a. What does the model predict the total U.S. personal income to be in 2014?

b. Graphically determine the year during which the model predicts that total U.S. personal income will reach $28.5 trillion.

SOLUTION

a. Evaluating the model at $x = 54$ gives total U.S. personal income in 2014 to be $19,013 billion, or $19.013 trillion.

b. Figure 5.10 shows the graph of the model and the graph of $y = 28{,}500$. Using Intersect gives $x = 59.98 \approx 60$, so the model predicts that total U.S. personal income will reach $28.5 trillion ($28,500 billion) in 2020.

Figure 5.10

Spreadsheet
▶ SOLUTION
Figure 5.11 shows the outputs of $y = 492.4(1.07^x)$ at several values of x, including $x = 60$. Using the data as we have to find other graphs, we get the graph of this exponential function, shown in Figure 5.11.*

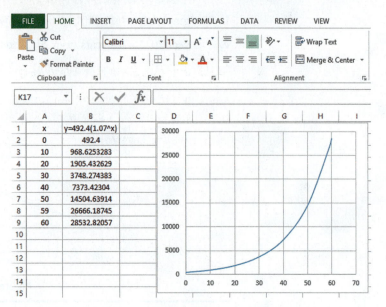

Figure 5.11

Exponential Decay Models

Functions of the form $y = a(b^{kx})$ with $b > 1$, $a > 0$, and $k < 0$ or $y = a(b^{kx})$ with $0 < b < 1$, $a > 0$, and $k > 0$ are decreasing and can be used to model decay. For example, suppose a couple retires with a fixed income of \$40,000 per year. If inflation averages 4% per year for each of the next 25 years, the value of what the couple can purchase with the \$40,000 will decrease each year, with its *purchasing power* in the xth year after they retire given by the *exponential decay model*

$$P(x) = 40{,}000(0.96^x)$$

Using this model gives the value of what the \$40,000 purchases in the 25th year after they retire to be

$$P(x) = 40{,}000(0.96^{25}) = 14{,}416 \text{ dollars}$$

Exponential Decay Models

Equation: $y = a(b^{kx})$, $b > 1$, $a > 0$, $k < 0$
or
$y = a(b^{kx})$, $0 < b < 1$, $a > 0$, $k > 0$

x-intercept: none
y-intercept: $(0, a)$
Domain: all real numbers
Range: $y > 0$
Horizontal asymptote: x-axis (the line $y = 0$)
Shape: decreasing on domain and concave up

---

* For details, see Appendix B, page 687.

EXAMPLE 6 ▶ Sales Decay

It pays to advertise, and it is frequently true that weekly sales will drop rapidly for many products after an advertising campaign ends. This decline in sales is called *sales decay*. Suppose that the decay in the sales of a product is given by

$$S = 1000(2^{-0.5x}) \text{ dollars}$$

where x is the number of weeks after the end of a sales campaign. Use this function to answer the following.

a. What is the level of sales when the advertising campaign ends?

b. What is the level of sales 1 week after the end of the campaign?

c. Use a graph of the function to estimate the week in which sales equal $500.

d. According to this model, will sales ever fall to zero?

SOLUTION

a. The campaign ends when $x = 0$, so $S = 1000(2^{-0.5(0)}) = 1000(2^0) = 1000(1) = \$1000$.

b. At 1 week after the end of the campaign, $x = 1$. Thus, $S = 1000(2^{-0.5(1)}) = \$707.11$.

c. The graph of this sales decay function is shown in Figure 5.12(a). One way to find the x-value for which $S = 500$ is to graph $y_1 = 1000(2^{-0.5x})$ and $y_2 = 500$ and find the point of intersection of the two graphs. See Figure 5.12(b), which shows that $y = 500$ when $x = 2$. Thus, sales fall to half their original amount after 2 weeks.

d. The graph of this sales decay function approaches the positive x-axis as x gets large, but it never reaches the x-axis. Thus, sales will never reach a value of $0.

(a)   (b)

Figure 5.12

The Number *e*

Many real applications involve exponential functions with the base e. The number e is an irrational number with decimal approximation 2.718281828 (to nine decimal places).

$$e \approx 2.718281828$$

The exponential function with base e occurs frequently in biology and in finance. We discuss the derivation of e and how it is used in finance in Section 5.5. Because e is close in value to 3, the graph of the exponential function $y = e^x$ has the same basic shape as the graph of $y = 3^x$. (See Figure 5.13, which compares the graphs of $y = e^x$, $y = 3^x$, and $y = 2^x$.) The graph of $y = e^x$ can be found with a graphing calculator by using the e^x (**2ND** **LN**) key, and it can be found with Excel by using the formula =exp(cell#).

Figure 5.13

The growth in the value of an investment is another important application that can be modeled by an exponential function. As we prove in Section 5.5, the future value S of an investment of P dollars for t years at interest rate r, compounded continuously, is given by

$$S = Pe^{rt} \text{ dollars}$$

EXAMPLE 7 ▶   **Growth of an Investment**

If \$10,000 is invested for 15 years at 10%, compounded continuously, what is the future value of the investment?

SOLUTION

The future value of this investment is

$$S = Pe^{rt} = 10{,}000e^{0.10(15)} = 10{,}000e^{1.5} = 44{,}816.89 \text{ dollars}$$

EXAMPLE 8 ▶   **Carbon-14 Dating**

During the life of an organism, the ratio of carbon-14 (C-14) atoms to carbon-12 (C-12) atoms remains about the same. When the organism dies, the number of C-14 atoms within its carcass begins to gradually decrease, but the number of C-12 atoms does not, so they can be measured to determine how many C-14 atoms were present at death. The number of C-14 atoms in an organism when it is discovered can then be compared to the number present at death to find how long ago the organism lived. The ratio of the present amount (y) of C-14 to the original amount (y_0) t years ago is given by

$$\frac{y}{y_0} = e^{-0.00012097t}$$

producing a function

$$y = f(t) = y_0 e^{-0.00012097t}$$

that gives the amount remaining after t years.
(Source: Thomas Jefferson National Accelerator Facility, http://www.jlab.org/)

a. If the original amount of carbon-14 present in an artifact is 100 grams, how much remains after 2000 years?

b. What percent of the original amount of carbon-14 remains after 4500 years?

SOLUTION

a. The initial amount y_0 is 100 grams and the time t is 2000 years, so we substitute these values in $f(t) = y_0 e^{-0.00012097t}$ and calculate:

$$f(2000) = 100e^{-0.00012097(2000)} \approx 78.51$$

Thus, the amount of carbon-14 present after 2000 years is approximately 78.5 grams.

b. If y_0 is the original amount, the amount that will remain after 4500 years is

$$f(4500) = y_0 e^{-0.00012097(4500)} = 0.58y_0$$

Thus, approximately 58% of the original amount will remain after 4500 years. (Note that you do not need to know the original amount of carbon-14 in order to answer this question.)

SKILLS CHECK 5.1

1. Which of the following functions are exponential functions?

    a. $y = 3x + 3$ **b.** $y = x^3$ **c.** $y = 3^x$

    d. $y = \left(\dfrac{2}{3}\right)^x$ **e.** $y = e^x$ **f.** $y = x^e$

2. Determine if each of the following functions models exponential growth or exponential decay.

    a. $y = 2^{0.1x}$ **b.** $y = 3^{-1.4x}$

    c. $y = 4e^{-5x}$ **d.** $y = 0.8^{3x}$

3. **a.** Graph the function $f(x) = e^x$ on the window $[-5, 5]$ by $[-10, 30]$.

    b. Find $f(1)$, $f(-1)$, and $f(4)$, rounded to three decimal places.

    c. What is the horizontal asymptote of the graph?

    d. What is the y-intercept?

4. **a.** Graph the function $f(x) = 5^x$ on the window $[-5, 5]$ by $[-10, 30]$.

    b. Find $f(1)$, $f(3)$, and $f(-2)$.

    c. What is the horizontal asymptote of the graph?

    d. What is the y-intercept?

5. Graph the function $y = 3^x$ on $[-5, 5]$ by $[-10, 30]$.

6. Graph $y = 5e^{-x}$ on $[-4, 4]$ by $[-1, 20]$.

In Exercises 7–12, graph each function.

7. $y = 5^{-x}$ 8. $y = 3^{-2x}$

9. $y = 3^x + 5$ 10. $y = 2(1.5)^{-x}$

11. $y = 3^{(x-2)} - 4$ 12. $y = 3^{(x-1)} + 2$

In Exercises 13–18, match the graph with its equation.

13. $y = 4^{x+3}$ **A.**

14. $y = 2 \cdot 4^x$ **B.**

15. $y = 4^x + 3$ **C.**

16. $y = 4^{-x}$ **D.**

17. $y = -4^x$ E.

18. $y = 4^{x-2} + 4$ F.

In Exercises 19–24, use your knowledge of transformations to compare the graph of the function with the graph of $f(x) = 4^x$. Then graph each function for $-4 \le x \le 4$.

19. $y = 4^x + 2$ **20.** $y = 4^{x-1}$

21. $y = 4^{-x}$ **22.** $y = -4^x$

23. $y = 3(4^x)$ **24.** $y = 3 \cdot 4^{(x-2)} - 3$

25. How does the graph of Exercise 24 compare with the graph of Exercise 23?

26. Which of the functions in Exercises 19–24 are increasing functions, and which are decreasing functions?

In Exercises 27–34, use your knowledge of transformations to compare the graph of each function with the graph of $f(x) = e^x$. Then graph each function.

27. $y = e^{x-3}$ **28.** $y = e^x - 4$

29. $y = \dfrac{1}{2}e^x$ **30.** $y = -e^x$

31. $y = e^{-x}$ **32.** $y = e^{x+2} + 1$

33. $y = e^{3x}$ **34.** $y = e^{(1/3)x}$

35. a. Graph $f(x) = 12e^{-0.2x}$ for $-5 \le x \le 15$.

   b. Find $f(10)$ and $f(-10)$.

   c. Does this function represent growth or decay?

36. If $y = 200(2^{-0.01x})$,

   a. Find the value of y (to two decimal places) when $x = 20$.

   b. Use graphical or numerical methods to determine the value of x that gives $y = 100$.

EXERCISES 5.1

37. *Sales Decay* At the end of an advertising campaign, weekly sales of smartphones declined according to the equation $y = 12{,}000(2^{-0.08x})$ dollars, where x is the number of weeks after the end of the campaign.

   a. Determine the sales at the end of the ad campaign.

   b. Determine the sales 6 weeks after the end of the campaign.

   c. Does this model indicate that sales will eventually reach \$0?

38. *Sales Decay* At the end of an advertising campaign, weekly sales of camcorders declined according to the equation $y = 10{,}000(3^{-0.05x})$ dollars, where x is the number of weeks after the campaign ended.

   a. Determine the sales at the end of the ad campaign.

   b. Determine the sales 8 weeks after the end of the campaign.

   c. How do we know, by inspecting the equation, that this function is decreasing?

39. *Investment*

   a. Graph the function $S = 80{,}000(1.05^t)$, which gives the future value of \$80,000 invested at 5%, compounded annually, for t years, $0 \le t \le 12$.

   b. Find the future value of \$80,000 invested for 10 years at 5%, compounded annually.

40. *Investment*

   a. Graph the function $S = 56{,}000(1.09^t)$, which gives the future value of \$56,000 invested at 9%, compounded annually, for t years, $0 \le t \le 15$.

   b. Find the future value of \$56,000 invested for 13 years at 9%, compounded annually.

41. *Continuous Compounding* If \$8000 is invested for t years at 8% interest, compounded continuously, the future value is given by $S = 8000e^{0.08t}$ dollars.

   a. Graph this function for $0 \le t \le 15$.

   b. Use the graph to estimate when the future value will be \$20,000.

c. Complete the following table.

t (year)	S ($)
10	
20	
	46,449.50

42. **Continuous Compounding** If $35,000 is invested for t years at 9% interest, compounded continuously, the future value is given by $S = 35,000e^{0.09t}$ dollars.

   a. Graph this function for $0 \leq t \leq 8$.

   b. Use the graph to estimate when the future value will be $60,000.

   c. Complete the following table.

t (year)	S ($)
10	
15	
	253,496

43. **Radioactive Decay** The amount of radioactive isotope thorium-234 present at time t is given by $A(t) = 500e^{-0.02828t}$ grams, where t is the time in years that the isotope decays. The initial amount present is 500 grams.

   a. How many grams remain after 10 years?

   b. Graph this function for $0 \leq t \leq 100$.

   c. If the half-life is the time it takes for half of the initial amount to decay, use graphical methods to estimate the half-life of this isotope.

44. **Radioactive Decay** A breeder reactor converts stable uranium-238 into the isotope plutonium-239. The decay of this isotope is given by $A(t) = 100e^{-0.00002876t}$, where $A(t)$ is the amount of the isotope at time t (in years) and 100 grams is the original amount.

   a. How many grams remain after 100 years?

   b. Graph this function for $0 \leq t \leq 50,000$.

   c. The half-life is the time it takes for half of the initial amount to decay; use graphical methods to estimate the half-life of this isotope.

45. **Sales Decay** At the end of an advertising campaign, weekly sales at an electronics store declined according to the equation $y = 2000(2^{-0.1x})$ dollars, where x is the number of weeks after the end of the campaign.

   a. Graph this function for $0 \leq x \leq 60$.

b. Use the graph to find the weekly sales 10 weeks after the campaign ended.

   c. Comment on "It pays to advertise" for this store.

46. **Sales Decay** At the end of an advertising campaign, weekly retail sales of a product declined according to the equation $y = 40,000(3^{-0.1x})$ dollars, where x is the number of weeks after the campaign ended.

   a. Graph this function for $0 \leq x \leq 50$.

   b. Find the weekly sales 10 weeks after the campaign ended.

   c. Should the retailers consider another advertising campaign even if it costs $5000?

47. **Purchasing Power** The purchasing power (real value of money) decreases if inflation is present in the economy. For example, the purchasing power of R dollars after t years of 5% inflation is given by the model

$$P = R(0.95^t) \text{ dollars}$$

   a. What will be the purchasing power of $40,000 after 20 years of 5% inflation?

   b. How does this affect people planning to retire at age 50?
   (Source: *Viewpoints*, VALIC, 1993)

48. **Purchasing Power** If a retired couple has a fixed income of $60,000 per year, the purchasing power (real value of money) after t years of 5% inflation is given by the equation $P = 60,000(0.95^t)$ dollars.

   a. What is the purchasing power of their income after 4 years?

   b. Test numerical values in this function to determine the years when their purchasing power will be less than $30,000.

49. **Real Estate Inflation** During a 5-year period of constant inflation, the value of a $100,000 property will increase according to the equation $v = 100,000e^{0.05t}$.

   a. What will be the value of this property in 4 years?

   b. Use a table or graph to estimate when this property will double in value.

50. **Inflation** An antique table increases in value according to the function $v(x) = 850(1.04^x)$ dollars, where x is the number of years after 1990.

   a. How much was the table worth in 1990?

   b. If the pattern indicated by the function remains valid, what was the value of the table in 2005?

   c. Use a table or graph to estimate the year when this table would reach double its 1990 value.

51. *Population* The population in a certain city was 53,000 in 2000, and its future size is predicted to be $P(t) = 53,000e^{0.015t}$ people, where t is the number of years after 2000.

   a. Does this model indicate that the population is increasing or decreasing?

   b. Use this function to estimate the population of the city in 2005.

   c. Use this function to estimate the population of the city in 2010.

   d. What is the average rate of growth between 2000 and 2010?

52. *Population* The population in a certain city was 800,000 in 2003, and its future size is predicted to be $P = 800,000e^{-0.020t}$ people, where t is the number of years after 2003.

   a. Does this model indicate that the population is increasing or decreasing?

   b. Use this model to estimate the population of the city in 2010.

   c. Use this model to predict the population of the city in 2020.

   d. What is the average rate of change in population between 2010 and 2020?

53. *Carbon-14 Dating* An exponential decay function can be used to model the number of grams of a radioactive material that remain after a period of time. Carbon-14 decays over time, with the amount remaining after t years given by $y = 100e^{-0.00012097t}$ if 100 grams is the original amount.

   a. How much remains after 1000 years?

   b. Use graphical methods to estimate the number of years until 10 grams of carbon-14 remain.

54. *Drugs in the Bloodstream* If a drug is injected into the bloodstream, the percent of the maximum dosage that is present at time t is given by

$$y = 100(1 - e^{-0.35(10-t)})$$

where t is in hours, with $0 \le t \le 10$.

   a. What percent of the drug is present after 2 hours?

   b. Graph this function.

   c. When is the drug totally gone from the bloodstream?

55. *Normal Curve* The "curve" on which many students like to be graded is the bell-shaped normal curve. The equation $y = \dfrac{1}{\sqrt{2\pi}}e^{-(x-50)^2/2}$ describes the normal curve for a standardized test, where x is the test score before curving.

   a. Graph this function for x between 47 and 53 and for y between 0 and 0.5.

   b. The average score for the test is the score that gives the largest output y. Use the graph to find the average score.

56. *IQ Measure* The frequency of IQ measures x follows a bell-shaped normal curve with equation $y = \dfrac{1}{\sqrt{2\pi}}e^{-(x-100)^2/20}$. Graph this function for x between 80 and 120 and for y between 0 and 0.4.

5.2   Logarithmic Functions; Properties of Logarithms

KEY OBJECTIVES

- Graph and evaluate logarithmic functions
- Convert equations to logarithmic and exponential forms
- Evaluate and apply common logarithms
- Evaluate and apply natural logarithms
- Apply logarithmic properties

SECTION PREVIEW   **Diabetes**

Projections from 2010 to 2050 indicate that the fraction of U.S. adults with diabetes (diagnosed and undiagnosed) will eventually reach one-third of the adult population. The percent can be modeled by the logarithmic function

$$p(x) = -12.975 + 11.851 \ln x$$

where x is the number of years after 2000. To find the year when the percent reaches 33%, we can graph the equations $y_1 = -12.975 + 11.851 \ln x$ and $y_2 = 33$ on the same axes and find the point of intersection. (See Example 8.) (Source: Centers for Disease Control and Prevention)

In this section, we introduce logarithmic functions, discuss the relationship between exponential and logarithmic functions, and use the definition of logarithmic functions to rewrite exponential equations in a new form, called the logarithmic form of the equation. We also use logarithmic functions in real applications. ■

Logarithmic Functions

Recall that we discussed inverse functions in Section 4.3. The inverse of a function is a second function that "undoes" what the original function does. For example, if the original function doubles each input, its inverse takes half of each input. If a function is defined by $y = f(x)$, the inverse of this function can be found by interchanging x and y in $y = f(x)$ and solving the new equation for y. As we saw in Section 4.3, a function has an inverse if it is a one-to-one function (that is, there is a one-to-one correspondence between the inputs and outputs defining the function).

Every exponential function of the form $y = b^x$, with $b > 0$ and $b \neq 1$, is one-to-one, so every exponential function of this form has an inverse function. The inverse function of

$$y = b^x$$

is found by interchanging x and y and solving the new equation for y. Interchanging x and y in $y = b^x$ gives

$$x = b^y$$

In this new function, y is the power to which we raise b to get the number x. To solve an expression like this for the exponent, we need new notation. We define y, the power to which we raise the base b to get the number x, as

$$y = \log_b x$$

This inverse function is called a **logarithmic function** with base b.

Logarithmic Function

For $x > 0$, $b > 0$, and $b \neq 1$, the logarithmic function with base b is

$$y = \log_b x$$

which is defined by $x = b^y$. That is, $\log_b x$ is the exponent to which we must raise the base b to get the number x.

This function is the inverse function of the exponential function $y = b^x$.

Note that, according to the definition of the logarithmic function, $y = \log_b x$ and $x = b^y$ are two different forms of the same equation. We call $y = \log_b x$ the *logarithmic form* of the equation and $x = b^y$ the *exponential form* of the equation. The number b is called the **base** in both $y = \log_b x$ and $x = b^y$, and y is the **logarithm** in $y = \log_b x$ and the **exponent** in $x = b^y$. Thus, a logarithm is an exponent.

EXAMPLE 1 ▶   Converting from Exponential to Logarithmic Form

Write each of the following exponential equations in logarithmic form.

a. $3^2 = 9$ **b.** $4^{-1} = \dfrac{1}{4}$

c. $5^1 = 5$ **d.** $x = 3^y$

SOLUTION

To write an exponential equation in its equivalent logarithmic form, remember that the base of the logarithm is the same as the base of the exponent in the exponential form. Also, the exponent in the exponential form is the value of the logarithm in the logarithmic form.

a. To write $3^2 = 9$ in logarithmic form, note that the base of the exponent is 3, so the base of the logarithm will also be 3. The exponent, 2, will be the value of the logarithm. Thus,

$$3^2 = 9 \quad \text{is equivalent to} \quad \log_3 9 = 2$$

We can refer to 9 as the **argument** of the logarithm.

b. To write $4^{-1} = \dfrac{1}{4}$ in logarithmic form, note that the base of the exponent is 4, so the base of the logarithm will also be 4. The exponent, -1, will be the value of the logarithm. Thus,

$$4^{-1} = \frac{1}{4} \quad \text{is equivalent to} \quad \log_4 \frac{1}{4} = -1$$

c. The base of the exponent in $5^1 = 5$ is 5, so the base of the logarithm will be 5, and the value of the logarithm will be 1. Thus,

$$5^1 = 5 \quad \text{is equivalent to} \quad \log_5 5 = 1$$

d. Similarly, the base of the exponent in $x = 3^y$ is 3, so the base of the logarithm will be 3, and the value of the logarithm will be y. Thus,

$$x = 3^y \quad \text{is equivalent to} \quad \log_3 x = y$$

EXAMPLE 2 ▶ **Converting from Logarithmic to Exponential Form**

Write each of the following logarithmic equations in exponential form.

a. $\log_2 16 = 4$ **b.** $\log_{10} 0.0001 = -4$ **c.** $\log_6 1 = 0$

SOLUTION

To write a logarithmic equation in its equivalent exponential form, remember that the base of the logarithm will be the base of the exponent in the exponential form. Also, the value of the logarithm will be the exponent in the exponential form. The argument of the logarithm will be the value of the exponential expression.

a. To write $\log_2 16 = 4$ in exponential form, note that the base of the exponent will be 2, the exponent will be 4, and the value of the exponential expression will be 16. Thus,

$$\log_2 16 = 4 \quad \text{is equivalent to} \quad 2^4 = 16$$

b. To write $\log_{10} 0.0001 = -4$ in exponential form, note that the base of the exponent will be 10, the exponent will be -4, and the value of the exponential expression will be 0.0001. Thus,

$$\log_{10} 0.0001 = -4 \quad \text{is equivalent to} \quad 10^{-4} = 0.0001$$

c. The base of the exponent in $\log_6 1 = 0$ will be 6, the exponent will be 0, and the value of the exponential expression will be 1. Thus,

$$\log_6 1 = 0 \quad \text{is equivalent to} \quad 6^0 = 1$$

We can sometimes more easily evaluate logarithmic functions for different inputs by changing the function from logarithmic form to exponential form.

EXAMPLE 3 ▶   **Evaluating Logarithms**

a. Write $y = \log_2 x$ in exponential form.

b. Use the exponential form from part (a) to find y when $x = 8$.

c. Evaluate $\log_2 8$.   **d.** Evaluate $\log_4 \dfrac{1}{16}$.   **e.** Evaluate $\log_{10} 0.001$.

SOLUTION

a. To rewrite $y = \log_2 x$ in exponential form, note that the base of the logarithm, 2, becomes the base of the exponential expression, and y equals the exponent in the equivalent exponential form. Thus, the exponential form is

$$2^y = x$$

b. To find y when x is 8, we solve $2^y = 8$. We can easily see that $y = 3$ because $2^3 = 8$.

c. To evaluate $\log_2 8$, we write $\log_2 8 = y$ in its equivalent exponential form $2^y = 8$. From part (b), $y = 3$. Thus, $\log_2 8 = 3$.

d. To evaluate $\log_4 \dfrac{1}{16}$, we write $\log_4 \dfrac{1}{16} = y$ in its equivalent exponential form $4^y = \dfrac{1}{16}$. Because $4^{-2} = \dfrac{1}{16}$, $y = -2$. Thus, $\log_4 \dfrac{1}{16} = -2$.

e. To evaluate $\log_{10} 0.001$, we write $\log_{10} 0.001 = y$ in its equivalent exponential form $10^y = 0.001$. Because $10^{-3} = 0.001$, $y = -3$. Thus, $\log_{10} 0.001 = -3$.

The relationship between the logarithmic and exponential forms of an equation is used in the following example.

EXAMPLE 4 ▶   **Graphing a Logarithmic Function**

Graph $y = \log_2 x$.

SOLUTION

We can graph $y = \log_2 x$ by graphing the equivalent equation $x = 2^y$. Table 5.4 gives the values (found by substituting values of y and calculating x), and the graph is shown in Figure 5.14.

Table 5.4

$x = 2^y$	y
$\dfrac{1}{8}$	-3
$\dfrac{1}{4}$	-2
$\dfrac{1}{2}$	-1
1	0
2	1
4	2
8	3

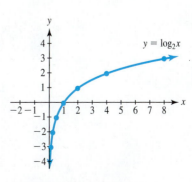

Figure 5.14

Figures 5.15(a) and (b) show graphs of the logarithmic function $y = \log_b x$ for a base $b > 1$ and a base $0 < b < 1$. Note that the graph of the function in Figure 5.15(a) is increasing and the graph of the function in Figure 5.15(b) is decreasing.

Logarithmic Function

Equation:	$y = \log_b x$
x-intercept:	$(1, 0)$
y-intercept:	none
Domain:	$x > 0$
Range:	all real numbers
Vertical asymptote:	*y*-axis (the line $x = 0$)

$y = \log_b x \ (b > 1)$

$y = \log_b x \ (0 < b < 1)$

(a)   (b)

Figure 5.15

Common Logarithms

We can use technology to evaluate logarithmic functions if the base of the logarithm is 10 or e. Because our number system uses base 10, logarithms with a base of 10 are called **common logarithms**, and, by convention, $\log_{10} x$ is simply denoted $\log x$. The graph of $y = \log x$ can be drawn by plotting points satisfying $x = 10^y$. Table 5.5 shows some points on the graph of $y = \log x$, using powers of 10 as *x*-values, and Figure 5.16 shows the graph.

Table 5.5

x	$y = \log x$
1000	3
100	2
10	1
1	0
0.1	−1
0.01	−2
0.001	−3

$y = \log x$

Figure 5.16

Graphing calculators can be used to evaluate and graph $y = \log x$, using the **LOG** key. Figure 5.17(a) shows the graph of $y = \log x$ on a graphing calculator.

(a)

Figure 5.17

Spreadsheet ▶ SOLUTION The graph of a logarithmic function can be created with Excel by entering the formula $= \log(x, b)$ where x is a cell entry and b is the base. The graph in Figure 5.17(b) results from the Excel command $= \log(x, 10)$.*

(b)

Figure 5.17

EXAMPLE 5 ▶ ## Evaluating Base 10 Logarithms

Using each of the indicated methods, find $f(10{,}000)$ if $f(x) = \log x$.

a. Write the equation $y = \log x$ in exponential form and find y when $x = 10{,}000$.

b. Use technology to evaluate $\log 10{,}000$.

SOLUTION

a. The exponential form of $y = \log x = \log_{10} x$ is $10^y = x$. Substituting 10,000 for x in this equation gives $10^y = 10{,}000$. Because $10^4 = 10{,}000$, we see that $y = 4$. Thus, $f(10{,}000) = 4$.

b. The value of $\log 10{,}000$ is shown to be 4 on the calculator screen in Figure 5.18, so $f(10{,}000) = 4$.

Figure 5.18

---

* For details, see Appendix B, page 687.

EXAMPLE 6 ▶ **pH Scale**

To simplify measurement of the acidity or basicity of a solution, the pH (hydrogen potential) scale was developed. The pH is given by the formula pH $= -\log[H^+]$, where $[H^+]$ is the concentration of hydrogen ions in moles per liter in the solution. The pH scale ranges from 0 to 14. A pH of 7 is neutral; a pH less than 7 is acidic; a pH greater than 7 is basic.

a. Using Table 5.6, find the pH level of bleach.

b. Using Table 5.6, find the pH level of lemon juice.

c. Lower levels of pH indicate a more acidic solution. How much more acidic is lemon juice than bleach?

Table 5.6

Concentration of Hydrogen Ions	Examples	Concentration of Hydrogen Ions	Examples
10^{-14}	Liquid drain cleaner	10^{-6}	Milk
10^{-13}	Bleach, oven cleaner	10^{-5}	Black coffee
10^{-12}	Soapy water	10^{-4}	Tomato juice
10^{-11}	Ammonia	10^{-3}	Orange juice
10^{-10}	Milk of magnesia	10^{-2}	Lemon juice
10^{-9}	Toothpaste	10^{-1}	Sulfuric acid
10^{-8}	Baking soda	1	Battery acid
10^{-7}	"Pure" water		

SOLUTION

a. The concentration of hydrogen ions for bleach is 10^{-13}. So the pH will be $-\log(10^{-13}) = 13$. See Figure 5.19(a).

(a) (b)

Figure 5.19

b. The concentration of hydrogen ions for lemon juice is 10^{-2}. So the pH will be $-\log(10^{-2}) = 2$. See Figure 5.19(b).

c. To find the ratio of the acidity of the two solutions, we divide the concentrations of hydrogen ions. Thus, the ratio is

$$\frac{10^{-2}}{10^{-13}} = \frac{10^{13}}{10^2} = 10^{11} = 100,000,000,000$$

indicating that lemon juice is 100,000,000,000 times as acidic as bleach.

EXAMPLE 7 ▶   **Transformations of Graphs of Logarithmic Functions**

Explain how the graph of each of the following functions compares with the graph of $y = \log x$, find the domain, and graph each function.

a. $y = \log(x + 3)$ **b.** $y = 4 + \log(x - 2)$ **c.** $y = \dfrac{1}{3}\log x$

SOLUTION

a. The graph of $y = \log(x + 3)$ has the same shape as the graph of $y = \log x$, but it is shifted 3 units to the left. Note that the vertical asymptote of the graph of $y = \log x$, $x = 0$, is also shifted 3 units left, so the vertical asymptote of $y = \log(x + 3)$ is $x = -3$. This means that, as the x-values approach -3, the y-values decrease without bound. See Figure 5.20(a). Because the graph has been shifted left 3 units, the domain is $(-3, \infty)$.

b. The graph of $y = 4 + \log(x - 2)$ has the same shape as the graph of $y = \log x$, but it is shifted 2 units to the right and 4 units up. The vertical asymptote is $x = 2$, and the graph is shown in Figure 5.20(b). The domain is $(2, \infty)$.

c. As we saw in Chapter 4, multiplication of a function by a constant less than 1 compresses the graph by a factor equal to that constant. So each of the y-values of $y = \dfrac{1}{3}\log x$ is $\dfrac{1}{3}$ times the corresponding y-value of $y = \log x$, and the graph is compressed. See Figure 5.20(c). The domain is $(0, \infty)$.

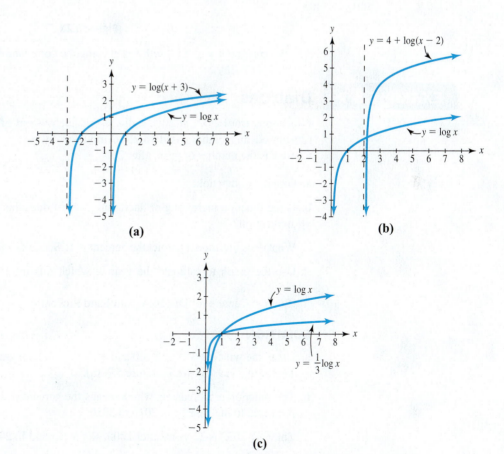

(a)

(b)

(c)

Figure 5.20

Natural Logarithms

As we have seen in the previous section, the number e is important in many business and life science applications. Likewise, logarithms to base e are important in many applications. Because they are used so frequently, especially in science, they are called **natural logarithms**, and we use the special notation $\ln x$ to denote $\log_e x$.

> ### Natural Logarithms
>
> The logarithmic function with base e is $y = \log_e x$, defined by $x = e^y$ for all positive numbers x and denoted by
>
> $$\ln x = \log_e x$$

Most graphing utilities and spreadsheets have the key or command **LN** to use in evaluating functions involving natural logarithms. Figure 5.21(a) shows the graph of $y = \ln x$, which is similar to the graph of $y = \log_3 x$ (Figure 5.21(b)) because the value of e, the base of $\ln x$, is close to 3.

(a)   (b)

Figure 5.21

We can graph $y = \ln x$ with Excel using the command =ln(x).*

EXAMPLE 8 ▶ **Diabetes**

Projections from 2010 to 2050 indicate that the percent of U.S. adults with diabetes (diagnosed and undiagnosed) can be modeled by $p(x) = -12.975 + 11.851 \ln x$, where x is the number of years after 2000.

a. Graph this function.

b. Is the function increasing or decreasing? What does this mean in the context of the application?

c. What does this model predict the percent of U.S. adults with diabetes will be in 2022?

d. Use the graph to estimate the year in which this model predicts the percent will reach 33%.
(Source: Centers for Disease Control and Prevention)

SOLUTION

a. Using the window $0 \le x \le 60$ and $0 \le y \le 50$, the graph of $p(x) = -12.975 + 11.851 \ln x$ is as shown in Figure 5.22(a).

b. The function is increasing, which means the percent of U.S. adults with diabetes is predicted to increase from 2010 to 2050.

c. The year 2022 is 22 years after 2000, so the percent in 2022 is estimated to be

$$p(22) = -12.975 + 11.851 \ln 22 \approx 23.7$$

---

* For details, see Appendix B, page 687.

d. To find the x-value for which $y = p(x) = 33$, we graph $y_1 = -12.975 + 11.851 \ln x$ and $y_2 = 33$ on the same axes and find the point of intersection. This shows that the percent is 33% when $x \approx 48.4$, during 2049. (See Figure 5.22(b).)

(a)

(b)

Figure 5.22

Logarithmic Properties

We have learned that logarithms are exponents and that a logarithmic function with base a is the inverse of an exponential function with base a. Thus, the properties of logarithms can be derived from the properties of exponents. These properties are useful in solving equations involving exponents and logarithms.

The basic properties of logarithms are easy to derive from the definition of a logarithm.

Basic Properties of Logarithms

For $b > 0, b \neq 1$,
1. $\log_b b = 1$
2. $\log_b 1 = 0$
3. $\log_b b^x = x$
4. $b^{\log_b x} = x$
5. For positive real numbers M and N, if $M = N$, then $\log_b M = \log_b N$.

These properties are easily justified by rewriting the logarithmic form as an exponential form or vice versa. In particular, $\log_b 1 = 0$ because $b^0 = 1$, $\log_b b = 1$ because $b^1 = b$, $\log_b b^x = x$ because $b^x = b^x$, and $b^{\log_b x} = x$ because $\log_b x = \log_b x$.

EXAMPLE 9 ▶ **Logarithmic Properties**

Use the basic properties of logarithms to simplify the following.

a. $\log_5 5^{10}$ **b.** $\log_4 4$ **c.** $\log_4 1$ **d.** $\log 10^7$ **e.** $\ln e^3$ **f.** $\log\left(\dfrac{1}{10^3}\right)$

SOLUTION

a. $\log_5 5^{10} = 10$ by Property 3   **b.** $\log_4 4 = 1$ by Property 1

c. $\log_4 1 = 0$ by Property 2   **d.** $\log 10^7 = \log_{10} 10^7 = 7$ by Property 3

e. $\ln e^3 = \log_e e^3 = 3$ by Property 3   **f.** $\log\left(\dfrac{1}{10^3}\right) = \log 10^{-3} = \log_{10} 10^{-3} = -3$

Other logarithmic properties are helpful in simplifying logarithmic expressions, which in turn help us solve some logarithmic equations. These properties are also directly related to the properties of exponents.

Additional Logarithmic Properties

For $b > 0, b \neq 1$, k a real number, and M and N positive real numbers,

6. $\log_b(MN) = \log_b M + \log_b N$ (Product Property)

7. $\log_b\left(\dfrac{M}{N}\right) = \log_b M - \log_b N$ (Quotient Property)

8. $\log_b M^k = k \log_b M$ (Power Property)

To prove Property 6, we let $x = \log_b M$ and $y = \log_b N$, so that $M = b^x$ and $N = b^y$. Then

$$\log_b(MN) = \log_b(b^x \cdot b^y) = \log_b(b^{x+y}) = x + y = \log_b M + \log_b N$$

To prove Property 8, we let $x = \log_b M$, so that $M = b^x$. Then

$$\log_b M^k = \log_b(b^x)^k = \log_b b^{kx} = kx = k \log_b M$$

EXAMPLE 10 ▶ **Rewriting Logarithms**

Rewrite each of the following expressions as a sum, difference, or product of logarithms, and simplify if possible.

a. $\log_4 5(x - 7)$ **b.** $\ln[e^2(e + 3)]$ **c.** $\log\left(\dfrac{x - 8}{x}\right)$

d. $\log_4 y^6$ **e.** $\ln\left(\dfrac{1}{x^5}\right)$

SOLUTION

a. By Property 6, $\log_4 5(x - 7) = \log_4 5 + \log_4(x - 7)$.

b. By Property 6, $\ln[e^2(e + 3)] = \ln e^2 + \ln(e + 3)$.
By Property 3, this equals $2 + \ln(e + 3)$.

c. By Property 7, $\log\left(\dfrac{x - 8}{x}\right) = \log(x - 8) - \log x$.

d. By Property 8, $\log_4 y^6 = 6 \log_4 y$.

e. By Property 7, Property 2, and Property 8,

$$\ln\left(\frac{1}{x^5}\right) = \ln 1 - \ln x^5 = 0 - 5 \ln x = -5 \ln x$$

Caution: There is no property of logarithms that allows us to rewrite a logarithm of a sum or a difference. That is why in Example 10(a) the expression $\log_4(x - 7)$ was **not** written as $\log_4 x - \log_4 7$.

EXAMPLE 11 ▶ **Rewriting Logarithms**

Rewrite each of the following expressions as a single logarithm.

a. $\log_3 x + 4 \log_3 y$ **b.** $\dfrac{1}{2} \log a - 3 \log b$ **c.** $\ln(5x) - 3 \ln z$

SOLUTION

a. $\log_3 x + 4 \log_3 y = \log_3 x + \log_3 y^4$ Use Logarithmic Property 8.

$\qquad\qquad\qquad\qquad = \log_3 xy^4$ Use Logarithmic Property 6.

b. $\dfrac{1}{2} \log a - 3 \log b = \log a^{1/2} - \log b^3$ Use Logarithmic Property 8.

$\qquad\qquad\qquad\qquad = \log\left(\dfrac{a^{1/2}}{b^3}\right)$ Use Logarithmic Property 7.

c. $\ln(5x) - 3 \ln z = \ln(5x) - \ln z^3$ Use Logarithmic Property 8.

$\qquad\qquad\qquad = \ln\left(\dfrac{5x}{z^3}\right)$ Use Logarithmic Property 7.

EXAMPLE 12 ▶ **Applying the Properties of Logarithms**

a. Use logarithmic properties to estimate $\ln(5e)$ if $\ln 5 \approx 1.61$.

b. Find $\log_a\left(\dfrac{15}{8}\right)$ if $\log_a 15 = 2.71$ and $\log_a 8 = 2.079$.

c. Find $\log_a\left(\dfrac{15^2}{\sqrt{8}}\right)^3$ if $\log_a 15 = 2.71$ and $\log_a 8 = 2.079$.

SOLUTION

a. $\ln(5e) = \ln 5 + \ln e \approx 1.61 + 1 = 2.61$

b. $\log_a\left(\dfrac{15}{8}\right) = \log_a 15 - \log_a 8 = 2.71 - 2.079 = 0.631$

c. $\log_a\left(\dfrac{15^2}{\sqrt{8}}\right)^3 = 3 \log_a\left(\dfrac{15^2}{\sqrt{8}}\right) = 3\left[\log_a 15^2 - \log_a 8^{1/2}\right]$

$\qquad\qquad\qquad = 3\left[2 \log_a 15 - \dfrac{1}{2} \log_a 8\right] = 3\left[2(2.71) - \dfrac{1}{2}(2.079)\right] = 13.1415$

Richter Scale

Because the intensities of earthquakes are so large, the Richter scale was developed to provide a "scaled down" measuring system that permits earthquakes to be more easily measured and compared. The Richter scale gives the magnitude R of an earthquake using the formula

$$R = \log\left(\frac{I}{I_0}\right)$$

where I is the intensity of the earthquake and I_0 is a certain minimum intensity used for comparison.

Table 5.7 describes the typical effects of earthquakes of various magnitudes near the epicenter.

Table 5.7

Description	Richter Magnitudes	Earthquake Effects
Micro	Less than 2.0	Microearthquakes, not felt
Very minor	2.0–2.9	Generally not felt, but recorded
Minor	3.0–3.9	Often felt, but rarely causes damage
Light	4.0–4.9	Noticeable shaking of indoor items, rattling noises; significant damage unlikely
Moderate	5.0–5.9	Can cause damage to poorly constructed buildings over small regions
Strong	6.0–6.9	Can be destructive in areas up to about 100 miles across in populated areas
Major	7.0–7.9	Can cause serious damage over larger areas
Great	8.0–8.9	Can cause serious damage in areas several hundred miles across
Rarely, great	9.0–9.9	Devastating in areas several thousand miles across
Meteoric	10.0+	Never recorded

(Source: U.S. Geological Survey)

EXAMPLE 13 ▶   **Richter Scale**

a. If an earthquake has an intensity of 10,000 times I_0, what is the magnitude of the earthquake? Describe the effects of this earthquake.

b. Show that if the Richter scale reading of an earthquake is k, the intensity of the earthquake is $I = 10^k I_0$.

c. An earthquake that measured 9.0 on the Richter scale occurred in the Indian Ocean in December 2004, causing a devastating tsunami that killed thousands of people. Express the intensity of this earthquake in terms of I_0.

d. If an earthquake measures 7.0 on the Richter scale, give the intensity of this earthquake in terms of I_0. How much more intense is the earthquake in part (c) than the one with the Richter scale measurement of 7.0?

e. If one earthquake has intensity 320,000 times I_0 and a second has intensity 3,200,000 times I_0, what is the difference in their Richter scale measurements?

SOLUTION

a. If the intensity of the earthquake is $10{,}000 I_0$, we substitute $10{,}000 I_0$ for I in the equation $R = \log\left(\dfrac{I}{I_0}\right)$, getting $R = \log \dfrac{10{,}000 I_0}{I_0} = \log 10{,}000 = 4$. Thus, the magnitude of the earthquake is 4, signifying a light earthquake that would not likely cause significant damage.

b. If $R = k$, $k = \log\left(\dfrac{I}{I_0}\right)$, and the exponential form of this equation is $10^k = \dfrac{I}{I_0}$, so $I = 10^k I_0$.

c. The Richter scale reading for this earthquake is 9.0, so its intensity is

$$.I = 10^{9.0} I_0 = 1{,}000{,}000{,}000 \, I_0$$

Thus, the earthquake had an intensity of 1 billion times I_0.

d. The Richter scale reading for this earthquake is 7.0, so its intensity is $I = 10^{7.0} I_0$. Thus, the ratio of intensities of the two earthquakes is

$$\frac{10^{9.0} I_0}{10^{7.0} I_0} = \frac{10^{9.0}}{10^{7.0}} = 10^2 = 100$$

Therefore, the Indian Ocean earthquake is 100 times as intense as the other earthquake.

e. Recall that the Richter scale measurement is given by $R = \log\left(\dfrac{I}{I_0}\right)$. The difference of the Richter scale measurements is

$$R_2 - R_1 = \log\frac{3{,}200{,}000 I_0}{I_0} - \log\frac{320{,}000 I_0}{I_0} = \log 3{,}200{,}000 - \log 320{,}000$$

Using Property 7 gives us

$$\log 3{,}200{,}000 - \log 320{,}000 = \log\frac{3{,}200{,}000}{320{,}000} = \log 10 = 1$$

Thus, the difference in the Richter scale measurements is 1.

Note from Example 13(d) that the Richter scale measurement of the Indian Ocean earthquake is 2 larger than the Richter scale measurement of the other earthquake, and that the Indian Ocean earthquake is 10^2 times as intense. In general, we have the following.

Richter Scale

1. If the intensity of an earthquake is I, its Richter scale measurement is

$$R = \log\frac{I}{I_0}$$

2. If the Richter scale reading of an earthquake is k, its intensity is

$$I = 10^k I_0$$

3. If the difference of the Richter scale measurements of two earthquakes is the positive number d, the intensity of the larger earthquake is 10^d times that of the smaller earthquake.

SKILLS CHECK 5.2

In Exercises 1–4, write the logarithmic equations in exponential form.

1. $y = \log_3 x$

2. $2y = \log_5 x$

3. $y = \ln(2x)$

4. $y = \log(-x)$

In Exercises 5–8, write the exponential equations in logarithmic form.

5. $x = 4^y$

6. $m = 3^p$

7. $32 = 2^5$

8. $9^{2x} = y$

In Exercises 9 and 10, evaluate the logarithms, if possible. Round each answer to three decimal places.

9. a. $\log 7$

   b. $\ln 86$

   c. $\log 63{,}980$

10. a. $\log 456$

   b. $\log(-12)$

   c. $\ln 10$

In Exercises 11–13, find the value of the logarithms without using a calculator.

11. a. $\log_2 32$

   b. $\log_9 81$

   c. $\log_3 27$

   d. $\log_4 64$

   e. $\log_5 625$

12. a. $\log_2 64$

   b. $\log_9 27$

   c. $\log_4 2$

   d. $\ln(e^3)$

   e. $\log 100$

13. a. $\log_3 \dfrac{1}{27}$

   b. $\ln 1$

   c. $\ln e$

   d. $\log 0.0001$

14. Graph the functions by changing to exponential form.

   a. $y = \log_3 x$

   b. $y = \log_5 x$

In Exercises 15–18, graph the functions.

15. $y = 2 \ln x$

16. $y = 4 \log x$

17. $y = \log(x + 1) + 2$

18. $y = \ln(x - 2) - 3$

19. a. Write the inverse of $y = 4^x$ in logarithmic form.

   b. Graph $y = 4^x$ and its inverse and discuss the symmetry of their graphs.

20. a. Write the inverse of $y = 3^x$ in logarithmic form.

   b. Graph $y = 3^x$ and its inverse and discuss the symmetry of their graphs.

21. Write $\log_a a = x$ in exponential form and find x to evaluate $\log_a a$ for any $a > 0, a \neq 1$.

22. Write $\log_a 1 = x$ in exponential form and find x to evaluate $\log_a 1$ for any $a > 0, a \neq 1$.

In Exercises 23–26, use the properties of logarithms to evaluate the expressions.

23. $\log 10^{14}$

24. $\ln e^5$

25. $10^{\log_{10} 12}$

26. $6^{\log_6 25}$

In Exercises 27–30, use $\log_a(20) = 1.4406$ and $\log_a(5) = 0.7740$ to evaluate each expression.

27. $\log_a(100)$

28. $\log_a(4)$

29. $\log_a 5^3$

30. $\log_a \sqrt{20}$

In Exercises 31–34, rewrite each expression as a sum, difference, or product of logarithms, and simplify if possible.

31. $\ln \dfrac{3x - 2}{x + 1}$

32. $\log[x^3(3x - 4)^5]$

33. $\log_3 \dfrac{\sqrt[3]{4x + 1}}{4x^2}$

34. $\log_3 \dfrac{\sqrt[3]{3x - 1}}{5x^2}$

In Exercises 35–38, rewrite each expression as a single logarithm.

35. $3 \log_2 x + \log_2 y$

36. $\log x - \dfrac{1}{3} \log y$

37. $4 \ln(2a) - \ln b$

38. $6 \ln(5y) + 2 \ln x$

EXERCISES 5.2

39. **Life Span** On the basis of data and projections for the years 1910 through 2020 the expected life span of people in the United States can be described by the function $f(x) = 11.027 + 14.304 \ln x$ years, where x is the number of years from 1900 to the person's birth year.

   a. What does this model estimate the life span to be for people born in 1925? In 2007? (Give each answer to the nearest year.)

   b. Explain why these numbers are so different.
   (Source: National Center for Health Statistics)

40. **U.S. Population** Using data for selected years from 1990 and projected to 2050, the total U.S. population can be modeled by $y = -363.3 + 166.9 \ln x$, with x equal to the number of years after 1950 and y in millions. Use the model to predict the U.S. population in 2026 and in 2033.

41. **Women in the Workforce** Using data from 1970 and projected to 2050, the number of millions of women in the civilian workforce can be modeled by $y = -35.700 + 27.063 \ln x$, where x is the number of years after 1960.

   a. What does this model predict the number of women in the workforce will be in 2025?

   b. Is this function increasing or decreasing?

   c. Graph this function from $x = 10$ to $x = 90$.

42. **Loan Repayment** The number of years t that it takes to pay off a $100,000 loan at 10% interest by making annual payments of R dollars is

   $$t = \frac{\ln R - \ln(R - 10{,}000)}{\ln 1.1}, R > 10{,}000$$

   If the annual payment is $16,274.54, in how many years will the loan be paid off?

43. **Supply** Suppose that the supply of a product is given by $p = 20 + 6 \ln(2q + 1)$, where p is the price per unit and q is the number of units supplied. What price will give a supply of 5200 units?

44. **Demand** Suppose that the demand function for a product is $p = \dfrac{500}{\ln(q + 1)}$, where p is the price per unit and q is the number of units demanded. What price will give a demand for 6400 units?

45. **Female Workers** For the years 1970 to 2040, the percent of females over 15 years old in the workforce is given or projected by $y = 27.4 + 5.02 \ln x$, where x is the number of years from 1960.

   a. What does the model predict the percent to be in 2011? In 2015?

   b. Is the percent of female workers increasing or decreasing?

46. **Obesity** The percent of Americas who are obese from 1990 projected through 2030 can be modeled by $y = -31.732 + 18.672 \ln x$, with x equal to the number of years after 1980.

   a. Use the model to predict the percent of Americas who will be obese in 2032.

   b. Does this model predict that obesity will get better or worse in the near future?

47. **Doubling Time** If $4000 is invested in an account earning 10% annual interest, compounded continuously, then the number of years that it takes for the amount to grow to $8000 is $n = \dfrac{\ln 2}{0.10}$. Find the number of years.

48. **Doubling Time** If $5400 is invested in an account earning 7% annual interest, compounded continuously, then the number of years that it takes for the amount to grow to $10,800 is $n = \dfrac{\ln 2}{0.07}$. Find the number of years.

49. **Doubling Time** The number of quarters (a quarter equals 3 months) needed to double an investment when a lump sum is invested at 8%, compounded quarterly, is given by $n = \dfrac{\log 2}{0.0086}$. In how many years will the investment double?

50. **Doubling Time** The number of periods needed to double an investment when a lump sum is invested at 12%, compounded semiannually, is $n = \dfrac{\log 2}{0.0253}$. How many years pass before the investment doubles in value?

The number of years t it takes for an investment to double if it earns r percent (as a decimal), compounded annually, is $t = \dfrac{\ln 2}{\ln(1 + r)}$. Use this formula in Exercises 51 and 52.

51. **Investing** In how many years will an investment double if it is invested at 8%, compounded annually?

52. *Investing* In how many years will an investment double if it is invested at 12.3%, compounded annually?

53. *Earthquakes* If an earthquake has an intensity of 25,000 times I_0, what is the magnitude of the earthquake?

54. *Earthquakes*
   a. If an earthquake has an intensity of 250,000 times I_0, what is the magnitude of the earthquake?

   b. Compare the magnitudes of two earthquakes when the intensity of one is 10 times the intensity of the other. (See Exercise 53.)

55. *Earthquakes* The Richter scale measurement for the southern Sumatra earthquake of 2007 was 6.4. Find the intensity of this earthquake as a multiple of I_0.

56. *Earthquakes* The Richter scale measurement for the San Francisco earthquake that occurred in 1906 was 8.25. Find the intensity of this earthquake as a multiple of I_0.

57. *Earthquakes* The Richter scale reading for the San Francisco earthquake of 1989 was 7.1. Find the intensity I of this earthquake as a multiple of I_0.

58. *Earthquakes* On January 9, 2008, an earthquake in northern Algeria measuring 4.81 on the Richter scale killed 1 person, and on May 12, 2008, an earthquake in China measuring 7.9 on the Richter scale killed 67,180 people. How many times more intense was the earthquake in China than the one in northern Algeria?

59. *Earthquakes* The largest earthquake ever to strike San Francisco (in 1906) measured 8.25 on the Richter scale, and the second largest (in 1989) measured 7.1. How many times more intense was the 1906 earthquake than the 1989 earthquake?

60. *Earthquakes* The largest earthquake ever recorded in Japan occurred in 2011 and measured 9.0 on the Richter scale. An earthquake measuring 8.25 occurred in Japan in 1983. Calculate the ratio of the intensities of these earthquakes.

61. *Japan Tsunami* In May of 2008, an earthquake measuring 6.8 on the Richter scale struck near the east coast of Honshu, Japan. In March of 2011, an earthquake measuring 9.0 struck that same region, causing a devastating tsunami that killed thousands of people. The 2011 earthquake was how many times as intense as the one in 2008?

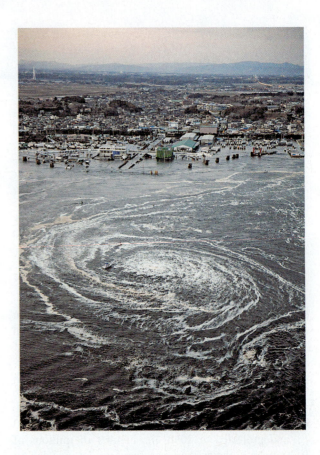

Use the following information to answer the questions in Exercises 62–67. To quantify the intensity of sound, the decibel scale (named for Alexander Graham Bell) was developed. The formula for loudness L on the decibel scale is $L = 10 \log \dfrac{I}{I_0}$, where I_0 is the intensity of sound just below the threshold of hearing, which is approximately 10^{-16} watt per square centimeter.

62. *Decibel Scale* Find the decibel reading for a sound with intensity 20,000 times I_0.

63. *Decibel Scale* Compare the decibel readings of two sounds if the intensity of the first sound is 100 times the intensity of the second sound.

64. *Decibel Scale* Use the exponential form of the function $L = 10 \log \dfrac{I}{I_0}$ to find the intensity of a sound if its decibel reading is 40.

65. *Decibel Scale* Use the exponential form of $L = 10 \log \dfrac{I}{I_0}$ to find the intensity of a sound if its decibel reading is 140. (Sound at this level causes pain.)

66. *Decibel Scale* The intensity of a whisper is $115I_0$, and the intensity of a busy street is $9,500,000I_0$. How do their decibel levels compare?

67. **Decibel Scale** If the decibel reading of a sound that is painful is 140 and the decibel reading of rock music is 120, how much more intense is the painful sound than the rock music?

To answer the questions in Exercises 68–72, use the fact that pH is given by the formula $\text{pH} = -\log[\text{H}^+]$.

68. **pH Levels** Find the pH value of beer for which $[\text{H}^+] = 0.0000631$.

69. **pH Levels** The pH value of eggs is 7.79. Find the hydrogen-ion concentration for eggs.

70. **pH Levels** The most common solutions have a pH range between 1 and 14. Use an exponential form of the pH formula to find the values of $[\text{H}^+]$ associated with these pH levels.

71. **pH Levels** Lower levels of pH indicate a more acidic solution. If the pH of ketchup is 3.9 and the pH of peanut butter is 6.3, how much more acidic is ketchup than peanut butter?

72. **pH Levels** Pure sea water in the middle of the ocean has a pH of 8.3. The pH of sea water in an aquarium must vary between 8 and 8.5; beyond these values, animals will experience certain physiological problems. If the pH of water in an aquarium is 8, how much more acidic is this water than pure sea water?

5.3 Exponential and Logarithmic Equations

KEY OBJECTIVES

- Solve an exponential equation by writing it in logarithmic form
- Convert logarithms using the change-of-base formula
- Solve an exponential equation by using properties of logarithms
- Solve logarithmic equations
- Solve exponential and logarithmic inequalities

SECTION PREVIEW **Carbon-14 Dating**

This is an artist's interpretation of the massive seabird *Pelagornis sandersi*, known from a 25-million-year-old fossil excavated in South Carolina. It had a wingspan of 24 feet.

Carbon-14 dating is one process by which scientists can tell the age of many fossils. To find the age of a fossil that originally contained 1000 grams of carbon-14 and now contains 1 gram, we solve the equation

$$1 = 1000e^{-0.00012097t}$$

An approximate solution of this equation can be found graphically, but sometimes a window that shows the solution is hard to find. If this is the case, it may be easier to solve an exponential equation like this one algebraically.

In this section, we consider two algebraic methods of solving exponential equations. The first method involves converting the equation to logarithmic form and then solving the logarithmic equation for the variable. The second method involves taking the logarithm of both sides of the equation and using the properties of logarithms to solve for the variable. ∎

Solving Exponential Equations Using Logarithmic Forms

When we wish to solve an equation for a variable that is contained in an exponent, we can remove the variable from the exponent by converting the equation to its logarithmic form. The steps in this solution method follow.

> **Solving Exponential Equations Using Logarithmic Forms**
>
> To solve an exponential equation using logarithmic form:
>
> 1. Rewrite the equation with the term containing the exponent by itself on one side.
> 2. Divide both sides by the coefficient of the term containing the exponent.
> 3. Change the new equation to logarithmic form.
> 4. Solve for the variable.

This method is illustrated in the following example.

EXAMPLE 1 ▶ **Solving a Base 10 Exponential Equation**

a. Solve the equation $3000 = 150(10^{4t})$ for t by converting it to logarithmic form.

b. Solve the equation graphically to confirm the solution.

SOLUTION

a. 1. The term containing the exponent is by itself on one side of the equation.

    2. We divide both sides of the equation by 150:

$$3000 = 150(10^{4t})$$
$$20 = 10^{4t}$$

    3. We rewrite this equation in logarithmic form:

$$4t = \log_{10} 20, \quad \text{or} \quad 4t = \log 20$$

    4. Solving for t gives the solution:

$$t = \frac{\log 20}{4} \approx 0.32526$$

b. We solve the equation graphically by using the intersection method. Entering $y_1 = 3000$ and $y_2 = 150(10^{4t})$, we graph using the window $[0, 0.4]$ by $[0, 3500]$ and find the point of intersection to be about $(0.32526, 3000)$ (Figure 5.23). Thus, the solution to the equation $3000 = 150(10^{4t})$ is $t \approx 0.32526$, as we found in part (a).

Figure 5.23

EXAMPLE 2 ▶   **Doubling Time**

a. Prove that the time it takes for an investment to double its value is $t = \dfrac{\ln 2}{r}$ if the interest rate is r, compounded continuously.

b. Suppose \$2500 is invested in an account earning 6% annual interest, compounded continuously. How long will it take for the amount to grow to \$5000?

SOLUTION

a. If P dollars are invested for t years at an annual interest rate r, compounded continuously, then the investment will grow to a future value S according to the equation

$$S = Pe^{rt}$$

and the investment will be doubled when $S = 2P$, giving

$$2P = Pe^{rt}, \quad \text{or} \quad 2 = e^{rt}$$

Converting this exponential equation to its equivalent logarithmic form gives

$$\log_e 2 = rt, \quad \text{or} \quad \ln 2 = rt$$

and solving this equation for t gives the doubling time,

$$t = \frac{\ln 2}{r}$$

b. We seek the time it takes to double the investment, and the doubling time is given by

$$t = \frac{\ln 2}{0.06} \approx 11.5525$$

Thus, it will take approximately 11.6 years for an investment of \$2500 to grow to \$5000 if it is invested at an annual rate of 6%, compounded continuously. We can confirm this solution graphically by finding the intersection of the graphs of $Y_1 = 2500e^{0.06x}$ and $Y_2 = 5000$, or numerically (see Figure 5.24). Note that this is the same time it would take any investment at 6%, compounded continuously, to double.

(a)

(b)

Figure 5.24

EXAMPLE 3 ▶   **Carbon-14 Dating**

Radioactive carbon-14 decays according to the equation

$$y = y_0 e^{-0.00012097t}$$

where y_0 is the original amount and y is the amount of carbon-14 at time t years. To find the age of a fossil if the original amount of carbon-14 was 1000 grams and the present amount is 1 gram, we solve the equation

$$1 = 1000e^{-0.00012097t}$$

a. Find the age of the fossil by converting the equation to logarithmic form.

b. Find the age of the fossil using graphical methods.

SOLUTION

a. To write the equation so that the coefficient of the exponential term is 1, we divide both sides by 1000, obtaining

$$0.001 = e^{-0.00012097t}$$

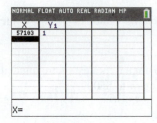

Figure 5.25

Writing this equation in logarithmic form gives the equation

$$-0.00012097t = \log_e 0.001, \quad \text{or} \quad -0.00012097t = \ln 0.001$$

Solving the equation for t gives

$$t = \frac{\ln 0.001}{-0.00012097} \approx 57{,}103$$

Thus, the age of the fossil is calculated to be about 57,103 years.

We can confirm that this is the solution to $1 = 1000e^{-0.00012097t}$ by using a table of values with $y_1 = 1000e^{-0.00012097t}$. The result of entering 57,103 for x is shown in Figure 5.25.

b. We can solve the equation graphically by using the intersection method. Entering

$$y_1 = 1000e^{-0.00012097t} \quad \text{and} \quad y_2 = 1$$

and investigating some y-values shows that we need very large x-values so that y is near 1 (see Figure 5.26 (a)). Graphing using the window $[0, 80000]$ by $[-1, 3]$, we find the x-coordinate of the intersection point to be 57,103.044 (see Figure 5.26(b).) Thus, as we found in part (a), the age of the fossil is about 57,103 years.

(a)

(b)

Figure 5.26

While finding this solution from a graph is easy, actually finding a window that contains the intersection point is not. It may be easier to solve the equation algebraically, as we did in part (a), and then use graphical or numerical methods to confirm the solution.

Change of Base

Most calculators and other graphing utilities can be used to evaluate logarithms and graph logarithmic functions if the base is 10 or e. Graphing logarithmic functions with other bases usually requires converting to exponential form to determine outputs and then graphing the functions by plotting points by hand. However, we can use a special formula called the **change-of-base formula** to rewrite logarithms so that the base is 10 or e. The formula for changing from base a to base b is developed below.

Suppose $y = \log_a x$. Then

$$a^y = x$$

$$\log_b a^y = \log_b x \qquad \text{Take logarithm, base } b, \text{ of both sides (Property 5 of Logarithms).}$$

$$y \log_b a = \log_b x \qquad \text{Use the Power Property of Logarithms.}$$

$$y = \frac{\log_b x}{\log_b a} \qquad \text{Solve for } y.$$

$$\log_a x = \frac{\log_b x}{\log_b a} \qquad \text{Substitute } \log_a x \text{ for } y.$$

The general change-of-base formula is summarized on the next page.

Change-of-Base Formula

If $b > 0$, $b \neq 1$, $a > 0$, $a \neq 1$, and $x > 0$, then

$$\log_a x = \frac{\log_b x}{\log_b a}$$

In particular, for base 10 and base e,

$$\log_a x = \frac{\log x}{\log a} \quad \text{and} \quad \log_a x = \frac{\ln x}{\ln a}$$

EXAMPLE 4 ▶   **Applying the Change-of-Base Formula**

a. Evaluate $\log_8 124$.

Graph the functions in parts (b) and (c) by changing each logarithm to a common logarithm and then by changing the logarithm to a natural logarithm.

b. $y = \log_3 x$ **c.** $y = \log_2(-3x)$

SOLUTION

a. The change-of-base formula (changing to base 10) can be used to evaluate this logarithm:

$$\log_8 124 = \frac{\log 124}{\log 8} = 2.318 \text{ approximately}$$

b. Changing $y = \log_3 x$ to base 10 gives $\log_3 x = \dfrac{\log x}{\log 3}$, so we graph $y = \dfrac{\log x}{\log 3}$

(Figure 5.27(a)). Changing to base e gives $\log_3 x = \dfrac{\ln x}{\ln 3}$, so we graph $y = \dfrac{\ln x}{\ln 3}$

(Figure 5.27(b)). Note that the graphs appear to be identical (as they should be because they both represent $y = \log_3 x$).

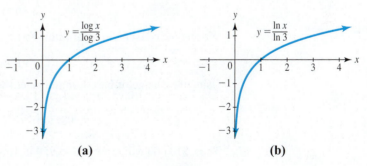

(a)   **(b)**

Figure 5.27

c. Changing $y = \log_2(-3x)$ to base 10 gives $\log_2(-3x) = \dfrac{\log(-3x)}{\log 2}$, so we graph

$y = \dfrac{\log(-3x)}{\log 2}$ (Figure 5.28(a)). Changing to base e gives $\log_2(-3x) = \dfrac{\ln(-3x)}{\ln 2}$,

so we graph $y = \dfrac{\ln(-3x)}{\ln 2}$ (Figure 5.28(b)). Note that the graphs are identical. Also notice that because logarithms are defined only for positive inputs, x must be negative so that $-3x$ is positive. Thus, the domain of $y = \log_2(-3x)$ is the interval $(-\infty, 0)$.

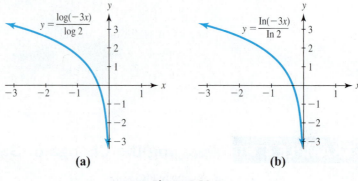

(a) **(b)**

Figure 5.28

The change-of-base formula is also useful in solving exponential equations whose base is neither 10 nor e.

EXAMPLE 5 ▶ **Investment**

If $10,000 is invested for t years at 10%, compounded annually, the future value is given by

$$S = 10,000(1.10^t)$$

In how many years will the investment grow to $45,950?

SOLUTION

We seek to solve the equation

$$45,950 = 10,000(1.10^t)$$

Dividing both sides by 10,000 gives

$$4.5950 = 1.10^t$$

Writing the equation in logarithmic form solves the equation for t:

$$t = \log_{1.10} 4.5950$$

Using the change-of-base formula gives the number of years:

$$t = \log_{1.10} 4.5950 = \frac{\log 4.5950}{\log 1.10} = 16$$

Thus, $10,000 will grow to $45,950 in 16 years.

Solving Exponential Equations Using Logarithmic Properties

With the properties of logarithms at our disposal, we consider a second method for solving exponential equations, which may be easier to use. This method of solving an exponential equation involves taking the logarithm of both sides of the equation (Property 5 of Logarithms) and then using properties of logarithms to write the equation in a form that we can solve.

> ## Solving Exponential Equations Using Logarithmic Properties
>
> To solve an exponential equation using logarithmic properties:
>
> 1. Rewrite the equation with a base raised to a power on one side.
> 2. Take the logarithm, base e or 10, of both sides of the equation.
> 3. Use a logarithmic property to remove the variable from the exponent.
> 4. Solve for the variable.

This method is illustrated in the following example.

EXAMPLE 6 ▶ Solution of Exponential Equations

Solve the following exponential equations.

a. $4096 = 8^{2x}$ **b.** $6(4^{3x-2}) = 120$

SOLUTION

a. 1. This equation has the base 8 raised to a variable power on one side.

2. Taking the logarithm, base 10, of both sides of the equation $4096 = 8^{2x}$ gives

$$\log 4096 = \log 8^{2x}$$

3. Using the Power Property of Logarithms removes the variable x from the exponent:

$$\log 4096 = 2x \log 8$$

4. Solving for x gives the solution:

$$\frac{\log 4096}{2 \log 8} = x$$

$$x = 2$$

b. We first isolate the exponential expression on one side of the equation by dividing both sides by 6:

$$\frac{6(4^{3x-2})}{6} = \frac{120}{6}$$

$$4^{3x-2} = 20$$

Taking the natural logarithm of both sides leads to the solution:

$$\ln 4^{3x-2} = \ln 20$$

$$(3x - 2) \ln 4 = \ln 20$$

$$3x - 2 = \frac{\ln 20}{\ln 4}$$

$$x = \frac{1}{3}\left(\frac{\ln 20}{\ln 4} + 2\right) \approx 1.387$$

An alternative method of solving the equation is to write $4^{3x-2} = 20$ in logarithmic form:

$$\log_4 20 = 3x - 2$$

$$x = \frac{\log_4 20 + 2}{3}$$

The change-of-base formula can be used to compute this value and verify that this solution is the same as above:

$$x = \frac{\dfrac{\ln 20}{\ln 4} + 2}{3} \approx 1.387$$

Solution of Logarithmic Equations

Some logarithmic equations can be solved by converting to exponential form.

EXAMPLE 7 ▶ Solving a Logarithmic Equation

Solve $4 \log_3 x = -8$ by converting to exponential form and verify the solution graphically.

SOLUTION

We first isolate the logarithm by dividing both sides of the equation by 4:

$$\log_3 x = -2$$

Writing $\log_3 x = -2$ in exponential form gives the solution

$$3^{-2} = x$$

$$x = \frac{1}{9}$$

Graphing $y_1 = 4 \log_3 x$ and $y_2 = -8$ with a window that contains $x = \dfrac{1}{9}$ and $y = -8$, we find the point of intersection to be $\left(\dfrac{1}{9}, -8 \right)$. See Figures 5.29(a) and (b).

(a)

(b)

Figure 5.29

EXAMPLE 8 ▶ Solving a Logarithmic Equation

a. Solve $6 + 3 \ln x = 12$ by writing the equation in exponential form.

b. Solve the equation graphically.

SOLUTION

a. We first solve the equation for $\ln x$:

$$6 + 3 \ln x = 12$$

$$3 \ln x = 6$$

$$\ln x = 2$$

Writing $\log_e x = 2$ in exponential form gives

$$x = e^2$$

b. We solve $6 + 3 \ln x = 12$ graphically by the intersection method. Entering $y_1 = 6 + 3 \ln x$ and $y_2 = 12$, we graph using the window $[-2, 10]$ by $[-10, 15]$ and find the point of intersection to be about $(7.38906, 12)$ (Figure 5.30). Thus, the solution to the equation $6 + 3 \ln x = 12$ is $x \approx 7.38906$. Because $e^2 \approx 7.38906$, the solutions agree.

Figure 5.30

The properties of logarithms are frequently useful in solving logarithmic equations. Consider the following example.

EXAMPLE 9 ▶ **Solving a Logarithmic Equation**

Solve $\ln x + 3 = \ln(x + 4)$ by converting the equation to exponential form and then using algebraic methods.

SOLUTION

We first write the equation with the logarithmic expressions on one side:

$$\ln x + 3 = \ln(x + 4)$$
$$3 = \ln(x + 4) - \ln x$$

Using the Quotient Property of Logarithms gives

$$3 = \ln \frac{x + 4}{x}$$

Writing the equation in exponential form gives

$$e^3 = \frac{x + 4}{x}$$

We can now solve for x:

$$e^3 x = x + 4$$
$$e^3 x - x = 4$$
$$x(e^3 - 1) = 4$$
$$x = \frac{4}{e^3 - 1} \approx 0.21$$

EXAMPLE 10 ▶   Global Warming

In an effort to reduce global warming, it has been proposed that a tax be levied based on the emissions of carbon dioxide into the atmosphere. The cost–benefit equation $\ln(1 - P) = -0.0034 - 0.0053t$ estimates the relationship between the percent reduction of emissions of carbon dioxide P (as a decimal) and the tax t in dollars per ton of carbon.
(Source: W. Clime, *The Economics of Global Warming*)

a. Solve the equation for P, the estimated percent reduction in emissions.

b. Determine the estimated percent reduction in emissions if a tax of $100 per ton is levied.

SOLUTION

a. We solve $\ln (1 - P) = -0.0034 - 0.0053t$ for P by writing the equation in exponential form.

$$(1 - P) = e^{-0.0034-0.0053t}$$

$$P = 1 - e^{-0.0034-0.0053t}$$

b. Substituting 100 for t gives $P = 0.4134$, so a $100 tax per ton of carbon is estimated to reduce carbon dioxide emissions by 41.3%.

Some exponential and logarithmic equations are difficult or impossible to solve algebraically, and finding approximate solutions to real data problems is frequently easier with graphical methods.

Exponential and Logarithmic Inequalities

Inequalities involving exponential and logarithmic functions can be solved by solving the related equation algebraically and then investigating the inequality graphically. Consider the following example.

EXAMPLE 11 ▶   Sales Decay

After the end of an advertising campaign, the daily sales of Genapet fell rapidly, with daily sales given by $S = 3200e^{-0.08x}$ dollars, where x is the number of days from the end of the campaign. For how many days after the campaign ended were sales at least $1980?

SOLUTION

To solve this problem, we find the solution to the inequality $3200e^{-0.08x} \geq 1980$. We begin our solution by solving the related equation:

$$3200e^{-0.08x} = 1980$$

$$e^{-0.08x} = 0.61875 \qquad \text{\color{blue}Divide both sides by 3200.}$$

$$\ln e^{-0.08x} = \ln 0.61875 \qquad \text{\color{blue}Take the logarithm, base } e \text{, of both sides.}$$

$$-0.08x \approx -0.4801 \qquad \text{\color{blue}Use the Power Property of Logarithms.}$$

$$x \approx 6$$

We now investigate the inequality graphically.

   To solve this inequality graphically, we graph $y_1 = 3200e^{-0.08x}$ and $y_2 = 1980$ on the same axes, with nonnegative x-values since x represents the number of days. The graph shows that $y_1 = 3200e^{-0.08x}$ is above $y_2 = 1980$ for $x < 6$ (Figure 5.31).

Thus, the daily revenue is at least $1980 for each of the first 6 days after the end of the advertising campaign.

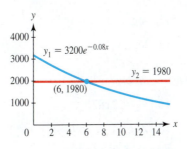

Figure 5.31

EXAMPLE 12 ▶   **Cost–Benefit**

The cost–benefit equation $\ln(1 - P) = -0.0034 - 0.0053t$ estimates the relationship between the percent reduction of emissions of carbon dioxide P (as a decimal) and the tax t in dollars per ton of carbon. What tax will give a reduction of at least 50%?

SOLUTION

To find the tax, we find the value of t that gives $P = 0.50$:

$$\ln(1 - 0.50) = -0.0034 - 0.0053t$$

$$-0.6931 = -0.0034 - 0.0053t$$

$$0.0053t = 0.6897$$

$$t = 130.14$$

In Example 10, we solved this cost–benefit for P, getting $P = 1 - e^{-0.0034-0.0053t}$. The graphs of this equation and $P = 0.50$ on the same axes are shown in Figure 5.32. The point of intersection of the graphs is (130.14, 0.5), and the graph shows that the percent reduction of emissions is more than 50% if the tax is above $130.14 per ton of carbon.

Figure 5.32

SKILLS CHECK 5.3

In Exercises 1–10, solve the equations algebraically and check graphically. Round to three decimal places.

1. $1600 = 10^x$

2. $4600 = 10^x$

3. $2500 = e^x$

4. $54.6 = e^x$

5. $8900 = e^{5x}$

6. $2400 = 10^{8x}$

7. $4000 = 200e^{8x}$

8. $5200 = 13e^{12x}$

9. $8000 = 500(10^x)$

10. $9000 = 400(10^x)$

In Exercises 11–14, use a change of base formula to evaluate each logarithm. Give your answers rounded to four decimal places.

11. $\log_6 18$

12. $\log_7 215$

13. $\log_8 \sqrt{2}$

14. $\log_4 \sqrt[3]{10}$

In Exercises 15–22, solve the equations.

15. $8^x = 1024$

16. $9^x = 2187$

17. $2(5^{3x}) = 31{,}250$

18. $2(6^{2x}) = 2592$

19. $5^{x-2} = 11.18$

20. $3^{x-4} = 140.3$

21. $18{,}000 = 30(2^{12x})$

22. $5880 = 21(2^{3x})$

In Exercises 23–36, solve the logarithmic equations.

23. $\log_2 x = 3$

24. $\log_4 x = -2$

25. $5 + 2 \ln x = 8$

26. $4 + 3 \log x = 10$

27. $5 + \ln(8x) = 23 - 2 \ln x$

28. $3 \ln x + 8 = \ln(3x) + 12.18$

29. $2 \log x - 2 = \log(x - 25)$

30. $\ln(x - 6) + 4 = \ln x + 3$

31. $\log_3 x + \log_3 9 = 1$

32. $\log_2 x + \log_2(x - 6) = 4$

33. $\log_2 x = \log_2 5 + 3$

34. $\log_2 x = 3 - \log_2 2x$

35. $\log 3x + \log 2x = \log 150$

36. $\ln(x + 2) + \ln x = \ln(x + 12)$

In Exercises 37–40, solve the inequalities.

37. $3^x < 243$

38. $7^x \geq 2401$

39. $5(2^x) \geq 2560$

40. $15(4^x) \leq 15{,}360$

EXERCISES 5.3

41. *Supply* The supply function for a certain size boat is given by $p = 340(2^q)$ boats, where p dollars is the price per boat and q is the quantity of boats supplied at that price. What quantity will be supplied if the price is $10,880 per boat?

42. *Demand* The demand function for a dining room table is given by $p = 4000(3^{-q})$ dollars per table, where p is the price and q is the quantity, in thousands of tables, demanded at that price. What quantity will be demanded if the price per table is $256.60?

43. *Sales Decay* After a television advertising campaign ended, the weekly sales of Korbel champagne fell rapidly. Weekly sales in a city were given by $S = 25{,}000e^{-0.072x}$ dollars, where x is the number of weeks after the campaign ended.

    a. Write the logarithmic form of this function.

    b. Use the logarithmic form of this function to find the number of weeks after the end of the campaign before weekly sales fell to $16,230.

44. *Sales Decay* After a television advertising campaign ended, sales of Genapet fell rapidly, with daily sales given by $S = 3200e^{-0.08x}$ dollars, where x is the number of days after the campaign ended.

    a. Write the logarithmic form of this function.

    b. Use the logarithmic form of this function to find the number of days after the end of the campaign before daily sales fell to $2145.

45. *Sales Decay* After the end of an advertising campaign, the daily sales of Genapet fell rapidly, with daily sales given by $S = 3200e^{-0.08x}$ dollars, where x is the number of days from the end of the campaign.

    a. What were daily sales when the campaign ended?

    b. How many days passed after the campaign ended before daily sales were below half of what they were at the end of the campaign?

46. *Sales Decay* After the end of a television advertising campaign, the weekly sales of Korbel champagne fell rapidly, with weekly sales given by $S = 25{,}000e^{-0.072x}$ dollars, where x is the number of weeks from the end of the campaign.

    a. What were weekly sales when the campaign ended?

    b. How many weeks passed after the campaign ended before weekly sales were below half of what they were at the end of the campaign?

47. **Super Bowl Ads** A minute ad during Super Bowl VII in 1973 cost $200,000. The price tag for a 30-second ad slot during the 2011 Super Bowl was $3 million. The cost of a 30-second slot of advertising for the Super Bowl can be modeled by $y = 0.0000966(1.101^x)$, where x is the number of years after 1900 and y is the cost in millions.

    a. According to the model, what was the cost of a 30-second Super Bowl ad in 2000?

    b. The cost of a 30-second ad during Super Bowl XLIX in 2015 was $4.5 million. Does the model underestimate or overestimate the cost of these ads?

48. **Population Growth** The population in a certain city was 53,000 in 2000, and its future size was predicted to be $P = 53,000e^{0.015t}$ people, where t is the number of years after 2000. Determine algebraically when the population was predicted to reach 60,000, and verify your solution graphically.

49. **Purchasing Power** The purchasing power (real value of money) decreases if inflation is present in the economy. For example, the purchasing power of $40,000 after t years of 5% inflation is given by the model

$$P = 40,000e^{-0.05t} \text{ dollars}$$

    How long will it take for the value of a $40,000 pension to have a purchasing power of $20,000 under 5% inflation?
    (Source: *Viewpoints*, VALIC)

50. **Purchasing Power** If a retired couple has a fixed income of $60,000 per year, the purchasing power of their income (adjusted value of the money) after t years of 5% inflation is given by the equation $P = 60,000e^{-0.05t}$. In how many years will the purchasing power of their income be half of their current income?

51. **Real Estate Inflation** During a 5-year period of constant inflation, the value of a $100,000 property increases according to the equation $v = 100,000e^{0.03t}$ dollars. In how many years will the value of this building be double its current value?

52. **Real Estate Inflation** During a 10-year period of constant inflation, the value of a $200,000 property is given by the equation $v = 200,000e^{0.05t}$ dollars. In how many years will the value of this building be $254,250?

53. **Radioactive Decay** The amount of radioactive isotope thorium-234 present in a certain sample at time t is given by $A(t) = 500e^{-0.02828t}$ grams, where t years is the time since the initial amount was measured.

    a. Find the initial amount of the isotope present in the sample.

b. Find the half-life of this isotope. That is, find the number of years until half of the original amount of the isotope remains.

54. **Radioactive Decay** The amount of radioactive isotope thorium-234 present in a certain sample at time t is given by $A(t) = 500e^{-0.02828t}$ grams, where t years is the time since the initial amount was measured. How long will it take for the amount of isotope to equal 318 grams?

55. **Drugs in the Bloodstream** The concentration of a drug in the bloodstream from the time the drug is injected until 8 hours later is given by

$$y = 100(1 - e^{-0.312(8-t)}) \text{ percent, } t \text{ in hours,}$$

    where the drug is administered at time $t = 0$. In how many hours will the drug concentration be 79% of the initial dose?

56. **Drugs in the Bloodstream** If a drug is injected into the bloodstream, the percent of the maximum dosage that is present at time t is given by $y = 100(1 - e^{-0.35(10-t)})$, where t is in hours, with $0 \le t \le 10$. In how many hours will the percent reach 65%?

57. **International Air Travel** The number of passengers using international air traffic into and out of the United States was projected by the Federal Aviation Administration to grow at 4% per year from 2012 until 2032. If the number was 172 million in 2012, the equation of the exponential function that models this projection of the number of millions of passengers y is $y = 172(1.04^x)$, where x is the number of years after 2012. When does the model project that the number of passengers using international air traffic into and out of the United States will be 335 million?

58. **Radioactive Decay** A breeder reactor converts stable uranium-238 into the isotope plutonium-239. The decay of this isotope is given by $A(t) = A_0e^{-0.00002876t}$, where $A(t)$ is the amount of the isotope at time t (in years) and A_0 is the original amount. If the original amount is 100 pounds, find the half-life of this isotope.

59. **Cost** Suppose the weekly cost for the production of x units of a product is given by $C(x) = 3452 + 50\ln(x + 1)$ dollars. Use graphical methods to estimate the number of units produced if the total cost is $3556.

60. **Supply** Suppose the daily supply function for a product is $p = 31 + \ln(x + 2)$, where p is in dollars and x is the number of units supplied. Use graphical methods to estimate the number of units that will be supplied if the price is $35.70.

61. *Doubling Time* If P dollars are invested at an annual interest rate r, compounded annually for t years, the future value of the investment is given by $S = P(1.07)^t$. Find a formula for the number of years it will take to double the initial investment.

62. *Doubling Time* The future value of a lump sum P that is invested for n years at 10%, compounded annually, is $S = P(1.10)^n$. Show that the number of years it would take for this investment to double is $n = \log_{1.10} 2$.

63. *Investment* At the end of t years, the future value of an investment of $20,000 at 7%, compounded annually, is given by $S = 20,000(1 + 0.07)^t$. In how many years will the investment grow to $48,196.90?

64. *Investment* At the end of t years, the future value of an investment of $30,000 at 9%, compounded annually, is given by $S = 30,000(1 + 0.09)^t$. In how many years will the investment grow to $129,829?

65. *Investing* Find in how many years $40,000 invested at 10%, compounded annually, will grow to $64,420.40.

66. *Investing* In how many years will $40,000 invested at 8%, compounded annually, grow to $86,357?

67. *Life Span* Based on data from 1920 and projected to 2020, the expected life span of people in the United States can be described by the function $f(x) = 11.027 + 14.304 \ln x$, where x is the number of years from 1900 to the person's birth year.

   a. Estimate the birth year for which the expected life span is 78 years.

   b. Use graphical methods to determine the birth year for which the expected life span is 78 years. Does this agree with the solution in part (a)?
   (Source: National Center for Health Statistics)

68. *Supply* Suppose that the supply of a product is given by $p = 20 + 6 \ln(2q + 1)$, where p is the price per unit and q is the number of units supplied. How many units will be supplied if the price per unit is $68.04?

69. *Alzheimer's Disease* The millions of U.S. citizens with Alzheimer's disease from 2000 and projected to 2050 can be modeled by $y = 3.294(1.025^x)$, with x equal to the number of years after 1990 and y equal to the millions of Americans with Alzheimer's disease. In what year does the model predict the number of U.S. citizens with Alzheimer's disease will be 8 million?
   (Source: National Academy on an Aging Society)

70. *Demand* Suppose that the demand function for a product is $p = \dfrac{500}{\ln(q + 1)}$, where p is the price per unit and q is the number of units demanded. How many units will be demanded if the price is $61.71 per unit?

71. *China's Shale-Natural Gas* The estimated annual production of shale-natural gas in China for the years 2013 through 2020, in billions of cubic feet, can be modeled by $y = 0.0131(1.725^x)$, with x equal to the number of years x after 2010. In what year does the model predict that the number of billions of cubic feet of shale-natural gas in China will be 1.77?
   (Source: Sanford C. Bernstein)

72. *Global Warming* In an effort to reduce global warming, it has been proposed that a tax be levied based on the emissions of carbon dioxide into the atmosphere. The cost–benefit equation $\ln(1 - P) = -0.0034 - 0.0053t$ estimates the relationship between the percent reduction of emissions of carbon dioxide P (as a decimal) and the tax t in dollars per ton of carbon dioxide.

   a. Solve the equation for t, giving t as a function of P. Graph the function.

   b. Use the equation in part (a) to find what tax will give a 30% reduction in emissions.
   (Source: W. Clime, *The Economics of Global Warming*)

73. *Rule of 72* The "Rule of 72" is a simplified way to determine how long an investment will take to double, given a fixed annual rate of interest. By dividing 72 by the annual interest rate, investors can get a rough estimate of how many years it will take for the initial investment to double. Algebraically we know that the time it takes an investment to double is $\dfrac{\ln 2}{r}$, when the interest is compounded continuously and r is written as a decimal.

   a. Complete the table to compare the exact time it takes for an investment to double to the "Rule of 72" time. (Round to two decimal places.)

Annual Interest Rate	Rule of 72 Years	Exact Years
2%		
3%		
4%		
5%		
6%		
7%		
8%		
9%		
10%		
11%		

   b. Compute the differences between the two sets of outputs. What conclusion can you reach about using the Rule of 72 estimate?

74. *Doubling Time* The number of quarters needed to double an investment when a lump sum is invested at 8%, compounded quarterly, is given by $n = \log_{1.02} 2$.

   a. Use the change of base formula to find n.

   b. In how many years will the investment double?

75. *Doubling Time* The number of periods needed to double an investment when a lump sum is invested at 12%, compounded semiannually, is given by $n = \log_{1.06} 2$.

   a. Use the change of base formula to find n.

   b. How many years pass before the investment doubles in value?

76. *Annuities* If $2000 is invested at the end of each year in an annuity that pays 5%, compounded annually, the number of years it takes for the future value to amount to $40,000 is given by $t = \log_{1.05} 2$. Use the change of base formula to find the number of years until the future value is $40,000.

77. *Annuities* If $1000 is invested at the end of each year in an annuity that pays 8%, compounded annually, the number of years it takes for the future value to amount to $30,000 is given by $t = \log_{1.08} 3.4$. Use the change of base formula to find the number of years until the future value is $30,000.

78. *Deforestation* One of the major causes of rain forest deforestation is agricultural and residential development. The number of hectares (2.47 acres) destroyed in a particular year t can be modeled by $y = -3.91435 + 2.62196 \ln t$, where $t = 0$ in 1950. When will more than 7 hectares be destroyed per year?

79. *Market Share* Suppose that after a company introduces a new product, the number of months before its market share is x percent is given by

$$m = 20 \ln \frac{50}{50 - x}, x < 50$$

After how many months is the market share more than 45%, according to this model?

80. *Drugs in the Bloodstream* The concentration of a drug in the bloodstream from the time the drug is administered until 8 hours later is given by $y = 100(1 - e^{-0.312(8-t)})$ percent, where the drug is administered at time $t = 0$. For what time period is the amount of drug present more than 60%?

81. *Carbon-14 Dating* An exponential decay function can be used to model the number of grams of a radioactive material that remain after a period of time. Carbon-14 decays over time, with the amount remaining after t years given by $y = y_0 e^{-0.00012097t}$, where y_0 is the original amount. If the original amount of carbon-14 is 200 grams, find the number of years until 155.6 grams of carbon-14 remain.

82. *Sales Decay* After a television advertising campaign ended, the weekly sales of Korbel champagne fell rapidly. Weekly sales in a city were given by $S = 25,000e^{-0.072x}$ dollars, where x is the number of weeks after the campaign ended.

   a. Use the logarithmic form of this function to find the number of weeks after the end of the campaign that passed before weekly sales fell below $16,230.

   b. Check your solution by graphical or numerical methods.

83. *Sales Decay* After a television advertising campaign ended, the weekly sales of Turtledove bars fell rapidly. Weekly sales are given by $S = 600e^{-0.05x}$ thousand dollars, where x is the number of weeks after the campaign ended.

   a. Use the logarithmic form of this function to find the number of weeks after the end of the campaign before weekly sales fell to $269.60.

   b. Check your solution by graphical or numerical methods.

5.4 Exponential and Logarithmic Models

KEY OBJECTIVES

- Model data with exponential functions
- Use constant percent change to determine whether data fit an exponential model
- Compare quadratic and exponential models of data
- Model data with logarithmic functions

SECTION PREVIEW Diabetes

In Example 8 of Section 5.2, we solved problems about diabetes by using the fact that the percent of U.S. adults with diabetes (diagnosed and undiagnosed) can be modeled by the logarithmic function

$$p(x) = -12.975 + 11.851 \ln x$$

where x is the number of years after 2000. In this section, we will use projections from the Centers for Disease Control and Prevention to create this logarithmic model using technology. (See Example 5.)

Many sources provide data that can be modeled by exponential growth and decay functions, and technology can be used to find exponential functions that model data. In this section, we model real data with exponential functions and determine when this type of model is appropriate. We also create exponential functions that model phenomena characterized by constant percent change, and we model real data with logarithmic functions when appropriate. ∎

Modeling with Exponential Functions

Exponential functions can be used to model real data if the data exhibit rapid growth or decay. When the scatter plot of data shows a very rapid increase or decrease, it is possible that an exponential function can be used to model the data. Consider the following examples.

EXAMPLE 1 ▶ Insurance Premiums

The monthly premiums for $250,000 in term-life insurance over a 10-year term period increase with the age of the men purchasing the insurance. The monthly premiums for nonsmoking males are shown in Table 5.8

Table 5.8

Age (years)	Monthly Premium for 10-Year Term Insurance ($)	Age (years)	Monthly Premium for 10-Year Term Insurance ($)
35	123	60	783
40	148	65	1330
45	225	70	2448
50	338	75	4400
55	500		

(Source: Quotesmith.com)

a. Graph the data in the table with x as age and y in dollars.

b. Create an exponential function that models these premiums as a function of age.

c. Graph the data and the exponential function that models the data on the same axes.

SOLUTION

a. The scatter plot of the data is shown in Figure 5.33(a). The plot shows that the outputs rise rapidly as the inputs increase.

b. Using technology gives the exponential model for the monthly premium, rounded to four decimal places, as

$$y = 4.0389\,(1.0946^x)\ \text{dollars}$$

where x is the age in years.

c. Figure 5.33(b) shows a scatter plot of the data and a graph of the (unrounded) exponential equation used to model it. The model appears to be a good, but not perfect, fit for the data.

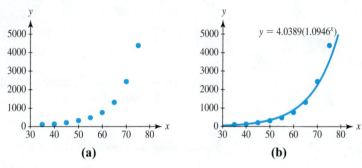

(a)   (b)

Figure 5.33

EXAMPLE 2 ▶ E-Commerce Sales

E-commerce is taking a bigger slice of the overall retail sales pie and is growing far faster than retail sales. E-commerce sales in the United States are projected to be $434 billion in 2017. This is more than double the 2010 sales, which were $171 billion. Table 5.9 gives retail trade sales by e-commerce in billions of dollars for the years 2000–2017.

a. Graph the data, with x as the number of years after 2000.

b. Find an exponential function that models these data, with x as the number of years after 2000.

c. Graph the data and the function on the same axes.

d. What does the model predict the online sales revenue to be in 2025?

Table 5.9

Year	Sales ($ billions)	Year	Sales ($ billions)
2000	29	2009	134
2001	34	2010	171
2002	45	2011	202
2003	58	2012	226
2004	74	2013	259
2005	87	2014	297
2006	106	2015	339
2007	126	2016	386
2008	132	2017	434

(Sources: U.S. Census Bureau, Forrester Research, Inc.)

SOLUTION

a. The graph of the data is shown in Figure 5.34(a).

Figure 5.34

b. The exponential equation that models the data, rounded to three decimal places, is
$f(x) = 35.590(1.166^x)$, where x is the number of years after 2000.

c. The graphs of the data and the exponential model are shown in Figure 5.34(b).

d. Evaluating the reported model at $x = 25$ for the year 2025 gives

$$f(25) = 35.590(1.166^{25}) \approx 1655$$

Thus, e-commerce retail sales are expected to be \$1655 billion in 2025, according to the model.

Spreadsheet ▸ SOLUTION   We can use software programs and spreadsheets to find the exponential function that is the best fit for the data. Figure 5.35 shows the Excel spreadsheet for the aligned data of Example 2. Selecting the cells containing the data, getting the scatter plot of the data, and selecting the Exponential Trendline

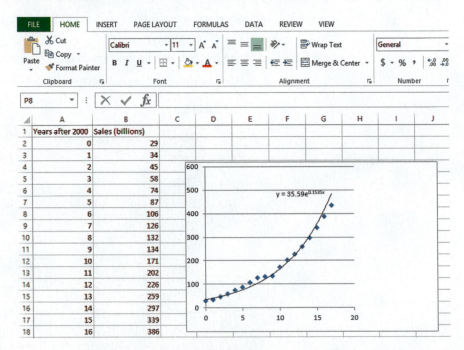

Figure 5.35

gives the equation of the exponential function that is the best fit for the data, along with the scatter plot and the graph of the best-fitting function. Checking Display equation gives the equation on the graph, shown in Figure 5.35. Although the function has a different form, it is equivalent to the function found in Example 2.

The function created by Excel,

$$y = 35.59e^{0.1535x},$$

is in a different form than the model found in Example 2(b), which is

$$f(x) = 35.59(1.166^x)$$

However, they are equivalent because $e^{0.1535} \approx 1.166$ which means that

$$y = 35.59e^{0.1535x} = 35.59(1.166^x)$$

Constant Percent Change in Exponential Models

How can we determine whether a set of data can be modeled by an exponential function? To investigate this, we look at Table 5.10, which gives the growth of paramecia for each hour up to 9 hours.

Table 5.10

x (hours)	0	1	2	3	4	5	6	7	8	9
y (paramecia)	1	2	4	8	16	32	64	128	256	512

If we look at the first differences in the output, we see that they are not constant (Table 5.11). But if we calculate the percent change of the outputs for equally spaced inputs, we see that they are constant. For example, from hour 2 to hour 3, the population grew by 4 units; this represents a 100% increase from the hour-2 population. From hour 3 to hour 4, the population grew by 8, which is a 100% increase over the hour-3 population. In fact, this population increases by 100% each hour. This means that the population 1 hour from now will be the present population plus 100% of the present population.

Table 5.11

Outputs	1	2	4	8	16	32	64	128	256	512
First Differences		1	2	4	8	16	32	64	128	256
Percent Change		100	100	100	100	100	100	100	100	100

Because the percent change in the outputs is constant for equally spaced inputs in this example, an exponential model $(y = 2^x)$ fits the data perfectly.

> ## Constant Percent Changes
>
> If the percent change of the outputs of a set of data is constant for equally spaced inputs, an exponential function will be a perfect fit for the data.
>
> If the percent change of the outputs is approximately constant for equally spaced inputs, an exponential function will be an approximate fit for the data.

EXAMPLE 3 ▶ **Sales Decay**

Suppose a company develops a product that is released with great expectations and extensive advertising, but sales suffer because of bad word of mouth from dissatisfied customers.

a. Use the monthly sales data shown in Table 5.12 to determine the percent change for each of the months given.

Table 5.12

Month	1	2	3	4	5	6	7	8
Sales (thousands of $)	780	608	475	370	289	225	176	137

b. Find the exponential function that models the data.

c. Graph the data and the model on the same axes.

SOLUTION

a. The inputs are equally spaced; the differences of outputs and the percent changes are shown in Table 5.13.

Table 5.13

Outputs	780	608	475	370	289	225	176	137
First Differences		−172	−133	−105	−81	−64	−49	−39
Percent Change		−22.1	−21.9	−22.1	−21.9	−22.1	−21.8	−22.2

The percent change of the outputs is approximately −22%. This means that the sales 1 month from now will be approximately 22% less than the sales now.

b. Because the percent change is nearly constant, an exponential function should fit these data well. Technology gives the model

$$y = 999.781(0.780^x)$$

where x is the month and y is the sales in thousands of dollars.

c. The graphs of the data and the (unrounded) model are shown in Figure 5.36.

Figure 5.36

Exponential Models

Most graphing utilities give the best-fitting exponential model in the form

$$y = a \cdot b^x$$

It can be shown that the constant percent change of this function is

$$(b - 1) \cdot 100\%$$

Thus, if $b > 1$, the function is increasing (growing), and if $0 < b < 1$, the function is decreasing (decaying).

For the function $y = 2^x$, $b = 2$, so the constant percent change is $r = (2 - 1)100\% = 100\%$ (which we found in Table 5.11). For the function $y = 1000(0.78^x)$, the base is $b = 0.78$, and the percent change is $r = (0.78 - 1) \cdot 100\% = -22\%$.

Because $a \cdot b^0 = a \cdot 1 = a$, the value of $y = a \cdot b^x$ is a when $x = 0$, so a is the y-intercept of the graph of the function. Because $y = a$ when $x = 0$, we say that the **initial value** of the function is a. If the constant percent change (as a decimal) is $r = b - 1$, then $b = 1 + r$, and we can write $y = a \cdot b^x$ as

$$y = a(1 + r)^x$$

Exponential Model

If a set of data has initial value a and a constant percent change r (written as a decimal) for equally spaced inputs x, the data can be modeled by the exponential function

$$y = a(1 + r)^x$$

for exponential growth and by the exponential function

$$y = a(1 - r)^x$$

for exponential decay.

To illustrate this, suppose we know that the present population of a city is 100,000 and that it will grow at a constant percent of 10% per year. Then we know that the future population can be modeled by an exponential function with initial value $a = 100,000$ and constant percent change $r = 10\% = 0.10$ per year. That is, the population can be modeled by

$$y = 100,000(1 + 0.10)^x = 100,000(1.10)^x$$

where x is the number of years from the present.

EXAMPLE 4 ▶ | **Inflation**

Suppose inflation averages 4% per year for each year from 2000 to 2010. This means that an item that costs \$1 one year will cost \$1.04 one year later. In the second year, the \$1.04 cost will increase by a factor of 1.04, to $(1.04)(1.04) = 1.04^2$.

a. Write an expression that gives the cost t years after 2000 of an item costing \$1 in 2000.

b. Write an exponential function that models the cost of an item t years from 2000 if its cost was \$100 in 2000.

c. Use the model to find the cost of the item from part (b) in 2010.

SOLUTION

a. The cost of an item is $1 in 2000, and the cost increases at a constant percent of 4% = 0.04 per year, so the cost after t years will be

$$1(1 + 0.04)^t = 1.04^t \text{ dollars}$$

b. The inflation rate is 4% = 0.04. Thus, if an item costs $100 in 2000, the function that gives the cost after t years is

$$f(t) = 100(1 + 0.04)^t = 100(1.04^t) \text{ dollars}$$

c. The year 2010 is 10 years after 2000, so the cost of an item costing $100 in 2000 is

$$f(10) = 100(1.04^{10}) = 148.02 \text{ dollars}$$

Comparison of Models

Sometimes it is hard to determine what type of model is the best fit for a set of data. When a scatter plot exhibits a single curvature without a visible high or low point, it is sometimes difficult to determine whether a power, quadratic, or exponential function should be used to model the data. Suppose after graphing the data points we are unsure which of these models is most appropriate. We can find the three models, graph them on the same axes as the data points, and inspect the graphs to see which is the best visual fit.*

Consider the exponential model for insurance premiums as a function of age that we found in Example 1. The curvature indicated by the scatter plot suggests that either a power, quadratic, or exponential model may fit the data. To compare the goodness of fit of the models, we find each model and compare its graph with the scatter plot (Figure 5.37).

Figure 5.37

The exponential function appears to provide a better fit than the quadratic function because the data points seem to be approaching the x-axis as the input becomes closer to zero from the right. (Recall that the x-axis is a horizontal asymptote for the basic exponential function.) The exponential function also appears to be a better fit than the power function, especially for larger values of x.

When fitting exponential functions to data using technology, it is important to know that problems arise if the input values are large. For instance, when the input values are years (like 2000 and 2010), a technology-determined model may appear to be of the form $y = 0 \cdot b^x$, a function that certainly does not make sense as a model for exponential data. At other times when large inputs are used, your calculator or

---

* Statistical measures of goodness of fit exist, but caution must be used in applying these measures. For example, the correlation coefficient r found for linear fits to data is different from the coefficient of determination R^2 found for quadratic fits to data.

computer software may give an error message and not even produce an equation. To avoid these problems, it is helpful to align the inputs by converting years to the number of years after a specified year (for example, inputting years from 1990 reduces the size of the inputs).

> **Technology Note**
>
> When using technology to fit an exponential model to data, you should align the inputs to reasonably small values.

Logarithmic Models

As with other functions we have studied, we can create models involving logarithms. Data that exhibit an initial rapid increase and then have a slow rate of growth can often be described by the function

$$f(x) = a + b \ln x \qquad (\text{for } b > 0)$$

Note that the parameter a is the vertical shift of the graph of $y = \ln x$ and the parameter b affects how much the graph of $y = \ln x$ is stretched. If $b < 0$, the graph of $f(x) = a + b \ln x$ decreases rather than increases. Most graphing utilities will create logarithmic models in the form $y = a + b \ln x$.

EXAMPLE 5 ▶   **Diabetes**

As Table 5.14 shows, projections indicate that the percent of U.S. adults with diabetes could dramatically increase.

a. Find a logarithmic model that fits the data in Table 5.14, with $x = 0$ in 2000.

b. Use the reported model to predict the percent of U.S. adults with diabetes in 2027.

c. In what year does this model predict the percent to be 26.9%?

Table 5.14

Year	Percent	Year	Percent	Year	Percent
2010	15.7	2025	24.2	2040	31.4
2015	18.9	2030	27.2	2045	32.1
2020	21.1	2035	29.0	2050	34.3

(Source: Centers for Disease Control and Prevention)

SOLUTION

a. Entering the aligned input data (number of years after 2000) as the x-values and the percents as the y-values, we use logarithmic regression on a graphing utility to find the function that models the data. This function, rounded to three decimal places, is

$$y = -12.975 + 11.851 \ln x$$

where x is the number of years after 2000. The graphs of the aligned data and the function are shown in Figure 5.38.

b. Evaluating the reported model at $x = 27$ gives $y = -12.975 + 11.851 \ln 27 \approx$ 26.1 percent in 2027.

Figure 5.38

Figure 5.39

c. Setting $y = 26.9$ gives

$$26.9 = -12.975 + 11.851 \ln x$$

$$39.875 = 11.851 \ln x$$

$$3.3647 = \ln x$$

$$x = e^{3.3647} = 28.9$$

We could also solve graphically by intersecting the graphs of $y_1 = -12.975 + 11.851 \ln x$ and $y_2 = 26.9$, as shown in Figure 5.39. Thus, the percent reaches 27% in 2029.

Note that the input data in Example 5 were not aligned as the number of years after 2010 because the first aligned input value would be 0 and the logarithm of 0 does not exist.

> **Technology Note**
>
> When using technology to fit a logarithmic model to data, you must align the data so that all input values are positive.

EXAMPLE 6 ▶   **Women in the Labor Force**

Table 5.15 contains the total percent of the labor force that is female for selected years from 1900 and projected to 2050.

Table 5.15

Year	Percent Female	Year	Percent Female
1940	24.3	1990	45.2
1950	29.6	2000	46.6
1960	33.4	2010	47.9
1970	38.1	2020	48.1
1980	42.5	2030	48.0

(Source: U.S. Bureau of Labor Statistics)

a. Find a logarithmic function that models these data. Align the input to be the number of years after 1900.

b. Graph the equation and the aligned data points. Comment on how the model fits the data.

c. Assuming that the model is valid in 2024, use it to estimate the percent of women in the labor force in 2024.

SOLUTION

a. A logarithmic function that models these data is

$$f(x) = -54.775 + 21.729 \ln x$$

where x is the number of years after 1900.

b. The scatter plot of the data and the graph of the logarithmic function that models the data are shown in Figure 5.40. The model appears to be a reasonably good fit for the data.

Figure 5.40

c. The percent of women in the labor force in 2024 can be estimated by evaluating $f(124) = 50\%$.

Spreadsheet ▸ SOLUTION We can use software programs and spreadsheets to find the logarithmic function that is the best fit for data. Figure 5.41 shows the Excel spreadsheet for the data of Example 6. Selecting the cells containing the data, getting the scatter plot of the data, and selecting Logarithmic Trendline gives the equation of the logarithmic function that is the best fit for the data, along with the scatter plot and the graph of the best-fitting function. Checking Display equation gives the equation on the graph, shown in Figure 5.41. The model found here agrees with that found in Example 6.

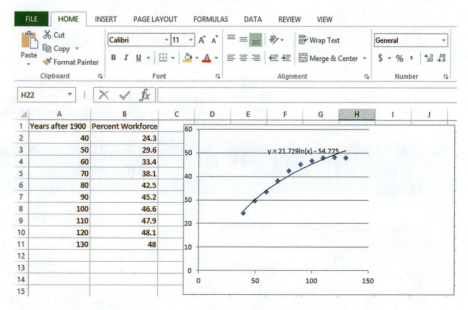

Figure 5.41

Exponents, Logarithms, and Linear Regression

Now that we have knowledge of the properties of logarithms and how exponential functions are related to logarithmic functions, we can show that linear regression can be used to create exponential models that fit data. (The development of linear regression is discussed in Section 2.2.)

Logarithmic Property 3 states that $\log_b b^x = x$ and, in particular, that $\ln e^x = x$, so if we have data that can be approximated by an exponential function, we can convert the data to a linear form by taking the logarithm, base e, of the outputs. We can use linear regression to find the linear function that is the best fit for the converted data and then use the linear function as the exponent of e, which gives the exponential function that is the best fit for the original data. Consider Table 5.16, which gives the number y of paramecia in a population after x hours. We showed in Section 5.1 that the population can be modeled by $y = 2^x$.

Table 5.16

x (hours)	0	1	2	3	4	5	6	7	8	9
y (paramecia)	1	2	4	8	16	32	64	128	256	512
ln y	0	.6931	1.3863	2.0794	2.7726	3.4657	4.1589	4.8520	5.5452	6.2383

The third row of Table 5.16 has the logarithms, base e, of the numbers (y-values) in the second row rounded to four decimal places, and the relationship between x and $\ln y$ is linear. The first differences of the $\ln y$ values are constant, with the difference approximately equal to 0.6931. Using linear regression on the x and $\ln y$ values (to ten decimal places) gives the equation of the linear model:*

$$\ln y = 0.6931471806x$$

Because we seek the equation solved for y, we can write this equation in its exponential form, getting

$$y = e^{0.6931471806x}$$

To show that this model for the data is equal to $y = 2^x$, which we found in Section 5.1, we use properties of exponents and note that $e^{0.6931471806} \approx 2$:

$$y = e^{0.6931471806x} = \left(e^{0.6931471806}\right)^x \approx 2^x$$

Fortunately, most graphing utilities have combined these steps to give the exponential model directly from the data.

---

* This equation could be found directly, in the form $\ln y = ax + b$, where $b = 0$ and a equals the constant difference. This is because the exponential function is a perfect fit for the data in this application.

SKILLS CHECK 5.4

1. Find the exponential function that models the data in the table below.

x	−2	−1	0	1	2	3	4	5
y	2/9	2/3	2	6	18	54	162	486

2. The following table has input x and output $f(x)$. Test the percent change of the outputs to determine

if the function is exactly exponential, approximately exponential, or not exponential.

x	1	2	3	4	5	6
f(x)	4	16	64	256	1024	4096

3. The following table has input x and output $g(x)$. Test the percent change of the outputs to determine if the function is exactly exponential, approximately exponential, or not exponential.

x	1	2	3	4	5	6
g(x)	2.5	6	8.5	10	8	6

4. The following table has input x and output $h(x)$. Test the percent change of the outputs to determine if the function is exactly exponential, approximately exponential, or not exponential.

x	1	2	3	4	5	6
h(x)	1.5	2.25	3.8	5	11	17

5. Find the exponential function that is the best fit for $f(x)$ defined by the table in Exercise 2.

6. Find the exponential function that is the best fit for $h(x)$ defined by the table in Exercise 4.

7. a. Make a scatter plot of the data in the table below.

   b. Does it appear that a linear model or an exponential model is the better fit for the data?

x	1	2	3	4	5	6
y	2	3.1	4.3	5.4	6.5	7.6

8. Find the exponential function that models the data in the table below. Round the model with three-decimal place accuracy.

x	−2	−1	0	1	2	3
y	5	30	150	1000	4000	20,000

9. Compare the first differences and the percent change of the outputs to determine if the data in the table below should be modeled by a linear or an exponential function.

x	1	2	3	4	5
y	2	6	14	34	81

10. Use a scatter plot to determine if a linear or exponential function is the better fit for the data in Exercise 9.

11. Find the linear *or* exponential function that is the better fit for the data in Exercise 9. Round the model to three-decimal-place accuracy.

12. Find the logarithmic function that models the data in the table below. Round the model to two-decimal-place accuracy.

x	1	2	3	4	5	6	7
y	2	4.08	5.3	6.16	6.83	7.38	7.84

13. a. Make a scatter plot of the data in the table below.

   b. Does it appear that a linear model or a logarithmic model is the better fit for the data?

x	1	3	5	7	9
y	−2	1	3	4	5

14. a. Find a logarithmic function that models the data in the table in Exercise 13. Round the model to three-decimal-place accuracy.

   b. Find a linear function that models the data.

   c. Visually determine which model is the better fit for the data.

15. a Make a scatter plot of the data in the table below.

   b. Find a power function that models the data. Round to three decimal places.

   c. Find a quadratic function that models the data. Round to three decimal places.

   d. Find a logarithmic function that models the data. Round to three decimal places.

x	1	2	3	4	5	6
y	3.5	5.5	6.8	7.2	8	9

16. Let $y = f(x)$ represent the power model found in Exercise 15(b), $y = g(x)$ represent the quadratic model found in Exercise 15(c), and $y = h(x)$ represent the logarithmic model found in Exercise 15(d). Graph each function on the same axes as the scatter plot, using the window $[0, 12]$ by $[0, 15]$. Which model appears to be the best fit?

EXERCISES 5.4

Report models accurate to three decimal places unless otherwise specified. Use the unrounded function to calculate and to graph the function.

Use the exponential form $y = a(1 + r)^x$ to model the information in Exercises 17–20.

17. Inflation Suppose that the retail price of an automobile is $30,000 in 2000 and that it increases at 4% per year.

   a. Write the equation of the exponential function that models the retail price of the automobile t years after 2000.

   b. Use the model to predict the retail price of the automobile in 2015.

18. Population Suppose that the population of a city is 190,000 in 2000 and that it grows at 3% per year.

   a. Write the equation of the exponential function that models the annual growth.

   b. Use the model to find the population of this city in 2010.

19. Sales Decay At the end of an advertising campaign, weekly sales amounted to $20,000. They then decreased by 2% each week after the end of the campaign.

   a. Write the equation of the exponential function that models the weekly sales.

   b. Find the sales 5 weeks after the end of the advertising campaign.

20. Inflation The average price of a house in a certain city was $220,000 in 2008, and it increases at 3% per year.

   a. Write the equation of the exponential function that models the average price of a house t years after 2008.

   b. Use the model to predict the average price of a house in 2013.

21. Personal Income Total personal income in the United States (in billions of dollars) for selected years from 1960 and projected to 2018 is given in the following table.

   a. The data can be modeled by an exponential function. Write the equation of this function, with x as the number of years after 1960.

Year	Personal Income ($ billions)
1960	411.5
1970	838.8
1980	2307.9
1990	4878.6
2000	8429.7
2008	12,100.7
2018	19,129.6

(Source: Bureau of Economic Analysis, U.S. Department of Commerce)

   b. If this model is accurate, what will be the total U.S. personal income in 2015?

   c. In what year does the model predict the total personal income will reach $19 trillion?

22. Spain's Budget Deficit During Spain's economic crisis in 2010, the International Monetary Fund released a forecast of Spain's budget deficit as a percent of its gross domestic product (GDP) from 2011 to 2015, as shown in the table.

   a. Find the exponential function that models these data, with x equal to the number of years after 2010 and y equal to the percent.

   b. Is the model an increasing or a decreasing function?

   c. Graph the model on the same set of axes as the data.

   d. What did the model project would be Spain's budget deficit, as a percent of its gross domestic product, in 2016?

Year	Deficit (% of GDP)
2011	9.0
2012	6.2
2013	4.9
2014	3.8
2015	3.2

(Source: *Wall Street Journal*)

23. *China's Shale Natural Gas* The estimated annual production of shale natural gas in China for the years 2013 through 2020, in millions of cubic feet, is shown in the table.

a. Find the equation of the exponential function that models these estimates of the number of millions of cubic feet of shale natural gas in China *y* as a function of the number of years *x* after 2010.

b. Graph the model and the data on the same axes and discuss the fit.

c. What does the model estimate that the number of millions of cubic feet of shale natural gas in China will be in 2021?

d. When does the model estimate that the number of millions of cubic feet of shale natural gas in China will be 9506?

Year	Natural Gas (millions of cubic feet)	Year	Natural Gas (millions of cubic feet)
2013	50	2017	600
2014	120	2018	1000
2015	200	2019	1700
2016	400	2020	2900

(Source: Sanford C. Bernstein)

24. *Insurance Premiums* The table below gives the annual premiums required for a $250,000 20-year term-life insurance policy on female nonsmokers of different ages.

a. Find an exponential function that models the monthly premium as a function of the age of the female nonsmoking policyholder.

b. Find the quadratic function that is the best fit for the data.

c. Graph each function on the same axes with the data points to determine visually which model is the better fit for the data. Use the window $[30, 80]$ by $[-10, 6800]$.

Age	Monthly Premium for a 20-Year Policy ($)	Age	Monthly Premium for a 20-Year Policy ($)
35	145	60	845
40	185	65	1593
45	253	70	2970
50	363	75	5820
55	550		

(Source: Quotesmith.com)

25. *Average Annual Wage* The following table shows the U.S. average annual wage, in thousands of dollars, for selected years from 1980 and projected to 2050.

a. Find an exponential function that models these data, with *x* equal to the number of years after 1980 and *y* equal to the average annual wage in thousands of dollars.

b. Graph the model and the data on the same axes.

c. What does the model predict that the average annual wage will be in 2040? Is this interpolation or extrapolation?

d. When does the model predict that the average annual wage will reach $225,000?

Year	Average Annual Wage ($ thousands)
1980	12.5
1990	21.0
2000	32.2
2010	41.7
2012	44.6
2014	48.6
2016	53.3
2018	58.7
2020	63.7
2025	76.8
2030	93.2
2035	113.2
2040	137.6
2045	167.1
2050	202.5

(Source: Social Security Administration)

26. *Alzheimer's Disease* Partially because people in the United States are living longer, the number with Alzheimer's disease and other forms of dementia is projected to grow each year. The table below gives the number of millions of U.S. citizens with Alzheimer's from 2000 and projected to 2050.

Year	2000	2010	2020	2030	2040	2050
Number	4.0	5.9	6.8	8.7	11.8	14.3

(Source: National Academy on an Aging Society)

a. Create a scatterplot of the data, with x equal to the number of years after 1990 and y equal to the number of millions of U.S. citizens with Alzheimer's disease.

b. Find an exponential function that models these data, with x equal to the number of years after 1990 and y equal to the millions of U.S. citizens with Alzheimer's disease.

c. Find a power function that models these data, with x equal to the number of years after 1990 and y equal to the millions of U.S. Citizens with Alzheimer's disease.

d. Graph each function on the same axes as the data points, and determine which model better fits the data.

27. *Life Span* The table below gives the life expectancy for people in the United States for the birth years from 1920 and projected to 2020.

a. Find the logarithmic function that models the data, with x equal to 0 in 1900.

b. Find the quadratic function that is the best fit for the data, with $x = 0$ in 1900. Round the quadratic coefficient to four decimal places.

c. Graph each of these functions on the same axes with the data points to determine visually which function is the better model for the data for the years 1920–2020.

d. Evaluate both models for the birth year 2016.

Year	Life Span (years)	Year	Life Span (years)
1920	54.1	1994	75.7
1930	59.7	1996	76.1
1940	62.9	1998	76.7
1950	68.2	1999	76.7
1960	69.7	2000	77.0
1970	70.8	2001	77.2
1975	72.6	2002	77.3
1980	73.7	2003	77.5
1987	75.0	2004	77.8
1988	74.9	2005	77.9
1989	75.2	2010	78.1
1990	75.4	2015	78.9
1992	75.8	2020	79.5

(Source: National Center for Health Statistics)

28. *Obesity* Obesity (BMI $\geq$ 30) is a serious problem in the United States and is expected to get worse. Being overweight increases the risk of diabetes, heart disease, and many other ailments. The table gives the percent of Americas who are obese from 1990 projected through 2030.

a. Find a logarithmic function that models the percent, y, as a function of the number of years after 1980, x.

b. Graph the model and the data on the same set of axes, and comment on the fit of the model to the data.

c. Use the model to predict what percent of Americas will be obese in 2027.

d. Graphically find the year in which the percent obese is projected to reach 38.

Year	1990	2000	2010	2015	2020	2025	2030
% Obese	12.7	22.1	30.9	34.5	37.4	39.9	42.2

(Source: *American Journal of Preventive Medicine*, Vol 42, (June 2012), 563–570. ajpmonline.org)

29. *Rate of Growth of Population* The table gives the annual rates of growth of the U.S. population for selected years from 1970 projected to 2049.

a. Find a logarithmic function that models the rate, y, as a function of the number of years after 1960.

b. Find an exponential function that models the rate, y, as a function of the number of years after 1960.

c. Graph each model on the same axes with the data to determine which model is the better fit to the data.

Year	Annual Growth Rate
1970	2.07
1980	1.69
1990	1.56
2000	1.25
2010	1.10
2020	0.90
2030	0.76
2040	0.62
2049	0.46

30. *Diabetes* The table shows the number of U.S. adults with diabetes (in millions), projected from 2015 through 2050.

Year	Millions of Adults with Diabetes
2015	37.3
2020	50.0
2025	59.5
2030	68.3
2035	77.2
2040	84.1
2045	91.7
2050	100.0

a. Create a scatter plot of the data, with x equal to the number of years after 2010.

b. Find a logarithmic function to model the data, with x equal to the number of years after 2010 and y equal to the number of millions of U.S. adults with diabetes.

c. When does the model predict that the number of millions of U.S. adults with diabetes will be 60.01?
(Source: Centers for Disease Control and Prevention)

31. *Vehicle Temperature* The temperature inside a vehicle can rise almost 20°F in just 10 minutes and almost 30°F in 20 minutes. The vehicle can quickly reach a temperature that puts a pet at risk of serious illness and even death. The estimated interior air temperature in a car after various numbers of minutes is shown in the table.

Estimated Vehicle Interior Air Temperature vs. Elapsed Time						
	Outside Air Temperature (°F)					
Elapsed time	**70**	**75**	**80**	**85**	**90**	**95**
0 minutes	70	75	80	85	90	95
10 minutes	89	94	99	104	109	114
20 minutes	99	104	109	114	119	124
30 minutes	104	109	114	119	124	129
40 minutes	108	113	118	123	128	133
50 minutes	111	116	121	126	131	136
60 minutes	113	118	123	128	133	138
> 1 hour	115	120	125	130	135	140

(Source: Courtesy Jan Null, CCM; Department of Geosciences, San Francisco State University)

a. Create a scatter plot of the vehicle interior air temperature as a function of the number of minutes in the first hour, when the outside air temperature is 80°F.

b. Omit the first data point and find a logarithmic function that models the data. (Why do we need to omit the first data point?)

c. Graph the model on the same axes as the data and comment on the fit.

d. What does the model predict that the interior air temperature will be after 45 minutes?

32. *Health Services Expenditures* The following table gives total U.S. expenditures (in billions of dollars) for health services and supplies for selected years from 2000 and projected to 2018.

Year	$ (billions)	Year	$ (billions)
2000	1264	2010	2458
2002	1498	2012	2746
2004	1733	2014	3107
2006	1976	2016	3556
2008	2227	2018	4086

(Source: U.S. Centers for Medicare and Medicaid Services)

a. Find an exponential function to model these data, with x equal to the number of years after 2000.

b. Use the model to estimate the U.S. expenditures for health services and supplies in 2020.

33. *Sexually Active Girls* The percents of girls age x or younger who have been sexually active are given in the table below.

a. Create a logarithmic function that models the data, using an input equal to the age of the girls.

b. Use the model to estimate the percent of girls age 17 or younger who have been sexually active.

c. Find the quadratic function that is the best fit for the data.

d. Graph each of these functions on the same axes with the data points to determine which function is the better model for the data.

Age	Cumulative Percent Sexually Active Girls	Cumulative Percent Sexually Active Boys
15	5.4	16.6
16	12.6	28.7
17	27.1	47.9
18	44.0	64.0
19	62.9	77.6
20	73.6	83.0

(Source: "National Longitudinal Survey of Youth," *Risking the Future*, Washington D.C.: National Academy Press)

34. *Sexually Active Boys* The percents of boys age x or younger who have been sexually active are given in the table in Exercise 33.

   a. Create a logarithmic function that models the data, using an input equal to the age of the boys.

   b. Use the model to estimate the percent of boys age 17 or younger who have been sexually active.

   c. Compare the percents that are sexually active for the two genders (see Exercise 33). What do you conclude?

35. *Fuel Economy* The lifetime gasoline use of light-duty vehicles is a function of the fuel economy, as shown in the figure below.

   a. Should the data be modeled by an exponential growth or an exponential decay function?

   b. Find an exponential function to model the data.

Light-Duty Vehicle Fuel Consumption

(Source: U.S. Department of Energy, Energy Efficiency, and Renewable Energy)

   c. Find a power function to model the data.

   d. Which function is the better fit to the data?

   e. If a vehicle had a fuel economy of 100 mpg, what would its lifetime gasoline use be, according to the power model?

36. *National Debt* The table gives the U.S. national debt (in $ billion) for selected years from 1945 and projected to 2015.

   a. Find an exponential function that models the debt, in billions of dollars, y, as a function of the number of years after 1940.

   b. Graph the function and the data on the same axes.

   c. Use the unrounded model to predict the national debt for 2021. Is this interpolation or extrapolation?

Year	U.S. Debt ($ billions)	Year	U.S. Debt ($ billions)
1945	259	2005	7933
1955	273	2009	11,957
1965	314	2010	13,529
1975	533	2011	15,476
1985	1823	2012	16,067
1990	3233	2013	16,856
1996	5225	2014	17,900
2000	5674	2015	18,700

(Source: usgovernmentspending.com)

5.5 Exponential Functions and Investing

KEY OBJECTIVES

- Find future value of investments when interest is compounded k times per year

- Find future value of investments when interest is compounded continuously

- Find the present value of an investment

- Use graphing utilities to model investment data

SECTION PREVIEW **Investments**

If $1000 is invested in an account that earns 6% interest, with the interest added to the account at the end of each year, the interest is said to be **compounded annually**. The amount to which an investment grows over a period of time is called its **future value**. In this section, we use general formulas to find the future value of money invested when the interest is compounded at regular intervals of time and when it is compounded continuously. We also find the lump sum that must be invested to have it grow to a specified amount in the future. This is called the **present value** of the investment. ■

Compound Interest

In Section 5.4, we found that if a set of data has initial value a and a constant percent change r for equally spaced inputs x, the data can be modeled by the exponential function

$$y = a(1 + r)^x$$

Thus, if an investment of $1000 earns 6% interest, compounded annually, the future value at the end of t years will be

$$S = \$1000(1.06)^t \text{ dollars}$$

EXAMPLE 1 ▶ Future Value of an Account

To numerically and graphically view how $1000 grows at 6% interest, compounded annually, do the following:

a. Construct a table that gives the future values $S = 1000(1 + 0.06)^t$ for $t = 0, 1, 2, 3, 4,$ and 5 years after the $1000 is invested.

b. Graph $S = 1000(1 + 0.06)^t$ for the values of t given in part (a).

c. Graph the function $S = 1000(1 + 0.06)^t$ as a continuous function of t for $0 \leq t \leq 5$.

d. Is there another graph that more accurately represents the amount that would be in the account at any time during the first 5 years that the money is invested?

SOLUTION

a. Substitute the values of t into the equation $S = 1000(1 + 0.06)^t$. Because the output units for S are dollars, we round to the nearest cent to obtain the values in Table 5.17.

b. Figure 5.42 shows the scatter plot of the data in Table 5.17.

Table 5.17

t (years)	Future Value, S ($)
0	1000.00
1	1060.00
2	1123.60
3	1191.02
4	1262.48
5	1338.23

Figure 5.42

c. Figure 5.43(a) shows a continuous graph of the function $S = 1000(1 + 0.06)^t$ for $0 \leq t \leq 5$.

d. The scatter plot in Figure 5.42 shows the amount in the account at the end of each year, but not at any other time during the 5-year period. The continuous graph in Figure 5.43(a) shows the amount continually increasing, but interest is paid only at

(a)   (b)

Figure 5.43

the end of each year and the amount in the account is constant at the previous level until more interest is added. So the graph in Figure 5.43(b) is a more accurate graph of the amount that would be in this account because it shows the amount remaining constant until the end of each year, when interest is added.

The function g graphed in Figure 5.43(b) is the step function defined by

$$g(t) = \begin{cases} 1000.00 & \text{if } 0 \le t < 1 \\ 1060.00 & \text{if } 1 \le t < 2 \\ 1123.60 & \text{if } 2 \le t < 3 \\ 1191.02 & \text{if } 3 \le t < 4 \\ 1262.48 & \text{if } 4 \le t < 5 \\ 1338.23 & \text{if } t = 5 \end{cases} \quad \text{dollars in } t \text{ years}$$

This step function is rather cumbersome to work with, so we will find the future value by using the graph in Figure 5.43(a) and the function $S = 1000(1 + 0.06)^t$ *interpreted discretely*. That is, we draw the graph of this function as it appears in Figure 5.43(a) and evaluate the formula for S with the understanding that the only inputs that make sense in this investment example are nonnegative integers. In general, we have the following.

Future Value of an Investment with Annual Compounding

If P dollars are invested at an interest rate r per year, compounded annually, the future value S at the end of t years is

$$S = P(1 + r)^t$$

The **annual interest rate** r is also called the **nominal interest rate**, or simply the **rate**. Remember that the interest rate is usually stated as a percent and that it is converted to a decimal when computing future values of investments.

If the interest is compounded more than once per year, then the additional compounding will result in a larger future value. For example, if the investment is compounded twice per year (semiannually), there will be twice as many compounding periods, with the interest rate in each period equal to

$$r\left(\frac{1}{2}\right) = \frac{r}{2}$$

Thus, the future value is found by doubling the number of periods and halving the interest rate if the compounding is done twice per year. In general, we use the following model, interpreted discretely, to find the amount of money that results when P dollars are invested and earn compound interest.

Future Value of an Investment with Periodic Compounding

If P dollars are invested for t years at the annual interest rate r, where the interest is compounded k times per year, then the interest rate per period is $\frac{r}{k}$, the number of compounding periods is kt, and the future value that results is given by

$$S = P\left(1 + \frac{r}{k}\right)^{kt} \text{ dollars}$$

For example, if $1000 is placed in an account that earns 6% interest per year, with the interest added to the account at the end of every month (compounded monthly), the future value of this investment in 5 years is given by

$$S = 1000\left(1 + \frac{0.06}{12}\right)^{12(5)} = \$1348.85$$

EXAMPLE 2 ▶ Future Value

Find the future value of $10,000 placed in an account earning 8% per year for 10 years, if the interest is compounded:

a. Annually

b. Daily

c. How much more is earned by compounding daily?

SOLUTION

a. Substituting $P = 10,000$, $r = 0.08$, $k = 1$, and $t = 10$ in $S = P\left(1 + \frac{r}{k}\right)^{kt}$ gives

$$S = 10,000\left(1 + \frac{0.08}{1}\right)^{1 \cdot 10} = S = 10,000(1.08)^{10} = 21,589.25$$

Thus the future value is $21,589.25.

b. Substituting $P = 10,000$, $r = 0.08$, $k = 365$ and $t = 10$ in $S = P\left(1 + \frac{r}{k}\right)^{kt}$ gives

$$S = 10,000\left(1 + \frac{0.08}{365}\right)^{365 \cdot 10} = 10,000\left(1 + \frac{0.08}{365}\right)^{3650} = 22,253.46$$

Thus the future value is $22,253.46.

c. Daily compounding gives $22,253.46 − $21,589.25 = $664.21 more than annual compounding.

Graphing calculators and spreadsheets such as Excel have built-in features that can be used to find future values and solve other financial problems. We can find the future value of investments such as those in Example 2, (called lump-sum investments) with a graphing calculator as follows:

Under the **APPS** menu, select Finance and then TMV Solver.

a. Set N = the total number of compounding periods.

b. Set I% = the annual percentage rate.

c. Set PV = the lump-sum investment (with a negative sign because the lump-sum is *leaving* the investor and going into the investment).

d. Set PMT = 0 because no additional money is being invested.

e. Set both P/Y (payments per year) and C/Y (compounding periods per year) equal to the number of compounding periods per year.

f. Highlight END, move the cursor to FV, and press **ALPHA** **ENTER** to get the future value.

Figure 5.44(a) and (b) show the data entered and the future value for part (a) of Example 2, and Figure 5.44(c) shows the solution for part (b).* The solutions agree with those in Example 2.

(a)   (b)   (c)

Figure 5.44

Spreadsheet
▶ SOLUTION   We can also use Excel to find the future value of a lump-sum investment by using the formula

$$= \text{fv}(\text{F2,B2,C2,A2, 0}), \text{where}$$

Cell F2 has the periodic interest rate as a decimal.
Cell B2 has the total number of periods.
Cell C2 has the payment amount, 0.
Cell A2 has the lump sum investment amount P.

Figure 5.45 shows the input for part (b) of Example 2 and the future value of $10,000 invested at 8% per year for 10 years, compounded daily, in Cell B4.†

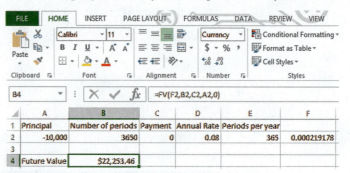

Figure 5.45

Continuous Compounding and the Number *e*

In Section 5.1, we stated that if P dollars are invested for t years at interest rate r, compounded continuously, the future value of the investment is

$$S = Pe^{rt}$$

To see how the number e becomes part of this formula, we consider the future value of $1 invested at an annual rate of 100%, compounded for 1 year with different compounding periods. If we denote the number of periods per year by k, the model that gives the future value is

$$S = \left(1 + \frac{1}{k}\right)^k \text{dollars}$$

---

* For details, see Appendix A, page 676.
† For details, see Appendix B, page 694.

Table 5.18 shows the future value of this investment for different compounding periods.

Table 5.18

Type of Compounding	Number of Compounding Periods per Year	Future Value ($)
Annually	1	$\left(1 + \dfrac{1}{1}\right)^1 = 2$
Quarterly	4	$\left(1 + \dfrac{1}{4}\right)^4 = 2.44140625$
Monthly	12	$\left(1 + \dfrac{1}{12}\right)^{12} \approx 2.61303529022$
Daily	365	$\left(1 + \dfrac{1}{365}\right)^{365} \approx 2.71456748202$
Hourly	8760	$\left(1 + \dfrac{1}{8760}\right)^{8760} \approx 2.71812669063$
Each minute	525,600	$\left(1 + \dfrac{1}{525,600}\right)^{525,600} \approx 2.7182792154$
x times per year	x	$\left(1 + \dfrac{1}{x}\right)^x$

As Table 5.18 indicates, the future value increases (but not very rapidly) as the number of compounding periods during the year increases. As x gets very large, the future value approaches the number e, which is 2.718281828 to nine decimal places. In general, the outputs that result from larger and larger inputs in the function

$$f(x) = \left(1 + \frac{1}{x}\right)^x$$

approach the number e. (In calculus, e is defined as the *limit* of $\left(1 + \dfrac{1}{x}\right)^x$ as x approaches ∞.)

This definition of e permits us to create a new model for the future value of an investment when interest is compounded continuously. We can let $m = \dfrac{k}{r}$ to rewrite the formula

$$S = P\left(1 + \frac{r}{k}\right)^{kt} \quad \text{as} \quad S = P\left(1 + \frac{1}{m}\right)^{mrt}, \quad \text{or} \quad S = P\left[\left(1 + \frac{1}{m}\right)^m\right]^{rt}$$

Now as the compounding periods k increase without bound, $m = \dfrac{k}{r}$ increases without bound and

$$\left(1 + \frac{1}{m}\right)^m$$

approaches e, so $S = P\left[\left(1 + \dfrac{1}{m}\right)^m\right]^{rt}$ approaches Pe^{rt}.

> **Future Value of an Investment with Continuous Compounding**
>
> If P dollars are invested for t years at an annual interest rate r, compounded continuously, then the future value S is given by
>
> $$S = Pe^{rt} \text{ dollars}$$

For a given principal and interest rate, the function $S = Pe^{rt}$ is a continuous function whose domain consists of real numbers greater than or equal to 0. Unlike the other compound interest functions, this one is not discretely interpreted because of the continuous compounding.

EXAMPLE 3 ▶ Future Value and Continuous Compounding

a. What is the future value of $2650 invested for 8 years at 12%, compounded continuously?

b. How much interest will be earned on this investment?

SOLUTION

a. The future value of this investment is $S = 2650e^{0.12(8)} = 6921.00$.

b. The interest earned on this investment is the future value minus the original investment:

$$\$6921 - \$2650 = \$4271$$

EXAMPLE 4 ▶ Continuous versus Annual Compounding of Interest

a. For each of 9 years, compare the future value of an investment of $1000 at 8%, compounded annually, and of $1000 at 8%, compounded continuously.

b. Graph the functions for annual compounding and for continuous compounding for $t = 30$ years on the same axes.

c. What conclusion can be made regarding compounding annually and compounding continuously?

SOLUTION

a. The future value of $1000, compounded annually, is given by the function $S = 1000(1 + 0.08)^t$, and the future value of $1000, compounded continuously, is given by $S = 1000e^{0.08t}$. By entering these formulas in an Excel spreadsheet and evaluating them for each of 9 years (see Figure 5.46), we can compare the future value for each of these 9 years. At the end of the 9 years, we see that compounding continuously results in $2054.43 - $1999.00 = $55.43 more than compounding annually yields.

b. The graphs are shown in Figure 5.46.

c. The value of the investment increases more under continuous compounding than under annual compounding.

Figure 5.46

Present Value of an Investment

We can write the formula for the future value S when a lump sum P is invested for $n = kt$ periods at interest rate $i = \dfrac{r}{k}$ per compounding period as

$$S = P(1 + i)^n$$

We sometimes need to know the lump sum investment P that will give an amount S in the future. This is called the **present value** P, and it can be found by solving $S = P(1 + i)^n$ for P. Solving for P gives the present value as follows.

Present Value

The lump sum that will give future value S in n compounding periods at rate i per period is the present value

$$P = \frac{S}{(1 + i)^n} \quad \text{or, equivalently,} \quad P = S(1 + i)^{-n}$$

EXAMPLE 5 ▶   Present Value

What lump sum must be invested at 10%, compounded semiannually, for the investment to grow to $15,000 in 7 years?

SOLUTION

We have the future value $S = 15{,}000$, $i = \dfrac{0.10}{2} = 0.05$, and $n = 2 \cdot 7 = 14$, so the present value of this investment is

$$P = \frac{15{,}000}{(1 + 0.05)^{14}} = 7576.02 \text{ dollars}$$

We can find the present value of the lump-sum investment in Example 5 with a graphing calculator by using the same steps used to find future value, with the following changes:

Set FV = the desired future value = 15000

Highlight END, move the cursor to PV, and press **ALPHA** **ENTER** to get the present value.*

Figures 5.47(a) and (b) show the data entered and the present value for Example 5. The present value shown agrees with that computed in Example 5. The present value is negative because it is the investment that *leaves* the investor.

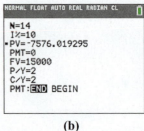

(a)   (b)

Figure 5.47

| **Spreadsheet** **▶ SOLUTION** | We can use Excel to find the present value (investment) that gives the given future value by using the formula |

$$= \text{pv}(\text{F2, B2, C2, A2, 0}), \text{ where}$$

Cell F2 has the periodic interest rate as a decimal = 0.05

Cell B2 has the total number of periods = 14

Cell C2 has the payment amount = 0

Cell A2 has the future value.

Figure 5.48 shows inputs for the investment in Example 5, which gives a future value of $15,000, with the lump-sum inverstment (present value) in Cell B4.† The negative value reflects the fact that $7576.02 is the necessary investment.

Figure 5.48

---

* For details, see Appendix A, page 677.

† For details, see Appendix B, page 695.

Investment Models

We have developed and used investment formulas. We can also use graphing utilities to model actual investment data. Consider the following example.

EXAMPLE 6 ▶ **Mutual Fund Growth**

The data in Table 5.19 give the annual return on an investment of $1.00 made on March 31, 1990, in the AIM Value Fund, Class A Shares. The values reflect reinvestment of all distributions and changes in net asset value but exclude sales charges.

Table 5.19

Time (years)	Average Annual Total Return ($)
1	1.23
3	2.24
5	3.11
10	6.82

(Source: FundStation)

a. Use an exponential function to model these data.

b. Use the model to find what the fund would amount to on March 31, 2001 if $10,000 was invested March 31, 1990 and the fund continued to follow this model after 2000.

c. Is it likely that this fund continued to grow at this rate in 2002?

SOLUTION

a. Using technology to create an exponential model for the data gives

$$A(x) = 1.1609(1.2004)^x$$

Thus, $1.00 invested in this fund grows to $A(x) = 1.1609(1.2004)^x$ dollars x years after March 31, 1990.

b. March 31, 2001 is 11 years after March 31, 1990, so using the equation from part (a) with $x = 11$ gives $A(11) = 1.1609(1.2004)^{11}$ as the future value of an investment of $1.00. Thus, the future value of an investment of $10,000 in 11 years is

$$\$10{,}000[1.1609(1.2004)^{11}] = \$86{,}572.64$$

c. No. Most stocks and mutual funds decreased in value in 2002 because of terrorist attacks in 2001.

SKILLS CHECK 5.5

Evaluate the expressions in Exercises 1–10. Write approximate answers rounded to two decimal places.

1. $15{,}000e^{0.06(20)}$

2. $8000e^{0.05(10)}$

3. $3000(1.06^x)$ for $x = 30$

4. $20{,}000(1.07^x)$ for $x = 20$

5. $12{,}000\left(1 + \dfrac{0.10}{k}\right)^{kn}$ for $k = 4$ and $n = 8$

6. $23{,}000\left(1 + \dfrac{0.08}{k}\right)^{kn}$ for $k = 12$ and $n = 20$

7. $P\left(1 + \dfrac{r}{k}\right)^{kn}$ for $P = 3000$, $r = 8\%$, $k = 2$, $n = 18$

8. $P\left(1 + \dfrac{r}{k}\right)^{kn}$ for $P = 8000$, $r = 12\%$, $k = 12$, $n = 8$

9. $300\left(\dfrac{1.02^n - 1}{0.02}\right)$ for $n = 240$

10. $2000\left(\dfrac{1.10^n - 1}{0.10}\right)$ for $n = 12$

11. Find $g(2.5)$, $g(3)$, and $g(3.5)$ if

$$g(x) = \begin{cases} 1000.00 & \text{if } 0 \le x < 1 \\ 1060.00 & \text{if } 1 \le x < 2 \\ 1123.60 & \text{if } 2 \le x < 3 \\ 1191.00 & \text{if } 3 \le x < 4 \\ 1262.50 & \text{if } 4 \le x < 5 \\ 1338.20 & \text{if } x = 5 \end{cases}$$

12. Find $f(2)$, $f(1.99)$, and $f(2.1)$ if

$$f(x) = \begin{cases} 100 & \text{if } 0 \leq x \leq 1 \\ 200 & \text{if } 1 < x < 2 \\ 300 & \text{if } 2 \leq x \leq 3 \end{cases}$$

13. Solve $S = P\left(1 + \dfrac{r}{k}\right)^{kn}$ for P.

14. Solve $S = P(1 + i)^n$ for P.

EXERCISES 5.5

15. *Future Value* If $8800 is invested for x years at 8% interest, compounded annually, find the future value that results in

  a. 8 years. **b.** 30 years.

16. *Investments* Suppose $6400 is invested for x years at 7% interest, compounded annually. Find the future value of this investment at the end of

  a. 10 years. **b.** 30 years.

17. *Future Value* If $3300 is invested for x years at 10% interest, compounded annually, the future value that results is $S = 3300(1.10)^x$ dollars.

  a. Graph the function for $x = 0$ to $x = 8$.

  b. Use the graph to estimate when the money in the account will double.

18. *Investments* If $5500 is invested for x years at 12% interest, compounded quarterly, the future value that results is $S = 5500(1.03)^{4x}$ dollars.

  a. Graph this function for $0 \leq x \leq 8$.

  b. Use the graph to estimate when the money in the account will double.

19. *Future Value* If $10,000 is invested at 12% interest, compounded quarterly, find the future value in 10 years.

20. *Future Value* If $8800 is invested at 6% interest, compounded semiannually, find the future value in 10 years.

21. *Future Value* An amount of $10,000 is invested at 12% interest, compounded daily.

  a. Find the future value in 10 years.

  b. How does this future value compare with the future value in Exercise 19? Why are they different?

22. *Future Value* A total of $8800 is invested at 6% interest, compounded daily.

  a. Find the future value in 10 years.

b. How does this future value compare with the future value in Exercise 20? Why are they different?

23. *Compound Interest* If $10,000 is invested at 12% interest, compounded monthly, find the interest earned in 15 years.

24. *Compound Interest* If $20,000 is invested at 8% interest, compounded quarterly, find the interest earned in 25 years.

25. *Continuous Compounding* Suppose $10,000 is invested for t years at 6% interest, compounded continuously. Give the future value at the end of

  a. 12 years.

  b. 18 years.

26. *Continuous Compounding* If $42,000 is invested for t years at 7% interest, compounded continuously, find the future value in

  a. 10 years.

  b. 20 years.

27. *Continuous versus Annual Compounding*
  a. If $10,000 is invested at 6%, compounded annually, find the future value in 18 years.

  b. How does this compare with the result from Exercise 25(b)?

28. *Continuous versus Annual Compounding*
  a. If $42,000 is invested at 7%, compounded annually, find the future value in 20 years.

  b. How does this compare with the result from Exercise 26(b)?

29. *Doubling Time* Use a spreadsheet, a table, or a graph to estimate how long it takes for an amount to double if it is invested at 10% interest

  a. Compounded annually.

  b. Compounded continuously.

30. Doubling Time Use a spreadsheet, a table, or a graph to estimate how long it takes for an amount to double if it is invested at 6% interest

   a. Compounded annually.

   b. Compounded continuously.

31. Future Value Suppose $2000 is invested in an account paying 5% interest, compounded annually. What is the future value of this investment

   a. After 8 years?

   b. After 18 years?

32. Future Value If $12,000 is invested in an account that pays 8% interest, compounded quarterly, find the future value of this investment

   a. After 2 quarters.

   b. After 10 years.

33. Investments Suppose $3000 is invested in an account that pays 6% interest, compounded monthly. What is the future value of this investment after 12 years?

34. Investments If $9000 is invested in an account that pays 8% interest, compounded quarterly, find the future value of this investment

   a. After 0.5 year.

   b. After 15 years.

35. Doubling Time If money is invested at 10% interest, compounded quarterly, the future value of the investment doubles approximately every 7 years.

   a. Use this information to complete the table below for an investment of $1000 at 10% interest, compounded quarterly.

Years	0	7	14	21	28
Future Value ($)	1000				

   b. Create an exponential function, rounded to three decimal places, that models the discrete function defined by the table.

   c. Because the interest is compounded quarterly, this model must be interpreted discretely. Use the rounded function to find the value of the investment in 5 years and in $10\frac{1}{2}$ years.

36. Doubling Time If money is invested at 11.6% interest, compounded monthly, the future value of the investment doubles approximately every 6 years.

   a. Use this information to complete the table below for an investment of $1000 at 11.6% interest, compounded monthly.

Years	0	6	12	18	24
Future Value ($)	1000				

   b. Create an exponential function, rounded to three decimal places, that models the discrete function defined by the table.

   c. Because the interest is compounded monthly, this model must be interpreted discretely. Use the rounded model to find the value of the investment in 2 months, in 4 years, and in $12\frac{1}{2}$ years.

37. Present Value What lump sum must be invested at 10%, compounded monthly, for the investment to grow to $65,000 in 8 years?

38. Present Value What lump sum must be invested at 8%, compounded quarterly, for the investment to grow to $30,000 in 12 years?

39. Present Value What lump sum investment will grow to $10,000 in 10 years if it is invested at 6%, compounded annually?

40. Present Value What lump sum investment will grow to $30,000 if it is invested for 15 years at 7%, compounded annually?

41. College Tuition New parents want to put a lump sum into a money market fund to provide $30,000 in 18 years to help pay for college tuition for their child. If the fund averages 10% per year, compounded monthly, how much should they invest?

42. Trust Fund Grandparents decide to put a lump sum of money into a trust fund on their granddaughter's 10th birthday so that she will have $1,000,000 on her 60th birthday. If the fund pays 11%, compounded monthly, how much money must they put in the account?

43. College Tuition The Toshes need to have $80,000 in 12 years for their son's college tuition. What amount must they invest to meet this goal if they can invest money at 10%, compounded monthly?

44. Retirement Hennie and Bob inherit $100,000 and plan to invest part of it for 25 years at 10%, compounded monthly. If they want it to grow to $1 million for their retirement, how much should they invest?

45. *Investment* At the end of t years, the future value of an investment of $10,000 in an account that pays 8%, compounded monthly, is

$$S = 10,000\left(1 + \frac{0.08}{12}\right)^{12t} \text{ dollars}$$

Assuming no withdrawals or additional deposits, how long will it take for the investment to amount to $40,000?

46. *Investment* At the end of t years, the future value of an investment of $25,000 in an account that pays 12%, compounded quarterly, is

$$S = 25,000\left(1 + \frac{0.12}{4}\right)^{4t} \text{ dollars}$$

In how many years will the investment amount to $60,000?

47. *Investment* At the end of t years, the future value of an investment of $38,500 in an account that pays 8%, compounded monthly, is

$$S = 38,500\left(1 + \frac{0.08}{12}\right)^{12t} \text{ dollars}$$

For what time period (in years) will the future value of the investment be more than $100,230?

48. *Investment* At the end of t years, the future value of an investment of $12,000 in an account that pays 8%, compounded monthly, is

$$S = 12,000\left(1 + \frac{0.08}{12}\right)^{12t} \text{ dollars}$$

Assuming no withdrawals or additional deposits, for what time period (in years) will the future value of the investment be less than $48,000?

49. *Interest Rate* Suppose that $15,000 is invested at continuous rate r and grows to $140,900 in 28 years. Solve the equation $140,900 = 15,000(e^{28r})$ to find the interest rate r.

50. *Interest Rate* Suppose that $10,000 is invested at continuous rate r and grows to $17,160 in 9 years. Find the interest rate r.

51. *Interest Rate* Suppose that $20,000 is invested at continuous rate r and grows to $81,104 in 28 years. Find the interest rate r.

52. *Investing* If an investment of P dollars earns r percent (as a decimal), compounded m times per year, its future value in t years is $A = P\left(1 + \dfrac{r}{m}\right)^{mt}$. Prove that the number of years it takes for this investment to double is $t = \dfrac{\ln 2}{m \ln(1 + r/m)}$.

5.6 Annuities; Loan Repayment

KEY OBJECTIVES

- Find the future value of an ordinary annuity
- Find the present value of an ordinary annuity
- Find the payments needed to amortize a loan

SECTION PREVIEW Annuities

To prepare for future retirement, people frequently invest a fixed amount of money at the end of each month into an account that pays interest that is compounded monthly. Such an investment plan or any other characterized by regular payments is called an annuity. In this section, we develop a model to find future values of annuities.

A person planning retirement may also want to know how much money is needed to provide regular payments to him or her during retirement, and a couple may want to know what lump sum to put in an investment to pay future college expenses for their child. The amount of money they need to invest to receive a series of payments in the future is the present value of an annuity.

We can also use the present value concept to determine the payments that are necessary to repay a loan with equal payments made on a regular schedule, which is called amortization. The formulas used to determine the future value and the present value of annuities and the payments necessary to repay loans are applications of exponential functions. ∎

Future Value of an Annuity

An **annuity** is a financial plan characterized by regular payments. We can view an annuity as a savings plan where regular payments are made to an account, and we can use an exponential function to determine what the future value of the account will be. One type of annuity, called an **ordinary annuity**, is a financial plan where equal payments are contributed at the end of each period to an account that pays a fixed rate of interest compounded at the same time as payments are made. For example, suppose $1000 is invested at the end of each year for 5 years in an account that pays interest at 10%, compounded annually. To find the future value of this annuity, we can think of it as the sum of 5 investments of $1000, one that draws interest for 4 years (from the end of the first year to the end of the fifth year), one that draws interest for 3 years, and so on (Table 5.20).

Table 5.20

Investment	Future Value
Invested at end of 1st year	$1000(1.10)^4$
Invested at end of 2nd year	$1000(1.10)^3$
Invested at end of 3rd year	$1000(1.10)^2$
Invested at end of 4th year	$1000(1.10)^1$
Invested at end of 5th year	1000

Note that the last investment is at the end of the fifth year, so it earned no interest. The future value of the annuity is the sum of the separate future values. It is

$$S = 1000 + 1000(1.10)^1 + 1000(1.10)^2 + 1000(1.10)^3 + 1000(1.10)^4$$
$$= 6105.10 \text{ dollars}$$

We can find the value of an ordinary annuity like the one above with the following formula.

Future Value of an Ordinary Annuity

If R dollars are contributed at the end of each period for n periods into an annuity that pays interest at rate i at the end of the period, the future value of the annuity is

$$S = R\left(\frac{(1 + i)^n - 1}{i}\right) \text{ dollars}$$

Note that the interest rate used in the above formula is the rate per period, not the rate for a year. We can use this formula to find the future value of an annuity in which $1000 is invested at the end of each year for 5 years into an account that pays interest at 10%, compounded annually. This future value, which we found earlier without the formula, is

$$S = R\left(\frac{(1 + i)^n - 1}{i}\right) = 1000\left(\frac{(1 + 0.10)^5 - 1}{0.10}\right) = 6105.10 \text{ dollars}$$

EXAMPLE 1 ▶   **Future Value of an Ordinary Annuity**

Find the 5-year future value of an ordinary annuity with a contribution of $500 per quarter into an account that pays 8% per year, compounded quarterly.

SOLUTION

The payments and interest compounding occur quarterly, so the interest rate per period is $\dfrac{0.08}{4} = 0.02$ and the number of compounding periods is $4(5) = 20$. Substituting the information into the formula for the future value of an annuity gives

$$S = 500\left(\frac{(1 + 0.02)^{20} - 1}{0.02}\right) = 12{,}148.68$$

Thus, the future value of this investment is $12,148.68.

We can find the future value of the ordinary annuity in Example 1 with a graphing calculator as follows:

Under the **APPS** menu, select Finance and then TMV Solver.

a. Set N = the total number of compounding periods = 5*4 = 20
b. Set I% = the annual percentage rate = 8
c. Set PV = lump-sum investment = 0
d. Set PMT = quarterly payment = −500
e. Set both P/Y (payments per year) and C/Y (compounding periods per year) equal to the number of compounding periods per year = 4
f. Highlight END, move the cursor to FV, and press **ALPHA** **ENTER** to get the future value.

Figures 5.49(a) and (b) show the data entered and the future value found in Example 1.*

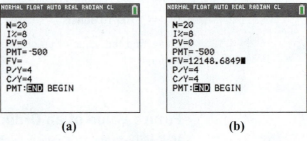

(a)   (b)

Figure 5.49

Spreadsheet ▶ SOLUTION   To find the future value of the ordinary annuity of Example 1 with Excel, we use the formula

$$= \text{fv}(\text{F2},\text{B2},\text{C2},\text{A2},0)$$

with the periodic rate in cell B4, and where

Cell F2 has the periodic interest rate as a decimal = 0.02
Cell B2 has the total number of periods = 20
Cell C2 has the payment amount = −500
Cell A2 has the future value = 0

Figure 5.50 shows the input for the investment of Example 1, with the future value in Cell B4.†

---

* For details, see Appendix A, page 677.
† For details, see Appendix B, page 695.

	A	B	C	D	E	F
1	Principal	Number Periods	Payment	Annual Rate	Periods per Year	Periodic Rate
2	0	20	-500	0.08	4	0.02
3						
4	Future Value	$12,148.68				

Figure 5.50

Future Value of an Annuity

Harry deposits $200 at the end of each month into an account that pays 12% interest per year, compounded monthly. Find the future value for every 4-month period, for up to 36 months.

SOLUTION

To find the future value of this annuity, we note that the interest rate per period (month) is $\dfrac{12\%}{12} = 0.01$, so the future value at the end of n months is given by the function

$$S = 200\left(\frac{1.01^n - 1}{0.01}\right) \text{ dollars}$$

This model for the future value of an ordinary annuity is a continuous function with discrete interpretation because the future value changes only at the end of each period. A graph of the function and the table of future values for every 4 months up to 36 months are shown in Figure 5.51 and Table 5.21, respectively. Keep in mind that the graph should be interpreted discretely; that is, only positive integer inputs and corresponding outputs make sense.

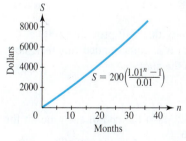

Figure 5.51

Table 5.21

Number of Months	Value of Annuity ($)	Number of Months	Value of Annuity ($)
4	812.08	24	5394.69
8	1657.13	28	6425.82
12	2536.50	32	7498.81
16	3451.57	36	8615.38
20	4403.80		

Present Value of an Annuity

Many retirees purchase annuities that give them regular payments through a period of years. Suppose a lump sum of money is invested to return payments of R dollars at the end of each of n periods, with the investment earning interest at rate i per period, after which $0 will remain in the account. This lump sum is the **present value** of this ordinary annuity, and it is equal to the sum of the present values of each of the future payments. Table 5.22 shows the present value of each payment of this annuity.

Table 5.22

Present	Present Value of This Payment ($)
Paid at end of 1st period	$R(1 + i)^{-1}$
Paid at end of 2nd period	$R(1 + i)^{-2}$
Paid at end of 3rd period	$R(1 + i)^{-3}$
$\vdots$	$\vdots$
Paid at end of $(n - 1)$st period	$R(1 + i)^{-(n-1)}$
Paid at end of nth period	$R(1 + i)^{-n}$

We must have enough money in the account to provide the present value of each of these payments, so the present value of the annuity is

$$A = R(1 + i)^{-1} + R(1 + i)^{-2} + R(1 + i)^{-3} + \cdots + R(1 + i)^{-(n-1)} + R(1 + i)^{-n}$$

This sum can be written in the form

$$A = R\left(\frac{1 - (1 + i)^{-n}}{i}\right)$$

Present Value of an Ordinary Annuity

If a payment of R dollars is to be made at the end of each period for n periods from an account that earns interest at a rate of i per period, then the account is an **ordinary annuity**, and the **present value** is

$$A = R\left(\frac{1 - (1 + i)^{-n}}{i}\right)$$

EXAMPLE 3 ▶ **Present Value of an Annuity**

Suppose a retiring couple wants to establish an annuity that will provide $2000 at the end of each month for 20 years. If the annuity earns 6%, compounded monthly, how much must the couple put in the account to establish the annuity?

SOLUTION

We seek the present value of an annuity that pays $2000 at the end of each month for $12(20) = 240$ months, with interest at $\dfrac{6\%}{12} = \dfrac{0.06}{12} = 0.005$ per month. The present value is

$$A = 2000\left(\frac{1 - (1 + 0.005)^{-240}}{0.005}\right) = 279{,}161.54 \text{ dollars}$$

Thus, the couple can receive a payment of $2000 at the end of each month for 20 years if they put a lump sum of $279,161.54 in an annuity. (Note that the sum of the money they will receive over the 20 years is $2000 \cdot 20 \cdot 12 = \$480{,}000$.)

We can find the present value of the annuity in Example 3 with a graphing calculator by using the same steps used to find future value, with the inputs for PV and FV changed.

Under the **APPS** menu, select Finance and then TMV Solver.
a. Set N = the total number of compounding periods = 240
b. Set I% = the annual percentage rate = 6
c. Set FV = lump-sum investment = 0
d. Set PMT = quarterly payment = 2000
e. Set both P/Y (payments per year) and C/Y (compounding periods per year) equal to the number of compounding periods per year = 12
f. Highlight END, move the cursor to PV, and press **ALPHA** **ENTER** to get the present value.

Figures 5.52(a) and (b) show the data entered and the present value for the investment in Example 3.*

(a)   (b)

Figure 5.52

Spreadsheet ▶ SOLUTION

To find the present value of the ordinary annuity of Example 3 with Excel, we use the formula

$$= \text{pv}(\text{F2,B2,C2,A2, 0})$$

with the periodic rate in cell B4, where

Cell F2 has the periodic interest rate as a decimal = 0.005
Cell B2 has the total number of periods = 240
Cell C2 has the payment amount = 2000
Cell A2 has the future value = 0

Figure 5.53 shows the input for the investment in Example 3, with the present value in Cell B4.† The parentheses indicate a negative value, because the money *leaves* the couple to start the annuity.

Figure 5.53

---

* For details, see Appendix A, page 678.
† For details, see Appendix B, page 696.

EXAMPLE 4 ▶   **Home Mortgage**

A couple who wants to purchase a home has $30,000 for a down payment and wants to make monthly payments of $2200. If the interest rate for a 25-year mortgage is 6% per year on the unpaid balance, what is the price of a house they can buy?

SOLUTION

The amount of money that they can pay for the house is the sum of the down payment and the present value of the payments that they can afford. The present value of the $12(25) = 300$ monthly payments of $2200 with interest at $\dfrac{6\%}{12} = \dfrac{0.06}{12} = 0.005$ per month is

$$A = 2200\left(\frac{1 - (1 + 0.005)^{-300}}{0.005}\right) = 341{,}455.10$$

They can buy a house costing $341,455.10 + \$30,000 = \$371,455.10$.

Loan Repayment

When money is borrowed, the borrower must repay the total amount that was borrowed (the debt) plus interest on that debt. Most loans (including those for houses, for cars, and for other consumer goods) require regular payments on the debt plus payment of interest on the unpaid balance of the loan. That is, loans are paid off by a series of partial payments with interest charged on the unpaid balance at the end of each period. The stated interest rate (the **nominal rate**) is the *annual rate*. The rate per period is the nominal rate divided by the number of payment periods per year.

There are two popular repayment plans for these loans. One plan applies an equal amount to the debt each payment period plus the interest for the period, which is the interest on the unpaid balance. For example, a loan of $240,000 for 10 years at 12% could be repaid with 120 monthly payments of $2000 plus 1% $\left(\dfrac{1}{12} \text{ of } 12\%\right)$ of the unpaid balance each month. When this payment method is used, the payments will decrease as the unpaid balance decreases. A few sample payments are shown in Table 5.23.

Table 5.23

Payment Number (month)	Unpaid Balance ($)	Interest on the Unpaid Balance ($)	Balance Reduction ($)	Monthly Payment ($)	New Balance ($)
1	240,000	2400	2000	4400	238,000
2	238,000	2380	2000	4380	236,000
⋮					
51	140,000	1400	2000	3400	138,000
⋮					
101	40,000	400	2000	2400	38,000
⋮					
118	6000	60	2000	2060	4000

A loan of this type can also be repaid by making all payments (including the payment on the balance and the interest) of equal size. The process of repaying the loan in this way is called **amortization**. If a bank makes a loan of this type, it uses an amortization formula that gives the size of the equal payments. This formula is developed by considering the loan as an annuity purchased from the borrower by the bank, with the annuity paying a fixed return to the bank each payment period. The lump sum that the bank gives to the borrower (the principal of the loan) is the present value of this annuity, and each payment that the bank receives from the borrower is a payment from this annuity. To find the size of these equal payments, we solve the formula for the present value of an annuity, $A = R\left(\dfrac{1 - (1 + i)^{-n}}{i}\right)$, for R. This gives the following formula.

Amortization Formula

If a debt of A dollars, with interest at a rate of i per period, is amortized by n equal periodic payments made at the end of each period, then the size of each payment is

$$R = A\left(\frac{i}{1 - (1 + i)^{-n}}\right)$$

We can use this formula to find the *equal* monthly payments that will amortize the loan of $240,000 for 10 years at 12%. The interest rate per month is 0.01, and there are 120 periods.

$$R = A\left(\frac{i}{1 - (1 + i)^{-n}}\right) = 240{,}000\left(\frac{0.01}{1 - (1 + 0.01)^{-120}}\right) \approx 3443.303$$

Thus, repaying this loan requires 120 equal payments of $3443.31. If we compare payment plans, we see that the first payment in Table 5.23 is nearly $1000 more than the equal amortization payment in Table 5.24 and that every payment after the 101st is $1000 less than the amortization payment. The advantage of the amortization method of repaying the loan is that the borrower can budget better by knowing what the payment will be each month. Of course, the balance of the loan will not be reduced as fast with this payment method, because most of the monthly payment will be needed to pay interest in the first few months. Table 5.24 shows the interest paid getting smaller, and balance reduction size increasing, as the payments increase.

Table 5.24

Payment Number	Unpaid Balance	Interest	Monthly Payment	Balance Reduction	New Balance
1	240,000	2400	3443.31	1043.31	238,956.69
2	238,956.69	2389.57	3443.31	1053.74	237,902.95
.					
.					
51	172,745.38	1727.45	3443.31	1715.88	171,029.52
.					
.					
101	63,136.30	631.36	3443.31	2811.95	60,324.35

EXAMPLE 5 ▶ Home Mortgage

A couple that wants to purchase a home with a price of $230,000 has $50,000 for a down payment. If they can get a 25-year mortgage at 9% per year on the unpaid balance,

a. What will be their equal monthly payments?

b. What is the total amount they will pay before they own the house outright?

c. How much interest will they pay?

SOLUTION

a. The amount of money that they must borrow is $230,000 - $50,000 = $180,000. The number of monthly payments is $12(25) = 300$, and the interest rate is $\dfrac{9\%}{12} = \dfrac{0.09}{12} = 0.0075$ per month. The monthly payment is

$$R = 180,000 \left(\frac{0.0075}{1 - (1.0075)^{-300}} \right) \approx 1510.553$$

so the required payment would be $1510.56.

b. The amount that they must pay before owning the house is the down payment plus the total of the 300 payments:

$$\$50,000 + 300(\$1510.56) = \$503,168$$

c. The total interest is the total amount paid minus the price paid for the house. The interest is

$$\$503,168 - \$230,000 = \$273,168$$

When the interest paid is based on the periodic interest rate on the unpaid balance of the loan, the nominal rate is also called the **annual percentage rate (APR)**. You will occasionally see an advertisement for a loan with a stated interest rate and a different, larger APR. This can happen if a lending institution charges fees (sometimes called points) in addition to the stated interest rate. Federal law states that all these charges must be included in computing the APR, which is the true interest rate that is being charged on the loan.

SKILLS CHECK 5.6

Give answers to two decimal places.

1. Solve $S = P(1 + i)^n$ for P with positive exponent n.

2. Solve $S = P(1 + i)^n$ for P, with no denominator.

3. Solve $Ai = R[1 - (1 + i)^{-n}]$ for A.

4. Evaluate $2000 \left(\dfrac{1 - (1 + 0.01)^{-240}}{0.01} \right)$.

5. Solve $A = R \left(\dfrac{1 - (1 + i)^{-n}}{i} \right)$ for R.

6. Evaluate $240,000 \left(\dfrac{0.01}{1 - (1 + 0.01)^{-120}} \right)$.

EXERCISES 5.6

7. *IRA* Anne Wright decides to invest $4000 in an IRA CD at the end of each year for 10 years. If she makes these payments and the certificates all pay 6%, compounded annually, how much will she have at the end of the 10 years?

8. *Future Value* Find the future value of an annuity of $5000 paid at the end of each year for 20 years, if interest is earned at 9%, compounded annually.

9. *Down Payment* To start a new business, Beth deposits $1000 at the end of each 6-month period in an account that pays 8%, compounded semiannually. How much will she have at the end of 8 years?

10. *Future Value* Find the future value of an annuity of $2600 paid at the end of each 3-month period for 5 years, if interest is earned at 6%, compounded quarterly.

11. *Retirement* Mr. Lawrence invests $600 at the end of each month in an account that pays 7%, compounded monthly. How much will be in the account in 25 years?

12. *Down Payment* To accumulate money for the down payment on a house, the Kings deposit $800 at the end of each month into an account paying 7%, compounded monthly. How much will they have at the end of 5 years?

13. *Dean's List* Parents agree to invest $1000 (at 10%, compounded semiannually) for their daughter on the December 31 or June 30 following each semester that she makes the Dean's List during her 4 years in college. If she makes the Dean's List in each of the 8 semesters, how much will the parents have to give her when she graduates?

14. *College Tuition* To help pay for tuition, parents deposited $1000 into an account on each of their child's birthdays, starting with the first birthday. If the account pays 6%, compounded annually, how much will they have for tuition on her 19th birthday?

15. *Annuities* Find the present value of an annuity that will pay $1000 at the end of each year for 10 years if the interest rate is 7%, compounded annually.

16. *Annuities* Find the present value of an annuity that will pay $500 at the end of each year for 20 years if the interest rate is 9%, compounded annually.

17. *Lottery Winnings* The winner of a "million dollar" lottery is to receive $50,000 plus $50,000 at the end of each year for 19 years or the present value of this annuity in cash. How much cash would she receive if money is worth 8%, compounded annually?

18. *College Tuition* A couple wants to establish a fund that will provide $3000 for tuition at the end of each 6-month period for 4 years. If a lump sum can be placed in an account that pays 8%, compounded semiannually, what lump sum is required?

19. *Insurance Payment* A man is disabled in an accident and wants to receive an insurance payment that will provide him with $3000 at the end of each month for 30 years. If the payment can be placed in an account that pays 9%, compounded monthly, what size payment should he seek?

20. *Auto Leasing* A woman wants to lease rather than buy a car but does not want to make monthly payments. A dealer has the car she wants, which leases for $400 at the end of each month for 48 months. If money is worth 8%, compounded monthly, what lump sum should she offer the dealer to keep the car for 48 months?

21. *Business Sale* A man can sell his Thrifty Electronics business for $800,000 cash or for $100,000 plus $122,000 at the end of each year for 9 years.

   a. Find the present value of the annuity that is offered if money is worth 10%, compounded annually.

   b. If he takes the $800,000, spends $100,000 of it, and invests the rest in a 9-year annuity at 10%, compounded annually, what size annuity payment will he receive at the end of each year?

   c. Which is better, taking the $100,000 and the annuity or taking the cash settlement? Discuss the advantages of your choice.

22. *Sale of a Practice* A physician can sell her practice for $1,200,000 cash or for $200,000 plus $250,000 at the end of each year for 5 years.

   a. Find the present value of the annuity that is offered if money is worth 7%, compounded annually.

   b. If she takes the $1,200,000, spends $200,000 of it, and invests the rest in a 5-year annuity at 7%, compounded annually, what size annuity payment will she receive at the end of each year?

   c. Which is better, taking the $200,000 and the annuity or taking the cash settlement? Discuss the advantages of your choice.

23. *Home Mortgage* A couple wants to buy a house and can afford to pay $1600 per month.

   a. If they can get a loan for 30 years with interest at 9% per year on the unpaid balance and make monthly payments, how much can they pay for a house?

   b. What is the total amount paid over the life of the loan?

   c. What is the total interest paid on the loan?

24. *Auto Loan* A man wants to buy a car and can afford to pay $400 per month.

   a. If he can get a loan for 48 months with interest at 12% per year on the unpaid balance and make monthly payments, how much can he pay for a car?

   b. What is the total amount paid over the life of the loan?

   c. What is the total interest paid on the loan?

25. *Loan Repayment* A loan of $10,000 is to be amortized with quarterly payments over 4 years. If the interest on the loan is 8% per year, paid on the unpaid balance,

   a. What is the interest rate charged each quarter on the unpaid balance?

b. How many payments are made to repay the loan?

c. What payment is required quarterly to amortize the loan?

26. **Loan Repayment** A loan of $36,000 is to be amortized with monthly payments over 6 years. If the interest on the loan is 6% per year, paid on the unpaid balance,

   a. What is the interest rate charged each month on the unpaid balance?

   b. How many payments are made to repay the loan?

   c. What payment is required each month to amortize the loan?

27. **Home Mortgage** A couple who wants to purchase a home with a price of $350,000 has $100,000 for a down payment. If they can get a 30-year mortgage at 6% per year on the unpaid balance,

a. What will be their monthly payments?

b. What is the total amount they will pay before they own the house outright?

c. How much interest will they pay over the life of the loan?

28. **Business Loan** Business partners want to purchase a restaurant that costs $750,000. They have $300,000 for a down payment, and they can get a 25-year business loan for the remaining funds at 8% per year on the unpaid balance, with quarterly payments.

   a. What will be the payments?

   b. What is the total amount that they will pay over the 25-year period?

   c. How much interest will they pay over the life of the loan?

5.7 Logistic and Gompertz Functions

KEY OBJECTIVES

- Graph and apply logistic functions
- Find limiting values of logistic functions
- Model with logistic functions
- Graph and apply Gompertz functions

SECTION PREVIEW Women in the Workforce

Figure 5.54 shows the total number of women in the workforce (in millions) for selected years from 1950 and projected to 2050. The S-shaped graph shows that the number of women in the workforce grew rapidly during the 20th century and that it is projected to grow more slowly in the first half of the 21st century. Very seldom does exponential growth continue indefinitely, so functions with this type of graph better represent many types of sales and organizational growth. Two functions that are characterized by rapid growth followed by leveling off are **logistic functions** and **Gompertz functions**. We consider applications of these functions in this section.

Figure 5.54

Logistic Functions

When growth begins slowly, then increases at a rapid rate, and finally slows over time to a rate that is almost zero, the amount (or number) present at any given time frequently fits on an S-shaped curve. Growth of business organizations and the spread of a virus or disease sometimes occur according to this pattern. For example, the total number y of people on a college campus infected by a virus can be modeled by

$$y = \frac{10,000}{1 + 9999e^{-0.99t}}$$

where t is the number of days after an infected student arrives on campus. Table 5.25 shows the number infected on selected days. Note that the number infected grows rapidly and then levels off near 10,000. The graph of this function is shown in Figure 5.55.

Table 5.25

Days	1	3	5	7	9	11	13	15	17
Number Infected	3	19	139	928	4255	8429	9749	9965	9995

A function that can be used to model growth of this type is called a **logistic function**; its graph is similar to the one shown in Figure 5.55.

Figure 5.55

Logistic Function

For real numbers a, b, and c, the function

$$f(x) = \frac{c}{1 + ae^{-bx}}$$

is a logistic function.
   If $a > 0$, a logistic function increases when $b > 0$ and decreases when $b < 0$.

The parameter c is often called the **limiting value**,* or **upper limit**, because the line $y = c$ is a horizontal asymptote for the logistic function. As seen in Figure 5.56, the line $y = 0$ is also a horizontal asymptote. Logistic growth begins at a rate that is near zero, rapidly increases in an exponential pattern, and then slows to a rate that is again near zero.

Logistic Growth Function

Equation: $y = \dfrac{c}{1 + ae^{-bx}}$ $(a > 0, b > 0)$

   x-intercept: none

   y-intercept: $\left(0, \dfrac{c}{1 + a}\right)$

   Horizontal asymptotes: $y = 0, y = c$

Figure 5.56

Even though the graph of an increasing logistic function has the elongated S-shape indicated in Figure 5.56, it may be the case that we have data showing only a portion of the S.

---

* The logistic function fit by most technologies is a best-fit (least-squares) logistic equation. There may therefore be data values that are larger than the value of c. To avoid confusion, we talk about the limiting value of the logistic function or the upper limit in the problem context.

EXAMPLE 1 ▶ Women in the Workforce

Table 5.26 gives the total number of women in the workforce (in millions) for selected years from 1950 and projected to 2050.

a. Create a scatter plot of the data, with x equal to the number of years after 1940 and y equal to the number of women in the workforce, in millions.

b. Find the logistic function that models these data, with x and y defined as in (a).

c. Graph the data and the model on the same axes, and comment on the fit.

d. What will be the number of women in the workforce in 2032, according to the model?

e. According to the model, what is the upper limit on the number of women in the workforce?

Table 5.26

Year	Women in Workforce (millions)
1950	18.4
1960	23.2
1970	31.5
1980	45.5
1990	56.8
2000	65.7
2010	75.5
2015	78.6
2020	79.3
2030	81.6
2040	86.5
2050	91.5

(Source: U.S. Bureau of Labor Statistics)

SOLUTION

a. A scatter plot of the data, with x equal to the number of years after 1940 and y equal to the number of women in the workforce, in millions, is shown in Figure 5.57(a).

b. The logistic function that models the data is $f(x) = \dfrac{92.227}{1 + 7.537e^{(-0.049x)}}$, rounded to three decimal places. (See Figure 5.57(b).)

c. The graphs of the data and the logistic model are shown in Figure 5.57(c). The fit is visually very good.

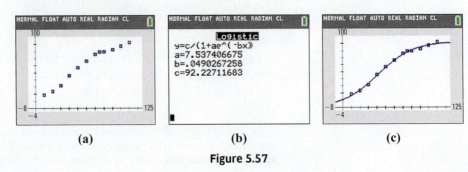

(a)   (b)   (c)

Figure 5.57

d. According to the model, the number of women in the workforce in 2032 will be $f(92) = 85.17$ million. See Figure 5.58.

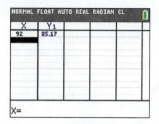

Figure 5.58

e. The value of c in the logistic function is 92.227, so this means the upper limit on the number of women in the workforce is 92.227 million according to this model. The model will probably become invalid eventually.

A logistic function can decrease; this behavior occurs when $b < 0$. Demand functions for products often follow such a pattern. Decreasing logistic functions can also be used to represent decay over time. Figure 5.59 shows a decreasing logistic function.

Logistic Decay Function

Equation: $y = \dfrac{c}{1 + ae^{-bx}} \ (a > 0, b < 0)$

x-intercept: none

y-intercept: $\left(0, \dfrac{c}{1 + a}\right)$

Horizontal asymptotes: $y = 0, y = c$

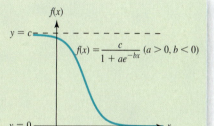

Figure 5.59

The properties of logarithms can also be used to solve for a variable in an exponent in a logistic function.

EXAMPLE 2 ▶   **Expected Life Span**

Using data from 1920 and projected to 2020, the expected life span at birth of people born in the United States can be modeled by the equation

$$y = \frac{82.488}{1 + 0.816e^{-0.024x}}$$

where x is the number of years from 1900.

Use this model to estimate the birth year after 1900 that gives an expected life span of 79 years with

a. Graphical methods.

b. Algebraic methods.

SOLUTION

a. To solve the equation

$$79 = \frac{82.488}{1 + 0.816e^{-0.024x}}$$

with the x-intercept method, we solve

$$0 = \frac{82.488}{1 + 0.816e^{-0.024x}} - 79$$

Graphing the function

$$y = \frac{82.488}{1 + 0.816e^{-0.024x}} - 79$$

on the window [100, 150] by [−4, 4] shows a point where the graph crosses the x-axis (Figure 5.60(a)). Finding the x-intercept gives the approximate solution

$$x \approx 121.53$$

Since x = 121.53 represents the number of years from 1900, we conclude that people born in 2022 have an expected life span of 79 years.

We can also use the intersection method to solve the equation graphically. To do this, we graph the equations $y_1 = \dfrac{82.488}{1 + 0.816e^{-0.024x}}$ and $y_2 = 79$ and find the point of intersection of the graphs (Figure 5.60(b)).

(a)   (b)

Figure 5.60

b. We can also solve the equation

$$79 = \frac{82.488}{1 + 0.816e^{-0.024x}}$$

algebraically. To do this, we first multiply both sides by the denominator and then solve for $e^{-0.024x}$:

$$79(1 + 0.816e^{-0.024x}) = 82.488$$

$$79 + 64.464e^{-0.024x} = 82.488$$

$$64.464e^{-0.024x} = 3.488$$

$$e^{-0.024x} = \frac{3.488}{64.464}$$

Taking the natural logarithm of both sides of the equation and using a logarithmic property are the next steps in the solution:

$$\ln e^{-0.024x} = \ln \frac{3.488}{64.464}$$

$$-0.024x = \ln \frac{3.488}{64.464}$$

$$x = \frac{\ln \dfrac{3.488}{64.464}}{-0.024} \approx 121.53$$

This solution is the same solution that was found (more easily) by using the graphical method in Figure 5.60(a); 2022 is the birth year for people with an expected life span of 79 years.

EXAMPLE 3 ▶ **Uninsured Americans**

The projected annual decrease in the number of uninsured people (in millions) under the Affordable Health Care law from 2014 and projected to 2022 is shown in Table 5.27.

a. Create a scatter plot of the data. Does a logistic function seem appropriate to model the data?

b. Find a logistic function that models the data, with y in millions and x in years after 2012.

c. Graph the model and the data on the same axes and comment on the fit.

d. What does this model project the decrease in the number of uninsured people to be in 2020?

e. What is the upper limit on the annual decrease in the number of uninsured Americans, according to the model?

Table 5.27

Years	Annual Decrease (millions)
2014	15
2015	20
2016	26
2017	28
2018	29
2019	29
2020	29
2021	31
2022	31

(Source: *Wall Street Journal*, July 25, 2012)

SOLUTION

a. The scatter plot of the data is shown in Figure 5.61(a), with y representing the annual decrease of uninsured people, in millions, and x representing the number of years after 2012. Because the scatter plot resembles an S-shape, with initial rapid growth and a slowing over time, a logistic function seems appropriate to model the data.

b. Technology gives the equation of the best-fitting logistic function as

$$y = \frac{30.298}{1 + 5.436e^{-0.822x}}$$

c. The graphs of the model and the data are shown in Figure 5.61(b). The fit appears to be excellent.

d. The model projects that the decrease in the number of uninsured people will be 30.1 million in 2020.

e. The parameter c in the logistic function is 30.298, which means that the upper limit on the annual decrease of uninsured Americans is 30.298 million. However, the actual data show that the annual decrease is expected to exceed 30.298 million in 2021 and 2022.

(a)   (b)

Figure 5.61

Gompertz Functions

Another type of function that models rapid growth that eventually levels off is called a **Gompertz function**. This type of function can also be used to describe human growth and development and the growth of organizations in a limited environment. These functions have equations of the form

$$N = Ca^{R^t}$$

where t represents the time, $R\,(0 < R < 1)$ is the expected rate of growth of the population, C is the maximum possible number of individuals, and a represents the proportion of C present when $t = 0$.

For example, the equation

$$N = 3000(0.2)^{0.6^t}$$

could be used to predict the number of employees t years after the opening of a new facility. Here the maximum number of employees C would be 3000, the proportion of 3000 present when $t = 0$ is $a = 0.2$, and R is 0.6. The graph of this function is shown in Figure 5.62. Observe that the initial number of employees, at $t = 0$, is

$$N = 3000(0.2)^{0.6^0} = 3000(0.2)^1 = 3000(0.2) = 600$$

To verify that the maximum possible number of employees is 3000, observe the following. Because $0.6 < 1$, higher powers of t make 0.6^t smaller, with 0.6^t approaching 0 as t approaches ∞. Thus, $N \to 3000(0.2)^0 = 3000(1) = 3000$ as $t \to \infty$.

Figure 5.62

EXAMPLE 4 ▶ **Deer Population**

The Gompertz equation

$$N = 1000(0.06)^{0.2^t}$$

predicts the size of a deer herd on a small island t decades from now.

a. What is the number of deer on the island now $(t = 0)$?

b. How many deer are predicted to be on the island 1 decade from now $(t = 1)$?

c. Graph the function.

d. What is the maximum number of deer predicted by this model?

e. In how many years will there be 840 deer on the island?

Figure 5.63

Figure 5.64

SOLUTION

a. If $t = 0$, $N = 1000(0.06)^{0.2^0} = 1000(0.06)^1 = 60$.

b. If $t = 1$, $N = 1000(0.06)^{0.2^1} = 1000(0.06)^{0.2} = 570$ (approximately).

c. The graph is shown in Figure 5.63.

d. According to the model, the maximum number of deer is predicted to be 1000. Evaluating N as t gets large, we see the values of N approach, but never reach, 1000. We say that $N = 1000$ is an asymptote for this graph.

e. To find the number of years from now when there will be 840 deer on the island, we graph $y = 840$ with $y = 1000(0.06)^{0.2^x}$ and find the point of intersection to be (1.728, 840). Thus, after about 17 years there will be 840 deer on the island. (See Figure 5.64.)

EXAMPLE 5 ▶   **Company Growth**

A new dotcom company starts with 3 owners and 5 employees but tells investors that it will grow rapidly, with the total number of people in the company given by the model

$$N = 2000(0.004)^{0.5^t}$$

where t is the number of years from the present. Use graphical methods to determine the year in which the number of employees is predicted to be 1000.

SOLUTION

If the company has 1000 employees plus the 3 owners, the total number in the company will be 1003; so we seek to solve the equation

$$1003 = 2000(0.004)^{0.5^t}$$

Using the equations $y_1 = 1003$ and $y_2 = 2000(0.004)^{0.5^t}$ and the intersection method gives $t = 3$ (Figure 5.65). Because t represents the number of years, the owners predict that they will have 1000 employees in 3 years.

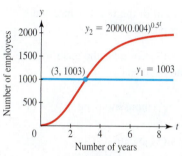

Figure 5.65

SKILLS CHECK 5.7

Give approximate answers to two decimal places.

1. Evaluate $\dfrac{79.514}{1 + 0.835e^{-0.0298(80)}}$.

2. If $y = \dfrac{79.514}{1 + 0.835e^{-0.0298x}}$ find

   a. y when $x = 10$.

   b. y when $x = 50$.

3. Evaluate $1000(0.06)^{0.2^t}$ for $t = 4$ and for $t = 6$.

4. Evaluate $2000(0.004)^{0.5^t}$ for $t = 5$ and for $t = 10$.

5. **a.** Graph the function $f(x) = \dfrac{100}{1 + 3e^{-x}}$ for $x = 0$ to $x = 15$.

   b. Find $f(0)$ and $f(10)$.

   c. Is this function increasing or decreasing?

   d. What is the limiting value of this function?

6. **a.** Graph $f(t) = \dfrac{1000}{1 + 9e^{-0.9t}}$ for $t = 0$ to $t = 15$.

   b. Find $f(2)$ and $f(5)$.

   c. What is the limiting value of this function?

7. **a.** Graph $y = 100(0.05)^{0.3^x}$ for $x = 0$ to $x = 10$.

   b. What is the initial value of this function (the y-value when $x = 0$)?

   c. What is the limiting value of this function?

8. **a.** Graph $N = 2000(0.004)^{0.5^t}$ for $t = 0$ to $t = 10$.

   b. What is the initial value of this function (the y-value when $x = 0$)?

   c. What is the limiting value of this function?

EXERCISES 5.7

9. *Spread of a Disease* The spread of a highly contagious virus in a high school can be described by the logistic function

$$y = \frac{5000}{1 + 999e^{-0.8x}}$$

where x is the number of days after the virus is identified in the school and y is the total number of people who are infected by the virus.

a. Graph the function for $0 \le x \le 15$.

b. How many students had the virus when it was first discovered?

c. What is the upper limit of the number infected by the virus during this period?

10. *Population Growth* Suppose that the size of the population of an island is given by

$$p(t) = \frac{98}{1 + 4e^{-0.1t}} \text{ thousand people}$$

where t is the number of years after 1988.

a. Graph this function for $0 \le x \le 30$.

b. Find and interpret $p(10)$.

c. Find and interpret $p(100)$.

d. What appears to be an upper limit for the size of this population?

11. *Sexually Active Boys* The percent of boys between ages 15 and 20 who have been sexually active at some time (the cumulative percent) can be modeled by the logistic function

$$y = \frac{89.786}{1 + 4.6531e^{-0.8256x}}$$

where x is the number of years after age 15.

a. Graph the function for $0 \le x \le 5$.

b. What does the model estimate the cumulative percent to be for boys whose age is 16?

c. What cumulative percent does the model estimate for boys of age 21, if it is valid after age 20?

d. What is the limiting value implied by this model?
(Source: "National Longitudinal Survey of Youth," riskingthefuture.com)

12. *Sexually Active Girls* The percent of girls between ages 15 and 20 who have been sexually active at

some time (the cumulative percent) can be modeled by the logistic function

$$y = \frac{83.84}{1 + 13.9233e^{-0.9248x}}$$

where x is the number of years after age 15.

a. Graph the function for $0 \le x \le 5$.

b. What does the model estimate the cumulative percent to be for girls of age 16?

c. What cumulative percent does the model estimate for girls of age 20?

d. What is the upper limit implied by the given logistic model?
(Source: "National Longitudinal Survey of Youth," riskingthefuture.com)

13. *Spread of a Rumor* The number of people in a small town who are reached by a rumor about the mayor and an intern is given by

$$N = \frac{10,000}{1 + 100e^{-0.8t}}$$

where t is the number of days after the rumor begins.

a. How many people will have heard the rumor by the end of the first day?

b. How many will have heard the rumor by the end of the fourth day?

c. Use graphical or numerical methods to find the day on which 7300 people in town have heard the rumor.

14. *Sexually Active Girls* The cumulative percent of sexually active girls ages 15 to 20 is given in the table below.

Age (years)	Cumulative Percent Sexually Active
15	5.4
16	12.6
17	27.1
18	44.0
19	62.9
20	73.6

(Source: "National Longitudinal Survey of Youth," riskingthefuture.com)

a. A logarithmic function was used to model this data in Exercises 5.4, problem 33. Find the logistic function that models the data, with *x* equal to the number of years after age 15.

b. Does this model agree with the model used in Exercise 12?

c. Find the linear function that is the best fit for the data.

d. Graph each function on the same axes with the data points to determine which model appears to be the better fit for the data.

15. *Sexually Active Boys* The cumulative percent of sexually active boys ages 15 to 20 is given in the table.

a. A logarithmic function was used to model this data in Exercises 5.4, problem 34. Create the logistic function that models the data, with *x* equal to the number of years after age 15.

b. Does this agree with the model used in Exercise 11?

c. Find the linear function that is the best fit for the data.

d. Determine which model appears to be the better fit for the data and graph this function on the same axes with the data points.

Age (years)	Cumulative Percent Sexually Active
15	16.6
16	28.7
17	47.9
18	64.0
19	77.6
20	83.0

(Source: "National Longitudinal Survey of Youth," riskingthefuture.com)

16. *Population Aged 20–64* The population of U.S. citizens aged 20–64, in millions, is shown in the table near the top of the next column.

a. Find a logistic function that is the best fit for the data, with *x* equal to the number of years after 1940. Report the best model as $f(x)$ with 3 significant digits.

b. Graph the data and the model on the same axes and comment on the fit.

c. Use the model to predict the population in 2023.

d. What does the model predict will be the maximum number in this population?

e. Is it realistic to assume this population will be limited to the number found in part (d) forever, or to assume this model ceases to be valid eventually?

Year	Population (millions)	Year	Population (millions)
1950	92.8	2010	189.1
1960	99.8	2015	194.6
1970	113.0	2020	198.2
1980	134.0	2030	202.8
1990	152.7	2040	213.3
2000	169.8	2050	224.3

(Source: Social Security Administration)

17. *Medicaid* The projected annual increases in the number of people receiving Medicaid (in millions) from 2013 and projected to 2022 are shown in the table.

a. Find a logistic function that models the data, with *y* in millions and *x* in years after 2012.

b. Graph the model and the data on the same axes, and comment on the fit.

c. What does this model predict that the maximum annual increase in the number of people receiving Medicaid will be? Does this number agree with the projections given in the table?

Years	Annual Increase (millions)	Years	Annual Increase (millions)
2013	1	2018	11
2014	7	2019	11
2015	9	2020	11
2016	10	2021	11
2017	10	2022	11

(Source: *Wall Street Journal*, July 25, 2012)

18. *World Population* The total population of the world, by decade, from 1950 and projected to 2050 is shown in the table on the following page.

a. Find the logistic function that is the best fit for the data, with *x* equal to the number of years after 1940 and *y* equal to the world population, in billions.

b. Graph the model and the data on the same axes, and comment on the fit.

c. What does the model predict the world population will be in 2028?

Year	Total World Population	Year	Total World Population
1950	2,556,000,053	2010	6,848,932,929
1960	3,039,451,023	2020	7,584,821,144
1970	3,706,618,163	2030	8,246,619,341
1980	4,453,831,714	2040	8,850,045,889
1990	5,278,639,789	2050	9,346,399,468
2000	6,082,966,429		

(Source: U.S. Census Bureau, International Database)

19. *Life Span* The expected life span of people in the United States for certain birth years is given in the table.

Birth Year	Life Span (years)	Birth Year	Life Span (years)
1920	54.1	1994	75.7
1930	59.7	1996	76.1
1940	62.9	1998	76.7
1950	68.2	1999	76.7
1960	69.7	2000	77.0
1970	70.8	2001	77.2
1975	72.6	2002	77.3
1980	73.7	2003	77.5
1987	75.0	2004	77.8
1988	74.9	2005	77.9
1989	75.2	2010	78.1
1990	75.4	2015	78.9
1992	75.8	2020	79.5

(Source: National Center for Health Statistics)

a. Find a logistic function to model the data, with x equal to how many years after 1900 the birth year is.

b. Estimate the expected life span of a person born in the United States in 1955 and a person born in 2006.

c. Find an upper limit for a person's expected life span in the United States, according to this model.

20. *World Internet Usage* The number of worldwide users (in millions) of the Internet for 1995–2014 is given in the following table.

a. Find an exponential function that models the data, with y equal to the number of users in millions and x equal to the number of years after 1990.

b. Find a logistic function that models the data, with y equal to the number of users in millions and x equal to the number of years after 1990.

c. Graph the aligned data with both models on the same axes. Which model appears to be a better fit?

Year	Users (millions)	Year	Users (millions)
1995	16	2005	1018
1996	36	2006	1093
1997	70	2007	1319
1998	147	2008	1574
1999	248	2009	1802
2000	361	2010	1971
2001	536	2011	2267
2002	598	2012	2497
2003	719	2013	2802
2004	817	2014	3035

(Source: www.internetworldstats.com)

d. Use each model to predict the number of users in 2020. Which estimate seems more reasonable?

e. What will be the maximum number of worldwide users, according to the logistic model?

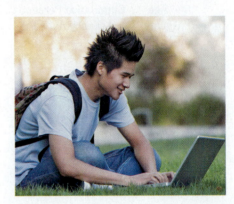

21. *Organizational Growth* A new community college predicts that its student body will grow rapidly at first and then begin to level off according to the Gompertz curve with equation $N = 10,000(0.4)^{0.2^t}$ students, where t is the number of years after the college opens.

a. What does this model predict the number of students to be when the college opens?

b. How many students are predicted to attend the college after 4 years?

c. Graph the equation for $0 \le t \le 10$ and estimate an upper limit on the number of students at the college.

22. *Organizational Growth* A new technology company started with 6 employees and predicted that its number of employees would grow rapidly at first and then begin to level off according to the Gompertz function

$$N = 150(0.04)^{0.5^t}$$

where t is the number of years after the company started.

   a. What does this model predict the number of employees to be in 8 years?

   b. Graph the function for $0 \leq t \leq 10$ and estimate the maximum predicted number of employees.

23. *Sales Growth* The president of a company predicts that sales will increase rapidly after a new product is brought to the market and that the number of units sold monthly can be modeled by $N = 40,000(0.2)^{0.4^t}$, where t represents the number of months after the product is introduced.

   a. How many units will be sold by the end of the first month?

   b. Graph the function for $0 \leq t \leq 10$.

   c. What is the predicted upper limit on sales?

24. *Company Growth* Because of a new research grant, the number of employees in a firm is expected to grow, with the number of employees modeled by $N = 1600(0.6)^{0.2^t}$, where t is the number of years after the grant was received.

   a. How many employees did the company have when the grant was received?

   b. How many employees did the company have at the end of 3 years after the grant was received?

   c. What is the expected upper limit on the number of employees?

   d. Graph the function.

25. *Company Growth* Suppose that the number of employees in a new company is expected to grow, with the number of employees modeled by $N = 1000(0.01)^{0.5^t}$, where t is the number of years after the company was formed.

   a. How many employees did the company have when it started?

   b. How many employees did the company have at the end of 1 year?

   c. What is the expected upper limit on the number of employees?

   d. Use graphical or numerical methods to find the year in which 930 people are employed.

26. *Sales Growth* The president of a company predicts that sales will increase rapidly after a new advertising campaign, with the number of units sold weekly modeled by $N = 8000(0.1)^{0.3^t}$, where t represents the number of weeks after the advertising campaign begins.

   a. How many units per week will be sold at the beginning of the campaign?

   b. How many units will be sold at the end of the first week?

   c. What is the expected upper limit on the number of units sold per week?

   d. Use graphical or numerical methods to find the first week in which 6500 units will be sold.

27. *Spread of Disease* An employee brings a contagious disease to an office with 150 employees. The number of employees infected by the disease t days after the employees are first exposed to it is given by

$$N = \frac{100}{1 + 79e^{-0.9t}}$$

Use graphical or numerical methods to find the number of days until 99 employees have been infected.

28. *Advertisement* The number of people in a community of 15,360 who are reached by a particular advertisement t weeks after it begins is given by

$$N(t) = \frac{14,000}{1 + 100e^{-0.6t}}$$

Use graphical or numerical methods to find the number of weeks until at least half of the community is reached by this advertisement.

29. *Population Growth* A pair of deer are introduced on a small island, and the population grows until the food supply and natural enemies of the deer on the island limit the population. If the number of deer is

$$N = \frac{180}{1 + 89e^{-0.5554t}}$$

where t is the number of years after the deer are introduced, how long does it take for the deer population to reach 150?

30. *Spread of Disease* A student brings a contagious disease to an elementary school of 1200 students. If the number of students infected by the disease t days after the students are first exposed to it is given by

$$N = \frac{800}{1 + 799e^{-0.9t}}$$

use numerical or graphical methods to find in how many days at least 500 students will be infected.

chapter 5 | SUMMARY

In this chapter, we discussed exponential and logarithmic functions and their applications. We explored the fact that logarithmic and exponential functions are inverses of each other. Many applications in today's world involve exponential and logarithmic functions, including the pH of substances, stellar magnitude, intensity of earthquakes, the growth of bacteria, the decay of radioactive isotopes, and compound interest. Logarithmic and exponential equations are solved by using both algebraic and graphical solution methods.

Key Concepts and Formulas

5.1 Exponential Functions

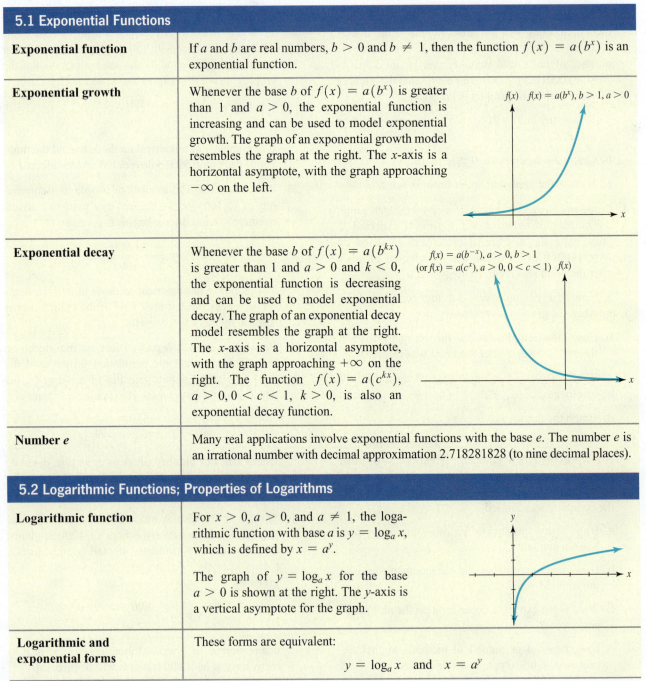

Exponential function	If a and b are real numbers, $b > 0$ and $b \neq 1$, then the function $f(x) = a(b^x)$ is an exponential function.
Exponential growth	Whenever the base b of $f(x) = a(b^x)$ is greater than 1 and $a > 0$, the exponential function is increasing and can be used to model exponential growth. The graph of an exponential growth model resembles the graph at the right. The x-axis is a horizontal asymptote, with the graph approaching $-\infty$ on the left.
Exponential decay	Whenever the base b of $f(x) = a(b^{kx})$ is greater than 1 and $a > 0$ and $k < 0$, the exponential function is decreasing and can be used to model exponential decay. The graph of an exponential decay model resembles the graph at the right. The x-axis is a horizontal asymptote, with the graph approaching $+\infty$ on the right. The function $f(x) = a(c^{kx})$, $a > 0, 0 < c < 1, k > 0$, is also an exponential decay function.
Number e	Many real applications involve exponential functions with the base e. The number e is an irrational number with decimal approximation 2.718281828 (to nine decimal places).

5.2 Logarithmic Functions; Properties of Logarithms

Logarithmic function	For $x > 0, a > 0$, and $a \neq 1$, the logarithmic function with base a is $y = \log_a x$, which is defined by $x = a^y$. The graph of $y = \log_a x$ for the base $a > 0$ is shown at the right. The y-axis is a vertical asymptote for the graph.
Logarithmic and exponential forms	These forms are equivalent: $$y = \log_a x \quad \text{and} \quad x = a^y$$

Common logarithms	Logarithms with a base of 10; $\log_{10} x$ is denoted $\log x$.
Natural logarithms	Logarithms with a base of e; $\log_e x$ is denoted $\ln x$.
Properties of logarithms	For $b > 0$, $b \neq 1$, k a real number, and M and N positive real numbers,

For $b > 0$, $b \neq 1$, k a real number, and M and N positive real numbers,

Property 1. $\log_b b = 1$

Property 2. $\log_b 1 = 0$

Property 3. $\log_b b^x = x$

Property 4. $b^{\log_b x} = x$

Property 5.   If $M = N$, then $\log_b M = \log_b N$

Property 6. $\log_b (MN) = \log_b M + \log_b N$

Property 7. $\log_b \left(\dfrac{M}{N} \right) = \log_b M - \log_b N$

Property 8. $\log_b M^k = k \log_b M$

5.3 Exponential and Logarithmic Equations

Solving exponential equations	Many exponential equations can be solved by converting to logarithmic form; that is, $x = a^y$ is equivalent to $y = \log_a x$.
Change of base	If $a > 0$, $a \neq 1$, $b > 0$, $b \neq 1$, and $x > 0$, then $\log_b x = \dfrac{\log_a x}{\log_a b}$.
Solving exponential equations	Some exponential equations can be solved by taking the logarithm of both sides and using logarithmic properties.
Solving logarithmic equations	Many logarithmic equations can be solved by converting to exponential form; that is, $y = \log_a x$ is equivalent to $x = a^y$. Sometimes properties of logarithms can be used to rewrite the equation in a form that can be more easily solved.
Exponential and logarithmic inequalities	Solving inequalities involving exponential and logarithmic functions involves rewriting and solving the related equation. Then graphical methods are used to find the values of the variable that satisfy the inequality.

5.4 Exponential and Logarithmic Models

Exponential models	Many sources provide data that can be modeled by exponential growth and decay functions according to the model $y = ab^x$ for real numbers a and b, $b > 0$ and $b \neq 1$.
Constant percent change	If the percent change of the outputs of a set of data is constant for equally spaced inputs, an exponential function will be a perfect fit for the data. If the percent change of the outputs is approximately constant for equally spaced inputs, an exponential function will be an approximate fit for the data.
Exponential model	If a set of data has initial value a and has a constant percent change r, the data can be modeled by the exponential function $$y = a(1 + r)^x$$
Comparison of models	To determine which type of model is the best fit for a set of data, we can look at the first differences and second differences of the outputs and the percent change of the outputs.

| Logarithmic models | Models can be created involving $\ln x$. Data that exhibit an initial rapid increase and then have a slow rate of growth can often be described by the function $f(x) = a + b \ln x$ for $b > 0$. |

5.5 Exponential Functions and Investing

Future value with annual compounding	If P dollars are invested for n years at an interest rate r, compounded annually, then the future value of P is given by $S = P(1 + r)^n$ dollars.
Future value of an investment	If P dollars are invested for t years at the annual interest rate r, where the interest is compounded k times per year, then the interest rate per period is $\dfrac{r}{k}$, the number of compounding periods is kt, and the future value that results is given by $S = P\left(1 + \dfrac{r}{k}\right)^{kt}$ dollars.
Future value with continuous compounding	If P dollars are invested for t years at an annual interest rate r, compounded continuously, then the future value S is given by $S = Pe^{rt}$ dollars.
Present value of an investment	The lump sum P that will give future value S if invested for n periods at interest rate i per compounding period is $$P = \frac{S}{(1 + i)^n}, \quad \text{or} \quad P = S(1 + i)^{-n}$$

5.6 Annuities; Loan Repayment

Future value of an ordinary annuity	If R dollars are contributed at the end of each period for n periods into an annuity that pays interest at rate i at the end of the period, the future value of the annuity is $$S = R\left(\frac{(1 + i)^n - 1}{i}\right)$$
Present value of an ordinary annuity	If a payment of R dollars is to be made at the end of each period for n periods from an account that earns interest at a rate of i per period, then the account is an ordinary annuity, and the present value is $$A = R\left(\frac{1 - (1 + i)^{-n}}{i}\right)$$
Amortization formula	If a debt of A dollars, with interest at a rate of i per period, is amortized by n equal periodic payments made at the end of each period, then the size of each payment is $$R = A\left(\frac{i}{1 - (1 + i)^{-n}}\right)$$

5.7 Logistic and Gompertz Functions

| Logistic function | For real numbers a, b, and c, the function $$f(x) = \frac{c}{1 + ae^{-bx}}$$ is a logistic function.

A logistic function increases when $a > 0$ and $b > 0$. The graph of a logistic function resembles the graph at right. | |

Logistic modeling	If data points indicate that growth begins slowly, then increases very rapidly, and finally slows over time to a rate that is almost zero, the data may fit the logistic model $$f(x) = \frac{c}{1 + ae^{-bx}}$$
Gompertz function	For real numbers C, a, and R $(0 < R < 1)$, the function $N = Ca^{R^t}$ is a Gompertz function. A Gompertz function increases rapidly, then levels off as it approaches a limiting value. The graph of a Gompertz function resembles the graph at right.

chapter 5 SKILLS CHECK

1. a. Graph the function $f(x) = 4e^{-0.3x}$.

   b. Find $f(-10)$ and $f(10)$.

2. Is $f(x) = 4e^{-0.3x}$ an increasing or a decreasing function?

3. Graph $f(x) = 3^x$ on $[-10, 10]$ by $[-10, 20]$.

4. Graph $y = 3^{(x-1)} + 4$ on $[-10, 10]$ by $[0, 15]$.

5. How does the graph of Exercise 4 compare with the graph of Exercise 3?

6. Is $y = 3^{(x-1)} + 4$ an increasing or a decreasing function?

7. a. If $y = 1000(2^{-0.1x})$, find the value of y when $x = 10$.

   b. Use graphical or numerical methods to determine the value of x that gives $y = 250$ if $y = 1000(2^{-0.1x})$.

In Exercises 8 and 9, write the exponential equations in logarithmic form.

8. $x = 6^y$

9. $y = 7^{3x}$

In Exercises 10–12, write the logarithmic equations in exponential form.

10. $y = \log_4 x$

11. $y = \log(x)$

12. $y = \ln x$

13. Write the inverse of $y = 4^x$ in logarithmic form.

In Exercises 14–16, evaluate the logarithms, if possible. Give approximate solutions to four decimal places.

14. $\log 22$

15. $\ln 56$

16. $\log 10$

In Exercises 17–19, find the value of the logarithms without using a calculator.

17. $\log_2 16$

18. $\ln(e^4)$

19. $\log 0.001$

In Exercises 20 and 21, use a change of base formula to evaluate each of the logarithms. Give approximate solutions to four decimal places.

20. $\log_3 54$

21. $\log_8 56$

In Exercises 22 and 23, graph the functions with technology.

22. $y = \ln(x - 3)$

23. $y = \log_3 x$

In Exercises 24–27, solve the exponential equations. Give approximate solutions to four decimal places.

24. $340 = e^x$

25. $1500 = 300e^{8x}$

26. $9200 = 23(2^{3x})$

27. $4(3^x) = 36$

28. Rewrite $\ln \dfrac{(2x - 5)^3}{x - 3}$ as the sum, difference, or product of logarithms, and simplify if possible.

29. Rewrite $6 \log_4 x - 2 \log_4 y$ as a single logarithm.

30. Determine whether a linear or an exponential function is the better fit for the data in the table. Then find the equation of the function with three-decimal-place accuracy.

x	1	2	3	4	5
y	2	5	12	30	75

31. Evaluate $P\left(1 + \dfrac{r}{k}\right)^{kn}$ for $P = 1000$, $r = 8\%$, $k = 12$, $n = 20$, to two decimal places.

32. Evaluate $1000\left(\dfrac{1 - 1.03^{-240+n}}{0.03}\right)$ for $n = 120$, to two decimal places.

33. a. Graph $f(t) = \dfrac{2000}{1 + 8e^{-0.8t}}$ for $t = 0$ to $t = 10$.

b. Find $f(0)$ and $f(8)$, to two decimal places.

c. What is the limiting value of this function?

34. a. Graph $y = 500(0.1^{0.2^x})$ for $x = 0$ to $x = 10$.

b. What is the initial value (the value of y when $x = 0$) of this function?

c. What is the limiting value of this function?

chapter 5 REVIEW

35. *Personal Consumption* Using data from the U.S. Bureau of Labor Statistics for selected years from 1998 and projected to 2018, the number of billions of dollars spent for personal consumption in the United States can be modeled by $P = 2969e^{0.051t}$, where t is the number of years after 1985.

a. Is this model one of exponential growth or exponential decay? Explain.

b. Graph this equation with a graphing utility to show the graph from 1985 through the year 2020.

36. *Sales Decay* At the end of an advertising campaign, the daily sales (in dollars) declined, with daily sales given by the equation $y = 2000(2^{-0.1x})$, where x is the number of weeks after the end of the campaign. According to this model, what will sales be 4 weeks after the end of the ad campaign?

37. *Sales Decay* At the end of an advertising campaign, the weekly sales (in dollars) declined, with weekly sales given by the equation $y = 2000(2^{-0.1x})$, where x is the number of weeks after the end of the campaign. In how many weeks will sales be half of the sales at the end of the ad campaign?

38. *Earthquakes*
a. If an earthquake has an intensity of 1000 times I_0, what is the magnitude of the earthquake?

b. An earthquake that measured 6.5 on the Richter scale occurred in Pakistan in February 1991. Express this intensity in terms of I_0.

39. *Earthquakes* On January 23, 2001, an earthquake registering 7.9 hit western India, killing more than 15,000 people. On the same day, an earthquake registering 4.8 hit the United States, killing no one. How much more intense was the Indian earthquake than the American earthquake?

40. *Investments* The number of years needed for an investment of $10,000 to grow to $30,000 when money is invested at 12%, compounded annually, is given by

$$t = \log_{1.12} \dfrac{30,000}{10,000} = \log_{1.12} 3$$

In how many years will this occur?

41. *Investments* If $1000 is invested at 10%, compounded quarterly, the future value of the investment is given by $S = 1000(2^{x/7})$, where x is the number of years after the investment is made.

a. Use the logarithmic form of this function to solve the equation for x.

b. Find when the future value of the investment is $19,504.

42. *Medicaid* The projected annual increases (in millions) in the number of people receiving Medicaid can be modeled by the function $y = 3.196 + 3.975 \ln x$, with y in millions and x in years after 2000. In what year does the model predict the annual increase will be 15.66 million?
(Source: *Wall Street Journal*, July 25, 2012)

43. *Hearing* High-pitched tones that young people can hear at 30 decibels must be produced at 90 decibels for elderly people to hear them. How many times louder must the tones be for the elderly to hear them?
(Source: *Discover*, October 2012

44. *Mobile Home Sales* A company that buys and sells used mobile homes estimates its cost to be given by $C(x) = 2x + 50$ thousand dollars when x mobile homes are purchased. The same company estimates that its revenue from the sale of x mobile homes is given by $R(x) = 10(1.26^x)$ thousand dollars.

    a. Combine C and R into a single function that gives the profit for the company when x used mobile homes are bought and sold.

    b. Use a graph to find how many mobile homes the company must sell for revenue to be at least $30,000 more than cost.

45. *Sales Decay* At the end of an advertising campaign, the sales (in dollars) declined, with weekly sales given by the equation $y = 40,000(3^{-0.1x})$, where x is the number of weeks after the end of the campaign. In how many weeks will sales be less than half of the sales at the end of the ad campaign?

46. *Carbon-14 Dating* An exponential decay function can be used to model the number of atoms of a radioactive material that remain after a period of time. Carbon-14 decays over time, with the amount remaining after t years given by

$$y = y_0 e^{-0.00012097t}$$

where y_0 is the original amount.

    a. If a sample of carbon-14 weighs 100 grams originally, how many grams will be present in 5000 years?

    b. If a sample of wood at an archaeological site contains 36% as much carbon-14 as living wood, determine when the wood was cut.

47. *Purchasing Power* If a retired couple has a fixed income of $60,000 per year, the purchasing power of their income (adjusted value of the money) after t years of 5% inflation is given by the equation $P = 60,000e^{-0.05t}$. In how many years will the purchasing power of their income fall below half of their current income?

48. *Investments* If $2000 is invested at 8%, compounded continuously, the future value of the investment after t years is given by $S = 2000e^{0.08t}$ dollars. What is the future value of this investment in 10 years?

49. *Investments* If $3300 is invested for x years at 10%, compounded annually, the future value that will result is $S = 3300(1.10)^x$ dollars. In how many years will the investment result in $13,784.92?

50. *Health Care Spending* In 2011, national health care spending was $2.7 trillion, and it was projected to grow at an average annual rate of 5.7% for 2011 through 2021.

    a. Write an exponential function that models the national health care spending, y, in trillions, as a function of the number of years after 2011, x.

    b. What is the projected national health care spending for 2021, according to this model?

51. *International Air Travel* The number of passengers using international air traffic into and out of the United States was projected by the Federal Aviation Administration to grow at 4% per year from 2012 until 2032.

    a. If the number was 172 million in 2012, write the equation of the exponential function that models this projection of the number of millions of passengers y as a function of the number of years x after 2012.

    b. What does this model project that the number of passengers using international air traffic into and out of the United States will be in 2032?

    c. When does the model project that the number of passengers using international air traffic into and out of the United States will be 254.6 million?

52. *Energy Use* Energy use per dollar of GDP is indexed to 1980, which means that energy use for any year is viewed as a percent of the use per dollar of GDP in 1980. The following data show the energy use per dollar of GDP, as a percent, for selected years from 1990 and projected to 2035.

    a. Find an exponential function that models these data, with x equal to the number of years after 1990 and y equal to the energy use per dollar of GDP, as a percent. Is this function increasing or decreasing?

    b. Graph the model and the data on the same axes.

    c. What does this model predict the energy use will be in 2032?

Energy Use per Dollar of 1980 GDP

Year	Percent	Year	Percent
1990	79	2015	51
1995	75	2020	45
2000	67	2025	41
2005	60	2030	37
2010	56	2035	34

(Source: U.S. Department of Energy)

53. *White Population* The table gives the number of millions of White, non-Hispanic individuals 16 years and older in the U.S. civilian non-institutional population.

a. Find the logarithmic function that is the best fit for the data, with y in millions of Americans and x equal to the number of years after 1990.

b. Graph the model and the data on the same axes.

c. Use the model to predict the number of this group in the U.S. civilian non-institutional population in 2019.

Year	2000	2010	2015	2020	2030	2040	2050
White non-Hispanic	153.1	162.1	164.6	166.3	168.8	169.7	169.4

(Source: U.S. Census Bureau)

54. *Corvette Acceleration* The following table shows the times that it takes a 2008 Corvette to reach speeds from 0 mph to 100 mph, in increments of 10 mph after 30 mph. Find a logarithmic function that gives the speed as a function of the time.

Time (sec)	Speed (mph)	Time (sec)	Speed (mph)
1.7	30	5.3	70
2.4	40	6.5	80
3.3	50	7.9	90
4.1	60	9.5	100

55. *Diabetes* The table gives the number of U.S. adults with diabetes (in millions) from 2010 and projected to 2050.

a. Find the logistic function that best models these data, with x equal to the number of years after 2000 and y equal to the number of millions of adults with diabetes. Report the model with three significant digits.

b. Use the model to estimate the number of U.S. adults who will have diabetes in 2028.

c. According to the model, what will be the maximum number of U.S. adults to have diabetes?

d. In what year does the model predict that the number of U.S. adults with diabetes will reach 80 million?

Year	Number (millions)	Year	Number (millions)
2010	32.2	2035	76.2
2015	37.3	2040	84.1
2020	50.0	2045	91.7
2025	59.5	2050	100.0
2030	68.3		

(Source: Centers for Disease Control and Prevention)

56. *Starbucks* The table shows the number of worldwide Starbucks stores from 2005 to 2013 and projected to 2017.

a. Find the logistic function that models the data, with x equal to the number of years after 2005.

b. What does the model predict the number of stores to be in 2020?

Year	Total Stores	Year	Total Stores
2005	10,241	2010	16,858
2006	12,440	2011	17,003
2007	15,011	2012	17,572
2008	16,680	2013	20,891
2009	16,635	2017	21,000

(Source: Starbucks annual report)

57. *Endangered Species* The following table gives the numbers of species that were endangered in various years from 1990 to 2012.

a. Find the logarithmic function that models these data. Use x as the number of years past 1980.

b. Use the model to predict the number of endangered species in 2017.

c. When does the model predict that 1303 species will be endangered?

Year	Endangered Species
1990	442
2003	987
2006	1009
2007	1037
2008	1049
2010	1061
2011	1058
2012	1086

(Source: U.S. Fish and Wildlife Service)

58. *Investments* Find the 7-year future value of an investment of $20,000 placed into an account that pays 6%, compounded annually.

59. *Annuities* At the end of each quarter, $1000 is placed into an account that pays 12%, compounded quarterly. What is the future value of this annuity in 6 years?

60. *Annuities* Find the 10-year future value of an ordinary annuity with a contribution of $1500 at the end of each month, placed into an account that pays 8%, compounded monthly.

61. *Present Value* Find the present value of an annuity that will pay $2000 at the end of each month for 15 years if the interest rate is 8%, compounded monthly.

62. *Present Value* Find the present value of an annuity that will pay $500 at the end of each 6-month period for 12 years if the interest rate is 10%, compounded semiannually.

63. *Loan Amortization* A debt of $2000 with interest at 12%, compounded monthly, is amortized by equal monthly payments for 36 months. What is the size of each payment?

64. *Loan Amortization* A debt of $120,000 with interest at 6%, compounded monthly, is amortized by equal monthly payments for 25 years. What is the size of each monthly payment?

65. *Births to Unmarried Mothers* The percent of live births to unmarried mothers for the years 1970–2007 can be modeled by the logistic function

$$y = \frac{44.742}{1 + 6.870e^{-0.0782x}}$$

where x is the number of years after 1960.

a. Use this model to estimate the percent in 1990 and in 1996.

b. What is the upper limit of the percent of teen mothers who were unmarried, according to this model?

66. *Spread of Disease* A student brings a contagious disease to an elementary school of 2000 students. The number of students infected by the disease is given by

$$n = \frac{1400}{1 + 200e^{-0.5x}}$$

where x is the number of days after the student brings the disease.

a. How many students will be infected in 14 days?

b. How many days will it take for 1312 students to be infected?

67. *Organizational Growth* The president of a new campus of a university predicts that the student body will grow rapidly after the campus is open, with the number of students at the beginning of year t given by

$$N = 4000 \left(0.06^{0.4^{t-1}}\right)$$

a. How many students does this model predict for the beginning of the second year $(t = 2)$?

b. How many students are predicted for the beginning of the tenth year?

c. What is the limit on the number of students that can attend this campus, according to the model?

68. *Sales Growth* The number of units of a new product that were sold each month after the product was introduced is given by

$$N = 18,000 \left(0.03^{0.4^t}\right)$$

where t is the number of months.

a. How many units were sold 10 months after the product was introduced?

b. What is the limit on sales if this model is accurate?

69. *Investing* In how many years will $50,000 invested at continuous rate 7% grow to $202,760?

70. *Interest Rate* Suppose that $30,000 is invested at continuous rate r and grows to $121,656 in 28 years. Find the interest rate r.

71. *Interest Rate* Suppose that $25,000 is invested at continuous rate r and grows to $45,552.97 in 15 years. Find the interest rate r.

Group Activities
▶ EXTENDED APPLICATIONS

1. Chain Letters

Suppose you receive a chain letter asking you to send $1 to the person at the top of the list of six names, then to add your name to the bottom of the list, and finally to send the revised letter to six people. The promise is that when the letters have been sent to each of the people above you on the list, you will be at the top of the list and you will receive a large amount of money. To investigate whether this is worth the dollar that you are asked to spend, create the requested models and answer the following questions.

1. Suppose each of the six people on the original list sent six letters. How much money would the person on the top of the list receive?
2. Consider the person who is second on the original list. How much money would this person receive?
3. Complete the partial table of amounts that will be sent to the person on the top of the list during each "cycle" of the chain letter.

Cycle Number (after original six names)	Money Sent to Person on Top of List ($)
1	6 × 6 = 36
2	6 × 36 = 216
3	
4	
5	

4. If you were going to start such a chain letter (don't, it's illegal), would you put your name at the top of the first 6 names or as number 5?
5. Find the best quadratic, power, and exponential models for the data in the table. Which of these function types gives the best model for the data?
6. Use the exponential function found in part (5) to determine the amount of money the person who was at the bottom of the original list would receive if all people contacted sent the $1 and mailed the letter to 6 additional people.
7. How many people will have to respond to the chain letter for the sixth person on the original list to receive all the money that was promised?
8. If you receive the letter in its tenth cycle, how many other people have been contacted, along with you, assuming that everyone who receives the letter cooperates with its suggestions?
9. Who remains in the United States for you to send your letter to, if no one sends a letter to someone who has already received it?
10. Why do you think the federal government has made it illegal to send chain letters in the U.S. mail?

2. Modeling

Your mission is to find real-world data for company sales, a stock price, biological growth, or some sociological trend over a period of years that fit an exponential, logistic, or logarithmic function. To do this, you can look at a statistical graph called a histogram, which describes the situation, or at a graph or table describing it. Once you have found data consisting of numerical values, you are to find the equation that is the best exponential, logistic, or logarithmic fit for the data and then write the model that describes the relationship. Some helpful steps for this process are given in the instructions below.

1. Look in newspapers, in periodicals, on the Internet, or in statistical abstracts for a graph or table of data containing at least six data points. The data must have numerical values for the independent and dependent variables.

2. If you decide to use a relation determined by a graph that you have found, read the graph very carefully to determine the data points or (for full credit) contact the source of the data to obtain the data from which the graph was drawn.
3. If you have a table of values for different years, create a graph of the data to determine the type of function that will be the best fit for the data. To do this, first align the independent variable by letting the input represent the number of years after some convenient year and then enter the data into your graphing calculator and draw a scatter plot. Check to see if the points on the scatter plot lie near some curve that could be described by an exponential, logistic, or logarithmic function. If not, save the data for possible later use and renew your search. You can also test to

see if the percent change of the outputs is nearly constant for equally spaced inputs. If so, the data can be modeled by an exponential function.

4. Use your calculator to create the equation of the function that is the best fit for the data. Graph this equation and the data points on the same axes to see if the equation is reasonable.

5. Write some statements about the data that you have been working with. For example, describe how the quantity is increasing or decreasing over periods of years, why it is changing as it is, or when this model is no longer appropriate and why.

6. Your completed project should include the following:

   a. A proper bibliographical reference for the source of your data.

   b. An original copy or photocopy of the data being used.

   c. A scatter plot of the data and reasons why you chose the model that you did to fit it.

   d. The equation that you have created, labeled with appropriate units of measure and variable descriptions.

   e. A graph of the scatter plot and the model on the same axes.

   f. One or more statements about how you think the model you have created could actually be used. Include any restrictions that should be placed on the model.

6

Higher-Degree Polynomial and Rational Functions

The amount of money spent by global tourists can be modeled by the function

$$y = 0.153x^3 - 2.914x^2 + 36.854x + 253.208$$

where y represents the amount spent by global travelers (in billions of dollars) and x represents the number of years after 1990. If the total cost of producing x units of a product is $C(x) = 100 + 30x + 0.01x^2$, the function

$$\overline{C}(x) = \frac{100 + 30x + 0.01x^2}{x}$$

models the average cost per unit. These are examples of polynomial and rational functions, which we will discuss in this chapter.

sections	objectives	applications
6.1 Higher-Degree Polynomial Functions	Identify graphs of higher-degree polynomial functions; graph cubic functions and quartic functions; find local extrema and absolute extrema	U.S. unemployment, elderly workers
6.2 Modeling with Cubic and Quartic Functions	Model data with cubic functions; model data with quartic functions	Aging workers, diabetes, Supplemental Nutrition Assistance Program, foreign-born population
6.3 Solution of Polynomial Equations	Solve polynomial equations by factoring and the root method; estimate solutions with technology	International travel, maximizing volume, photosynthesis, cost, future value of an investment, China's labor pool
6.4 Polynomial Equations Continued; Fundamental Theorem of Algebra	Divide polynomials with synthetic division; solve cubic equations; combine graphical and algebraic methods to solve polynomial equations; solve quartic equations; use the Fundamental Theorem of Algebra to find the number of complex solutions of a polynomial equation; solve polynomial equations with complex solutions	Break-even
6.5 Rational Functions and Rational Equations	Graph rational functions; solve rational equations algebraically and graphically	Average cost of production, cost-benefit, advertising and sales, illumination
6.6 Polynomial and Rational Inequalities	Solve polynomial inequalities; solve rational inequalities	Average cost, box construction

Algebra TOOLBOX

Polynomials

Recall from Chapter 1 that an expression containing a finite number of additions, subtractions, and multiplications of constants and positive integer powers of variables is called a **polynomial**. The general form of a polynomial in x is

$$a_nx^n + a_{n-1}x^{n-1} + \cdots + a_1x + a_0$$

where a_0 and each coefficient of x are real numbers and each power of x is a positive integer. If $a_n \neq 0$, n is the highest power of x, a_n is called the **leading coefficient**, and n is the **degree** of the polynomial. Thus, $5x^4 + 3x^2 - 6$ is a fourth-degree (quartic) polynomial with leading coefficient 5.

EXAMPLE 1 ▶ **Polynomials**

What are the degree and the leading coefficient of each of the following polynomials?

a. $4x^2 + 5x^3 - 17$ **b.** $30 - 6x^5 + 3x^2 - 7$

SOLUTION

a. This is a third-degree polynomial. The leading coefficient is 5, the coefficient of the highest-degree term.

b. This is a fifth-degree polynomial. The leading coefficient is -6, the coefficient of the highest-degree term.

Factoring Higher-Degree Polynomials

Some higher-degree polynomials can be factored by expressing them as the product of a monomial and another polynomial, which sometimes can be factored to complete the factorization.

EXAMPLE 2 ▶ **Factoring**

Factor: **a.** $3x^3 - 21x^2 + 36x$ **b.** $-6x^4 - 10x^3 + 4x^2$

SOLUTION

a. $3x^3 - 21x^2 + 36x = 3x(x^2 - 7x + 12) = 3x(x - 3)(x - 4)$

b. $-6x^4 - 10x^3 + 4x^2 = -2x^2(3x^2 + 5x - 2) = -2x^2(3x - 1)(x + 2)$

419

Some polynomials can be written in *quadratic form* by using substitution. Then quadratic factoring methods can be used to begin the factorization.

EXAMPLE 3 ▶   **Factoring Polynomials in Quadratic Form**

Factor: **a.** $x^4 - 6x^2 + 8$ **b.** $x^4 - 18x^2 + 81$

SOLUTION

a. The expression $x^4 - 6x^2 + 8$ is not a quadratic polynomial, but substituting u for x^2 converts it into the quadratic expression $u^2 - 6u + 8$, so we say that the original expression is in *quadratic form*. We factor it using quadratic methods, as follows:

$$u^2 - 6u + 8 = (u - 2)(u - 4)$$

Replacing u with x^2 gives

$$(x^2 - 2)(x^2 - 4)$$

Completing the factorization by factoring $x^2 - 4$ gives

$$x^4 - 6x^2 + 8 = (x^2 - 2)(x - 2)(x + 2)$$

b. Substituting u for x^2 converts $x^4 - 18x^2 + 81$ into the quadratic expression $u^2 - 18u + 81$. Factoring this expression gives

$$u^2 - 18u + 81 = (u - 9)(u - 9)$$

Replacing u with x^2 and completing the factorization gives

$$\begin{aligned} x^4 - 18x^2 + 81 &= (x^2 - 9)(x^2 - 9) \\ &= (x - 3)(x + 3)(x - 3)(x + 3) = (x - 3)^2(x + 3)^2 \end{aligned}$$

Rational Expressions

An expression that is the quotient of two nonzero polynomials is called a **rational expression**. For example, $\dfrac{3x + 1}{x^2 - 2}$ is a rational expression. A rational expression is undefined when the denominator equals 0.

> ### Fundamental Principle of Rational Expressions
>
> The **Fundamental Principle of Rational Expressions** provides the means to simplify them.
>
> $$\frac{ac}{bc} = \frac{a}{b} \quad \text{for } b \neq 0, c \neq 0$$

We simplify a rational expression by factoring the numerator and denominator and dividing both the numerator and denominator by any common factors. We will assume that all rational expressions are defined for those values of the variables that do not make any denominator 0.

EXAMPLE 4 ▶ **Rational Expressions**

Simplify the rational expression.

a. $\dfrac{2x^2 - 8}{x + 2}$ **b.** $\dfrac{3x}{3x + 6}$ **c.** $\dfrac{3x^2 - 14x + 8}{x^2 - 16}$

SOLUTION

a. $\dfrac{2x^2 - 8}{x + 2} = \dfrac{2(x^2 - 4)}{x + 2} = \dfrac{2\overset{1}{\cancel{(x + 2)}}(x - 2)}{\underset{1}{\cancel{(x + 2)}}} = 2(x - 2)$

b. Note that we cannot divide both numerator and denominator by $3x$, because $3x$ is not a factor of $3x + 6$. Instead, we factor the denominator.

$$\dfrac{3x}{3x + 6} = \dfrac{\overset{1}{\cancel{3}}x}{\underset{1}{\cancel{3}}(x + 2)} = \dfrac{x}{x + 2}$$

c. $\dfrac{3x^2 - 14x + 8}{x^2 - 16} = \dfrac{(3x - 2)\overset{1}{\cancel{(x - 4)}}}{(x + 4)\underset{1}{\cancel{(x - 4)}}} = \dfrac{3x - 2}{x + 4}$

Multiplying and Dividing Rational Expressions

To multiply rational expressions, we write the product of the numerators divided by the product of the denominators and then simplify. We may also simplify prior to multiplying; this is usually easier than multiplying first.

EXAMPLE 5 ▶ **Products of Rational Expressions**

Multiply and simplify.

a. $\dfrac{4x^2}{5y^2} \cdot \dfrac{25y}{12x}$ **b.** $\dfrac{6 - 3x}{2x + 4} \cdot \dfrac{4x - 20}{2 - x}$

SOLUTION

a. $\dfrac{4x^2}{5y^2} \cdot \dfrac{25y}{12x} = \dfrac{4x^2 \cdot 25y}{5y^2 \cdot 12x} = \left(\dfrac{4 \cdot 25}{5 \cdot 12}\right)\left(\dfrac{x^2}{x}\right)\left(\dfrac{y}{y^2}\right) = \left(\dfrac{5}{3}\right)\left(\dfrac{x}{1}\right)\left(\dfrac{1}{y}\right) = \dfrac{5x}{3y}$

b. $\dfrac{6 - 3x}{2x + 4} \cdot \dfrac{4x - 20}{2 - x} = \dfrac{(6 - 3x)(4x - 20)}{(2x + 4)(2 - x)} = \dfrac{3\overset{1}{\cancel{(2 - x)}} \cdot \overset{2}{\cancel{4}}(x - 5)}{\underset{1}{\cancel{2}}(x + 2) \cdot \underset{1}{\cancel{(2 - x)}}} = \dfrac{6x - 30}{x + 2}$

To divide rational expressions, we multiply by the reciprocal of the divisor.

EXAMPLE 6 ▶ **Quotients of Rational Expressions**

Divide and simplify.

a. $\dfrac{a^2 b}{c} \div \dfrac{ab^3}{c^2}$ **b.** $\dfrac{x^2 + 7x + 12}{2 - x} \div \dfrac{x^2 - 9}{x^2 - x - 2}$

SOLUTION

a. $\dfrac{a^2 b}{c} \div \dfrac{ab^3}{c^2} = \dfrac{a^2 b}{c} \cdot \dfrac{c^2}{ab^3} = \left(\dfrac{a^2}{a}\right)\left(\dfrac{b}{b^3}\right)\left(\dfrac{c^2}{c}\right) = \dfrac{ac}{b^2}$

b. $\dfrac{x^2 + 7x + 12}{2 - x} \div \dfrac{x^2 - 9}{x^2 - x - 2} = \dfrac{x^2 + 7x + 12}{2 - x} \cdot \dfrac{x^2 - x - 2}{x^2 - 9}$

$= \dfrac{(x + 3)(x + 4)(x - 2)(x + 1)}{-(x - 2)(x - 3)(x + 3)} = \dfrac{(x + 4)(x + 1)}{-(x - 3)} = \dfrac{x^2 + 5x + 4}{3 - x}$

Adding and Subtracting Rational Expressions

We can add or subtract two fractions with the same denominator by adding or subtracting the numerators over their common denominator. If the denominators are not the same, we can write each fraction as an equivalent fraction with the common denominator. Our work is easier if we use the **least common denominator (LCD)**. The LCD is the lowest-degree polynomial that all denominators will divide into. For example, if several denominators are $3x$, $6x^2y$, and $9y^3$, the lowest-degree polynomial that all three denominators will divide into is $18x^2y^3$. We can find the LCD as follows.

> ### Finding the Least Common Denominator of a Set of Fractions
> 1. Completely factor all the denominators.
> 2. The LCD is the product of each factor used the maximum number of times it occurs in any one denominator.

EXAMPLE 7 ▶ Least Common Denominator

Find the LCD of the fractions $\dfrac{3x}{x^2 - 4x - 5}$ and $\dfrac{5x - 1}{x^2 - 2x - 3}$.

SOLUTION

The factored denominators are $(x - 5)(x + 1)$ and $(x - 3)(x + 1)$. The factors $(x - 5)$, $(x + 1)$, and $(x - 3)$ each occur a maximum of one time in any one denominator. Thus, the LCD is $(x - 5)(x + 1)(x - 3)$.

The procedure for adding or subtracting rational expressions follows.

> ### Adding or Subtracting Rational Expressions
> 1. Find the LCD of all the rational expressions.
> 2. Write the equivalent of each rational expression with the LCD as its denominator.
> 3. Combine the like terms in the numerators and write this expression in the numerator with the LCD in the denominator.
> 4. Simplify the rational expression, if possible.

EXAMPLE 8 ▶ Adding Rational Expressions

Add $\dfrac{3}{x^2y} + \dfrac{5}{xy^2}$.

SOLUTION

The factor x occurs a maximum of two times in any denominator and the factor y occurs a maximum of two times in any denominator, so the LCD is x^2y^2. We rewrite the fractions with the common denominator by multiplying the numerator and denominator of the first fraction by y and multiplying the numerator and denominator of the second fraction by x, as follows:

$$\frac{3}{x^2y} + \frac{5}{xy^2} = \frac{3\cdot y}{x^2y\cdot y} + \frac{5\cdot x}{xy^2\cdot x} = \frac{3y}{x^2y^2} + \frac{5x}{x^2y^2}$$

We can now add the fractions by adding the numerators over the common denominator:

$$\frac{3y}{x^2y^2} + \frac{5x}{x^2y^2} = \frac{3y + 5x}{x^2y^2}$$

This result cannot be simplified because there is no common factor of the numerator and denominator.

EXAMPLE 9 ▶

Subtracting Rational Expressions

Subtract $\dfrac{x + 1}{x^2 + x - 6} - \dfrac{2x - 1}{x - 2}$.

SOLUTION

We factor the first denominator to get the LCD:

$$x^2 + x - 6 = (x + 3)(x - 2)$$

so the LCD is $(x + 3)(x - 2)$. We rewrite the second fraction so that it has the common denominator, subtract the second numerator, then combine the like terms of the numerators.

$$\frac{x + 1}{x^2 + x - 6} - \frac{2x - 1}{x - 2} = \frac{x + 1}{(x - 2)(x + 3)} - \frac{(2x - 1)(x + 3)}{(x - 2)(x + 3)}$$

$$= \frac{x + 1}{(x - 2)(x + 3)} - \frac{2x^2 + 5x - 3}{(x - 2)(x + 3)}$$

$$= \frac{x + 1 - (2x^2 + 5x - 3)}{(x - 2)(x + 3)}$$

$$= \frac{x + 1 - 2x^2 - 5x + 3}{(x - 2)(x + 3)}$$

$$= \frac{-2x^2 - 4x + 4}{(x - 2)(x + 3)}$$

This result cannot be simplified.

Simplifying Complex Fractions

A complex fraction has a fraction in the numerator, in the denominator, or in both the numerator and the denominator. To simplify a complex fraction, find the LCD of all fractions contained in the numerator and denominator of the complex fraction. Multiply both the numerator and the denominator of the complex fraction by the LCD. (By the distributive property, this means that every term of the complex fraction will be multiplied by the LCD.) Simplify each term of the complex fraction either by dividing out common factors or by multiplying. Factor the resulting numerator and denominator, and simplify if possible.

EXAMPLE 10 ▶ **Simplifying a Complex Fraction—Method 1**

Simplify $\dfrac{\dfrac{1}{3} + \dfrac{4}{x}}{3 - \dfrac{1}{xy}}$.

SOLUTION

To simplify $\dfrac{\dfrac{1}{3} + \dfrac{4}{x}}{3 - \dfrac{1}{xy}}$, note that the LCD of the numerator and denominator of the complex

fraction is $3xy$. We multiply the numerator and denominator (and each term) of the complex fraction by $3xy$.

$$\frac{\dfrac{1}{3} + \dfrac{4}{x}}{3 - \dfrac{1}{xy}} = \left[\frac{\dfrac{1}{3} + \dfrac{4}{x}}{3 - \dfrac{1}{xy}}\right] \cdot 3xy = \frac{\left[\dfrac{1}{3} + \dfrac{4}{x}\right] \cdot 3xy}{\left[3 - \dfrac{1}{xy}\right] \cdot 3xy}$$

$$= \frac{\dfrac{1}{3} \cdot 3xy + \dfrac{4}{x} \cdot 3xy}{3 \cdot 3xy - \dfrac{1}{xy} \cdot 3xy} = \frac{xy + 12y}{9xy - 3} = \frac{y(x + 12)}{3(3xy - 1)}$$

Another way to simplify a complex fraction involves treating the fraction as a division. Rewrite the complex fraction with the main fraction bar written as a division symbol. If more than one fraction appears in the dividend (the numerator of the complex fraction) or divisor (the denominator of the complex fraction), find the LCD for each. Write the equivalent of each fraction with the desired LCD as its denominator. Add or subtract the fractions in the numerator and denominator so that the numerator contains a single fraction and the denominator contains a single fraction. Divide the fractions by inverting the divisor and multiplying. Write the indicated product of the numerators and the indicated product of the denominators. Simplify the fraction if possible.

EXAMPLE 11 ▶ **Simplifying a Complex Fraction—Method 2**

Simplify $\dfrac{\dfrac{1}{3} + \dfrac{4}{x}}{3 - \dfrac{1}{xy}}$.

SOLUTION

To simplify $\dfrac{\dfrac{1}{3} + \dfrac{4}{x}}{3 - \dfrac{1}{xy}}$, we first rewrite the fraction as a division:

$$\frac{\dfrac{1}{3} + \dfrac{4}{x}}{3 - \dfrac{1}{xy}} = \left[\frac{1}{3} + \frac{4}{x}\right] \div \left[3 - \frac{1}{xy}\right]$$

Next we get the LCD of the dividend and the LCD of the divisor and combine the fractions in the resulting dividend and divisor.

$$= \left[\frac{1}{3} \cdot \frac{x}{x} + \frac{4}{x} \cdot \frac{3}{3}\right] \div \left[3 \cdot \frac{xy}{xy} - \frac{1}{xy}\right] = \left[\frac{x}{3x} + \frac{12}{3x}\right] \div \left[\frac{3xy}{xy} - \frac{1}{xy}\right]$$

$$= \frac{x + 12}{3x} \div \frac{3xy - 1}{xy}$$

We then invert the divisor, multiply, and simplify the result.

$$= \frac{x + 12}{3x} \cdot \frac{xy}{3xy - 1} = \frac{(x + 12)xy}{3x(3xy - 1)} = \frac{y(x + 12)}{3(3xy - 1)}$$

Division of Polynomials

When the divisor has a degree less than the dividend, the division of one polynomial by another is done in a manner similar to long division in arithmetic. The degree of the quotient polynomial will be less than the degree of the dividend polynomial, and the degree of the remainder will be less than the degree of the divisor. (If the remainder is 0, the divisor is a factor of the dividend.) The procedure follows.

Long Division of Polynomials

1. Write both the divisor and the dividend in descending order of a variable. Include missing terms with a 0 coefficient in the dividend.

   Divide $(4x^3 + 4x^2 + 5)$ by $(2x^2 + 1)$.

   $$2x^2 + 1 \overline{)4x^3 + 4x^2 + 0x + 5}$$

2. Divide the highest power of the divisor into the highest power of the dividend, and write this partial quotient above the dividend. Multiply the partial quotient times the divisor, write the product under the dividend, subtract, and bring down the next term, getting a new dividend.

   $$\begin{array}{r} 2x \\ 2x^2 + 1 \overline{)4x^3 + 4x^2 + 0x + 5} \\ \underline{4x^3 \qquad\quad + 2x} \\ 4x^2 - 2x + 5 \end{array}$$

3. Repeat until the degree of the new dividend is less than the degree of the divisor. If a nonzero expression remains, it is the remainder from the division.

   $$\begin{array}{r} 2x + 2 \\ 2x^2 + 1 \overline{)4x^3 + 4x^2 + 0x + 5} \\ \underline{4x^3 \qquad\quad + 2x} \\ 4x^2 - 2x + 5 \\ \underline{4x^2 \qquad\quad + 2} \\ -2x + 3 \end{array}$$

Thus, the quotient is $2x + 2$, with remainder $-2x + 3$.

EXAMPLE 12 ▶   **Division of Polynomials**

Perform the indicated division: $\dfrac{x^4 + 6x^3 - 4x + 2}{x + 2}$.

SOLUTION

Begin by placing $0x^2$ in the dividend.

$$
\begin{array}{r}
x^3 + 4x^2 - 8x + 12 \\
x + 2\overline{)x^4 + 6x^3 + 0x^2 - 4x + 2} \\
\underline{x^4 + 2x^3} \\
4x^3 + 0x^2 \\
\underline{4x^3 + 8x^2} \\
-8x^2 - 4x \\
\underline{-8x^2 - 16x} \\
12x + 2 \\
\underline{12x + 24} \\
-22
\end{array}
$$

Thus, the quotient is $x^3 + 4x^2 - 8x + 12$, with remainder -22. We can write the result as

$$
\frac{x^4 + 6x^3 - 4x + 2}{x + 2} = x^3 + 4x^2 - 8x + 12 - \frac{22}{x + 2}
$$

Toolbox EXERCISES

For Exercises 1–4, (a) give the degree of the polynomial and (b) give the leading coefficient.

1. $3x^4 - 5x^2 + \dfrac{2}{3}$ **2.** $5x^3 - 4x + 7$

3. $7x^2 - 14x^5 + 16$ **4.** $2y^5 + 7y - 8y^6$

In Exercises 5–10, factor the polynomials completely.

5. $4x^3 - 8x^2 - 140x$

6. $4x^2 + 7x^3 - 2x^4$

7. $x^4 - 13x^2 + 36$

8. $x^4 - 21x^2 + 80$

9. $2x^4 - 8x^2 + 8$

10. $3x^5 - 24x^3 + 48x$

In Exercises 11–16, simplify each rational expression.

11. $\dfrac{x - 3y}{3x - 9y}$ **12.** $\dfrac{x^2 - 9}{4x + 12}$

13. $\dfrac{2y^3 - 2y}{y^2 - y}$ **14.** $\dfrac{4x^3 - 3x}{x^2 - x}$

15. $\dfrac{x^2 - 6x + 8}{x^2 - 16}$ **16.** $\dfrac{3x^2 - 7x - 6}{x^2 - 4x + 3}$

In Exercises 17–33, perform the indicated operations and simplify.

17. $\dfrac{6x^3}{8y^3} \cdot \dfrac{16x}{9y^2} \cdot \dfrac{15y^4}{x^3}$

18. $\dfrac{x - 3}{x^3} \cdot \dfrac{x^2 - 4x}{x^2 - 7x + 12}$

19. $(x^2 - 4) \cdot \dfrac{2x - 3}{x + 2}$

20. $(x^2 - x - 6) \div \dfrac{9 - x^2}{x^2 + 3x}$

21. $\dfrac{4x + 4}{x - 4} \div \dfrac{8x^2 + 8x}{x^2 - 6x + 8}$

22. $\dfrac{6x^2}{4x^2y - 12xy} \div \dfrac{3x^2 + 12x}{x^2 + x - 12}$

23. $\dfrac{x^2 + x}{x^2 - 5x + 6} \cdot \dfrac{x^2 - 2x - 3}{2x + 4} \div \dfrac{x^3 - 3x^2}{4 - x^2}$

24. $\dfrac{6x - 2}{3xy} + \dfrac{3x + 2}{3xy}$

25. $\dfrac{2x + 3}{x^2 - 1} + \dfrac{4x + 3}{x^2 - 1}$ ____

26. $3 + \dfrac{1}{x^2} - \dfrac{2}{x^3}$

27. $\dfrac{5}{x} - \dfrac{x - 2}{x^2} + \dfrac{4}{x^3}$

28. $\dfrac{a}{a^2 - 2a} - \dfrac{a - 2}{a^2}$

29. $\dfrac{5x}{x^4 - 16} + \dfrac{8x}{x + 2}$

30. $\dfrac{x - 1}{x + 1} - \dfrac{2}{x^2 + x}$

31. $1 + \dfrac{1}{x - 2} - \dfrac{2}{x^2}$

32. $\dfrac{x - 7}{x^2 - 9x + 20} + \dfrac{x + 2}{x^2 - 5x + 4}$

33. $\dfrac{2x + 1}{4x - 2} + \dfrac{5}{2x} - \dfrac{x + 1}{2x^2 - x}$

In Exercises 34 – 37, simplify the complex fraction.

34. $\dfrac{\dfrac{1}{x} + \dfrac{1}{y}}{\dfrac{1}{x} - \dfrac{1}{y}}$

35. $\dfrac{\dfrac{5}{2y} + \dfrac{3}{y}}{\dfrac{1}{4} + \dfrac{1}{3y}}$

36. $\dfrac{2 - \dfrac{1}{x}}{2x - \dfrac{3x}{x + 1}}$

37. $\dfrac{1 - \dfrac{2}{x - 2}}{x - 6 + \dfrac{10}{x + 1}}$

In Exercises 38–41, perform the long division.

38. $(x^5 + x^3 - 1) \div (x + 1)$

39. $(a^4 + 3a^3 + 2a^2) \div (a + 2)$

40. $(3x^5 - x^4 + 5x - 1) \div (x^2 - 2)$

41. $\dfrac{3x^4 + 2x^2 + 1}{3x^2 - 1}$

6.1 Higher-Degree Polynomial Functions

KEY OBJECTIVES

- Identify the graphs of higher-degree polynomial functions
- Graph cubic functions
- Graph quartic functions
- Find local minima, local maxima, absolute minima, and absolute maxima of polynomial functions

SECTION PREVIEW   U.S. Unemployment

About 25 million people in the world's 30 richest countries lost their jobs between the end of 2007 and the end of 2010 as the economic downturn pushed most countries into recession. The unemployment rate in the United States was 8.9% in February of 2011. From 1948 to 2010, the U.S. unemployment rate averaged 5.70%, reaching a historic high of 10.80% in November of 1982 and a record low of 2.50% in May of 1953.

Table 6.1 shows the U.S. unemployment rates from 2000 and projected to 2015. The data can be used to find a polynomial function

$$y = -0.002x^4 + 0.051x^3 - 0.371x^2 + 1.055x + 4.234$$

that models the U.S. unemployment rate as a function of the number of years after 2000. (See Example 4.)

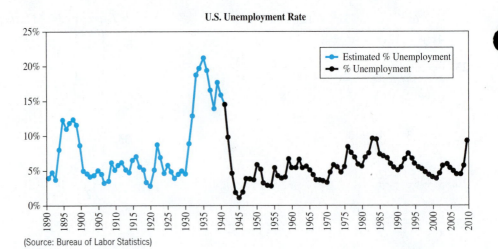

U.S. Unemployment Rate

(Source: Bureau of Labor Statistics)

Table 6.1

Year	Unemployment Rate (%)	Year	Unemployment Rate (%)
2000	3.97	2008	5.82
2001	4.74	2009	9.28
2002	5.78	2010	9.73
2003	5.99	2011	9.59
2004	5.54	2012	8.85
2005	5.08	2013	8.05
2006	4.61	2014	7.38
2007	4.61	2015	6.81

(Source: Bureau of Labor Statistics)

In Chapters 2–4, we solved many real problems by modeling data with linear functions and with quadratic functions. (Recall that $ax + b$ is a first-degree polynomial and $ax^2 + bx + c$ is a second-degree polynomial.) Both of these functions are types of **polynomial functions**. **Higher-degree polynomial functions** are functions with degree higher than 2. Examples of higher-degree polynomial functions are

$$y = x^3 - 16x^2 - 2x \qquad f(x) = 3x^4 - x^3 + 2 \qquad y = 2x^5 - x^3$$

In this section, we graph higher-degree polynomial functions and use technology to find local maxima and minima. ■

Cubic Functions

Using data from 1986 to 2010, the number of international visitors to the United States can be modeled by

$$y = 0.010x^3 - 0.575x^2 + 10.814x - 20.752$$

where y is in millions and x is the number of years after 1980. This function is a third-degree function, or **cubic function**, because the highest-degree term in the function has degree 3. The graph of this function is shown in Figure 6.1.

Figure 6.1

(a)

(b)

(c)

Figure 6.2

A **cubic function** in the variable x has the form

$$f(x) = ax^3 + bx^2 + cx + d \quad (a \neq 0)$$

The basic cubic function $f(x) = x^3$ was discussed in Section 3.3. The graph of this function, shown in Figure 6.2(a), is one of the possible shapes of the graph of a cubic function. The graphs of two other cubic functions are shown in Figures 6.2(b) and (c).

The graph of a cubic function will have 0 or 2 turning points, as shown in Figures 6.1 and 6.2. Also, it is characteristic of a cubic polynomial function to have a graph that has one of the end behaviors shown in these figures. This end behavior can be described as "one end opening up and one end opening down." The specific end behavior is determined by the leading coefficient (the coefficient of the third-degree term); the curve opens up on the right if the leading coefficient is positive, and it opens down on the right if the leading coefficient is negative.

Notice that the graph of the cubic function shown in Figure 6.2(a) has one x-intercept, the graph of the cubic function shown in Figure 6.2(b) has three x-intercepts, and the graph of the cubic function in Figure 6.2(c) has two x-intercepts.

In general, the graph of a polynomial function of degree n has at most n x-intercepts.

Recall that the graph of a quadratic (second-degree) function is a parabola with 1 turning point, called a vertex, and that the vertex is a maximum point or a minimum point. Notice that in each of the graphs of the cubic (third-degree) functions shown in Figures 6.2(b) and (c), there are 2 points where the function changes from increasing to decreasing or from decreasing to increasing. (Recall that a curve is increasing on an interval if the curve rises as it moves from left to right on that interval.) These *turning points* are called the **local extrema points** of the graph of the function. In particular, a point where the curve changes from decreasing to increasing is called a **local minimum point**, and a point where the curve changes from increasing to decreasing is called a **local maximum point**. For example, in Figure 6.2(c), the point $(0, 0)$ is a local minimum point, and the point $(2, 8)$ is a local maximum point. Note that at the point $(0, 0)$ the function changes from decreasing to increasing, and at the point $(2, 8)$ the function changes from increasing to decreasing. Observe that the graph in Figure 6.2(a) has no local extrema points. In general, a cubic equation will have 0 or 2 local extrema points.

The highest point on the graph over an interval is called the **absolute maximum point** on the interval, and the lowest point on the graph over an interval is called the **absolute minimum point** on that interval. If the graph in Figure 6.2(c) were limited to the x-interval $[-0.5, 3.2]$, the absolute maximum point on this interval would be $(2, 8)$, and the absolute minimum point on this interval would be $(3.2, -4.096)$.

EXAMPLE 1 ▶   ## Cubic Graph

a. Using an appropriate window, graph $y = x^3 - 27x$.

b. Find the local maximum and local minimum, if possible.

c. Where is the absolute maximum of this function on the interval $[0, 3]$?

SOLUTION

a. This function is a third-degree (cubic) function, so it has 0 or 2 local extrema. We want to graph the function in a window that allows us to see any turning points that exist and recognize the end behavior of the function. Using the window $[-40, 30]$ by $[-100, 100]$, we get the graph shown in Figure 6.3(a). This graph has a shape like Figure 6.2(b); it has three x-intercepts and 2 local extrema, so it is a complete graph. A different viewing window can be used to provide a more detailed view of the turning points of the graph (Figure 6.3(b)).

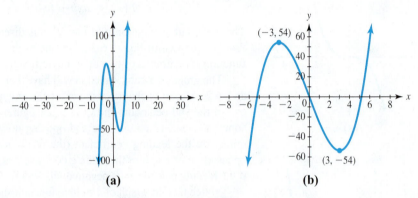

(a)   (b)

Figure 6.3

b. To find the local maximum of the function graphed in Figure 6.3:

1. Under the CALC menu (**2ND** **TRACE**), choose 4: maximum.

2. Respond to the prompts, by choosing a point to the left and right of the suspected local maximum, and then pressing **ENTER** after each response and after Guess.*

   These steps give a local maximum at $(-3, 54)$. See Figure 6.4(a).

Using the same steps while choosing 3: minimum gives a local minimum at $(3, -54)$. See Figure 6.4(b).

(a)   (b)

Figure 6.4

c. The highest point on the graph on the interval $[0, 3]$ is $(0, 0)$, so the absolute maximum is $y = 0$ at $x = 0$.

---

* For detailed steps, see Appendix A, page 669.

EXAMPLE 2 ▶ **Elderly Workers**

Table 6.2 shows the number of millions of elderly in the workforce for the years from 1970 and projected to 2050. These data can be modeled by the cubic function $y = -0.0000597x^3 + 0.00934x^2 - 0.308x + 5.63$, where x is the number of years after 1960 and y is in millions.

Table 6.2

Years	Elderly Workforce (millions)
1970	3.22
1980	3.05
1990	3.45
2000	4.20
2010	5.44
2020	8.24
2030	10.1
2040	9.60
2050	10.2

(Source: U.S. Bureau of Labor Statistics)

a. Graph the data points and the function on the same axes, using the window $[0, 100]$ by $[0, 12]$. Comment on the fit.

b. Use the model to predict the number of elderly workers in 2025.

c. Use technology and the model to find the year in the period from 1960 to 2060 during which the number of elderly workers is expected to reach its maximum.

d. A local minimum occurs at $(20.5, 2.73)$. What does the function do on the interval $[20.5, 83.8]$?

SOLUTION

a. Figure 6.5(a) shows a graph of the model and the data on the same axes. The model is a good fit.

b. Evaluating the function at $x = 65$ gives 8.68, so the model predicts that the number of elderly in the workforce will be 8.68 million in 2025.

c. Using the "maximum" feature under the CALC menu (see Figure 6.5(b)) shows that the local maximum is at approximately $(83.8, 10.3)$. This indicates that the model predicts that the number of elderly in the workforce will reach its maximum, 10.3 million, during 2044.

d. The function increases over this interval.

(a)

(b)

Figure 6.5

Quartic Functions

A polynomial function of the form $f(x) = ax^4 + bx^3 + cx^2 + dx + e, a \neq 0$, is a fourth-degree, or **quartic**, function. The basic quartic function $f(x) = x^4$ has the graph shown in Figure 6.6(a). Notice that this graph resembles a parabola, although it is not actually a parabola (recall the shape of a parabola, which was discussed in

Section 3.1). The graph in Figure 6.6(a) is one of the possible shapes of the graph of a quartic function. The graphs of two additional quartic functions are shown in Figures 6.6(b) and (c).

Figure 6.6

Notice that in the graph of the quartic function shown in Figure 6.6(b) there are 3 turning points and that in the graphs shown in Figures 6.6(a) and 6.6(c) there is 1 turning point each. The end behavior of a quartic function can be described as "both ends opening up" (if the leading coefficient is positive) or "both ends opening down" (if the leading coefficient is negative). Variations in the curvature may occur for different quartic functions, but these are the only possible end behaviors.

Table 6.3 summarizes the information about turning points and end behavior for graphs of cubic and quartic functions as well as for linear and quadratic functions.

Table 6.3

Function	Possible Graphs	Degree	Number of Turning Points	End Behavior — Positive leading coefficient	End Behavior — Negative leading coefficient	Degree: Even or Odd
Linear	Positive slope Negative slope	1	0			odd
Quadratic	Positive leading coefficient Negative leading coefficient	2	1			even
Cubic Positive leading coefficient Negative leading coefficient		3	2 or 0			odd
Quartic Positive leading coefficient Negative leading coefficient		4	3 or 1			even

Table 6.3 helps us observe the following.

> ## Polynomial Graphs
>
> 1. The graph of a polynomial function of degree n has at most $n - 1$ turning points.
>
> 2. The graph of a polynomial function of degree n has at most n x-intercepts.
>
> 3. The end behavior of the graph of a polynomial function with odd degree can be described as "one end opening up and one end opening down."
>
> 4. The end behavior of the graph of a polynomial function with even degree can be described as "both ends opening up" or "both ends opening down."

EXAMPLE 3 ▶   Graphs of Polynomial Functions

Figures 6.7(a)–(d) show the complete graphs of several polynomial functions. For each function, determine

i. the number of x-intercepts.

ii. the number of turning points.

iii. whether the leading coefficient is positive or negative.

iv. whether the degree of the polynomial is even or odd.

v. the minimum possible degree.

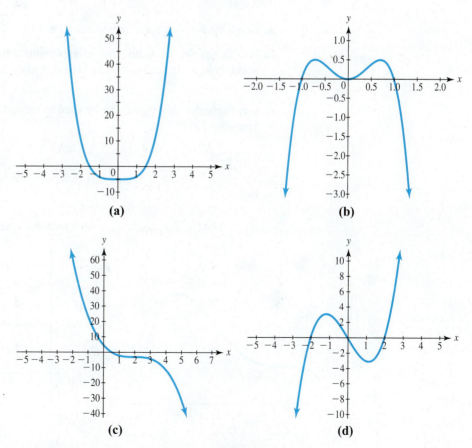

(a)

(b)

(c)

(d)

Figure 6.7

SOLUTION

a. The graph clearly shows two x-intercepts and 1 turning point. Because the end behavior is "both ends opening up," the degree of the polynomial function is even and the leading coefficient is positive. Because there is only 1 turning point, the function could be quadratic or possibly quartic, so the minimum degree of the polynomial is 2.

b. The graph shows 3 turning points and three x-intercepts. The end behavior is "both ends opening down," so the degree of the polynomial function is even and the leading coefficient is negative. The minimum possible degree is 4 because there is more than 1 turning point, and thus the function cannot be quadratic.

c. The graph shows one x-intercept and 0 turning points. The degree of the function is odd because the end behavior is "one end up and one end down." The leading coefficient is negative because the left end is up. Because the graph is not a line, the minimum degree is 3.

d. The graph shows three x-intercepts and 2 turning points. The end behavior is "one end up and one end down," with the right end opening up, so the function has odd degree and the leading coefficient is positive. Because the graph has three x-intercepts, the minimum degree is 3.

EXAMPLE 4 ▶ **U.S. Unemployment**

Using data from 2000 and projected to 2015, the function

$$y = -0.00201x^4 + 0.05076x^3 - 0.37132x^2 + 1.05544x + 4.23362$$

can be used to model the U.S. unemployment rate, with x equal to the number of years after 2000.

a. Graph the function for x-values from 0 to 15.

b. Use the graph and technology to approximate the year after 2000 in which U.S. unemployment is a maximum and the year after 2002 and before 2015 in which it is a minimum.

c. Is it likely that this model can be used to estimate the U.S. unemployment rate for long after 2015?

SOLUTION

a. The graph is shown in Figure 6.8(a), with two local maxima and one local minimum indicated.

Figure 6.8

b. The graph in Figure 6.8(a) shows the absolute maximum occurring at (12.30, 9.49). This indicates that U.S. unemployment is predicted to be highest in 2013, at 9.49%. The minimum of this function after $x = 2$ (for 2002) occurs at (3.92, 5.25), which indicates that the lowest unemployment rate is 5.25% in 2004.

c. Looking at the graph in Figure 6.8(b), we see that it drops very rapidly after $x = 13$ (which corresponds to 2013), with outputs becoming negative before $x = 17$ (which corresponds to 2017). Thus, the model ceases to be valid eventually, certainly before 2017.

SKILLS CHECK 6.1

1. Graph the function $h(x) = 3x^3 + 5x^2 - x - 10$ on the windows given in parts (a) and (b). Which window gives a complete graph?

   a. $[-5, 5]$ by $[-5, 5]$ **b.** $[-5, 5]$ by $[-20, 20]$

2. Graph the function $f(x) = 2x^3 - 3x^2 - 6x$ on the windows given in parts (a) and (b). Which window gives a complete graph?

   a. $[-5, 5]$ by $[-5, 5]$

   b. $[-10, 10]$ by $[-10, 10]$

3. Graph the function $g(x) = 3x^4 - 12x^2$ on the windows given in parts (a) and (b). Which window gives a complete graph?

   a. $[-5, 5]$ by $[-5, 5]$ **b.** $[-3, 3]$ by $[-12, 10]$

4. Graph the function $g(x) = 3x^4 - 4x^2 + 10$ on the windows given in parts (a) and (b). Which window gives a complete graph?

   a. $[-10, 10]$ by $[-10, 10]$

   b. $[-5, 5]$ by $[-20, 20]$

For Exercises 5–10, use the given graph of the polynomial function to (a) estimate the x-intercept(s), (b) state whether the leading coefficient is positive or negative, and (c) determine whether the polynomial function is cubic or quartic.

5.

6.

7.

8.

9.

B.

10.

C.

D.

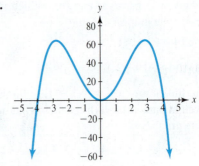

For Exercises 11–16, match the polynomial function with its graph (below or on the next page).

11. $y = 2x^3 + 3x^2 - 23x - 12$

12. $y = -2x^3 - 8x^2 + 0.5x + 2$

13. $y = -6x^3 + 5x^2 + 17x - 6$

14. $y = x^4 - 0.5x^3 - 14.5x^2 + 17x + 12$

15. $y = x^4 + 3$

16. $y = -x^4 + 16x^2$

E.

A.

F.

For Exercises 17–20, use the equation of the polynomial function to (a) state the degree and the leading coefficient and (b) describe the end behavior of the graph of the function. (c) Support your answer by graphing the function.

17. $f(x) = 2x^3 - x$

18. $g(x) = 0.3x^4 - 6x^2 + 17x$

19. $f(x) = -2(x - 1)(x^2 - 4)$

20. $g(x) = -3(x - 3)^2(x - 1)^2$

21. a. Graph the function $y = x^3 - 3x^2 - x + 3$ using the window $[-10, 10]$ by $[-10, 10]$.

    b. Is the graph complete?

22. a. Graph $y = x^3 + 6x^2 - 4x$ using the window $[-10, 10]$ by $[-10, 10]$.

    b. Is the graph complete?

23. Graph $y = 25x - x^3$ using

    a. the window $[-10, 10]$ by $[-10, 10]$.

    b. a window that shows 2 turning points.

24. Graph the function $y = x^3 - 16x$ using

    a. the window $[-10, 10]$ by $[-10, 10]$.

    b. a window that shows 2 turning points.

25. Graph the function $y = x^4 - 4x^3 + 4x^2$ using

    a. the window $[-10, 10]$ by $[-10, 10]$.

    b. the window $[-4, 4]$ by $[-4, 4]$.

    c. Which window gives a more detailed view of the graph near the turning points?

26. a. Graph $y = x^4 - 4x^2$ using the window $[-10, 10]$ by $[-10, 10]$.

    b. Is the graph complete?

27. a. Graph $y = x^4 - 4x^2 - 12$ on $[-8, 8]$ by $[-20, 10]$.

    b. How many turning points does the graph have?

c. Could the graph of this function have more turning points?

28. a. Graph $y = x^4 + 6x^2$ on $[-10, 10]$ by $[-10, 50]$.

    b. Is it possible for the graph of this function to have only 1 turning point?

29. Sketch a graph of any cubic polynomial function that has a negative leading coefficient and one x-intercept.

30. Sketch a graph of any polynomial function that has degree 3, a positive leading coefficient, and three x-intercepts.

31. Sketch a graph of any polynomial function that has degree 4, a positive leading coefficient, and two x-intercepts.

32. Sketch a graph of any cubic polynomial function that has a negative leading coefficient and three x-intercepts.

33. a. Graph $y = x^3 + 4x^2 + 5$ on a window that shows a local maximum and a local minimum.

    b. A local maximum occurs at what point?

    c. A local minimum occurs at what point?

34. a. Graph $y = x^4 - 8x^2$ on a window that shows 2 local minima and 1 local maximum.

    b. A local maximum occurs at what point?

    c. The local minima occur at what points?

35. Use technology to find a local maximum and 2 local minima of the graph of the function $y = x^4 - 4x^3 + 4x^2$.

36. a. Graph the function $y = -x^3 - x^2 + 9x$ using the window $[-5, 5]$ by $[-15, 10]$.

    b. Graph the function on an interval with $x \geq 0$ and with $y \geq 0$.

    c. The graph of what other type of function resembles the *piece* of the graph of $y = -x^3 - x^2 + 9x$ shown in part (b)?

EXERCISES 6.1

37. *Daily Revenue* The daily revenue in dollars from the sale of a product is given by $R = -0.1x^3 + 11x^2 - 100x$, where x is the number of units sold.

    a. Graph this function on the window $[-100, 100]$ by $[-5000, 25,000]$. How many turning points do you see?

b. Because x represents the number of units sold, what restriction should be placed on x in the context of the problem? What restriction should be placed on R?

c. Using your result from part (b), graph the function on a new window that makes sense for the problem.

d. What does this function give as the revenue if 50 units are produced?

38. **Weekly Revenue** A firm has total weekly revenue in dollars for its product given by $R(x) = 2000x + 30x^2 - 0.3x^3$, where x is the number of units sold.

a. Graph this function on the window $[-100, 150]$ by $[-30,000, 220,000]$.

b. Because x represents the number of units sold, what restrictions should be placed on x in the context of the problem? What restrictions should be placed on y?

c. Using your result from part (b), graph the function on a new window that makes sense for the problem.

d. What does this function give as the revenue if 60 units are produced?

39. **Daily Revenue** The daily revenue in dollars from the sale of a product is given by $R = 600x - 0.1x^3 + 4x^2$, where x is the number of units sold.

a. Graph this function on the window $[0, 100]$ by $[0, 30,000]$.

b. Use the graph of this function to estimate the number of units that will give the maximum daily revenue and the maximum possible daily revenue.

c. Find a window that will show a complete graph—that is, will show all of the turning points and intercepts. Graph the function using this window.

d. Does the graph in part (a) or the graph in part (c) better represent the revenue from the sale of x units of a product?

e. Over what interval of x-values is the revenue increasing, if $x \geq 0$?

40. **Weekly Revenue** A firm has total weekly revenue in dollars for its product given by $R(x) = 2800x - 8x^2 - x^3$, where x is the number of units sold.

a. Graph this function on the window $[0, 50]$ by $[0, 51,000]$.

b. Use technology to find the maximum possible revenue and the number of units that gives the maximum revenue.

c. Find a window that will show a complete graph—that is, will show all of the turning points and intercepts. Graph the function using this window.

d. Does the graph in part (a) or the graph in part (c) better represent the revenue from the sale of x units of a product?

e. Over what interval of x-values is the revenue increasing, if $x \geq 0$?

41. **Investment** If $2000 is invested for 3 years at rate r, compounded annually, the future value of this investment is given by $S = 2000(1 + r)^3$, where r is the rate written as a decimal.

a. Complete the following table to see how increasing the interest rate affects the future value of this investment.

Rate, r	Future Value, S ($)
0.00	
0.05	
0.10	
0.15	
0.20	

b. Graph this function for $0 \leq r \leq 0.24$.

c. Use the table and/or graph to compare the future value if $r = 10\%$ and if $r = 20\%$. How much more money is earned at 20%?

d. Which interest rate, 10% or 20%, is more realistic for an investment?

42. **Investment** The future value of $10,000 invested for 5 years at rate r, compounded annually, is given by $S = 10,000(1 + r)^5$, where r is the rate written as a decimal.

a. Complete the following table to see how increasing the interest rate affects the future value of this investment.

Rate, r	Future Value, S ($)
0.00	
0.05	
0.07	
0.12	
0.18	

b. Graph this function for $0 \leq r \leq 0.24$.

c. Use the graph to compare the future value if $r = 10\%$ and if $r = 24\%$. How much more money is earned at 24%?

d. Is it more likely that you can get an investment paying 10% or 24%?

43. *Starbucks* Using data from 2005 and projected to 2017, the total number of Starbucks stores in the world can be modeled by the cubic function $y = 10.3x^3 - 400x^2 + 5590x - 8694$, where x is the number of years after 2000 and y is the total number of stores.

a. Graph the function on the window $[2, 20]$ by $[8000, 23,000]$.

b. What does the model predict the number of stores will be in 2020?

c. Does the number of stores ever decrease over this period of years, according to the model?
(Source: Starbucks)

44. *U.S. Population* Using data from 1980 and projected to 2050, the U.S. population, in millions, can be modeled by $y = -0.000233x^3 + 0.0186x^2 + 2.32x + 235$, with x equal to the number of years after 1980.

a. Graph the function on an interval representing 1980 through 2050.

b. What does the model predict the population will be in 2022? Round to the nearest million.

c. What is the maximum predicted population over this period of time, according to the model?
(Source: Social Security Administration)

45. *World Cell Phone Subscribership* Using International Telecommunications Union data from 1990 with projections to 2020, the function

$$f(x) = -0.000728x^3 + 0.0414x^2 - 0.296x + 0.340$$

can be used to model the world cell phone subscriberships, y, in billions, as a function of the number of years after 1990, x.

a. Graph the function on the window $[10, 30]$ by $[0, 10]$.

b. What does the model give as the number of subscriberships in 2020?

c. In what year before 2025 does the number of subscriberships reach a maximum, according to the model?
(Source: International Telecommunications Union, Key Global Telecom Indicators)

46. *Aging Workers* The number of millions of Americans who are working full time at selected ages from 27 to 62 can be modeled by the cubic function $f(x) = -0.000362x^3 + 0.0401x^2 - 1.39x + 21.7$, where x is their age.
(Source: Wall Street Journal, June 18, 2012)

a. Use the model to predict the number of Americans age 50 who are working full time.

b. Use technology to find the local maximum on the interval $[27, 62]$. What does this point indicate?

c. Find the local minimum on the interval $[27, 62]$. What does this indicate?

47. *Risky Drivers* Both inexperienced young drivers and very old drivers are involved in far more accidents than other age groups. The quartic function that models the police-reported crashes for different age groups is $y = 0.00001828x^4 - 0.003925x^3 + 0.3031x^2 - 9.907x + 118.2$, where x is the age of the driver and y equals the number of crashes per million miles traveled.

a. Graph the function for the ages 15 to 86.

b. What does the model estimate the number of crashes per million miles to be for 22-year-old drivers?

c. At what age among older drivers does the model predict the number of crashes will reach 8 per million miles?

d. At what age among younger drivers does the model predict the number of crashes will be the minimum number?
(Source: Consumer Reports, October 2012)

48. *Salaries* The median salary for male workers, in thousands of dollars, for the years 1975–2006 can be modeled by the function

$$y = -0.0000929x^4 + 0.00784x^3 - 0.226x^2 + 2.508x + 34.347$$

where x is the number of years after 1970.

a. Find the minimum on the graph of this function between $x = 10$ and $x = 30$ and interpret it.

b. Is the median salary found in part (a) the lowest median salary in the period 1975–2006?

49. *Profit* The weekly revenue for a product is given by $R(x) = 120x - 0.015x^2$, and the weekly cost is $C(x) = 10,000 + 60x - 0.03x^2 + 0.00001x^3$, where x is the number of units produced and sold.

a. How many units will give maximum profit?

b. What is the maximum possible profit?

50. *Profit* The annual revenue for a product is given by $R(x) = 60,000x - 50x^2$, and the annual cost is $C(x) = 800 + 100x^2 + x^3$, where x is the number of thousands of units produced and sold.

a. How many units will give maximum profit?

b. What is the maximum possible profit?

Modeling with Cubic and Quartic Functions

KEY OBJECTIVES

- Model and apply data with cubic functions
- Model and apply data with quartic functions

SECTION PREVIEW **Aging Workers**

It is reasonable to assume that the numbers of Americans of differing ages working full time will vary considerably. Table 6.4 shows the number of millions of Americans who are working full time at selected ages from 27 to 62. The scatter plot of the data in Figure 6.9 shows that neither a line nor a parabola fits the data well, and it appears that a cubic function would be a better fit.

Table 6.4

Age	Millions Working Full Time
27	6.30
32	6.33
37	6.96
42	7.07
47	7.52
52	7.10
57	5.58
62	3.54

(Source: *Wall Street Journal*, June 18, 2012)

Figure 6.9

The use of graphing calculators and spreadsheets enables us to find cubic and quartic functions that model nonlinear data. In this section, we will model and apply these functions.

Modeling with Cubic Functions

The scatter plot in Figure 6.9 indicates that the data may fit close to one of the possible shapes of the graph of a cubic function, so it is possible to find a cubic function that models the data.

EXAMPLE 1 ▶ **Aging Workers**

Use the data in Table 6.4 to do the following:

a. Find a cubic model that gives the number of Americans working full time, y (in millions), as a function of their age, x. Report the model with three significant digits in the coefficients.

b. Graph the unrounded model from part (a) on the same axes with the data, and comment on the fit.

c. Use the model to estimate the number of Americans working at age 59.

SOLUTION

a. A cubic function that models the number of Americans working full time, y (in millions), as a function of their age, x, is

$$f(x) = -0.000362x^3 + 0.0401x^2 - 1.39x + 21.7$$

b. Figure 6.10 shows a graph of the model and the data on the same axes. The model is an excellent visual fit.

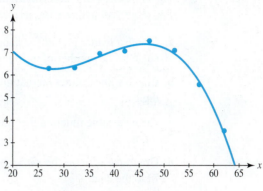

Figure 6.10

c. The number of Americans age 59 who are working is estimated from the model to be $y = f(59) \approx 5$ million.

Spreadsheet ► SOLUTION   We can use graphing calculators, software programs, and spread-sheets to find the polynomial function that is the best fit for a set of data. Figure 6.11 includes an Excel spreadsheet for the data of Example 1. Selecting the cells containing the data, getting the scatter plot of the data, and selecting Polynomial Trendline with the order (degree) of the polynomial gives the equation of the polynomial function that is the best fit for the data, along with the scatter plot and the graph of the best-fitting curve. Figure 6.11 shows an Excel worksheet with the data from Example 1 and the graph of the cubic function that fits the data.*

Figure 6.11

Sometimes a set of data points can be modeled reasonably well by both a linear function and a cubic function. Consider the following example.

---

* For detailed steps, see Appendix B, page 691.

EXAMPLE 2 ▶   ## Diabetes

The number of U.S. adults with diabetes is projected to increase rapidly in the future. The projected numbers for selected years, in millions, are shown in Table 6.5.

a. Create a scatter plot of the data, with x equal to the number of years after 2015 and y equal to the number of millions of U.S. adults with diabetes.

b. Create a linear function that models the data. Report the model with three significant digits.

c. Create a cubic function that models the data. Report the model with three significant digits.

d. Graph the data on the same axes as each model, and determine which is the better fit to the data.

Table 6.5

Year	Millions	Year	Millions
2015	37.3	2035	76.2
2020	50.0	2040	84.1
2025	59.5	2045	91.7
2030	68.3	2050	100.0

(Source: Centers for Disease Control and Prevention)

SOLUTION

a. The scatter plot of the data is shown in Figure 6.12(a).

(a)

(b)

(c)

Figure 6.12

b. The linear function that is the best fit for the data is
$$y = 1.74x + 40.5$$

c. The cubic function that is the best fit for the data is
$$y = 0.000697x^3 - 0.0486x^2 + 2.64x + 37.5$$

d. The graphs of the data on the same axes as the linear function and the cubic function are shown in Figure 6.12(b) and Figure 6.12(c). The fit of the linear function to the data is excellent, but the cubic function is nearly an exact fit to the data, so the cubic function is the better fit.

Modeling with Quartic Functions

If a scatter plot of data indicates that the data may fit close to one of the possible shapes that are graphs of quartic functions, it is possible to find a quartic function that models the data. Consider the following example.

EXAMPLE 3. ▶ **Supplemental Nutrition Assistance Program**

The Supplemental Nutrition Assistance Program (SNAP) caseloads grew significantly between 2007 and 2011 as the recession and lagging economic recovery led more low-income households to qualify and apply for help. The Congressional Budget Office (CBO) expects that as the economy improves, the number of participants will fall by about 2–4% each year over the next decade: from 46.5 million in fiscal year 2014 to 46.0 million in 2015, 44.3 million in 2016, and 32.8 million by 2025. The CBO forecasts that by 2025, the share of the population receiving SNAP will return to close to 2007 levels (about 9%). Table 6.6 shows the number of SNAP participants, in thousands, from 1990 to 2015 and projected to 2025.

Table 6.6

Year	SNAP Participants (thousands)	Year	SNAP Participants (thousands)
1990	20,049	2004	23,811
1991	22,625	2005	25,628
1992	25,407	2006	26,549
1993	26,987	2007	26,316
1994	27,474	2008	28,223
1995	26,619	2009	33,490
1996	25,543	2010	40,302
1997	22,858	2011	44,709
1998	19,791	2012	46,609
1999	18,183	2013	47,636
2000	17,194	2014	46,536
2001	17,318	2015	46,000
2002	19,096	2016	44,300
2003	21,250	2025	32,800

(Source: Center on Budget and Policy Priorities)

a. Create a scatter plot of the data, with $x = 0$ in 1990 and y equal to the number of thousands of SNAP participants, to determine whether the data can be modeled by a quartic function.

b. Find the quartic function that models the data, with $x = 0$ in 1990. Round the model to four decimal places.

c. Graph the unrounded model and the scatter plot of the data on the same axes.

d. Use the unrounded model to determine whether the CBO's projection for SNAP participants in 2025 is accurate.

SOLUTION

a. The scatter plot of the data is shown in Figure 6.13(a). The scatter plot indicates that a function whose graph has end behavior such that both ends open down would be a good fit. Thus, a quartic model is appropriate.

b. The quartic function that models the data is

$$y = -0.4445x^4 + 24.3876x^3 - 333.3264x^2 + 1086.8474 + 23441.2847$$

where x is the number of years after 1990.

c. The graph of the model and the scatter plot of the data are shown in Figure 6.13(b).

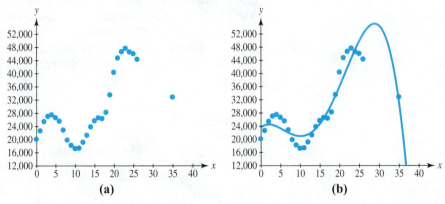

Figure 6.13

d. The year 2025 corresponds to an x-value of 35, and when $x = 35$, $y = 31,731$. This number is close to the number projected by the CBO: 32,800.

Model Comparisons

Even when a scatter plot does not show all the characteristics of the graph of a quartic function, a quartic function may be a good fit for the data. Consider the following example, which shows a comparison of a cubic and a quartic model for a set of data. It also shows that many decimal places are sometimes required in reporting a model so that the leading coefficient is not written as zero.

EXAMPLE 4 ▶ **Foreign-Born Population**

The percents of the U.S. population that were foreign born in selected years from 1900 and projected to 2020 are shown in Table 6.7 and Figure 6.14.

a. Create a scatter plot for the data, using the number of years after 1900 as the input x and the percent as the output y.

b. Find a cubic function to model the data and graph the function on the same axes with the data points. Report the model by rounding coefficients to seven decimal places.

c. Find a quartic function to model the data and graph the function on the same axes with the data points. Report the model by rounding coefficients to seven decimal places.

d. Use the graphs of the unrounded functions in parts (b) and (c) to determine which function is the better fit for the data.

Table 6.7

Year	Percent	Year	Percent
1900	13.6	2000	10.4
1910	14.7	2005	11.7
1920	13.2	2009	12.2
1930	11.6	2010	12.4
1940	8.8	2016	11.6
1950	6.9	2017	11.6
1960	5.4	2018	11.7
1970	4.8	2019	11.7
1980	6.2	2020	11.8
1990	8.0		

(Source: U.S. Census Bureau)

Figure 6.14

SOLUTION

a. The scatter plot of the data is shown in Figure 6.15.

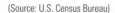

Figure 6.15

b. A cubic function that models the data is

$$y = 0.0000059x^3 + 0.0009376x^2 - 0.2194844x + 15.5962344$$

with $x = 0$ representing 1900. Figure 6.16(a) shows the graph of the unrounded model on the same axes with the data points.

Cubic
(a)

Quartic
(b)

Figure 6.16

c. A quartic function that models the data is

$$y = -0.0000011x^4 + 0.0002840x^3 - 0.0206528x^2 + 0.3253713x + 13.3935645$$

with $x = 0$ representing 1900. Figure 6.16(b) shows the graph of the unrounded model on the same axes as the data points.

d. The quartic model is a much better fit.

Third and Fourth Differences

Recall that for equally spaced inputs, the first differences of outputs are constant for data modeled by linear functions, and the second differences are constant for data modeled by quadratic functions. In a like manner, the third differences of outputs are constant for data modeled by cubic functions, and the fourth differences are constant for data modeled by quartic functions, if the inputs are equally spaced.

EXAMPLE 5 ▶ **Model Comparisons**

Table 6.8 has a set of equally spaced inputs, x, in column 1 and three sets of outputs in columns 2, 3, and 4. Use third and/or fourth differences with each set of outputs to determine if each set of data can be modeled by a cubic or quartic function. Then find the function that is the best fit for each set of outputs and the input, using

a. Output Set I.   **b.** Output Set II.   **c.** Output Set III.

Table 6.8

Inputs, x	Output Set I	Output Set II	Output Set III
1	3	4	−16
2	14	4	−13
3	47	18	1
4	114	88	33
5	227	280	87
6	398	684	173

SOLUTION

a. The inputs (x-values in column 1) are equally spaced. The third differences for Output Set I data in column 2 follow.

Outputs	3	14	47	114	227	398
First Differences		11	33	67	113	171
Second Differences			22	34	46	58
Third Differences				12	12	12

The third differences are constant, so the data points fit exactly on the graph of a cubic function. The function that models the data is $f(x) = 2x^3 - x^2 + 2$. Figure 6.17 shows the graph of this function for $0 \le x \le 6$.

$f(x)$

$f(x) = 2x^3 - x^2 + 2$

Figure 6.17

b. The third and fourth differences for Output Set II data in column 3 follow.

The fourth differences are constant, so the data points fit exactly on the graph of a quartic function. The function that models the data is $g(x) = x^4 - 3x^3 + 6x$ (see Figure 6.18).

Figure 6.18

c. The third differences for Output Set III data in column 4 follow.

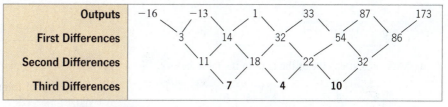

The third differences are not constant, but are relatively close, so the data points can be approximately fitted by a cubic function. The cubic function that is the best fit for the data is $y = h(x) = 1.083x^3 - 1.107x^2 - 1.048x - 15$. Figure 6.19 shows that the graph is a good fit for the data points.

Figure 6.19

Note that the third (or fourth) differences will not be constant unless the data points fit on the graph of the cubic (or quartic) function exactly, and real data points will rarely fit on the graph of a function exactly. As with other models, statistical measures exist to measure the goodness of fit of the model to the data.

SKILLS CHECK 6.2

1. Find the cubic function that models the data in the table below.

x	−2	−1	0	1	2	3	4
y	−16	−3	0	−1	0	9	32

2. Find the cubic function that is the best fit for the data in the table below.

x	0	5	10	15
y	1	270	2600	9220

3. Find the quartic function that models the data in the table below.

x	−2	−1	0	1	2	3	4
y	0	−3	0	−3	0	45	192

4. Find the quartic function that is the best fit for the data in the table below.

x	−2	−1	0	1	4
y	17	−0.25	0	−0.25	356

5. **a.** Make a scatter plot of the data in the table below.

x	1	2	3	4	5	6
y	−2	−1	0	4	8	16

   b. Does it appear that a linear model or a cubic model is the better fit for the data?

6. **a.** Make a scatter plot of the data in the table below.

x	−2	−1	0	1	2	3
y	−8	7	1	2	9	57

   b. Does it appear that a cubic model or a quartic model is the better fit for the data?

7. **a.** Find a cubic function that models the data in the table in Exercise 5.

   b. Find a linear function that models the data.

   c. Visually determine which model is the better fit for the data.

8. **a.** Find a cubic function that models the data in the table in Exercise 6.

   b. Find a quartic function that models the data.

   c. Visually determine which model is the better fit for the data.

9. **a.** Find the cubic function that is the best fit for the data in the table below.

x	1	2	3	4	5	6	7
y	−2	0	4	16	54	192	1500

   b. Find the quartic function that is the best fit for the data in the table.

10. **a.** Graph each of the functions found in Exercise 9 on the same axes with the data in the table.

   b. Does it appear that a cubic model or a quartic model is the better fit for the data?

11. Find the quartic function that is the best fit for the data in the table below.

x	−3	−2	−1	0	1	2	3
y	55	7	1	1	−5	−5	37

12. Is the model found in Exercise 11 an exact fit to the data?

13. The following table has the inputs, x, and the outputs for two functions, f and g. Use third differences to determine whether a cubic function exactly fits the data with input x and output $f(x)$.

x	0	1	2	3	4	5
f(x)	0	1	5	24	60	110
g(x)	0	0.5	4	13.5	32	62.5

14. Use third differences with the data in the table in Exercise 13 to determine whether a cubic function exactly fits the data with input x and output $g(x)$.

15. Find the cubic function that is the best fit for $f(x)$ defined by the table in Exercise 13.

16. Find the cubic function that is the best fit for $g(x)$ defined by the table in Exercise 13.

EXERCISES 6.2

Use unrounded models for graphing and calculations unless otherwise stated. Report models with coefficients to three decimal places unless otherwise instructed.

17. *Internet Access* The table shows the percent of households with Internet access for selected years from 2009 and projected through 2015.

  a. Find the cubic function that is the best model for the data, with x equal to the number of years after 2005 and y equal to the percent of households with Internet access.

  b. Use the model to predict the percent of households with Internet access in 2022.

  c. Compare the graphs of the model and the data to determine whether the model is a good or a poor fit for the data.

Year	Percent of Households
2009	67
2010	70
2011	72.5
2012	75
2013	76.5
2014	77.2
2015	78

(Source: Jupiter Media Metrix)

18. *Social Security Beneficiaries* The table that follows gives the numbers of Social Security beneficiaries (in millions) for selected years from 1950 to 2000, with projections through 2030.

  a. Find the cubic equation that models the data, with x equal to the number of years after 1950 and $B(x)$ equal to the number of millions of beneficiaries. Report the model with four decimal places.

Year	Beneficiaries (millions)	Year	Beneficiaries (millions)
1950	2.9	2000	44.8
1960	14.3	2010	53.3*
1970	25.2	2020	68.8*
1980	35.1	2030	82.7*
1990	39.5		

(Source: 2000 Social Security Trustees Report)   * Projected.

  b. Graph the data points and the model on the same axes.

  c. Is the model a good fit for the data?

19. *Population of the World* The table gives the total population of the world from 1950 and projected to 2050.

  a. Create a scatter plot of the data, with y equal to the world population in billions and x equal to the number of years after 1940.

  b. Find a cubic function that models the data, with y equal to the world population in billions and x equal to the number of years after 1940. Report the model with three significant digits.

  c. Graph the data points and the model on the same axes. Is the model a good fit?

  d. What does the model predict the world population will be in 2022?

  e. In what year does the model estimate the world population will reach 8 billion?

  f. Is the model also a good predictor of the population after 2050? Why or why not?

Year	Total World Population (billions)	Year	Total World Population (billions)
1950	2.556	2010	6.849
1960	3.039	2020	7.585
1970	3.707	2030	8.247
1980	4.454	2040	8.850
1990	5.279	2050	9.346
2000	6.083		

(Source: U.S. Census Bureau, International Database)

20. *Income by Age* The median income for workers age 20 to 62 is shown in the table on the next page.

  a. Find a cubic function that models the data with x equal to age and y equal to median salary in thousands.

  b. Graph the function on the same axes as the scatter plot.

  c. At what age does the model estimate the median income to be $56,520?

Age	Median Income ($)	Age	Median Income ($)
20	25,404	47	54,260
27	37,896	52	54,405
32	45,770	57	56,427
37	50,740	62	59,222
42	51,613		

(Source: *Wall Street Journal*, June 18, 2012)

The table below shows the U.S. median age at first marriage for men and women for selected years from 1900 to 2015. Use the table in Exercises 21 and 22.

Median Age at First Marriage

Year	Men	Women	Year	Men	Women
1900	25.9	21.9	1990	26.1	23.9
1910	25.1	21.6	2000	26.8	25.1
1920	24.6	21.2	2004	27.1	25.8
1930	24.3	21.3	2009	28.1	25.9
1940	24.3	21.5	2010	28.2	26.1
1950	22.8	20.3	2011	28.7	26.5
1960	22.8	20.3	2012	28.6	26.6
1970	23.2	20.8	2013	29.0	26.6
1980	24.7	22.0	2015	29.2	27.1

(Source: U.S. Census Bureau)

21. Age at First Marriage
   a. Find the cubic function $y = f(x)$ that is the best fit for the data, with y equal to the age at first marriage for men and x equal to the number of years after 1900. Round coefficients to three significant digits.

   b. What does the unrounded model predict to be the age at first marriage for men in 2020?

22. Age at First Marriage
   a. Find the quartic function $y = f(x)$ that is the best fit for the data, with y equal to the age at first marriage for women and x equal to the number of years after 1900. Round coefficients to three significant digits.

   b. What does the unrounded model predict to be the age at first marriage for women in 2020?

23. China's Labor Pool The following table shows UN data from 1975 and projections to 2050 of the size China's labor pool (ages 15–59).

   a. Find the cubic function that best models this population as a function of the number of years after 1970. Let x represent the number of years after 1970 and y represent the number of millions of people in this labor pool. Report the model with coefficients rounded to three significant digits.

   b. Graph the model and the data on the same axes, and comment on the fit.

   c. Use the model to predict the size of the labor pool in 2022.

   d. According to the model, when is the number in the labor pool projected to be at a maximum? What is the maximum?

Year	Labor Pool (millions)	Year	Labor Pool (millions)
1975	490	2015	920
1980	560	2020	920
1985	650	2025	905
1990	730	2030	875
1995	760	2035	830
2000	800	2040	820
2005	875	2045	800
2010	910	2050	670

(Source: United Nations; *Wall Street Journal*, January 19, 2013)

24. Inflation Rate The annual changes in the consumer price index (CPI) for 1990–2008 are shown in the following table.

Year	CPI Annual Change (%)	Year	CPI Annual Change (%)	Year	CPI Annual Change (%)
1990	5.4	1997	2.3	2003	2.3
1991	4.2	1998	1.6	2004	2.7
1992	3.0	1999	2.2	2005	3.4
1993	3.0	2000	3.4	2006	3.2
1994	2.6	2001	2.8	2007	2.8
1995	2.8	2002	1.6	2008	3.8
1996	3.0				

(Source: Bureau of Labor Statistics)

a. Find the quartic function that models the data, with $x = 0$ in 1990. Report the model to 5 decimal places.

b. Is the fit of this model to the data best described as fair, good, or exact?

25. *Elderly Men in the Workforce* The table below gives the percent of men 65 years or older in the workforce for selected years from 1920 and projected to 2030.

Year	Percent	Year	Percent
1920	55.6	1980	19.0
1930	54.0	1990	16.3
1940	41.8	2000	17.7
1950	45.8	2010	22.6
1960	33.1	2020	27.2
1970	21.8	2030	27.6

(Source: U.S. Bureau of the Census)

a. With $x = 0$ representing 1900, find the cubic function that models the data.

b. Use the model to determine when the percent of elderly men in the workforce reached its minimum.

26. *World Cell Phone Subscribership* The table shows the number of billions of world cell phone subscriberships from 1990 with projections to 2020.

Year	Subscriberships (billions)	Year	Subscriberships (billions)
1990	0.011	2006	2.76
1991	0.016	2007	3.30
1992	0.023	2008	4.10
1993	0.034	2009	4.72
1994	0.056	2010	5.40
1995	0.091	2011	6.05
1996	0.145	2012	6.65
1997	0.215	2013	7.19
1998	0.318	2014	7.65
1999	0.490	2015	8.03
2000	0.738	2016	8.35
2001	0.961	2017	8.60
2002	1.16	2018	8.79
2003	1.42	2019	8.95
2004	1.76	2020	9.07
2005	2.22		

(Source: International Telecommunications Union)

a. Find a cubic function that models the world cell phone subscriberships, y, in billions, as a function of the number of years after 1990, x.

b. Graph the function on the window $[10, 30]$ by $[0, 10]$.

c. What does the model give as the number of subscriberships in 2024?

27. *U.S. Population* The table gives the U.S. population, in millions, for selected years from 1980 and projected to 2050.

Year	Millions	Year	Millions
1980	235.1	2020	342.9
1990	259.6	2030	369.5
2000	287.9	2040	390.8
2010	315.2	2050	409.1
2015	328.4		

(Source: Social Security Administration)

a. Create a scatter plot for the data in the table. Use x as the number of years after 1980 and y as the number of millions of Americans.

b. Find a cubic function that models the data with x equal to the number of years after 1980. Report the model with three significant digits.

c. Graph the model and the data on the same axes, and comment on the fit.

d. Change the view of the graph by increasing the x-max to 140. Does the model seem appropriate for this increased time period?

28. *Consumer Prices* The data in the table give the consumer price index (CPI) for selected years from 1945 to 2009.

a. Create a scatter plot of the data, with x equal to the number of years after 1945.

Year	CPI	Year	CPI
1945	18.0	1980	82.4
1950	24.1	1985	107.6
1955	26.8	1990	130.7
1960	29.6	1995	152.4
1965	31.5	2000	172.2
1970	38.8	2005	185.3
1975	53.8	2009	214.5

(Source: Bureau of Labor Statistics)

b. What type of function would be best to model the data? Find the equation of that function. Round the coefficient of the cubic term to five decimal places.

c. Use the function to estimate the CPI in 2015.

29. *Federal Tax Per Capita* The table gives the amount of federal tax per capita (per person) for selected years from 2000 and projected to 2018.

  a. Create a scatter plot of the data, with x equal to the number of years after 2000.

  b. Find a cubic model that fits the data points, with x equal to the number of years after 2000.

  c. Graph the data and the model on the same axes.

  d. What does the model predict the federal tax per capita will be in 2023?

  e. Use technology to find the year when the model predicts the federal tax per capita will be $16,226.

Year	Tax ($)	Year	Tax ($)
2000	5110	2013	8559
2005	7177	2015	10,335
2010	7228	2018	12,029

(Source: Internal Revenue Service)

30. *Japanese Economy* The economy of Japan is the third largest in the world, after those of the United States and the People's Republic of China. With average growth rates of 10% in the 1960s, 5% in the 1970s, and 4% in the 1980s, Japan was able to establish and maintain itself as the world's second largest economy from 1968 until 2010. However, Japan's economy was disrupted in early 2011 because of the earthquake and tsunami. The percent changes in gross domestic product (GDP) for each year since 1997 are given in the figure.

(Source: tradingeconomics.com)

  a. Find a quartic function that models the data, where x equals the number of years after 1990 and y is the percent change.

b. Use the model to estimate the percent change in GDP for 2012.

c. Does the model appear to be accurate after 2013? Why or why not?

31. *Women in the Workforce* The following table gives the number of women age 16 years and older (in millions) in the U.S. civilian workforce for selected years from 1950 and projected to 2050.

  a. Use x as the number of years past 1950 to create a cubic model using these data.

  b. During what year does the model indicate that the number of women in the workforce will reach its maximum?

Year	Women in the Workforce (millions)	Year	Women in the Workforce (millions)
1950	18.4	2010	75.5
1960	23.2	2015	78.6
1970	31.5	2020	79.2
1980	45.5	2030	81.6
1990	56.8	2040	86.5
2000	65.6	2050	91.5

(Source: Bureau of Labor Statistics)

32. *Homicide Rates* The table gives the U.S. homicide rates per 100,000 people for the years from 1990 through 2009.

  a. Find the quartic function that models the data, with x representing the number of years after 1990. Report the model with four decimal places.

Year	Homicide Rate	Year	Homicide Rate
1990	9.4	2000	5.5
1991	9.8	2001	5.6
1992	9.3	2002	5.6
1993	9.5	2003	5.7
1994	9.0	2004	5.5
1995	8.2	2005	5.6
1996	7.4	2006	5.8
1997	6.8	2007	5.7
1998	6.3	2008	5.4
1999	5.7	2009	5.0

(Source: www.ojp.usdog.gov/bjs)

b. Graph the model on the same axes with the data.

c. What does the rounded model estimate for the homicide rate in 2011? Is this possible?

33. *Gross Domestic Product* The table gives the U.S. gross domestic product (GDP), in trillions of dollars, for selected years from 2005 and projected to 2070.

a. Create a scatter plot of the data, with y representing GDP in billions of dollars and x representing the number of years after 2000.

b. Find the quadratic function that best fits the data, with x equal to the number of years after 2000. Report the model with three significant digits.

c. Find the cubic function that best fits the data, with x equal to the number of years after 2000. Report the model with three significant digits.

d. Graph each of these functions on the same axes as the data points, and visually determine which model is the better fit for the data.

Year	GDP ($ trillions)	Year	GDP ($ trillions)
2005	12.145	2040	79.680
2010	16.174	2045	103.444
2015	21.270	2050	133.925
2020	27.683	2055	173.175
2025	35.919	2060	224.044
2030	46.765	2065	290.042
2035	61.100	2070	375.219

(Source: Bureau of Economic Analysis)

34. *U.S. Newspaper Industry* The following graph shows ad revenue from help wanted classifieds dropping 92% in the last 10 years, hitting $723 million last year, down from $8.7 billion in 2000.

U.S. Newspaper Industry's Help Wanted Classified Ad Revenue

(Source: Silicon Alley Insider Chart of the Day; data from naa.org and The Ladders, March 2011)

a. Create a scatter plot of the data, with x equal to the number of years after 2000.

b. Find a cubic function that models the data, with x equal to the number of years after 2000. Round coefficients to four decimal places.

c. Find a quartic function that models the data, with x equal to the number of years after 2000. Round coefficients to four decimal places.

d. Graph each model on the same axes as the data.

e. Which model appears to be the better choice for prediction after 2010? Why?

35. *Carbon Dioxide Emissions* For selected years from 2010 and projected to 2032, the following table shows the number of millions of metric tons of carbon dioxide (CO_2) emissions from biomass energy combustion in the United States.

a. Create a scatter plot of the data with x equal to the number of years after 2000 and y equal to the millions of metric tons of CO_2.

b. Create a cubic function that models the data, with x equal to the number of years after 2000. Report the model with four significant digits.

c. Graph the data and the model on the same axes, and comment on the fit of the model to the data.

d. Use the model to predict the CO_2 emissions in 2026.

e. Use technology to estimate the year in which CO_2 emissions are projected to be 781 million metric tons with the model.

Year	CO₂ Emissions (millions of metric tons)	Year	CO₂ Emissions (millions of metric tons)
2010	338.5	2022	556.2
2012	364.5	2024	590.2
2014	396.1	2026	629.7
2016	425.8	2028	663.1
2018	453.1	2030	701.1
2020	498.4	2032	743.7

(Source: U.S. Department of Energy)

36. *Union Membership* The table on the next page gives the numbers of U.S. workers (in thousands) who were union members for selected years from 1930 to 2009.

Year	Members	Year	Members
1930	3401	1975	19,611
1935	3584	1980	19,843
1940	8717	1985	16,996
1945	14,322	1990	16,740
1950	14,267	1995	16,360
1955	16,802	2000	16,258
1960	17,049	2005	15,685
1965	17,299	2009	16,904
1970	19,381		

(Source U.S. Department of Labor)

a. With $x = 0$ representing 1900, find a cubic function that models the data. Round coefficients to four decimal places.

b. Use the model and a graphing utility to estimate the year in which union membership was maximized.

c. How well does the relative maximum from the model agree with the data?

37. U.S. Pay TV The table shows the projected number of millions of U.S. households with pay TV for 2010 through 2017.

a. Find the quadratic function that models the data for the years 2010 through 2017, with x equal to the number of years after 2010 and y equal to the projected number of millions of U.S. households with pay TV.

b. Find the cubic function that models the data for the years 2010 through 2017, with x equal to the number of years after 2010 and y equal to the projected number of millions of U.S. households with pay TV.

Years	Pay-TV Households (millions)
2010	99.0
2011	100.9
2012	100.8
2013	99.3
2014	98.0
2015	97.2
2016	96.0
2017	94.6

(Source: Data from gigaom.com)

c. Graph each of the models on the same axes as the data to decide which model best fits the data.

d. Describe what happens to each model after 2017. Which model seems more appropriate for these data?

38. Medicaid The numbers of thousands of people covered by Medicaid for the years 2000–2014 are given in the table that follows.

Year	People Covered by Medicaid (thousands)
2000	32,652
2001	35,878
2002	38,644
2003	40,606
2004	41,938
2005	42,632
2006	42,301
2007	42,808
2008	44,903
2009	48,918
2010	51,804
2011	53,535
2012	54,828
2013	55,413
2014	57,888

(Source: Kaiser Family Foundation)

a. Create a scatter plot of the data to determine what type of function would best model the data.

b. Find the equation of the best-fit cubic model. Align the data so that $x = 0$ is 2000, and round your answer to three decimal places.

c. Graph the model with the data to see how well the model fits the data.

d. Find a quartic function that models the data, rounded to three decimal places, and graph this function with the data.

e. Which function is the better fit for extrapolation after 2014? Why?

Solution of Polynomial Equations

KEY OBJECTIVES

- Solve polynomial equations using factoring, factoring by grouping, and the root method

- Find factors, zeros, *x*-intercepts, and solutions

- Estimate solutions with technology

- Solve polynomial equations using the intersection method and the *x*-intercept method

SECTION PREVIEW **International Travel to the United States**

Tourism has become a popular global leisure activity. International travel to the United States has seen increases and decreases since 1986, with the decreases resulting from recessions and terrorist attacks. The number of millions of international visitors to the United States can be modeled by the equation

$$y = 0.0101x^3 - 0.575x^2 + 10.814x - 20.753$$

where *x* is the number of years after 1980.

The graph of this model, in Figure 6.20, shows that in approximately 2001, the number of international visitors reached a minimum and then began to increase. If the model is accurate, when will the number of international visitors be 75 million?

In Example 8, we answer this question by solving the equation

$$0.0101x^3 - 0.575x^2 + 10.814x - 20.753 = 75$$

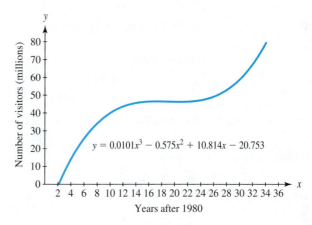

Figure 6.20

The technology of graphing utilities allows us to find approximate solutions of equations such as this one.

Although many equations derived from real data require technology to find or to approximate solutions, some higher-degree equations can be solved by factoring, and some can be solved by the root method. In this section, we discuss these methods for solving polynomial equations; in the next section, we discuss additional methods of solving higher-degree polynomial equations. ■

Solving Polynomial Equations by Factoring

Some polynomial equations can be solved by factoring and using the zero-product property, much like the method used to solve quadratic equations described in Chapter 3. From our discussion in Chapter 3, we know that the following statements regarding a function *f* are equivalent:

- $(x - a)$ is a factor of $f(x)$.

- *a* is a real zero of the function *f*.

- *a* is a real solution to the equation $f(x) = 0$.

- *a* is an *x*-intercept of the graph of $y = f(x)$.

- The graph crosses the *x*-axis at the point $(a, 0)$. Note that if the graph of $y = f(x)$ touches, but does not cross, the *x*-axis at $(a, 0)$, then $(x - a)^2$ is a factor of $f(x)$, and $(x - a)$ is a factor with even multiplicity.

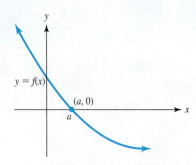

EXAMPLE 1 ▶ Factors, Zeros, Intercepts, and Solutions

If $P(x) = (x - 1)(x + 2)(x + 5)$, find

a. The zeros of $P(x)$.

b. The solutions of $P(x) = 0$.

c. The *x*-intercepts of the graph of $y = P(x)$.

SOLUTION

a. The factors of $P(x)$ are $(x - 1)$, $(x + 2)$, and $(x + 5)$; thus, the zeros of $P(x)$ are $x = 1, x = -2$, and $x = -5$.

b. The solutions to $P(x) = 0$ are the zeros of $P(x)$: $x = 1, x = -2$, and $x = -5$.

c. The *x*-intercepts of the graph of $y = P(x)$ are $1, -2$, and -5 (Figure 6.21).

Figure 6.21

EXAMPLE 2 ▶ Finding the Maximum Volume

A box is to be formed by cutting a square of *x* inches per side from each corner of a square piece of cardboard that is 24 inches on each side and folding up the sides. This will give a box whose height is *x* inches, with the length of each side of the bottom $2x$ less than the original length of the cardboard (see Figure 6.22). Using $V = lwh$, the volume of the box is given by

$$V = (24 - 2x)(24 - 2x)x$$

where *x* is the length of the side of the square that is cut out.

Figure 6.22

a. Use the factors of V to find the values of x that give volume 0.

b. Graph the function that gives the volume as a function of the side of the square that is cut out over an x-interval that includes the x-values found in part (a).

c. What input values make sense for this problem (that is, actually result in a box)?

d. Graph the function using the input values that make sense for the problem (found in part (c)).

e. Use technology to determine the size of the square that should be cut out to give the maximum volume.

SOLUTION

a. We use the zero-product property to solve the equation $0 = (24 - 2x)(24 - 2x)x$.

$$0 = (24 - 2x)(24 - 2x)x$$

$$x = 0 \quad \text{or} \quad 24 - 2x = 0$$

$$-2x = -24$$

$$x = 0 \quad \text{or} \qquad x = 12$$

b. A graph of $V(x)$ from $x = -1$ to $x = 16$ is shown in Figure 6.23(a).

c. The box can exist only if each of the dimensions is positive, resulting in a volume that is positive. Thus, $24 - 2x > 0$, or $x < 12$, so the box can exist only if the value of x is greater than 0 and less than 12. Note that if squares 12 inches on a side are removed, no material remains to make a box, and no squares larger than 12 inches on a side can be removed from all four corners.

d. The graph of $V(x)$ over the x-interval $0 < x < 12$ is shown in Figure 6.23(b).

Figure 6.23

e. By observing the graph, we see that the function has a maximum value, which gives the maximum volume over the interval $0 < x < 12$. Using the "maximum" feature on a graphing utility, we see that the maximum occurs at or near $x = 4$

(Figure 6.23(c)). Cutting a square with side 4 inches from each corner and folding up the sides gives a box that is 16 inches by 16 inches by 4 inches. The volume of this box is $16 \cdot 16 \cdot 4 = 1024$ cubic inches, and testing values near 4 shows that this is the maximum possible volume.

EXAMPLE 3 ▶ Photosynthesis

The amount y of photosynthesis that takes place in a certain plant depends on the intensity x of the light (in lumens) present, according to the function $y = 120x^2 - 20x^3$. The model is valid only for nonnegative x-values that produce a positive amount of photosynthesis.

a. Graph this function with a viewing window large enough to see 2 turning points.

b. Use factoring to find the x-intercepts of the graph.

c. What nonnegative values of x will give positive photosynthesis?

d. Graph the function with a viewing window that contains only the values of x found in part (c) and nonnegative values of y.

SOLUTION

a. The graph in Figure 6.24 shows the function with 2 turning points.

b. We can find the x-intercepts of the graph by solving $120x^2 - 20x^3 = 0$ for x. Factoring gives

$$20x^2(6 - x) = 0$$
$$20x^2 = 0 \quad \text{or} \quad 6 - x = 0$$
$$x = 0 \quad \text{or} \quad x = 6$$

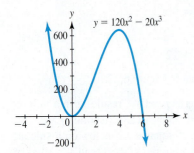

Figure 6.24

Thus, the x-intercepts are 0 and 6.

c. From the graph of $y = 120x^2 - 20x^3$ in Figure 6.24, we see that positive output values result when the input values are between 0 and 6. Thus, the nonnegative x-values that produce positive amounts of photosynthesis are $0 < x < 6$.

d. Graphing the function in the viewing window $[0, 6]$ by $[0, 700]$ will show the region that is appropriate for the model (see Figure 6.25).

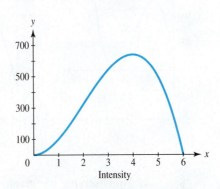

Figure 6.25

Solution Using Factoring by Grouping

Some higher-degree equations can be solved by using the method of factoring by grouping.* Consider the following examples.

EXAMPLE 4 ▶   **Cost**

The total cost of producing a product is given by the function

$$C(x) = x^3 - 12x^2 + 3x + 9 \text{ thousand dollars}$$

where x is the number of hundreds of units produced. How many units must be produced to give a total cost of $45,000?

SOLUTION

Because $C(x)$ is the total cost in thousands of dollars, a total cost of $45,000 is represented by $C(x) = 45$. So we seek to solve

$$45 = x^3 - 12x^2 + 3x + 9, \quad \text{or} \quad 0 = x^3 - 12x^2 + 3x - 36$$

This equation appears to have a form that permits factoring by grouping. The solution steps follow.

$0 = (x^3 - 12x^2) + (3x - 36)$ Separate the terms into two groups, each having a common factor.

$0 = x^2(x - 12) + 3(x - 12)$ Factor x^2 from the first group, then 3 from the second group.

$0 = (x - 12)(x^2 + 3)$ Factor the common factor $(x - 12)$ from each of the two terms.

$x - 12 = 0 \quad \text{or} \quad x^2 + 3 = 0$ Set each factor equal to 0 and solve for x.

$x = 12 \quad \text{or} \quad x^2 = -3$

$x = 12$

Because no real number x can satisfy $x^2 = -3$, the only solution is 12 hundred units. Thus, producing 1200 units gives a total cost of $45,000.

The Root Method

We solved quadratic equations of the form $x^2 = C$ in Chapter 3 by taking the square root of both sides and using a $\pm$ symbol to indicate that we get both the positive and the negative root: $x = \pm \sqrt{C}$. In the same manner, we can solve $x^3 = C$ by taking the cube root of both sides. There will not be a $\pm$ sign because there is only one real cube root of a number. In general, there is only one real nth root if n is odd, and there are zero or two real roots if n is even and $C \geq 0$.

> ### Root Method
>
> The real solutions of the equation $x^n = C$ are found by taking the nth root of both sides:
>
> $$x = \sqrt[n]{C} \text{ if } n \text{ is odd} \quad \text{and} \quad x = \pm\sqrt[n]{C} \text{ if } n \text{ is even and } C \geq 0$$

---

* Factoring by grouping was discussed on page 164, in the Chapter 3 Algebra Toolbox.

EXAMPLE 5 ▶ **Root Method**

Solve the following equations.

a. $x^3 = 125$ **b.** $5x^4 = 80$ **c.** $4x^2 = 18$

SOLUTION

a. Taking the cube root of both sides gives

$$x^3 = 125$$

$$x = \sqrt[3]{125} = 5$$

No $\pm$ sign is needed because the root is odd.

b. We divide both sides by 5 to isolate the x^4 term:

$$5x^4 = 80$$

$$x^4 = 16$$

Then we take the fourth root of both sides, using a $\pm$ sign (because the root is even):

$$x = \pm\sqrt[4]{16} = \pm 2$$

The solutions are 2 and -2.

c. We divide both sides by 4 to isolate the x^2 term:

$$4x^2 = 18$$

$$x^2 = \frac{9}{2}$$

Then we take the square root of both sides, using a $\pm$ sign (because the root is even):

$$x = \pm\sqrt{\frac{9}{2}} = \pm\frac{3}{\sqrt{2}}$$

The *exact* solutions are $\pm\dfrac{3}{\sqrt{2}}$; the solutions are *approximately* 2.121 and -2.121.

EXAMPLE 6 ▶ **Future Value of an Investment**

The future value of $10,000 invested for 4 years at interest rate r, compounded annually, is given by $S = 10,000(1 + r)^4$. Find the rate r, as a percent, for which the future value is $14,641.

SOLUTION

We seek to solve $14,641 = 10,000(1 + r)^4$ for r. We can rewrite this equation with $(1 + r)^4$ on one side by dividing both sides by 10,000:

$$14,641 = 10,000(1 + r)^4$$

$$1.4641 = (1 + r)^4$$

We can now remove the exponent by taking the fourth root of both sides of the equation, using a $\pm$ sign:

$$\pm\sqrt[4]{1.4641} = 1 + r$$

$$\pm 1.1 = 1 + r$$

Solving for r gives

$$1.1 = 1 + r \quad \text{or} \quad -1.1 = 1 + r$$
$$0.1 = r \qquad \text{or} \quad -2.1 = r$$

The negative value cannot represent an interest rate, so the interest rate that gives this future value is

$$r = 0.1 = 10\%$$

The graphs of $y_1 = 10{,}000(1 + x)^4$ and $y_2 = 14{,}641$ shown in Figure 6.26 support the conclusion that $r = 0.10$ gives the future value 14,641.*

Figure 6.26

Estimating Solutions with Technology

Although we can solve some higher-degree equations by using factoring, many real problems do not have "nice" solutions, and factoring is not an appropriate solution method. Thus, many equations resulting from real data applications require graphical or numerical methods of solution.

EXAMPLE 7 ▶   **China's Labor Pool**

Using United Nations data from 1975 with projections to 2050, the size of China's labor pool (age 15 to 59) can be modeled by the function

$$y = -0.000487x^3 - 0.162x^2 + 19.8x + 388$$

where x is the number of years after 1970 and y represents the number of millions of people in this labor pool. Use this model to estimate graphically the years after 1970 when the population is and will be 900 million. (Source: United Nations; Wall Street Journal Jan 19, 2013)

SOLUTION

To find the years when China's labor pool is or will be 900 million, we solve

$$900 = -0.000487x^3 - 0.162x^2 + 19.8x + 388$$

To solve this equation graphically, we can use either the intersection method or the x-intercept method. To use the intersection method, we graph

$$y_1 = -0.000487x^3 - 0.162x^2 + 19.8x + 388 \quad \text{and} \quad y_2 = 900$$

for $x \geq 0$ and find the x-coordinates of the points of intersection.

---

* The graph of the quartic function $y = 10{,}000(1 + x)^4$ is actually curved, but within this small window it appears to be nearly linear.

Figure 6.27(a) shows that the graph of $y_1 = -0.000487x^3 - 0.162x^2 + 19.8x + 388$ crosses the graph of $y_2 = 900$ in two places. Using 5: intersect under the CALC menu gives the x-coordinates of the two points of intersection. From these graphs, the model estimates that China's labor pool reached 900 million in 2013 ($x = 42.6$) and will fall back to 900 million in 2027 ($x = 57$).

(a)   (b)   (c)

Figure 6.27

EXAMPLE 8 ▶ ## International Visitors to the United States

The number of international visitors to the United States, in millions, can be modeled by the equation

$$y = 0.0101x^3 - 0.575x^2 + 10.814x - 20.753$$

where x is the number of years after 1980.

If the model remains accurate, when will the number of international visitors be 75 million?

SOLUTION

To answer this question, we solve the equation

$$0.0101x^3 - 0.575x^2 + 10.814x - 20.753 = 75, \text{ or}$$
$$0.0101x^3 - 0.575x^2 + 10.814x - 95.753 = 0$$

To solve this equation graphically with the intersection method, we graph

$$y_1 = 0.0101x^3 - 0.575x^2 + 10.814x - 20.753 \quad \text{and} \quad y_2 = 75$$

and find the point of intersection (Figure 6.28).

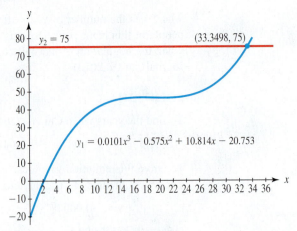

Figure 6.28

Thus, the number of international visitors will be 75 million in $1980 + 34$, or 2014.

SKILLS CHECK 6.3

In Exercises 1–4, solve the polynomial equation.

1. $(2x - 3)(x + 1)(x - 6) = 0$

2. $(3x + 1)(2x - 1)(x + 5) = 0$

3. $(x + 1)^2(x - 4)(2x - 5) = 0$

4. $(2x + 3)^2(5 - x)^2 = 0$

In Exercises 5–10, solve the polynomial equations by factoring, and check the solutions graphically.

5. $x^3 - 16x = 0$ **6.** $2x^3 - 8x = 0$

7. $x^4 - 4x^3 + 4x^2 = 0$

8. $x^4 - 6x^3 + 9x^2 = 0$

9. $4x^3 - 4x = 0$ **10.** $x^4 - 3x^3 + 2x^2 = 0$

In Exercises 11–14, use factoring by grouping to solve the equations.

11. $x^3 - 4x^2 - 9x + 36 = 0$

12. $x^3 + 5x^2 - 4x - 20 = 0$

13. $3x^3 - 4x^2 - 12x + 16 = 0$

14. $4x^3 + 8x^2 - 36x - 72 = 0$

In Exercises 15–18, solve the polynomial equations by using the root method, and check the solutions graphically.

15. $2x^3 - 16 = 0$ **16.** $3x^3 - 81 = 0$

17. $\dfrac{1}{2}x^4 - 8 = 0$ **18.** $2x^4 - 162 = 0$

In Exercises 19–24, use factoring and the root method to solve the polynomial equations.

19. $4x^4 - 8x^2 = 0$ **20.** $3x^4 - 24x^2 = 0$

21. $0.5x^3 - 12.5x = 0$ **22.** $0.2x^3 - 24x = 0$

23. $x^4 - 6x^2 + 9 = 0$ **24.** $x^4 - 10x^2 + 25 = 0$

In Exercises 25–30, use the graph of the polynomial function $f(x)$ to (a) solve $f(x) = 0$, and (b) find the factorization of $f(x)$.

25. $f(x) = x^3 - 2x^2 - 11x + 12$

26. $f(x) = -x^3 + 6.5x^2 + 13x - 8$

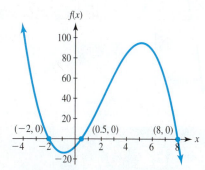

27. $y = -2x^4 + 6x^3 + 6x^2 - 14x - 12$

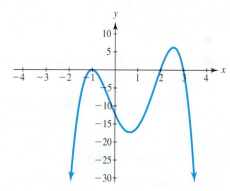

28. $y = x^4 - 5x^3 - 3x^2 + 13x + 10$

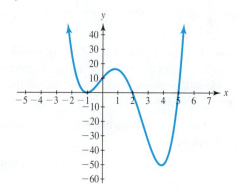

29. $y = -0.5x^3 + 2.5x^2 + 0.5x - 2.5$

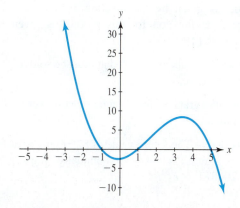

30. $y = x^3 - x^2 - 4x + 4$

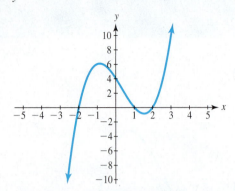

In Exercises 31 and 32, find the solutions graphically.

31. $4x^3 - 15x^2 - 31x + 30 = 0$

32. $2x^3 - 15x^2 - 62x + 120 = 0$

EXERCISES 6.3

33. *Revenue* The revenue from the sale of a product is given by the function $R = 400x - x^3$.

   a. Use factoring to find the numbers of units that must be sold to give zero revenue.

   b. Does the graph of the revenue function verify this solution?

34. *Revenue* The revenue from the sale of a product is given by the function $R = 12{,}000x - 0.003x^3$.

   a. Use factoring and the root method to find the numbers of units that must be sold to give zero revenue.

   b. Does the graph of the revenue function verify this solution?

35. *Revenue* The price for a product is given by $p = 100{,}000 - 0.1x^2$, where x is the number of units sold, so the revenue function for the product is $R = px = (100{,}000 - 0.1x^2)x$.

   a. Find the numbers of units that must be sold to give zero revenue.

   b. Does the graph of the revenue function verify this solution?

36. *Revenue* If the price from the sale of x units of a product is given by the function $p = 100x - x^2$, the revenue function for this product is given by $R = px = (100x - x^2)x$.

   a. Find the numbers of units that must be sold to give zero revenue.

   b. Does the graph of the revenue function verify this solution?

37. *Future Value* The future value of $2000 invested for 3 years at rate r, compounded annually, is given by $S = 2000(1 + r)^3$.

   a. Complete the table below to determine the future value of $2000 at certain interest rates.

Rate	Future Value
4%	
5%	
7.25%	
10.5%	

   b. Graph this function on the window $[0, 0.24]$ by $[0, 5000]$.

   c. Use the root method to find the rate r, as a percent, for which the future value is $2662.

   d. Use the root method to find the rate r, as a percent, for which the future value is $3456.

38. *Future Value* The future value of $5000 invested for 4 years at rate r, compounded annually, is given by $S = 5000(1 + r)^4$.

   a. Graph this function on the window $[0, 0.24]$ by $[0, 12{,}000]$.

   b. Use the root method to find the rate r, as a percent, for which the future value is $10{,}368.

   c. What rate, as a percent, gives $2320.50 in interest on this investment?

39. *Constructing a Box* A box can be formed by cutting a square out of each corner of a piece of tin and folding the sides up. Suppose the piece of tin is 18 inches by 18 inches and each side of the square that is cut out has length x.

a. Write an expression for the height of the box that is constructed.

b. Write an expression for the dimensions of the base of the box that is constructed.

c. Use the formula $V = lwh$ to find an equation that represents the volume of the box.

d. Use the equation that you constructed to find the values of x that make $V = 0$.

e. For which of these values of x does a box exist if squares of length x are cut out and the sides are folded up?

40. *Constructing a Box* A box can be formed by cutting a square out of each corner of a piece of cardboard and folding the sides up. If the piece of cardboard is 12 inches by 12 inches and each side of the square that is cut out has length x, the function that gives the volume of the box is $V = 144x - 48x^2 + 4x^3$.

a. Find the values of x that make $V = 0$.

b. For each of the values of x in part (a), discuss what happens to the box if squares of length x are cut out.

c. For what values of x does a box exist?

d. Graph the function that gives the volume as a function of the side of the square that is cut out, using an x-interval that makes sense for the problem (see part (c)).

41. *Profit* The profit function for a product is given by $P(x) = -x^3 + 2x^2 + 400x - 400$, where x is the number of units produced and sold and P is in hundreds of dollars. Use factoring by grouping to find the numbers of units that will give a profit of $40,000.

42. *Cost* The total cost function for a product is given by $C(x) = 3x^3 - 6x^2 - 300x + 1800$, where x is the number of units produced and C is the cost in hundreds of dollars. Use factoring by grouping to find the numbers of units that will give a total cost of $120,000.

43. *Ballistics* Ballistics experts are able to identify the weapon that fired a certain bullet by studying the markings on the bullet after it is fired. They test the rifling (the grooves in the bullet as it travels down the barrel of the gun) by comparing it to that of a second bullet fired into a bale of paper. The speed, s, in centimeters

per second, that the bullet travels through the paper is given by $s = 30(3 - 10t)^3$, where t is the time after the bullet strikes the bale and $t \leq 0.3$ second.

a. Complete the following table to find the speed for given values of t.

t	0	0.1	0.2	0.3
s (cm/sec)				

b. Use the root method to find the number of seconds to give $s = 0$. Does this agree with the data in the table?

44. *Federal Tax Per Capita* The amount of federal tax per capita (per person) for selected years from 2000 and projected to 2018 can be modeled by the function $y = 2.481x^3 - 52.25x^2 + 528.7x + 5193$, with x equal to the number of years after 2000. In what year does the model indicate the federal tax per capita will be $41,016?
(Source: Internal Revenue Service)

45. *Civilian Labor Force* The size of the civilian labor force (in millions) for selected years from 1950 and projected to 2050 can be modeled by

$$y = -0.0001x^3 + 0.0088x^2 + 1.43x + 57.9$$

where x is the number of years after 1950.

a. What does this model project the civilian labor force to be in 2015?

b. In what year does the model project the civilian labor force to be 163 million?

46. *Japanese Economy* The percent change y in the Japanese gross domestic product (GDP) for years since 1997 is given by $y = 0.0041x^4 - 0.222x^3 + 4.287x^2 - 34.8398x + 101.33417$, where x is the number of years after 1990.

a. Use graphical or numerical methods with this model to find the year before 1999 when the change in the GDP was approximately 1.3 percentage points.

b. According to this model, would the change in the GDP ever be 1.3% again?
(Source: tradingeconomics.com)

47. *Women in the Labor Force* Using data and projections from 1950 to 2050, the number of women age 16 years and older (in millions) in the U.S. civilian labor force can be modeled by

$$y = -0.00007x^3 + 0.00567x^2 + 0.863x + 16.009$$

where x is the number of years after 1950. Use the model and a graphing utility to estimate when the number of women in the labor force reached 76 million.

48. *Foreign-Born Population* The percent of the U.S. population that was foreign born during the years 1900–2010 is given by the function

$$y = 0.0000384x^3 - 0.00397x^2 - 0.03829x + 14.58102$$

where x is the number of years after 1900. Use graphical methods to determine the year after 1900 when 16% of the population would be foreign born.

49. *Gross Domestic Product* The U.S. gross domestic product (GDP), in trillions of dollars, for selected years from 2005 and projected to 2070 can be modeled by

$$y = 0.00188x^3 - 0.0954x^2 + 2.79x - 2.70,$$

with x equal to the number of years after 2000. In what year does the model indicate the GDP will be $85.48 trillion?
(Source: Bureau of Economic Analysis)

50. *Personal Savings* The following table gives Americans' personal savings rate for selected years from 2000 to 2015.

a. Find a quartic function that models the personal savings rate as a function of the number of years after 1980.

b. In what year would the personal savings rate be 6.0, according to the model?

c. Is the result in part (b) interpolation or extrapolation?

Year	Personal Savings Rate (%)	Year	Personal Savings Rate (%)
2000	3.4	2010	4.9
2002	3.26	2011	3.4
2004	3.23	2012	4.1
2006	3.39	2013	4.3
2008	3.76	2014	5.0
2009	4.03	2015	5.5

(Source: Trading Economics)

6.4 Polynomial Equations Continued; Fundamental Theorem of Algebra

KEY OBJECTIVES

- Divide polynomials with synthetic division
- Solve cubic equations using division
- Combine graphical and algebraic methods to solve polynomial equations
- Solve quartic equations
- Use the Fundamental Theorem of Algebra to find the number of complex solutions of a polynomial equation
- Find the complex solutions of polynomial equations

SECTION PREVIEW   Break-Even

Suppose the profit function for a product is given by

$$P(x) = -x^3 + 98x^2 - 700x - 1800 \text{ dollars}$$

where x is the number of units produced and sold. To find the number of units that gives break-even for the product, we must solve $P(x) = 0$. Solving this equation by factoring would be difficult. Finding all the zeros of $P(x)$ by using a graphical method may also be difficult without knowing an appropriate viewing window. In general, if we seek to find the exact solutions to a polynomial equation, a number of steps are involved if the polynomial is not easily factored.

In this section, we investigate methods of finding additional solutions to polynomial equations after one or more solutions have been found. Specifically, if we know that a is one of the solutions to a polynomial equation, we know that $(x - a)$ is a factor of $P(x)$, and we can divide $P(x)$ by $(x - a)$ to find a second factor of $P(x)$ that may give additional solutions. ■

Division of Polynomials; Synthetic Division

Suppose $f(x)$ is a cubic function and we know that a is a solution of $f(x) = 0$. We can write $(x - a)$ as a factor of $f(x)$, and if we divide this factor into the cubic function, the quotient will be a quadratic factor of $f(x)$. If there are additional real solutions to $f(x) = 0$, this quadratic factor can be used to find the remaining solutions.

Division of a polynomial by a binomial was discussed in the Algebra Toolbox for this chapter. When we divide a polynomial by a binomial of the form

$$x - a$$

the division process can be simplified by using a technique called **synthetic division**. Note in the division of $x^4 + 6x^3 + 5x^2 - 4x + 2$ by $x + 2$ on the left below, many terms are recopied. Because the divisor is linear, the quotient has degree one less than the degree of the dividend, and the degree decreases by one in each remaining term. This is true in general, so we can simplify the division process by omitting all variables and writing only the coefficients. If we rewrite the work without the terms that were recopied in the division and without the x's, the process looks like the one on the right below.

$$
\begin{array}{r}
x^3 + 4x^2 - 3x + 2 \\
x + 2\overline{)x^4 + 6x^3 + 5x^2 - 4x + 2} \\
\underline{x^4 + 2x^3} \\
4x^3 + 5x^2 \\
\underline{4x^3 + 8x^2} \\
-3x^2 - 4x \\
\underline{-3x^2 - 6x} \\
2x + 2 \\
\underline{2x + 4} \\
-2
\end{array}
\qquad
\begin{array}{r}
1 + 4 - 3 + 2 \\
+2\overline{)1 + 6 + 5 - 4 + 2} \\
\underline{2} \\
4 \\
\underline{8} \\
-3 \\
\underline{-6} \\
2 \\
\underline{4} \\
-2
\end{array}
$$

Look at the process on the right above and observe that the coefficient of the first term of the dividend and the differences found in each step are identical to the coefficients of the quotient; the last difference, -2, is the remainder. Because of this, we can compress the division process as follows:

$$
\begin{array}{r}
+2\overline{)1 + 6 + 5 - 4 + 2} \\
\underline{2 + 8 - 6 + 4} \\
1 + 4 - 3 + 2 - 2
\end{array}
$$

Note that the last line in this process gives the coefficients of the quotient, with the last number equal to the remainder. Changing the sign in the divisor allows us to use addition instead of subtraction in the process, as follows:

$$
\begin{array}{r}
-2\overline{)1 + 6 + 5 - 4 + 2} \\
\underline{-2 - 8 + 6 - 4} \\
1 + 4 - 3 + 2 - 2
\end{array}
$$

The bottom line of this process gives the coefficients of the quotient, and the degree of the quotient is one less than the degree of the dividend, so the quotient is

$$x^3 + 4x^2 - 3x + 2, \quad \text{with remainder } -2$$

just as it was in the original division.

The synthetic division process uses the following steps.

Synthetic Division Steps to Divide a Polynomial by $x - a$

To divide a polynomial by $x - a$:

Divide $2x^3 - 9x - 27$ by $x - 3$.

1. Arrange the coefficients in descending powers of x, with a 0 for any missing power. Place a from $x - a$ to the left of the coefficients.

$$3\overline{)2 + 0 - 9 - 27}$$

2. Bring down the first coefficient to the third line. Multiply the last number in the third line by a and write the product in the second line under the next term.

$$3 \overline{)2 + 0 - 9 - 27}$$
$$\underline{ 6 }$$
$$2$$

3. Add the last number in the second line to the number above it in the first line. Continue this process until all numbers in the first line are used.

$$3 \overline{)2 + 0 - 9 - 27}$$
$$\underline{ 6 + 18 + 27}$$
$$2 + 6 + 9 + 0$$

4. The third line represents the coefficients of the quotient, with the last number the remainder. The quotient is a polynomial of degree one less than the dividend.

The quotient is
$$2x^2 + 6x + 9$$

If the remainder is 0, $x - a$ is a factor of the polynomial, and the polynomial can be written as the product of the divisor $x - a$ and the quotient.

The remainder is 0, so $x - 3$ is a factor of $2x^3 - 9x - 27$.
$$2x^3 - 9x - 27$$
$$= (x - 3)(2x^2 + 6x + 9)$$

Using Synthetic Division to Solve Cubic Equations

If we know one solution of a cubic equation $P(x) = 0$, we can divide by the corresponding factor using synthetic division to obtain a second quadratic factor that can be used to find any additional solutions of the polynomial.

EXAMPLE 1 ▶ Solving an Equation with Synthetic Division

If $x = -2$ is a solution of $x^3 + 8x^2 + 21x + 18 = 0$, find the remaining solutions.

SOLUTION

If -2 is a solution of the equation $x^3 + 8x^2 + 21x + 18 = 0$, then $x + 2$ is a factor of

$$f(x) = x^3 + 8x^2 + 21x + 18$$

Synthetically dividing $f(x)$ by $x + 2$ gives

$$-2 \overline{)1 + 8 + 21 + 18}$$
$$\underline{ - 2 - 12 - 18}$$
$$1 + 6 + 9 + 0$$

Thus, the quotient is $x^2 + 6x + 9$, with remainder 0, so $x + 2$ is a factor of $x^3 + 8x^2 + 21x + 18$ and

$$x^3 + 8x^2 + 21x + 18 = (x + 2)(x^2 + 6x + 9)$$

Because $x^2 + 6x + 9 = (x + 3)(x + 3)$, we have

$$x^3 + 8x^2 + 21x + 18 = 0$$
$$(x + 2)(x^2 + 6x + 9) = 0$$
$$(x + 2)(x + 3)(x + 3) = 0$$

and the solutions to the equation are $x = -2$, $x = -3$, and $x = -3$.

Because one of the solutions of this equation, -3, occurs *twice*, we say that it is a double solution, or a solution of **multiplicity** 2.

To check the solutions, we can graph the function $f(x) = x^3 + 8x^2 + 21x + 18$ and observe x-intercepts at -2 and -3 (Figure 6.29). Observe that the graph touches but does not cross the x-axis at $x = -3$, where a *double* solution to the equation (and zero of the function) occurs.

Figure 6.29

A graph touches but does not cross the x-axis at a zero of even multiplicity, and it crosses the x-axis at a zero of odd multiplicity.

Graphs and Solutions

Sometimes a combination of graphical methods and factoring can be used to solve a polynomial equation. If we are seeking exact solutions to a cubic equation and one solution can be found graphically, we can use that solution and division to find the remaining quadratic factor and to find the remaining two solutions. Consider the following example.

EXAMPLE 2 ▶ **Combining Graphical and Algebraic Methods**

Solve the equation $x^3 + 9x^2 - 610x + 600 = 0$.

SOLUTION

If we graph $P(x) = x^3 + 9x^2 - 610x + 600$ in a standard window, shown in Figure 6.30, we see that the graph appears to cross the x-axis at $x = 1$, so one solution to $P(x) = 0$ may be $x = 1$. (Notice that this window does not give a complete graph of the cubic polynomial function.)

Figure 6.30

To verify that $x = 1$ is a solution to $P(x) = 0$, we use synthetic division to divide $x^3 + 9x^2 - 610x + 600$ by $x - 1$.

$$
\begin{array}{r}
1)\overline{1 + 9 - 610 + 600} \\
\underline{1 + 10 - 600} \\
1 + 10 - 600 + 0
\end{array}
$$

The fact that the division results in a 0 remainder verifies that 1 is a solution to the equation $P(x) = 0$. The quotient is $x^2 + 10x - 600$, so the remaining solutions to $P(x) = 0$ can be found by factoring or by using the quadratic formula to solve $x^2 + 10x - 600 = 0$.

$$x^3 + 9x^2 - 610x + 600 = 0$$
$$(x - 1)(x^2 + 10x - 600) = 0$$
$$x - 1 = 0 \quad \text{or} \quad x^2 + 10x - 600 = 0$$
$$(x + 30)(x - 20) = 0$$
$$x = 1 \quad \text{or} \quad x = -30 \quad \text{or} \quad x = 20$$

Graphing the function $P(x) = x^3 + 9x^2 - 610x + 600$ using a viewing window that includes x-values of -30, 1, and 20 verifies that -30, 1, and 20 are zeros of the function and thus solutions to the equation $x^3 + 9x^2 - 610x + 600 = 0$ (Figure 6.31).

Figure 6.31

EXAMPLE 3 ▶   **Break-Even**

The weekly profit for a product is $P(x) = -0.1x^3 + 11x^2 - 80x - 2000$ thousand dollars, where x is the number of thousands of units produced and sold. To find the number of units that gives break-even,

a. Graph the function using a window representing up to 50 thousand units and find one x-intercept of the graph.

b. Use synthetic division to find a quadratic factor of $P(x)$.

c. Find all of the zeros of $P(x)$.

d. Determine the levels of production that give break-even.

SOLUTION

a. Break-even occurs where the profit is 0, so we seek the solution to

$$0 = -0.1x^3 + 11x^2 - 80x - 2000$$

for $x \geq 0$ (because x represents the number of units). Because $x = 50$ represents 50 thousand units and we seek an x-value where $P(x) = 0$, we use the viewing window $[0, 50]$ by $[-2500, 8000]$. The graph of

$$P(x) = -0.1x^3 + 11x^2 - 80x - 2000$$

is shown in Figure 6.32. (Note that this is not a complete graph of the polynomial function.)

Figure 6.32

An x-intercept of this graph appears to be $x = 20$ (see Figure 6.32).

b. To verify that $P(x) = 0$ at $x = 20$ and to determine whether any other x-values give break-even, we divide

$$P(x) = -0.1x^3 + 11x^2 - 80x - 2000$$

by the factor $x - 20$:

$$
\begin{array}{r}
20\overline{)\ -0.1 + 11 -\ \ 80 - 2000} \\
-\ \ 2 + 180 + 2000 \\
\hline
-0.1 +\ \ 9 + 100 +\ \ \ \ 0
\end{array}
$$

Thus, $P(x) = 0$ at $x = 20$ and the quotient is the quadratic factor $-0.1x^2 + 9x + 100$, so

$$-0.1x^3 + 11x^2 - 80x - 2000 = (x - 20)(-0.1x^2 + 9x + 100)$$

c. The solution to $P(x) = 0$ follows.

$$-0.1x^3 + 11x^2 - 80x - 2000 = 0$$
$$(x - 20)(-0.1x^2 + 9x + 100) = 0$$
$$(x - 20)[-0.1(x^2 - 90x - 1000)] = 0$$
$$-0.1(x - 20)(x - 100)(x + 10) = 0$$

$$x - 20 = 0 \quad \text{or} \quad x - 100 = 0 \quad \text{or} \quad x + 10 = 0$$
$$x = 20 \quad \text{or} \quad x = 100 \quad \text{or} \quad x = -10$$

Thus, the zeros are 20, 100, and -10. Figure 6.33 shows the graph of $P(x) = -0.1x^3 + 11x^2 - 80x - 2000$ on a window that shows all three x-intercepts.

Figure 6.33

d. The solutions indicate that break-even occurs at $x = 20$ and at $x = 100$. The value $x = -10$ also makes $P(x) = 0$, but a negative number of units does not make sense in the context of this problem. Thus, producing and selling 20,000 or 100,000 units gives break-even.

Combining graphical methods and synthetic division, as we did in Example 3, is especially useful in finding exact solutions to polynomial equations. If some of the solutions are irrational or nonreal solutions, the quadratic formula can be used rather than factoring. Remember that the quadratic formula can always be used to solve a quadratic equation, and factoring cannot always be used.

Rational Solutions Test

The Rational Solutions Test provides information about the rational solutions of a polynomial equation with integer coefficients.

> ### Rational Solutions Test
>
> The rational solutions of the polynomial equation
>
> $$a_n x^n + a_{n-1} x^{n-1} + \cdots + a_1 x + a_0 = 0$$
>
> with integer coefficients must be of the form $\dfrac{p}{q}$, where p is a factor of the constant term a_0 and q is a factor of a_n, the leading coefficient.

Recall that if a is a solution to $f(x) = 0$, then a is also a zero of the function $f(x)$. The Rational Solutions Test provides a list of *possible* rational zeros of a polynomial function. For a polynomial function with integer coefficients, if there is a rational zero, then it will be in this list of possible rational zeros. Keep in mind that there may be *no* rational zeros of a polynomial function. Graphing a polynomial function $f(x)$ will give us a better sense of the location of the zeros, and we can also use our list of possible zeros to find an appropriate viewing window for the function.

For example, the rational solutions of $x^3 + 8x^2 + 21x + 18 = 0$ must be numbers that are factors of 18 divided by factors of 1. Thus, the possible rational solutions are ± 1, ± 18, ± 2, ± 9, ± 3, and ± 6. A viewing window containing x-values from -18 to 18 will show all possible rational solutions. Note that the solutions to this equation, found in Example 1 to be -2 and -3, are contained in this list.

Many cubic and quartic equations can be solved by using the following steps.

> ### Solving Cubic and Quartic Equations of the Form $f(x) = 0$
>
> 1. Determine the possible rational solutions of $f(x) = 0$.
>
> 2. Graph $y = f(x)$ to see if any of the values from Step 1 are x-intercepts. Those values that are x-intercepts are rational solutions to $f(x) = 0$.
>
> 3. Find the factors associated with the x-intercepts from Step 2.
>
> 4. Use synthetic division to divide $f(x)$ by the factors from Step 3 to confirm the graphical solutions and find additional factors. Continue until a quadratic factor remains.
>
> 5. Use factoring or the quadratic formula to find the solutions associated with the quadratic factor. These solutions are also solutions to $f(x) = 0$.

Example 4 shows how the Rational Solutions Test is used when the leading coefficient of the polynomial is not 1.

EXAMPLE 4 ▶ **Solving a Quartic Equation**

Solve the equation $2x^4 + 10x^3 + 13x^2 - x - 6 = 0$.

SOLUTION

The rational solutions of this equation must be factors of the constant -6 divided by factors of 2, the leading coefficient. Thus, the possible rational solutions are

$$\pm 1, \ \pm 2, \ \pm 3, \ \pm 6 \quad \text{and} \quad \pm \frac{1}{2}, \ \pm \frac{2}{2}, \ \pm \frac{3}{2}, \ \pm \frac{6}{2}*$$

All of these possible rational solutions are between -6 and 6. Using a window with x-values from -4 to 4, Figure 6.34 shows that the graph of

$$y = P(x) = 2x^4 + 10x^3 + 13x^2 - x - 6$$

appears to cross the x-axis at $x = -1$ and at $x = -2$, so two of the solutions to $y = 0$ appear to be $x = -1$ and $x = -2$. We will use synthetic division to confirm that these values are solutions to the equation and to find the two remaining solutions. (We know there are two additional solutions, which may be irrational, because the graph crosses the x-axis at two other points.)

$$y = 2x^4 + 10x^3 + 13x^2 - x - 6$$

Figure 6.34

We can confirm that $x = -1$ is a solution and find the remaining factor of $P(x)$ by using synthetic division to divide $2x^4 + 10x^3 + 13x^2 - x - 6$ by $x + 1$.

$$
\begin{array}{r}
-1)\overline{2 + 10 + 13 - 1 - 6} \\
\underline{-\ 2 -\ 8 - 5 + 6} \\
2 +\ 8 +\ 5 - 6 + 0
\end{array}
$$

Thus, we have $2x^4 + 10x^3 + 13x^2 - x - 6 = (x + 1)(2x^3 + 8x^2 + 5x - 6)$.

We can also confirm that $x = -2$ is a solution of $P(x)$ by showing that $x + 2$ is a factor of $2x^3 + 8x^2 + 5x - 6$ and thus a factor of $P(x)$. Dividing $x + 2$ into $2x^3 + 8x^2 + 5x - 6$ using synthetic division gives a quadratic factor of $P(x)$:

$$
\begin{array}{r}
-2)\overline{2 + 8 + 5 - 6} \\
\underline{-\ 4 - 8 + 6} \\
2 + 4 - 3 + 0
\end{array}
$$

So the quotient is the quadratic factor $2x^2 + 4x - 3$, and $P(x)$ has the following factorization:

$$2x^4 + 10x^3 + 13x^2 - x - 6 = (x + 1)(x + 2)(2x^2 + 4x - 3)$$

Thus, $P(x) = 0$ is equivalent to $(x + 1)(x + 2)(2x^2 + 4x - 3) = 0$, so the solutions can be found by solving $x + 1 = 0$, $x + 2 = 0$, and $2x^2 + 4x - 3 = 0$.

The two remaining x-intercepts of the graph of $y = 2x^2 + 4x - 3$ do not appear to cross the x-axis at integer values, and $2x^2 + 4x - 3$ does not appear to be factorable.

---

* Some of these possible solutions are duplicates.

Using the quadratic formula is the obvious method to solve $2x^2 + 4x - 3 = 0$ to find the two remaining solutions:

$$x = \frac{-4 \pm \sqrt{16 - 4(2)(-3)}}{2(2)} = \frac{-4 \pm \sqrt{40}}{4} = \frac{-2 \pm \sqrt{10}}{2}$$

So the solutions are $-1, -2, \dfrac{-2 + \sqrt{10}}{2}$, and $\dfrac{-2 - \sqrt{10}}{2}$. Note that two of these solutions are rational and two are irrational, approximated by 0.58114 and -2.58114. The two irrational solutions can also be verified graphically (Figure 6.35).

Figure 6.35

Note that some graphing utilities and spreadsheets have commands that can be used to find or to approximate solutions to polynomial equations.

Fundamental Theorem of Algebra

As we saw in Section 3.2, not all quadratic equations have real solutions. However, every quadratic equation has complex solutions. If the complex number a is a solution to $P(x) = 0$, then a is a complex zero of P. The **Fundamental Theorem of Algebra** states that every polynomial function has a complex zero.

> ### Fundamental Theorem of Algebra
>
> If $f(x)$ is a polynomial function of degree $n \geq 1$, then f has at least one complex zero.

As a result of this theorem, it is possible to prove the following.

> ### Complex Zeros
>
> Every polynomial function $f(x)$ of degree $n \geq 1$ has exactly n complex zeros. Some of these zeros may be imaginary, and some may be repeated. Nonreal zeros occur in conjugate pairs, $a + bi$ and $a - bi$.

EXAMPLE 5 ▶   **Solution in the Complex Number System**

Find the complex solutions to $x^3 + 2x^2 - 3 = 0$.

SOLUTION

Because the function is cubic, there are three solutions, and because any nonreal solutions occur in conjugate pairs, at least one of the solutions must be real. To estimate that solution, we graph

$$y = x^3 + 2x^2 - 3$$

on a window that includes the possible rational solutions, which are ± 1 and ± 3 (see Figure 6.36).

Figure 6.36

The graph appears to cross the x-axis at 1, so we (synthetically) divide $x^3 + 2x^2 - 3$ by $x - 1$:

$$\begin{array}{r} 1)\overline{1 + 2 + 0 - 3} \\ 1 + 3 + 3 \\ \hline 1 + 3 + 3 + 0 \end{array}$$

The division confirms that 1 is a solution to $x^3 + 2x^2 - 3 = 0$ and shows that

$$x^2 + 3x + 3$$

is a factor of $x^3 + 2x^2 - 3$. Thus, any remaining solutions to the equation satisfy

$$x^2 + 3x + 3 = 0$$

Using the quadratic formula gives the remaining solutions:

$$x = \frac{-3 \pm \sqrt{3^2 - 4(1)(3)}}{2(1)} = \frac{-3 \pm \sqrt{-3}}{2} = \frac{-3 \pm i\sqrt{3}}{2}$$

Thus, the solutions are 1, $-\dfrac{3}{2} + \dfrac{\sqrt{3}}{2}i$, and $-\dfrac{3}{2} - \dfrac{\sqrt{3}}{2}i$. Because there can only be three solutions to a cubic equation, we have all of the solutions.

SKILLS CHECK 6.4

In Exercises 1–4, use synthetic division to find the quotient and the remainder.

1. $(x^4 - 4x^3 + 3x + 10) \div (x - 3)$

2. $(x^4 + 2x^3 - 3x^2 + 1) \div (x + 4)$

3. $(2x^4 - 3x^3 + x - 7) \div (x - 1)$

4. $(x^4 - 1) \div (x + 1)$

In Exercises 5 and 6, determine whether the given constant is a solution to the given polynomial equation.

5. $2x^4 - 4x^3 + 3x + 18 = 0$; 3

6. $x^4 + 3x^3 - 10x^2 + 8x + 40 = 0$; -5

In Exercises 7 and 8, determine whether the second polynomial is a factor of the first polynomial.

7. $(-x^4 - 9x^2 + 3x)$; $(x + 3)$

8. $(2x^4 + 5x^3 - 6x - 4)$; $(x + 2)$

In Exercises 9–12, one or more solutions of a polynomial equation are given. Use synthetic division to find any remaining solutions.

9. $-x^3 + x^2 + x - 1 = 0$; -1

10. $x^3 + 4x^2 - x - 4 = 0$; 1

11. $x^4 + 2x^3 - 21x^2 - 22x = -40$; $-5, 1$

12. $2x^4 - 17x^3 + 51x^2 - 63x + 27 = 0;\ 3,\ 1$

In Exercises 13–16, find one solution graphically and then find the remaining solutions using synthetic division.

13. $x^3 + 3x^2 - 18x - 40 = 0$

14. $x^3 - 3x^2 - 9x - 5 = 0$

15. $3x^3 + 2x^2 - 7x + 2 = 0$

16. $4x^3 + x^2 - 27x + 18 = 0$

In Exercises 17–20, determine all possible rational solutions of the polynomial equation.

17. $x^3 - 6x^2 + 5x + 12 = 0$

18. $4x^3 + 3x^2 - 9x + 2 = 0$

19. $9x^3 + 18x^2 + 5x - 4 = 0$

20. $6x^4 - x^3 - 42x^2 - 29x + 6 = 0$

In Exercises 21–24, find all rational zeros of the polynomial function.

21. $f(x) = x^3 - 6x^2 + 5x + 12$

22. $f(x) = 4x^3 + 3x^2 - 9x + 2$

23. $g(x) = 9x^3 + 18x^2 + 5x - 4$

24. $g(x) = 6x^3 + 19x^2 - 19x + 4$

Solve each of the equations in Exercises 25–30 exactly in the complex number system. (Find one solution graphically and then use the quadratic formula.)

25. $x^3 = 10x - 7x^2$

26. $t^3 - 2t^2 + 3t = 0$

27. $w^3 - 5w^2 + 6w - 2 = 0$

28. $2w^3 + 3w^2 + 3w + 2 = 0$

29. $z^3 - 8 = 0$ **30.** $x^3 + 1 = 0$

EXERCISES 6.4

In Exercises 31–36, use synthetic division and factoring to solve the problems.

31. *Break-Even* The profit function, in dollars, for a product is given by $P(x) = -0.2x^3 + 66x^2 - 1600x - 60{,}000$, where x is the number of units produced and sold. If break-even occurs when 50 units are produced and sold,

   a. Use synthetic division to find a quadratic factor of $P(x)$.

   b. Use factoring to find a number of units other than 50 that gives break-even for the product, and verify your answer graphically.

32. *Break-Even* The profit function, in dollars, for a product is given by $P(x) = -x^3 + 98x^2 - 700x - 1800$, where x is the number of units produced and sold. If break-even occurs when 10 units are produced and sold,

   a. Use synthetic division to find a quadratic factor of $P(x)$.

   b. Find a number of units other than 10 that gives break-even for the product, and verify your answer graphically.

33. *Break-Even* The weekly profit for a product, in thousand of dollars, is $P(x) = -0.1x^3 + 50.7x^2 - 349.2x - 400$, where x is the number of thousands of units produced and sold. To find the number of units that gives break-even,

   a. Graph the function on the window $[-8, 10]$ by $[-60, 100]$.

   b. Graphically find one x-intercept of the graph.

   c. Use synthetic division to find a quadratic factor of $P(x)$.

   d. Find all of the zeros of $P(x)$.

   e. Determine the levels of production and sale that give break-even.

34. *Break-Even* The weekly profit for a product, in thousands of dollars, is $P(x) = -0.1x^3 + 10.9x^2 - 97.9x - 108.9$, where x is the number of thousands of units produced and sold. To find the number of units that gives break-even,

   a. Graph the function on the window $[-10, 20]$ by $[-100, 100]$.

   b. Graphically find one x-intercept of the graph.

   c. Use synthetic division to find a quadratic factor of $P(x)$.

   d. Find all of the zeros of $P(x)$.

   e. Determine the levels of production and sale that give break-even.

35. Revenue The revenue from the sale of a product is given by $R = 1810x - 81x^2 - x^3$. If the sale of 9 units gives a total revenue of $9000, use synthetic division to find another number of units that will give $9000 in revenue.

36. Revenue The revenue from the sale of a product is given by $R = 250x - 5x^2 - x^3$. If the sale of 5 units gives a total revenue of $1000, use synthetic division to find another number of units that will give $1000 in revenue.

37. Drunk Driving Crashes Suppose the total number of fatalities in drunk driving crashes in South Carolina is modeled by the function $y = 0.4566x^3 - 14.3085x^2 + 117.2978x + 107.8456$, where x is the number of years after 1980. This model indicates that the number of fatalities was 244 in 1992. To find any other years that the number was 244,

 a. Set the function equal to 244 and rewrite the resulting equation with 0 on one side.

 b. Graph the function from part (a) on the window $[-10, 25]$ by $[-75, 200]$ and use the graph to find an integer solution to this equation.

 c. Use synthetic division to find a quadratic factor.

 d. Use the quadratic formula to find any other solutions to this equation.

 e. Use this information to determine the years in which the number of fatalities in drunk driving crashes was 244.
 (Source: Anheuser-Busch, www.beeresponsible.com)

38. College Enrollment The number of college enrollments, in thousands, by recent high school graduates is given by $f(x) = 0.20x^3 - 13.71x^2 + 265.06x + 1612.56$, where x is the number of years after 1960. To find any years after 1960 when the number of enrollments was 2862 thousand,

 a. Set the function equal to 2862, and rewrite the resulting equation with 0 on one side.

 b. Set the nonzero side of the equation in part (a) equal to y and graph this function on the window $[25, 50]$ by $[-50, 50]$. Use the graph to find an integer solution to this equation.

 c. Use synthetic division to find a quadratic factor.

 d. Use the quadratic formula to find any other solutions to this equation.

 e. Use this information to determine the years in which the number of enrollments was 2,862,000.
 (Source: U.S. Census Bureau)

39. Per Capita Health Care Costs For the years from 2000 and projected to 2018, the U.S. per capita out-of-pocket cost for health care C (in dollars) can be modeled by the function

$$C(t) = 0.118t^3 - 2.51t^2 + 40.2t + 677$$

where t is the number of years after 2000.

 a. Verify that one solution to $946 = 0.118t^3 - 2.51t^2 + 40.2t + 677$ is $t = 10$. Interpret this result.

 b. Use the quadratic formula to determine if the model gives any other years when the U.S. per capita out-of-pocket cost for healthcare is estimated to be $946.
 (Source: U.S. Centers for Medicare)

40. Births The number of births to females in the United States under 15 years of age can be modeled by the function $y = -0.0014x^3 + 0.1025x^2 - 2.2687x + 25.5704$, where y is the number of births in thousands and x is the number of years after 1970. The model indicates that 10,808 births occurred in 1994; find two other years when the model indicates that 10,808 births occurred.
 (Source: guttmacher.org)

41. Births The number of births to females in the United States between 15 and 17 years of age can be modeled by the cubic function $y = 0.0790x^3 - 2.103x^2 + 10.695x + 179.504$, where y is the number of thousands of births and x is the number of years after 1990. The model indicates that 155,154 births occurred in 2000; find another year after 1990 when the model indicates that 155,154 births occurred.
 (Source: Child Trends, Inc., MSNBC)

42. Personal Savings The model $y = 0.000447x^3 - 0.0361x^2 + 0.615x + 7.966$, where $x = 0$ in 1960, gives Americans' personal savings rates for selected years from 1960 to 2009. In what year after 1970 is the personal savings rate 9.4, according to the model?

SECTION PREVIEW   Average Cost

The function that gives the daily average cost (in hundreds of dollars) for the production of Stanley golf carts is formed by dividing the total cost function for the production of the golf carts

$$C(x) = 25 + 13x + x^2$$

by x, the number of golf carts produced. Thus, the average cost per golf cart is given by

$$\overline{C}(x) = \frac{25 + 13x + x^2}{x} \qquad (x > 0)$$

where x is the number of golf carts produced. The graph of this function with x restricted to positive values is shown in Figure 6.37(a). The graph of the function

$$\overline{C}(x) = \frac{25 + 13x + x^2}{x}$$

with x not restricted is shown in Figure 6.37(b). (See Example 3.)

    This function, as well as any other function that is formed by taking the quotient of two polynomials, is called a **rational function**. In this section, we discuss the characteristics of the graphs of rational functions.

(a)   **(b)**

Figure 6.37

Graphs of Rational Functions

A rational function is defined as follows.

> ### Rational function
>
> The function f is a rational function if
>
> $$f(x) = \frac{P(x)}{Q(x)}$$
>
> where $P(x)$ and $Q(x)$ are polynomials and $Q(x) \neq 0$.

Figure 6.38

In Section 3.3, we discussed the simple rational function $y = \dfrac{1}{x}$, whose graph is shown in Figure 6.38. We now expand this discussion.

Because division by 0 is not possible, those real values of x for which $Q(x) = 0$ are not in the domain of the rational function $f(x) = \dfrac{P(x)}{Q(x)}$. Observe that the graph of $y = \dfrac{1}{x}$ approaches, but does not touch, the y-axis. Thus, the y-axis is a **vertical asymptote**. The graph of a rational function frequently has vertical asymptotes.

To find where the vertical asymptote(s) occur, set the denominator equal to 0 and solve for x. If any of these values of x do not also make the numerator equal to 0, then a vertical asymptote occurs at those values.

> ### Vertical Asymptote
>
> A vertical asymptote occurs in the graph of $f(x) = \dfrac{P(x)}{Q(x)}$ at those values of x where $Q(x) = 0$ and $P(x) \neq 0$—that is, at values of x where the denominator equals 0 but the numerator does not equal 0.

EXAMPLE 1 ▶ Rational Function

Find the vertical asymptote and sketch the graph of $y = \dfrac{x + 1}{(x + 2)^2}$.

Figure 6.39

SOLUTION

Setting the denominator equal to 0, taking the square root of both sides of the equation, and solving gives

$$(x + 2)^2 = 0$$
$$(x + 2) = 0$$
$$x = -2$$

The value $x = -2$ makes the denominator of the function equal to 0 and does not make the numerator equal to 0, so a vertical asymptote occurs at $x = -2$. The graph of the function $y = \dfrac{x + 1}{(x + 2)^2}$ is shown in Figure 6.39.

EXAMPLE 2 ▶ Cost-Benefit

Suppose that for specified values of p, the function

$$C(p) = \frac{800p}{100 - p}$$

can be used to model the cost of removing $p\%$ of the particulate pollution from the exhaust gases at an industrial site.

a. Graph this function on the window $[-100, 200]$ by $[-4000, 4000]$.

b. Does the graph of this function have a vertical asymptote on this window? Where?

c. For what values of p does this function serve as a model for the cost of removing particulate pollution?

d. Use the information determined in part (c) to graph the model.

e. What does the part of the graph near the vertical asymptote tell us about the cost of removing particulate pollution?

SOLUTION

a. The graph is shown in Figure 6.40(a).

b. The denominator is equal to 0 at $p = 100$, and the numerator does not equal 0 at $p = 100$, so a vertical asymptote occurs at $p = 100$.

c. Because p represents the percent of pollution, p must be limited to values from 0 to 100. The domain cannot include 100, so the p-interval is $[0, 100)$.

d. The graph of the function on the window [0, 100] by $[-1000, 10,000]$ is shown in Figure 6.40(b).

(a)   (b)

Figure 6.40

e. From Figure 6.40(b), we see that the graph is increasing rapidly as it approaches the vertical asymptote at $p = 100$, which tells us that the cost of removing pollution gets extremely high as the amount of pollution removed approaches 100%. It is impossible to remove 100% of the particulate pollution.

Figure 6.41

We see in Figure 6.41 that the graph of

$$y = \frac{3}{x + 1}$$

approaches the x-axis asymptotically on the left and on the right. Because the x-axis is a horizontal line, we say that the x-axis is a **horizontal asymptote** for this graph.

We can study the graph of a rational function to determine if the curve approaches a horizontal asymptote. If the graph approaches the horizontal line $y = a$ as $|x|$ gets very large, the graph has a horizontal asymptote at $y = a$. We can denote this as follows.

> ### Horizontal Asymptote
>
> If y approaches a as x approaches $+\infty$ or as x approaches $-\infty$, the graph of $y = f(x)$ has a horizontal asymptote at $y = a$.

We can also compare the degrees of the numerator and denominator of the rational function to determine if the curve approaches a horizontal asymptote, and to find the horizontal asymptote if it exists.

> ### Determining the Horizontal Asymptotes of a Rational Function
>
> Consider the rational function
>
> $$f(x) = \frac{P(x)}{Q(x)} = \frac{a_n x^n + \cdots + a_1 x + a_0}{b_m x^m + \cdots + b_1 x + b_0}, \quad a_n \neq 0 \quad \text{and} \quad b_m \neq 0$$

1. If $n < m$ (that is, if the degree of the numerator is less than the degree of the denominator), a horizontal asymptote occurs at $y = 0$ (the x-axis).

2. If $n = m$ (that is, if the degree of the numerator is equal to the degree of the denominator), a horizontal asymptote occurs at $y = \dfrac{a_n}{b_m}$. (This is the ratio of the leading coefficients.)

3. If $n > m$ (that is, if the degree of the numerator is greater than the degree of the denominator), there is no horizontal asymptote.

In the function $y = \dfrac{3}{x + 1}$, the degree of the numerator is less than the degree of the denominator, so a horizontal asymptote occurs at $y = 0$, the x-axis (Figure 6.41).

In the function $C(p) = \dfrac{800p}{100 - p}$ of Example 2, the degrees of the numerator and the denominator are equal, so the graph has a horizontal asymptote at $y = C(p) = \dfrac{800}{-1} = -800$. Observing the graph of the function, shown in Figure 6.40(a) on the previous page, we see that this horizontal asymptote is reasonable for the graph of $C(p) = \dfrac{800p}{100 - p}$.

Although it is impossible for a graph to cross a vertical asymptote (because the function is undefined at that value of x), the curve *may* cross a horizontal asymptote because the horizontal asymptote describes the end behavior (as x approaches ∞ or x approaches $-\infty$) of a graph. Figure 6.42 shows the graph of $y = \dfrac{6 - 5x}{x^2}$; it crosses the x-axis but approaches the line $y = 0$ (the x-axis) as x approaches ∞ and as x approaches $-\infty$.

Figure 6.42

EXAMPLE 3 ▶ Average Cost

The function $\overline{C}(x) = \dfrac{25 + 13x + x^2}{x}$ represents the daily average cost (in hundreds of dollars) for the production of Stanley golf carts, with x equal to the number of golf carts produced.

a. Graph the function on the window $[-20, 20]$ by $[-30, 50]$.

b. Does the graph in part (a) have a horizontal asymptote?

c. Graph the function on the window $[0, 20]$ by $[0, 50]$.

d. Does the graph of the function using the window in part (a) or part (b) better model the average cost function? Why?

e. Use technology to find the minimum daily average cost and the number of golf carts that gives the minimum daily average cost.

SOLUTION

a. The graph on the window $[-20, 20]$ by $[-30, 50]$ is shown in Figure 6.43(a). Note that the graph has a vertical asymptote at $x = 0$ (the $\overline{C}(x)$-axis).

b. The graph in part (a) does not have a horizontal asymptote, because the degree of the numerator is greater than the degree of the denominator.

c. The graph on the window $[0, 20]$ by $[0, 50]$ is shown in Figure 6.43(b).

(a)   (b)

Figure 6.43

Figure 6.44

d. Because the number of golf carts produced cannot be negative, the window used in part (b) gives a better representation of the graph of the average cost function.

e. It appears that the graph reaches a low point somewhere between $x = 4$ and $x = 6$. By using technology, we can find the minimum point. Figure 6.44 shows that the minimum value of y is 23 at $x = 5$. Thus, the minimum daily average cost is $2300 per golf cart when 5 golf carts are produced.

Slant Asymptotes and Missing Points

The average cost function from Example 3,

$$\overline{C}(x) = \frac{25 + 13x + x^2}{x}$$

does not have a horizontal asymptote because the degree of the numerator is greater than the degree of the denominator. However, if the degree of the numerator of a rational function is *one* more than the degree of the denominator, its graph approaches a **slant asymptote**, which we can find by using the following method:

1. Divide the numerator by the denominator, getting a linear function plus a rational expression with a constant numerator (the remainder).

2. Observe that the rational expression found in Step 1 above will approach 0 as x approaches ∞ or $-\infty$, so the graph of the original function will have a slant asymptote, which is the graph of the linear function.

The average cost function $\overline{C}(x) = \dfrac{25 + 13x + x^2}{x}$ can be written as

$$\overline{C}(x) = \frac{25}{x} + 13 + x$$

so its graph approaches the graph of the slant asymptote $y = 13 + x$ (Figure 6.45).

Figure 6.45

EXAMPLE 4 ▶ **Slant Asymptote**

Find the equations of the vertical, horizontal, and slant asymptotes (if any) for the graph of the function

$$f(x) = \frac{x^2 + 4}{x - 2}$$

Verify your results graphically.

SOLUTION

The denominator of the rational function is equal to 0 at $x = 2$, and the numerator does not equal 0 at $x = 2$, so there is a vertical asymptote at $x = 2$. The degree of the numerator of the rational function, 2, is greater than the degree of the denominator, 1, so there is no horizontal asymptote. Because the degree of the numerator is exactly one more than the degree of the denominator, there is a slant asymptote. To find the equation of the slant asymptote, we synthetically divide the numerator by the denominator:

$$\begin{array}{r} 2)\overline{1 \quad 0 \quad 4} \\ \underline{2 \quad 4} \\ 1 \quad 2 \quad 8 \end{array}$$

Thus, when $x^2 + 4$ is divided by $x - 2$, the quotient is $x + 2$ and the remainder is 8. So the equation of the slant asymptote is $y = x + 2$. Figure 6.46 shows the graph of

$$f(x) = \frac{x^2 + 4}{x - 2}$$ and its asymptotes.

Figure 6.46

There is another type of rational function that is undefined at a value of x but does not have a vertical asymptote at that value. Note that if there is a value a such that $Q(a) = 0$ and $P(a) = 0$, then the function

$$f(x) = \frac{P(x)}{Q(x)}$$

is undefined at $x = a$, but its graph may have a "hole" in it at $x = a$ rather than an asymptote. Figure 6.47 shows the graph of such a function,

$$y = \frac{x^2 - 4}{x - 2}$$

Figure 6.47

Observe that the function is undefined at $x = 2$, but the graph does not have an asymptote at this value of x. Notice that for all values of x except 2,

$$y = \frac{x^2 - 4}{x - 2} = \frac{(x + 2)(x - 2)}{x - 2} = x + 2$$

Thus, the graph of this function looks like the graph of $y = x + 2$, except that it has a missing point (hole) at $x = 2$ (Figure 6.47).

Algebraic and Graphical Solution of Rational Equations

To solve an equation involving rational expressions algebraically, we multiply both sides of the equation by the least common denominator (LCD) of the fractions in the equation. This will result in a linear or polynomial equation which can be solved by using the methods discussed earlier in the text. For these types of equations it is essential that all solutions be checked in the original equation, because some solutions to the resulting polynomial equation may not be solutions to the original equation. In particular, some solutions to the polynomial equation may result in a zero in the denominator of the original equation and thus cannot be solutions to the equation. Such solutions are called **extraneous solutions**.

The steps used to solve a rational equation follow.

Solving a Rational Equation Algebraically

To solve a rational equation algebraically:

1. Multiply both sides of the equation by the LCD of the fractions in the equation.

2. Solve the resulting polynomial equation for the variable.

3. Check each solution in the original equation. Some solutions to the polynomial equation may not be solutions to the original rational equation (these are called extraneous solutions).

EXAMPLE 5 ▶   **Solving a Rational Equation**

Solve the equation $x^2 + \dfrac{x}{x-1} = x + \dfrac{x^3}{x-1}$ for x.

SOLUTION

We first multiply both sides of the equation by $x - 1$, the LCD of the fractions in the equation, and then we solve the resulting polynomial equation:

$$x^2 + \frac{x}{x-1} = x + \frac{x^3}{x-1}$$

$$(x-1)\left(x^2 + \frac{x}{x-1}\right) = (x-1)\left(x + \frac{x^3}{x-1}\right)$$

$$(x-1)x^2 + (x-1)\frac{x}{x-1} = (x-1)x + (x-1)\frac{x^3}{x-1}$$

$$x^3 - x^2 + x = x^2 - x + x^3$$

$$0 = 2x^2 - 2x$$

$$0 = 2x(x-1)$$

$$x = 0 \quad \text{or} \quad x = 1$$

We must check these answers in the original equation.

Check $x = 0$:

$$0^2 + \frac{0}{0-1} = 0 + \frac{0^3}{0-1} \quad \text{checks}$$

Check $x = 1$:

$$1^2 + \frac{1}{1-1} = 1 + \frac{1^3}{1-1}, \quad \text{or} \quad 1 + \frac{1}{0} = 1 + \frac{1}{0}$$

But $1 + \dfrac{1}{0}$ cannot be evaluated, so $x = 1$ does not check in the original equation. This means that 1 is an extraneous solution of the equation. Thus, the only solution is $x = 0$.

As with polynomial equations derived from real data, some equations involving rational expressions may require technology to find or to approximate their solutions. The following example shows how applied problems involving rational functions can be solved algebraically and graphically.

EXAMPLE 6 ▶   **Advertising and Sales**

Monthly sales y (in thousands of dollars) for Yang products are related to monthly advertising expenses x (in thousands of dollars) according to the function

$$y = \frac{300x}{15 + x}$$

Determine the amount of money that must be spent on advertising to generate $100,000 in sales

a. Algebraically.

b. Graphically.

SOLUTION

a. To solve the equation $100 = \dfrac{300x}{15 + x}$ algebraically, we multiply both sides of the equation by $15 + x$, getting

$$100\,(15 + x) = 300x$$

Solving this equation gives

$$1500 + 100x = 300x$$
$$1500 = 200x$$
$$x = 7.5$$

Checking this solution in the original equation gives

$$100 = \frac{300\,(7.5)}{15 + 7.5}, \quad \text{or} \quad 100 = \frac{2250}{22.5}$$

so the solution checks. Thus, spending 7.5 thousand dollars ($7500) per month on advertising results in monthly sales of $100,000.

b. To solve the equation graphically, we graph the rational functions $y_1 = \dfrac{300x}{15 + x}$ and $y_2 = 100$ on the same axes on an interval with $x \geq 0$ (Figure 6.48). We then find the point of intersection $(7.5, 100)$, which gives the same solution as in part (a).

Figure 6.48

Inverse Variation

Recall from Section 2.1 that two variables are directly proportional (or vary directly) if their quotient is constant. When two variables x and y are **inversely proportional** (or vary inversely), an increase in one variable results in a decrease in the other. That is, y is inversely proportional to x (or y varies inversely as x) if there exists a nonzero number k such that

$$y = \frac{k}{x}$$

Also, y is inversely proportional to the nth power of x if there exists a nonzero number k such that

$$y = \frac{k}{x^n}$$

We can also say that y varies inversely as the nth power of x.

EXAMPLE 7 ▶ Illumination

The illumination produced by a light varies inversely as the square of the distance from the source of the light. If the illumination 30 feet from a light source is 60 candela, what is the illumination 20 feet from the source?

SOLUTION

If L represents the illumination and d represents the distance, the relation is

$$L = \frac{k}{d^2}$$

Substituting for L and d and solving for k gives

$$60 = \frac{k}{30^2}, \text{ or } k = 54{,}000$$

Thus, the relation is $L = \dfrac{54{,}000}{d^2}$, and when $d = 20$ feet,

$$L = \frac{54{,}000}{20^2} = 135 \text{ candela}$$

So the illumination 20 feet from the source of the light is 135 candela.

SKILLS CHECK 6.5

Give the equations of any (a) vertical and (b) horizontal asymptotes for the graphs of the rational functions $y = f(x)$ in Exercises 1–6.

1. $f(x) = \dfrac{3}{x - 5}$

2. $f(x) = \dfrac{7}{x - 4}$

3. $f(x) = \dfrac{x - 4}{5 - 2x}$

4. $f(x) = \dfrac{2x - 5}{3 - x}$

5. $f(x) = \dfrac{x^3 + 4}{x^2 - 1}$

6. $f(x) = \dfrac{x^2 + 6}{x^2 + 3}$

7. Which of the following functions has a graph that does not have a vertical asymptote? Why?

a. $f(x) = \dfrac{x - 3}{x^2 - 4}$

b. $f(x) = \dfrac{3}{x^2 - 4x}$

c. $f(x) = \dfrac{3x^2 - 6}{x^2 + 6}$

d. $f(x) = \dfrac{x + 4}{(x - 6)(x + 2)}$

8. Which of the following functions has a graph that does not have a horizontal asymptote? Why?

a. $f(x) = \dfrac{2x + 3}{x - 4}$

b. $f(x) = \dfrac{5x}{x^2 - 16}$

c. $f(x) = \dfrac{x^3}{3x^2 + 2}$

d. $f(x) = \dfrac{5}{(x + 2)(x - 3)}$

In Exercises 9–14, match each function with its graph below.

9. $y = \dfrac{3}{x - 1}$

10. $y = \dfrac{x + 1}{x - 2}$

11. $y = \dfrac{x - 3}{x^2 - 2x - 8}$

12. $y = \dfrac{3}{x^2 - 4}$

13. $y = \dfrac{-3x^2 + 2x}{x - 3}$

14. $y = \dfrac{2x^2 - 3}{x - 4}$

A.

B.

C.

D.

E.

F.

In Exercises 15–18, (a) find the horizontal asymptotes, (b) find the vertical asymptotes, and (c) sketch a graph of the function.

15. $f(x) = \dfrac{x + 1}{x - 2}$

16. $f(x) = \dfrac{5x}{x - 3}$

17. $f(x) = \dfrac{5 + 2x}{1 - x^2}$

18. $f(x) = \dfrac{2x^2 + 1}{2 - x}$

19. Graph the function $y = \dfrac{x^2 - 9}{x - 3}$. What happens at $x = 3$?

20. Graph the function $y = \dfrac{x^2 - 16}{x + 4}$. What happens at $x = -4$?

In Exercises 21–24, find the equations of any vertical, horizontal, and slant asymptotes. Then graph the function.

21. $f(x) = \dfrac{x^2 - 4}{x + 1}$

22. $f(x) = \dfrac{x^2 - x - 6}{x + 3}$

23. $f(x) = \dfrac{x^2 - x + 2}{x - 1}$

24. $f(x) = \dfrac{x^3 - 1}{x^2 - 9}$

In Exercises 25–28, use graphical methods to find any turning points of the graph of the function.

25. $f(x) = \dfrac{x^2}{x-1}$ **26.** $f(x) = \dfrac{(x-1)^2}{x}$

27. $g(x) = \dfrac{x^2+4}{x}$ **28.** $g(x) = \dfrac{x}{x^2+1}$

29. a. Graph $y = \dfrac{1-x^2}{x-2}$ on the window $[-10, 10]$ by $[-10, 10]$.

b. Use the graph to find y when $x = 1$ and when $x = 3$.

c. Use the graph to find the value(s) of x that give $y = -7.5$.

d. Use algebraic methods to solve $-7.5 = \dfrac{1-x^2}{x-2}$.

30. a. Graph $y = \dfrac{2+4x}{x^2+1}$ on the window $[-10, 10]$ by $[-5, 5]$.

b. Use the graph to find y when $x = -3$ and when $x = 3$.

c. Use the graph to find the value(s) of x that give $y = 2$.

d. Use algebraic methods to solve $2 = \dfrac{2+4x}{x^2+1}$.

31. a. Graph $y = \dfrac{3-2x}{x}$ on the window $[-10, 10]$ by $[-10, 10]$.

b. Use the graph to find y when $x = -3$ and when $x = 3$.

c. Use the graph to find the value(s) of x that give $y = -5$.

d. Use algebraic methods to solve $-5 = \dfrac{3-2x}{x}$.

32. a. Graph $y = \dfrac{x^2}{(x+1)^2}$ on the window $[-10, 10]$ by $[-6, 6]$.

b. Use the graph to find y when $x = 0$ and when $x = -2$.

c. Use the graph to find the value(s) of x that give $y = \dfrac{1}{4}$.

d. Use algebraic methods to solve $\dfrac{1}{4} = \dfrac{x^2}{(x+1)^2}$.

33. Use algebraic methods to solve $\dfrac{x^2+1}{x-1} + x = 2 + \dfrac{2}{x-1}$.

34. Use algebraic methods to solve $\dfrac{x}{x-2} - x = 1 + \dfrac{2}{x-2}$.

35. If y varies inversely as the 4th power of x and $y = 5$ when $x = -1$, what is y when $x = 0.5$?

36. If S varies inversely as the square root of T and $S = 4$ when $T = 4$, what is S when $T = 16$?

EXERCISES 6.5

37. *Average Cost* The average cost per unit for the production of a certain brand of DVD players is given by

$$\overline{C} = \dfrac{400 + 50x + 0.01x^2}{x}$$

where x is the number of units produced.

a. What is the average cost per unit when 500 units are produced?

b. What is the average cost per unit when 60 units are produced?

c. What is the average cost per unit when 100 units are produced?

d. Is it reasonable to say that the average cost continues to fall as the number of units produced rises?

38. *Average Cost* The average cost per set for the production of a certain brand of television sets is given by

$$\overline{C}(x) = \dfrac{1000 + 30x + 0.1x^2}{x}$$

where x is the number of units produced.

a. What is the average cost per set when 30 sets are produced?

b. What is the average cost per set when 300 sets are produced?

c. What happens to the function when $x = 0$? What does this tell you about the average cost when 0 units are produced?

39. *Advertising and Sales* The monthly sales volume y (in thousands of dollars) of a product is related to monthly advertising expenditures x (in thousands of dollars) according to the equation

$$y = \frac{400x}{x + 20}$$

a. What monthly sales will result if \$5000 is spent monthly on advertising?

b. What value of x makes the denominator of this function 0? Will this ever happen in the context of this problem?

40. **Productivity** As an 8-hour day shift progresses, the rate at which workers produce picture frames, in units per hour, changes according to the equation

$$f(t) = \frac{100(t^2 + 3t)}{(t^2 + 3t + 12)^2} \qquad (0 \le t \le 8)$$

where t is the number of hours after the beginning of the shift.

a. Graph this function using the window $[-10, 10]$ by $[-5, 5]$.

b. Graph this function using the window $[0, 8]$ by $[0, 5]$.

c. Does the graph in part (a) or part (b) better represent the rate of change of productivity function? Why?

d. Is the rate of productivity higher near lunch ($t = 4$) or near quitting time ($t = 8$)?

41. **Average Cost** The average cost per set for the production of television sets is given by

$$\overline{C}(x) = \frac{1000 + 30x + 0.1x^2}{x}$$

where x is the number of hundreds of units produced.

a. Graph this function using the window $[-10, 10]$ by $[-200, 300]$.

b. Graph this function using the window $[0, 50]$ by $[0, 300]$.

c. Which window makes sense for this application?

d. Use the graph with the window $[0, 250]$ by $[0, 300]$ to find the minimum average cost and the number of units that gives the minimum average cost.

42. **Average Cost** The average cost per unit for the production of a certain brand of DVD players is given by

$$\overline{C}(x) = \frac{400 + 50x + 0.01x^2}{x}$$

where x is the number of hundreds of units produced.

a. Graph this function using the window $[-100, 100]$ by $[-100, 400]$.

b. Graph this function using the window $[0, 300]$ by $[0, 400]$.

c. Which window is more appropriate for this problem?

d. Use the graph to find the minimum average cost and the number of units that gives the minimum average cost.

43. **Population** Suppose the number of employees of a start-up company is given by

$$f(t) = \frac{30 + 40t}{5 + 2t}$$

where t is the number of months after the company is organized.

a. Graph this function using the window $[0, 20]$ by $[0, 20]$.

b. Use the graph to find $f(0)$. What does this represent?

c. Use the graph to find $f(12)$. What does this represent? If $f(t)$ represents a number of individuals, how should you report your answer?

44. **Drug Concentration** Suppose the concentration of a drug (as a percent) in a patient's bloodstream t hours after injection is given by

$$C(t) = \frac{200t}{2t^2 + 32}$$

a. Graph the function using the window $[0, 20]$ by $[0, 20]$.

b. What is the drug concentration 1 hour after injection? 5 hours?

c. What is the highest percent concentration? In how many hours will it occur?

d. Describe how the end behavior of the graph of this function relates to the drug concentration.

45. **Cost-Benefit** Suppose the cost C of removing $p\%$ of the impurities from the waste water in a manufacturing process is given by

$$C(p) = \frac{3600p}{100 - p}$$

a. Where does the graph of this function have a vertical asymptote?

b. What does this tell us about removing the impurities from this process?

46. **Cost-Benefit** The percent p of particulate pollution that can be removed from the smokestacks of an industrial plant by spending C dollars is given by

$$p = \frac{100C}{8300 + C}$$

a. Find the percent of pollution that could be removed if spending were allowed to increase without bound.

b. Can 100% of the pollution be removed? Explain.

47. **Sales Volume** Suppose the weekly sales volume (in thousands of units) for a product is given by

$$V = \frac{640}{(p + 2)^2}$$

where p is the price in dollars per unit.

a. Graph this function on the p-interval $[-10, 10]$. Does the graph of this function have a vertical asymptote on this interval? Where?

b. Complete the table below.

Price/Unit ($)	5	20	50	100	200	500
Weekly Sales Volume						

c. What values of p give a weekly sales volume that makes sense? Does the graph of this function have a vertical asymptote on this interval?

d. What is the horizontal asymptote of the graph of the weekly sales volume? Explain what this means.

48. **Population** Suppose the number of employees of a start-up company is given by

$$N = \frac{30 + 40t}{5 + 2t}$$

where t is the number of months after the company is organized.

a. Does the graph of this function have a vertical asymptote on $[-10, 10]$? Where?

b. Does the graph of this function have a vertical asymptote for values of t in this application?

c. Does the graph of this function have a horizontal asymptote? Where?

d. What is the maximum number of employees predicted by this model?

49. **Demand** The quantity of a product demanded by consumers is defined by the function

$$p = \frac{100,000}{(q + 1)^2}$$

where p is the price and q is the quantity demanded.

a. Graph this function using the window $[0, 100]$ by $[0, 1000]$.

b. What is the horizontal asymptote that this graph approaches?

c. Use the graph to explain what happens to quantity demanded as price becomes lower.

50. **Advertising and Sales** Weekly sales y (in hundreds of dollars) are related to weekly advertising expenses x (in hundreds of dollars) according to the equation

$$y = \frac{800x}{20 + 5x}$$

a. Graph this function using the window $[0, 100]$ by $[0, 300]$.

b. Find the horizontal asymptote for this function.

c. Complete the table below.

Weekly Expenses ($)	0	50	100	200	300	500
Weekly Sales						

d. Use your results from parts (a) and (b) to determine the maximum weekly sales even if an unlimited amount of money is spent on advertising.

51. **Sales and Training** During the first 3 months of employment, the monthly sales S (in thousands of dollars) for an average new salesperson depend on the number of hours of training x, according to

$$S = \frac{40}{x} + \frac{x}{4} + 10 \quad \text{for} \quad x \geq 4$$

a. Combine the terms of this function over a common denominator to create a rational function.

b. How many hours of training should result in monthly sales of $21,000?

52. **Sales and Training** The average monthly sales volume (in thousands of dollars) for a company depends on the number of hours of training x of its sales staff, according to

$$S(x) = \frac{20}{x} + 40 + \frac{x}{2} \quad \text{for} \quad 4 \leq x \leq 120$$

a. Graph this function.

b. How many hours of training will give average monthly sales of $51,000?

53. **Worker Productivity** Suppose the average time (in hours) that a new production team takes to assemble 1 unit of a product is given by

$$H = \frac{5 + 3t}{2t + 1}$$

where t is the number of days of training for the team.

a. Graph this function using the window $[0, 20]$ by $[0, 8]$.

b. What is the horizontal asymptote? What does this mean in this application?

c. Use a graphical or numerical solution method to find the number of days of training necessary to reduce the production time to 1.6 hours.

54. *Advertising and Sales* Weekly sales y (in hundreds of dollars) are related to weekly advertising expenses x (in hundreds of dollars) according to the equation

$$y = \frac{800x}{20 + 5x}$$

Use graphical or numerical methods to find the amount of weekly advertising expenses that will result in weekly sales of $14,000, according to this model.

55. *Age at First Marriage* The table that follows shows the U.S. median age at first marriage for men and women for selected years from 1900 to 2015.

Median Age at First Marriage

Year	Men	Women	Year	Men	Women
1900	25.9	21.9	1990	26.1	23.9
1910	25.1	21.6	2000	26.8	25.1
1920	24.6	21.2	2009	28.1	25.9
1930	24.3	21.3	2010	28.2	26.1
1940	24.3	21.5	2011	28.7	26.5
1950	22.8	20.3	2012	28.6	26.6
1960	22.8	20.3	2013	29.0	26.6
1970	23.2	20.8	2015	29.2	27.1
1980	24.7	22.0			

(Source: U.S. Census Bureau)

Using these data, the ratio of age at first marriage for women to age at first marriage for men can be modeled by

$$y = \frac{0.00117x^2 - 0.0948x + 22.4}{0.00132x^2 - 0.127x + 26.3}$$

where x is the number of years after 1900.

a. Graph the function on a window representing the years from 1900 to 2020.

b. What is the predicted ratio of women's age to men's age at first marriage in 2020, according to the model?

c. Evaluate the function for large values of x to determine whether the function continues to increase indefinitely.

56. *Fences* Suppose that a rectangular field is to have an area of 51,200 square feet and that it needs to be enclosed by fence on three of its four sides (see the figure). What is the minimum length of fence needed? What dimensions should this field have to minimize the amount of fence needed? To solve this problem,

a. Write an equation that describes the area of the proposed field, with x representing the length and y representing the width of the field.

b. Write an equation that gives the length of fence L as a function of the dimensions of the field, remembering that only three sides are fenced in.

c. Use the equation from part (a) to write the length L as a function of one of the dimensions.

d. Graph the function L on a window that applies to the context of the problem: $[0, 400]$ by $[0, 1000]$.

e. Use graphical or numerical methods to find the minimum length of fence needed and the dimensions required to give this length.

57. *Females in the Workforce* For selected years from 1950 and projected to 2050, the following table shows the percent of total U.S. workers who were female.

Year	% Female	Year	% Female
1950	29.6	2010	47.9
1960	33.4	2015	48.3
1970	38.1	2020	48.1
1980	42.5	2030	48.0
1990	45.2	2040	47.9
2000	46.6	2050	47.7

(Source: U.S. Bureau of Labor Statistics)

Assume these data can be modeled with the function

$$p(t) = \frac{78.6t + 2090}{1.38t + 64.1}$$

where t is the number of years after 1950.

a. What does this model project the percent to be in 2030? Does this agree with the data?

b. What is the maximum possible percent of women in the workforce, according to this model?

c. In what year does the model indicate that the percent will reach 47.4%?

58. *Wind Chill* If x is the wind speed in miles per hour and is greater than or equal to 5, then the wind chill (in degrees Fahrenheit) for an air temperature of 0°F can be approximated by the function

$$f(x) = \frac{289.173 - 58.5731x}{x + 1}, \quad x \geq 5.$$

a. Ignoring the restriction $x \geq 5$, does $f(x)$ have a vertical asymptote? If so, what is it?

b. Does $f(x)$ have a vertical asymptote within its domain?

c. Does $f(x)$ have a horizontal asymptote? If so, what is it?

6.6 Polynomial and Rational Inequalities

KEY OBJECTIVES

- Solve polynomial inequalities using graphical and algebraic solution methods
- Solve rational inequalities using graphical and algebraic solution methods

SECTION PREVIEW   **Average Cost**

The average cost per set for the production of 42-inch plasma televisions is given by

$$\overline{C}(x) = \frac{5000 + 80x + x^2}{x}$$

where x is the number of hundreds of units produced. To find the number of televisions that must be produced to keep the average cost to at most $590 per TV, we solve the rational inequality

$$\frac{5000 + 80x + x^2}{x} \leq 590$$

(See Example 2.) In this section, we use algebraic and graphical methods similar to those used in Section 4.4 to solve inequalities involving polynomial functions and rational functions. ■

Polynomial Inequalities

The quadratic inequalities studied in Section 4.4 are examples of polynomial inequalities. Other examples of polynomial inequalities are

$$3x^3 + 3x^2 - 4x \geq 2, \qquad x - 3 < 8x^4, \qquad x \geq (x - 2)^3$$

We can solve the inequality $144x - 48x^2 + 4x^3 > 0$ with a graphing utility. The graph of the function $y = 144x - 48x^2 + 4x^3$ is shown in Figure 6.49.

Figure 6.49

We can determine that the x-intercepts of the graph are $x = 0$ and $x = 6$. The graph touches but does not cross the x-axis at $x = 6$, and the graph of the function is above the x-axis for values of $x > 0$ and $x \neq 6$. The solution to $144x - 48x^2 + 4x^3 > 0$ is the set of all x that give positive values for this function: $0 < x < 6$ and $x > 6$, or x in the intervals $(0, 6)$ and $(6, \infty)$.

The steps used to solve polynomial inequalities algebraically are similar to those used to solve quadratic inequalities.

Solving Polynomial Inequalities

To solve a polynomial inequality algebraically:

1. Write an equivalent inequality with 0 on one side and with the function $f(x)$ on the other side.

2. Solve $f(x) = 0$.

3. Create a sign diagram that uses the solutions from Step 2 to divide the number line into intervals. Pick a test value in each interval to determine whether $f(x)$ is positive or negative in that interval.*

4. Identify the intervals that satisfy the inequality in Step 1. The values of x that define these intervals are the solutions to the original inequality.

We illustrate the use of this method in the following example.

EXAMPLE 1 ▶   Constructing a Box

A box can be formed by cutting a square out of each corner of a piece of tin and folding the sides up. If the piece of tin is 12 inches by 12 inches and each side of the square that is cut out has length x, the function that gives the volume of the box is $V(x) = 144x - 48x^2 + 4x^3$.

a. Use factoring and then find the values of x that give positive values for $V(x)$.

b. Which of the values of x that give positive values for $V(x)$ result in a box?

SOLUTION

a. We seek those values of x that give positive values for $V(x)$—that is, for which

$$144x - 48x^2 + 4x^3 > 0$$

Writing the related equation $V(x) = 0$ and solving for x gives

$$144x - 48x^2 + 4x^3 = 0$$
$$4x(x^2 - 12x + 36) = 0$$
$$4x(x - 6)(x - 6) = 0$$
$$x = 0 \quad \text{or} \quad x = 6$$

These two values of x, 0 and 6, divide the number line into three intervals. Testing a value in each of the intervals (Figure 6.50(a)) determines the intervals where $V(x) > 0$ and thus determines the solutions to the inequality. The values of x that give positive values for V satisfy $0 < x < 6$ and $x > 6$. We found the same solutions to this inequality using graphical methods (see Figure 6.49) on the previous page.

---

* The numerical feature of your graphing utility can be used to test the x-values.

sign of $4x(x-6)(x-6)$ $----------$ $+++++++++$ $+++++++++$
sign of $(x-6)$ $----------$ $----------$ $+++++++++$
sign of $4x$ $----------$ $+++++++++$ $+++++++++$

(a)   **(b)**

Figure 6.50

The interval notation for this solution set is $(0, 6) \cup (6, \infty)$, and the graph of the solution set is shown in Figure 6.50(b).

b. The box is formed by cutting 4 squares of length x inches from a piece of tin that is 12 inches by 12 inches, so it is impossible to cut squares longer than 6 inches. Thus, the solution $x > 6$ does not apply to the physical building of the box, and so the values of x that result in a box satisfy $0 < x < 6$.

Rational Inequalities

To solve a **rational inequality** using an algebraic method, we first get 0 on the right side of the inequality; then, if necessary, we get a common denominator and combine the rational expressions on the left side. We find the numbers that make the numerator of the rational expression equal to zero and those that make the denominator equal to zero. These numbers are used to create a sign diagram much like the one we used when solving polynomial inequalities. Note that we should avoid multiplying both sides of an inequality by any term containing a variable, because we cannot easily determine when the term is positive or negative.

EXAMPLE 2 ▶ Average Cost

The average cost per set for the production of 42-inch plasma televisions is given by

$$\overline{C}(x) = \frac{5000 + 80x + x^2}{x}$$

where x is the number of hundreds of units produced. Find the number of televisions that must be produced to keep the average cost to at most $590 per TV.

SOLUTION

To find the number of televisions that must be produced to keep the average cost to at most $590 per TV, we solve the rational inequality

$$\frac{5000 + 80x + x^2}{x} \le 590$$

To solve this inequality with algebraic methods, we rewrite the inequality with 0 on the right side of the inequality, getting

$$\frac{5000 + 80x + x^2}{x} - 590 \le 0$$

Combining the terms over a common denominator and factoring gives

$$\frac{5000 + 80x + x^2 - 590x}{x} \le 0$$

$$\frac{x^2 - 510x + 5000}{x} \le 0$$

$$\frac{(x - 10)(x - 500)}{x} \le 0$$

The values of x that make the numerator equal to 0 are 10 and 500, and the value $x = 0$ makes the denominator equal to 0. We cannot have a negative number of units, so the average cost is defined only for values of $x > 0$. The values 0, 10, and 500 divide a number line into three intervals. Next we find the sign of $f(x) = \dfrac{(x - 10)(x - 500)}{x}$ in each interval. Testing a value in each interval determines which interval(s) satisfy the inequality. We do this by using the sign diagram in Figure 6.51.

Figure 6.51

The function $f(x)$ is negative on $10 < x < 500$, so the solution to

$$\frac{5000 + 80x + x^2}{x} - 590 \leq 0$$

and thus to the original inequality, is $10 \leq x \leq 500$. Because x represents the number of hundreds, between 1000 and 50,000 televisions must be produced to keep the average cost to at most $590 per TV.

We can also use graphical methods to solve rational inequalities. For example, we can solve the inequality

$$\frac{5000 + 80x + x^2}{x} \leq 590$$

for $x > 0$ by graphing

$$y_1 = \frac{5000 + 80x + x^2}{x} \quad \text{and} \quad y_2 = 590$$

on the same axes. We are only interested in the graph for positive values of x, so we restrict our graph to the first quadrant. The points of intersection can be found by the intersection method (Figure 6.52). The inequality is satisfied by the x-interval for which the graph of y_1 is below (or on) the graph of y_2—that is, for $10 \leq x \leq 500$. Recall that this is the solution that we found using algebraic methods in Example 2.

Figure 6.52

SKILLS CHECK 6.6

In Exercises 1–16, use algebraic and/or graphical methods to solve the inequalities.

1. $16x^2 - x^4 \geq 0$ **2.** $x^4 - 4x^2 \leq 0$

3. $2x^3 - x^4 < 0$ **4.** $3x^3 \geq x^4$

5. $(x - 1)(x - 3)(x + 1) \geq 0$

6. $(x - 3)^2(x + 1) < 0$

7. $\dfrac{4 - 2x}{x} > 2$ **8.** $\dfrac{x - 3}{x + 1} \geq 3$

9. $\dfrac{x}{2} + \dfrac{x - 2}{x + 1} \leq 1$ **10.** $\dfrac{x}{x - 1} \leq 2x + \dfrac{1}{x - 1}$

11. $(x - 1)^3 > 27$ **12.** $(2x + 3)^3 \leq 8$

13. $(x - 1)^3 < 64$ **14.** $(x + 4)^3 - 125 \geq 0$

15. $-x^3 - 10x^2 - 25x \leq 0$

16. $x^3 + 10x^2 + 25x < 0$

For Exercises 17–20, use the graph of $y = f(x)$ to solve the requested inequality.

17.

    a. Solve $f(x) < 0$.

    b. Solve $f(x) \geq 0$.

18.

    a. Solve $f(x) < 0$.

    b. Solve $f(x) \geq 0$.

19.

    Solve $f(x) \geq 2$.

20.

    a. Solve $f(x) < 0$.

    b. Solve $f(x) \geq 0$.

EXERCISES 6.6

Combine factoring with graphical and/or numerical methods to solve Exercises 21–26.

21. *Revenue* The revenue from the sale of a product is given by the function $R = 400x - x^3$. Selling how many units will give positive revenue?

22. *Revenue* The price for a product is given by $p = 1000 - 0.1x^2$, where x is the number of units sold.

    a. Form the revenue function for the product.

    b. Selling how many units gives positive revenue?

23. *Constructing a Box* A box can be formed by cutting a square out of each corner of a piece of cardboard and folding the sides up. If the piece of cardboard is 36 cm by 36 cm and each side of the square that is cut out has length x cm, the function that gives the volume of the box is $V = 1296x - 144x^2 + 4x^3$.

a. Find the values of x that make $V > 0$.

b. What size squares can be cut out to construct a box?

24. **Constructing a Box** A box can be formed by cutting a square out of each corner of a piece of cardboard and folding the sides up. If the piece of cardboard is 12 inches by 16 inches and each side of the square that is cut out has length x inches, the function that gives the volume of the box is $V = 192x - 56x^2 + 4x^3$. What size squares can be cut out to construct a box?

25. **Cost** The total cost function for a product is given by $C(x) = 3x^3 - 6x^2 - 300x + 1800$, where x is the number of units produced and C is the cost in hundreds of dollars. Use factoring by grouping and then find the number of units that will give a total cost of at least $120,000. Verify your conclusion with a graphing utility.

26. **Profit** The profit function for a product is given by $P(x) = -x^3 + 2x^2 + 400x - 400$, where x is the number of units produced and sold and P is in hundreds of dollars. Use factoring by grouping to find the number of units that will give a profit of at least $40,000. Verify your conclusion with a graphing utility.

27. **Advertising and Sales** The monthly sales volume y (in thousands of dollars) is related to monthly advertising expenditures x (in thousands of dollars) according to the equation

$$y = \frac{400x}{x + 20}$$

Spending how much money on advertising will result in sales of at least $200,000 per month?

28. **Average Cost** The average cost per set for the production of a portable stereo system is given by

$$\overline{C} = \frac{100 + 30x + 0.1x^2}{x}$$

where x is the number of hundreds of units produced. What number of units can be produced while keeping the average cost to at most $41?

29. **Future Value** The future value of $2000 invested for 3 years at rate r, compounded annually, is given

by $S = 2000(1 + r)^3$. Find the rate r that gives a future value from $2662 to $3456, inclusive.

30. **Future Value** The future value of $5000 invested for 4 years at rate r, compounded annually, is given by $S = 5000(1 + r)^4$. Find the rate r that gives a future value of at least $6553.98.

31. **Revenue** The revenue from the sale of a product is given by the function $R = 4000x - 0.1x^3$ dollars. Use graphical or numerical methods to determine how many units should be sold to give a revenue of at least $39,990.

32. **Revenue** The price for a product is given by $p = 1000 - 0.1x^2$, where x is the number of units sold.

a. Form the revenue function for the product.

b. Use graphical or numerical methods to determine how many units should be sold to give a revenue of at most $37,500.

33. **Supply and Demand** Suppose the supply function for a product is given by $6p - q = 180$ and the demand is given by $(p + 20)q = 30,000$. Over what meaningful price interval does supply exceed demand?

34. **Drug Concentration** A pharmaceutical company claims that the concentration of a drug in a patient's bloodstream will be at least 10% for 8 hours. Suppose clinical tests show that the concentration of a drug (as percent) t hours after injection is given by

$$C(t) = \frac{200t}{2t^2 + 32}$$

During what time period is the concentration at least 10%? Is the company's claim supported by the evidence?

35. **Population** Suppose the number of employees of a start-up company is given by

$$f(t) = \frac{30 + 40t}{5 + 2t}$$

where t is the number of months after the company is organized.

a. For what values of t is $f(t) \le 18$?

b. During what months is the number of employees below 18?

chapter 6   SUMMARY

In this chapter, we studied higher-degree polynomial functions, their graphs, and the solution of polynomial equations by factoring, by the root method, by graphical methods, and by synthetic division. We also studied rational functions, their graphs, vertical asymptotes, horizontal asymptotes, slant asymptotes, and the solution of equations and inequalities containing rational functions.

Key Concepts and Formulas

6.1 Higher-Degree Polynomial Functions

Polynomial function	A polynomial function of degree n with independent variable x is a function in the form $P(x) = a_n x^n + a_{n-1} x^{n-1} + \cdots + a_1 x + a_0$ with $a_n \neq 0$. Each a_i represents a real number and each n is a positive integer.
Cubic function	A polynomial function of the form $f(x) = ax^3 + bx^2 + cx + d$ with $a \neq 0$.
Quartic function	A polynomial function of the form $f(x) = ax^4 + bx^3 + cx^2 + dx + e$ with $a \neq 0$.
Extrema points	Turning points on the graph of a polynomial function: local minimum, local maximum, absolute minimum, absolute maximum.
Graphs of cubic functions	End behavior: one end opening up and one end opening down. Turning points: 0 or 2
Graphs of quartic functions	End behavior: Both ends opening up or both ends opening down. Turning points: 1 or 3

6.2 Modeling with Cubic and Quartic Functions

Cubic models	If a scatter plot of data indicates that the data may fit close to one of the possible shapes of the graph of a cubic function, it is possible to find a cubic function that is an approximate model for the data.
Quartic models	If a scatter plot of data indicates that the data may fit close to one of the possible shapes of the graph of a quartic function, it is possible to find a quartic function that is an approximate model for the data.
Third differences	For equally spaced inputs, the third differences are constant for data fitting cubic functions exactly.
Fourth differences	For equally spaced inputs, the fourth differences are constant for data fitting quartic functions exactly.

6.3 Solution of Polynomial Equations

Solving polynomial equations	
By factoring	Some polynomial equations can be solved by factoring and using the zero-product property.
By the root method	The real solutions of the equation $x^n = C$ are $x = \pm \sqrt[n]{C}$ if n is even and $C \geq 0$, or $x = \sqrt[n]{C}$ if n is odd.
Factors, zeros, intercepts, and solutions	If $(x - a)$ is a factor of $f(x)$, then a is an x-intercept of the graph of $y = f(x)$, a is a solution to the equation $f(x) = 0$, and a is a zero of the function f.
Estimating solutions	Solving equations derived from real data requires graphical or numerical methods using technology.

6.4 Polynomial Equations Continued; Fundamental Theorem of Algebra

Division of polynomials	If $f(x)$ is a function and we know that $x = a$ is a solution of $f(x) = 0$, then we can write $x - a$ as a factor of $f(x)$. If we divide this factor into the function, the quotient will be another factor of $f(x)$.
Synthetic division	To divide a polynomial by a binomial of the form $x - a$ (or $x + a$), we can use synthetic division to simplify the division process by omitting the variables and writing only the coefficients.
Solving cubic equations	If we know one exact solution of a cubic equation $P(x) = 0$, we can divide by the corresponding factor using synthetic division, obtaining a second quadratic factor that can be used to find any additional solutions of the equation.
Combining graphical and algebraic methods	To find exact solutions to a polynomial equation when one solution can be found graphically, we can use that solution and synthetic division to find a second factor.
Rational Solutions Test	The rational solutions of the polynomial equation $a_n x^n + a_{n-1} x^{n-1} + \cdots + a_1 x + a_0 = 0$ with integer coefficients must be of the form $\dfrac{p}{q}$, where p is a factor of the constant term a_0 and q is a factor of the leading coefficient a_n.
Fundamental Theorem of Algebra	If $f(x)$ is a polynomial function of degree $n \geq 1$, then f has at least one complex zero.
Complex solutions of polynomial equations	Every polynomial function $f(x)$ of degree $n \geq 1$ has exactly n complex zeros. Some of these zeros may be imaginary, and some may be repeated.

6.5 Rational Functions and Rational Equations

Rational function	The function f is a rational function if $f(x) = \dfrac{P(x)}{Q(x)}$, where $P(x)$ and $Q(x)$ are polynomials and $Q(x) \neq 0$.
Graphs of rational functions	The graph of a rational function may have vertical asymptotes and/or a horizontal asymptote. The graph may have a "hole" at $x = a$ rather than a vertical asymptote.
Vertical asymptotes	A vertical asymptote occurs in the graph of $f(x) = \dfrac{P(x)}{Q(x)}$ at those values of x where $Q(x) = 0$ and $P(x) \neq 0$, that is, where the denominator equals 0 but the numerator does not equal 0.

Horizontal asymptotes	Consider the rational function $$f(x) = \frac{P(x)}{Q(x)} = \frac{a_n x^n + \cdots + a_1 x + a_0}{b_m x^m + \cdots + b_1 x + b_0}$$ **1.** If $n < m$, a horizontal asymptote occurs at $y = 0$ (the x-axis). **2.** If $n = m$, a horizontal asymptote occurs at $y = \dfrac{a_n}{b_m}$. **3.** If $n > m$, there is no horizontal asymptote.
Slant asymptote	If the degree of the numerator of a rational function is *one* more than the degree of the denominator, its graph approaches a slant asymptote. The equation of the slant asymptote can be found by dividing the numerator by the denominator, getting a linear function plus a rational expression with a constant numerator. The graph of the original function will have a slant asymptote, which is the graph of the linear function.
Missing point (hole)	If there is a value a such that $Q(a) = 0$ and $P(a) = 0$, then the graph of the function is undefined at $x = a$ but may have a "hole" at $x = a$ rather than an asymptote.
Solution of rational equations	
Algebraically	Multiplying both sides of the equation by the LCD of any fractions in the equation converts the equation to a polynomial equation that can be solved. All solutions must be checked in the original equation because some may be extraneous.
Graphically	Solutions to rational equations can be found or approximated by graphical methods.

6.6 Polynomial and Rational Inequalities

Polynomial inequalities	The steps used to solve polynomial inequalities graphically are similar to those used to solve quadratic inequalities. To solve a polynomial inequality algebraically, solve the corresponding polynomial equation and use a sign diagram to complete the solution.
Rational inequalities	To solve a rational inequality, get 0 on the right side of the inequality; then, if necessary, get a common denominator and combine the rational expressions on the left side. Find the numbers that make the numerator of the rational expression equal to 0 and those that make the denominator equal to 0. Use these numbers to create a sign diagram and complete the solution. Graphical methods can be used to find or to verify the solutions.

chapter 6   SKILLS CHECK

1. What is the degree of the polynomial $5x^4 - 2x^3 + 4x^2 + 5$?

2. What is the type of the polynomial function $y = 5x^4 - 2x^3 + 4x^2 + 5$?

3. Graph $y = -4x^3 + 4x^2 + 1$ using the window $[-10, 10]$ by $[-10, 10]$. Is the graph complete on this window?

4. Graph the function $y = x^4 - 4x^2 - 20$ using

   a. The window $[-8, 8]$ by $[-8, 8]$.

   b. A window that gives a complete graph.

5. a. Graph $y = x^3 - 3x^2 - 4$ using a window that shows a local maximum and local minimum.

b. A local maximum occurs at what point?

c. A local minimum occurs at what point?

6. a. Graph $y = x^3 + 11x^2 - 16x + 40$ using the window $[-10, 10]$ by $[0, 100]$.

b. Is the graph complete?

c. Sketch a complete graph of this function.

d. Find the points where a local maximum and a local minimum occur.

7. Solve $x^3 - 16x = 0$.

8. Solve $2x^4 - 8x^2 = 0$.

9. Solve $x^4 - x^3 - 20x^2 = 0$.

10. Solve $0 = x^3 - 15x^2 + 56x$ by factoring.

11. Use factoring by grouping to solve $4x^3 - 20x^2 - 4x + 20 = 0$.

12. Solve $12x^3 - 9x^2 - 48x + 36 = 0$ using factoring by grouping.

13. Use technology to find the solutions to $x^4 - 3x^3 - 3x^2 + 7x + 6 = 0$.

14. Use technology to solve $6x^3 - 59x^2 - 161x + 60 = 0$.

15. Use the root method to solve $(x - 4)^3 = 8$.

16. Use the root method to solve $5(x - 3)^4 = 80$.

17. Use synthetic division to divide $4x^4 - 3x^3 + 2x - 8$ by $x - 2$.

18. Use synthetic division to solve $2x^3 + 5x^2 - 11x + 4 = 0$, given that $x = 1$ is one solution.

19. Find one solution of $3x^3 - x^2 - 12x + 4 = 0$ graphically, and use synthetic division to find any additional solutions.

20. Find one solution of $2x^3 + 5x^2 - 4x - 3 = 0$ graphically, and use synthetic division to find any additional solutions.

21. For the function $y = \dfrac{1 - x^2}{x + 2}$,

a. Find the x-intercepts and y-intercept if they exist.

b. Find any horizontal asymptotes and vertical asymptotes that exist.

c. Find any slant asymptote that exists.

d. Sketch the graph of the function.

22. For the function $y = \dfrac{3x - 2}{x - 3}$,

a. Find the x-intercepts and y-intercept if they exist.

b. Find any horizontal asymptotes and vertical asymptotes that exist.

c. Sketch the graph of the function.

23. Use a graph and technology to find any local maxima and/or minima of the function $y = \dfrac{x^2}{x - 4}$.

24. Use a graph and technology to find any local maxima and/or minima of the function $y = \dfrac{x^2 + 3}{1 - x}$.

25. a. Graph $y = \dfrac{1 + 2x^2}{x + 2}$ using the window $[-10, 10]$ by $[-30, 10]$.

b. Use the graph to find y when $x = 1$ and when $x = 3$.

c. Use the graph to find the value(s) of x that give $y = \dfrac{9}{4}$.

d. Use an algebraic method to solve $\dfrac{9}{4} = \dfrac{1 + 2x^2}{x + 2}$.

26. Solve $x^4 - 13x^2 + 36 = 0$.

In Exercises 27 and 28, find the exact solutions to $f(x) = 0$ in the complex number system.

27. $x^3 + x^2 + 2x - 4 = 0$

28. $4x^3 + 10x^2 + 5x + 2 = 0$

29. Solve $x^3 - 5x^2 \geq 0$.

30. Use a graphing utility to solve $x^3 - 5x^2 + 2x + 8 \geq 0$.

31. Solve $2 < \dfrac{4x - 6}{x}$.

32. Solve $\dfrac{5x - 10}{x + 1} \geq 20$.

33. *Revenue* The monthly revenue for a product is given by $R = -0.1x^3 + 15x^2 - 25x$, where x is the number of thousands of units sold.

   a. Graph this function using the window $[-100, 200]$ by $[-2000, 60,000]$.

   b. Graph the function on a window that makes sense for the problem—that is, with nonnegative x and R.

   c. What is the revenue when 50,000 units are produced?

34. *Revenue* The monthly revenue for a product is given by $R = -0.1x^3 + 13.5x^2 - 150x$, where x is the number of thousands of units sold. Use technology to find the number of units that gives the maximum possible revenue.

35. *Investment* If \$5000 is invested for 6 years at interest rate r (as a decimal), compounded annually, the future value of the investment is given by $S = 5000(1 + r)^6$ dollars.

   a. Find the future value of this investment for selected interest rates by completing the following table.

Rate, r	Future Value, S (\$)
0.01	
0.05	
0.10	
0.15	

   b. Graph this function for $0 \le r \le 0.20$.

   c. Compute the future value if the rate is 10% and 20%. How much more money is earned at 20%?

36. *Photosynthesis* The amount y of photosynthesis that takes place in a certain plant depends on the intensity x of the light present, according to the function $y = 120x^2 - 20x^3$. The model is valid only for x-values that are nonnegative and that produce a positive amount of photosynthesis. Use graphical or numerical methods to find what intensity allows the maximum amount of photosynthesis.

37. *Poverty Threshold* Using data from selected years from 1990 and projected to 2013, the yearly income poverty thresholds for a single person can be modeled by $y = -0.200x^3 + 7.494x^2 + 159.305x + 6685.097$, where x is the number of years after 1990 and y is in dollars.

   a. Graph this function on the window representing the years 1990 through 2020.

   b. What does the model predict the poverty threshold for a single person will be in 2017?

   c. Use technology to find the year in which the poverty threshold for a single person will be a maximum during 1990–2025, according to this model.

38. *Elderly in the Workforce* The table gives the number of millions of elderly Americans in the workforce for selected years from 1970 and projected to 2050.

   a. Create a scatter plot of the data on the window $[0, 100]$ by $[0, 12]$, with x equal to the number of years after 1960 and y equal to the number of millions of elderly workers.

   b. Find a cubic model for these data, with x equal to the number of years after 1960 and y equal to the number of elderly in the workforce, in millions. Report the model with three significant digits.

   c. Graph the model and the data on the window representing 1960 through 2060.

   d. According to the model, how many elderly Americans will be in the workforce in 2060?

Year	Workforce (millions)
1970	3.22
1980	3.05
1990	3.45
2000	4.20
2010	5.44
2015	6.99
2020	8.24
2030	10.10
2040	9.60
2050	10.20

(Source: U.S. Bureau of Labor Statistics)

39. *Foreign Born Population* The table on the following page gives the percent of the U.S. population that is foreign born for selected years from 1900 and projected to 2020.

   a. Create a scatter plot of the data, with x equal to the number of years after 1900 and y equal to the percent of the U.S. population that is foreign born.

   b. Find the quartic function that models the data, with x equal to the number of years after 1900 and y equal to the percent of the U.S. population that is foreign born. Report the model with three significant digits.

c. Graph the function and the data on the same axes, and comment on the fit.

d. According to the model, in what year was the percent foreign born a minimum?

Year	Percent Foreign Born	Year	Percent Foreign Born
1900	13.6	1980	6.2
1910	14.7	1990	8.0
1920	13.2	2000	10.4
1930	11.6	2005	11.7
1940	8.8	2010	12.4
1950	6.9	2016	11.6
1960	5.4	2020	11.8
1970	4.7		

(Source: U.S. Census Bureau)

40. Investment The future value of an investment of $8000 at interest rate r (as a decimal), compounded annually, for 3 years is given by $S = 8000(1 + r)^3$. Use the root method to find the rate r that gives a future value of $9261.

41. Constructing a Box A box can be formed by cutting a square out of each corner of a piece of tin and folding the sides up. If the piece of tin is 18 inches by 18 inches and each side of the square that is cut out has length x, the function that gives the volume of the box is $V = 324x - 72x^2 + 4x^3$.

a. Use factoring to find the values of x that make $V = 0$.

b. For these values of x, discuss what happens to the box if squares of length x are cut out.

c. What values of x are reasonable for the squares that can be cut out to make a box?

42. Break-Even The daily profit in dollars for a product is given by $P(x) = -0.2x^3 + 20.5x^2 - 48.8x - 120$, where x is the number of hundreds of units produced. To find the number of units that gives break-even,

a. Graph the function on the window [0, 40] by [−15, 20], and graphically find an x-intercept of the graph.

b. Use synthetic division to find a quadratic factor of $P(x)$.

c. Find all of the zeros of $P(x)$.

d. Determine the levels of production that give break-even.

43. Starbucks The table gives the total number of Starbucks stores in the world and in the United States for selected years from 2005 through 2017.

a. Create a scatter plot of the data, with x representing the number of years after 2000 and y representing the total number of stores in the world.

b. Find a cubic model for the data, with x equal to the number of years after 2000 and y equal to the total number of stores in the world. Report the model with three significant digits.

c. Graph the data and the cubic model on the same axes, with the window [2, 20] by [8000, 23,000].

d. What does the model give as the projected total number of worldwide Starbucks stores for 2019?

Year	Total	U.S.
2005	10,241	7353
2006	12,440	8896
2007	15,011	10,684
2008	16,680	11,567
2009	16,635	11,128
2010	16,858	11,131
2011	17,003	10,787
2012	17,572	10,924
2013	20,891	13,279
2017	21,000	12,543

(Source: Starbucks)

44. Drugs in the Bloodstream The concentration of a drug in a patient's bloodstream is given by

$$C = \frac{0.3t}{t^2 + 1}$$

where t is the number of hours after the drug was ingested and C is in mg/cm^3.

a. What is the horizontal asymptote for the graph of this function?

b. What does this say about the concentration of the drug?

c. Use technology to find the maximum concentration of the drug, and when it occurs, for $0 \le t \le 4$.

45. **Average Cost** The average cost for production of a product is given by

$$\overline{C}(x) = \frac{50x + 5600}{x}$$

where x is the number of units produced.

a. Find $\overline{C}(0)$, if it exists. What does this tell us about average cost?

b. Find the horizontal asymptote for the graph of this function. What does this tell us about the average cost of this product?

c. Does this function increase or decrease for $x > 0$?

46. **Average Cost** The average cost for production of a product is given by

$$\overline{C}(x) = \frac{30x^2 + 12{,}000}{x}$$

where x is the number of units produced.

a. Graph the function for $x > 0$.

b. Use technology to find the number of units that gives the minimum average cost. What is the minimum average cost?

47. **Demand** The quantity of a certain product that is demanded is related to the price per unit by the equation $p = \dfrac{30{,}000 - 20q}{q}$, where p is the price per unit. Graph the function on a window that fits the context of the problem, with $0 \le q \le 1000$, with q on the horizontal axis.

48. **Printing** A printer has a contract to run 20,000 posters for a state fair. He can use any number of plates from 1 to 8 to run the posters; each stamp of the press will produce as many posters as there are plates. The cost of printing all the posters is given by

$$C(x) = 200 + 20x + \frac{180}{x}, \text{ where } x \text{ is the number}$$

of plates.

a. Combine the terms of this function over a common denominator to create a rational function.

b. Use numerical or graphical methods to find the number of plates that will make the cost $336.

c. How many plates should the printer create to produce the posters at the minimum cost?

49. **U.S. Smartphones** Using data from 2010 through 2016, the fraction of cell phone users who are smartphone users can be modeled by the rational function

$$f(t) = \frac{21.35t + 69.59}{4.590t + 233.1}$$

where t is equal to the number of years after 2010.

a. Graph the function on a window that represents the years 2010 through 2020.

b. What fraction of cell phone users in the United States used smartphones in 2015, according to this model?

c. What is $f(10)$? What does this indicate about the validity of the function?
(Source: emarketer.com)

50. **Sales and Training** During the first month of employment, the monthly sales S (in thousands of dollars) for an average new salesperson depend on the number of hours of training x, according to

$$S = \frac{40}{x} + \frac{x}{4} + 10 \quad \text{for} \quad x \ge 4$$

How many hours of training should result in monthly sales greater than $23,300?

51. **Risky Drivers** The following table gives the police-reported crashes per million miles traveled for different age groups.

a. Create a scatter plot of the data, with x equal to the age of the drivers and y equal to the number of crashes per million miles travel.

b. Find the quartic function that is the best fit for the data, where x is the age of the drivers and y equals the number of crashes per million miles traveled. Report the model with four significant digits.

c. Graph the function and the data on the same set of axes, for $x = 15$ to $x = 85$.

d. At what ages is the number of crashes less than 3, according to the model?

Age	Crashes	Age	Crashes
16	32	45	3.5
17	21	50	3.5
18	10	55	3
19	9	60	2
20	9.5	65	2.5
25	8	70	3
30	5	75	4
35	3.5	80	5
40	3	85	8

(Source: *Consumer Reports*, October 2012)

52. **Revenue** The revenue from the sale of x units of a product is given by the function $R = 1200x - 0.003x^3$. Selling how many units will give a revenue of at least $59,625?

53. *Average Cost* The average cost per set for the production of television sets is given by

$$\overline{C} = \frac{100 + 30x + 0.1x^2}{x}$$

where x is the number of hundreds of units produced. For how many units is the average cost at most $37?

54. *Cost-Benefit* The percent p of particulate pollution that can be removed from the smokestacks of an industrial plant by spending C dollars per week is given by $p = \dfrac{100C}{9600 + C}$. How much would it cost to remove at least 30.1% of the particulate pollution?

Group Activities
▶ EXTENDED APPLICATIONS

1. Global Climate Change

1. Suppose the annual cost C (in hundreds of dollars) of removing $p\%$ of the particulate pollution from the smokestack of a power plant is given by

$$C(p) = \frac{242{,}000}{100 - p} - 2420$$

  a. What is the domain of the function described by this equation?

  b Taking into account the meaning of the variables, give the domain and range of the function.

2. Find $C(60)$ and $C(80)$ and interpret $C(80)$.

3. Find the asymptotes of the function. What does the vertical asymptote tell you about the cost of removing pollution?

4. Find and interpret the y-intercept.

5. Suppose the annual fine for removing less than 80% of the particulate pollution from the smokestack is $700,000. If you were the company's chief financial officer (CFO), would you advise the company to remove the pollution or pay the fine? Why?

6. If the company has already paid to remove 60% of the particulate pollution, would you advise the company to remove another 20% or pay the fine?

7. If you were a lawmaker, to what amount would you raise the fine in order to encourage power plants to comply with the pollution regulations?

8. Compare the costs of removing 80% and 90% of the pollution from the smokestack. Do you think the benefit of removing 90% of particulate pollution rather than 80% is worth the cost?

2. Printing

A printer has a contract to print 100,000 invitations for a political candidate. He can run the invitations by using any number of metal printing plates from 1 to 20 on his press. Each stamp of the press will produce as many invitations as there are plates. For example, if he prepares 10 plates, each impression of his press will make 10 invitations. Preparing each plate costs $8, and he can make 1000 impressions per plate per hour. If it costs $128 per hour to run the press, how many metal printing plates should the printer use to minimize the cost of printing the 100,000 invitations?

  A. To understand the problem:

    1. Determine how many hours it would take to print the 100,000 invitations with 10 plates.

    2. How much would it cost to print the invitations with 10 plates?

    3. Determine how much it would cost to prepare 10 printing plates.

    4. What is the total cost to make the 10 plates and print 100,000 invitations?

  B. To find the number of plates that will minimize the total cost of printing the invitations:

    1. Find the cost of preparing x plates.

    2. Find the number of invitations that can be printed with one impression of x plates.

    3. Find the number of invitations that can be made per hour with x plates and the number of hours it would take to print 100,000 invitations.

    4. Write a function of x that gives the total cost of producing the invitations.

    5. Use technology to graph the cost function and to estimate the number of plates that will give the minimum cost of producing 100,000 invitations.

    6. Use numerical methods to find the number of plates that will give the minimum cost and the minimum possible cost of producing 100,000 invitations.

7

Systems of Equations and Matrices

To improve the highway infrastructure, departments of transportation can use systems of equations to analyze traffic flow. Solving systems of equations is also useful in finding break-even and market equilibrium. We can represent systems with matrices and use matrix reduction to solve the systems and related problems. We can also use matrices to make comparisons of data, to solve pricing problems and advertising problems, and to encode and decode messages.

sections	objectives	applications
7.1 Systems of Linear Equations in Three Variables	Solve systems of linear equations using left-to-right elimination; find solutions of dependent systems; determine that a system is inconsistent	Manufacturing, purchasing
7.2 Matrix Solution of Systems of Linear Equations	Represent systems of equations with matrices; solve systems with matrices; solve systems of equations with Gauss–Jordan elimination; find nonunique solutions of dependent systems	Transportation, investment
7.3 Matrix Operations	Add and subtract two matrices; multiply a matrix by a constant; multiply two matrices	Life expectancy, pricing, advertising, manufacturing
7.4 Inverse Matrices; Matrix Equations	Find the inverse of a matrix; find inverses with technology; encode and decode messages; solve matrix equations; solve matrix equations with technology	Credit card security, encoding messages, decoding messages, manufacturing
7.5 Determinants and Cramer's Rule	Find determinants of matrices; solve systems of equations using Cramer's Rule; Cramer's Rule with inconsistent and dependent systems	Business properties
7.6 Systems of Nonlinear Equations	Solve systems of nonlinear equations algebraically and graphically	Market equilibrium, constructing a box

Algebra TOOLBOX

KEY OBJECTIVES

- Solve problems involving proportions
- Determine whether sets of triples are proportional
- Determine whether a pair of equations in three variables represent the same plane, parallel planes, or neither
- Determine whether given values of three variables satisfy a system of three equations in the three variables

Proportion

Often in mathematics we need to express relationships between quantities. One relationship that is frequently used in applied mathematics occurs when two quantities are proportional. Two variables x and y are proportional to each other (or vary directly) if their quotient is constant. That is, y is **directly proportional** (or simply **proportional**) to x if y and x are related by the equation

$$\frac{y}{x} = k \quad \text{or, equivalently,} \quad y = kx$$

where $k \neq 0$ is called the **constant of proportionality**.

EXAMPLE 1 ▶ Circles

Is the circumference of a circle proportional to the radius of the circle?

SOLUTION

The circumference C of a circle is proportional to the radius r of the circle because $C = 2\pi r$. In this case, C and r are the variables, and 2π is the constant of proportionality. We could write an equivalent equation,

$$\frac{C}{r} = 2\pi$$

which says that the quotient of the circumference divided by the radius is constant, and that constant is 2π.

Proportionality

One pair of numbers is said to be **proportional** to another pair of numbers if the ratio of one pair is equal to the ratio of the other pair. In general, the pair a, b is proportional to the pair c, d if and only if

$$\frac{a}{b} = \frac{c}{d} \qquad (b \neq 0, d \neq 0)$$

This proportion can be written $a:b$ as $c:d$.

If $b \neq 0$, $d \neq 0$, we can verify that $\frac{a}{b} = \frac{c}{d}$ if $ad = bc$—that is, if the cross-products are equal.

EXAMPLE 2 ▶ Proportional Pairs

Is the pair 4, 3 proportional to the pair 12, 9?

SOLUTION

It is true that $\frac{4}{3} = \frac{12}{9}$ because we can reduce $\frac{12}{9}$ to $\frac{4}{3}$; or we can show that $\frac{4}{3} = \frac{12}{9}$ is true by showing that the cross-products are equal: $4 \cdot 9 = 3 \cdot 12$. Thus, the pair 4, 3 is proportional to 12, 9.

Proportional Triples

For two sets of triples a, b, c and d, e, f, if there exists some number k such that multiplying each number in the first set of triples by k results in the second set of triples, then the two sets of triples are proportional. That is, a, b, c is proportional to d, e, f if there exists a number k such that $ak = d$, $bk = e$, and $ck = f$.

EXAMPLE 3 ▶

Proportional Triples

Determine whether the sets of triples are proportional.

a. 2, 3, 5 and 4, 6, 10   **b.** 4, 6, 10 and 6, 9, 12

SOLUTION

a. If the sets of triples are proportional, then there is a number k such that $2k = 4$, $3k = 6$, and $5k = 10$. Because $k = 2$ is the solution to all three equations, we conclude that the sets of triples are proportional.

b. If the sets of triples are proportional, then there is a number k such that $4k = 6$, $6k = 9$, and $10k = 12$. Because $k = 1.5$ is the solution to $4k = 6$ but not a solution to $10k = 12$, the sets of triples are not proportional.

Linear Equations in Three Variables

If a, b, c, and d represent nonzero constants, then

$$ax + by + cz = d$$

is a first-degree equation in three variables. We also call these equations linear equations in three variables, but unlike linear equations in two variables, their graphs are not lines. When equations of this form are graphed in a three-dimensional coordinate system, their graphs are planes. Similar to the graphs of linear equations in two variables, the graphs of two linear equations in three variables will be parallel planes if the respective coefficients of x, y, and z are proportional and the constants are not in the same proportion. If the respective coefficients of x, y, and z in two equations are not proportional, then the two planes will intersect along a line. And if the two equations are equivalent, the equations represent the same plane.

EXAMPLE 4 ▶

Planes

Determine whether the planes represented by the following pairs of equations represent the same plane, parallel planes, or neither.

a. $3x + 4y + 5z = 1$ and $6x + 8y + 10z = 2$

b. $2x - 5y + 3z = 4$ and $-6x + 10y - 6z = -12$

c. $3x - 6y + 9z = 12$ and $x - 2y + 3z = 26$

SOLUTION

a. If we multiply both sides of the first equation by 2, the second equation results, so the equations are equivalent, and thus the graphs of both equations are the same plane.

b. The respective coefficients of the variables are not in proportion, so the graphs are not the same plane and are not parallel planes, and thus the planes representing these equations must intersect in a line.

c. The respective coefficients of the variables are in proportion and the constants are not in the same proportion, so the graphs are parallel planes.

Three different planes may intersect in a single point (Figure 7.1), may intersect in a line (as in a paddle wheel; see Figure 7.2), or may not have a common intersection (Figure 7.3). Thus, three linear equations in three variables may have a unique solution, infinitely many solutions, or no solution. For example, the solution of the system

$$\begin{cases} 3x + 2y + z = 6 \\ x - y - z = 0 \\ x + y - z = 4 \end{cases}$$

is the single point with coordinates $x = 1, y = 2$, and $z = -1$, because these three values satisfy all three equations, and these are the only values that satisfy them.

Unique solution	Infinitely many solutions	No solution
Figure 7.1	**Figure 7.2**	**Figure 7.3**

EXAMPLE 5 ▶ Solutions of Systems of Equations

Determine whether the given values of x, y, and z satisfy the system of equations.

a. $\begin{cases} 2x + y + z = -3 \\ x - y - 2z = -4; \quad x = -2, y = 0, z = 1 \\ x + y + 3z = 1 \end{cases}$

b. $\begin{cases} x + y + z = 4 \\ x - y - z = 2; \quad x = 3, y = -1, z = 2 \\ 2x + y - 2z = 3 \end{cases}$

SOLUTION

a. Substituting the values of the variables into each equation gives

$$2(-2) + 0 + 1 = -4 + 1 = -3 \quad \checkmark$$
$$-2 - 0 - 2(1) = -2 - 2 = -4 \quad \checkmark$$
$$-2 + 0 + 3(1) = -2 + 3 = \quad 1 \quad \checkmark$$

The values of x, y, and z satisfy all three equations, so we conclude that $x = -2$, $y = 0, z = 1$ is a solution to the system.

b. Substituting the values of the variables into each equation gives

$$3 + (-1) + 2 = 2 + 2 = 4 \quad \checkmark$$
$$3 - (-1) - 2 = 4 - 2 = 2 \quad \checkmark$$
$$2(3) + (-1) - 2(2) = 5 - 4 = 1 \neq 3$$

Although the values of x, y, and z satisfy the first two equations, they do not satisfy the third equation, so we conclude that $x = 3, y = -1, z = 2$ is not a solution to the system.

Toolbox EXERCISES

1. If y is proportional to x, and $y = 12x$, what is the constant of proportionality?

2. If x is proportional to y, and $y = 10x$, what is the constant of proportionality?

3. Is the pair of numbers 5, 6 proportional to the pair 10, 18?

4. Is the pair of numbers 8, 3 proportional to the pair 24, 9?

5. Is the pair of numbers 4, 6 proportional to the pair 2, 3?

6. Is the pair of numbers 5.1, 2.3 proportional to the pair 51, 23?

7. Suppose y is proportional to x, and when $x = 2$, $y = 18$. Find y when $x = 11$.

8. Suppose x is proportional to y, and when $x = 3$, $y = 18$. Find y when $x = 13$.

In Exercises 9–12, determine whether the pairs of triples are proportional.

9. 8, 6, 2 and 20, 12.5, 5

10. 4.1, 6.8, 9.3 and 8.2, 13.6, 18.6

11. −1, 6, 4 and −3.4, 10.2, 13.6

12. 5, −12, 16 and 2, −3, 4

In Exercises 13–16, determine whether the given values of x, y, and z satisfy the system of equations.

13. $\begin{cases} x - 2y + z = 8 \\ 2x - y + 2z = 10; \quad x = 1, y = -2, z = 3 \\ 3x - 2y + z = 5 \end{cases}$

14. $\begin{cases} x - 2y + 2z = -9 \\ 2x - 3y + z = -14; \quad x = -1, y = 4, z = 0 \\ x + 2y - 3z = 7 \end{cases}$

15. $\begin{cases} x + y - z = 4 \\ 2x + 3y - z = 5; \quad x = 5, y = -2, z = -1 \\ 3x - 2y + 5z = 14 \end{cases}$

16. $\begin{cases} x - 2y - 3z = 2 \\ 2x + 3y + 3z = 19; \quad x = 2, y = 3, z = -2 \\ -x - 2y + 2z = -12 \end{cases}$

17. Determine whether the planes represented by these equations represent the same plane, parallel planes, or neither:

$$x + 3y - 7z = 12 \quad \text{and} \quad 2x + 6y - 14z = 18$$

18. Determine whether the planes represented by these equations represent the same plane, parallel planes, or neither:

$$x - 3y - 2z = 15 \quad \text{and} \quad 2x - 6y - 4z = 30$$

19. Determine whether the planes represented by these equations represent the same plane, parallel planes, or neither:

$$2x - y - z = 2 \quad \text{and} \quad 2x + y - 4z = 8$$

20. Determine whether the planes represented by these following equations represent the same plane, parallel planes, or neither:

$$3x - y + 2z = 9 \quad \text{and} \quad 9x - 3y + 6z = 27$$

7.1 Systems of Linear Equations in Three Variables

KEY OBJECTIVES

- Solve systems of linear equations in three variables using left-to-right elimination

- Find solutions of dependent systems

- Determine when a system of equations is inconsistent

SECTION PREVIEW Manufacturing

A manufacturer of furniture has three models of chairs: Anderson, Blake, and Colonial. The numbers of hours required for framing, upholstery, and finishing for each type of chair are given in Table 7.1. The company has 1500 hours per week for framing, 2100 hours for upholstery, and 850 hours for finishing.

Finding how many of each type of chair can be produced under these conditions involves solving a system with three linear equations in three variables. It is not possible to solve this equation graphically, and new algebraic methods are needed. In this section, we discuss how to solve a system of equations by using the left-to-right elimination method.

Table 7.1

	Anderson	Blake	Colonial
Framing	2	3	1
Upholstery	1	2	3
Finishing	1	2	1/2

Systems in Three Variables

In Section 2.3, we used the elimination method to reduce a system of two linear equations in two variables to one equation in one variable. If a system of three linear equations in three variables has a unique solution, we can reduce the system to an equivalent system where at least one of the equations contains only one variable. We can then easily solve that equation and use the resulting value to find the values of the remaining variables. For example, we will show later that the system

$$\begin{cases} x + y + z = 2 \\ x - y - z = 0 \\ x + y - z = 4 \end{cases}$$

can be reduced to the equivalent system

$$\begin{cases} x + y + z = 2 & (1) \\ \phantom{x + {}} y + z = 1 & (2) \\ \phantom{x + y + {}} z = -1 & (3) \end{cases}$$

Clearly, Equation (3) in the reduced system has solution $z = -1$. Using this value of z in Equation (2) gives

$$y + (-1) = 1, \quad \text{or} \quad y = 2$$

Using $z = -1$ and $y = 2$ in Equation (1) gives

$$x + 2 + (-1) = 2, \quad \text{or} \quad x = 1$$

Thus, the solution to the reduced system is $x = 1, y = 2, z = -1$. The process of using the value of the third variable to work back to the values of the other variables is called **back substitution**.

We say that the reduced system is equivalent to the original system because both systems have the same solutions. We verify this by showing that these values of x, y, and z satisfy the equations of the original system.

$$\begin{cases} 1 + 2 - 1 = 2 \\ 1 - 2 + 1 = 0 \\ 1 + 2 + 1 = 4 \end{cases}$$

Left-to-Right Elimination

The reduced system of equations that we discussed above is in a special form called the **echelon form**, which, in general, looks like

$$\begin{cases} x + ay + bz = d \\ \phantom{x + {}} y + cz = e \\ \phantom{x + ay + {}} z = f \end{cases}$$

where a, b, c, d, e, and f are constants. Notice that in this form, each equation begins with a new (leading) variable with coefficient 1. Any system of linear equations with a unique solution can be reduced to this form. By using the following operations, we can be assured that the new system will be equivalent to the original system.

Operations on a System of Equations

The following operations result in an equivalent system.

1. Interchange any two equations.

2. Multiply both sides of any equation by the same nonzero number.

3. Multiply any equation by a number, add the result to a second equation, and then replace the second equation with the sum.

The following process, which uses these operations to reduce a system of linear equations, is called the **left-to-right elimination method**.

Left-to-Right Elimination Method of Solving Systems of Linear Equations in Three Variables x, y, and z

1. If necessary, interchange two equations or use multiplication to make the coefficient of x in the first equation a 1.

2. Add a multiple of the first equation to each of the following equations so that the coefficients of x in the second and third equations become 0.

3. Multiply (or divide) both sides of the second equation by a number that makes the coefficient of y in the second equation equal to 1.

4. Add a multiple of the (new) second equation to the (new) third equation so that the coefficient of y in the newest third equation becomes 0.

5. Multiply (or divide) both sides of the third equation by a number that makes the coefficient of z in the third equation equal to 1. This gives the solution for z in the system of equations.

6. Use the solution for z to solve for y in the second equation. Then substitute values for y and z to solve for x in the first equation. (This is called back substitution.)

EXAMPLE 1 ▶ | **Solution by Elimination**

Solve the system

$$\begin{cases} 2x - 3y + z = -1 & (1) \\ x - y + 2z = -3 & (2) \\ 3x + y - z = 9 & (3) \end{cases}$$

SOLUTION

1. To write the system with x-coefficient 1 in the first equation, interchange equations (1) and (2).

$$\begin{cases} x - y + 2z = -3 \\ 2x - 3y + z = -1 \\ 3x + y - z = 9 \end{cases}$$

2. To eliminate x in the second equation, multiply the (new) first equation by -2 and add it to the second equation. To eliminate x in the third equation, multiply the (new) first equation by -3 and add it to the third equation. This gives the equivalent system

$$\begin{cases} x - y + 2z = -3 \\ -y - 3z = 5 \\ 4y - 7z = 18 \end{cases}$$

3. To get 1 as the coefficient of y in the second equation, multiply the (new) second equation by -1.

$$\begin{cases} x - y + 2z = -3 \\ y + 3z = -5 \\ 4y - 7z = 18 \end{cases}$$

4. To eliminate y from the (new) third equation, multiply the (new) second equation by -4 and add it to the third equation.

$$\begin{cases} x - y + 2z = -3 \\ y + 3z = -5 \\ -19z = 38 \end{cases}$$

5. To solve for z, divide both sides of the (new) third equation by -19.

$$\begin{cases} x - y + 2z = -3 \\ y + 3z = -5 \\ z = -2 \end{cases}$$

This gives the solution $z = -2$.

6. Substituting -2 for z in the second equation and solving for y gives $y + 3(-2) = -5$, so $y = 1$. Substituting -2 for z and 1 for y in the first equation and solving for x gives $x - 1 + 2(-2) = -3$, so $x = 2$, and the solution is $(2, 1, -2)$.

The process used in step 6 is called *back substitution* because we use the solution for the third variable to work back to the solutions of the first and second variables. Checking in the original system verifies that the solution satisfies all the equations:

$$\begin{cases} 2(2) - 3(1) + (-2) = -1 \\ 2 - 1 + 2(-2) = -3 \\ 3(2) + 1 - (-2) = 9 \end{cases}$$

Modeling Systems of Equations

Solution of real problems sometimes requires us to create three or more equations whose simultaneous solution is the solution to the problem.

EXAMPLE 2 ▶ **Manufacturing**

A manufacturer of furniture has three models of chairs: Anderson, Blake, and Colonial. The numbers of hours required for framing, upholstery, and finishing for each type of chair are given in Table 7.2. The company has 1500 hours per week for framing, 2100 hours for upholstery, and 850 hours for finishing. How many of each type of chair can be produced under these conditions?

Table 7.2

	Anderson	Blake	Colonial
Framing	2	3	1
Upholstery	1	2	3
Finishing	1	2	$\frac{1}{2}$

SOLUTION

If we represent the number of units of the Anderson model by x, the number of units of the Blake model by y, and the number of units of the Colonial model by z, then we have the following equations:

$$\begin{aligned}
\text{Framing:} \quad & 2x + 3y + z = 1500 \\
\text{Upholstery:} \quad & x + 2y + 3z = 2100 \\
\text{Finishing:} \quad & x + 2y + \tfrac{1}{2}z = 850
\end{aligned}$$

We seek the values of x, y, and z that satisfy all three equations above. That is, we seek the solution to the system

$$\begin{cases}
2x + 3y + z = 1500 & (1) \\
x + 2y + 3z = 2100 & (2) \\
x + 2y + \tfrac{1}{2}z = 850 & (3)
\end{cases}$$

We use the following steps to solve the system by left-to-right elimination. The notation that we use to represent each step is shown beside the system. We use this notation in subsequent examples.

To write the system with x-coefficient 1 in the first equation, interchange Equations (1) and (2).

$$\begin{cases}
x + 2y + 3z = 2100 \\
2x + 3y + z = 1500 \quad \text{Eq1} \leftrightarrow \text{Eq2} \\
x + 2y + 0.5z = 850
\end{cases}$$

To eliminate x in the second equation, multiply the first equation by -2 and add it to the second equation. To eliminate x in the third equation, multiply the first equation by -1 and add it to the third equation.

$$\begin{cases}
x + 2y + 3z = 2100 \\
-y - 5z = -2700 \quad (-2)\text{Eq1} + \text{Eq2} \rightarrow \text{Eq2} \\
-2.5z = -1250 \quad (-1)\text{Eq1} + \text{Eq3} \rightarrow \text{Eq3}
\end{cases}$$

Observe that we can solve Equation (3) for z, getting $z = -1250/(-2.5) = 500$. Substituting $z = 500$ in Equation (2) gives $-y - 5(500) = -2700$, or $y = 200$. Substituting $z = 500$ and $y = 200$ in Equation (1) gives $x + 2(200) + 3(500) = 2100$, or $x = 200$.

Thus 200 Anderson, 200 Blake, and 500 Colonial chairs should be made under the given conditions.

Nonunique Solutions

It is not always possible to reduce a system of three equations in three variables to a system in which the third equation contains one variable. For example, the system

$$\begin{cases}
x + 2y + z = 2 & (1) \\
x - y - 2z = -4 & (2) \\
2x + y - z = 4 & (3)
\end{cases}$$

reduces to

$$\begin{cases}
x + 2y + z = 2 \\
y + z = 2 \\
0 + 0 = 6
\end{cases}$$

No solution

Figure 7.4

The third equation in the reduced system is $0 = 6$, which is impossible. Thus, this system and the original system have no solution. Such a system is called an **inconsistent system**.

When a system of three linear equations in three variables has no solution, the graphs of the equations will not have a common intersection, as shown in Figure 7.4.

The system

$$\begin{cases} x + 2y + z = 2 & (1) \\ x - y - 2z = -4 & (2) \\ 2x + y - z = -2 & (3) \end{cases}$$

looks similar but is slightly different from the inconsistent system just discussed. Reducing this system with left-to-right elimination results in the system

$$\begin{cases} x + 2y + z = 2 \\ y + z = 2 \\ 0 + 0 = 0 \end{cases}$$

The third equation in this system is $0 = 0$, and we will see that the system has infinitely many solutions. Notice that the z is not a leading variable in any of the equations, so we can solve for the other variables in terms of z. Back substitution gives

$$y = 2 - z$$
$$x = 2 - 2(2 - z) - z = -2 + z$$

Thus, the solutions have the form $x = -2 + z, y = 2 - z$, and $z =$ any number.

By selecting different values of z, we can find some particular solutions of this system.

For example,

$$z = 1 \quad \text{gives} \quad x = -1, \quad y = 1, \quad z = 1$$
$$z = 2 \quad \text{gives} \quad x = 0, \quad y = 0, \quad z = 2$$
$$z = 5 \quad \text{gives} \quad x = 3, \quad y = -3, \quad z = 5$$

Infinitely many solutions

Figure 7.5

A system of this type, with infinitely many solutions, is called a **dependent system**. When a system of three equations in three variables has infinitely many solutions, the graphs of the equations may be planes that intersect in a line, as in the paddle wheel shown in Figure 7.5.

EXAMPLE 3 ▶   Find the solution to each of the following systems, if it exists.

a. $\begin{cases} x - y + 4z = -5 \\ 3x \qquad + z = 0 \\ -x + y - 4z = 20 \end{cases}$ **b.** $\begin{cases} -x + y + 2z = 0 \\ x + 2y + z = 6 \\ -2x - y + z = -6 \end{cases}$

SOLUTION

a. The system

$$\begin{cases} x - y + 4z = -5 & (1) \\ 3x \qquad + z = 0 & (2) \\ -x + y - 4z = 20 & (3) \end{cases}$$

reduces as follows:

$$\begin{cases} x - y + 4z = -5 \\ 3y - 11z = 15 \qquad \text{(−3)Eq1 + Eq2 → Eq2} \\ 0 = 15 \qquad \text{Eq1 + Eq3 → Eq3} \end{cases}$$

The third equation in the reduced system is $0 = 15$, which is impossible. Thus, the system is inconsistent and has no solution.

b. The system

$$\begin{cases} -x + y + 2z = 0 & (1) \\ x + 2y + z = 6 & (2) \\ -2x - y + z = -6 & (3) \end{cases}$$

reduces as follows:

$$\begin{cases} x - y - 2z = 0 & (-1)\text{Eq1} \rightarrow \text{Eq1} \\ x + 2y + z = 6 \\ -2x - y + z = -6 \end{cases}$$

$$\begin{cases} x - y - 2z = 0 \\ 3y + 3z = 6 & (-1)\text{Eq1} + \text{Eq2} \rightarrow \text{Eq2} \\ -3y - 3z = -6 & (2)\text{Eq1} + \text{Eq3} \rightarrow \text{Eq3} \end{cases}$$

$$\begin{cases} x - y - 2z = 0 \\ y + z = 2 & (1/3)\text{Eq2} \rightarrow \text{Eq2} \\ -3y - 3z = -6 \end{cases}$$

$$\begin{cases} x - y - 2z = 0 \\ y + z = 2 \\ 0 = 0 & (3)\text{Eq2} + \text{Eq3} \rightarrow \text{Eq3} \end{cases}$$

Because the third equation in the reduced system is $0 = 0$, the system has infinitely many solutions. We let z be any number and use back substitution to solve the other variables in terms of z.

$$y = 2 - z$$
$$x = 2 - z + 2z = 2 + z$$

Thus, the solution has the form

$$x = 2 + z$$
$$y = 2 - z$$
$$z = \text{any number}$$

If a system of equations has fewer equations than variables, the system must be dependent or inconsistent. Consider the following example.

EXAMPLE 4 ▶ Purchasing

A young man wins $200,000 and (foolishly) decides to buy a new car for each of the seven days of the week. He wants to choose from cars that are priced at $40,000, $30,000, and $25,000. How many cars of each price can he buy with the $200,000?

SOLUTION

If he buys x cars costing $40,000 each, y cars costing $30,000 each, and z cars costing $25,000 each, then

$$x + y + z = 7 \quad \text{and} \quad 40{,}000x + 30{,}000y + 25{,}000z = 200{,}000$$

Writing these two equations as a system of equations gives

$$\begin{cases} x + y + z = 7 & (1) \\ 40{,}000x + 30{,}000y + 25{,}000z = 200{,}000 & (2) \end{cases}$$

Because there are three variables and only two equations, the solution cannot be unique. The solution follows.

$$\begin{cases} x + y + z = 7 \\ -10{,}000y - 15{,}000z = -80{,}000 \end{cases} \qquad -40{,}000\text{Eq1} + \text{Eq2} \rightarrow \text{Eq2}$$

$$\begin{cases} x + y + z = 7 \\ y + 1.5z = 8 \end{cases} \qquad -0.0001\text{Eq2} \rightarrow \text{Eq2}$$

In the solution to this system, we see that z can be any value and that x and y depend on z.

$$y = 8 - 1.5z$$
$$x = -1 + 0.5z$$

In the context of this application, the solutions must be nonnegative integers (none of $x, y,$ or z can be negative), so z must be at least 2 (or x will be negative) and z cannot be more than 5 (or y will be negative). In addition, z must be an even number (to get integer solutions). Thus, the only possible selections are

$$z = 2, x = 0, y = 5 \quad \text{or} \quad z = 4, x = 1, y = 2$$

Thus, the man can buy two \$25,000 cars, zero \$40,000 cars, and five \$30,000 cars, or he can buy four \$25,000 cars, one \$40,000 car, and two \$30,000 cars. So even though there are an infinite number of solutions to this system of equations, there are only two solutions in the context of this application.

SKILLS CHECK 7.1

In Exercises 1–4, solve the systems of equations.

1. $\begin{cases} x + 2y - z = 3 \\ y + 3z = 11 \\ z = 3 \end{cases}$

2. $\begin{cases} x - 4y - 3z = 3 \\ y - 2z = 11 \\ z = 4 \end{cases}$

3. $\begin{cases} x + 2y - z = 6 \\ y + 3z = 3 \\ z = -2 \end{cases}$

4. $\begin{cases} x + 2y - z = 22 \\ y + 3z = 21 \\ z = -5 \end{cases}$

In Exercises 5–16, use left-to-right elimination to solve the systems of equations.

5. $\begin{cases} x - y - 4z = 0 \\ y + 2z = 4 \\ 3y + 7z = 22 \end{cases}$

6. $\begin{cases} x + 4y - 11z = 33 \\ y - 3z = 11 \\ 2y + 7z = -4 \end{cases}$

7. $\begin{cases} x - 2y + 3z = 0 \\ y - 2z = -1 \\ y + 5z = 6 \end{cases}$

8. $\begin{cases} x - 2y + 3z = -10 \\ y - 2z = 7 \\ y - 3z = 6 \end{cases}$

9. $\begin{cases} x + 2y - 2z = 0 \\ x - y + 4z = 3 \\ x + 2y + 2z = 3 \end{cases}$

10. $\begin{cases} x + 4y - 14z = 22 \\ x + 5y + z = -2 \\ x + 4y - z = 9 \end{cases}$

11. $\begin{cases} x + 3y - z = 0 \\ x - 2y + z = 8 \\ x - 6y + 2z = 6 \end{cases}$

12. $\begin{cases} x - 5y - 2z = 7 \\ x - 3y + 4z = 21 \\ x - 5y + 2z = 19 \end{cases}$

13. $\begin{cases} 2x + 4y - 14z = 0 \\ 3x + 5y + z = 19 \\ x + 4y - z = 12 \end{cases}$

14. $\begin{cases} 3x - 6y + 6z = 24 \\ 2x - 6y + 2z = 6 \\ x - 3y + 2z = 4 \end{cases}$

15. $\begin{cases} 3x - 4y + 6z = 10 \\ 2x - 4y - 5z = -14 \\ x + 2y - 3z = 0 \end{cases}$

16. $\begin{cases} x - 3y + 4z = 7 \\ 2x + 2y - 3z = -3 \\ x - 3y + z = -2 \end{cases}$

The systems in Exercises 17–26 do not have unique solutions. Solve each system, if possible.

17. $\begin{cases} x - 3y + z = -2 \\ y - 2z = -4 \end{cases}$ **18.** $\begin{cases} x + 2y + 2z = 4 \\ y + 3z = 6 \end{cases}$

19. $\begin{cases} x - y + 2z = 4 \\ y + 8z = 16 \end{cases}$ **20.** $\begin{cases} x - 3y - 2z = 3 \\ y + 4z = 5 \\ 0 = 4 \end{cases}$

21. $\begin{cases} x - y + 2z = -1 \\ 2x + 2y + 3z = -3 \\ x + 3y + z = -2 \end{cases}$

22. $\begin{cases} x + 5y - 2z = 5 \\ x + 3y - 2z = 23 \\ x + 4y - 2z = 14 \end{cases}$

23. $\begin{cases} 3x - 4y - 6z = 10 \\ 2x - 4y - 5z = -14 \\ x - z = 0 \end{cases}$

24. $\begin{cases} 3x - 9y + 4z = 24 \\ 2x - 6y + 2z = 6 \\ x - 3y + 2z = 20 \end{cases}$

25. $\begin{cases} 2x + 3y - 14z = 16 \\ 4x + 5y - 30z = 34 \\ x + y - 8z = 9 \end{cases}$

26. $\begin{cases} 2x + 4y - 4z = 6 \\ x + 5y - 3z = 3 \\ 3x + 3y - 5z = 9 \end{cases}$

EXERCISES 7.1

27. Rental Cars A car rental agency rents compact, midsize, and luxury cars. Its goal is to purchase 60 cars with a total of $1,400,000 and to earn a daily rental of $2200 from all the cars. The compact cars cost $15,000 and earn $30 per day in rental, the midsize cars cost $25,000 and earn $40 per day, and the luxury cars cost $45,000 and earn $50 per day. To find the number of each type of car the agency should purchase, solve the system of equations

$$\begin{cases} x + y + z = 60 \\ 15{,}000x + 25{,}000y + 45{,}000z = 1{,}400{,}000 \\ 30x + 40y + 50z = 2200 \end{cases}$$

where x, y, and z are the numbers of compact cars, midsize cars, and luxury cars, respectively.

28. Ticket Pricing A theater owner wants to divide an 1800-seat theater into three sections, with tickets costing $20, $35, and $50, depending on the section. He wants to have twice as many $20 tickets as the sum of the other kinds of tickets, and he wants to earn $48,000 from a full house. To find how many seats he should have in each section, solve the system of equations

$$\begin{cases} x + y + z = 1800 \\ x = 2(y + z) \\ 20x + 35y + 50z = 48{,}000 \end{cases}$$

where x, y, and z are the numbers of $20 tickets, $35 tickets, and $50 tickets, respectively.

29. Pricing A concert promoter needs to make $120,000 from the sale of 2600 tickets. The promoter charges $40 for some tickets and $60 for the others.

a. If there are x of the $40 tickets and y of the $60 tickets, write an equation that states that the total number of the tickets sold is 2600.

b. How much money is made from the sale of x tickets for $40 each?

c. How much money is made from the sale of y tickets for $60 each?

d. Write an equation that states that the total amount made from the sale is $120,000.

e. Solve the equations simultaneously to find how many tickets of each type must be sold to yield the $120,000.

30. Investment A trust account manager has $500,000 to be invested in three different accounts. The accounts pay 8%, 10%, and 14%, respectively, and the goal is to earn $49,000, with the amount invested at 8% equal to the sum of the other two investments. To accomplish this, assume that x dollars are invested at 8%, y dollars are invested at 10%, and z dollars are invested at 14%.

a. Write an equation that describes the sum of money in the three investments.

b. Write an equation that describes the total amount of money earned by the three investments.

c. Write an equation that describes the relationship among the three investments.

d. Solve the system of equations to find how much should be invested in each account to satisfy the conditions.

31. *Investment* A man has $400,000 invested in three rental properties. One property earns 7.5% per year on the investment, the second earns 8%, and the third earns 9%. The total annual earnings from the three properties is $33,700, and the amount invested at 9% equals the sum of the first two investments. Let x equal the investment at 7.5%, y equal the investment at 8%, and z represent the investment at 9%.

a. Write an equation that represents the sum of the three investments.

b. Write an equation that states that the sum of the returns from all three investments is $33,700.

c. Write an equation that states that the amount invested at 9% equals the sum of the other two investments.

d. Solve the system of equations to find how much is invested in each property.

32. *Loans* A bank loans $285,000 to a development company to purchase three business properties. One of the properties costs $45,000 more than another, and the third costs twice the sum of these two properties.

a. Write an equation that represents the total loaned as the sum of the costs of the three properties, if the costs are x, y, and z, respectively.

b. Write an equation that states that one cost is $45,000 more than another.

c. Write an equation that states that the third cost is equal to twice the sum of the other two.

d. Solve the system of equations to find the cost of each property.

33. *Transportation* Ace Trucking Company has an order for delivery of three products, A, B, and C. The table below gives the volume in cubic feet, the weight in pounds, and the value for insurance in dollars for a unit of each of the products. If the carrier can carry 8000 cubic feet and 12,400 pounds and is insured for $52,600, how many units of each product can be carried?

	Product A	Product B	Product C
Volume (cu ft)	24	20	40
Weight (lb)	40	30	60
Value ($)	150	180	200

34. *Nutrition* The following table gives the calories, fat, and carbohydrates per ounce for three brands of cereal, as well as the total amount of calories, fat, and carbohydrates required for a special diet.

	Calories	Fat	Carbohydrates
All Bran	50	0	22.0
Sugar Frosted Flakes	108	0.1	25.5
Natural Mixed Grain	127	5.5	18.0
Total Required	393	5.7	91.0

To find the number of ounces of each brand that should be combined to give the required totals of calories, fat, and carbohydrates, let x represent the number of ounces of All Bran, y represent the number of ounces of Sugar Frosted Flakes, and z represent the number of ounces of Natural Mixed Grain.

a. Use the information in the table to write a system of three equations.

b. Solve the system to find the amount of each brand of cereal that will give the necessary nutrition.

Exercises 35–40 have nonunique solutions.

35. *Investment* A trust account manager has $500,000 to invest in three different accounts. The accounts pay 8%, 10%, and 14%, and the goal is to earn $49,000. To solve this problem, assume that x dollars are invested at 8%, y dollars are invested at 10%, and z dollars are invested at 14%. Then x, y, and z must satisfy the equations $x + y + z = 500,000$ and $0.08x + 0.10y + 0.14z = 49,000$.

a. Explain the meaning of each of these equations.

b. Solve the systems containing these two equations. Find x and y in terms of z.

c. What limits must there be on z so that all investment values are nonnegative?

36. *Loans* A bank gives three loans totaling $200,000 to a development company for the purchase of three business properties. The largest loan is $45,000 more than the sum of the other two. Represent the amount of money in each loan as x, y, and z, respectively.

a. Write a system of two equations in three variables to represent the problem.

b. Does the system have a unique solution?

c. Find the amount of the largest loan.

d. What is the relationship between the remaining two loans?

37. *Manufacturing* A company manufactures three types of air conditioners: the Acclaim, the Bestfrig, and the

Cool King. The hours needed for assembly, testing, and packing for each product are given in the table below. The numbers of hours that are available daily are 300 for assembly, 120 for testing, and 210 for packing. Can all of the available hours be used to produce these three types of air conditioners? If so, how many of each type can be produced?

	Assembly	Testing	Packing
Acclaim	5	2	1.4
Bestfrig	4	1.4	1.2
Cool King	4.5	1.7	1.3

38. *Transportation* Hohman Trucking Company has an order for delivery of three products, A, B, and C. The following table gives the volume in cubic feet, the weight in pounds, and the value for insurance in dollars per unit for each of the products. If one of the company's trucks can carry 4000 cubic feet and 6000 pounds and is insured to carry $24,450, how many units of each product can be carried on the truck?

	Product A	Product B	Product C
Volume (cu ft)	24	20	50
Weight (lb)	36	30	75
Value ($)	150	180	120

39. *Nutrition* A psychologist, studying the effects of good nutrition on the behavior of rabbits, feeds one group a combination of three foods: I, II, and III. Each of these foods contains three additives: A, B, and C. The following table gives the percent of each additive that is present in each food. If the diet being used requires 6.88 g per day of A, 6.72 g of B, and 6.8 g of C, find the number of grams of each food that should be used each day.

	Food I	Food II	Food III
Additive A (%)	12	14	8
Additive B (%)	8	6	16
Additive C (%)	10	10	12

40. *Investment* A woman has $1,400,000 to purchase three investments. One investment earns 8%, a second earns 7.5%, and the third earns 10%.

   a. What investment strategies will give annual earnings of $120,000 from these three investments?

   b. What one investment strategy will minimize her risk and return $120,000 if the 10% investment has the most risk?

7.2 Matrix Solution of Systems of Linear Equations

KEY OBJECTIVES

- Represent systems of equations with matrices
- Find dimensions of matrices
- Identify square matrices
- Identify an identity matrix
- Form an augmented matrix
- Identify a coefficient matrix
- Reduce a matrix with row operations
- Reduce a matrix to its row-echelon form
- Reduce a matrix to its reduced row-echelon form
- Solve systems of equations using the Gauss–Jordan elimination method

SECTION PREVIEW Transportation

Ace Trucking Company has an order for delivery of three products: A, B, and C. Table 7.3 gives the volume in cubic feet, the weight in pounds, and the value for insurance in dollars per unit of each of the products. If the carrier can carry 30,000 cubic feet and 62,000 pounds and is insured for $276,000, how many units of each product can be carried?

Table 7.3

	Product A	Product B	Product C
Volume (cu ft)	25	22	30
Weight (lb)	25	38	70
Value ($)	150	180	300

If the numbers of units of products A, B, and C are x, y, and z, respectively, the system of linear equations that satisfies these conditions is

$$\begin{cases} 25x + 22y + 30z = 30,000 & \text{Volume} \\ 25x + 38y + 70z = 62,000 & \text{Weight} \\ 150x + 180y + 300z = 276,000 & \text{Value} \end{cases}$$

As we will see in Example 5, there is a solution, but not a unique solution. In this section, we use matrices to solve systems with unique solutions and with infinitely many solutions, and we see how to determine when a system has no solution. ∎

Matrix Representation of Systems of Equations

Recall from Section 7.1 that in solving a system of three linear equations in three variables, our goal in each step was to use elimination to remove each variable, in turn, from the equations below it in the system. This process can be simplified if we represent the system in a new way without including the symbols for the variables. We can do this by creating a rectangular array containing the coefficients and constants from the system. This array, called a **matrix** (the plural is **matrices**), uses each row (horizontal) to represent an equation and each column (vertical) to represent a variable. For example, the system of equations that was solved in Example 1 of Section 7.1,

$$\begin{cases} 2x - 3y + z = -1 \\ x - y + 2z = -3 \\ 3x + y - z = 9 \end{cases}$$

can be represented in a matrix as follows:

$$\left[\begin{array}{rrr|r} 2 & -3 & 1 & -1 \\ 1 & -1 & 2 & -3 \\ 3 & 1 & -1 & 9 \end{array}\right]$$

In this matrix, the first column contains the coefficient of x for each of the equations, the second column contains the coefficients of y, and the third column contains the coefficients of z. The column to the right of the vertical line, containing the constants of the equations, is called the **augment** of the matrix, and a matrix containing an augment is called an **augmented matrix**. The matrix that contains only the coefficients of the variables is called a **coefficient matrix**. The coefficient matrix for this system is

$$\left[\begin{array}{rrr} 2 & -3 & 1 \\ 1 & -1 & 2 \\ 3 & 1 & -1 \end{array}\right]$$

The augmented matrix above has 3 rows and 4 columns, so its **dimension** is 3×4 (three by four). The coefficient matrix above is a 3×3 matrix. Matrices like this one are called **square matrices**. A square matrix that has 1's down its diagonal and 0's everywhere else, like matrix I below, is called an **identity matrix**. Matrix I is a 3×3 identity matrix.

$$I = \left[\begin{array}{rrr} 1 & 0 & 0 \\ 0 & 1 & 0 \\ 0 & 0 & 1 \end{array}\right]$$

Echelon Forms of Matrices; Solving Systems with Matrices

The matrix that represents the system of equations

$$\begin{cases} x - y + 2z = -3 \\ y + 3z = -5 \quad \text{is} \\ z = -2 \end{cases} \left[\begin{array}{rrr|r} 1 & -1 & 2 & -3 \\ 0 & 1 & 3 & -5 \\ 0 & 0 & 1 & -2 \end{array}\right]$$

The coefficient part of this augmented matrix has all 1's on its diagonal and all 0's below its diagonal.

Any augmented matrix that has 1's or 0's on the diagonal of its coefficient part and 0's below the diagonal is said to be in **row-echelon form**.

If the rows of an augmented matrix represent the equations of a system of linear equations, we can perform **row operations** on the matrices that are equivalent to operations that we performed on equations to solve a system of equations. These operations are as follows.

Matrix Row Operations

Row Operation	**Corresponding Equation Operation**
1. Interchange two rows of the matrix.	1. Interchange two equations.
2. Multiply a row by any nonzero constant.	2. Multiply an equation by any nonzero constant.
3. Add a multiple of one row to another row.	3. Add a multiple of one equation to another.

When a new matrix results from one or more of these row operations performed on a matrix, the new matrix is **equivalent** to the original matrix.

EXAMPLE 1 ▶ Matrix Solution of Linear Systems

Use matrix row operations to solve the system of equations

$$\begin{cases} 2x - 3y + z = -1 & (1) \\ x - y + 2z = -3 & (2) \\ 3x + y - z = 9 & (3) \end{cases}$$

SOLUTION

We begin by writing the augmented matrix that represents the system.

$$\left[\begin{array}{ccc|c} 2 & -3 & 1 & -1 \\ 1 & -1 & 2 & -3 \\ 3 & 1 & -1 & 9 \end{array}\right]$$

To get a first row that has 1 in the first column, we interchange row 1 and row 2. (Compare this to the equivalent system of equations, which is to the right.)

$$R_1 \leftrightarrow R_2 \quad \left[\begin{array}{ccc|c} 1 & -1 & 2 & -3 \\ 2 & -3 & 1 & -1 \\ 3 & 1 & -1 & 9 \end{array}\right] \qquad \begin{cases} x - y + 2z = -3 \\ 2x - 3y + z = -1 \\ 3x + y - z = 9 \end{cases}$$

To get 0 as the first entry in the second row, we multiply the first row by -2 and add it to the second row. To get 0 as the first entry in the third row, we multiply the first row by -3 and add it to the third row. (Compare this to the equivalent system of equations, which is to the right.)

$$\begin{array}{c} -2R_1 + R_2 \rightarrow R_2 \\ -3R_1 + R_3 \rightarrow R_3 \end{array} \quad \left[\begin{array}{ccc|c} 1 & -1 & 2 & -3 \\ 0 & -1 & -3 & 5 \\ 0 & 4 & -7 & 18 \end{array}\right] \qquad \begin{cases} x - y + 2z = -3 \\ -y - 3z = 5 \\ 4y - 7z = 18 \end{cases}$$

To get 1 as the second entry in the second row, we multiply the second row by -1. (Compare this to the equivalent system of equations, which is to the right.)

$$-1R_2 \rightarrow R_2 \quad \begin{bmatrix} 1 & -1 & 2 & | & -3 \\ 0 & 1 & 3 & | & -5 \\ 0 & 4 & -7 & | & 18 \end{bmatrix} \qquad \begin{cases} x - y + 2z = -3 \\ \phantom{x - {}} y + 3z = -5 \\ \phantom{x - {}} 4y - 7z = 18 \end{cases}$$

To get 0 as the second entry in the third row, we multiply the new second row by -4 and add it to the third row. (Compare this to the equivalent system of equations, which is to the right.)

$$-4R_2 + R_3 \rightarrow R_3 \quad \begin{bmatrix} 1 & -1 & 2 & | & -3 \\ 0 & 1 & 3 & | & -5 \\ 0 & 0 & -19 & | & 38 \end{bmatrix} \qquad \begin{cases} x - y + 2z = -3 \\ \phantom{x - {}} y + 3z = -5 \\ -19z = 38 \end{cases}$$

To get 1 as the third entry in the third row, we multiply the third row by $-\dfrac{1}{19}$. (Compare this to the equivalent system of equations, which is to the right.)

$$(-1/19)\, R_3 \rightarrow R_3 \quad \begin{bmatrix} 1 & -1 & 2 & | & -3 \\ 0 & 1 & 3 & | & -5 \\ 0 & 0 & 1 & | & -2 \end{bmatrix} \qquad \begin{cases} x - y + 2z = -3 \\ \phantom{x - {}} y + 3z = -5 \\ z = -2 \end{cases}$$

The matrix is now in row-echelon form. The equivalent system can be solved by back substitution, giving the solution $(2, 1, -2)$, the same solution that was found in Example 1 of Section 7.1.

Gauss–Jordan Elimination

One method that can be used to solve a system of n equations in n variables is called the **Gauss–Jordan elimination method**. To use this method, we attempt to reduce the $n \times n$ coefficient matrix to one that contains 1's on its diagonal and 0's everywhere else.

> The augmented matrix representing n equations in n variables is said to be in **reduced row-echelon form** if it has 1's or 0's on the diagonal of its coefficient part and 0's everywhere else.

When the original augmented matrix represents a system of n equations in n variables with a unique solution, this method will reduce the coefficient matrix of the augmented matrix to an identity matrix. When this happens, we can easily find the solution to the system, as we will see below.

Rather than use back substitution to complete the solution to the system of Example 1, we can apply several additional row operations to reduce the matrix to reduced row-echelon form. That is, we can reduce the matrix

$$\begin{bmatrix} 1 & -1 & 2 & | & -3 \\ 0 & 1 & 3 & | & -5 \\ 0 & 0 & 1 & | & -2 \end{bmatrix}$$

to reduced row-echelon form with the following steps:

1. Add row 2 to row 1:

$$R_2 + R_1 \rightarrow R_1 \quad \begin{bmatrix} 1 & 0 & 5 & | & -8 \\ 0 & 1 & 3 & | & -5 \\ 0 & 0 & 1 & | & -2 \end{bmatrix}$$

2. Add (-5) times row 3 to row 1 and (-3) times row 3 to row 2:

$$\begin{matrix} -5R_3 + R_1 \to R_1 \\ -3R_3 + R_2 \to R_2 \end{matrix} \longrightarrow \left[\begin{array}{ccc|c} 1 & 0 & 0 & 2 \\ 0 & 1 & 0 & 1 \\ 0 & 0 & 1 & -2 \end{array}\right]$$

3. The coefficient matrix part of this reduced row-echelon matrix is the identity matrix, so the solutions to the system can be easily "read" from the reduced augmented matrix. The rows of this reduced row-echelon matrix translate into the equations

$$\begin{cases} x + 0y + 0z = 2 \\ 0x + y + 0z = 1 \\ 0x + 0y + z = -2 \end{cases} \quad \text{or} \quad \begin{cases} x = 2 \\ y = 1 \\ z = -2 \end{cases}$$

This solution agrees with the solution to the system found in Example 1.

EXAMPLE 2 ▶ Investment

The Trust Department of Century Bank divided a $150,000 investment among three mutual funds with different levels of risk and return. The Potus Fund returns 10% per year, the Stong Fund returns 8%, and the Franklin Fund returns 7%. If the annual return from the combined investments is $12,900 and if the investment in the Potus Fund has $20,000 less than the sum of the investments in the other two funds, how much is invested in each fund?

SOLUTION

If we represent the amount invested in the Potus Fund by x, the amount invested in the Stong Fund by y, and the amount invested in the Franklin Fund by z, the equations that represent this situation are

$$x + y + z = 150,000$$
$$0.10x + 0.08y + 0.07z = 12,900$$
$$x = y + z - 20,000$$

We can write these equations in a system, create an augmented matrix, and reduce the matrix representing the system to reduced row-echelon form.

$$\begin{cases} x + y + z = 150,000 \\ 0.10x + 0.08y + 0.07z = 12,900 \\ x - y - z = -20,000 \end{cases}$$

$$\left[\begin{array}{ccc|c} 1 & 1 & 1 & 150,000 \\ 0.10 & 0.08 & 0.07 & 12,900 \\ 1 & -1 & -1 & -20,000 \end{array}\right] \begin{matrix} -0.10R_1 + R_2 \to R_2 \\ -R_1 + R_3 \to R_3 \end{matrix} \left[\begin{array}{ccc|c} 1 & 1 & 1 & 150,000 \\ 0 & -0.02 & -0.03 & -2,100 \\ 0 & -2 & -2 & -170,000 \end{array}\right]$$

$$\xrightarrow{-50R_2 \to R_2} \left[\begin{array}{ccc|c} 1 & 1 & 1 & 150,000 \\ 0 & 1 & 1.5 & 105,000 \\ 0 & -2 & -2 & -170,000 \end{array}\right] \begin{matrix} -R_2 + R_1 \to R_1 \\ 2R_2 + R_3 \to R_3 \end{matrix} \left[\begin{array}{ccc|c} 1 & 0 & -0.5 & 45,000 \\ 0 & 1 & 1.5 & 105,000 \\ 0 & 0 & 1 & 40,000 \end{array}\right]$$

$$\begin{matrix} 0.5R_3 + R_1 \to R_1 \\ -1.5R_3 + R_2 \to R_2 \end{matrix} \left[\begin{array}{ccc|c} 1 & 0 & 0 & 65,000 \\ 0 & 1 & 0 & 45,000 \\ 0 & 0 & 1 & 40,000 \end{array}\right]$$

From the reduced row-echelon matrix, we see that

$$x = 65,000, \quad y = 45,000, \quad \text{and} \quad z = 40,000$$

Thus, the investments are $65,000 in the Potus Fund, $45,000 in the Stong Fund, and $40,000 in the Franklin Fund.

(a)

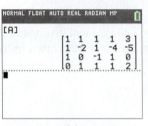

(b)

Figure 7.6

Solution with Technology

Graphing calculators are very useful in reducing an augmented matrix to solve a system of linear equations. To solve a system of linear equations with a graphing calculator, we use the following steps.

1. Enter the coefficients and constants in a matrix: Select MATRIX (accessed by **2ND** **x⁻¹**), move to EDIT, select a matrix name, enter the number of rows and columns of the matrix, and enter the coefficients of x, y, and z and the constants into columns 1, 2, 3, and 4, respectively. See Figure 7.6(a), which shows the augmented matrix for the system of equations in Example 2.*

2. Reduce the augmented matrix: Select MATRIX, move to MATH, choose "B:rref(" on the display, and select MATRIX and [A] under NAMES (getting rref([A]) on the screen). Pressing **ENTER** gives the reduced matrix, from which the solution can be read. See Figure 7.6(b), which shows the same solution matrix as the one found in Example 2.[†]

Solving systems of linear equations in more than three variables can also be done using Gauss–Jordan elimination. Because of the increased number of steps involved, technology is especially useful in solving larger systems.[††]

EXAMPLE 3 ▶ **Four Equations in Four Variables**

Solve the system

$$\begin{cases} x + y + z + w = 3 \\ x - 2y + z - 4w = -5 \\ x \quad\quad - z + w = 0 \\ y + z + w = 2 \end{cases}$$

(a)

(b)

Figure 7.7

SOLUTION

This system can be represented by the augmented matrix

$$\begin{bmatrix} 1 & 1 & 1 & 1 & | & 3 \\ 1 & -2 & 1 & -4 & | & -5 \\ 1 & 0 & -1 & 1 & | & 0 \\ 0 & 1 & 1 & 1 & | & 2 \end{bmatrix}$$

We can enter this augmented matrix into a graphing calculator (Figure 7.7(a)) and reduce the matrix to its reduced row-echelon form (Figure 7.7(b)). The solution to the system can be "read" from the reduced matrix:

$$x = 1, \quad y = 11, \quad z = -4, \quad w = -5, \quad \text{or} \quad (1, 11, -4, -5)$$

Nonunique Solution

If we represent a system of three equations in three variables by an augmented matrix and the coefficient matrix reduces to a 3×3 identity matrix, then the system has a unique solution. If the coefficient matrix representing a system of three equations in three variables does not reduce to an identity matrix, then either there is no solution to the system or there are an infinite number of solutions.

If a row of the reduced row-echelon coefficient matrix associated with a system contains all 0's and the augment of that row contains a nonzero number, the system has no solution and is an **inconsistent system**.

---

* For detailed steps, see Appendix A, page 678.

[†] For detailed steps, see Appendix A, page 680.

[††] For details of this solution method, see Appendix A, page 680.

If a row of the reduced 3×3 row-echelon coefficient matrix associated with a system contains all 0's and the augment of that row also contains 0, then there are infinitely many solutions. If any column of the reduced matrix does not contain a leading 1, that variable can assume any value, and the variables corresponding to columns containing leading 1's will have values dependent on that variable. (There may be more than one variable that can assume any value, with other variables depending on them.) This is a **dependent system**, and it has infinitely many solutions.

Of course, any system with two equations in three variables cannot have a unique solution.

> A system with fewer equations than variables has either infinitely many solutions or no solutions.

Dependent Systems

The following example illustrates the solution of a dependent system.

EXAMPLE 4 ▶   ## A Dependent System

Solve the system

$$\begin{cases} 5x + 10y + 12z = 6{,}140 \\ 10x + 18y + 30z = 13{,}400 \\ 300x + 480y + 1080z = 435{,}600 \end{cases}$$

SOLUTION

The augmented matrix that represents the system is

$$\left[\begin{array}{ccc|c} 5 & 10 & 12 & 6{,}140 \\ 10 & 18 & 30 & 13{,}400 \\ 300 & 480 & 1080 & 435{,}600 \end{array} \right]$$

The procedure follows:

$$\left[\begin{array}{ccc|c} 5 & 10 & 12 & 6{,}140 \\ 10 & 18 & 30 & 13{,}400 \\ 300 & 480 & 1080 & 435{,}600 \end{array} \right] \xrightarrow{(1/5)R_1 \to R_1} \left[\begin{array}{ccc|c} 1 & 2 & 2.4 & 1{,}228 \\ 10 & 18 & 30 & 13{,}400 \\ 300 & 480 & 1080 & 435{,}600 \end{array} \right]$$

$$\begin{array}{c} -10R_1 + R_2 \to R_2 \\ -300R_1 + R_3 \to R_3 \end{array} \left[\begin{array}{ccc|c} 1 & 2 & 2.4 & 1{,}228 \\ 0 & -2 & 6 & 1{,}120 \\ 0 & -120 & 360 & 67{,}200 \end{array} \right]$$

$$\xrightarrow{(-1/2)R_2 \to R_2} \left[\begin{array}{ccc|c} 1 & 2 & 2.4 & 1{,}228 \\ 0 & 1 & -3 & -560 \\ 0 & -120 & 360 & 67{,}200 \end{array} \right]$$

$$\begin{array}{c} 120R_2 + R_3 \to R_3 \\ -2R_2 + R_1 \to R_1 \end{array} \left[\begin{array}{ccc|c} 1 & 0 & 8.4 & 2348 \\ 0 & 1 & -3 & -560 \\ 0 & 0 & 0 & 0 \end{array} \right]$$

Note that the third row contains only 0's and there is no leading 1 in the "z" column of the reduced matrix, and that z occurs in the equations which correspond to the first two rows of the matrix. Thus, we can solve for x in terms of z in the first equation and for y in terms of z in the second equation. These solutions are

$$x = -8.4z + 2348$$
$$y = 3z - 560$$
$$z = \text{any real number}$$

Different values of z give different solutions to the system. Two sample solutions are $z = 0$, $y = -560$, $x = 2348$ and $z = 10$, $y = -530$, $x = 2264$. If $z = a$, the solution is $(-8.4a + 2348, 3a - 560, a)$.

Notice that the row-reduction technique is the same for systems that have nonunique solutions as it is for systems that have unique solutions. To solve a system of n linear equations in n variables, we attempt to reduce the coefficient matrix to the identity matrix. If we succeed, the system has a unique solution. If we are unable to reduce the coefficient matrix to the identity matrix, either there is no solution or the system is dependent and we can solve it in terms of one or more of the variables.

Figure 7.8(a) shows the original augmented matrix for the system of Example 4 on a calculator screen, and Figure 7.8(b) shows the reduced row-echelon form that is equivalent to the original matrix. This is the same as the reduced matrix found in Example 4 and yields the same solution.

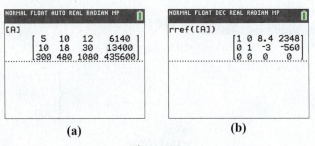

(a)   (b)

Figure 7.8

EXAMPLE 5 ▶ Transportation

Ace Trucking Company has an order for delivery of three products: A, B, and C. Table 7.4 gives the volume in cubic feet, the weight in pounds, and the value for insurance in dollars per unit of each of the products. If the company can carry 30,000 cubic feet and 62,000 pounds and is insured for $276,000, how many units of each product can be carried if the shipment of product C is as large as possible?

Table 7.4

	Product A	Product B	Product C
Volume (cu ft)	25	22	30
Weight (lb)	25	38	70
Value ($)	150	180	300

SOLUTION

If we represent the number of units of product A by x, the number of units of product B by y, and the number of units of product C by z, then we can write a system of equations to represent the problem.

$$\begin{cases} 25x + 22y + 30z = 30{,}000 & \text{Volume} \\ 25x + 38y + 70z = 62{,}000 & \text{Weight} \\ 150x + 180y + 300z = 276{,}000 & \text{Value} \end{cases}$$

The Gauss–Jordan elimination method gives

$$
\begin{bmatrix}
25 & 22 & 30 & | & 30{,}000 \\
25 & 38 & 70 & | & 62{,}000 \\
150 & 180 & 300 & | & 276{,}000
\end{bmatrix}
\xrightarrow{(1/25)R_1 \to R_1}
\begin{bmatrix}
1 & \dfrac{22}{25} & \dfrac{6}{5} & | & 1{,}200 \\
25 & 38 & 70 & | & 62{,}000 \\
150 & 180 & 300 & | & 276{,}000
\end{bmatrix}
$$

$$
\xrightarrow[\substack{-25R_1 + R_2 \to R_2 \\ -150R_1 + R_3 \to R_3}]{}
\begin{bmatrix}
1 & \dfrac{22}{25} & \dfrac{6}{5} & | & 1{,}200 \\
0 & 16 & 40 & | & 32{,}000 \\
0 & 48 & 120 & | & 96{,}000
\end{bmatrix}
$$

$$
\xrightarrow{(1/16)R_2 \to R_2}
\begin{bmatrix}
1 & \dfrac{22}{25} & \dfrac{6}{5} & | & 1{,}200 \\
0 & 1 & \dfrac{5}{2} & | & 2{,}000 \\
0 & 48 & 120 & | & 96{,}000
\end{bmatrix}
$$

$$
\xrightarrow[\substack{(-22/25)R_2 + R_1 \to R_1 \\ -48R_2 + R_3 \to R_3}]{}
\begin{bmatrix}
1 & 0 & -1 & | & -560 \\
0 & 1 & \dfrac{5}{2} & | & 2000 \\
0 & 0 & 0 & | & 0
\end{bmatrix}
$$

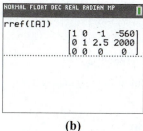

(a)

(b)

Figure 7.9

Reducing this augmented matrix by algebraic methods (above) is quite time consuming. Much time and effort can be saved if we use technology to obtain the reduced row-echelon form (Figure 7.9).

The solution to this system is $x = -560 + z$, $y = 2000 - 2.5z$, with the values of z limited so that all values are nonnegative integers. Because x must be nonnegative, z must be at least 560, and because y must be a nonnegative integer, z must be an even number that is no more than 800. This also means that the size of the shipments of products A and B is limited by the size of the shipments of product C, because the limits on z put limits on x and y. We can write the solution as follows.

$$
\begin{aligned}
&\text{Product A:} \quad x = -560 + z \\
&\text{Product B:} \quad y = 2000 - 2.5z \\
&\text{Product C:} \quad 560 \le z \le 800 \quad (z \text{ is even})
\end{aligned}
$$

The maximum shipment of product C is 800 units, making the shipment of products A and B equal to 240 units and 0 units, respectively.

Inconsistent Systems

The following example shows how we can use matrices to determine that a system has no solution.

EXAMPLE 6 ▶   An Inconsistent System

Solve the following system of equations if a solution exists.

$$
\begin{cases}
5x + 10y + 12z = 6{,}140 \\
10x + 18y + 30z = 13{,}400 \\
300x + 480y + 1080z = 214{,}800
\end{cases}
$$

SOLUTION

The augmented matrix for this system is $\begin{bmatrix} 5 & 10 & 12 & | & 6{,}140 \\ 10 & 18 & 30 & | & 13{,}400 \\ 300 & 480 & 1080 & | & 214{,}800 \end{bmatrix}$.

We reduce this matrix as follows.

$$\begin{bmatrix} 5 & 10 & 12 & | & 6{,}140 \\ 10 & 18 & 30 & | & 13{,}400 \\ 300 & 480 & 1080 & | & 214{,}800 \end{bmatrix} \xrightarrow{(1/5)R_1 \to R_1} \begin{bmatrix} 1 & 2 & 2.4 & | & 1{,}228 \\ 10 & 18 & 30 & | & 13{,}400 \\ 300 & 480 & 1080 & | & 214{,}800 \end{bmatrix}$$

$$\xrightarrow[\substack{-10R_1 + R_2 \to R_2 \\ -300R_1 + R_3 \to R_3}]{} \begin{bmatrix} 1 & 2 & 2.4 & | & 1{,}228 \\ 0 & -2 & 6 & | & 1{,}120 \\ 0 & -120 & 360 & | & -153{,}600 \end{bmatrix}$$

$$\xrightarrow{(-1/2)R_2 \to R_2} \begin{bmatrix} 1 & 2 & 2.4 & | & 1{,}228 \\ 0 & 1 & -3 & | & -560 \\ 0 & -120 & 360 & | & -153{,}600 \end{bmatrix}$$

$$\xrightarrow[\substack{-2R_2 + R_1 \to R_1 \\ 120R_2 + R_3 \to R_3}]{} \begin{bmatrix} 1 & 0 & 8.4 & | & 2{,}348 \\ 0 & 1 & -3 & | & -560 \\ 0 & 0 & 0 & | & -220{,}800 \end{bmatrix}$$

This matrix represents the reduced system that is equivalent to the original system.

$$\begin{cases} x + 0y + 8.4z = & 2{,}348 \\ 0x + y - 3z = & -560 \\ 0x + 0y + 0z = -220{,}800 \end{cases}$$

The third equation is $0 = -220{,}800$, which is impossible. Thus, the system has no solution.

Figure 7.10(a) shows the augmented matrix and Figure 7.10(b) gives the resulting reduced row-echelon form for the system of equations in Example 6.

         (a)   (b)

Figure 7.10

Finally, we consider systems with more equations than variables. Such systems of equations may have zero, one, or many solutions, and they are solved in the same manner as other systems.

1. If the reduced augmented matrix contains a row of 0's in the coefficient matrix with a nonzero number in the augment, the system has no solution.

2. If the coefficient matrix in the reduced augmented matrix contains an identity matrix and all remaining rows of the reduced augmented matrix contain all 0's, there is a unique solution to the system.

3. Otherwise, the system has many solutions.

SKILLS CHECK 7.2

In Exercises 1–4, write the augmented matrix associated with the given system.

1. $\begin{cases} x + y - z = 4 \\ x - 2y - z = -2 \\ 2x + 2y + z = 11 \end{cases}$ **2.** $\begin{cases} 3x - 4y + 6z = 10 \\ 2x - 4y - 5z = -14 \\ x + 2y - 3z = 0 \end{cases}$

3. $\begin{cases} 5x - 3y + 2z = 12 \\ 3x + 6y - 9z = 4 \\ 2x + 3y - 4z = 9 \end{cases}$ **4.** $\begin{cases} x - 3y + 4z = 7 \\ 2x + 2y - 3z = -3 \\ x - 3y + z = -2 \end{cases}$

In Exercises 5–14, the matrix associated with the solution to a system of linear equations in x, y, and z is given. Write the solution to the system, if it exists.

5. $\begin{bmatrix} 1 & 0 & 0 & | & -1 \\ 0 & 1 & 0 & | & 4 \\ 0 & 0 & 1 & | & -2 \end{bmatrix}$ **6.** $\begin{bmatrix} 1 & 0 & 0 & | & -2 \\ 0 & 1 & 0 & | & 4 \\ 0 & 0 & 1 & | & 8 \end{bmatrix}$

7. $\begin{bmatrix} 1 & 1 & -1 & | & 4 \\ 1 & -2 & -1 & | & -2 \\ 2 & 2 & 1 & | & 11 \end{bmatrix}$

8. $\begin{bmatrix} 1 & -1 & 3 & | & 4 \\ 2 & 3 & -1 & | & 11 \\ 4 & 2 & -4 & | & 12 \end{bmatrix}$

9. $\begin{bmatrix} 2 & -3 & 4 & | & 13 \\ 1 & -2 & 1 & | & 3 \\ 2 & -3 & 1 & | & 4 \end{bmatrix}$ **10.** $\begin{bmatrix} 3 & 1 & 2 & | & 1 \\ 2 & 3 & -4 & | & -20 \\ 2 & 4 & 8 & | & 14 \end{bmatrix}$

11. $\begin{bmatrix} 1 & 0 & 4 & | & 3 \\ 0 & 1 & 2 & | & 2 \\ 0 & 0 & 0 & | & 1 \end{bmatrix}$ **12.** $\begin{bmatrix} 1 & 0 & 2 & | & 1 \\ 0 & 1 & 3 & | & 5 \\ 0 & 0 & 0 & | & 0 \end{bmatrix}$

13. $\begin{bmatrix} 1 & 0 & 3 & | & 2 \\ 0 & 1 & -5 & | & 5 \\ 0 & 0 & 0 & | & 0 \end{bmatrix}$ **14.** $\begin{bmatrix} 1 & 0 & -1 & | & 3 \\ 0 & 1 & 2 & | & -2 \\ 0 & 0 & 0 & | & 0 \end{bmatrix}$

Solve the systems in Exercises 15–22.

15. $\begin{cases} x + y - z = 0 \\ x - 2y - z = 6 \\ 2x + 2y + z = 3 \end{cases}$

16. $\begin{cases} x - 2y + z = -5 \\ 2x - y + 2z = 6 \\ 3x + 2y - z = 1 \end{cases}$

17. $\begin{cases} 3x - 2y + 5z = 15 \\ x - 2y - 2z = -1 \\ 2x - 2y = 0 \end{cases}$

18. $\begin{cases} x + 3y + 5z = 5 \\ 2x + 4y + 3z = 9 \\ 2x + 3y + z = 1 \end{cases}$ **19.** $\begin{cases} 2x + 3y + 4z = 5 \\ 6x + 7y + 8z = 9 \\ 2x + y + z = 1 \end{cases}$

20. $\begin{cases} 4x + 3y + 8z = 1 \\ 2x + 3y + 8z = 5 \\ 2x + 5y + 5z = 6 \end{cases}$

21. $\begin{cases} x - y + z - w = -2 \\ 2x + 4z + w = 5 \\ 2x - 3y + z = -5 \\ y + 2z + 20w = 4 \end{cases}$

22. $\begin{cases} x - 2y + z - 3w = 10 \\ 2x - 3y + 4z + w = 12 \\ 2x - 3y + z - 4w = 7 \\ x - y + z + w = 4 \end{cases}$

In Exercises 23–32, find the solutions, if any exist, to the systems.

23. $\begin{cases} -2x + 3y + 2z = 13 \\ -2x - 2y + 3z = 0 \\ 4x + y + 4z = 11 \end{cases}$

24. $\begin{cases} 2x + 3y + 4z = 5 \\ x + y + z = 1 \\ 6x + 7y + 8z = 9 \end{cases}$ **25.** $\begin{cases} 2x + 5y + 6z = 6 \\ 3x - 2y + 2z = 4 \\ 5x + 3y + 8z = 10 \end{cases}$

26. $\begin{cases} -x + 5y - 3z = 10 \\ 3x + 7y + 2z = 5 \\ 4x + 12y - z = 15 \end{cases}$

27. $\begin{cases} -x - 5y + 3z = -2 \\ 3x + 7y + 2z = 5 \\ 4x + 12y - z = 7 \end{cases}$

28. $\begin{cases} x - 3y + 2z = 12 \\ 2x - 6y + z = 7 \end{cases}$

29. $\begin{cases} 2x - 3y + 2z = 5 \\ 4x + y - 3z = 6 \end{cases}$ **30.** $\begin{cases} 3x + 2y - z = 4 \\ 2x - 3y + z = 3 \end{cases}$

31. $\begin{cases} x - 3z - 3w = -2 \\ x + y + z + 3w = 2 \\ 2x + y - 2z - 2w = 0 \\ 3x + 2y - z + w = 2 \end{cases}$

32. $\begin{cases} x + y + z + 5w = 10 \\ x + 2y + z + 6w = 16 \\ x + y + 2z + 7w = 11 \\ 2x + 3y + 3z + 13w = 27 \end{cases}$

EXERCISES 7.2

33. Ticket Pricing A theater owner wants to divide a 3600-seat theater into three sections, with tickets costing $40, $70, and $100, depending on the section. He wants to have twice as many $40 tickets as the sum of the other types of tickets, and he wants to earn $192,000 from a full house. Find how many seats he should have in each section.

34. Rental Cars A car rental agency rents compact, mid-size, and luxury cars. Its goal is to purchase 90 cars with a total of $2,270,000 and to earn a daily rental of $3150 from all the cars. The compact cars cost $18,000 each and earn $25 per day in rental, the mid-size cars cost $25,000 each and earn $35 per day, and the luxury cars cost $40,000 each and earn $55 per day. Find the number of each type of car the agency should purchase to meet its goal.

35. Testing A professor wants to create a test that has 15 true-false questions, 10 multiple-choice questions, and 5 essay questions. She wants the test to be worth 100 points, with each multiple-choice question worth twice as many points as a true-false question and with each essay question equal to three times the number of points of a true-false question.

a. Write a system of equations to represent this problem, with x, y, and z equal to the number of points for a true-false, multiple-choice, and essay question, respectively.

b. How many points should be assigned to each type of problem? Note that only positive integer answers are useful.

36. Testing A professor wants to create a test that has true-false questions, multiple-choice questions, and essay questions. She wants the test to have 35 questions, with twice as many multiple-choice questions as essay questions and with twice as many true-false questions as multiple-choice questions.

a. Write a system of equations to represent this problem.

b. How many of each type question should the professor put on the test? Note that only positive integer answers are useful.

37. Investment A company offers three mutual fund plans for its employees. Plan I consists of 4 blocks of common stock and 2 municipal bonds. Plan II consists of 8 blocks of common stock, 4 municipal bonds, and 6 blocks of preferred stock. Plan III consists of 14 blocks of common stock, 6 municipal bonds, and 6 blocks of preferred stock. If an employee wants to combine these

plans so that she has 42 blocks of common stock, 20 municipal bonds, and 18 blocks of preferred stock, how many units of each plan does she need?

38. Manufacturing A manufacturer of swing sets has three models, Deluxe, Premium, and Ultimate, which must be painted, assembled, and packaged for shipping. The following table gives the number of hours required for each of these operations for each type of swing set. If the manufacturer has 55 hours available per day for painting, 75 hours for assembly, and 40 hours for packaging, how many of each type swing set can be produced each day?

	Deluxe	Premium	Ultimate
Painting	0.8	1	1.4
Assembly	1	1.5	2
Packaging	0.6	0.75	1

39. Nutrition A psychologist, studying the effects of good nutrition on the behavior of rabbits, feeds one group a combination of three foods: I, II, and III. Each of these foods contains three additives: A, B, and C. The table below gives the percent of each additive that is present in each food. If the diet being used requires 3.74 grams per day of A, 2.04 grams of B, and 1.35 grams of C, find the number of grams of each food that should be used each day.

	Food I	Food II	Food III
Additive A	12%	15%	28%
Additive B	8%	6%	16%
Additive C	15%	2%	6%

40. Manufacturing To expand its manufacturing capacity, Krug Industries borrowed $440,000, part at 6%, part at 8%, and part at 10%. The sum of the money borrowed at 6% and 8% was three times that borrowed at 10%. The loan was repaid in full at the end of 5 years and the annual interest paid was $34,400. How much was borrowed at each rate?

Some of the following exercises have nonunique solutions.

41. Social Services A social agency is charged with providing services to three types of clients: A, B, and C. A total of 500 clients are to be served, with $300,000 available for counseling and $200,000 available for emergency food and shelter. Type A clients require an average of $400 for counseling and $600 for emergencies. Type B clients require an average of $1000 for counseling and $400 for emergencies.

Type C clients require an average of $600 for counseling and $200 for emergencies. How many of each type of client can be served?

42. *Investment* A man has $235,000 invested in three rental properties. One property earns 7.5% per year on the investment, a second earns 10%, and the third earns 8%. The annual earnings from the properties total $18,000.

    a. Write a system of two equations to represent the problem, with x, y, and z representing the 7.5%, 10%, and 8% investments, respectively.

    b. Solve this system.

    c. If $60,000 is invested at 8%, how much is invested in each of the other properties?

43. *Investment* A brokerage house offers three stock portfolios for its clients. Portfolio I consists of 10 blocks of common stock, 2 municipal bonds, and 3 blocks of preferred stock. Portfolio II consists of 12 blocks of common stock, 8 municipal bonds, and 5 blocks of preferred stock. Portfolio III consists of 10 blocks of common stock, 6 municipal bonds, and 4 blocks of preferred stock. A client wants to combine these portfolios so that she has 180 blocks of common stock, 140 municipal bonds, and 110 blocks of preferred stock. Can she do this? To answer this question, let x equal the number of units of portfolio I, y equal the number of units of portfolio II, and z equal the number of units of portfolio III, so that the equation $10x + 12y + 10z = 180$ represents the total number of blocks of common stock.

    a. Write the remaining two equations to create a system of three equations.

    b. Solve the system of equations, if possible.

44. *Investment* A company offers three mutual fund plans for its employees. Plan I consists of 14 blocks of common stock, 4 municipal bonds, and 6 blocks of preferred stock. Plan II consists of 4 blocks of common stock and 2 municipal bonds. Plan III consists of 18 blocks of common stock, 6 municipal bonds, and 6 blocks of preferred stock. Suppose an employee wants to combine these plans so that she has 58 blocks of common stock, 20 municipal bonds, and 18 blocks of preferred stock. How many units of each plan does she need if she wants 1 unit of Plan III?

45. *Purchasing* A young man wins $100,000 and decides to buy four new cars. He wants to choose from cars that are priced at $40,000, $30,000, and $20,000 and spend all of the money.

    a. Write a system of two equations in three variables to represent the problem.

    b. Can this system have a unique solution?

c. Solve the system.

d. Use the context of the problem to find how many cars of each price he can buy with the $100,000.

46. *Investment* A trust account manager has $400,000 to invest in three different accounts. The accounts pay 8%, 10%, and 12%, respectively, and the goal is to earn $42,400 with minimum risk. To solve this problem, assume that x dollars are invested at 8%, y dollars are invested at 10%, and z dollars are invested at 12%.

    a. Write a system of two equations in three variables to represent the problem.

    b. How much can be invested in each account with the largest possible amount invested at 8%?

47. *Investments* An investor wants to place a total of $500,000 in two or more of three investments with different degrees of risk. One investment pays 4%, one pays 8%, and one pays 10%, and the investor wants to earn $35,000 from the investments.

    a. Write a system of two equations to represent the problem, representing the amount in the 4% investment by x, the amount in the 8% investment by y, and the amount in the 10% investment by z.

    b. Solve the system.

    c. If the investor wants to put $225,000 in the 8% investment, how much should she put in each of the other investments?

48. *Investing* An investor wants to place a total of $700,000 in two or more of three investments with different degrees of risk. One investment pays 4%, one pays 8%, and one pays 12%, and the investor wants to earn $42,000 from the investments.

    a. Write a system of two equations to represent the problem, representing the amount in the 4% investment by x, the amount in the 8% investment by y, and the amount in the 12% investment by z.

    b. Solve the system.

    c. If the investor wants to put the maximum amount in the 4% investment because of lower risk, how much should he put in each investment?

49. *Traffic Flow* In an analysis of traffic, a certain city estimates the traffic flow as illustrated in the following figure, where the arrows indicate the flow of the traffic. If x_1 represents the number of cars traveling from intersection A to intersection B, x_2 represents the number of cars traveling from intersection B to intersection C, and so on, we can formulate equations based on the principle that the number of vehicles entering the intersection equals the number leaving it.

The arrows indicate that $3050 + x_1$ cars are entering intersection B and that $2500 + x_2$ cars are leaving intersection B. Thus, the equation that describes the number of cars entering and leaving intersection B is $3050 + x_1 = 2500 + x_2$. The equations that describe the numbers of cars entering and leaving all intersections are

B: $3050 + x_1 = 2500 + x_2$
C: $4100 + x_2 = x_3 + 2800$
D: $x_3 + 1800 = 3000 + x_4$
A: $x_4 + 2250 = x_1 + 2900$

Solve the system of these four equations to find how traffic between the other intersections is related to the traffic from intersection D to intersection A.

50. *Traffic Flow* In an analysis of traffic, a retirement community estimates the traffic flow on its "town square" at 6 P.M. to be as illustrated in the figure. If x_1 illustrates the number of cars moving from intersection A to intersection B, x_2 represents the number of cars traveling from intersection B to intersection C, and so on, we can formulate equations based on the principle that the number of vehicles entering the intersection equals the number leaving it. For example, the equation that represents the traffic through A is $x_4 + 470 = x_1 + 340$.

a. Formulate an equation for the traffic at each of the four intersections.

b. Solve the system of these four equations to find how traffic between the other intersections is related to the traffic from intersection D to intersection A.

51. *Irrigation* An irrigation system allows water to flow in the pattern shown in the figure below. Water flows into the system at A and exits at B, C, and D, with amounts shown. If x_1 represents the number of gallons of water moving from A to B, x_2 represents the number of gallons moving from A to C, and so on, we can formulate equations using the fact that at each point the amount of water entering the system equals the amount exiting. For example, the equation that represents the water flow through C is $x_2 = x_3 + 200,000$.

a. Formulate an equation for the water flow at each of the other three points.

b. Solve the system of these four equations.

7.3 Matrix Operations

KEY OBJECTIVES

- Add two matrices
- Subtract two matrices
- Multiply a matrix by a constant
- Multiply two matrices

SECTION PREVIEW **Expected Life Span**

Table 7.5 gives the years of life expected at birth for male and female blacks and whites born in the United States in the years 1920, 1940, 1960, 1980, 2000, 2010, and 2020. To determine what the data tell us about the relationships among race, sex, and life expectancy, we can make a matrix W containing the information for whites and a matrix B for blacks, and use these matrices to find additional information. For example, we can find a matrix D that shows how many more years whites in each category are expected to live than blacks.

Table 7.5   Life Expectancy

Year	Whites		Blacks	
	Males	Females	Males	Females
1920	54.4	55.6	45.5	54.9
1940	62.1	66.6	51.1	45.2
1960	67.4	74.1	61.1	67.4
1980	70.7	78.1	63.8	72.5
2000	74.7	79.9	68.2	75.1
2010	76.5	81.3	70.2	77.2
2020	77.7	82.4	72.6	79.2

(Source: National Center for Health Statistics)

We used matrices to solve systems of equations in the last section. In this section, we consider the operations of addition, subtraction, and multiplication of matrices, and we use these operations in applications like the one above. ∎

Addition and Subtraction of Matrices

If two matrices have the same numbers of rows and columns (the same dimensions), we can add them by adding the corresponding entries of the two matrices.

> ### Matrix Addition
>
> The sum of two matrices with the same dimensions is the matrix that is formed by adding the corresponding entries of the two matrices. Addition is not defined if the matrices do not have the same number of rows and the same number of columns.

EXAMPLE 1 ▶   Matrix Addition

Find the following sums of matrices.

a. $\begin{bmatrix} a & b \\ c & d \end{bmatrix} + \begin{bmatrix} w & y \\ x & z \end{bmatrix}$

b. $\begin{bmatrix} 1 & 3 & -8 & 0 \\ 3 & 6 & 1 & -3 \\ -4 & 5 & 3 & 2 \end{bmatrix} + \begin{bmatrix} -1 & 4 & 2 & 4 \\ 5 & -2 & 4 & 2 \\ -5 & 1 & 4 & -1 \end{bmatrix}$

c. $A + B$ if $A = \begin{bmatrix} 2 & -3 \\ -1 & 4 \end{bmatrix}$ and $B = \begin{bmatrix} -2 & 3 \\ 1 & -4 \end{bmatrix}$

SOLUTION

a. $\begin{bmatrix} a & b \\ c & d \end{bmatrix} + \begin{bmatrix} w & y \\ x & z \end{bmatrix} = \begin{bmatrix} a + w & b + y \\ c + x & d + z \end{bmatrix}$

b. $\begin{bmatrix} 1 & 3 & -8 & 0 \\ 3 & 6 & 1 & -3 \\ -4 & 5 & 3 & 2 \end{bmatrix} + \begin{bmatrix} -1 & 4 & 2 & 4 \\ 5 & -2 & 4 & 2 \\ -5 & 1 & 4 & -1 \end{bmatrix}$

$= \begin{bmatrix} 1 + (-1) & 3 + 4 & -8 + 2 & 0 + 4 \\ 3 + 5 & 6 + (-2) & 1 + 4 & -3 + 2 \\ -4 + (-5) & 5 + 1 & 3 + 4 & 2 + (-1) \end{bmatrix} = \begin{bmatrix} 0 & 7 & -6 & 4 \\ 8 & 4 & 5 & -1 \\ -9 & 6 & 7 & 1 \end{bmatrix}$

c. $A + B = \begin{bmatrix} 2 & -3 \\ -1 & 4 \end{bmatrix} + \begin{bmatrix} -2 & 3 \\ 1 & -4 \end{bmatrix} = \begin{bmatrix} 0 & 0 \\ 0 & 0 \end{bmatrix}$

The matrix that is the sum in Example 1(c) is called a **zero matrix** because each of its elements is 0. Matrix B in Example 1(c) is called the **negative** of matrix A, denoted $-A$, because the sum of matrices A and B is a zero matrix. Similarly, matrix A is the negative of matrix B and can be denoted by $-B$.

Matrix Subtraction

If matrices M and N have the same dimension, the difference $M - N$ is found by subtracting the elements of N from the corresponding elements of M. This difference can also be defined as

$$M - N = M + (-N)$$

EXAMPLE 2 ▶ **Matrix Subtraction**

Complete the following matrix operations.

a. $\begin{bmatrix} 2 & -1 & 3 \\ 5 & -4 & 2 \\ 1 & 4 & -2 \end{bmatrix} - \begin{bmatrix} 1 & 2 & 4 \\ 5 & -2 & 3 \\ 6 & 2 & -3 \end{bmatrix}$

b. $\begin{bmatrix} 1 & 4 \\ 3 & 6 \\ -2 & 1 \end{bmatrix} - \begin{bmatrix} -5 & 2 \\ 3 & 5 \\ 12 & 3 \end{bmatrix} + \begin{bmatrix} 2 & 3 \\ -4 & -8 \\ 2 & 0 \end{bmatrix}$

SOLUTION

a. $\begin{bmatrix} 2 & -1 & 3 \\ 5 & -4 & 2 \\ 1 & 4 & -2 \end{bmatrix} - \begin{bmatrix} 1 & 2 & 4 \\ 5 & -2 & 3 \\ 6 & 2 & -3 \end{bmatrix}$

$$= \begin{bmatrix} 2-1 & -1-2 & 3-4 \\ 5-5 & -4-(-2) & 2-3 \\ 1-6 & 4-2 & -2-(-3) \end{bmatrix} = \begin{bmatrix} 1 & -3 & -1 \\ 0 & -2 & -1 \\ -5 & 2 & 1 \end{bmatrix}$$

b. $\begin{bmatrix} 1 & 4 \\ 3 & 6 \\ -2 & 1 \end{bmatrix} - \begin{bmatrix} -5 & 2 \\ 3 & 5 \\ 12 & 3 \end{bmatrix} + \begin{bmatrix} 2 & 3 \\ -4 & -8 \\ 2 & 0 \end{bmatrix}$

$$= \begin{bmatrix} 6 & 2 \\ 0 & 1 \\ -14 & -2 \end{bmatrix} + \begin{bmatrix} 2 & 3 \\ -4 & -8 \\ 2 & 0 \end{bmatrix} = \begin{bmatrix} 8 & 5 \\ -4 & -7 \\ -12 & -2 \end{bmatrix}$$

We can use technology to perform the operations of addition and subtraction of matrices, such as those in Examples 1 and 2. Figure 7.11 shows the computations for Example 2(b).*

---

* For more details, see Appendix A, page 679.

(a)   (b)   (c)   (d)

Figure 7.11

EXAMPLE 3 ▶ Life Expectancy

Table 7.6 gives the years of life expected at birth for male and female blacks and whites born in the United States in the years 1920, 1940, 1960, 1980, 2000, 2010, and 2020.

a. Make a matrix W containing the life expectancy data for whites and a matrix B for blacks.

b. Use these matrices to find matrix $D = W - B$, which represents the difference between white and black life expectancy.

c. What does this tell us about race and life expectancy?

Table 7.6

Year	Whites Males	Whites Females	Blacks Males	Blacks Females
1920	54.4	55.6	45.5	54.9
1940	62.1	66.6	51.1	45.2
1960	67.4	74.1	61.1	67.4
1980	70.7	78.1	63.8	72.5
2000	74.7	79.9	68.2	75.1
2010	76.5	81.3	70.2	77.2
2020	77.7	82.4	72.6	79.2

(Source: National Center for Health Statistics)

SOLUTION

a. $W = \begin{bmatrix} 54.4 & 55.6 \\ 62.1 & 66.6 \\ 67.4 & 74.1 \\ 70.7 & 78.1 \\ 74.7 & 79.9 \\ 76.5 & 81.3 \\ 77.7 & 82.4 \end{bmatrix}$ $B = \begin{bmatrix} 45.5 & 54.9 \\ 51.1 & 45.2 \\ 61.1 & 67.4 \\ 63.8 & 72.5 \\ 68.2 & 75.1 \\ 70.2 & 77.2 \\ 72.6 & 79.2 \end{bmatrix}$ **b.** $D = W - B = \begin{bmatrix} 8.9 & 0.7 \\ 11.0 & 21.4 \\ 6.3 & 6.7 \\ 6.9 & 5.6 \\ 6.5 & 4.8 \\ 6.3 & 4.1 \\ 5.1 & 3.2 \end{bmatrix}$

c. Because each element in the difference matrix D is positive, we conclude that the life expectancy for whites is longer than that for blacks for all birth years and for both sexes.

Spreadsheet ▸ SOLUTION   We can use spreadsheets as well as calculators to perform operations with matrices. Figure 7.12 Shows an Excel spreadsheet with W, B, and D from Example 3. The matrix W is entered in cells that we denote as B1:C7, and the matrix B is entered in cells B9:C15. The first entry of the difference matrix $D = W - B$ is found by entering the formula "$= B1 - B9$" in cell $B17$ and pressing ENTER. Using Fill Across to C17 and then Fill Down to copy the new row B17:C17 to B23:C23 gives the matrix containing the differences.*

▲	A	B	C
1	W	54.4	55.6
2		62.1	66.6
3		67.4	74.1
4		70.7	78.1
5		74.7	79.9
6		76.5	81.3
7		77.7	82.4
8			
9	B	45.5	54.9
10		51.1	45.2
11		61.1	67.4
12		63.8	72.5
13		68.2	75.1
14		70.2	77.2
15		72.6	79.2
16			
17	D=W - B	8.9	0.7
18		11	21.4
19		6.3	6.7
20		6.9	5.6
21		6.5	4.8
22		6.3	4.1
23		5.1	3.2

Figure 7.12

Multiplication of a Matrix by a Number

As in operations with real numbers, we can multiply a matrix by a positive integer to find the sum of repeated additions. For example, if

$$A = \begin{bmatrix} 1 & -2 & 4 \\ 6 & 3 & -3 \end{bmatrix}$$

then we can find $2A$ in two ways:

$$2A = A + A = \begin{bmatrix} 1 & -2 & 4 \\ 6 & 3 & -3 \end{bmatrix} + \begin{bmatrix} 1 & -2 & 4 \\ 6 & 3 & -3 \end{bmatrix} = \begin{bmatrix} 2 & -4 & 8 \\ 12 & 6 & -6 \end{bmatrix}$$

and

$$2A = \begin{bmatrix} 2 \cdot 1 & 2(-2) & 2 \cdot 4 \\ 2 \cdot 6 & 2 \cdot 3 & 2(-3) \end{bmatrix} = \begin{bmatrix} 2 & -4 & 8 \\ 12 & 6 & -6 \end{bmatrix}$$

Product of a Number and a Matrix

Multiplying a matrix A by a real number c results in a matrix in which each entry of matrix A is multiplied by the number c.

---

* For details, see Appendix B, page 697.

In general, we define multiplication of a matrix by a real number as follows. For example, if $A = \begin{bmatrix} a & b & c & d \\ e & f & g & h \end{bmatrix}$, then $nA = \begin{bmatrix} na & nb & nc & nd \\ ne & nf & ng & nh \end{bmatrix}$.

Note that $-A = (-1)A = \begin{bmatrix} -a & -b & -c & -d \\ -e & -f & -g & -h \end{bmatrix}$.

EXAMPLE 4 ▶   Price Increases

Table 7.7 contains the purchase prices and delivery costs (per unit) for plywood, siding, and 2×4 lumber. If the supplier announces a 5% increase in all of these prices and in delivery costs, find the matrix that gives the new prices and costs.

Table 7.7

	Plywood	Siding	2 × 4's
Purchase Price ($)	32	23	2.60
Delivery Cost ($)	3	1	0.60

SOLUTION

The matrix that represents the original prices and costs is

$$\begin{bmatrix} 32 & 23 & 2.60 \\ 3 & 1 & 0.60 \end{bmatrix}$$

To find the prices and costs after a 5% increase, we multiply the matrix by 1.05 (100% of the old prices plus the 5% increase).

$$1.05\begin{bmatrix} 32 & 23 & 2.60 \\ 3 & 1 & 0.60 \end{bmatrix} = \begin{bmatrix} 33.60 & 24.15 & 2.73 \\ 3.15 & 1.05 & 0.63 \end{bmatrix}$$

Matrix Multiplication

Suppose Circuitown made a special purchase for one of its stores, consisting of 22 televisions, 15 washers, and 12 dryers. If the value of each television is $550, each washer is $435, and each dryer is $325, then the value of this purchase is

$$550 \cdot 22 + 435 \cdot 15 + 325 \cdot 12 = 22{,}525 \text{ dollars}$$

If we write the value of each item in a 1×3 *row matrix*

$$A = [550 \quad 435 \quad 325]$$

and the number of each of the items in the special purchase in a 3×1 *column matrix*

$$B = \begin{bmatrix} 22 \\ 15 \\ 12 \end{bmatrix}$$

then the value of the special purchase can be represented by the **matrix product**

$$AB = [550 \quad 435 \quad 325]\begin{bmatrix} 22 \\ 15 \\ 12 \end{bmatrix}$$

$$= [550 \cdot 22 + 435 \cdot 15 + 325 \cdot 12] = [22{,}525]$$

In general, we have the following.

Product of a Row Matrix and a Column Matrix

The product of a $1 \times n$ row matrix and an $n \times 1$ column matrix is a 1×1 matrix given by

$$[a_1 \quad a_2 \quad \cdots \quad a_n] \begin{bmatrix} b_1 \\ b_2 \\ \vdots \\ b_n \end{bmatrix} = [a_1 b_1 + a_2 b_2 + \cdots + a_n b_n]$$

We can expand the multiplication to larger matrices. For example, suppose Circuitown has a second store and purchases 28 televisions, 21 washers, and 26 dryers for it. Rather than writing two matrices to represent the two stores, we can use a two-column matrix C to represent the purchases for the two stores.

$$C = \begin{matrix} & \text{Store I} & \text{Store II} & \\ & \begin{bmatrix} 22 & 28 \\ 15 & 21 \\ 12 & 26 \end{bmatrix} & \begin{matrix} \text{TVs} \\ \text{Washers} \\ \text{Dryers} \end{matrix} \end{matrix}$$

If these products have the same values as given above, we can find the value of the purchases for each store by multiplying the row matrix times each of the column matrices. The value of the store I purchases is found by multiplying the row matrix A times the first column of matrix C, and the value of the store II purchases is found by multiplying matrix A times the second column of matrix C. The result is

$$AC = [550 \quad 435 \quad 325] \begin{bmatrix} 22 & 28 \\ 15 & 21 \\ 12 & 26 \end{bmatrix}$$

$$= [550 \cdot 22 + 435 \cdot 15 + 325 \cdot 12 \quad 550 \cdot 28 + 435 \cdot 21 + 325 \cdot 26]$$

$$= [22{,}525 \quad 32{,}985]$$

This indicates that the value of the store I purchase is \$22,525 (the value found before) and the value of the store II purchase is \$32,985.

The matrix AC is the product of the 1×3 matrix A and the 3×2 matrix C. This product is a 1×2 matrix. In general, the product of an $m \times n$ matrix and an $n \times k$ matrix is an $m \times k$ matrix, and the product is undefined if the number of columns in the first matrix does not equal the number of rows in the second matrix.

In general, we can define the product of two matrices by defining how each element of the product is formed.

Product of Two Matrices

The product of an $m \times n$ matrix A and an $n \times k$ matrix B is the $m \times k$ matrix $C = AB$. The element in the ith row and jth column of matrix C has the form

$$c_{ij} = [a_{i1} \quad a_{i2} \quad \cdots \quad a_{in}] \begin{bmatrix} b_{1j} \\ b_{2j} \\ \vdots \\ b_{nj} \end{bmatrix} = [a_{i1}b_{1j} + a_{i2}b_{2j} + \cdots + a_{in}b_{nj}]$$

We illustrate below the product AB, with each of the c_{ij} elements found as shown in the box above:

$$C = AB = \begin{bmatrix} a_{11} & a_{12} & \cdots & a_{ln} \\ a_{21} & a_{22} & \cdots & a_{2n} \\ \vdots & \vdots & & \vdots \\ a_{i1} & a_{i2} & \cdots & a_{in} \\ \vdots & \vdots & & \vdots \\ a_{m1} & a_{m2} & \cdots & a_{mn} \end{bmatrix} \begin{bmatrix} b_{11} & b_{12} & \cdots & b_{1j} & \cdots & b_{1k} \\ b_{21} & b_{22} & \cdots & b_{2j} & \cdots & b_{2k} \\ \vdots & \vdots & & \vdots & & \vdots \\ b_{n1} & b_{n2} & \cdots & b_{nj} & \cdots & b_{nk} \end{bmatrix} = \begin{bmatrix} c_{11} & c_{12} & \cdots & c_{1j} & \cdots & c_{1k} \\ c_{21} & c_{22} & \cdots & c_{2j} & \cdots & c_{2k} \\ \vdots & \vdots & & \vdots & & \vdots \\ c_{i1} & c_{i2} & \cdots & c_{ij} & \cdots & c_{ik} \\ \vdots & \vdots & & \vdots & & \vdots \\ c_{m1} & c_{m2} & \cdots & c_{mj} & \cdots & c_{mk} \end{bmatrix}$$

$$m \times n \qquad\qquad n \times k \qquad\qquad n \times k$$

EXAMPLE 5 ▶ Matrix Product

Compute the products AB and BA for the matrices $A = \begin{bmatrix} 1 & 2 \\ 3 & 4 \end{bmatrix}$ and $B = \begin{bmatrix} a & b \\ c & d \end{bmatrix}$.

SOLUTION

$$AB = \begin{bmatrix} 1 & 2 \\ 3 & 4 \end{bmatrix}\begin{bmatrix} a & b \\ c & d \end{bmatrix} = \begin{bmatrix} 1a + 2c & 1b + 2d \\ 3a + 4c & 3b + 4d \end{bmatrix}$$

$$BA = \begin{bmatrix} a & b \\ c & d \end{bmatrix}\begin{bmatrix} 1 & 2 \\ 3 & 4 \end{bmatrix} = \begin{bmatrix} 1a + 3b & 2a + 4b \\ 1c + 3d & 2c + 4d \end{bmatrix}$$

Note that in Example 5 the product AB is quite different from the product BA. That is, $AB \neq BA$. For some matrices, but not all, $AB \neq BA$. We indicate this by saying that *matrix multiplication is not commutative.*

EXAMPLE 6 ▶ Advertising

A business plans to use three methods of advertising—newspapers, radio, and cable TV—in each of its two markets, I and II. The cost per ad type in each market (in thousands of dollars) is given by matrix A. The business has three target groups: teenagers, single women, and men aged 35 to 50. Matrix B gives the number of ads per week directed at each of these groups.

$$A = \begin{matrix} & \text{Mkt I} & \text{Mkt II} & \\ & \begin{bmatrix} 12 & 10 \\ 10 & 8 \\ 5 & 9 \end{bmatrix} & \begin{matrix} \text{Paper} \\ \text{Radio} \\ \text{TV} \end{matrix} \end{matrix} \qquad B = \begin{matrix} & \text{Paper} & \text{Radio} & \text{TV} & \\ & \begin{bmatrix} 3 & 12 & 15 \\ 5 & 16 & 10 \\ 10 & 11 & 6 \end{bmatrix} & \begin{matrix} \text{Teens} \\ \text{Single females} \\ \text{Men 35–50} \end{matrix} \end{matrix}$$

a. Does AB or BA give the matrix that represents the cost of ads for each target group in each market?

b. Find this matrix.

c. For what group of people is the most money spent on advertising?

SOLUTION

a. Multiplying matrix B times matrix A gives the total cost of ads for each target group in each market. Note that the product AB is undefined, because multiplying a 3×2 matrix times a 3×3 matrix is not possible.

b. $BA = \begin{bmatrix} 3 & 12 & 15 \\ 5 & 16 & 10 \\ 10 & 11 & 6 \end{bmatrix} \begin{bmatrix} 12 & 10 \\ 10 & 8 \\ 5 & 9 \end{bmatrix}$

$= \begin{bmatrix} 36 + 120 + 75 & 30 + 96 + 135 \\ 60 + 160 + 50 & 50 + 128 + 90 \\ 120 + 110 + 30 & 100 + 88 + 54 \end{bmatrix} = \begin{bmatrix} 231 & 261 \\ 270 & 268 \\ 260 & 242 \end{bmatrix}$

The columns of this matrix represent the markets and the rows represent the target groups.

$\begin{array}{cc} \text{Mkt I} & \text{Mkt II} \end{array}$
$\begin{bmatrix} 231 & 261 \\ 270 & 268 \\ 260 & 242 \end{bmatrix} \begin{array}{l} \text{Teens} \\ \text{Single females} \\ \text{Men 35–50} \end{array}$

c. The largest amount is spent on single females, \$270,000 in market I and \$268,000 in market II.

Multiplication with Technology

We can multiply two matrices by using technology. We can find the product of matrix B times matrix A on a graphing utility by entering $[B] * [A]$ (or $[B][A]$) and pressing **ENTER**.* Figure 7.13 shows the product BA from Example 6.

(a)   (b)   (c)

Figure 7.13

EXAMPLE 7 ▶ **Multiplying with Technology**

Use technology to compute BA and AB if $A = \begin{bmatrix} 1 & 2 \\ 0 & -1 \\ 3 & -2 \end{bmatrix}$ and $B = \begin{bmatrix} 2 & 4 & -1 \\ 3 & -2 & 1 \\ 2 & 0 & 2 \\ 1 & -3 & 0 \end{bmatrix}$.

SOLUTION

Figure 7.14 shows displays with matrix A, matrix B, and the product BA.

(a)   (b)   (c)

Figure 7.14

* For more details, see Appendix A, page 679.

Figure 7.15 shows that the matrix product AB does not exist because the dimensions do not match (that is, the number of columns of A does not equal the number of rows of B).

(a)   (b)

Figure 7.15

Spreadsheet ▶ SOLUTION   Like graphing calculators and software programs, spreadsheets can be used to find products of matrices. Consider the following example, which is solved with Excel.

EXAMPLE 8 ▶   Manufacturing

A furniture company manufactures two products, A and B, which are constructed using steel, plastic, and fabric. The number of units of each raw material that is required for each product is given in Table 7.8.

Table 7.8

	Steel	Plastic	Fabric
Product A	2	3	8
Product B	3	1	10

Because of transportation costs to the company's two plants, the unit costs for some of the raw materials differ. Table 7.9 gives the unit costs for each of the raw materials at the two plants. Create two matrices from the information in the two tables and use matrix multiplication with a spreadsheet to find the cost of manufacturing each product at each plant.

Table 7.9

	Plant I	Plant II
Steel	$15	$16
Plastic	$11	$10
Fabric	$ 6	$ 7

SOLUTION

We represent the number of units of raw materials for each product by matrix P and the costs of the raw materials at each plant by matrix C. The cost of manufacturing each product at each plant is given by the matrix product PC. The spreadsheet shown in Figure 7.16 gives matrix P, matrix C, and the matrix product PC.

The product *PC* is found by selecting cells B8:C9, entering "=mmult(", then selecting the cells B1:D2, typing a comma, selecting the cells B4:C6, and closing the parentheses, and then holding the CTRL and SHIFT keys down and pressing ENTER.*

Figure 7.16

The rows of matrix *PC* represent Product A and Product B, respectively, and the columns represent Plant I and Plant II. The entries give the cost of each product at each plant.

$$\begin{array}{cc} \text{Plant I} & \text{Plant II} \\ \begin{bmatrix} 111 & 118 \\ 116 & 128 \end{bmatrix} & \begin{array}{l} \text{Product A} \\ \text{Product B} \end{array} \end{array}$$

Recall that a square ($n \times n$) matrix with 1's on the diagonal and 0's elsewhere is an identity matrix. For any $n \times k$ matrix *A* and the $n \times n$ matrix *I*, $IA = A$, and for any $m \times n$ matrix *B* and the $n \times n$ matrix *I*, $BI = B$. If matrix *C* is an $n \times n$ matrix, then for the $n \times n$ matrix *I*, $IC = C$ and $CI = C$, so the product of an identity matrix and another square matrix is *commutative*. (Recall that multiplication of matrices, in general, is not commutative.)

EXAMPLE 9 ▶ **The Identity Matrix**

a. Write the 2×2 identity matrix *I*.

b. For the matrix $A = \begin{bmatrix} 3 & -2 \\ -1 & 5 \end{bmatrix}$, find *IA* and *AI*.

SOLUTION

a. $\begin{bmatrix} 1 & 0 \\ 0 & 1 \end{bmatrix}$

b. $\begin{bmatrix} 1 & 0 \\ 0 & 1 \end{bmatrix}\begin{bmatrix} 3 & -2 \\ -1 & 5 \end{bmatrix} = \begin{bmatrix} 3 & -2 \\ -1 & 5 \end{bmatrix}$ and $\begin{bmatrix} 3 & -2 \\ -1 & 5 \end{bmatrix}\begin{bmatrix} 1 & 0 \\ 0 & 1 \end{bmatrix} = \begin{bmatrix} 3 & -2 \\ -1 & 5 \end{bmatrix}$

Thus, $IA = A$ and $AI = A$.

* For detailed steps, see Appendix B, page 698.

SKILLS CHECK 7.3

Use the following matrices for Exercises 1–12.

$$A = \begin{bmatrix} 1 & 3 & -2 \\ 3 & 1 & 4 \\ -5 & 3 & 6 \end{bmatrix} \quad B = \begin{bmatrix} 2 & 1 & -1 \\ 3 & 2 & 4 \end{bmatrix}$$

$$C = \begin{bmatrix} 1 & 3 \\ 2 & 1 \\ 3 & -1 \end{bmatrix} \quad D = \begin{bmatrix} 2 & 3 & 1 \\ 3 & 4 & -1 \\ 2 & 5 & 1 \end{bmatrix}$$

$$E = \begin{bmatrix} 9 & 2 & -7 \\ -5 & 0 & 5 \\ 7 & -4 & -1 \end{bmatrix} \quad F = \begin{bmatrix} 2 & 1 & 3 \\ 4 & 0 & 1 \end{bmatrix}$$

1. Which pairs of the matrices can be added?

2. Use letters to represent the matrix products that are defined.

3. Find the sum of A and D if it is defined.

4. **a.** Find $D + E$ and $E + D$.

   b. Are the sums equal?

5. Find $3A$.   6. Find $-4F$.

7. Find $2D - 4A$.   8. Find $2B - 4F$.

9. **a.** Find AD and DA if these products exist.

   b. Are these products equal?

   c. Do these products have the same dimension?

10. **a.** Find BC and CB if these products exist.

    b. Are these products equal?

    c. Do these products have the same dimension?

11. **a.** Compute DE and ED.

    b. What is the name of $\dfrac{1}{10} DE$?

12. If I is a 3×3 identity matrix, find ID and DI.

13. Compute the sum of $A = \begin{bmatrix} 1 & 5 \\ 3 & 2 \end{bmatrix}$ and $B = \begin{bmatrix} 2a & 3b \\ -c & -2d \end{bmatrix}$.

14. Compute the difference $A - B$ if $A = \begin{bmatrix} a & b \\ c & d \\ f & g \end{bmatrix}$ and $B = \begin{bmatrix} 1 & 2 \\ 3 & 4 \\ 5 & 6 \end{bmatrix}$.

15. Compute $3A - 2B$ if $A = \begin{bmatrix} a & b \\ c & d \end{bmatrix}$ and $B = \begin{bmatrix} 1 & 2 \\ 3 & 4 \end{bmatrix}$.

16. Compute $2A - 3B$ if $A = \begin{bmatrix} 1 & -3 & 2 \\ 2 & 2 & -1 \\ 3 & 4 & 2 \end{bmatrix}$ and $B = \begin{bmatrix} 2 & 2 & 2 \\ 3 & -2 & -1 \\ 1 & 1 & 2 \end{bmatrix}$.

17. If an $m \times n$ matrix A is multiplied by an $n \times k$ matrix B, what is the dimension of the matrix that is the product AB?

18. If A and B are any two matrices, does $AB = BA$ always, sometimes, or never?

19. If A is a 2×3 matrix and B is a 4×2 matrix:

    a. Which product is defined, AB or BA?

    b. What is the dimension of the product that is defined?

20. If C is a 3×4 matrix and D is a 4×3 matrix, what are the dimensions of CD and DC?

21. Find AB and BA if $A = \begin{bmatrix} a & b & c \\ d & e & f \end{bmatrix}$ and $B = \begin{bmatrix} 1 & 2 \\ 3 & 4 \\ 5 & 6 \end{bmatrix}$.

22. Find EF and FE if $E = \begin{bmatrix} a & b \\ c & d \end{bmatrix}$ and $F = \begin{bmatrix} e & f \\ g & h \end{bmatrix}$.

23. If $A = \begin{bmatrix} 1 & 5 \\ 3 & 2 \end{bmatrix}$ and $B = \begin{bmatrix} 2 & 3 \\ -1 & -2 \end{bmatrix}$, compute AB and BA, if possible.

24. If $A = \begin{bmatrix} 1 & 4 \\ 3 & -1 \\ -2 & 2 \end{bmatrix}$ and $B = \begin{bmatrix} 4 & 2 & 2 \\ -1 & 3 & 1 \end{bmatrix}$, compute AB and BA, if possible.

25. If $A = \begin{bmatrix} 1 & -1 & 2 \\ 3 & 4 & 4 \end{bmatrix}$ and $B = \begin{bmatrix} 3 & 1 \\ 1 & 3 \\ -2 & 1 \end{bmatrix}$, compute AB and BA, if possible.

26. Suppose $A = \begin{bmatrix} 1 & -\dfrac{1}{2} & -\dfrac{1}{4} \\ -\dfrac{1}{2} & 0 & \dfrac{1}{2} \\ 0 & \dfrac{1}{2} & -\dfrac{1}{4} \end{bmatrix}$ and $B = \begin{bmatrix} 2 & 2 & 2 \\ 1 & 2 & 3 \\ 2 & 4 & 2 \end{bmatrix}$.

    a. Compute AB and BA, if possible.

    b. Are the products equal?

EXERCISES 7.3

27. *Future Demographics* The tables below show important demographics for China, Bangladesh, and the Philippines for 2012 and projected for 2062.

  a. Create matrix B for the 2012 table and matrix A for the 2062 table.

  b. Find a matrix C that shows the changes of demographics from 2012 to 2062 for these countries.

  c. Which negative entries in matrix C definitely indicate a positive change for the countries?

2012

	China	Bangladesh	The Philippines
Population (millions)	1341.3	148.7	93.3
Life Expectancy	73.8	69.4	69.2
Fertility Rate*	1.56	2.16	3.05
Infant Mortality Rate**	19.6	41.8	20.9

2062

	China	Bangladesh	The Philippines
Population (millions)	1211.5	192.4	165.5
Life Expectancy	80.8	79.0	78.4
Fertility Rate*	1.88	1.68	1.95
Infant Mortality Rate**	8.3	12.5	10.2

Source: *Discover Almanac*

28. *Life Expectancy* The table gives the years of expected life at birth for whites and blacks in the United States for selected years from 2000 and projected to 2020.

  a. Create a matrix W containing the information for whites and a matrix B containing the corresponding information for blacks.

  b. Find a matrix $D = W - B$.

  c. Use matrix D to find the categories in which life expectancy for whites is greater than that for blacks.

  d. For which of these birth years is the life expectancy difference greatest?

---

* The fertility rate is the number of children an average woman will produce in her lifetime.

** Infant mortality is the number of deaths per thousand births.

	Whites		Blacks	
	Males	Females	Males	Females
2000	74.7	79.9	67.8	74.7
2006	75.7	80.6	69.7	76.5
2010	76.5	81.3	70.2	77.2
2015	77.1	81.8	71.4	78.2
2020	77.7	82.4	72.6	79.2

(Source: National Center for Health Statistics)

29. *Population* The following tables give U.S. Census Bureau data and projections of the population by selected age groups and gender in the United States, in thousands.

  a. Matrix M gives the data for males, and matrix F gives the data for females. Find a matrix $D = M - F$ and discuss what it means.

  b. For what years and ages is the female population projected to be greater than the male population?

Males

	Under Age 5	5–13	14–17	18–64	Over Age 64
2015	10,763	18,784	8541	99,232	21,041
2020	11,150	19,309	8478	100,904	24,970
2025	11,307	20,210	8472	102,004	29,204
2030	11,377	20,649	9078	103,510	32,709

Females

	Under Age 5	5–13	14–17	18–64	Over Age 64
2015	10,288	17,988	8153	99,918	26,654
2020	10,658	18,640	8104	100,863	30,999
2025	10,807	19,301	8093	101,162	35,848
2030	10,875	19,717	8653	101,839	40,066

30. *Endangered Species* The tables give the numbers of some species of threatened and endangered wildlife in the United States and in foreign countries in 2011.

United States Wildlife

	Mammals	Birds	Reptiles	Amphibians	Fishes
Endangered	71	76	13	14	72
Threatened	14	16	24	10	68

Foreign Wildlife

	Mammals	Birds	Reptiles	Amphibians	Fishes
Endangered	256	198	66	8	11
Threatened	20	14	16	1	1

(Source: U.S. Fish and Wildlife Service)

a. Form the matrix A that contains the number of each of these species in the United States in 2011 and matrix B that contains the number of each of these species outside the United States in 2011.

b. Write a matrix containing the total number of each of these species, assuming that the U.S. and foreign species are different.

c. Find the matrix $C = B - A$ and tell what the elements of this matrix signify.

d. What do the negative elements of matrix C mean?

31. **Exchange Rates** The table below gives the exchange rates from U.S. dollars to euros and Japanese yen for the years 2000 and 2015.

a. Assuming the fee for exchanging dollars to these units reduces the national currency returned per U.S. dollar by 5%, place the data in a matrix and use a matrix operation to find the currency units actually returned per dollar.

National Currency Units per U.S. Dollar

Year	Euros	Yen
2000	0.94	107.91
2015	0.90	119.74

(Source: International Monetary Fund)

b. Which of these two currencies has increased in value during this period?

32. **Income** The table on the top of the next column gives the median weekly earnings for different ages and sexes in 2012. Create a matrix containing the data, and use a matrix operation to form a matrix that contains the weekly earnings if the 2012 weekly earnings are increased by 12% in all categories. Round to the nearest dollar.

Median Weekly Earnings, 2012 ($)

Age	16–24	25–54	55–64	65+
Men	460	902	1015	780
Women	409	722	760	661

(Source: Bureau of Labor Statistics)

33. **Advertising** A political candidate plans to use three methods of advertising: newspapers, radio, and cable TV. The cost per ad (in thousands of dollars) for each type of media is given by matrix A. Matrix B shows the number of ads per month in these three media that are targeted to single people, to married males aged 35 to 55, and to married females over 65 years of age. Find the matrix that gives the cost of ads for each target group.

$$A = \begin{bmatrix} 12 \\ 15 \\ 5 \end{bmatrix} \begin{matrix} \text{TV} \\ \text{Radio} \\ \text{Papers} \end{matrix}$$

$$B = \begin{bmatrix} 30 & 45 & 35 \\ 25 & 32 & 40 \\ 22 & 12 & 30 \end{bmatrix} \begin{matrix} \text{Singles} \\ \text{Males 35–55} \\ \text{Females 65+} \end{matrix}$$

34. **Cost** Men and women in a church choir wear choir robes in the sizes shown in matrix A. Matrix B contains the prices (in dollars) of new robes and hoods according to size.

$$A = \begin{bmatrix} 10 & 24 \\ 22 & 10 \\ 33 & 3 \end{bmatrix} \begin{matrix} \text{Small} \\ \text{Medium} \\ \text{Large} \end{matrix}$$

$$B = \begin{bmatrix} 45 & 50 & 55 \\ 20 & 20 & 20 \end{bmatrix} \begin{matrix} \text{Robes} \\ \text{Hoods} \end{matrix}$$

a. Find the product BA and label the rows and columns to show what each row represents.

b. What is the cost of the robes for all the men? For all the women?

35. **Manufacturing** Two departments, A and B, of a firm need differing amounts of steel, wood, and plastic. The following table gives the amounts of the products needed by the departments.

	Steel	Wood	Plastic
Department A	60	40	20
Department B	40	20	40

These three products are supplied by two suppliers, DeTuris and Marriott, with the unit prices (in dollars) given in the following table.

	DeTuris	Marriott
Steel	600	560
Wood	300	200
Plastic	300	400

a. Use matrix multiplication to determine how much these orders will cost each department at each of the two suppliers.

b. From which supplier should each department make its purchase?

36. Manufacturing A furniture manufacturer produces three styles of chairs, with the number of units of each type of raw material needed for each style given in the table below.

	Wood	Nylon	Velvet	Springs
Style A	5	20	0	10
Style B	10	9	0	0
Style C	5	10	10	10

The cost in dollars per unit for each of the raw materials is given in the table below.

Wood	15
Nylon	12
Velvet	14
Springs	30

Create two matrices to represent the data and use matrix multiplication to find the price of manufacturing each style of chair.

37. Politics In a midwestern state, it is determined that 90% of all Republicans vote for Republican candidates and the remainder for Democratic candidates, while 80% of all Democrats vote for Democratic candidates and the remainder for Republican candidates. The percent of each party predicted to win the next election is given by

$$\begin{bmatrix} R \\ D \end{bmatrix} = \begin{bmatrix} 0.90 & 0.20 \\ 0.10 & 0.80 \end{bmatrix} \begin{bmatrix} a \\ b \end{bmatrix}$$

where a is the percent of Republicans and b is the percent of Democrats who won the last election. If 50% of those winning the election last time were Republicans and 50% were Democrats, what is the percent of each party that is predicted to win the next election?

38. Competition Two phone companies compete for customers in the southeastern region of a state. Company X retains $\frac{3}{5}$ of its customers and loses $\frac{2}{5}$ of its customers to company Y; company Y retains $\frac{2}{3}$ of its customers and loses $\frac{1}{3}$ of its customers to company X. If we represent the fraction of the market held last year by

$$\begin{bmatrix} a \\ b \end{bmatrix}$$

where a is the number of customers that company X had and b is the number of customers that company Y had, then the fraction that each company will have this year can be found from

$$\begin{bmatrix} x \\ y \end{bmatrix} = \begin{bmatrix} 3/5 & 1/3 \\ 2/5 & 2/3 \end{bmatrix} \begin{bmatrix} a \\ b \end{bmatrix}$$

If company X had 120,000 customers and company Y had 90,000 customers last year, how many customers did each have this year?

39. Wages The graph below gives the median weekly earnings for male and female full-time wage and salary workers in selected age groups in 2010.

a. Make a 2 × 6 matrix containing the data in this graph.

b. Suppose the median weekly earnings of men are found to be 10% more than the earnings given in the graph, and the median weekly earnings of women are found to be 25% more than the earnings given in the graph. Use matrix multiplication by a 2 × 2 matrix to find the new median weekly earnings by age for the male and female workers.

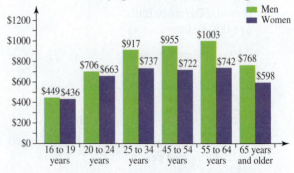

Median Usual Weekly Earnings of Full-time Wage and Salary Workers by Age and Sex, 2010 Annual Averages

(Source: U.S. Dept. of Labor, Bureau of Labor Statistics)

40. Libraries The number of libraries and the operating income of public libraries in selected states for 2000 and for 2008 are shown in the tables on the next page.

a. Form the matrix A that contains the numbers of libraries and operating incomes for 2000.

b. Form the matrix B that contains the numbers of libraries and operating incomes for 2008.

2000

	Number of Libraries	Operating Income ($ millions)
Illinois	606	358
Iowa	518	49
Kansas	324	50
Alabama	207	48
California	170	566

c. Use these two matrices to find the increase in the number of libraries and operating income from 2000 through 2008.

d. For which state is the increase in the number of libraries largest?

e. For which state is the increase in operating income the largest?

2008

	Number of Libraries	Operating Income ($ millions)
Illinois	634	714
Iowa	539	99
Kansas	327	108
Alabama	210	99
California	181	1326

(Source: Institute of Museum and Library Services.)

7.4 Inverse Matrices; Matrix Equations

KEY OBJECTIVES

- Find the inverse of a matrix
- Find matrix inverses with technology
- Encode and decode messages
- Solve matrix equations
- Solve matrix equations with technology

SECTION PREVIEW Encryption

Security of credit card numbers on the Internet depends on encryption of the data. Encryption involves providing a way for the sender to encode a message so that the message is not apparent and a way for the receiver to decode the message so that it can be read. Throughout history, different military units have used encoding and decoding systems of varying sophistication. In this section, we find the inverse of a matrix, use it to decode messages, and use it to solve matrix equations that have unique solutions. ■

Inverse Matrices

If the product of matrices A and B is an identity matrix, I, we say that B is the inverse of A (and A is the inverse of B). B is called the **inverse matrix** of A and is denoted A^{-1}.

> ### Inverse Matrices
>
> Two square matrices, A and B, are called **inverses** of each other if
>
> $$AB = I \quad \text{and} \quad BA = I$$
>
> where I is the identity matrix. We denote this by $B = A^{-1}$ and $A = B^{-1}$.

EXAMPLE 1 ▶ **Inverse Matrices**

Show that A and B are inverse matrices if

$$A = \begin{bmatrix} 1 & -0.6 & -0.2 \\ 0 & 0.4 & -0.2 \\ -1 & 0.4 & 0.8 \end{bmatrix} \quad \text{and} \quad B = \begin{bmatrix} 2 & 2 & 1 \\ 1 & 3 & 1 \\ 2 & 1 & 2 \end{bmatrix}$$

(a)

(b)

Figure 7.17

SOLUTION

$$AB = \begin{bmatrix} 1 & -0.6 & -0.2 \\ 0 & 0.4 & -0.2 \\ -1 & 0.4 & 0.8 \end{bmatrix} \begin{bmatrix} 2 & 2 & 1 \\ 1 & 3 & 1 \\ 2 & 1 & 2 \end{bmatrix}$$

$$= \begin{bmatrix} 2 - 0.6 - 0.4 & 2 - 1.8 - 0.2 & 1 - 0.6 - 0.4 \\ 0 + 0.4 - 0.4 & 0 + 1.2 - 0.2 & 0 + 0.4 - 0.4 \\ -2 + 0.4 + 1.6 & -2 + 1.2 + 0.8 & -1 + 0.4 + 1.6 \end{bmatrix}$$

$$= \begin{bmatrix} 1 & 0 & 0 \\ 0 & 1 & 0 \\ 0 & 0 & 1 \end{bmatrix}$$

The product AB is the 3×3 identity matrix; we can use technology (Figure 7.17) to see that the product BA is also the identity matrix. Thus, A and B are inverse matrices.

We have used elementary row operations on augmented matrices to solve systems of equations. We can also find the inverse of a matrix A, if it exists, by using elementary row operations. Note that if a matrix is not square, then it does not have an inverse, and that not all square matrices have inverses.

Finding the Inverse of a Square Matrix

1. Write the matrix with the same dimension identity matrix in its augment, getting a matrix of the form $[A\,|\,I]$.

2. Use elementary row operations on $[A\,|\,I]$ to attempt to transform A into an identity matrix, giving a new matrix of the form $[I\,|\,B]$. The matrix B is the inverse of A.

3. If A does not have an inverse, the reduction process will yield a row of zeros in the left half (representing the original matrix) of the augmented matrix.

EXAMPLE 2 ▶ **Finding an Inverse Matrix**

Find the inverse of $A = \begin{bmatrix} 2 & 2 \\ 2 & 1 \end{bmatrix}$.

SOLUTION

Creating the matrix $[A\,|\,I]$ and performing the operations to convert A to I gives

$$\begin{bmatrix} 2 & 2 & | & 1 & 0 \\ 2 & 1 & | & 0 & 1 \end{bmatrix} \xrightarrow{(1/2)R_1 \rightarrow R_1} \begin{bmatrix} 1 & 1 & | & \dfrac{1}{2} & 0 \\ 2 & 1 & | & 0 & 1 \end{bmatrix}$$

$$\xrightarrow{-2R_1 + R_2 \rightarrow R_2} \begin{bmatrix} 1 & 1 & | & \dfrac{1}{2} & 0 \\ 0 & -1 & | & -1 & 1 \end{bmatrix}$$

$$\xrightarrow{-R_2 \rightarrow R_2} \begin{bmatrix} 1 & 1 & | & \dfrac{1}{2} & 0 \\ 0 & 1 & | & 1 & -1 \end{bmatrix} \xrightarrow{-R_2 + R_1 \rightarrow R_1} \begin{bmatrix} 1 & 0 & | & -\dfrac{1}{2} & 1 \\ 0 & 1 & | & 1 & -1 \end{bmatrix}$$

Thus, the inverse matrix of A is $A^{-1} = \begin{bmatrix} -\dfrac{1}{2} & 1 \\ 1 & -1 \end{bmatrix}$. We can verify this by observing

that $\begin{bmatrix} 2 & 2 \\ 2 & 1 \end{bmatrix} \begin{bmatrix} -\dfrac{1}{2} & 1 \\ 1 & -1 \end{bmatrix} = \begin{bmatrix} 1 & 0 \\ 0 & 1 \end{bmatrix}$.

EXAMPLE 3 ▶ **Inverse of a 3 × 3 Matrix**

Find the inverse of $A = \begin{bmatrix} -2 & 1 & 2 \\ 1 & 0 & -1 \\ 4 & -2 & -3 \end{bmatrix}$.

SOLUTION

Creating the matrix $[A\,|\,I]$ and performing the operations to convert A to I gives

$$\begin{bmatrix} -2 & 1 & 2 & | & 1 & 0 & 0 \\ 1 & 0 & -1 & | & 0 & 1 & 0 \\ 4 & -2 & -3 & | & 0 & 0 & 1 \end{bmatrix} \xrightarrow{R_1 \leftrightarrow R_2} \begin{bmatrix} 1 & 0 & -1 & | & 0 & 1 & 0 \\ -2 & 1 & 2 & | & 1 & 0 & 0 \\ 4 & -2 & -3 & | & 0 & 0 & 1 \end{bmatrix}$$

$$\xrightarrow[\substack{2R_1 + R_2 \rightarrow R_2 \\ -4R_1 + R_3 \rightarrow R_3}]{} \begin{bmatrix} 1 & 0 & -1 & | & 0 & 1 & 0 \\ 0 & 1 & 0 & | & 1 & 2 & 0 \\ 0 & -2 & 1 & | & 0 & -4 & 1 \end{bmatrix}$$

$$\xrightarrow{2R_2 + R_3 \rightarrow R_3} \begin{bmatrix} 1 & 0 & -1 & | & 0 & 1 & 0 \\ 0 & 1 & 0 & | & 1 & 2 & 0 \\ 0 & 0 & 1 & | & 2 & 0 & 1 \end{bmatrix}$$

$$\xrightarrow{R_3 + R_1 \rightarrow R_1} \begin{bmatrix} 1 & 0 & 0 & | & 2 & 1 & 1 \\ 0 & 1 & 0 & | & 1 & 2 & 0 \\ 0 & 0 & 1 & | & 2 & 0 & 1 \end{bmatrix}$$

Thus, we have the inverse of A.

$$A^{-1} = \begin{bmatrix} 2 & 1 & 1 \\ 1 & 2 & 0 \\ 2 & 0 & 1 \end{bmatrix}$$

Inverses and Technology

Computer software, spreadsheets, and calculators can be used to find the inverse of a matrix. We will see that if the inverse of a square matrix exists, it can be found easily with a calculator.* Figure 7.18(a) shows matrix A from Example 3, and Figure 7.18(b) shows the inverse of A found with a graphing calculator. We can confirm that these matrices are inverses by computing AA^{-1} (Figure 7.19).

(a)

(b)

Figure 7.18

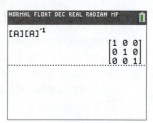

Figure 7.19

_____
* For more details. see Appendix A. page 680.

Figure 7.20

Spreadsheet ▶ SOLUTION The inverse of matrix A is found with Excel by entering the matrix in cells B1:D3, selecting cells B5:D7 to contain the inverse, entering "=minverse(", and then selecting the cells B1:D3, closing the parentheses, and holding the CTRL and SHIFT keys down while pressing ENTER. Figure 7.20 shows the spreadsheet with matrix A and its inverse.*

EXAMPLE 4 ▶ Does the Inverse Exist?

Find the inverse of $A = \begin{bmatrix} 1 & 2 & -2 \\ 2 & 0 & 2 \\ 6 & 4 & 0 \end{bmatrix}$, if it exists.

SOLUTION

Attempting to find the inverse of matrix A with technology results in an error statement, indicating that the inverse does not exist (Figure 7.21).

Recall that if A does not have an inverse, the reduction process using elementary row operations will yield a row of zeros in the left half of the augmented matrix.[†]

$$\begin{bmatrix} 1 & 2 & -2 & | & 1 & 0 & 0 \\ 2 & 0 & 2 & | & 0 & 1 & 0 \\ 6 & 4 & 0 & | & 0 & 0 & 1 \end{bmatrix} \xrightarrow[\substack{(-6)R_1 + R_3 \rightarrow R_3}]{(-2)R_1 + R_2 \rightarrow R_2} \begin{bmatrix} 1 & 2 & -2 & | & 1 & 0 & 0 \\ 0 & -4 & 6 & | & -2 & 1 & 0 \\ 0 & -8 & 12 & | & -6 & 0 & 1 \end{bmatrix}$$

$$\xrightarrow{(-1/4)R_2 \rightarrow R_2} \begin{bmatrix} 1 & 2 & -2 & | & 1 & 0 & 0 \\ 0 & 1 & -\frac{3}{2} & | & \frac{1}{2} & -\frac{1}{4} & 0 \\ 0 & -8 & 12 & | & -6 & 0 & 1 \end{bmatrix}$$

$$\xrightarrow{(8)R_2 + R_3 \rightarrow R_3} \begin{bmatrix} 1 & 2 & -2 & | & 1 & 0 & 0 \\ 0 & 1 & -\frac{3}{2} & | & \frac{1}{2} & -\frac{1}{4} & 0 \\ 0 & 0 & 0 & | & -2 & -2 & 1 \end{bmatrix}$$

We see that the left half of the bottom row of the reduced matrix contains all 0's, so it is not possible to reduce the original matrix to the identity matrix, and thus the matrix A does not have an inverse.

```
NORMAL FLOAT DEC REAL RADIAN MP
[A]
                    [1 2 -2]
                    [2 0 2]
                    [6 4 0]
[A]⁻¹
```
(a)

```
NORMAL FLOAT DEC REAL RADIAN MP
  ERROR: SINGULAR MATRIX
1:Quit
2:Goto

No inverse matrix exists.
Calculated result not
  found if dependent on a
  matrix inverse.
```
(b)

Figure 7.21

Encoding and Decoding Messages

In sending messages during military maneuvers, in business transactions, and in sending secure data on the Internet, encoding (or encryption) of messages is important. Suppose we want to encode the message "Cheer up." The following simple code could be used to change letters of the alphabet to the numbers 1 to 26, respectively, with the number 27 representing a blank space.

---

* For detailed steps, see Appendix B, page 698.

† The methods used by technology will sometimes yield approximations for numbers that should be zeros.

$$a \ b \ c \ d \ e \ f \ g \ h \ i \ j \ k \ l \ m \ n \ o \ p \ q \ r \ s \ t \ u \ v \ w \ x \ y \ z$$
$$1 \ 2 \ 3 \ 4 \ 5 \ 6 \ 7 \ 8 \ 9 \ 10 \ 11 \ 12 \ 13 \ 14 \ 15 \ 16 \ 17 \ 18 \ 19 \ 20 \ 21 \ 22 \ 23 \ 24 \ 25 \ 26 \ 27$$

Then "Cheer up" can be represented by the numbers

$$3 \ 8 \ 5 \ 5 \ 18 \ 27 \ 21 \ 16$$

To further encode the message, we put these numbers in pairs and then create a 2×1 matrix for each pair of numbers. Next we choose an *encoding matrix*, like

$$A = \begin{bmatrix} 3 & -2 \\ -1 & 1 \end{bmatrix}$$

and multiply each pair of numbers (in order) by the encoding matrix, as follows.

$$\begin{bmatrix} 3 & -2 \\ -1 & 1 \end{bmatrix}\begin{bmatrix} 3 \\ 8 \end{bmatrix} = \begin{bmatrix} -7 \\ 5 \end{bmatrix} \qquad \begin{bmatrix} 3 & -2 \\ -1 & 1 \end{bmatrix}\begin{bmatrix} 5 \\ 5 \end{bmatrix} = \begin{bmatrix} 5 \\ 0 \end{bmatrix}$$

$$\begin{bmatrix} 3 & -2 \\ -1 & 1 \end{bmatrix}\begin{bmatrix} 18 \\ 27 \end{bmatrix} = \begin{bmatrix} 0 \\ 9 \end{bmatrix} \qquad \begin{bmatrix} 3 & -2 \\ -1 & 1 \end{bmatrix}\begin{bmatrix} 21 \\ 16 \end{bmatrix} = \begin{bmatrix} 31 \\ -5 \end{bmatrix}$$

Because multiplying a 2×2 matrix by a 2×1 matrix gives a 2×1 matrix, the resulting products are pairs of numbers. Combining the pairs of numbers gives the encoded numerical message

$$-7 \ 5 \ 5 \ 0 \ 0 \ 9 \ 31 \ -5$$

Note that another encoding matrix could be used rather than the one used above. We can also encode a message by putting triples of numbers in 3×1 matrices and multiplying each 3×1 matrix by a 3×3 encoding matrix.

EXAMPLE 5 ▶ Encoding Messages

Use the encoding matrix $A = \begin{bmatrix} 1 & -3 & 2 \\ 2 & -2 & 2 \\ 3 & -1 & 1 \end{bmatrix}$ to encode the message "Meet me for lunch."

SOLUTION

Converting the letters of the message to triples of numbers gives

$$13 \ 5 \ 5 \qquad 20 \ 27 \ 13 \qquad 5 \ 27 \ 6 \qquad 15 \ 18 \ 27 \qquad 12 \ 21 \ 14 \qquad 3 \ 8 \ 27$$

with 27 used to complete the last triple.

Placing these triples of numbers in 3×1 matrices and multiplying by matrix A gives

$$\begin{bmatrix} 1 & -3 & 2 \\ 2 & -2 & 2 \\ 3 & -1 & 1 \end{bmatrix}\begin{bmatrix} 13 \\ 5 \\ 5 \end{bmatrix} = \begin{bmatrix} 8 \\ 26 \\ 39 \end{bmatrix} \qquad \begin{bmatrix} 1 & -3 & 2 \\ 2 & -2 & 2 \\ 3 & -1 & 1 \end{bmatrix}\begin{bmatrix} 20 \\ 27 \\ 13 \end{bmatrix} = \begin{bmatrix} -35 \\ 12 \\ 46 \end{bmatrix}$$

$$\begin{bmatrix} 1 & -3 & 2 \\ 2 & -2 & 2 \\ 3 & -1 & 1 \end{bmatrix}\begin{bmatrix} 5 \\ 27 \\ 6 \end{bmatrix} = \begin{bmatrix} -64 \\ -32 \\ -6 \end{bmatrix} \qquad \begin{bmatrix} 1 & -3 & 2 \\ 2 & -2 & 2 \\ 3 & -1 & 1 \end{bmatrix}\begin{bmatrix} 15 \\ 18 \\ 27 \end{bmatrix} = \begin{bmatrix} 15 \\ 48 \\ 54 \end{bmatrix}$$

$$\begin{bmatrix} 1 & -3 & 2 \\ 2 & -2 & 2 \\ 3 & -1 & 1 \end{bmatrix}\begin{bmatrix} 12 \\ 21 \\ 14 \end{bmatrix} = \begin{bmatrix} -23 \\ 10 \\ 29 \end{bmatrix} \qquad \begin{bmatrix} 1 & -3 & 2 \\ 2 & -2 & 2 \\ 3 & -1 & 1 \end{bmatrix}\begin{bmatrix} 3 \\ 8 \\ 27 \end{bmatrix} = \begin{bmatrix} 33 \\ 44 \\ 28 \end{bmatrix}$$

The products are triples of numbers. Combining the triples of numbers gives the encoded message

$$8 \ 26 \ 39 \ -35 \ 12 \ 46 \ -64 \ -32 \ -6 \ 15 \ 48 \ 54 \ -23 \ 10 \ 29 \ 33 \ 44 \ 28$$

We can also find the encoded triples of numbers for the message of Example 5 by writing the original triples of numbers as columns in one matrix and then multiplying this matrix by the encoding matrix. The columns of the product will be the encoded triples of numbers.

Figure 7.22

$$\begin{bmatrix} 1 & -3 & 2 \\ 2 & -2 & 2 \\ 3 & -1 & 1 \end{bmatrix} \begin{bmatrix} 13 & 20 & 5 & 15 & 12 & 3 \\ 5 & 27 & 27 & 18 & 21 & 8 \\ 5 & 13 & 6 & 27 & 14 & 27 \end{bmatrix} = \begin{bmatrix} 8 & -35 & -64 & 15 & -23 & 33 \\ 26 & 12 & -32 & 48 & 10 & 44 \\ 39 & 46 & -6 & 54 & 29 & 28 \end{bmatrix}$$

Observe that the columns of this product have the same triples, respectively, as the individual products in Example 5. Figure 7.22 shows a calculator display of the product.

Sending an encoded message is of little value if the recipient is not able to decode it. If the message is encoded with a matrix, then the message can be decoded with the inverse of the matrix.

Recall that the message "Cheer up" was represented by the numbers 3 8 5 5 18 27 21 16 and encoded to −7 5 5 0 0 9 31 −5 with the encoding matrix

$$A = \begin{bmatrix} 3 & -2 \\ -1 & 1 \end{bmatrix}$$

To decode the encoded message, we multiply each pair of numbers in the encoded message on the left by the inverse of matrix A.

$$A^{-1} = \begin{bmatrix} 1 & 2 \\ 1 & 3 \end{bmatrix}$$

Multiplying the pairs of the coded message by A^{-1} gives us back the original message.

$$\begin{bmatrix} 1 & 2 \\ 1 & 3 \end{bmatrix}\begin{bmatrix} -7 \\ 5 \end{bmatrix} = \begin{bmatrix} 3 \\ 8 \end{bmatrix} \qquad \begin{bmatrix} 1 & 2 \\ 1 & 3 \end{bmatrix}\begin{bmatrix} 5 \\ 0 \end{bmatrix} = \begin{bmatrix} 5 \\ 5 \end{bmatrix}$$

$$\begin{bmatrix} 1 & 2 \\ 1 & 3 \end{bmatrix}\begin{bmatrix} 0 \\ 9 \end{bmatrix} = \begin{bmatrix} 18 \\ 27 \end{bmatrix} \qquad \begin{bmatrix} 1 & 2 \\ 1 & 3 \end{bmatrix}\begin{bmatrix} 31 \\ -5 \end{bmatrix} = \begin{bmatrix} 21 \\ 16 \end{bmatrix}$$

Thus, the decoded numbers are 3 8 5 5 18 27 21 16, which correspond to the letters in the message "Cheer up."

EXAMPLE 6 ▶ Decoding Messages

Suppose we encoded messages using a simple code that changed letters of the alphabet to numbers, with the number 27 representing a blank space, and the encoding matrix

$$A = \begin{bmatrix} 1 & 2 & -1 \\ 2 & 1 & 2 \\ 3 & 2 & -3 \end{bmatrix}$$

Decode the following coded message:

18 41 34 61 93 75 39 77 9 34 62 46 3 62 −23

SOLUTION

To find the numbers representing the message, we place triples of numbers from the coded message into columns of a 3×5 matrix C and multiply that matrix by A^{-1}. The inverse of A is

$$A^{-1} = \begin{bmatrix} \dfrac{-7}{16} & \dfrac{1}{4} & \dfrac{5}{16} \\[2mm] \dfrac{3}{4} & 0 & \dfrac{-1}{4} \\[2mm] \dfrac{1}{16} & \dfrac{1}{4} & \dfrac{-3}{16} \end{bmatrix}$$

Multiplying by A^{-1} on the left gives

$$A^{-1}\begin{bmatrix} 18 & 61 & 39 & 34 & 3 \\ 41 & 93 & 77 & 62 & 62 \\ 34 & 75 & 9 & 46 & -23 \end{bmatrix} = \begin{bmatrix} \dfrac{-7}{16} & \dfrac{1}{4} & \dfrac{5}{16} \\[2mm] \dfrac{3}{4} & 0 & \dfrac{-1}{4} \\[2mm] \dfrac{1}{16} & \dfrac{1}{4} & \dfrac{-3}{16} \end{bmatrix}\begin{bmatrix} 18 & 61 & 39 & 34 & 3 \\ 41 & 93 & 77 & 62 & 62 \\ 34 & 75 & 9 & 46 & -23 \end{bmatrix}$$

$$= \begin{bmatrix} 13 & 20 & 5 & 15 & 7 \\ 5 & 27 & 27 & 14 & 8 \\ 5 & 13 & 20 & 9 & 20 \end{bmatrix}$$

Figure 7.23

If we use technology to perform this multiplication, we do not need to display the elements of A^{-1} in the computation. We enter $[A]$ and $[C]$ and compute $[A]^{-1}[C]$ (Figure 7.23).

Reading down the columns of the product gives the numbers representing the message:

13 5 5 20 27 13 5 27 20 15 14 9 7 8 20

The message is "Meet me tonight."

Matrix Equations

The system of equations

$$\begin{cases} -2x + y + 2z = 5 \\ x \qquad - z = 2 \\ 4x - 2y - 3z = 4 \end{cases}$$

can be written as the matrix equation

$$\begin{bmatrix} -2x + y + 2z \\ x \qquad - z \\ 4x - 2y - 3z \end{bmatrix} = \begin{bmatrix} 5 \\ 2 \\ 4 \end{bmatrix}$$

Because

$$\begin{bmatrix} -2 & 1 & 2 \\ 1 & 0 & -1 \\ 4 & -2 & -3 \end{bmatrix}\begin{bmatrix} x \\ y \\ z \end{bmatrix} = \begin{bmatrix} -2x + y + 2z \\ x \qquad - z \\ 4x - 2y - 3z \end{bmatrix}$$

we can write the system of equations as the matrix equation in the form

$$\begin{bmatrix} -2 & 1 & 2 \\ 1 & 0 & -1 \\ 4 & -2 & -3 \end{bmatrix}\begin{bmatrix} x \\ y \\ z \end{bmatrix} = \begin{bmatrix} 5 \\ 2 \\ 4 \end{bmatrix}$$

Note that the 3×3 matrix on the left side of the matrix equation is the coefficient matrix for the original system. We will call the coefficient matrix A, the matrix containing the variables X, and the matrix containing the constants C, giving the form

$$AX = C$$

If A^{-1} exists and we multiply both sides of this equation on the left by A^{-1}, the product is as follows:*

$$A^{-1}AX = A^{-1}C$$

Because $A^{-1}A = I$ and because multiplication of matrix X by an identity matrix gives matrix X, we have

$$IX = A^{-1}C \quad \text{or} \quad X = A^{-1}C$$

In general, multiplying both sides of the matrix equation $AX = C$ on the left by A^{-1} gives the solution to the system that the matrix equation represents, if the solution is unique.

EXAMPLE 7 ▶ **Solution of Matrix Equations with Inverses**

Solve the system

$$\begin{cases} -2x + y + 2z = 5 \\ x \quad\quad - z = 2 \\ 4x - 2y - 3z = 4 \end{cases}$$

by writing a matrix equation and using an inverse matrix.

SOLUTION

We can write this system of equations as $AX = C$, where

$$A = \begin{bmatrix} -2 & 1 & 2 \\ 1 & 0 & -1 \\ 4 & -2 & -3 \end{bmatrix}, \quad X = \begin{bmatrix} x \\ y \\ z \end{bmatrix} \quad \text{and} \quad C = \begin{bmatrix} 5 \\ 2 \\ 4 \end{bmatrix}$$

That is,

$$\begin{bmatrix} -2 & 1 & 2 \\ 1 & 0 & -1 \\ 4 & -2 & -3 \end{bmatrix} \begin{bmatrix} x \\ y \\ z \end{bmatrix} = \begin{bmatrix} 5 \\ 2 \\ 4 \end{bmatrix}$$

As we saw in Example 3, the inverse of matrix A is

$$A^{-1} = \begin{bmatrix} 2 & 1 & 1 \\ 1 & 2 & 0 \\ 2 & 0 & 1 \end{bmatrix}$$

Thus, we can solve the system by multiplying both sides of the matrix equation on the left by A^{-1} as follows:

$$A^{-1}A \begin{bmatrix} x \\ y \\ z \end{bmatrix} = A^{-1} \begin{bmatrix} 5 \\ 2 \\ 4 \end{bmatrix}$$

$$I \begin{bmatrix} x \\ y \\ z \end{bmatrix} = \begin{bmatrix} 2 & 1 & 1 \\ 1 & 2 & 0 \\ 2 & 0 & 1 \end{bmatrix} \begin{bmatrix} 5 \\ 2 \\ 4 \end{bmatrix}$$

_____

* Recall that multiplication on the right may give a different product than multiplication on the left.

$$\begin{bmatrix} x \\ y \\ z \end{bmatrix} = \begin{bmatrix} 16 \\ 9 \\ 14 \end{bmatrix}$$

Thus, we have the solution to the system, $x = 16$, $y = 9$, and $z = 14$, or (16, 9, 14).

Matrix Equations and Technology

We can use technology to multiply both sides of the matrix equation $AX = C$ by the inverse of A and get the solution to the system that the matrix equation represents. If we use technology to solve a system of linear equations, it is not necessary to display the inverse of the coefficient matrix.

EXAMPLE 8 ▶ **Manufacturing**

Sharper Technology Company manufactures three types of calculators: a business calculator, a scientific calculator, and a graphing calculator. The production requirements are given in Table 7.10. If during each month the company has 134,000 circuit components, 56,000 hours for assembly, and 14,000 cases, how many of each type of calculator can it produce each month?

Table 7.10

	Business Calculator	Scientific Calculator	Graphing Calculator
Circuit Components	5	7	12
Assembly Time (hours)	2	3	5
Cases	1	1	1

(a)

(b)

Figure 7.24

SOLUTION

If the company produces x business calculators, y scientific calculators, and z graphing calculators, then the problem can be represented using the following system of equations.

$$\begin{cases} 5x + 7y + 12z = 134{,}000 \\ 2x + 3y + 5z = 56{,}000 \\ x + y + z = 14{,}000 \end{cases}$$

This system can be represented by the matrix equation $AX = C$, where A is the coefficient matrix and C is the constant matrix.

$$\begin{bmatrix} 5 & 7 & 12 \\ 2 & 3 & 5 \\ 1 & 1 & 1 \end{bmatrix} \begin{bmatrix} x \\ y \\ z \end{bmatrix} = \begin{bmatrix} 134{,}000 \\ 56{,}000 \\ 14{,}000 \end{bmatrix}$$

Multiplying both sides of this equation by the inverse of the coefficient matrix gives the solution (Figure 7.24):

$$\begin{bmatrix} x \\ y \\ z \end{bmatrix} = A^{-1} \begin{bmatrix} 134{,}000 \\ 56{,}000 \\ 14{,}000 \end{bmatrix} = \begin{bmatrix} 2000 \\ 4000 \\ 8000 \end{bmatrix}$$

This shows that the company can produce 2000 business calculators, 4000 scientific calculators, and 8000 graphing calculators each month. (Recall that we could also solve this system of linear equations by using elementary row operations with an augmented matrix, as discussed in Section 7.2.)

**Spreadsheet
▸ SOLUTION**   We can also use inverse matrices with Excel to solve a system of linear equations if a unique solution exists. We can solve such a system by finding the inverse of the coefficient matrix and multiplying this inverse times the matrix containing the constants. The spreadsheet in Figure 7.25 shows the solution of Example 8 found with Excel.

We can find this solution as follows:

1. Enter the coefficient matrix A in cells B1:D3, compute the inverse of A in B5:D7, enter matrix C containing the constants in B9:B11, and select cells b13:B15.

2. Type "=mmult(" in the formula bar, select B5:D7, type a comma, select B9:B11, and close the parentheses.

3. Hold the CTRL and SHIFT keys down and press ENTER, getting the solution in B13:B15. The inverse is shown in Figure 7.25.*

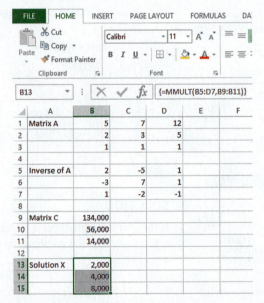

Figure 7.25

We can use the inverse of the coefficient matrix to solve a system of linear equations only if the system has a unique solution. Otherwise, the inverse of the coefficient matrix will not exist, and an attempt to use technology to solve the system with an inverse will give an error message.

---

* For detailed steps, see Appendix B, page 699.

SKILLS CHECK 7.4

1. Suppose $A = \begin{bmatrix} 3 & 1 \\ 4 & 2 \end{bmatrix}$ and $B = \begin{bmatrix} 1 & -0.5 \\ -2 & 1.5 \end{bmatrix}$.

  a. Compute AB and BA, if possible.

  b. What is the relationship between A and B?

2. Suppose $A = \begin{bmatrix} 1 & -\dfrac{1}{2} & -\dfrac{1}{4} \\ -\dfrac{1}{2} & 0 & \dfrac{1}{2} \\ 0 & \dfrac{1}{2} & -\dfrac{1}{4} \end{bmatrix}$ and

$B = \begin{bmatrix} 2 & 2 & 2 \\ 1 & 2 & 3 \\ 2 & 4 & 2 \end{bmatrix}$.

  a. Compute AB and BA, if possible.

  b. What is the relationship between A and B?

3. Show that A and B are inverse matrices if

$A = \begin{bmatrix} 1 & -1 & 1 \\ 2 & -1 & 0 \\ -2 & 2 & -1 \end{bmatrix}$ and $B = \begin{bmatrix} 1 & 1 & 1 \\ 2 & 1 & 2 \\ 2 & 0 & 1 \end{bmatrix}$.

4. Show that C and D are inverse matrices if

$C = \begin{bmatrix} -\dfrac{1}{2} & \dfrac{1}{2} & 1 \\ \dfrac{1}{2} & \dfrac{1}{2} & 0 \\ -\dfrac{1}{2} & \dfrac{3}{2} & 1 \end{bmatrix}$ and $D = \begin{bmatrix} 1 & 2 & -1 \\ -1 & 0 & 1 \\ 2 & 1 & -1 \end{bmatrix}$.

In Exercises 5–14, find the inverse of the given matrix.

5. $A = \begin{bmatrix} 1 & 3 \\ 2 & 7 \end{bmatrix}$ **6.** $A = \begin{bmatrix} 2 & 4 \\ 2 & 5 \end{bmatrix}$

7. $A = \begin{bmatrix} 2 & -2 & 2 \\ 2 & 1 & 2 \\ 2 & 0 & 1 \end{bmatrix}$

8. $A = \begin{bmatrix} \dfrac{1}{2} & -1 & 1 \\ 1 & -2 & 0 \\ 1 & 2 & -1 \end{bmatrix}$ **9.** $A = \begin{bmatrix} 1 & 1 & -1 \\ -1 & 0 & 1 \\ 1 & 1 & 2 \end{bmatrix}$

10. $C = \begin{bmatrix} 0 & 1 & -1 \\ 1 & 0 & 0 \\ 0 & 0 & 1 \end{bmatrix}$ **11.** $A = \begin{bmatrix} 2 & 3 & 1 \\ 3 & 4 & -1 \\ 2 & 5 & 1 \end{bmatrix}$

12. $B = \begin{bmatrix} 9 & 2 & -7 \\ -5 & 0 & 5 \\ 7 & -4 & -1 \end{bmatrix}$

13. $C = \begin{bmatrix} 1 & 0 & -1 & -1 \\ 0 & 1 & -1 & -1 \\ 0 & 0 & 1 & 0 \\ 0 & 0 & 0 & 1 \end{bmatrix}$

14. $A = \begin{bmatrix} -1 & 0 & 1 & 1 \\ 0 & 1 & 1 & 1 \\ 1 & -1 & 0 & 0 \\ 0 & 0 & 0 & 1 \end{bmatrix}$

15. If $A^{-1} = \begin{bmatrix} 1 & 2 \\ 4 & 3 \end{bmatrix}$, solve $AX = \begin{bmatrix} 2 \\ 4 \end{bmatrix}$ for X.

16. If $A^{-1} = \begin{bmatrix} 1 & 2 & -1 \\ 0 & 2 & 1 \\ 2 & 0 & -2 \end{bmatrix}$, solve $AX = \begin{bmatrix} 1 \\ 2 \\ -2 \end{bmatrix}$ for X.

17. Solve the matrix equation

$$\begin{bmatrix} -1 & 1 & 0 \\ -2 & 3 & -2 \\ 2 & -2 & 1 \end{bmatrix} \begin{bmatrix} x \\ y \\ z \end{bmatrix} = \begin{bmatrix} 3 \\ 5 \\ 8 \end{bmatrix}$$

18. Solve the matrix equation

$$\begin{bmatrix} -2 & 1 & 0 \\ -1 & 3 & -2 \\ 2 & -2 & 1 \end{bmatrix} \begin{bmatrix} x \\ y \\ z \end{bmatrix} = \begin{bmatrix} 2 \\ 4 \\ 8 \end{bmatrix}$$

In Exercises 19–24, use an inverse matrix to find the solution to the systems.

19. $\begin{cases} 4x - 3y + z = 2 \\ -6x + 5y - 2z = -3 \\ x - y + z = 1 \end{cases}$

20. $\begin{cases} 5x + 3y + z = 12 \\ 4x + 3y + 2z = 9 \\ x + y + 2z = 2 \end{cases}$

21. $\begin{cases} 2x + y + z = 4 \\ x + 4y + 2z = 4 \\ 2x + y + 2z = 3 \end{cases}$

22. $\begin{cases} x + 2y - z = 2 \\ -x + z = 4 \\ 2x + y - z = 6 \end{cases}$

23. $\begin{cases} x_1 & + x_3 + x_4 = 90 \\ x_2 + x_3 + x_4 = 72 \\ 2x_1 + 5x_2 + x_3 \quad = 108 \\ 3x_2 \quad\quad + x_4 = 144 \end{cases}$

24. $\begin{cases} 2x_1 + x_2 + 3x_3 + 4x_4 = 1 \\ x_1 + x_2 + x_3 - x_4 = 2 \\ x_1 + x_2 + x_3 + x_4 = 10 \\ 2x_1 - x_2 + 3x_3 - x_4 = 5 \end{cases}$

EXERCISES 7.4

25. **Competition** Two phone companies compete for customers in the southeastern region of a state. Each year, company X retains $\frac{3}{5}$ of its customers and loses $\frac{2}{5}$ of its customers to company Y, while company Y retains $\frac{2}{3}$ of its customers and loses $\frac{1}{3}$ of its customers to company X. If we represent the fraction of the market held last year by

$$\begin{bmatrix} a \\ b \end{bmatrix}$$

where a is the number that company X had last year and b is the number that company Y had last year, then the number that each company will have this year can be found by

$$\begin{bmatrix} A \\ B \end{bmatrix} = \begin{bmatrix} \frac{3}{5} & \frac{1}{3} \\ \frac{2}{5} & \frac{2}{3} \end{bmatrix} \begin{bmatrix} a \\ b \end{bmatrix}$$

If company X has $A = 150,000$ customers and company Y has $B = 120,000$ customers this year, how many customers did each have last year?

26. **Competition** A satellite company and a cable company compete for customers in a city. The satellite company retains $\frac{2}{3}$ of its customers and loses $\frac{1}{3}$ of its customers to the cable company, while the cable company retains $\frac{3}{5}$ of its customers and loses $\frac{2}{5}$ of its customers to the satellite company. If we represent the fraction of the market held last year by

$$\begin{bmatrix} a \\ b \end{bmatrix}$$

where a is the number of customers that the satellite company had last year and b is the number that the cable company had last year, then the number that each company will have this year can be found by

$$\begin{bmatrix} A \\ B \end{bmatrix} = \begin{bmatrix} \frac{2}{3} & \frac{2}{5} \\ \frac{1}{3} & \frac{3}{5} \end{bmatrix} \begin{bmatrix} a \\ b \end{bmatrix}$$

If the satellite company has $A = 90,000$ customers and the cable company has $B = 85,000$ customers this year, how many customers did each have last year?

27. **Politics** In a midwestern state, it is observed that 90% of all Republicans vote for the Republican candidate for governor and the remainder vote for the Democratic candidate, while 80% of all Democrats vote for the Democratic candidate and the remainder vote for the Republican candidate. The percent voting for each party in the next election is given by

$$\begin{bmatrix} R \\ D \end{bmatrix} = \begin{bmatrix} 0.90 & 0.20 \\ 0.10 & 0.80 \end{bmatrix} \begin{bmatrix} r \\ d \end{bmatrix}$$

where r and d are the respective percents voting for each party in the last election. If 55% of the votes in this election were for a Republican and 45% were for a Democrat, what are the percents for each party candidate in the last election?

28. **Politics** Suppose that in a certain city, the Democratic, Republican, and Consumer parties always nominate candidates for mayor. The percent of people voting for each party candidate in the next election depends on the percent voting for that party in the last election. The percent voting for each party in the next election is given by

$$\begin{bmatrix} D \\ R \\ C \end{bmatrix} = \begin{bmatrix} 50 & 40 & 30 \\ 40 & 50 & 30 \\ 10 & 10 & 40 \end{bmatrix} \begin{bmatrix} d \\ r \\ c \end{bmatrix}$$

where d, r, and c are the respective percents (in decimals) voting for each party in the last election. If the percent voting for a Democrat in this election is $D = 42\%$, the percent voting for a Republican in this election is $R = 42\%$, and the percent voting for the Consumer party candidate in this election is $C = 16\%$, what were the respective percents of people voting for these parties in the last election?

29. **Loans** A bank gives three loans totaling $400,000 to a development company for the purchase of three business properties. The largest loan is $100,000 more than the sum of the other two, and the smallest loan is one-half of the next larger loan. Represent

the amount of money in each loan as x, y, and z, respectively.

a. Write a system of three equations in three variables to represent the problem.

b. Use a matrix equation to solve the system and find the amount of each loan.

30. *Transportation* Ross Freightline has an order to deliver two products, A and B, to a store. The table gives the volume and weight for one unit of each of the two products.

	Product A	Product B
Volume (cu ft)	40	65
Weight (lb)	320	360

If the truck that can deliver the products can carry 5200 cubic feet and 31,360 pounds, how many units of each product can the truck carry?

31. *Investment* An investor has $400,000 in three accounts, paying 6%, 8%, and 10%, respectively. If she has twice as much invested at 8% as she has at 6%, how much does she have invested in each account if she earns a total of $36,000 in interest?

32. *Pricing* A theater has 1000 seats divided into orchestra, main, and balcony. The orchestra seats cost $80, the main seats cost $50, and the balcony seats cost $40. If all the seats are sold, the revenue is $50,000. If all the orchestra and balcony seats are sold and $\frac{3}{4}$ of the main seats are sold, the revenue is $42,500. How many of each type of seat does the theater have?

33. *Venture Capital* Suppose a bank draws its venture-capital funds annually from three sources of income: business loans, auto loans, and home mortgages. One spreadsheet shows the income from each of these sources for the years 2001, 2002, and 2003, and the second spreadsheet shows the venture capital for these years. If the bank uses a fixed percent of its income from each of the business loans, find the percent of income from each of these loans.

	A	B	C	D
1		Income	from Loans	(millions)
2	Years	Business	Auto	Home
3	2001	532	58	682
4	2002	562	62	695
5	2003	578	69	722

	A	B
1	Venture	Capital (millions)
2	2001	483.94
3	2002	503.28
4	2003	521.33

34. *Bookcases* A company produces three types of bookcases: 4-shelf metal, 6-shelf metal, and 4-shelf wooden. The 4-shelf metal bookcase requires 2 hours for fabrication, 1 hour to paint, and $\frac{1}{2}$ hour to package for shipping. The 6-shelf bookcase requires 3 hours for fabrication, 1.5 hours to paint, and $\frac{1}{2}$ hour to package. The wooden bookcase requires 3 hours to fabricate, 2 hours to paint, and $\frac{1}{2}$ hour to package. The daily amount of time available is 124 hours for fabrication, 68 hours for painting, and 24 hours for packaging. How many of each type of bookcase can be produced each day?

Encoding Messages We have encoded messages by assigning the numbers 1 to 26 to the letters a to z of the alphabet, respectively, and assigning 27 to a blank space.

a	b	c	d	e	f	g	h	i	j	k	l	m	n	o	p
1	2	3	4	5	6	7	8	9	10	11	12	13	14	15	16

q	r	s	t	u	v	w	x	y	z	blank
17	18	19	20	21	22	23	24	25	26	27

To further encode the messages, we can use an encoding matrix A to convert these numbers into new pairs or triples of numbers. In Exercises 35–38, use the given matrix A to encode the message.

35. *Encoding Messages*
   a. Convert "Just do it" from letters to numbers.

   b. Multiply $A = \begin{bmatrix} 4 & 4 \\ 1 & 2 \end{bmatrix}$ times 2×1 matrices created with columns containing pairs of numbers from part (a) to encode the message "Just do it" into pairs of coded numbers.

36. *Encoding Messages*
   a. Convert "Call home" from letters to numbers.

   b. Multiply $A = \begin{bmatrix} 2 & 3 \\ 2 & 2 \end{bmatrix}$ times 2×1 matrices created with columns containing pairs of numbers from part (a) to encode the message "Call home" into pairs of coded numbers.

37. *Encoding Messages*

Use the matrix $\begin{bmatrix} 4 & 4 & 4 \\ 1 & 2 & 3 \\ 2 & 4 & 2 \end{bmatrix}$ to encode the message "Neatness counts" into triples of numbers.

38. Encoding Messages

Use the matrix $\begin{bmatrix} 4 & 4 & 4 \\ 1 & 2 & 3 \\ 2 & 4 & 2 \end{bmatrix}$ to encode the message

"Meet for lunch" into triples of numbers.

Decoding Messages We have encoded messages by assigning the numbers 1 to 26 to the letters a to z of the alphabet, respectively, and assigning 27 to a blank space. We can decode messages of this type by finding the inverse of the encoding matrix and multiplying it times the coded message. Use A^{-1} and the conversion table below to decode the messages in Exercises 39–44.

$$
\begin{array}{cccccccccccccccc}
a & b & c & d & e & f & g & h & i & j & k & l & m & n & o & p \\
1 & 2 & 3 & 4 & 5 & 6 & 7 & 8 & 9 & 10 & 11 & 12 & 13 & 14 & 15 & 16
\end{array}
$$
$$
\begin{array}{ccccccccccc}
q & r & s & t & u & v & w & x & y & z & blank \\
17 & 18 & 19 & 20 & 21 & 22 & 23 & 24 & 25 & 26 & 27
\end{array}
$$

39. Decoding Messages The encoding matrix is $A = \begin{bmatrix} 3 & -1 \\ -2 & 1 \end{bmatrix}$, and the encoded message is 51, −29, 55, −35, 76, −49, −15, 16, 11, 1.

40. Decoding Messages The encoding matrix is $A = \begin{bmatrix} 2 & 3 \\ 2 & 4 \end{bmatrix}$, and the encoded message is 59, 74, 72, 78, 84, 102, 81, 90, 121, 148.

41. Decoding Messages The encoding matrix is $A = \begin{bmatrix} -1 & 0 & 1 \\ -1 & 1 & 0 \\ 3 & -1 & -1 \end{bmatrix}$, and the encoded message is 1, −4, 16, 21, 23, −40, 3, 6, 6, −26, −14, 67, −9, 0, 23, 9, 1, 8.

42. Decoding Messages The encoding matrix is $A = \begin{bmatrix} 1 & 1 & 0 \\ 1 & 3 & 0 \\ 5 & 3 & 3 \end{bmatrix}$, and the encoded message is 16, 34, 128, 32, 86, 145, 32, 86, 109, 29, 33, 195, 6, 8, 61.

43. Decoding Messages Use the encoding matrix $A = \begin{bmatrix} 1 & 2 & -1 \\ 2 & 1 & -3 \\ -1 & 4 & 2 \end{bmatrix}$ to decode the message 29, −1, 75, −19, −66, 50, 46, 41, 47, 3, −38, 65.

44. Decoding Messages Use the encoding matrix $A = \begin{bmatrix} 2 & 3 & -2 \\ 3 & 2 & -4 \\ -2 & 5 & 3 \end{bmatrix}$ to decode the message 20, −33, 142, 74, 51, 107, 67, 87, −26, 22, −22, 99.

45. Coded message

a. Choose a partner and create a message to send to the partner. Change the message from letters to numbers with the code

$$
\begin{array}{cccccccccccccccc}
a & b & c & d & e & f & g & h & i & j & k & l & m & n & o & p \\
1 & 2 & 3 & 4 & 5 & 6 & 7 & 8 & 9 & 10 & 11 & 12 & 13 & 14 & 15 & 16
\end{array}
$$
$$
\begin{array}{ccccccccccc}
q & r & s & t & u & v & w & x & y & z & blank \\
17 & 18 & 19 & 20 & 21 & 22 & 23 & 24 & 25 & 26 & 27
\end{array}
$$

b. Encode your message with encoding matrix

$$
A = \begin{bmatrix} 1 & 2 & 1 \\ 2 & 1 & -3 \\ -1 & 4 & 2 \end{bmatrix}
$$

c. Exchange coded messages.

d. Decode the message that you receive from your partner, and report your original message and your partner's message as you have decoded it.

7.5 Determinants and Cramer's Rule

KEY OBJECTIVES

- Find determinants of matrices
- Solve systems of equations with Cramer's Rule
- Use Cramer's Rule with inconsistent and dependent systems

SECTION PREVIEW **Business Properties**

Suppose a bank lends $285,000 to a development company to purchase three business properties, where one of the properties costs $45,000 more than another, and the third costs twice the sum of these two properties. To find the cost of each property, we can solve the system of linear equations defining this problem using one of several methods. In this section, we consider the use of determinants to solve certain systems of equations in two variables and three variables. After we have defined determinants, we will use them to solve systems of this type with Cramer's Rule. (See Example 5.)

Development of the computer can be traced back to the 19th century, when the English inventor Charles Babbage designed a machine to perform calculations involved with Cramer's Rule. Babbage's machine, the "difference engine," demonstrated how complex calculations could be handled mechanically. In 1944 scientists at IBM used the lessons of his difference engine to create the world's first computer. ∎

The Determinant of a 2 × 2 Matrix

Every square matrix has a number associated with it called the **determinant.** The determinant for a 2 × 2 matrix is defined as follows:

> The determinant of the matrix
> $$\begin{bmatrix} a & b \\ c & d \end{bmatrix} \text{ is } D = \det\begin{bmatrix} a & b \\ c & d \end{bmatrix} = \begin{vmatrix} a & b \\ c & d \end{vmatrix} = ad - bc.$$

Note that the difference of the two diagonal products gives the determinant.

$$\begin{vmatrix} a & b \\ c & d \end{vmatrix} = ad - bc$$

EXAMPLE 1 ▶ **Evaluating the Determinant of a 2 × 2 Matrix**

Evaluate the determinants of the matrices

a. $\begin{bmatrix} 2 & 4 \\ -1 & -4 \end{bmatrix}$ **b.** $\begin{bmatrix} -3 & 2 \\ -1 & 4 \end{bmatrix}$

SOLUTION

a. $\begin{vmatrix} 2 & 4 \\ -1 & -4 \end{vmatrix} = (2)(-4) - (4)(-1) = -8 + 4 = -4$

b. $\begin{vmatrix} -3 & 2 \\ -1 & 4 \end{vmatrix} = (-3)(4) - (2)(-1) = -12 + 2 = -10$

Cramer's Rule

Determinants can be used to solve systems of equations if the number of variables equals the number of equations. Let's solve the system

$$\begin{cases} ax + by = u \\ cx + dy = w \end{cases}$$

using the elimination method. Assuming $d \neq 0$ and $b \neq 0$,

$$\begin{cases} ax + by = u & \to \\ cx + dy = w & \to \end{cases} \begin{cases} adx + bdy = du & \text{\color{blue}Multiply by } d. \\ -bcx - bdy = -bw & \text{\color{blue}Multiply by } -b. \end{cases}$$

$$\begin{aligned} adx - bcx &= du - bw && \text{\color{blue}Add these equations.} \\ (ad - bc)x &= du - bw && \text{\color{blue}Factor out } x. \\ x &= \frac{du - bw}{ad - bc} && \text{\color{blue}Divide by the coefficient of } x. \end{aligned}$$

Note that, by the definition of determinants, $\begin{vmatrix} a & b \\ c & d \end{vmatrix} = ad - bc$ and $\begin{vmatrix} u & b \\ w & d \end{vmatrix} =$
$du - bw$. Thus, the solution x can be expressed as the quotient of two determinants:

$$x = \frac{\begin{vmatrix} u & b \\ w & d \end{vmatrix}}{\begin{vmatrix} a & b \\ c & d \end{vmatrix}}$$

Similarly, we could use the elimination method to solve the original system for y as the quotient of two determinants.

$$\begin{cases} ax + by = u \\ cx + dy = w \end{cases} \rightarrow \begin{cases} -acx - bcy = -cu \quad\quad \text{Multiply by } -c. \\ acx + ady = aw \quad\quad\;\; \text{Multiply by } a. \end{cases}$$

$$-bcy + ady = -cu + aw \quad\quad \text{Add these equations.}$$

$$(-bc + ad)y = -cu + aw \quad\quad \text{Factor out } y.$$

$$y = \frac{aw - cu}{ad - bc} \quad\quad\quad\quad\quad\;\; \text{Divide by the coefficient of } y.$$

Because $\begin{vmatrix} a & b \\ c & d \end{vmatrix} = ad - bc$ and $\begin{vmatrix} a & u \\ c & w \end{vmatrix} = aw - cu$, y can be expressed as the quotient of two determinants:

$$y = \frac{\begin{vmatrix} a & u \\ c & w \end{vmatrix}}{\begin{vmatrix} a & b \\ c & d \end{vmatrix}}$$

These solutions can be summarized in the form below, which is called **Cramer's Rule**.

Solving a System of Two Equations in Two Variables Using Cramer's Rule

If $\begin{cases} ax + by = u \\ cx + dy = w \end{cases}$, then $x = \dfrac{\begin{vmatrix} u & b \\ w & d \end{vmatrix}}{\begin{vmatrix} a & b \\ c & d \end{vmatrix}}$ and $y = \dfrac{\begin{vmatrix} a & u \\ c & w \end{vmatrix}}{\begin{vmatrix} a & b \\ c & d \end{vmatrix}}$,

provided $D = \begin{vmatrix} a & b \\ c & d \end{vmatrix} = ad - bc \neq 0$.

Look carefully at the pattern when using Cramer's Rule. The denominator of the solution for both x and y above is the determinant for the coefficients of the variables,

$$D = \begin{vmatrix} a & b \\ c & d \end{vmatrix}.$$

In the solution for x, the numerator is the determinant formed by replacing the entries in the first column (the coefficients of x) by the constants on the right side of the equation. We write this as $D_x = \begin{vmatrix} u & b \\ w & d \end{vmatrix}.$

In the solution for y, the numerator is the determinant formed by replacing the entries in the second column (the coefficients of y) by the constants on the right side of the equation. We write this as $D_y = \begin{vmatrix} a & u \\ c & w \end{vmatrix}$.

Thus, Cramer's Rule states that if $D \neq 0$, then

$$x = \frac{D_x}{D} \quad \text{and} \quad y = \frac{D_y}{D}$$

EXAMPLE 2 ▶ Using Cramer's Rule to Solve a Linear System

Use Cramer's Rule to solve the system

$$\begin{cases} x - 2y = 5 \\ 5x - y = -2 \end{cases}$$

SOLUTION

We need to evaluate three determinants.
D, the determinant of the coefficients of the variables:

$$D = \begin{vmatrix} 1 & -2 \\ 5 & -1 \end{vmatrix} = (1)(-1) - (-2)(5) = -1 + 10 = 9$$

D_x, the determinant obtained by replacing the coefficients of x with the constants on the right side of the equations:

$$D_x = \begin{vmatrix} 5 & -2 \\ -2 & -1 \end{vmatrix} = (5)(-1) - (-2)(-2) = -5 - 4 = -9$$

D_y, the determinant obtained by replacing the coefficients of y with the constants on the right side of the equations:

$$D_y = \begin{vmatrix} 1 & 5 \\ 5 & -2 \end{vmatrix} = (1)(-2) - (5)(5) = -2 - 25 = -27$$

Then, by Cramer's Rule,

$$x = \frac{D_x}{D} = \frac{-9}{9} = -1 \quad \text{and} \quad y = \frac{D_y}{D} = \frac{-27}{9} = -3$$

Thus the solution to the system is $(-1, -3)$. The solution can be checked by substituting these values into the original equations.

Remember, if the determinant of the coefficients of the variables, D, equals 0, a solution cannot be found with Cramer's Rule.

The Determinant of a 3 × 3 Matrix

The determinant of a 3 × 3 matrix is defined as follows:

$$\begin{vmatrix} a_1 & b_1 & c_1 \\ a_2 & b_2 & c_2 \\ a_3 & b_3 & c_3 \end{vmatrix} = a_1b_2c_3 + b_1c_2a_3 + c_1a_2b_3 - a_3b_2c_1 - b_3c_2a_1 - c_3a_2b_1$$

We can rearrange the terms and factor this expression to obtain

$$a_1b_2c_3 - b_3c_2a_1 - c_3a_2b_1 + c_1a_2b_3 + b_1c_2a_3 - a_3b_2c_1$$
$$= a_1(b_2c_3 - b_3c_2) - a_2(b_1c_3 - b_3c_1) + a_3(b_1c_2 - b_2c_1)$$
$$= a_1\begin{vmatrix} b_2 & c_2 \\ b_3 & c_3 \end{vmatrix} - a_2\begin{vmatrix} b_1 & c_1 \\ b_3 & c_3 \end{vmatrix} + a_3\begin{vmatrix} b_1 & c_1 \\ b_2 & c_2 \end{vmatrix}$$

Thus, the determinant of a 3×3 determinant is defined by

$$\begin{vmatrix} a_1 & b_1 & c_1 \\ a_2 & b_2 & c_2 \\ a_3 & b_3 & c_3 \end{vmatrix} = a_1 \begin{vmatrix} b_2 & c_2 \\ b_3 & c_3 \end{vmatrix} - a_2 \begin{vmatrix} b_1 & c_1 \\ b_3 & c_3 \end{vmatrix} + a_3 \begin{vmatrix} b_1 & c_1 \\ b_2 & c_2 \end{vmatrix}$$

Note that each of the three terms in the definition contains a numerical factor that is an element from the first column of the 3×3 determinant, and a second-order determinant obtained by crossing out the row and column containing the numerical factor. That is,

To obtain $a_1 \begin{vmatrix} b_2 & c_2 \\ b_3 & c_3 \end{vmatrix}$, remove the first row and first column of $\begin{vmatrix} a_1 & b_1 & c_1 \\ a_2 & b_2 & c_2 \\ a_3 & b_3 & c_3 \end{vmatrix}$ and multiply by a_1.

To obtain $a_2 \begin{vmatrix} b_1 & c_1 \\ b_3 & c_3 \end{vmatrix}$, remove the second row and first column of $\begin{vmatrix} a_1 & b_1 & c_1 \\ a_2 & b_2 & c_2 \\ a_3 & b_3 & c_3 \end{vmatrix}$ and multiply by a_2.

To obtain $a_3 \begin{vmatrix} b_1 & c_1 \\ b_2 & c_2 \end{vmatrix}$, remove the third row and first column of $\begin{vmatrix} a_1 & b_1 & c_1 \\ a_2 & b_2 & c_2 \\ a_3 & b_3 & c_3 \end{vmatrix}$ and multiply by a_3.

The 2×2 determinants in the definition above are called **minors** of the 3×3 determinant.

Note that the signs alternate on the expressions involving the different minors, with a positive sign on the first expression, which corresponds to the element in the first row and first column.

EXAMPLE 3 ▶   Evaluating a 3 × 3 Determinant

Find the determinant of $\begin{bmatrix} -1 & 2 & -3 \\ 3 & 0 & -2 \\ 1 & 2 & -2 \end{bmatrix}$.

SOLUTION

For the first minor, we cross out row 1 and column 1: $\begin{vmatrix} -1 & 2 & -3 \\ 3 & 0 & -2 \\ 1 & 2 & -2 \end{vmatrix}$

For the second minor, we cross out row 2 and column 1: $\begin{vmatrix} -1 & 2 & -3 \\ 3 & 0 & -2 \\ 1 & 2 & -2 \end{vmatrix}$

For the third minor, we cross out row 3 and column 1: $\begin{vmatrix} -1 & 2 & -3 \\ 3 & 0 & -2 \\ 1 & 2 & -2 \end{vmatrix}$

To evaluate the determinant of $\begin{vmatrix} -1 & 2 & -3 \\ 3 & 0 & -2 \\ 1 & 2 & -2 \end{vmatrix}$, we multiply -1 by the first minor, multiply 3 by the second minor, and multiply 1 by the third minor. Then we add the first and third products and subtract the second product.

$$\begin{vmatrix} -1 & 2 & -3 \\ 3 & 0 & -2 \\ 1 & 2 & -2 \end{vmatrix} = (-1)\begin{vmatrix} 0 & -2 \\ 2 & -2 \end{vmatrix} - 3\begin{vmatrix} 2 & -3 \\ 2 & -2 \end{vmatrix} + 1\begin{vmatrix} 2 & -3 \\ 0 & -2 \end{vmatrix}$$

$$= (-1)((0)(-2) - (2)(-2)) - 3((2)(-2) - (2)(-3)) +$$
$$\quad 1((2)(-2) - (0)(-3))$$
$$= (-1)(0 + 4) - 3(-4 + 6) + 1(-4 - 0)$$
$$= (-1)(4) - 3(2) + 1(-4)$$
$$= -4 - 6 - 4$$
$$= -14$$

A graphing utility can be used to evaluate the determinant of a matrix.

Using a Graphing Calculator to Evaluate a Determinant

1. Enter the matrix and name it A.

2. Select "det(" under MATRIX, MATH.

3. Enter MATRIX, A, and a right parenthesis, and press **ENTER**.

NORMAL FLOAT AUTO REAL RADIAN MP		NORMAL FLOAT DEC REAL RADIAN MP	
[A]		det([A])	
	$\begin{bmatrix} -1 & 2 & -3 \\ 3 & 0 & -2 \\ 1 & 2 & -2 \end{bmatrix}$		-14

Figure 7.26

The determinant of $\begin{bmatrix} -1 & 2 & -3 \\ 3 & 0 & -2 \\ 1 & 2 & -2 \end{bmatrix}$, from Example 3, is shown in Figure 7.26.

Solving Systems of Equations in Three Variables with Cramer's Rule

Cramer's Rule can be used to solve systems of equations in three variables if the determinant D of the coefficients of the variables does not equal 0.

Cramer's Rule

The solution to $\begin{cases} a_1x + b_1y + c_1z = d_1 \\ a_2x + b_2y + c_2z = d_2 \\ a_3x + b_3y + c_3z = d_3 \end{cases}$ is $x = \dfrac{D_x}{D}, y = \dfrac{D_y}{D}$, and $z = \dfrac{D_z}{D}$, where

$$D = \begin{vmatrix} a_1 & b_1 & c_1 \\ a_2 & b_2 & c_2 \\ a_3 & b_3 & c_3 \end{vmatrix} \qquad \text{The coefficients of the variables } x, y, \text{ and } z$$

$$D_x = \begin{vmatrix} d_1 & b_1 & c_1 \\ d_2 & b_2 & c_2 \\ d_3 & b_3 & c_3 \end{vmatrix} \qquad \text{Replace } x\text{-coefficients with the constants at the right of the three equations.}$$

(continued)

$$D_y = \begin{vmatrix} a_1 & d_1 & c_1 \\ a_2 & d_2 & c_2 \\ a_3 & d_3 & c_3 \end{vmatrix} \qquad \text{Replace } y\text{-coefficients with the constants at the right of the three equations.}$$

$$D_z = \begin{vmatrix} a_1 & b_1 & d_1 \\ a_2 & b_2 & d_2 \\ a_3 & b_3 & d_3 \end{vmatrix} \qquad \text{Replace } z\text{-coefficients with the constants at the right of the three equations.}$$

EXAMPLE 4 ▶ Using Cramer's Rule to Solve a System in Three Variables

Use Cramer's Rule to solve the system

$$\begin{cases} 2x - y - z = 2 \\ 3x - y + z = -3 \\ x + y - z = 7 \end{cases}$$

SOLUTION

First we will evaluate the determinant of the coefficients.

$$D = \begin{vmatrix} 2 & -1 & -1 \\ 3 & -1 & 1 \\ 1 & 1 & -1 \end{vmatrix} = 2\begin{vmatrix} -1 & 1 \\ 1 & -1 \end{vmatrix} - 3\begin{vmatrix} -1 & -1 \\ 1 & -1 \end{vmatrix} + 1\begin{vmatrix} -1 & -1 \\ -1 & 1 \end{vmatrix}$$

$$= 2((-1)(-1) - (1)(1)) - 3((-1)(-1) - (1)(-1)) + 1((-1)(1) - (-1)(-1))$$

$$= 2(1 - 1) - 3(1 + 1) + 1(-1 - 1)$$

$$= 2(0) - 3(2) + 1(-2)$$

$$= -8$$

$$D_x = \begin{vmatrix} 2 & -1 & -1 \\ -3 & -1 & 1 \\ 7 & 1 & -1 \end{vmatrix} = 2\begin{vmatrix} -1 & 1 \\ 1 & -1 \end{vmatrix} - (-3)\begin{vmatrix} -1 & -1 \\ 1 & -1 \end{vmatrix} + 7\begin{vmatrix} -1 & -1 \\ -1 & 1 \end{vmatrix}$$

$$= 2(0) + 3(2) + 7(-2)$$

$$= -8$$

$$D_y = \begin{vmatrix} 2 & 2 & -1 \\ 3 & -3 & 1 \\ 1 & 7 & -1 \end{vmatrix} = 2\begin{vmatrix} -3 & 1 \\ 7 & -1 \end{vmatrix} - 3\begin{vmatrix} 2 & -1 \\ 7 & -1 \end{vmatrix} + 1\begin{vmatrix} 2 & -1 \\ -3 & 1 \end{vmatrix}$$

$$= 2(-4) - 3(5) + 1(-1)$$

$$= -24$$

$$D_z = \begin{vmatrix} 2 & -1 & 2 \\ 3 & -1 & -3 \\ 1 & 1 & 7 \end{vmatrix} = 2\begin{vmatrix} -1 & -3 \\ 1 & 7 \end{vmatrix} - 3\begin{vmatrix} -1 & 2 \\ 1 & 7 \end{vmatrix} + 1\begin{vmatrix} -1 & 2 \\ -1 & -3 \end{vmatrix}$$

$$= 2(-4) - 3(-9) + 1(5)$$

$$= 24$$

Thus,

$$x = \frac{D_x}{D} = \frac{-8}{-8} = 1$$

$$y = \frac{D_y}{D} = \frac{-24}{-8} = 3$$

$$z = \frac{D_z}{D} = \frac{24}{-8} = -3$$

EXAMPLE 5 ▶ Business Properties

A bank lends $285,000 to a development company to purchase three business properties. One of the properties costs $45,000 more than another, and the third costs twice the sum of these two properties. What is the cost of each property?

SOLUTION

If we let x, y, and z represent the cost of the three properties, respectively, then we have the following equations:

$$x + y + z = 285,000$$
$$x = 45,000 + y$$
$$z = 2(x + y)$$

The system for these equations is

$$\begin{cases} x + y + z = 285,000 \\ x - y = 45,000 \\ 2x + 2y - z = 0 \end{cases}$$

The determinant of the coefficients is

$$\begin{vmatrix} 1 & 1 & 1 \\ 1 & -1 & 0 \\ 2 & 2 & -1 \end{vmatrix} = 1\begin{vmatrix} -1 & 0 \\ 2 & -1 \end{vmatrix} - 1\begin{vmatrix} 1 & 1 \\ 2 & -1 \end{vmatrix} + 2\begin{vmatrix} 1 & 1 \\ -1 & 0 \end{vmatrix}$$

$$= 1(1 - 0) - 1(-1 - 2) + 2(0 + 1)$$
$$= 1 + 3 + 2$$
$$= 6$$

Then

$$D_x = \begin{vmatrix} 285,000 & 1 & 1 \\ 45,000 & -1 & 0 \\ 0 & 2 & -1 \end{vmatrix} = 285,000\begin{vmatrix} -1 & 0 \\ 2 & -1 \end{vmatrix} - 45,000\begin{vmatrix} 1 & 1 \\ 2 & -1 \end{vmatrix} + 0\begin{vmatrix} 1 & 1 \\ -1 & 0 \end{vmatrix}$$

$$= 285,000(1) - 45,000(-3) + 0$$
$$= 420,000$$

$$D_y = \begin{vmatrix} 1 & 285,000 & 1 \\ 1 & 45,000 & 0 \\ 2 & 0 & -1 \end{vmatrix} = 1\begin{vmatrix} 45,000 & 0 \\ 0 & -1 \end{vmatrix} - 1\begin{vmatrix} 285,000 & 1 \\ 0 & -1 \end{vmatrix} + 2\begin{vmatrix} 285,000 & 1 \\ 45,000 & 0 \end{vmatrix}$$

$$= 1(-45,000) - 1(-285,000) + 2(-45,000)$$
$$= 150,000$$

$$D_z = \begin{vmatrix} 1 & 1 & 285{,}000 \\ 1 & -1 & 45{,}000 \\ 2 & 2 & 0 \end{vmatrix} = 1\begin{vmatrix} -1 & 45{,}000 \\ 2 & 0 \end{vmatrix} - 1\begin{vmatrix} 1 & 285{,}000 \\ 2 & 0 \end{vmatrix} + 2\begin{vmatrix} 1 & 285{,}000 \\ -1 & 45{,}000 \end{vmatrix}$$

$$= 1(-90{,}000) - 1(-570{,}000) + 2(330{,}000)$$

$$= 1{,}140{,}000$$

So

$$x = \frac{D_x}{D} = \frac{420{,}000}{6} = 70{,}000$$

$$y = \frac{D_y}{D} = \frac{150{,}000}{6} = 25{,}000$$

$$z = \frac{D_z}{D} = \frac{1{,}140{,}000}{6} = 190{,}000$$

Thus, the first property costs $70,000, the second costs $25,000, and the third costs $190,000.

Cramer's Rule with Inconsistent and Dependent Systems

If D, the determinant of the coefficients of the variables of a system of three equations in three variables is 0, then the system either is inconsistent or contains dependent equations. In particular,

1. If $D = 0$ and at least one of the determinants in the numerator $(D_x, D_y, \text{ or } D_z)$ is not 0, then the system is inconsistent, and there is no solution.

2. If $D = 0$ and all the determinants in the numerators $(D_x, D_y, \text{ and } D_z)$ are 0, then the equations in the system are dependent, and there are infinitely many solutions. These solutions can be found using different methods.

SKILLS CHECK 7.5

In Exercises 1–8, evaluate each determinant.

1. $\begin{vmatrix} 3 & 2 \\ 4 & 1 \end{vmatrix}$

2. $\begin{vmatrix} -2 & -1 \\ -3 & 4 \end{vmatrix}$

3. $\begin{vmatrix} 5 & 1 \\ 5 & 1 \end{vmatrix}$

4. $\begin{vmatrix} \dfrac{1}{2} & \dfrac{2}{3} \\ \dfrac{3}{4} & \dfrac{1}{6} \end{vmatrix}$

5. $\begin{vmatrix} 3 & -2 & 1 \\ -1 & 0 & 2 \\ 0 & 1 & 1 \end{vmatrix}$

6. $\begin{vmatrix} 1 & 2 & 2 \\ 3 & 0 & 1 \\ -1 & 0 & -2 \end{vmatrix}$

7. $\begin{vmatrix} 0 & -1 & 2 \\ 3 & 1 & -1 \\ 4 & -1 & 3 \end{vmatrix}$

8. $\begin{vmatrix} 1 & 2 & 3 \\ -1 & 5 & 6 \\ -1 & 3 & 3 \end{vmatrix}$

A matrix has an inverse if and only if its determinant does not equal 0. In Exercises 9–12, determine whether the matrix has an inverse.

9. $\begin{bmatrix} 2 & 3 \\ -1 & 4 \end{bmatrix}$

10. $\begin{bmatrix} 3 & 3 \\ 1 & -1 \end{bmatrix}$

11. $\begin{bmatrix} 1 & 3 & -2 \\ 2 & -1 & 5 \\ 3 & 2 & 3 \end{bmatrix}$

12. $\begin{bmatrix} 4 & 1 & 0 \\ 0 & 5 & 0 \\ 0 & 0 & -1 \end{bmatrix}$

In Exercises 13–24, use Cramer's Rule to solve each system, if possible. If it is not possible, state whether the system is inconsistent or has infinitely many solutions.

13. $\begin{cases} x + 2y = 4 \\ 3x + 4y = 10 \end{cases}$

14. $\begin{cases} 3x - 4y = 11 \\ 2x + 3y = -4 \end{cases}$

15. $\begin{cases} 2x - y = 4 \\ 3x + y = 5 \end{cases}$ **16.** $\begin{cases} 5x - 2y = 6 \\ 3x + 3y = 12 \end{cases}$

17. $\begin{cases} 2x - y = 2 \\ -4x + 2y = 1 \end{cases}$ **18.** $\begin{cases} 3x + 2y = 4 \\ 6x + 4y = 3 \end{cases}$

19. $\begin{cases} 2x - 4y = 5 \\ 3x - 6y = 7.5 \end{cases}$ **20.** $\begin{cases} x - 2y = 3 \\ -2x + 4y = -6 \end{cases}$

21. $\begin{cases} x + y + z = 3 \\ 2x + y + z = 4 \\ 2x + 2y + z = 5 \end{cases}$ **22.** $\begin{cases} 2x - y - 2z = 2 \\ 3x - y + z = -3 \\ x + y - z = 7 \end{cases}$

23. $\begin{cases} x + y + 2z = 8 \\ 2x + y + z = 7 \\ 2x + 2y + z = 10 \end{cases}$ **24.** $\begin{cases} x - 2y + z = 0 \\ 2x + y - 2z = 2 \\ 3x + 2y - 3z = 2 \end{cases}$

In Exercises 25–28, solve for x.

25. $\begin{vmatrix} -2 & -11 \\ x & 6 \end{vmatrix} = 32$ **26.** $\begin{vmatrix} 5 & x \\ -1 & -3 \end{vmatrix} = 5$

27. $\begin{vmatrix} x + 3 & -6 \\ x - 2 & -4 \end{vmatrix} = 18$

28. $\begin{vmatrix} 2 & x & 1 \\ -3 & 1 & 0 \\ 2 & 1 & 4 \end{vmatrix} = 39$

EXERCISES 7.5

In Exercises 29–38, set up each system of equations and then solve it using Cramer's Rule.

29. *Medication* Medication A is given every 4 hours, and medication B is given twice a day. For patient I, the total intake of the two medications is required to be 50.6 mg per day, and for patient II, the total intake is required to be 92 mg per day. The ratio of the dosage of A to the dosage of B is always 5 to 8.

    a. Find the dosage for each administration of each medication for patient I.

    b. Find the dosage for each administration of each medication for patient II.

30. *Investment* One safe investment pays 10% per year, and a more risky investment pays 18% per year. A woman has $145,000 to invest and would like to have an income of $20,000 per year from her investments. How much should she invest at each rate?

31. *Manufacturing* A manufacturer of table saws has three models—Deluxe, Premium, and Ultimate—that must be painted, assembled, and packaged for shipping. The table gives the number of hours required for each of these operations for each type of table saw. If the manufacturer has 96 hours available for painting, 156 hours for assembling, and 37 hours for packaging, how many of each type of saw can be produced each day?

	Deluxe	Premium	Ultimate
Painting	1.6	2	2.4
Assembling	2	3	4
Packaging	0.5	0.5	1

32. *Ticket Pricing* A theater has 3600 seats, with tickets costing $40, $70, and $100, for three different seating sections: C, B, and A. The owner wants to earn $192,000 from a full house and wants to have twice as many $40 seats as the sum of all other seats. How many seats should he have in each section?

33. *Rental Cars* A car rental company has $3,160,000 available to purchase large, midsize, and compact cars. It can purchase compact cars for $18,000 each, midsize cars for $28,000 each, and large cars for $48,000 each and can rent them for $50, $70, and $90 per day, respectively. If the company's goal is to buy 120 cars and earn $7600 in daily rental, how many of each car type should it buy?

34. *Investment* A trust account manager has $1,000,000 to be invested in three different accounts. The accounts pay 6%, 8%, and 10%, respectively, and the goal is to earn $86,000 with the amount invested at 10% equal to the sum of the other two investments. To accomplish this, assume that x dollars are invested at 6%, y dollars at 8%, and z dollars at 10%. Find out how much should be invested in each account to satisfy the conditions.

35. *Nutrition* Three different bacterial species are cultured in one dish and feed on three nutrients. Each individual of species I consumes 1 unit of each of the first and second nutrients and 2 units of the third nutrient. Each individual of species II consumes 2 units of the first nutrient and 2 units of the third nutrient. Each individual of species III consumes 2 units of the first nutrient, 3 units of the second nutrient, and 5 units of the third nutrient. If the culture is given 5100 units of the first nutrient, 6900 units of the second nutrient, and 12,000 units of the third nutrient, how many of each species can be supported so all of the nutrients are consumed?

36. *Investment* The members of an investment club set a goal of earning 15% on the money they earn in stocks. The members are considering purchasing four possible stocks. Their cost per share (in dollars) and their projected growth per share (in dollars) are summarized in the following table.

Stocks				
	Oil	Bank	Computer	Retail
Cost/share	100	40	30	20
Growth/share	10.00	3.60	6.00	2.40

If they have $102,300 to invest, how many shares of each stock should they buy to meet their goal?

37. *Investment* A woman has $800,000 to invest, and she is considering buying shares in three investment properties. One property earns 9%, one earns 8%, and one earns 7.5%. The total annual earnings on the three properties is $67,400. If she invests as much in the 9% property as the sum of her investments in the other two properties, how much did she invest in each property?

38. *Social Services* A social services agency provides services to three types of clients: A, B, and C. The agency has $800,000 available for emergency food and shelter and has $660,000 available for counseling. Type A clients require an average of $1200 for emergencies and $800 for counseling, type B clients require an average of $800 for emergencies and $500 for counseling, and type C clients require an average of $400 for emergencies and $600 for counseling. If a total of 1000 clients can be served, how many of each type of client can be served?

7.6 Systems of Nonlinear Equations

KEY OBJECTIVES

- Solve systems of nonlinear equations algebraically
- Solve systems of nonlinear equations graphically

SECTION PREVIEW   Market Equilibrium

The supply function for a product is given by $p = q^2 + 804$, and the demand function for this product is $p = 1200 - 60q$, where p is the price in dollars and q is the number of hundreds of units. Suppose we are interested in finding the price that gives market equilibrium.

To answer this question, we create a system of equations, one of which is nonlinear, and solve the system. In this section, we use algebraic and graphical methods to solve systems of nonlinear equations. ■

We have solved systems of linear equations in two variables by substitution, by elimination, and by graphing. To solve systems of nonlinear equations, we can use these same methods. A system of nonlinear equations can have any number of solutions or no solution.

Algebraic Solution of Nonlinear Systems

EXAMPLE 1 ▶   Solving a Nonlinear System Algebraically

Use substitution to solve the system

$$\begin{cases} x^2 - x - 3y = 0 \\ x - y = -4 \end{cases}$$

and check the solution.

SOLUTION

To solve the system

$$\begin{cases} x^2 - x - 3y = 0 & (1) \\ x - y = -4 & (2) \end{cases}$$

by substitution, notice that Equation (2) can easily be solved for y:

$$x - y = -4$$
$$y = x + 4$$

Substituting this expression for y in Equation (1) gives

$$x^2 - x - 3y = 0$$
$$x^2 - x - 3(x + 4) = 0$$
$$x^2 - x - 3x - 12 = 0$$
$$x^2 - 4x - 12 = 0$$
$$(x - 6)(x + 2) = 0$$
$$x = 6 \quad \text{or} \quad x = -2$$

Using these values for x in $y = x + 4$ gives the solutions $(6, 10)$ and $(-2, 2)$.

To check the solution, we substitute each ordered pair into both of the original equations:

Check $(6, 10)$:
$$
\begin{array}{ll}
x^2 - x - 3y = 0 & x - y = -4 \\
(6)^2 - (6) - 3(10) = 0 & 6 - 10 = -4 \\
36 - 6 - 30 = 0 & -4 = -4 \\
0 = 0 &
\end{array}
$$

Check $(-2, 2)$:
$$
\begin{array}{ll}
x^2 - x - 3y = 0 & x - y = -4 \\
(-2)^2 - (-2) - 3(2) = 0 & -2 - 2 = -4 \\
4 + 2 - 6 = 0 & -4 = -4 \\
0 = 0 &
\end{array}
$$

Both solutions check in each equation of the original system.

EXAMPLE 2 ▶   ## Solving a Nonlinear System Algebraically

Solve the system algebraically, if $y \neq 0$:

$$
\begin{cases}
x + \dfrac{x}{y} = 6 \\
xy + y = 10
\end{cases}
$$

SOLUTION

To solve the system

$$
\begin{cases}
x + \dfrac{x}{y} = 6 & (1) \\
xy + y = 10 & (2)
\end{cases}
$$

algebraically, we begin by multiplying Equation (1) by y:

$$xy + x = 6y \qquad (3)$$

Next we solve Equation (2) for x:

$$xy + y = 10$$
$$xy = 10 - y$$
$$x = \frac{10 - y}{y}$$

Substituting this expression for x in Equation (3) gives

$$\frac{10 - y}{y}(y) + \frac{10 - y}{y} = 6y$$

$$10 - y + \frac{10 - y}{y} = 6y$$

Now we multiply both sides of this equation by y and solve the quadratic equation that results:

$$10 - y + \frac{10 - y}{y} = 6y$$

$$10y - y^2 + 10 - y = 6y^2$$

$$0 = 7y^2 - 9y - 10$$

$$0 = (y - 2)(7y + 5)$$

$$y = 2 \quad \text{or} \quad y = -\frac{5}{7}$$

If $y = 2$, then

$$x = \frac{10 - 2}{2} = \frac{8}{2} = 4$$

and if $y = -\dfrac{5}{7}$, then

$$x = \frac{10 + \dfrac{5}{7}}{-\dfrac{5}{7}} = -\frac{75}{5} = -15$$

Thus, the solutions are $(4, 2)$ and $\left(-15, -\dfrac{5}{7}\right)$.

Both solutions check in the original system of equations.

EXAMPLE 3 ▶ **Market Equilibrium**

The supply function for a product is given by $p = q^2 + 804$, and the demand function for this product is $p = 1200 - 60q$, where p is the price in dollars and q is the number of hundreds of units. Find the price that gives market equilibrium.

SOLUTION

Market equilibrium occurs when the supply price (and quantity) equals the demand price (and quantity). Thus, the solution to the system

$$\begin{cases} p = q^2 + 804 & (1) \\ p = 1200 - 60q & (2) \end{cases}$$

gives the equilibrium quantity and price.

Substituting for p (from Equation (1)) into Equation (2) and solving gives

$$q^2 + 804 = 1200 - 60q$$

$$q^2 + 60q - 396 = 0$$

$$(q - 6)(q + 66) = 0$$

$$q = 6 \quad \text{or} \quad q = -66$$

The negative value of q has no meaning in this application, so we have $q = 6$ and $p = 6^2 + 804 = 840$. Thus, market equilibrium occurs when the price is \$840 and the quantity is 600 units.

Graphical Solution of Nonlinear Systems

As with systems of linear equations in two variables, we can use graphical methods to check and to solve systems of nonlinear equations.

EXAMPLE 4 ▶ **Solving a Nonlinear System Graphically**

Solve the system

$$\begin{cases} y = -x^2 - 2x + 8 \\ 2x + y = 4 \end{cases}$$

graphically.

SOLUTION

To solve the system

$$\begin{cases} y = -x^2 - 2x + 8 & (1) \\ 2x + y = 4 & (2) \end{cases}$$

graphically, we first solve Equation (2) for y, getting $y = -2x + 4$. Graphing the two equations on the same axes gives the graph in Figure 7.27.

Figure 7.27

The points of intersection are $(2, 0)$ and $(-2, 8)$, as shown in Figures 7.28(a) and (b).

(a)   (b)

Figure 7.28

Thus, the solutions to the system are $(2, 0)$ and $(-2, 8)$.

EXAMPLE 5 ▶ **Constructing a Box**

An open box with a square base is to be constructed so that the area of the material used is 300 cm^2 and the volume of the box is 500 cm^3. What are the dimensions of the box?

SOLUTION

Let x represent the length of a side of the base of the box and let y represent the height of the box (Figure 7.29). The area of the base of the box is x^2, and the area of each of the four sides is xy, so the area of the material is

$$x^2 + 4xy = 300$$

Figure 7.29

The volume of the box is $x^2 y = 500$. So the system whose solution we seek is

$$\begin{cases} x^2 + 4xy = 300 & (1) \\ x^2y = 500 & (2) \end{cases}$$

Solving Equation (2) for y and substituting in Equation (1) gives

$$y = \frac{500}{x^2}$$

$$x^2 + 4x\left(\frac{500}{x^2}\right) = 300$$

$$x^2 + \left(\frac{2000}{x}\right) = 300$$

Figure 7.30

The resulting equation is difficult to solve algebraically, so we solve the original system by graphing. Solving Equation (1) and Equation (2) for y gives

$$y = \frac{300 - x^2}{4x} \quad \text{and} \quad y = \frac{500}{x^2}$$

Figure 7.30 shows the graphs of these equations on the same axes and the point of intersection.

Thus, $x = 10$ cm gives the length of each side of the base of the box, and $y = 5$ cm is the height of the box.

EXAMPLE 6 ▶   **Oil Production**

The following functions, with x equal to the number of years after 2010 and y equal to the number of millions of barrels of oil per day, give forecasts of oil production for the United States and Saudi Arabia from 2011 to 2025. Use technology to find the years in which U.S. production will equal that of Saudi Arabia, according to this model.

$$\text{United States:} \quad y = -0.0332x^2 + 0.795x + 7.130$$

$$\text{Saudi Arabia:} \quad y = 0.00194x^3 - 0.0283x^2 - 0.0153x + 11.512$$

(Source: International Energy Agency)

SOLUTION

Entering the equations of the forecast models and intersecting the graphs gives two points of intersection, $x = 6.22$ and $x = 15.18$. See Figure 7.31(a) and (b).

(a)   (b)

Figure 7.31

These values of x correspond to the number of years after 2010. Thus, according to this model, U.S. oil production will equal that of Saudi Arabia in 2017 (with 10.8 million barrels per day) and in 2026 (with 11.5 million barrels per day).

SKILLS CHECK 7.6

In Exercises 1–8, solve the system algebraically.

1. $\begin{cases} x^2 - y = 0 \\ 3x + y = 0 \end{cases}$ **2.** $\begin{cases} x^2 - 2y = 0 \\ 2y - x = 0 \end{cases}$

3. $\begin{cases} x^2 - 3y = 4 \\ 2x + 3y = 4 \end{cases}$ **4.** $\begin{cases} x^2 - 5y = 2 \\ 3x + 5y = 8 \end{cases}$

5. $\begin{cases} x^2 + y^2 = 80 \\ y = 2x \end{cases}$ **6.** $\begin{cases} x^2 + y^2 = 72 \\ x + y = 0 \end{cases}$

7. $\begin{cases} x + y = 8 \\ xy = 12 \end{cases}$ **8.** $\begin{cases} 2xy + y = 36 \\ 2x - 3y = 2 \end{cases}$

In Exercises 9–16, solve the system algebraically or graphically.

9. $\begin{cases} x^2 + 5x - y = 6 \\ 2x - y + 4 = 0 \end{cases}$ **10.** $\begin{cases} x^2 - y - 8x = 6 \\ y + 10x = 18 \end{cases}$

11. $\begin{cases} 2x^2 - 2y + 7x = 19 \\ 2y - x = 61 \end{cases}$

12. $\begin{cases} 2x^2 + 4x - 2y = 24 \\ 2x - y + 37 = 0 \end{cases}$

13. $\begin{cases} x^2 + 4y = 28 \\ \sqrt{x} - y = -1 \end{cases}$ **14.** $\begin{cases} x^2 - y = 0 \\ 2x + \sqrt{y} = 4 \end{cases}$

15. $\begin{cases} 4xy + x = 10 \\ x - 4y = -8 \end{cases}$ **16.** $\begin{cases} xy + y = 4 \\ 2x - 3y = -4 \end{cases}$

In Exercises 17–20, solve the system graphically.

17. $\begin{cases} y = x^3 - 7x^2 + 10x \\ y = 70x - 10x^2 - 100 \end{cases}$

18. $\begin{cases} y = x^3 - 14x^2 + 40x \\ y = 70x - 5x^2 - 200 \end{cases}$

19. $\begin{cases} x^3 + x^2 + y = 15 \\ 2x^2 + y = 11 \end{cases}$ **20.** $\begin{cases} x^3 - 3x + 2y = 2 \\ x^2 + x + y = 4 \end{cases}$

21. a. Algebraically eliminate y from the system in Exercise 19.

b. Use synthetic division to confirm that there is only one real solution to the system.

22. a. Algebraically eliminate y from the system in Exercise 20.

b. Use synthetic division to confirm that the system has three real solutions.

23. Use graphical methods to find two solutions to the system

$$\begin{cases} x^2 + 4y = 17 \\ x^2 y - 5x = 3 \end{cases}$$

24. Use graphical or algebraic methods to find two solutions to the system

$$\begin{cases} xy + 4x = 10 \\ \dfrac{y}{x} + y = \dfrac{3}{2} \end{cases}$$

EXERCISES 7.6

25. *Supply and Demand* The supply function for a product is given by $p = q^2 + 2q + 122$ and the demand function for this product is $p = 650 - 30q$, where p is the price in dollars and q is the number of hundreds of units. Find the price that gives market equilibrium and the equilibrium quantity.

26. *Supply and Demand* The supply function for a product is given by $p = q^2 + 500$ and the demand function for this product is $p = 1124 - 40q$, where p is the price in dollars and q is the number of hundreds of units. Find the price that gives market equilibrium and the equilibrium quantity.

27. *Supply and Demand* The supply function for a product is given by $p = 0.1q^2 + 50q + 1027.50$ and the demand function for this product is $p = 6000 - 20q$, where p is the price in dollars and q is the number of hundreds of units. Find the price that gives market equilibrium and the equilibrium quantity.

28. *Supply and Demand* The supply function for a product is given by $p = q^3 + 800q + 6000$ and the demand function for this product is $p = 12,200 - 10q - q^2$, where p is the price in dollars and q is the number of hundreds of units. Find the price that gives market equilibrium and the equilibrium quantity.

29. *Break-Even* The weekly total cost function for a piece of equipment is $C(x) = 2000x + 18,000 + 60x^2$ dollars, and the weekly revenue is $R(x) = 4620x - 12x^2 - x^3$, where x is the number of thousands of units. The x-values that are the solutions to the system

$$\begin{cases} y = 2000x + 18,000 + 60x^2 \\ y = 4620x - 12x^2 - x^3 \end{cases}$$

are the quantities that give break-even. Find these quantities.

30. *Break-Even* The daily total cost function for a product is $C(x) = 400x + 4800$ dollars and the daily revenue is $R(x) = 600x + 4x^2 - 0.1x^3$, where x is the number of units. Use graphical methods to find the numbers of units that give break-even.

31. *Break-Even* The monthly total cost function for a product is $C(x) = 400x + 0.1x^2 + 17,901$ dollars and the monthly revenue is $R(x) = 1000x - 0.01x^2$, where x is the number of units. Find the numbers of units that give break-even for the product. (*Hint*: Use graphical methods to find one solution less than 100 units and one greater than 5000 units.)

32. *Break-Even* The weekly total cost function for a product is $C(x) = 3x^2 + 1228$ dollars and the weekly revenue is $R(x) = 177.50x$, where x is the number of hundreds of units. Use graphical methods to find the numbers of units that give break-even.

33. *Constructing a Box* A rectangular piece of cardboard with area 180 square inches is made into an open box by cutting a 2-inch square from each corner and turning up the sides. If the box is to have a volume of 176 cubic inches, what are the dimensions of the original piece of cardboard? (See the figure below.)

34. *Constructing a Box* An open box is constructed from a rectangular piece of tin by cutting a 4-centimeter square from each corner and turning up the sides. If the box has a volume of 768 cubic centimeters and the area of the original piece of tin was 512 square centimeters, what are the dimensions of the original piece of tin?

35. *Dimensions of a Box* An open box has a rectangular base and its height is equal to the length of the shortest side of the base. What dimensions give a volume of 2000 cubic centimeters using 800 square centimeters of material?

36. *Constructing a Box* A box with a top is to be constructed with the width of its base equal to half of the length of its base. If it is to have volume 6000 cubic centimeters and uses 2200 square centimeters of material, what are the dimensions of the box?

37. *Trust Fund* A father gives a trust fund to his daughter by depositing \$50,000 into an account paying 10%, compounded annually. This account grows according to $y = 50,000(1.10^t)$. He will give her \$12,968.72 per year from another account until she has received in total the value to which the trust fund has grown. For how many years will she receive the \$12,968.72?

38. *Constructing a Box* An open box with a square base is to be constructed so that the area of the material used is 500 cm^2 and the volume of the box is 500 cm^3. What are the dimensions of the box?

---

chapter 7   **Summary**

In this chapter, we studied systems of equations and their applications. We first solved systems of linear equations in three variables by using left-to-right elimination. Matrices were introduced to help solve systems of equations, and other applications of matrices were investigated. We also solved systems of nonlinear equations by algebraic and by graphical methods.

Key Concepts and Formulas

7.1 Systems of Linear Equations in Three Variables

Left-to-right elimination method	An extension of the elimination method used to solve systems of two equations in two variables.
Possible solutions to a system of linear equations in three variables	
• Unique solution	Planes intersect in a single point.
• No solution	Planes have no common intersection.
• Many solutions	Planes intersect in a line.
Modeling systems of equations	Solution of real problems sometimes requires us to create three or more equations whose simultaneous solution is the solution to the problem.

7.2 Matrix Solution of Systems of Linear Equations

Matrix representation of systems of equations	Systems of linear equations can be represented using matrices, and the solution can be found easily.
Matrices	
• Dimension	The numbers of rows and columns of a matrix.
• Square matrix	Equal number of rows and columns.
• Coefficient matrix	Coefficients of the variables in a linear system placed in a matrix.
• Augmented matrix	A constant column is added to a coefficient matrix.
• Identity matrix	A square matrix that has 1's down its diagonal and 0's everywhere else.
• Row-echelon form	The coefficient part has all 1's or 0's on its diagonal and all 0's below its diagonal.
• Reduced row-echelon form	The coefficient part has 1's or 0's on its diagonal and 0's elsewhere.
Matrix row operations	1. Interchange two rows of the matrix.
	2. Multiply a row by a nonzero constant.
	3. Add a multiple of one row to another row.
Gauss–Jordan elimination method	Row operations are used to reduce a matrix to reduced row-echelon form. If the coefficient part reduces to the identity matrix, the system has a unique solution.
Using technology to find reduced row-echelon form	Graphing utilities and spreadsheets can be used to find the reduced row-echelon form.
Unique/nonunique solutions	Systems of equations can have a unique solution, no solution, or infinitely many solutions.
• Unique solution	The reduced coefficient matrix contains the identity matrix.
• No solution	A row of the reduced coefficient matrix contains all 0's and the augment of that row contains a nonzero number.
• Infinite number of solutions	If neither of the above occurs.

7.3 Matrix Operations

Addition of matrices	Matrices of the same dimension are added by adding the corresponding entries of the two matrices.
Zero matrix	Each element is zero.
Negative of a matrix	The sum of a matrix and its negative is the zero matrix.
Subtraction of matrices	If the matrices M and N are the same dimension, the difference $M - N$ is found by subtracting the elements of N from the corresponding elements of M.
Multiplication of matrices by a number	Multiplying a matrix A by a real number c results in a matrix in which each entry in matrix A is multiplied by the number c.
Matrix multiplication	The product of an $m \times n$ matrix A and an $n \times k$ matrix B is an $m \times k$ matrix $C = AB$ with the element in the ith row and jth column of matrix C given by the product of the ith row of A and the jth column of B.

7.4 Inverse Matrices; Matrix Equations

Inverse matrices	Two square matrices, A and B, are called inverses of each other if $AB = I$ and $BA = I$.
Finding the inverse of a square matrix	Placing the same dimension identity matrix in its augment and reducing the matrix to its reduced row-echelon form will give the inverse matrix in the augment if the inverse exists.
Inverses and technology	Computer software, spreadsheets, and calculators can be used to find the inverse of a matrix, if it exists.
Matrix equations	Multiplying both sides of the matrix equation $AX = C$ on the left by A^{-1}, if it exists, gives the solution to the system that the matrix equation represents.
Encoding messages and decoding messages	Matrix multiplication can be used to encode messages. The inverse of an encoding matrix can be used to decode messages.

7.5 Determinants and Cramer's Rule

Finding determinants of matrices	Every square matrix has a number associated with it called the **determinant.** The determinant for a 2×2 matrix is defined as the difference of the diagonal products. $$\begin{vmatrix} a & b \\ c & d \end{vmatrix} = ad - bc$$
Solving systems of matrices with Cramer's Rule	Cramer's Rule for a 2×2 system $$\text{If } \begin{cases} ax + by = u \\ cx + dy = w \end{cases}, \text{ then } x = \frac{\begin{vmatrix} u & b \\ w & d \end{vmatrix}}{\begin{vmatrix} a & b \\ c & d \end{vmatrix}} \text{ and } y = \frac{\begin{vmatrix} a & u \\ c & w \end{vmatrix}}{\begin{vmatrix} a & b \\ c & d \end{vmatrix}}, \text{ provided}$$ $$D = \begin{vmatrix} a & b \\ c & d \end{vmatrix} = ad - bc \neq 0.$$

Cramer's Rule with inconsistent and dependent systems	If D, the determinant of the coefficients of the variables of a system of three equations in three variables, is 0, then the system either is inconsistent or contains dependent equations. In particular,
	1. If $D = 0$ and at least one of the determinants in the numerator $(D_x, D_y, \text{ or } D_z)$ is not 0, then the system is inconsistent, and there is no solution.
	2. If $D = 0$ and all the determinants in the numerators $(D_x, D_y, \text{ and } D_z)$ are 0, then the equations in the system are dependent, and there are infinitely many solutions. These solutions can be found using different methods.

7.6 Systems of Nonlinear Equations

Systems of nonlinear equations	A system of nonlinear equations contains at least one equation that is nonlinear. These systems can have any number of solutions or no solution.
Algebraic solution	Systems of nonlinear equations can be solved algebraically by substitution or elimination.
Graphical solution	Solutions to systems of nonlinear equations can be checked with graphical methods, or, if solving the system algebraically is difficult, graphical methods may be used to solve the system.

chapter 7 SKILLS CHECK

In Exercises 1–10, solve the systems of linear equations, if possible, by (a) left-to-right elimination and (b) using technology.

1. $\begin{cases} 2x - 3y + z = 2 \\ 3x + 2y - z = 6 \\ x - 4y + 2z = 2 \end{cases}$ **2.** $\begin{cases} 3x - 2y - 4z = 9 \\ x + 3y + 2z = -1 \\ 2x + 4y + 4z = 2 \end{cases}$

3. $\begin{cases} 3x + 2y - z = 6 \\ 2x - 4y - 2z = 0 \\ 5x + 3y + 6z = 2 \end{cases}$

4. $\begin{cases} 3x + 6y + 9z = 27 \\ 2x + 3y - z = -2 \\ 4x + 5y + z = 6 \end{cases}$

5. $\begin{cases} x + 2y - 2z = 1 \\ 2x - y + 5z = 15 \\ 3x - 4y + z = 7 \end{cases}$

6. $\begin{cases} -6x + 4y - 2z = 4 \\ 3x - 2y + 5z = -6 \\ x - 4y + z = -8 \end{cases}$

7. $\begin{cases} 2x + 5y + 8z = 30 \\ 18x + 42y + 18z = 60 \end{cases}$

8. $\begin{cases} 9x + 21y + 15z = 60 \\ 2x + 5y + 8z = 30 \\ x + 2y - 3z = -10 \end{cases}$

9. $\begin{cases} x + 3y + 2z = 5 \\ 9x + 12y + 15z = 6 \\ 2x + y + 3z = -10 \end{cases}$

10. $\begin{cases} 5x + 7y + 10z = 6 \\ 2x + 5y + 6z = 1 \\ 3x + 2y + 4z = 6 \end{cases}$

In Exercises 11 and 12, solve the systems of linear equations using technology.

11. $\begin{cases} 3x + 2y - z + w = 12 \\ x - 4y + 3z - w = -18 \\ x + y + 3z + 2w = 0 \\ 2x - y + 3z - 3w = -10 \end{cases}$

12. $\begin{cases} 2x + y - 3z + 4w = 7 \\ x - 2y + z - 2w = 0 \\ 3x + y + 4z + w = -2 \\ x + 3y + 2z + 2w = -1 \end{cases}$

Perform the matrix operations, if possible, in Exercises 13–21, with

$$A = \begin{bmatrix} 1 & 3 & -3 \\ 2 & 4 & 1 \\ -1 & 3 & 2 \end{bmatrix} \quad B = \begin{bmatrix} 1 & 2 & 1 \\ 2 & -1 & 3 \end{bmatrix}$$

$$C = \begin{bmatrix} 2 & 3 \\ -1 & 2 \\ 3 & -2 \end{bmatrix} \quad D = \begin{bmatrix} -2 & 3 & 1 \\ -3 & 2 & 2 \end{bmatrix}$$

13. $B + D$ **14.** $D - B$ **15.** $5C$

16. AB **17.** BA **18.** CD

19. DC **20.** $A^2 = A \cdot A$ **21.** A^{-1}

Find the inverse of each matrix in Exercises 22–24.

22. $\begin{bmatrix} 2 & 1 & 1 \\ 1 & 2 & -1 \\ 2 & 2 & 1 \end{bmatrix}$ **23.** $\begin{bmatrix} 1 & 2 & 1 \\ 1 & 1 & 0 \\ 0 & 2 & 1 \end{bmatrix}$

24. $\begin{bmatrix} -1 & 0 & 1 & 1 \\ 1 & 1 & 2 & 1 \\ 2 & -1 & 0 & 1 \\ 0 & 0 & 0 & 1 \end{bmatrix}$

In Exercises 25–28, solve the system of equations by using the inverse of the coefficient matrix.

25. $\begin{cases} x + y - 3z = 8 \\ 2x + 4y + z = 15 \\ -x + 3y + 2z = 5 \end{cases}$

26. $\begin{cases} 2x + 3y + z = 20 \\ 3x + 4y - z = 40 \\ 2x + 5y + z = 60 \end{cases}$

27. $\begin{cases} -x_1 + x_3 + x_4 = 6 \\ x_2 + x_3 + x_4 = 12 \\ 2x_1 + 5x_2 + x_3 = 20 \\ 3x_2 + x_4 = 24 \end{cases}$

28. $\begin{cases} 2x_1 + 2x_2 + x_4 = 4 \\ 2x_1 + x_2 + 2x_4 = 12 \\ x_1 + x_2 + x_3 + x_4 = 4 \\ x_2 + x_4 = 8 \end{cases}$

In Exercises 29 and 30, find the determinant.

29. $\begin{vmatrix} -1 & 4 \\ 2 & -3 \end{vmatrix}$ **30.** $\begin{vmatrix} 2 & 0 & -3 \\ -1 & 4 & 3 \\ 0 & 1 & -2 \end{vmatrix}$

In Exercises 31 and 32, solve the system of equations using Cramer's Rule, if possible.

31. $\begin{cases} 2x - 3y = 7 \\ -x + y = 4 \end{cases}$

32. $\begin{cases} -x + 3y - 2z = 4 \\ -x + 2y + 2z = 5 \\ x - 2y + z = 6 \end{cases}$

33. Solve the nonlinear system

$$\begin{cases} x^2 - y = x \\ 4x - y = 4 \end{cases}$$

34. Find two solutions to the system

$$\begin{cases} x^2 y = 2000 \\ x^2 + 2y = 140 \end{cases}$$

chapter 7 REVIEW

35. **Pricing** A concert promoter needs to make $200,000 from the sale of 4000 tickets. The promoter charges $40 for some tickets, $60 for some tickets, and $100 for the others. If the number of $60 tickets is $\dfrac{1}{4}$ the sum of the remaining tickets, how many tickets of each type should be sold?

36. **Medication** Medication A is given six times per day, and medication B is given twice per day. Medication C is given once every two days, and its dosage is $\dfrac{1}{2}$ the sum of the dosages of the other two medications. For a certain patient, the total intake of the three medications is limited to 28.7 milligrams on days when all three medications are given. If the ratio of the dosage of medication A to the dosage of medication B is 2 to 3, how many milligrams are in each dosage?

37. Rental Income A woman has $750,000 invested in three rental properties. One yields an annual return of 12% on her investment, and the second returns 15% per year on her investment. The third investment returns 10% and is limited to $\frac{1}{2}$ the sum of the other two investments. Her total annual return from the three investments is $89,500. How much is invested in each property?

38. Investment A woman invests $360,000 in three different mutual funds. One averages 12% per year, and another averages 16% per year. A third fund averages 8% per year, and because it has much less risk, she invests twice as much in it as in the sum of the other two funds. To realize an annual return of $35,200, how much should she invest in each fund?

39. Loans A bank lends $1,180,000 to a development company to purchase three business properties. If one of the properties costs $75,000 more than the other and the third costs three times the sum of these two properties, find the cost of each property.

40. Investment A brokerage house offers three stock portfolios for its clients. Portfolio I consists of 10 blocks of common stock, 2 municipal bonds, and 3 blocks of preferred stock. Portfolio II consists of 12 blocks of common stock, 8 municipal bonds, and 5 blocks of preferred stock. Portfolio III consists of 10 blocks of common stock, 4 municipal bonds, and 8 blocks of preferred stock. If a client wants to combine these portfolios so that she has 290 blocks of common stock, 138 municipal bonds, and 161 blocks of preferred stock, how many units of each portfolio does she need? To answer this question, let x equal the number of units of portfolio I, y equal the number of units of portfolio II, and z equal the number of units of portfolio III.

41. Property A realty partnership purchases three business properties at a total cost of $375,000. One property costs $50,000 more than the second, and the third property costs half the sum of these two properties. Find the cost of each property.

42. Nutrition A nutritionist wants to create a diet that uses a combination of three foods: I, II, and III. Each of these foods contains three additives: A, B, and C. The following table gives the percent of each additive that is present in each food. If the diet being used requires 12.5 grams per day of A, 9.1 grams of B, and 9.6 grams of C, find the number of grams of each food that should be used each day for this diet.

	Food I	Food II	Food III
Additive A (%)	10	11	18
Additive B (%)	12	9	10
Additive C (%)	14	12	8

43. Transportation A delivery service has three types of aircraft, each of which carries three types of cargo. The payload of each type is summarized in the table below. Suppose on a given day the airline must move 2200 next-day delivery letters, 3860 two-day delivery letters, and 920 units of air freight. How many aircraft of each type should be scheduled?

	Passenger	Transport	Jumbo
Next-Day Letters	200	200	200
Two-Day Letters	300	40	700
Air Freight	40	130	70

44. Finance A financial planner promises a long-term return of 12% from a combination of three mutual funds. The cost per share and average annual return per share for each mutual fund are given in the table below. If the annual returns are accurate, how many shares of each mutual fund will return an average of 12% on a total investment of $210,000?

	Tech Fund	Balanced Fund	Utility Fund
Cost per Share ($)	180	210	120
Annual Return per Share ($)	18	42	18

45. Transportation Marshall Trucking Company has an order for delivery of three products: A, B, and C. The following table gives the volume in cubic feet, the weight in pounds, and the value for insurance in dollars per unit of each of the products. If one of the company's trucks can carry 9260 cubic feet and 12,000 pounds and is insured to carry $52,600, how many units of each product can be carried on the truck?

	Product A	Product B	Product C
Volume (cu ft)	25	30	40
Weight (lb)	30	36	60
Value ($)	150	180	200

46. Nutrition A biologist is growing three types of slugs (A, B, and C) in the same laboratory environment. Each day, the slugs are given different nutrients

(I, II, and III). Each type A slug requires 2 units of I, 6 units of II, and 2 units of III per day. Each type B slug requires 2 units of I, 8 units of II, and 4 units of III per day. Each type C slug requires 4 units of I, 20 units of II, and 12 units of III per day. If the daily mixture contains 4000 units of I, 16,000 units of II, and 8000 units of III, find the number of slugs of each type that can be supported.

47. *Trade Balances* The tables below give the values of U.S. exports and imports with Mexico and with Canada for selected years from 2009 to 2014.

a. Form the matrix *A* that contains the values of exports to Mexico and to Canada for these years.

b. Form the matrix *B* that contains the values of imports from Mexico and from Canada for these years.

c. Use these two matrices to find the trade balances for the United States for these years.

d. Are the trade balances with these countries improving for the United States?

e. With which country was the trade balance worse in 2014?

Mexico

Year	Exports to ($ millions)	Imports from ($ millions)
2014	240,326	294,158
2013	226,079	280,529
2012	215,907	277,594
2011	198,289	262,874
2010	163,665	229,986
2009	128,892	176,654

Canada

Year	Exports to ($ millions)	Imports from ($ millions)
2014	312,032	346,063
2013	301,610	332,553
2012	292,651	324,264
2011	281,292	315,325
2010	249,257	277,637
2009	204,658	226,248

(Source: census.gov/foreign-trade/balance)

48. *Investment* An investor wants to place a total of $800,000 in two or more of three investments with different degrees of risk. One investment pays 6%, one pays 7%, and one pays 10%, and the investor wants to earn $71,600 from the investments.

a. Write a system of two equations to represent the problem, representing the amount in the 6% investment by x, the amount in the 7% investment by y, and the amount in the 10% investment by z.

b. Solve the system.

c. How much should the investor put in each investment if she wants to put more in the 6% investment than in the 7% investment, because the 6% investment has a lower risk?

d. How much should she put in each investment if she wants to put more in the 7% investment than in the 6% investment?

49. *Income and SAT Scores* The following table gives the average critical reading and math SAT scores for certain family income levels.

a. Use linear regression to find a linear equation that gives the critical reading SAT score as a function of family income.

b. Use linear regression to find a linear equation that gives the math SAT score as a function of family income.

c. Interpret the slopes of the equations from parts (a) and (b) as rates of change.

Family Income ($ thousands)	Critical Reading SAT	Math SAT
10	434	457
30	462	475
50	488	497
70	503	512
90	517	528
110	525	538
130	529	542
150	536	550
180	542	554
200	563	579

(Source: collegeboard.com)

50. *Medication* Suppose combining x cubic centimeters (cc) of a 20% solution of a medication and y cc of a 5% solution of the medication gives 10 cc of a 15.5% solution.

a. Write the equation that gives the total amount of solution in terms of x and y.

b. Write the equation that gives the total amount of medication in the solution.

c. Use substitution to find the amount of each solution that is needed.

51. *Supply and Demand* The supply function for a product is given by $p = q + 578$ and the demand function for this product is $p = 396 + q^2$, where p is the price in dollars and q is the number of units. Find the price that gives market equilibrium and the equilibrium quantity.

52. *Break-Even* The daily total cost function for a product is $C(x) = 2500x + x^2 + 27{,}540$ dollars and the daily revenue is $R(x) = 3899x - 0.1x^2$, where x is the number of units. Find the numbers of units that give break-even.

Group Activities
▶ EXTENDED APPLICATIONS

1. Salaries

Average Salaries of College Professors

Academic Rank		Men			Women		
		Public	Private-Independent	Church-Related	Public	Private-Independent	Church-Related
Doctoral	Professor	108,481	138,921	121,312	98,552	127,542	109,720
	Associate	76,030	89,936	82,941	70,764	83,148	76,814
	Assistant	65,498	78,079	70,227	60,155	71,207	65,882
	Instructor	43,352	50,418	55,513	42,012	49,246	53,549
	Lecturer	51,182	59,443	52,257	46,340	52,047	46,476
Master's	Professor	82,834	93,020	84,493	79,493	86,476	78,355
	Associate	66,102	71,130	66,448	63,630	66,881	63,338
	Assistant	56,225	57,610	55,107	53,907	55,065	52,467
	Instructor	41,495	46,444	45,997	40,820	44,315	44,174
	Lecturer	46,030	51,037	49,064	44,018	43,907	45,631
Baccalaureate	Professor	78,390	91,563	69,969	73,372	87,570	66,427
	Associate	63,658	67,210	56,752	61,363	66,444	55,315
	Assistant	52,960	55,180	47,853	50,714	54,011	47,085
	Instructor	40,844	45,139	40,966	41,183	40,705	40,181
	Lecturer	47,743	58,064	42,095	45,565	54,343	39,233

(Source: American Association of University Professors)

The table above gives the average salaries of male and female college professors who teach at three types of institutions at three different levels of instruction and at five job designations. To investigate the relationship between salaries of males and females, do the following:

1. Create matrix A and matrix B, which contain the salaries of male and female professors, respectively, in each type of school and at each level of instruction.
2. Subtract the matrix of male professors' salaries from the matrix of female professors' salaries.

3. What does this show about female salaries versus male salaries in every category? Based on the data, discuss gender bias in educational institutions.
4. Create a scatter plot with male professors' salaries as the input and female professors' salaries as the output.
5. Find a function that gives a relationship between the salaries of male and female professors from part 4.
6. Graph the scatter plot from part 4 and the function from part 5 on the same axes.

7. The graph of the function $y = x$ would represent the relationship between male and female professors' salaries if they were equal. Graph the function from part 5 and the function $y = x$ on the same axes.
8. Use the two graphs in part 7 to determine how female professors' salaries compare to male professors' salaries as male salaries rise.
9. Solve the two equations whose graphs are discussed in part 8 to determine at what salary level male salaries become greater than female salaries.

2. Parts Listing

Wingo Playgrounds must maintain numerous parts in inventory for its Fun-in-the-Sun swing set. The swing set has 4 legs and a top connecting the legs. Each of the legs is made from 1 nine-foot pipe connected with 1 brace and 2 bolts, and the top is made of 1 nine-foot pipe, 2 clamps, and 6 bolts. The parts listing for these swing sets can be described by the following matrix.

$$P = \begin{bmatrix} 0 & 0 & 0 & 0 & 0 & 0 & 0 \\ 4 & 0 & 0 & 0 & 0 & 0 & 0 \\ 1 & 0 & 0 & 0 & 0 & 0 & 0 \\ 0 & 1 & 1 & 0 & 0 & 0 & 0 \\ 0 & 0 & 2 & 0 & 0 & 0 & 0 \\ 0 & 1 & 0 & 0 & 0 & 0 & 0 \\ 0 & 2 & 6 & 0 & 0 & 0 & 0 \end{bmatrix} \begin{matrix} \text{Swing sets} \\ \text{Legs} \\ \text{Top} \\ \text{Pipes} \\ \text{Clamps} \\ \text{Braces} \\ \text{Bolts} \end{matrix}$$

with column headings SS, L, T, P, C, Bc, B.

Each column indicates how many of each part are required to produce the part shown at the top of the column. For example, column 1 indicates that to produce a swing set requires 4 legs and 1 top, and the second column indicates that each leg (L) is constructed from 1 pipe, 1 brace, and 2 bolts.

Suppose an order is received for 8 complete swing sets plus the following spare parts: 2 legs, 1 top, 2 pipes, 4 clamps, 4 braces, and 8 bolts. We indicate the number of each part type that must be supplied to fill this order by the matrix

$$X = \begin{bmatrix} x_1 \\ x_2 \\ x_3 \\ x_4 \\ x_5 \\ x_6 \\ x_7 \end{bmatrix}$$

To find the numbers of each part that are needed, do the following:

1. Form the 7×1 matrix D that contains the numbers of items to be delivered to fill the order for the 8 swing sets and the spare parts. Some parts of each type are needed to produce other parts of the swing set. The matrix PX gives the number of parts of each type used to produce other parts, and the matrix $X - PX$ gives the number of parts of each type that are available to be delivered. That is,

$$X - PX = D,$$

where D is the number of parts of each type that are available for delivery.
2. The matrix equation $X - PX = D$ can be rewritten in the form $(I - P)X = D$, where I is the 7×7 identity matrix. Find the matrix $I - P$ that satisfies this equation.
3. Find the inverse of the matrix $I - P$.
4. Multiply both sides of the matrix equation $(I - P)X = D$ by the inverse of $I - P$. This gives the matrix X that contains the numbers of parts of each type that are needed.
5. The primary assembly parts are the pipes, clamps, braces, and bolts. What is the total number of primary parts of each type needed to fill the order?

8

Special Topics in Algebra

To determine how many cars of different types it must buy and rent to maximize its profit, a car rental agency must consider the constraints on its purchases. The constraints include how much money it has to invest in cars and how many cars it needs. These constraints can be written as inequalities and "pictured" by graphs of systems of inequalities. The process of finding the optimal solution to a problem subject to the constraint inequalities, called linear programming, is examined in this chapter. In addition, we use sequences and series to solve several types of financial problems, including future values of lump sum investments and annuities. We also review numerous algebra skills in the context of calculus applications in this chapter, discuss the Binomial Theorem, and introduce the geometric properties of conic sections.

sections	objectives	applications
8.1 Systems of Inequalities	Solve linear inequalities in two variables; solve systems of linear inequalities in two variables; solve systems of nonlinear inequalities in two variables	Auto purchases, car rental, advertising
8.2 Linear Programming: Graphical Methods	Optimize a function subject to constraints with linear programming	Maximizing car rental profit, cost minimization, profit maximization, manufacturing
8.3 Sequences and Discrete Functions	Find terms of sequences, arithmetic sequences, and geometric sequences	Football contracts, depreciation, rebounding ball, compound interest
8.4 Series	Find the sum of a finite series; find the sum of the first n terms of an infinite series, an arithmetic series, and a geometric series; find the sum of an infinite geometric series	Football contracts, depreciation, annuities
8.5 Preparing for Calculus	Simplify expressions; evaluate functions; simplify fractions; rewrite radicals; solve equations; find the decomposition of functions; rewrite expressions using negative exponents; rewrite expressions using positive exponents; use logarithmic properties; reduce fractions	Slope of a tangent line, evaluating the difference quotient
8.6 The Binomial Theorem	Evaluate factorials; use Pascal's triangle; find coefficients of terms in a binomial expansion; use the Binomial Theorem to expand binomials; find a particular term of a binomial expansion	
8.7 Conic Sections: Circles and Parabolas	Use the distance formula; use the midpoint formula; write the equation of a circle; locate the center and radius of a circle; graph circles; write the equation of a parabola given the vertex and focus; graph parabolas	Parabolic cables on a bridge, satellite dishes
8.8 Conic Sections: Ellipses and Hyperbolas	Graph horizontal ellipses; graph vertical ellipses; write the standard form of ellipses; graph horizontal hyperbolas; graph vertical hyperbolas; write the standard form of hyperbolas; identify the asymptotes of hyperbolas; graph rectangular hyperbolas	Planets, cones, bridge arches, Roman Colosseum, Ohm's Law, Boyle's Law

KEY OBJECTIVES

- Solve linear inequalities in two variables
- Graph *solution regions* to linear inequalities in two variables
- Solve systems of nonlinear inequalities in two variables

SECTION PREVIEW   Car Rental

A rental agency has a maximum of $1,260,000 to invest in the purchase of at most 71 new cars of two different types: compact and midsize. The cost per compact car is $15,000, and the cost per midsize car is $28,000. The number of cars of each type is limited (constrained) by the budget available and the number of cars needed. To get a "picture" of the constraints on this rental agency, we can graph the inequalities determined by the maximum amount of money and the limit on the number of cars purchased. Because these inequalities deal with two types of cars, they will contain two variables. For example, if we denote the number of compact cars by x and the number of midsize cars by y, the statement that the agency wants to purchase at most 71 cars can be written as the inequality

$$x + y \leq 71, \quad \text{where } x \text{ and } y \text{ are integers}$$

The collection of all **constraint** inequalities can be expressed by a **system of inequalities** in two variables. Finding the values that satisfy all these constraints at the same time is called solving the system of inequalities. In this section, we will graph solutions of linear and nonlinear inequalities in two variables. ■

Linear Inequalities in Two Variables

We solved linear inequalities in one variable in Section 2.4. We now consider linear inequalities in two variables. For example, consider the inequality $y > x$. The solutions to this inequality are the ordered pairs (x, y) that satisfy the statement $y > x$. The graph of $y > x$ consists of all points above the line $y = x$ (the shaded area in Figure 8.1). The line $y = x$ is dashed in Figure 8.1 because the inequality does not include $y = x$. This (dashed) line divides the plane into two **half-planes**, $y > x$ and $y < x$. We can determine which half-plane is the solution to the inequality by selecting any point not on the line. If the coordinates of this **test point** satisfy the inequality, then the half-plane containing that point is the graph of the solution. If the test point does not satisfy the inequality, then the other half-plane is the graph of the solution. For example, the point $(2, 4)$ is above the line $y = x$, and $x = 2$, $y = 4$ satisfies the inequality $y > x$, so the region above the line is the graph of the solution of the inequality. Note that any other point on the half-plane above the line also satisfies the inequality.

Figure 8.1

EXAMPLE 1 ▶   **Graph of an Inequality**

Graph the solution of the inequality $6x - 3y \leq 15$.

SOLUTION

First we graph the line $6x - 3y = 15$ (which we can also write as $y = 2x - 5$) as a solid line, because the inequality symbol "≤" indicates that points on the line are part of the solution (Figure 8.2(a)). Next we pick a test point that is not on the line. If we use

(a) $6x - 3y = 15$ (b) $6x - 3y \leq 15$

Figure 8.2

(a)

(b)

Figure 8.3

$(0, 0)$, we get $6(0) - 3(0) \leq 15$, or $0 \leq 15$, which is true, so the inequality is satisfied. Thus, the half-plane that contains the point $(0, 0)$ and the line $6x - 3y = 15$ is the graph of the solution, which we call the **solution region** of the inequality.

Technology Note

Graphing calculators can also be used to create shaded regions representing inequalities. To graph the solution region of the inequality $y \geq 2x - 5$, we enter $2x - 5$ to the right of Y1= in the equation editor, move to the icon at the left of Y1=, change the icon from \ to the icon showing a shaded area above a line (see Figure 8.3(a)), and press **ENTER**. Figure 8.3(b) shows the line $y = 2x - 5$ and the region above it.

EXAMPLE 2 ▶ Car Rental Agency

One of the constraint inequalities for the car rental company application given in the Section Preview is that the sum of x and y (the numbers of compact and midsize cars, respectively) is at most 71. This constraint can be described by the inequality

$$x + y \leq 71$$

where x and y are integers. Find the graphical solution of this inequality.

SOLUTION

(a)

(b)

Figure 8.4

The inequality $x + y \leq 71$, where x and y are integers, has many integer solutions. The inequality is satisfied if x is any integer from 0 to 71 and y is any integer less than or equal to $71 - x$. For example, some of the solutions are $(0, 71)$, $(1, 70)$, $(60, 5)$, and $(43, 20)$. As we stated in Chapter 2, problems are frequently easier to solve if we treat the data as continuous rather than discrete. With this in mind, we graph this inequality as a region on a coordinate plane rather than attempting to write all of the possible integer solutions.

We create the graph of the inequality by first graphing the equation

$$x + y = 71, \quad \text{or, equivalently,} \quad y = 71 - x$$

(Figure 8.4(a)). The graph of this line divides the coordinate plane into two half-planes, $y > 71 - x$ and $y < 71 - x$. We can find the half-plane that represents the graph of $y < 71 - x$ by testing the coordinates of a point and determining whether it satisfies $y < 71 - x$. For example, the point $(0, 0)$ is in the half-plane below the line, and $x = 0, y = 0$ satisfies the inequality $y < 71 - x$ because $0 < 71 - 0$. Thus, the half-plane below the line satisfies the inequality. If we shade the half-plane that satisfies $y < 71 - x$, the line and the shaded region constitute the graph of $x + y \leq 71$ (Figure 8.4(b)). This is the solution region for the inequality.

Note that the number of cars satisfying this constraint has other limitations. In the context of this application, the number of cars cannot be negative, so the values of x and y must be nonnegative. We can show the solution to the inequality for nonnegative x and y by using a window with $x \geq 0$ and $y \geq 0$. Figure 8.5 shows the solution of $x + y \leq 71$ for $x \geq 0$ and $y \geq 0$.

All points in this region with integer coordinates satisfy the stated conditions of this example.

Figure 8.5

We now investigate the second inequality in the car rental problem.

EXAMPLE 3 ▶   **Rental Cars**

The car rental agency in the Section Preview has a maximum of $1,260,000 to invest, with x compact cars costing $15,000 each and y midsized cars costing $28,000 each.

a. Write the inequality representing this information.

b. Graph this inequality on a coordinate plane.

SOLUTION

a. If x represents the number of compact cars, the cost of all the compact cars purchased is $15,000x$ dollars. Similarly, the cost of all the midsized cars purchased is $28,000y$ dollars. Because the total amount available to spend is $1,260,000$ dollars, x and y satisfy

$$15,000x + 28,000y \leq 1,260,000$$

b. To graph the solutions to this inequality, we first graph the equation

$$15,000x + 28,000y = 1,260,000$$

Figure 8.6

or, equivalently,

$$y = \frac{1,260,000 - 15,000x}{28,000} = \frac{1260 - 15x}{28}$$

The graph is shown in Figure 8.6.

To decide which of the two half-planes determined by the line satisfies the inequality, we test points in one half-plane. The point $(0, 0)$ is in the half-plane below the line, and it satisfies the inequality because $15,000(0) + 28,000(0) < 1,260,000$. Thus, the half-plane containing $(0, 0)$, along with the line, forms the solution region. Because x and y must be nonnegative in this application, the graph of the inequality is shown only in the first quadrant (Figure 8.7). All points in this region with integer coordinates satisfy the stated conditions of this example.

Figure 8.7

Systems of Inequalities in Two Variables

We now have pictures of the limitations (constraints) on the car purchases of the car rental agency, but how are they related? We know that the number of cars the rental agency can buy is limited by *both* inequalities in Examples 2 and 3, so we seek the solution to both inequalities *simultaneously*. That is, we seek the solution to a **system of inequalities**. In general, if we have two or more inequalities in two variables and seek the values of the variables that satisfy all inequalities, we are solving a system of inequalities. The solution to the system can be found by finding the intersection of the solution sets of the inequalities.

EXAMPLE 4 ▶   **System of Inequalities**

Solve the system of inequalities $\begin{cases} 2x - 4y \geq 12 \\ x + 3y > -4 \end{cases}$.

SOLUTION

To find the solution, we first solve the inequalities for y.

$$y \leq \frac{1}{2}x - 3 \qquad (1)$$

$$y > -\frac{1}{3}x - \frac{4}{3} \qquad (2)$$

The *borders* of the inequality region are graphed as the solid line with equation $y = \frac{1}{2}x - 3$ and the dashed line with equation $y = -\frac{1}{3}x - \frac{4}{3}$. One of these borders is a solid line because $y = \frac{1}{2}x - 3$ is part of Inequality (1), and one is a dashed line because $y = -\frac{1}{3}x - \frac{4}{3}$ is not part of Inequality (2) (Figure 8.8(a)). We find a *corner* of the solution region by finding the intersection of the two lines. We can find this point of intersection by algebraically or graphically solving the equations simultaneously. Using substitution and solving gives the point of intersection:

$$\frac{1}{2}x - 3 = -\frac{1}{3}x - \frac{4}{3}$$
$$3x - 18 = -2x - 8$$
$$5x = 10$$
$$x = 2$$
$$y = \frac{1}{2}(2) - 3 = -2$$

Figure 8.8

The two border lines divide the plane into four regions, one of which is the solution region. We can determine what region satisfies both inequalities by choosing a test point in each of the four regions. For example:

- $(0, 0)$ and $(1, -2)$ do not satisfy Inequality (1).

- $(2, -4)$ does not satisfy Inequality (2).

- $(6, -1)$ satisfies both inequalities, so the region that contains this point is the solution region.

Figure 8.8(b) shows the solution region for the system of inequalities, with its corner at $(2, -2)$. Note that the graph of $y = -\frac{1}{3}x - \frac{4}{3}$ is not part of the solution.

We now turn our attention to finding the values of x and y that satisfy both the constraints on the car rental agency problem in the Section Preview.

EXAMPLE 5 ▶   Car Rental Agency

The inequalities that satisfy the conditions given in the car rental problem in the Section Preview are called the **constraint inequalities** for the problem. They form the system

$$\begin{cases} y \le 71 - x \\ y \le \dfrac{1260 - 15x}{28} \\ x \ge 0 \\ y \ge 0 \end{cases}$$

a. Graph the solution to the system.

b. Find the coordinates of the points where the borders intersect and the points where the region intersects the x- and y-axes.

SOLUTION

a. We graphed the solutions to the first two inequalities discussed for the car rental agency application in Examples 2 and 3. If we graph them both in the first quadrant, the region where the two solution regions overlap is where both inequalities are true simultaneously (the coordinates of the points satisfy both inequalities) (Figure 8.9).

Figure 8.9

Any point in the solution region with integer coordinates is a solution to the application. For example, $x = 50, y = 8$ is a solution, as is $x = 6, y = 40$.

b. The region is bordered on one side by the *line*

$$y = 71 - x$$

and on the other side by the *line*

$$y = \frac{1260 - 15x}{28}$$

A **corner** of the graph of the solution set occurs at the *point* where these two lines intersect. We find this point of intersection graphically as shown in Figure 8.9 or by solving the two equations simultaneously.

$$\begin{cases} y = 71 - x \\ y = \dfrac{1260 - 15x}{28} \end{cases}$$

Substitution gives

$$71 - x = \frac{1260 - 15x}{28}$$
$$1988 - 28x = 1260 - 15x$$
$$728 = 13x$$
$$x = 56$$
$$y = 71 - 56 = 15$$

Thus, this point of intersection of the borders of the solution region, or the *corner*, occurs at $x = 56, y = 15$. This corresponds to purchasing 56 compact cars and 15 midsized cars and is one of many possible solutions to the rental application.

Other corners occur where the region intersects the axes:

- At the origin ($x = 0, y = 0$, representing no cars purchased)

- Where $y = 71 - x$ intersects the x-axis ($x = 71, y = 0$, representing 71 compact and no midsize cars)

- Where $y = \dfrac{1260 - 15x}{28}$ intersects the y-axis ($x = 0, y = 45$, representing 0 compact and 45 midsize cars)

(a)

(b)

Figure 8.10

Technology Note

We can also use a graphing calculator to graph the solution to the system in Example 5. We enter the inequalities as equations in Y1 and Y2, with the icons to the left of Y1 and Y2 changed from \ to an icon denoting a shaded area below a line (see Figure 8.10(a)). The intersection of the two regions is the solution region, which is the area shaded by the intersection of blue lines and red lines in Figure 8.10(b). In Section 8.2, we will discuss how to find the corners of this region with a graphing calculator.

EXAMPLE 6 ▶   **Advertising**

A candidate for mayor of a city wishes to use a combination of radio and television advertisements in her campaign. Research has shown that each 1-minute spot on television reaches 0.09 million people and each 1-minute spot on radio reaches 0.006 million. The candidate believes that she must reach at least 2.16 million people, and she can buy a total of no more than 80 minutes of advertising time.

a. Write the inequalities that describe her advertising needs.

b. Graph the region determined by these constraint inequalities.

c. Interpret the solution region in the context of this problem.

SOLUTION

a. If we represent by x the number of minutes of television time, the total number of people reached by television is $0.09x$ million, and if we represent by y the number of minutes of radio time, the total number reached by radio is $0.006y$ million. Thus, one condition that must be satisfied is given by the inequality

$$0.09x + 0.006y \geq 2.16$$

In addition, the statement that the total number of minutes of advertising can be no more than 80 minutes can be written as the inequality

$$x + y \leq 80$$

Neither x nor y can be negative in this application, so we also have

$$x \geq 0$$
$$y \geq 0$$

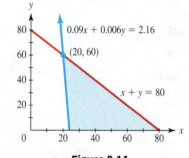

Figure 8.11

b. Graphing the equations corresponding to these inequalities and shading the regions satisfying the inequalities give the graph of the solution (Figure 8.11).

One corner of the region occurs where the boundary lines intersect. We can write the equations of the boundary lines as

$$0.09x + 0.006y = 2.16 \quad \text{and} \quad y = 80 - x$$

Their y-values are equal where the lines intersect, so we substitute for y in the first equation and solve.

$$0.09x + 0.006(80 - x) = 2.16$$
$$0.09x + 0.48 - 0.006x = 2.16$$
$$0.084x = 1.68$$
$$x = 20$$
$$y = 80 - 20 = 60$$

The other corners occur where the region intersects the axes:

- $0.09x + 0.006y = 2.16$ intersects the x-axis at $x = 24, y = 0$.

- $x + y = 80$ intersects the x-axis at $x = 80, y = 0$.

c. Each point in this region represents a combination of television and radio time that the candidate for mayor can use in her campaign. For example, she could use 20 minutes of television time and 60 minutes of radio time or 40 minutes of television time and 20 minutes of radio time. No point outside this region can represent a number of minutes of television and radio advertising for this candidate because of the constraints given in the problem.

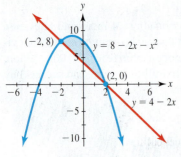

Figure 8.12

Systems of Nonlinear Inequalities

As with systems of linear inequalities in two variables, the solutions of systems of non-linear inequalities in two variables are regions in a plane. For example, the solution of the system

$$\begin{cases} y \le 8 - 2x - x^2 \\ y \ge 4 - 2x \end{cases}$$

is the area below the graph of $y = 8 - 2x - x^2$ and above the graph of $y = 4 - 2x$, as well as any points on the graphs of $y = 4 - 2x$ and $y = 8 - 2x - x^2$ between $(-2, 8)$ and $(2, 0)$ (Figure 8.12). The points of intersection of the two border curves are $(2, 0)$ and $(-2, 8)$.

EXAMPLE 7 ▶ Nonlinear System of Inequalities

Graphically solve the system

$$\begin{cases} x^2 - 10x + y \le 0 \\ -x^2 + 8x + y \ge 28 \end{cases}$$

SOLUTION

To solve the system

$$\begin{cases} x^2 - 10x + y \le 0 \\ -x^2 + 8x + y \ge 28 \end{cases}$$

graphically, we first solve the inequalities for y:

$$\begin{cases} y \le 10x - x^2 \\ y \ge x^2 - 8x + 28 \end{cases}$$

Graphing the associated equations on the same axes gives the graphs in Figure 8.13(a).

(a)   **(b)**

Figure 8.13

The points of intersection of the two border curves are $(2, 16)$ and $(7, 21)$. We can show this algebraically as follows:

$$x^2 - 8x + 28 = 10x - x^2$$
$$2x^2 - 18x + 28 = 0$$
$$2(x - 2)(x - 7) = 0$$
$$x = 2 \qquad \text{or} \quad x = 7$$
$$y = 16 \qquad\qquad y = 21$$

The region that we seek is the portion of the graph that is above the graph of $y = x^2 - 8x + 28$ and below the graph of $y = 10x - x^2$, as well as any points on the graphs of these two functions between $(2, 16)$ and $(7, 21)$, as shown in Figure 8.13(b).

EXAMPLE 8 ▶ **Nonlinear System of Inequalities**

Graphically solve the system

$$\begin{cases} y > x^3 - 3x \\ y < x \\ x > -2.5 \\ x < 2.5 \end{cases}$$

SOLUTION

To solve the system

$$\begin{cases} y > x^3 - 3x \\ y < x \\ x > -2.5 \\ x < 2.5 \end{cases}$$

we first graph the related equations (Figure 8.14(a)).

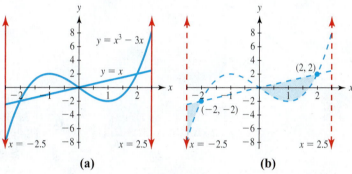

Figure 8.14

One region is bounded on the left by $x = -2.5$ and the two curves, with a corner at $(-2, -2)$. The second region is bounded by the two curves, with corners at $(0, 0)$ and $(2, 2)$. These corner points can also be found algebraically as follows.

$$x^3 - 3x = x$$
$$x^3 - 4x = 0$$
$$x(x - 2)(x + 2) = 0$$
$$x = 0 \quad \text{or} \quad x = 2 \quad \text{or} \quad x = -2$$
$$y = 0 \qquad\quad y = 2 \qquad\quad y = -2$$

The solution to the system of inequalities consists of the two regions that are above the graph of $y = x^3 - 3x$ and below the graph of $y = x$ in the interval $[-2.5, 2.5]$. This solution is shown in Figure 8.14(b).

SKILLS CHECK 8.1

In Exercises 1–6, graph each inequality.

1. $y \le 5x - 4$

2. $y > 3x + 2$

3. $6x - 3y \ge 12$

4. $\dfrac{x}{2} + \dfrac{y}{3} \le 6$

5. $4x + 5y \le 20$

6. $3x - 5y \ge 30$

In Exercises 7–10, match each solution of a system of inequalities with the correct graph.

7. $\begin{cases} y > 3x - 4 \\ y < 2x + 3 \end{cases}$

8. $\begin{cases} y \ge x + 1 \\ y \le 2x + 1 \\ x \ge 0 \\ y \ge 0 \end{cases}$

9. $\begin{cases} 3x + y \geq 5 \\ x + 3y \geq 6 \\ x \geq 0 \\ y \geq 0 \end{cases}$ **10.** $\begin{cases} x + 2y \leq 20 \\ 3x + 2y \leq 42 \\ x \geq 0 \\ y \geq 0 \end{cases}$ **13.** $\begin{cases} 4x + 2y > 8 \\ 3x + y > 5 \\ x \geq 0, y \geq 0 \end{cases}$

A.

B.

14. $\begin{cases} 2x + 4y \geq 12 \\ x + 3y \geq 8 \\ x \geq 0, y \geq 0 \end{cases}$

C.

D.

15. $\begin{cases} 2x + 6y \geq 12 \\ 3x + y \geq 5 \\ x + 2y \geq 5 \\ x \geq 0, y \geq 0 \end{cases}$

In Exercises 11–16, the graph of the boundary equations for each system of inequalities is shown with the system. Locate the solution region, and identify it by finding the corners.

11. $\begin{cases} x + y \leq 5 \\ 2x + y \leq 8 \\ x \geq 0, y \geq 0 \end{cases}$

16. $\begin{cases} 2x + y \geq 12 \\ x + y \leq 8 \\ 2x + y \leq 14 \\ x \geq 0, y \geq 0 \end{cases}$

12. $\begin{cases} 2x + y \leq 12 \\ x + 5y \leq 15 \\ x \geq 0, y \geq 0 \end{cases}$

For each system of inequalities in Exercises 17–22, graph the solution region and identify the corners of the region.

17. $\begin{cases} y \leq 8 - 3x \\ y \leq 2x + 3 \\ y > 3 \end{cases}$ **18.** $\begin{cases} 2x + y \geq 10 \\ 3x + 2y \geq 17 \\ x + 2y \geq 7 \end{cases}$

19. $\begin{cases} 2x + y < 5 \\ 2x - y > -1 \\ x \geq 0, y \geq 0 \end{cases}$ **20.** $\begin{cases} y < 2x \\ y > x + 2 \\ x \geq 0, y \geq 0 \end{cases}$

21. $\begin{cases} x + 2y \geq 4 \\ x + y \leq 5 \\ 2x + y \leq 8 \\ x \geq 0, y \geq 0 \end{cases}$ **22.** $\begin{cases} x + y < 4 \\ x + 2y < 6 \\ 2x + y < 7 \\ x \geq 0, y \geq 0 \end{cases}$

In Exercises 23–26, graph the solution of the system of nonlinear inequalities.

23. $\begin{cases} x^2 - 3y < 4 \\ 2x + 3y < 4 \end{cases}$ **24.** $\begin{cases} x + y \leq 8 \\ xy \geq 12 \end{cases}$

25. $\begin{cases} x^2 - y - 8x \leq -6 \\ y + 9x \leq 18 \end{cases}$ **26.** $\begin{cases} x^2 - y > 0 \\ y > (4 - 2x)^2 \end{cases}$

EXERCISES 8.1

27. *Manufacturing* A company manufactures two types of leaf blowers: an electric Turbo model and a gas-powered Tornado model. The company's production plan calls for the production of at least 780 blowers per month.

  a. Write the inequality that describes the production plan, if x represents the number of Turbo blowers and y represents the number of Tornado blowers.

  b. Graph the region determined by this inequality in the context of the application.

28. *Manufacturing* A company manufactures two types of leaf blowers: an electric Turbo model and a gas-powered Tornado model. It costs $78 to produce each of the x Turbo models and $117 to produce each of the y Tornado models, and the company has at most $76,050 per month to use for production.

  a. Write the inequality that describes this constraint on production.

  b. Graph the region determined by this inequality in the context of the application.

29. *Sales* Trix Auto Sales sells used cars. To promote its cars to target groups, it advertises with x 1-minute spots on cable television, at a cost of $240 per minute, and y 1-minute spots on radio, at a cost of $150 per minute. Suppose the company has at most $36,000 to spend on advertising.

  a. Write the inequality that describes this constraint on advertising.

  b. Graph the region determined by this inequality in the context of the application.

30. *Sales* Trix Auto Sales sells used cars. To promote its cars to target groups, it advertises with x 1-minute spots on cable television and y 1-minute spots on radio. Research shows that it sells one vehicle for every 4 minutes of cable television advertising and it sells one vehicle for each 10 minutes of radio advertising.

  a. Write the inequality that describes the requirement that at least 33 cars be sold from this advertising.

  b. Graph the region determined by this inequality in the context of the application.

31. *Politics* A political candidate wishes to use a combination of x television and y radio advertisements in his campaign. Each 1-minute ad on television reaches 0.12 million eligible voters, and each 1-minute ad on radio reaches 0.009 million eligible voters. The candidate feels that he must reach at least 7.56 million eligible voters and that he must buy at least 100 minutes of advertising.

  a. Write the inequalities that describe these advertising requirements.

  b. Graph the region determined by these inequalities in the context of the application.

32. *Housing* A contractor builds two models of homes: the Van Buren and the Jefferson. The Van Buren requires 200 worker-days of labor and $240,000 in capital, and the Jefferson requires 500 worker-days of labor and $300,000 in capital. The contractor has a total of 5000 worker-days and $3,600,000 in capital available per month. Let x represent the number of Van Buren models and y represent the number of Jefferson models, and graph the region that satisfies these inequalities.

33. *Manufacturing* A company manufactures two types of leaf blowers: an electric Turbo model and a gas-powered Tornado model. The company's production plan calls for the production of at least 780 blowers per month. It costs $78 to produce each Turbo model and $117 to manufacture each Tornado model, and the company has at most $76,050 per month to use for production. Let x represent the number of Turbo models and y represent the number of Tornado models, and graph the region that satisfies these constraints.

34. *Sales* Trix Auto Sales sells used cars. To promote its cars to target groups, it advertises with x 1-minute spots on cable television, at a cost of $240 per minute, and y 1-minute spots on radio, at a cost of $150 per minute. Research shows that it sells one vehicle for every 4 minutes of cable television advertising (that is, $\frac{1}{4}$ car per minute) and it sells one vehicle for each 10 minutes of radio advertising. Suppose the company has at most $36,000 to spend on advertising and sells at least 33 cars per month from this advertising. Let x represent the number of cable TV minutes and y represent the number of radio minutes, and graph the region that satisfies these constraints. That is, graph the solution to this system of inequalities.

35. *Production* A firm produces three different-size television sets on two assembly lines. The following table summarizes the production capacity of each assembly line and the minimum number of each size TV needed to fill orders.

    a. Write the inequalities that are the constraints.

    b. Graph the region that satisfies the inequalities, and identify the corners.

	Assembly Line 1	Assembly Line 2	Number Ordered
19-in. TV	80 per day	40 per day	3200
25-in. TV	20 per day	20 per day	1000
35-in. TV	100 per day	40 per day	3400

36. *Advertising* Tire Town is developing an advertising campaign. The following table indicates the cost per ad package in newspapers and the cost per ad package on radio, as well as the number of ads in each type of ad package.

	Newspaper	Radio
Cost per Ad Package ($)	1000	3000
Ads per Package	18	36

The owner of the company can spend no more than $18,000 per month, and he wants to have at least 252 ads per month.

    a. Write the inequalities that are the constraints.

    b. Graph the region that satisfies the inequalities, and identify the corners.

37. *Nutrition* A privately owned lake contains two types of fish: bass and trout. The owner provides two types of food, A and B, for these fish. Trout require 4 units of food A and 6 units of food B, and bass require 10 units of food A and 7 units of food B. If the owner has 1600 units of each food, graph the region that satisfies the constraints.

38. *Manufacturing* Evergreen Company produces two types of printers: the inkjet and the laserjet. It takes 2 hours to make the inkjet and 6 hours to make the laserjet. The company can make at most 120 printers per day and has 400 labor-hours available per day.

    a. Write the inequalities that describe this application.

    b. Graph the region that satisfies the inequalities, and identify the corners of the region.

39. *Manufacturing* Easyboy manufactures two types of chairs: Standard and Deluxe. Each Standard chair requires 4 hours to construct and finish, and each Deluxe chair requires 6 hours to construct and finish. Upholstering takes 2 hours for a Standard chair and 6 hours for a Deluxe chair. There are 480 hours available each day for construction and finishing, and there are 300 hours available per day for upholstering.

    a. Write the inequalities that describe the application.

    b. Graph the solution of the system of inequalities, and identify the corners of the region.

40. *Manufacturing* Safeco Company produces two types of chainsaws: the Safecut and the Safecut Deluxe. The Safecut model requires 2 hours to assemble and 1 hour to paint, and the Deluxe model requires 3 hours to assemble and $\frac{1}{2}$ hour to paint. The daily maximum number of hours available for assembly is 36, and the daily maximum number of hours available for painting is 12.

    a. Write the inequalities that describe the application.

    b. Graph the solution of the system of inequalities, and identify the corners of the region.

41. *Manufacturing* A company manufactures commercial and domestic heating systems at two plant sites. It can produce no more than 1400 units per month, and it needs to fill orders of at least 500 commercial units and 750 domestic units.

    a. Write the system of inequalities that describes the constraints on production for these orders.

    b. Graph the solution set of this system of inequalities.

42. *Constructing a Box* An open box with a square base is to be constructed so that the area of the material used is at most 500 cm². What dimensions of the base will yield a box that gives a volume of at least 500 cm³?

Linear Programming: Graphical Methods

SECTION PREVIEW Maximizing Profit

A rental agency has a maximum of $1,260,000 to invest in the purchase of at most 71 new cars of two different types, compact and midsize. The cost per compact car is $15,000, and the cost per midsize car is $28,000. The number of cars of each type is limited (constrained) by the budget available and the number of cars needed. We saw in the previous section that purchases of the rental agency were constrained by the number of cars needed and the amount of money available for purchasing the cars. The constraints were given by a system of inequalities, and we found the solution region determined by the constraints.

If the average yearly profit is $6000 for each compact car and $9000 for each midsize car, buying a different number of each type will affect the profit. How do we find the correct number of each type so that the profit is maximized under these conditions? To answer questions of this type without knowledge of special techniques, people must use the "guess-and-check" method repeatedly to arrive at a solution. And when the number of conditions that enter into a problem of this type increases, guessing is a very unsatisfactory solution method. We can answer these questions with a mathematical technique called **linear programming**. Linear programming is a valuable tool in management; in fact, 85% of all Fortune 500 firms use linear programming. In this section, we use graphical methods to find the **optimal solutions** to linear programming problems such as the one above. ∎

Linear Programming

In any linear programming problem, we seek the point or points in the region determined by the constraint inequalities that give optimal (*maximum* or *minimum*) values for the **objective function** (the function we seek to optimize).* Any point in the region determined by the constraints is called a **feasible solution** to the linear programming problem, and the region itself is called the **feasible region**.

Note that the constraints in our car rental application above can be written as the system of inequalities

$$\begin{cases} x + y \le 71 \\ 15x + 28y \le 1260 \\ x \ge 0 \\ y \ge 0 \end{cases}$$

The solution set of the inequalities for the problem, the feasible region, is shown as the shaded region in Figure 8.15. Any point inside the shaded region or on its boundary satisfies the constraints, so it is a feasible solution. In the context of this application, only feasible solutions with integer coordinates can be solutions to the problem.

Figure 8.15

---

* The region determined by the constraints must be convex for the optimal solution to exist. A convex region is one such that for any two points in the region, the segment joining those points lies entirely within the region. We restrict our discussion to convex regions.

We will see that it is important to find the corners of feasible regions.

> ### Technology Note
>
> We can use graphing calculators to graph the feasible region and then use the (TRACE), zero (accessed by **2ND** (TRACE)), and intersect (accessed by **2ND** (TRACE)) commands to find the corners. Figure 8.16 shows the calculator graph of the feasible region shown in Figure 8.15, with the corners (0, 45), (71, 0), and (56, 15) identified.

Figure 8.16

Because the profit P is 6 thousand dollars for each of the x compact cars and 9 thousand dollars for each of the y midsize cars, we can model the profit function as

$$P = 6x + 9y, \quad \text{where } P \text{ is in thousands of dollars}$$

In this application, we seek to find the maximum profit subject to the constraints listed above. Because we seek to maximize the profit subject to the constraints, we evaluate the objective function

$$P = 6x + 9y$$

for points (x, y) in the feasible region. For example, at (20, 19), $P = 291$, at (50, 10), $P = 390$, and at (50, 15), $P = 435$. But there are a large number of points to test, so we look at a different way to find the maximum value of P. We can graph $P = 6x + 9y$ for different values of P by first writing the equation in the form

$$y = -\frac{2}{3}x + \frac{P}{9}$$

Then, for each possible value of P, the graph is a line with slope $-\frac{2}{3}$ and y-intercept $\frac{P}{9}$. Figure 8.17 shows the graphs of the feasible region and objective functions for $P = 360, 471,$ and 495. Note that the graphs are parallel $\left(\text{with slope } -\frac{2}{3}\right)$, and that larger values of P coincide with larger y-intercepts $\left(\frac{P}{9}\right)$. Observe that if $P = 471$, the graph of $y = -\frac{2}{3}x + \frac{P}{9}$ intersects the feasible region and that values of P less than 471 will give lines that pass through the feasible region, but each of these values of P

Figure 8.17

represents a smaller profit. Note also that the value $P = 495$ (or any P-value greater than 471) results in a line that "misses" the feasible region, so this value of P is not a solution to the problem. Thus, the maximum value for P, subject to the constraints, is 471 (thousand dollars). We can verify that the graph of $471 = 6x + 9y$ intersects the region at $(56, 15)$ and that $P = 6x + 9y$ will not intersect the feasible region if $P > 471$. Thus, the profit is maximized at \$471,000, when $x = 56$ and $y = 15$—that is, when 56 compact and 15 midsize cars are purchased.

Note that the values of x and y that correspond to the maximum value of P occur at $(56, 15)$, which is a corner of the feasible region. In fact, the maximum value of an objective function always occurs at a corner of the feasible region if the feasible region is closed and bounded. (A feasible region is closed and bounded if it is enclosed by and includes the lines associated with the constraints.) This gives us the basis for finding the solutions to linear programming problems using the graphical method.

Solutions to Linear Programming Problems

1. If a linear programming problem has an optimal solution, then the optimal value (maximum or minimum) of an objective function occurs at a corner of the feasible region determined by the constraints.

2. When a feasible region for a linear programming problem is closed and bounded, the objective function has a maximum and a minimum value.

3. If the objective function has the same optimal value at two corners, then it also has that optimal value at any point on the boundary line segment connecting the two corners.

4. When the feasible region is not closed and bounded, the objective function may have a maximum only, a minimum only, or neither.

Thus, for a closed and bounded region, we can find the maximum and minimum values of an objective function by evaluating the function at each of the corners of the feasible region formed by the graphical solution of the constraint inequalities.

EXAMPLE 1 ▶ Finding the Optimal Values of a Function Subject to Constraints

Find (a) the maximum value and (b) the minimum value of $C = 5x + 3y$ subject to the constraints

$$\begin{cases} x + y \leq 5 \\ 2x + y \leq 8 \\ x \geq 0, y \geq 0 \end{cases}$$

SOLUTION

The feasible region is in the first quadrant because it is bounded by $x \geq 0$ and $y \geq 0$. We can graph the remaining boundary lines with technology if we solve the first two inequalities for y.

$$\begin{cases} y \leq 5 - x \\ y \leq -2x + 8 \\ x \geq 0, y \geq 0 \end{cases}$$

The solution set for the system of inequalities (that is, the feasible region) is bounded by the lines $y = 5 - x$ and $y = -2x + 8$ and the x- and y-axes. The corners of the feasible region occur at the origin $(0, 0)$, at the x-intercept of one of the two boundary lines, at the y-intercept of one of the two boundary lines, and at the point of intersection of the lines $y = 5 - x$ and $y = -2x + 8$ (Figure 8.18).

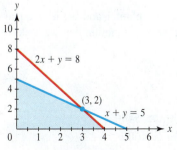

Figure 8.18

There are two y-intercept points, $(0, 5)$ and $(0, 8)$, and the one that satisfies both inequalities is $(0, 5)$. The x-intercept points are $(4, 0)$ and $(5, 0)$; the one that satisfies both inequalities is $(4, 0)$. We can find the point of intersection of the two boundary lines by solving the system of equations

$$\begin{cases} y = 5 - x \\ y = -2x + 8 \end{cases}$$

Substitution gives

$$5 - x = -2x + 8$$

Solving gives $x = 3$, and substituting gives $y = 2$. Thus, the point of intersection of these two boundary lines is $(3, 2)$. This point of intersection can also be found by using a graphing utility.

a. Any point inside the shaded region or on the boundary is a feasible solution to the problem, but the maximum possible value of

$$C = 5x + 3y$$

occurs at a corner of the feasible region. We evaluate the objective function at each of the corner points:

$$\begin{aligned} \text{At } (0, 0): \quad & C = 5x + 3y = 5(0) + 3(0) = 0 \\ \text{At } (0, 5): \quad & C = 5x + 3y = 5(0) + 3(5) = 15 \\ \text{At } (4, 0): \quad & C = 5x + 3y = 5(4) + 3(0) = 20 \\ \text{At } (3, 2): \quad & C = 5x + 3y = 5(3) + 3(2) = 21 \end{aligned}$$

Thus, the maximum value of the objective function is 21 when $x = 3$, $y = 2$.

b. Observing the values of $C = 5x + 3y$ at each corner point of the feasible region, we see that the minimum value of the objective function is 0 at $(0, 0)$.

EXAMPLE 2 ▶   **Cost Minimization**

Star Manufacturing Company produces two types of blu-ray players, which are assembled at two different locations. Plant 1 can assemble 60 units of the Star model and 80 units of the Prostar model per hour, and plant 2 can assemble 300 units of the Star model and 80 units of the Prostar model per hour. The company needs to produce at least 5400 units of the Star model and 4000 units of the Prostar model to fill an order. If it costs \$2000 per hour to run plant 1 and \$3000 per hour to run plant 2, how many hours should each plant spend on manufacturing blu-ray players to minimize its cost for this order? What is the minimum cost for this order?

SOLUTION

If we let x equal the number of hours of assembly time at plant 1 and we let y equal the number of hours of assembly time at plant 2, then the cost function that we seek to minimize is

$$C = 2000x + 3000y$$

The constraints on the assembly hours follow:

Star model units: $60x + 300y \geq 5400$

Prostar model units: $80x + 80y \geq 4000$

Nonnegative hours: $x \geq 0, y \geq 0$

Figure 8.19

We can solve the first two inequalities for y, write these constraints as a system of inequalities, and graph the feasible region (Figure 8.19):

$$\begin{cases} y \geq 18 - 0.2x \\ y \geq 50 - x \\ x \geq 0, y \geq 0 \end{cases}$$

The two boundary lines intersect where $18 - 0.2x = 50 - x$, at $x = 40$, which then gives $y = 10$. The boundary lines of the feasible region intersect the axes at $(0, 50)$ and at $(90, 0)$.

Thus, the corners of the feasible region are

- $(0, 50)$, the y-intercept that satisfies all the inequalities.

- $(90, 0)$, the x-intercept that satisfies all the inequalities.

- $(40, 10)$, the intersection of the two boundary lines.

To find the hours of assembly time at each plant that will minimize the cost, we test the corner points:

$$\begin{aligned} \text{At } (0, 50): & \quad C = 2000(0) + 3000(50) = 150,000 \\ \text{At } (90, 0): & \quad C = 2000(90) + 3000(0) = 180,000 \\ \text{At } (40, 10): & \quad C = 2000(40) + 3000(10) = 110,000 \end{aligned}$$

Thus, the cost is minimized when assembly time is 40 hours at plant 1 and 10 hours at plant 2. The minimum cost is \$110,000.

Note that the feasible region in Example 2 is not closed and bounded, and that there is no maximum value for the objective function even though there is a minimum value. It should also be noted that some applied linear programming problems require discrete solutions; for example, the optimal solution to the earlier car rental problem required a number of cars, so only nonnegative integer solutions were possible.

EXAMPLE 3 ▶   **Maximizing Profit**

Smoker Meat Packing Company makes two different types of hot dogs: regular and all-beef. Each pound of all-beef hot dogs requires 0.8 pound of beef and 0.2 pound of spices, and each pound of regular hot dogs requires 0.3 pound of beef and 0.2 pound of spices, with the remainder nonbeef meat products. The company has at most 1020 pounds of beef and has at most 500 pounds of spices available for hot dogs. If the profit is \$0.90 on each pound of all-beef hot dogs and \$1.20 on each pound of regular hot dogs, how many of each type should be produced to maximize the profit?

SOLUTION

Letting x equal the number of pounds of all-beef hot dogs and y equal the number of pounds of regular hot dogs, we seek to maximize the function $P = 0.90x + 1.20y$ subject to the constraints

$$\begin{cases} 0.8x + 0.3y \leq 1020 \\ 0.2x + 0.2y \leq 500 \\ x \geq 0, y \geq 0 \end{cases}$$

The graph of this system of inequalities is more easily found if the first two inequalities are solved for y.

$$\begin{cases} y \leq 3400 - \dfrac{8}{3}x \\ y \leq 2500 - x \\ x \geq 0, y \geq 0 \end{cases}$$

Graphing the boundary lines and testing the four regions determines the feasible region (Figure 8.20).

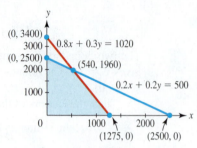

Figure 8.20

The boundary lines intersect where

$$3400 - \frac{8}{3}x = 2500 - x$$

or at $x = 540$. Substituting for x gives $y = 1960$.

The x-intercept $(1275, 0)$ and the y-intercept $(0, 2500)$ satisfy the inequalities. Thus, the corners of this region are $(0, 0)$, $(0, 2500)$, $(1275, 0)$, and $(540, 1960)$. $P = 0$ at $(0, 0)$. Testing the objective function at the other corner points gives

At $(0, 2500)$: $P = 0.90x + 1.20y = 0.90(0) + 1.20(2500) = 3000$

At $(1275, 0)$: $P = 0.90x + 1.20y = 0.90(1275) + 1.20(0) = 1147.50$

At $(540, 1960)$: $P = 0.90x + 1.20y = 0.90(540) + 1.20(1960) = 2838$

This indicates that profit will be maximized if 2500 pounds of regular hot dogs and no all-beef hot dogs are produced.

EXAMPLE 4 ▶   **Manufacturing**

A company manufactures air conditioning units and heat pumps at its factories in Atlanta, Georgia, and Newark, New Jersey. The Atlanta plant can produce no more than 1000 items per day, and the number of air conditioning units cannot exceed 100 more than half the number of heat pumps. The Newark plant can produce no more than 850 units per day. The profit on each air conditioning unit is $400 at the Atlanta plant and $390 at the Newark plant. The profit on each heat pump is $200 at the Atlanta plant and $215 at the Newark plant. Suppose there is an order for 500 air conditioning units and 750 heat pumps.

a. Graph the feasible region and identify the corners.

b. Find the maximum profit that can be made on this order and what production distribution will give the maximum profit.

SOLUTION

a. Because the total number of air conditioning units needed is 500, we can represent the number produced at Atlanta by x and the number produced at Newark by $(500 - x)$. Similarly, we can represent the number of heat pumps produced at Atlanta by y and the number produced at Newark by $(750 - y)$. Table 8.1 summarizes the constraints.

Table 8.1

	Air Conditioning Units	Heat Pumps	Total
Atlanta	x	$+ y$	≤ 1000
Newark	$500 - x$	$+ (750 - y)$	≤ 850
Other Constraints	x		$\leq 0.5y + 100$
	x		≥ 0
		y	≥ 0
Profit	$400x + 390(500 - x)$	$+ 200y + 215(750 - y)$	

Using this information gives the following system:

$$\begin{cases} x + y \leq 1000 & (1) \\ (500 - x) + (750 - y) \leq 850 & (2) \\ x \leq 0.5y + 100 & (3) \\ x \geq 0, y \geq 0 \end{cases}$$

We solve each of these inequalities for y and graph the solution set (Figure 8.21):

$$\begin{cases} y \leq 1000 - x & (1) \\ y \geq 400 - x & (2) \\ y \geq 2x - 200 & (3) \\ x \geq 0, y \geq 0 \end{cases}$$

Figure 8.21

The intersections of the boundary lines associated with the inequalities follow.

• Inequalities (1) and (3): where $1000 - x = 2x - 200$, at $x = 400$, $y = 600$.

• Inequalities (2) and (3): where $400 - x = 2x - 200$, at $x = 200$, $y = 200$.

• Inequalities (1) and (2): where $1000 - x = 400 - x$, which has no solution, so they do not intersect.

The other corners are on the y-axis, at $(0, 400)$ and $(0, 1000)$.

b. The objective function gives the profit for the products.

$$P = 400x + 390(500 - x) + 200y + 215(750 - y)$$
$$= 10x - 15y + 356{,}250$$

Testing the profit function at the corners of the feasible region determines where the profit is maximized.

At $(0, 400)$: $P = 350{,}250$

At $(200, 200)$: $P = 355{,}250$

At $(0, 1000)$: $P = 341{,}250$

At $(400, 600)$: $P = 351{,}250$

Thus, the profit is maximized at \$355,250 when 200 air conditioning units and 200 heat pumps are manufactured in Atlanta. The remainder are manufactured in Newark, so 300 air conditioning units and 550 heat pumps are manufactured in Newark.

Solution with Technology

As we have seen, graphing utilities can be used to graph the feasible region and to find the corners of the region satisfying the constraint inequalities. The graphical method that we have been using cannot be used if the problem involves more than two variables, and most real linear programming applications involve more than two variables. In fact, some real applications involve as many as 40 variables. Problems involving more than two variables can be solved with spreadsheet programs such as Lotus 1-2-3, Microsoft Excel, and Quatro Pro. Use of these programs to solve linear programming problems is beyond the scope of this text.

SKILLS CHECK 8.2

In Exercises 1–4, use the given feasible region determined by the constraint inequalities to find the maximum possible value and the minimum possible value of the objective function.

1. $f = 4x + 9y$ subject to the constraints

$$\begin{cases} 2x + 3y \le 134 \\ x + 5y \le 200 \\ x \ge 0, y \ge 0 \end{cases}$$

2. $f = 2x + 3y$ subject to the constraints

$$\begin{cases} 6x + 5y \le 170 \\ x + 5y \le 100 \\ x \ge 0, y \ge 0 \end{cases}$$

3. $f = 4x + 2y$ subject to the constraints

$$\begin{cases} 3x + y \le 15 \\ x + 2y \le 10 \\ -x + y \le 2 \\ x \ge 0, y \ge 0 \end{cases}$$

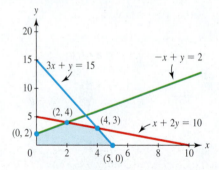

4. $f = 3x + 9y$ subject to the constraints

$$\begin{cases} 2x + y \le 12 \\ x + 3y \le 15 \\ x + y \le 7 \\ x \ge 0, y \ge 0 \end{cases}$$

5. Perform the following steps to maximize $f = 3x + 5y$ subject to the constraints

$$\begin{cases} y \le \dfrac{54 - 2x}{3} \\ y \le 22 - x \\ x \ge 0, y \ge 0 \end{cases}$$

a. Graph the region that satisfies the system of inequalities, and identify the corners of the region.

b. Test the objective function $f = 3x + 5y$ at each of the corners of the feasible region to determine which corner gives the maximum value. Give the maximum possible value of f and the values of x and y that give the value.

6. Perform the following steps to maximize $f = 3x + 4y$ subject to the constraints

$$\begin{cases} x + 2y \le 16 \\ x + y \le 10 \\ x \ge 0, y \ge 0 \end{cases}$$

a. Graph the region that satisfies the system of inequalities, and identify the corners of the region.

b. Test the objective function $f = 3x + 4y$ at each of the corners of the feasible region to determine which corner gives the maximum value. Give the maximum possible value of f and the values of x and y that give that value.

7. Perform the following steps to minimize $g = 4x + 2y$ subject to the constraints

$$\begin{cases} x + 2y \ge 15 \\ x + y \ge 10 \\ x \ge 0, y \ge 0 \end{cases}$$

a. Graph the region that satisfies the system of inequalities, and identify the corners of the region.

b. Test the objective function $g = 4x + 2y$ at each of the corners of the feasible region to determine which corner gives the minimum value. Give the minimum possible value of g and the values of x and y that give that value.

8. Find the maximum possible value of $f = 400x + 300y$ and the values of x and y that give that value, subject to the constraints

$$\begin{cases} 3x + 2y \ge 6 \\ 2x + y \le 7 \\ x \ge 0, y \ge 0 \end{cases}$$

9. Find the maximum possible value of $f = 20x + 30y$ and the values of x and y that give that value, subject to the constraints

$$\begin{cases} y \le -\dfrac{1}{2}x + 4 \\ y \le -x + 6 \\ y \le -\dfrac{1}{3}x + 4 \\ x \ge 0, y \ge 0 \end{cases}$$

10. Find the maximum possible value of $f = 100x + 100y$ and the values of x and y that give that value, subject to the constraints

$$\begin{cases} x + 2y \le 6 \\ x + 4y \le 10 \\ 2x + y \le 9 \\ x \ge 0, y \ge 0 \end{cases}$$

11. Find the maximum possible value of $f = 80x + 160y$ and the values of x and y that give that value, subject to the constraints

$$\begin{cases} 3x + 4y \le 32 \\ x + 2y \le 15 \\ 2x + y \le 18 \\ x \ge 0, y \ge 0 \end{cases}$$

12. Find the minimum possible value of $g = 30x + 40y$ and the values of x and y that give that value, subject to the constraints

$$\begin{cases} x + 2y \ge 16 \\ x + y \ge 10 \\ x \ge 0, y \ge 0 \end{cases}$$

13. Find the minimum possible value of $g = 40x + 30y$ and the values of x and y that give that value, subject to the constraints

$$\begin{cases} 2x + y \ge 6 \\ 4x + y \ge 8 \\ x + 2y \ge 6 \\ x \ge 0, y \ge 0 \end{cases}$$

14. Find the minimum possible value of $g = 30x + 40y$ and the values of x and y that give that value, subject to the constraints

$$\begin{cases} x + 2y \ge 5 \\ x + y \ge 4 \\ 2x + y \ge 6 \\ x \ge 0, y \ge 0 \end{cases}$$

15. Find the minimum possible value of $g = 46x + 23y$ and the values of x and y that give that value, subject to the constraints

$$\begin{cases} 3x + y \ge 6 \\ x + y \ge 4 \\ x + 5y \le 8 \\ x \ge 0, y \ge 0 \end{cases}$$

16. Consider the constraints

$$\begin{cases} 2x + y \le 9 \\ 3x + y \ge 11 \\ x + y \ge 5 \\ x \ge 0, y \ge 0 \end{cases}$$

a. Find the maximum possible value of $f = 4x + 5y$ and the values of x and y that give that value.

b. Find the minimum possible value of $g = 3x + 2y$ and the values of x and y that give that value.

17. Find the minimum possible value of $g = 60x + 10y$ and the values of x and y that give that value, subject to the constraints

$$\begin{cases} 3x + 2y \ge 12 \\ 2x + y \ge 7 \\ x \ge 0, y \ge 0 \end{cases}$$

18. Find the maximum possible value of $f = 10x + 10y$ and the values of x and y that give that value, subject to the constraints

$$\begin{cases} x + 2y \le 10 \\ x + y \le 6 \\ 2x + y \le 10 \\ x \ge 0, y \ge 0 \end{cases}$$

EXERCISES 8.2

19. *Manufacturing* A company manufactures two types of leaf blowers: an electric Turbo model and a gas-powered Tornado model. The company's production plan calls for the production of at least 780 blowers per month. It costs $78 to produce each Turbo model and $117 to manufacture each Tornado model, and the company has at most $76,050 per month to use for production. Find the number of units that should be produced to maximize profit for the company, and the maximum profit, if the profit on each Turbo model is $32 and the profit on each Tornado model is $45. (See Exercise 27 in Section 8.1.)

20. *Manufacturing* Evergreen Company produces two types of printers: the Inkjet and the Laserjet. The company can make at most 120 printers per day and has 400 labor-hours available per day. It takes 2 hours to make the Inkjet and 6 hours to make the Laserjet. If the profit on the Inkjet is $80 and the profit on the Laserjet is $120, find the maximum possible daily profit and the number of each type of printer that gives it. (See Exercise 38 in Section 8.1.)

21. *Manufacturing* Safeco Company produces two types of chainsaws: the Safecut and the Safecut Deluxe. The Safecut model requires 2 hours to assemble and 1 hour to paint, and the Deluxe model requires 3 hours to assemble and $\frac{1}{2}$ hour to paint. The daily maximum number of hours available for assembly is 36, and the daily maximum number of hours available for painting is 12. If the profit is $24 per unit

on the Safecut model and $30 per unit on the Deluxe model, how many units of each type will maximize the daily profit and what will that profit be? (See Exercise 40 in Section 8.1.)

22. *Production* Two models of riding mowers, the Lawn King and the Lawn Master, are produced on two assembly lines. Producing the Lawn King requires 2 hours on line I and 1 hour on line II. Producing the Lawn Master requires 1 hour on line I and 3 hours on line II. The number of hours available for production is limited to 60 on line I and 40 on line II. If there is $150 profit per mower on the Lawn King and $200 profit per mower on the Lawn Master, producing how many of each model maximizes the profit? What is the maximum possible profit?

23. *Sales* Trix Auto Sales sells used cars. To promote its cars to target groups, it advertises with x 1-minute spots per month on cable television, at a cost of $240 per minute, and y 1-minute spots per month on radio, at a cost of $150 per minute. Research shows that it sells one vehicle for every 4 minutes of cable television advertising (that is, $\frac{1}{4}$ car per minute) and it sells one vehicle for each 10 minutes of radio advertising. Suppose the company has at most $36,000 per month to spend on advertising and sells at least 33 cars per month from this advertising. Graph the region that satisfies these two inequalities. That is, graph the solution to this system of inequalities. (See Exercise 34 in Section 8.1.)

a. If the profit on cars advertised on television averages $500 and the profit on cars advertised on radio averages $550, how many minutes per month of advertising should be spent on television advertising and how many should be spent on radio advertising?

b. What is the maximum possible profit?

24. *Advertising* Tire Town is developing an advertising campaign. The table below indicates the cost per ad package in newspapers and the cost per ad package on radio, as well as the number of ads in each type of ad package.

	Newspaper	Radio
Cost per Ad Package ($)	1000	3000
Ads per Package	18	36

   The owner of the company can spend no more than $18,000 per month, and he wants to have at least 252 ads per month. If each newspaper ad package reaches 6000 people over 20 years of age and each radio ad reaches 8000 of these people, how many newspaper ad packages and radio ad packages should he buy to maximize the number of people over age 20 reached? How many people will these ads reach? (See Exercise 36 in Section 8.1.)

25. *Production* A firm produces three different-size television sets on two assembly lines. The table below summarizes the production capacity of each assembly line, the number of each size TV ordered by a retailer, and the daily operating costs for each assembly line. How many days should each assembly line run to fill this order with minimum cost? What is the minimum cost? (See Exercise 35 in Section 8.1.)

	Assembly Line 1	Assembly Line 2	Number Ordered
19-in. TV	80 per day	40 per day	3200
27-in. TV	20 per day	20 per day	1000
35-in. TV	100 per day	40 per day	3400
Daily Cost ($)	20,000	40,000	

26. *Nutrition* A privately owned lake contains two types of fish: bass and trout. The owner provides two types of food, A and B, for these fish. Trout require 4 units of food A and 5 units of food B, and bass require 10 units of food A and 4 units of food B. If the owner has 1600 units of food A and 1000 units of food B, find the maximum number of fish that the lake can support.

27. *Housing* A contractor builds two models of homes: the Van Buren and the Jefferson. The Van Buren requires 200 worker-days of labor and $240,000 in capital, and the Jefferson requires 500 worker-days of labor and $300,000 in capital. The contractor has a total of 5000 worker-days and $3,600,000 in capital available per month. The profit is $60,000 on the Van Buren and $75,000 on the Jefferson. Building how many of each model will maximize the monthly profit? What is the maximum possible profit? (See Exercise 32 in Section 8.1.)

28. *Manufacturing* Easyboy manufactures two types of chairs: Standard and Deluxe. Each Standard chair requires 4 hours to construct and finish, and each Deluxe chair requires 6 hours to construct and finish. Upholstering takes 2 hours for a Standard chair and 6 hours for a Deluxe chair. There are 480 hours available each day for construction and finishing, and there are 300 hours available per day for upholstering. Suppose the revenue is $178 for each Standard chair and $267 for each Deluxe chair. (See Exercise 39 in Section 8.1.)

a. What is the maximum possible daily revenue?

b. How many of each type should be produced each day to maximize the daily revenue?

29. *Production* Ace Company produces three models of DVRs, the Ace, the Ace Plus, and the Ace Deluxe, at two facilities, A and B. The company has orders for at least 4000 of the Ace, at least 1800 of the Ace Plus, and at least 2400 of the Ace Deluxe. The weekly production capacity of each model at each facility and the cost per week to operate each facility are given in the table that follows. For how many weeks should each facility operate to minimize the cost of filling the orders? What is the minimum cost?

	A	B
Ace	400	400
Ace Plus	300	100
Ace Deluxe	200	400
Cost per Week ($)	15,000	20,000

30. *Politics* A political candidate wishes to use a combination of television and radio advertisements in his campaign. Each 1-minute ad on television reaches 0.12 million eligible voters and each 1-minute ad on radio reaches 0.009 million eligible voters. The candidate believes that he must reach at least 7.56 million eligible voters and that he must buy at least 100 minutes of advertisements. If television ads cost $1000 per minute and radio ads cost $200 per minute, how many minutes of radio and television advertising does he need to minimize costs? What is the minimum cost? (See Exercise 31 in Section 8.1.)

31. *Nutrition* In a hospital ward, patients are grouped into two general nutritional categories depending on the amount of solid food in their diet, and are provided two different diets with different amounts of solid foods and detrimental substances. The following table gives the patient groups, the weekly diet requirements for each group, and the amount of detrimental substance in each diet. How many servings of each diet will satisfy the nutritional requirements and minimize detrimental substances? What is the minimum amount of detrimental substance?

	Diet A (oz/serving)	Diet B (oz/serving)	Minimum Daily Requirements (oz)
Group 1	2	1	18
Group 2	4	1	26
Detrimental Substance	0.09	0.035	

32. *Manufacturing* Kitchen Pride manufactures toasters and can openers, which are assembled on two different lines. Line 1 can assemble 15 toasters and 20 can openers per hour, and line 2 can assemble 75 toasters and 20 can openers per hour. The company needs to produce at least 540 toasters and 400 can openers. If it costs $300 per hour to run line 1 and $600 per hour to run line 2, how many hours should each line be run to fill all the orders at the minimum cost? What is the minimum cost?

8.3 Sequences and Discrete Functions

KEY OBJECTIVES

- Find terms of sequences
- Find terms of arithmetic sequences
- Find terms of geometric sequences

SECTION PREVIEW **Football Contracts**

Suppose a football player is offered a chance to sign a contract for 18 games with one of the following salary plans:

Plan A: $10,000 for the first game with a $10,000 raise for each game thereafter

Plan B: $2 for the first game with his salary doubled for each game thereafter

Which salary plan should he accept if he wants to make the most money for his last game?

The payments for the games form the sequences

Plan A: 10,000, 20,000, 30,000, . . .

Plan B: 2, 4, 8, 16, . . .

In this section, we answer this question after we have developed formulas that apply to sequences, and we use sequences to solve applied problems. ■

Sequences

In Chapter 5 we found the future values of investments at simple and compound interest. Consider a $5000 investment that pays 1% simple interest for each of 6 months, shown in Table 8.2.

These future values are outputs that result when the inputs are positive integers that correspond to the number of months of the investment. Outputs (such as these future values) that result uniquely from the positive integer inputs define a function whose domain is a set of positive integers. Such a function is called a **sequence**, and the ordered outputs corresponding to the integer inputs are called the **terms of the sequence**.

Table 8.2

Month	Interest, $I = Prt$ ($)	Future Value of the Investment ($)
1	$(5000)(0.01)(1) = 50$	$5000 + 50 = 5050$
2	$(5000)(0.01)(1) = 50$	$5050 + 50 = 5100$
3	$(5000)(0.01)(1) = 50$	$5100 + 50 = 5150$
4	$(5000)(0.01)(1) = 50$	$5150 + 50 = 5200$
5	$(5000)(0.01)(1) = 50$	$5200 + 50 = 5250$
6	$(5000)(0.01)(1) = 50$	$5250 + 50 = 5300$

Sequences have the same properties as other functions that we have studied, *except that the domains of sequences are positive integers, and so sequences are* **discrete functions**. Rather than denoting a functional output with y, we denote the output of the sequence f that corresponds to input n with $f(n) = a_n$, where n is a positive integer.

> ### Sequence
>
> The functional values $a_1, a_2, a_3, \ldots$ of a sequence are called the terms of the sequence, with a_1 the first term, a_2 the second term, and so on. The general term (or nth term) is denoted by a_n.

If the domain of a sequence is the set of all positive integers, the outputs form an **infinite sequence**. If the domain is the set of positive integers from 1 to n, the outputs form a **finite sequence**. Sequences are important because they permit us to apply discrete functions to real problems instead of having to approximate them with continuous functions.

By looking at the future values in Table 8.2, we can see that the future value of $5000 at the end of each month is given by

$$a_1 = f(1) = 5000 + 50(1) = 5050$$
$$a_2 = f(2) = 5000 + 50(2) = 5100$$
$$a_3 = f(3) = 5000 + 50(3) = 5150$$
$$a_4 = f(4) = 5000 + 50(4) = 5200$$
$$a_5 = f(5) = 5000 + 50(5) = 5250$$
$$a_6 = f(6) = 5000 + 50(6) = 5300$$

These six terms can be written in the form 5050, 5100, 5150, 5200, 5250, 5300. By observing this pattern, we write the general term as $a_n = 5000 + 50n$, which represents

$$f(n) = 5000 + 50n$$

As with other functions, we can represent sequences in tables and on graphs. Figure 8.22 shows the graph of the function for these values of n.

Figure 8.22

EXAMPLE 1 ▶ Finding Terms of Sequences

Find the first four terms of the sequence (for $n = 1, 2, 3, 4$) defined by

a. $a_n = \dfrac{8}{n}$

b. $b_n = (-1)^n n(n+1)$

SOLUTION

a. $a_1 = \dfrac{8}{1} = 8, a_2 = \dfrac{8}{2} = 4, a_3 = \dfrac{8}{3}, a_4 = \dfrac{8}{4} = 2$

We write these terms in the form $8, 4, \dfrac{8}{3}, 2$.

b. $b_1 = (-1)^1 (1)(1+1) = -2, b_2 = (-1)^2(2)(2+1) = 6,$

$b_3 = (-1)^3(3)(3+1) = -12, b_4 = (-1)^4(4)(4+1) = 20$

Notice that $(-1)^n$ causes the signs of the terms to alternate. The terms are -2, $6, -12, 20$.

> ## Technology Note
>
> We can find n terms of a sequence with a calculator if we set the calculator to sequence mode. Figure 8.23(a) shows the first four terms defined in Example 1(a), in decimal and fractional form. The graph of the sequence and the table of values for the sequence are shown in Figures 8.23(b) and 8.23(c), respectively.*

| (a) | (b) | (c) |

Figure 8.23

EXAMPLE 2 ▶   **Depreciation**

For tax purposes, a firm depreciates its $900,000 building over 30 years by the straight-line method, which depreciates the value of the building by $\dfrac{900,000}{30} = 30,000$ dollars each year. Write a sequence that gives the value of the building at the end of each of the first 5 years.

SOLUTION

The value would be reduced each year by 30,000 dollars, so the first five terms of the sequence are

$$870,000, 840,000, 810,000, 780,000, 750,000$$

Arithmetic Sequences

The future values of the investment described in Table 8.2 are the first six terms of the sequence

$$5050, 5100, 5150, 5200, 5250, 5300, \dots$$

---

* For more details, see Appendix A, page 682.

We can define this sequence **recursively**, with each term after the first defined from the previous term, as

$$a_1 = 5050, a_n = a_{n-1} + 50, \quad \text{for } n > 1$$

This sequence is an example of a special sequence called an **arithmetic sequence**. In such a sequence, each term after the first term is found by adding a constant to the preceding term. Thus, we have the following definition.

Arithmetic Sequence

A sequence is called an arithmetic sequence if there exists a number d, called the **common difference**, such that

$$a_n = a_{n-1} + d, \quad \text{for } n > 1$$

EXAMPLE 3 ▶   **Arithmetic Sequences**

Write the next three terms of the arithmetic sequences:

a. $1, 4, 7, 10, \ldots$ **b.** $11, 9, 7, \ldots$ **c.** $\dfrac{1}{2}, \dfrac{2}{3}, \dfrac{5}{6}, \ldots$

SOLUTION

a. The common difference that gives each term from the previous one is 3, so the next three terms are 13, 16, 19.

b. The common difference is -2, so the next three terms are 5, 3, 1.

c. The common difference is $\dfrac{1}{6}$, so the next three terms are $1, \dfrac{7}{6}, \dfrac{4}{3}$.

Note that the differences of terms of an arithmetic sequence are constant, so the function defining the sequence is linear, with its rate of change equal to the common difference. That is, an arithmetic sequence is really a linear function whose domain is restricted to a subset of the positive integers. Because each term after the first in an arithmetic sequence is obtained by adding d to the preceding term, the second term is $a_1 + d$, the third term is $(a_1 + d) + d = a_1 + 2d$, the fourth term is $a_1 + 3d, \ldots,$ and the nth term is $a_1 + (n - 1)d$. Thus, we have a formula for the nth term of an arithmetic sequence.

nth Term of an Arithmetic Sequence

The nth term of an arithmetic sequence is given by

$$a_n = a_1 + (n - 1)d$$

where a_1 is the first term of the sequence, n is the number of the term, and d is the common difference between consecutive terms.

Thus, the 25th term of the arithmetic sequence $1, 4, 7, 10, \ldots$ (see Example 3) is

$$a_{25} = 1 + (25 - 1)3 = 73$$

EXAMPLE 4 ▶ **Depreciation**

For tax purposes, a firm depreciates its $900,000 building over 30 years by the straight-line method, which depreciates the value of the building by $\dfrac{900,000}{30} = 30,000$ dollars each year. What is the value of the building after 12 years?

SOLUTION

The description indicates that the value of the building is 900,000 dollars at the beginning, that its value is $900,000 - 30,000 = 870,000$ dollars after 1 year, and that it decreases by 30,000 dollars each year for 29 additional years. Because the value of the building is reduced by the same amount each year, after n years the value is given by the arithmetic sequence with first term 870,000, common difference $-30,000$, and nth term

$$a_n = 870,000 + (n - 1)(-30,000), \quad \text{for } n = 1, 2, 3, \ldots, 30$$

Thus, the value of the building at the end of 12 years is

$$a_{12} = 870,000 + (12 - 1)(-30,000) = 540,000 \text{ dollars}$$

EXAMPLE 5 ▶ **Football Contract Plan A**

Recall from the Section Preview that the first payment plan offered to the football player was

Plan A: 10,000, 20,000, 30,000, ...

a. Do the payments form an arithmetic sequence?

b. What is the 18th payment under this payment plan?

c. If the contract is extended into the postseason, what is the 20th payment under this payment plan?

SOLUTION

a. The payments form a finite sequence with 18 terms. The common difference between terms is 10,000, so the sequence is an arithmetic sequence.

b. Because this is an arithmetic sequence with first term 10,000 and common difference $d = 10,000$, the 18th payment is

$$a_{18} = 10,000 + (18 - 1)10,000 = 180,000 \text{ dollars}$$

c. The 20th payment is

$$a_{20} = 10,000 + (20 - 1)10,000 = 200,000 \text{ dollars}$$

Geometric Sequences

Notice in the Section Preview that Plan B pays the football player $2, $4, $8, ... It is not an arithmetic sequence, because there is no constant difference between the terms. Each successive term of this sequence is doubled (that is, multiplied by 2). This is an example of another special sequence, where each term is found by multiplying the previous term by the same number. This sequence is called a **geometric sequence**.

> ### Geometric Sequence
>
> A sequence is called a geometric sequence if there exists a number r, called the **common ratio**, such that
> $$a_n = ra_{n-1}, \quad \text{for } n > 1$$

EXAMPLE 6 ▶ **Geometric Sequences**

Write the next three terms of the geometric sequences:

a. $1, 3, 9, \ldots$ **b.** $64, 16, 4, \ldots$ **c.** $2, -4, 8, \ldots$

SOLUTION

a. The common ratio that gives each term from the previous one is 3, so the next three terms are $27, 81, 243$.

b. The common ratio is $\dfrac{1}{4}$, so the next three terms are $1, \dfrac{1}{4}, \dfrac{1}{16}$.

c. The common ratio is -2, so the next three terms are $-16, 32, -64$.

Note that there is a **constant percent change** in the terms of a geometric sequence, so the function defining the sequence is exponential, and a geometric sequence is really an exponential function with its domain restricted to the positive integers. Because each term after the first in a geometric sequence is obtained by multiplying r times the preceding term, the second term is $a_1 r$, the third term is $a_1 r \cdot r = a_1 r^2$, the fourth term is $a_1 r^3, \ldots$, and the nth term is $a_1 r^{n-1}$. Thus, we have a formula for the nth term of a geometric sequence.

> ### *n*th Term of a Geometric Sequence
> The nth term of a geometric sequence is given by
> $$a_n = a_1 r^{n-1}$$
> where a_1 is the first term of the sequence, n is the number of the term, and r is the common ratio of consecutive terms.

Thus, the 25th term of the geometric sequence $2, -4, 8, \ldots$ (see Example 6) is

$$a_{25} = 2(-2)^{25-1} = 33{,}554{,}432$$

EXAMPLE 7 ▶ **Rebounding**

A ball is dropped from a height of 100 feet and rebounds $\dfrac{2}{5}$ of the height from which it falls every time it hits the ground. How high will it bounce after it hits the ground the fourth time?

SOLUTION

The first rebound is $\dfrac{2}{5}$ of $100 = 40$ feet, the second is $\dfrac{2}{5}$ of $40 = 16$ feet, and the rebounds form a geometric sequence with first term 40 and common ratio $\dfrac{2}{5}$.

   Thus, the fourth rebound is

$$40\left(\frac{2}{5}\right)^{4-1} = 2.56 \text{ feet}$$

We found the future value of money invested at compound interest in Chapter 5 by using exponential functions. Because geometric sequences are really exponential functions with domains restricted to the positive integers, the future value of an investment with interest compounded over a number of discrete periods can be found using a geometric sequence.

EXAMPLE 8 ▶ **Compound Interest**

The future value of $1000 invested for 3 years at 6%, compounded annually, can be found using the simple interest formula $S = P + Prt$, as follows.

$$
\begin{aligned}
\text{Year 1:} \quad S &= 1000 + 1000(0.06)(1) \\
&= 1000(1 + 0.06) \\
&= 1000(1.06) \text{ dollars} \\
\text{Year 2:} \quad S &= 1000(1.06) + [1000(1.06)](0.06)(1) \\
&= 1000(1.06)(1 + 0.06) \\
&= 1000(1.06)^2 \text{ dollars} \\
\text{Year 3:} \quad S &= 1000(1.06)^2 + [1000(1.06)^2](0.06)(1) \\
&= 1000(1.06)^2(1 + 0.06) \\
&= 1000(1.06)^3 \text{ dollars}
\end{aligned}
$$

a. Do the future values of this investment form a geometric sequence?

b. What is the value of this investment in 25 years?

SOLUTION

a. The future values of the investment form a sequence with first term $1000(1.06) = 1060$ and common ratio 1.06, so the sequence is a geometric sequence.

b. The future value of the investment is the 25th term of the geometric sequence:

$$1060(1.06)^{25-1} = 4291.87 \text{ dollars}$$

EXAMPLE 9 ▶ **Football Contract Plan B**

Recall from the Section Preview that the second payment plan offered to the football player was

Plan B: 2, 4, 8, ...

a. Do the payments form a geometric sequence?

b. What is the 18th payment under this payment plan?

SOLUTION

a. The payments form a finite sequence with 18 terms. The common ratio between terms is 2 because each payment is doubled. Thus, the sequence is a geometric sequence.

b. Because this is a geometric sequence with first term 2 and common ratio $r = 2$, the 18th payment is

$$a_{18} = 2 \cdot 2^{18-1} = 262{,}144 \text{ dollars}$$

Thus, the 18th payment ($262,144) from Plan B is significantly larger than the 18th payment ($180,000) from Plan A.

But does the fact that the 18th payment from Plan B is much larger than that from Plan A mean that the total payment from Plan B is larger than that from Plan A? We answer this question in the next section.

SKILLS CHECK 8.3

1. Find the first 6 terms of the sequence defined by $f(n) = 2n + 3$.

2. Find the first 4 terms of the sequence defined by $f(n) = \dfrac{1}{2n} + n$.

3. Find the first 5 terms of the sequence defined by $a_n = \dfrac{10}{n}$.

4. Find the first 5 terms of the sequence defined by $a_n = (-1)^n(2n)$.

5. Find the next 3 terms of the arithmetic sequence 1, 3, 5, 7, ...

6. Find the next 3 terms of the arithmetic sequence 2, 5, 8, ...

7. Find the eighth term of the arithmetic sequence with first term -3 and common difference 4.

8. Find the 40th term of the arithmetic sequence with first term 5 and common difference 15.

9. Write 4 additional terms of the geometric sequence 3, 6, 12, ...

10. Write 4 additional terms of the geometric sequence 8, 20, 50, ...

11. Find the sixth term of the geometric sequence with first term 10 and common ratio 3.

12. Find the tenth term of the geometric sequence with first term 48 and common ratio $-\dfrac{1}{2}$.

13. Find the first 4 terms of the sequence with first term 5 and nth term $a_n = a_{n-1} - 2$.

14. Find the first 6 terms of the sequence with first term 8 and nth term $a_n = a_{n-1} + 3$.

15. Find the first 4 terms of the sequence with first term 2 and nth term $a_n = 2a_{n-1} + 3$.

16. Find the first 5 terms of the sequence with first term 26 and nth term $a_n = \dfrac{a_{n-1} + 4}{2}$.

EXERCISES 8.3

17. *Salaries* Suppose you are offered a job with a relatively low starting salary but with a $1500 raise for each of the next 7 years. How much more than your starting salary would you be making in the eighth year?

18. *Depreciation* A new car costing $35,000 is purchased for business and is depreciated with the straight-line depreciation method over a 5-year period, which means that it is depreciated by the same amount each year. Write a sequence that gives the value after depreciation for each of the 5 years.

19. *Landscaping* Grading equipment used for landscaping costs $300 plus $60 per hour or part of an hour thereafter.

    a. Write an expression for the cost of a job lasting n hours.

    b. If the answer to part (a) is the nth term of a sequence, write the first 6 terms of this sequence.

20. *Profit* A new firm loses $2000 in its first month, but its profit increases by $400 in each succeeding month for the rest of the year. What is its profit in the 12th month?

21. *Salaries* Suppose you are offered two identical jobs, one paying a starting salary of $40,000 with yearly raises of $2000 and a second one paying a starting salary of $36,000 with yearly raises of $2400.

    a. Which job will be paying more in 6 years? By how much?

    b. Which job will be paying more in 11 years? By how much?

    c. How much more will the second job pay in the 13th year?

22. *Phone Calls* Suppose a long-distance call costs 99¢ for the first minute plus 25¢ for each additional minute.

    a. Write the cost of a call lasting n minutes.

    b. If the answer to part (a) is the nth term of a sequence, write the first 6 terms of this sequence.

23. *Interest* If $1000 is invested at 5% interest, compounded annually, write a sequence that gives the amount in the account at the end of each of the first 4 years.

24. *Interest* If $10,000 is invested at 6% interest, compounded annually, write a sequence that gives the amount in the account at the end of each of the first 3 years.

25. *Depreciation* A new car costing $50,000 depreciates by 20% of its original value each year.

    a. What is the value of the car at the end of the third year?

    b. Write an expression that gives the value at the end of the nth year.

    c. Write the first 5 terms of the sequence of the values after depreciation.

26. *Salaries* If you accept a job paying $32,000 for the first year, with a guaranteed 8% raise each year, how much will you earn in your fifth year?

27. *Bacteria* The size of a certain bacteria culture doubles each hour. If the number of bacteria present initially is 5000, how many will be present at the end of 6 hours?

28. *Bacteria* If a bacteria culture increases by 20% every hour and 2000 are present initially, how many will be present at the end of 10 hours?

29. *Animal Fossils* In the 1980s, a number of spherical fossils that were segmented on their surfaces were found in China. Some scientists believe these fossils were embryos of early ancestors of the first animals on earth, because they had a key characteristic of animal embryos: They "remained the same size while undergoing successive rounds of cell division." That is, different fossils consisted of different numbers of cells: one, two, four, eight, and so on.
(Source: *Discover,* December 2012)

    a. What type of sequence do the numbers of cells in these fossils follow? What is the common ratio?

    b. Some of the fossils had more than 1000 cells. What is the smallest number in the sequence from part (a) that is larger than 1000?

    c. How many cell divisions are necessary to get more than 500 cells?

30. *Bouncing Ball* A ball is dropped from 64 feet. If it rebounds $\frac{3}{4}$ of the height from which it falls every time it hits the ground, how high will it bounce after it hits the ground for the fourth time?

31. *Profit* If changing market conditions cause a company earning a profit of $8,000,000 this year to project decreases of 2% of its profit in each of the next 5 years, what profit does it project 5 years from now?

32. *Pumps* A pump removes $\frac{1}{3}$ of the water in a container with every stroke. What amount of water is removed on the fifth stroke if the container originally had 81 cm^3 of water?

33. *Salaries* If you begin a job making $54,000 and are given a $3600 raise each year, use numerical methods to find in how many years your salary will double.

34. *Interest* If $2500 is invested at 8% interest, compounded annually, how long will it take the account to reach $5829.10?

35. *Compound Interest* Find the future value of $10,000 invested for 10 years at 8%, compounded daily.

36. *Compound Interest* Find the future value of $10,000 invested for 10 years at 8%, compounded annually.

37. *Rabbit Breeding* The number of pairs of rabbits in the population during the first 6 months can be written in the Fibonacci sequence

$$1, 1, 2, 3, 5, 8, \ldots$$

in which each term after the second is the sum of the two previous terms. Find the number of pairs of rabbits for the next 4 months.

38. *Bee Ancestry* A female bee hatches from a fertilized egg, whereas a male bee hatches from an unfertilized egg. Thus, a female bee has a male parent and a female parent, whereas a male bee has only a female parent. Therefore, the number of ancestors of a male bee follow the Fibonacci sequence

$$1, 2, 3, 5, 8, 13, \ldots$$

    a. Observe the pattern and write three more terms of the sequence.

    b. What do the 1, the 2, and the 3, respectively, represent for a given male bee?

39. *Credit Card Debt* Kirsten has a $10,000 credit card debt. Each month she pays 1% interest plus a payment of 10% of the monthly balance. If she makes no new purchases, write a sequence that gives the payments for the first 4 months.

40. *Bouncing Ball* A ball is dropped from 128 feet. It rebounds $\frac{1}{4}$ of the height from which it falls every time it hits the ground. How high will it bounce after it hits the ground for the fourth time?

8.4 Series

KEY OBJECTIVES

- Find the sum of a finite series
- Evaluate sums written in sigma notation
- Find the sum of the first *n* terms of an arithmetic sequence
- Find the sum of the first *n* terms of a geometric sequence
- Find the sum of an infinite geometric series

SECTION PREVIEW **Football Contracts**

In Section 8.3, we discussed a football player's opportunity to sign a contract for 18 games for either

Plan A: $10,000 for the first game with a $10,000 raise for each game thereafter

Plan B: $2 for the first game with his salary doubled for each game thereafter

Which salary plan should he accept if he wants to make the most money?

The payments for the games form the sequences:

Plan A: 10,000, 20,000, 30,000, . . .

Plan B: 2, 4, 8, 16, . . .

We found that Plan B paid more than Plan A for the 18th game, but we did not determine which plan pays more for all 18 games. To answer this question, we need to find the *sum* of the first 18 terms of these sequences. The sum of the terms of a sequence is called a series. Series are useful in computing finite and infinite sums and in finding depreciation and future values of annuities. ■

Finite and Infinite Series

Suppose a firm loses $2000 in its first month of operation, but its profit increases by $500 in each succeeding month for the remainder of its first year. What is its profit for the year? The description indicates that the profit is −$2000 for the first month and that it increases by $500 each month for 11 months. The monthly profits form the sequence that begins

$$-2000, -1500, -1000, \ldots$$

The profit for the 12 months is the sum of these terms:

$$(-2000) + (-1500) + (-1000) + (-500) + 0 + 500 + 1000 + 1500 + 2000$$
$$+ 2500 + 3000 + 3500 = 9000$$

The total profit for the year is this sum, $9000.

The sum of the 12 terms of this sequence can be referred to as a **finite series**. In general, a **series** is defined as the sum of the terms of a sequence. If there are infinitely many terms, the series is an **infinite series**.

> ### Series
> A finite series is defined by
> $$a_1 + a_2 + a_3 + \cdots + a_n$$
> where $a_1, a_2, \ldots a_n$ are terms of a sequence.
> An infinite series is defined by
> $$a_1 + a_2 + a_3 + \cdots + a_n + \cdots$$

We can use the Greek letter Σ (sigma) to express the sum of numbers or expressions. For example, we can write

$$\sum_{i=1}^{3} i \quad \text{to denote} \quad 1 + 2 + 3$$

and we can write the general finite series in the form

$$\sum_{i=1}^{n} a_i = a_1 + a_2 + a_3 + \cdots + a_n$$

The symbol $\displaystyle\sum_{i=1}^{n} a_i$ may be read as "The sum of a_i as i goes from 1 to n one unit at a time."

The general infinite series can be written in the form

$$\sum_{i=1}^{\infty} a_i = a_1 + a_2 + a_3 + \cdots + a_n + \cdots$$

where ∞ indicates that there is no end to the number of terms.

EXAMPLE 1 ▶ **Finite Series**

Find the sum of the first six terms of the sequence $6, 3, \dfrac{3}{2}, \ldots$.

SOLUTION

Each succeeding term of this sequence is found by multiplying the preceding term by $\dfrac{1}{2}$, so the sequence is

$$6, 3, \frac{3}{2}, \frac{3}{4}, \frac{3}{8}, \frac{3}{16}, \cdots$$

and the sum of the first six terms is $6 + 3 + \dfrac{3}{2} + \dfrac{3}{4} + \dfrac{3}{8} + \dfrac{3}{16} = \dfrac{189}{16}$.

Arithmetic Series

The sum of the payments to the football player in the Section Preview for the first three games under Plan A is

$$10{,}000 + 20{,}000 + 30{,}000 = 60{,}000 \text{ dollars}$$

and the sum of the payments for the first three games under Plan B is

$$2 + 4 + 8 = 14 \text{ dollars}$$

so the player would be well advised to take Plan A if there were only three games. To find the sum of the payments for 18 games and to answer other questions, it would be more efficient to have formulas to use in finding the sums of terms of sequences. We begin by finding a formula for the sum of n terms of an arithmetic sequence.

Given the first term a_1 in an arithmetic sequence with common difference d, we use s_n to represent the sum of the first n terms and write it as follows:

$$s_n = a_1 + (a_1 + d) + (a_1 + 2d) + \cdots + (a_n - 2d) + (a_n - d) + a_n$$

Writing this sum in reverse order gives

$$s_n = a_n + (a_n - d) + (a_n - 2d) + \cdots + (a_1 + 2d) + (a_1 + d) + a_1$$

Adding these two equations term by term gives

$$s_n + s_n = (a_1 + a_n) + (a_1 + a_n) + (a_1 + a_n) + \cdots + (a_1 + a_n)$$
$$+ (a_1 + a_n) + (a_1 + a_n)$$

so $2s_n = n(a_1 + a_n)$ because there are n terms, and we have $s_n = \dfrac{n(a_1 + a_n)}{2}$.

Sum of *n* Terms of an Arithmetic Sequence

The sum of the first n terms of an arithmetic sequence is given by

$$s_n = \frac{n(a_1 + a_n)}{2}$$

where a_1 is the first term of the sequence and a_n is the nth term.

Recall that Plan A of the football player's contract began with $a_1 = 10{,}000$ and had 18th payment $a_{18} = 180{,}000$. Thus, the sum of the 18 payments is

$$s_{18} = \frac{18(10{,}000 + 180{,}000)}{2} = 1{,}710{,}000 \text{ dollars}$$

EXAMPLE 2 ▶ **Depreciation**

An automobile valued at \$24,000 is depreciated over 5 years with the *sum-of-the-years'-digits* depreciation method. Under this method, annual depreciation is found by multiplying the value of the property by a fraction whose denominator is the sum of the 5 years' digits and whose numerator is 5 for the first year's depreciation, 4 for the second year's, 3 for the third, and so on.

a. Write a sequence that represents the parts of the automobile value that are depreciated in each year, and show that the sum of the terms is 100%.

b. Find the yearly depreciation and show that the sum of the depreciations is \$24,000.

SOLUTION

a. The sum of the 5 years' digits is $1 + 2 + 3 + 4 + 5 = 15$, so the parts of the value that are depreciated for each of the 5 years are the numbers

$$\frac{5}{15}, \frac{4}{15}, \frac{3}{15}, \frac{2}{15}, \frac{1}{15}$$

This is an arithmetic sequence with first term $\dfrac{5}{15}$, common difference $-\dfrac{1}{15}$, and fifth term $\dfrac{1}{15}$. Thus, the sum of the terms is

$$s_5 = \frac{5\left(\dfrac{5}{15} + \dfrac{1}{15}\right)}{2} = \frac{5}{2} \cdot \frac{6}{15} = 1 = 100\%$$

b. The respective annual depreciations for the automobile are

$$\frac{5}{15} \cdot 24{,}000 = 8000 \qquad \frac{4}{15} \cdot 24{,}000 = 6400 \qquad \frac{3}{15} \cdot 24{,}000 = 4800$$

$$\frac{2}{15} \cdot 24{,}000 = 3200 \qquad \frac{1}{15} \cdot 24{,}000 = 1600$$

The depreciations form an arithmetic sequence with difference -1600, first term 8000, and fifth term 1600, so the sum of these depreciations is

$$s_n = \frac{5(8000 + 1600)}{2} = 24{,}000 \text{ dollars}$$

Note that the sum is the original value of the automobile, so it is totally depreciated in 5 years.

Geometric Series

In Section 5.6, we found the future value of the annuity with payments of $1000 at the end of each of 5 years, with interest at 10%, compounded annually, by finding the sum

$$S = 1000 + 1000(1.10)^1 + 1000(1.10)^2 + 1000(1.10)^3 + 1000(1.10)^4$$
$$= 6105.10 \text{ dollars}$$

We can see that this is a finite geometric series with first term 1000 and common ratio 1.10. We can develop a formula to find the sum of this and any other finite geometric series.

The sum of the first n terms of a geometric sequence with first term a_1 and common ratio r is

$$s_n = a_1 + a_1 r + a_1 r^2 + \cdots + a_1 r^{n-1}$$

To find a formula for this sum, we multiply it by r to get a new expression and then find the difference of the two expressions.

$$s_n = a_1 + a_1 r + a_1 r^2 + a_1 r^3 + \cdots + a_1 r^{n-1}$$
$$rs_n = a_1 r + a_1 r^2 + a_1 r^3 + \cdots + a_1 r^{n-1} + a_1 r^n$$
$$s_n - rs_n = a_1 - a_1 r^n$$

Rewriting this difference and solving for s_n gives a formula for the sum.

$$s_n(1 - r) = a_1(1 - r^n)$$
$$s_n = \frac{a_1(1 - r^n)}{(1 - r)}$$

Sum of n Terms of a Geometric Sequence

The sum of the first n terms of a geometric sequence is

$$s_n = \frac{a_1(1 - r^n)}{1 - r}, \quad r \neq 1$$

where a_1 is the first term of the sequence and r is the common ratio.

EXAMPLE 3 ▶ Sums of Terms of Geometric Sequences

a. Find the sum of the first 10 terms of the geometric sequence with first term 4 and common ratio -3.

b. Find the sum of the first 8 terms of the sequence $2, 1, \frac{1}{2}, \ldots$.

SOLUTION

a. The sum of the first 10 terms of the geometric sequence with first term 4 and common ratio -3 is given by

$$s_{10} = \frac{a_1(1 - r^{10})}{1 - r} = \frac{4(1 - (-3)^{10})}{1 - (-3)} = -59{,}048$$

b. This sequence is geometric, with first term 2 and common ratio $\frac{1}{2}$, so the sum of the first 8 terms is

$$s_8 = \frac{a_1(1 - r^8)}{1 - r} = \frac{2\left(1 - \left(\frac{1}{2}\right)^8\right)}{1 - \left(\frac{1}{2}\right)} = 4\left(1 - \frac{1}{256}\right) = \frac{255}{64}$$

EXAMPLE 4 ▶   **Annuities**

Show that if regular payments of $1000 are made at the end of each year for 5 years into an account that pays interest at 10% per year, compounded annually, the future value of this annuity is $6105.10.

SOLUTION

As we saw on the previous page, the future values of the individual payments into this annuity form a geometric sequence with first term 1000 and common ratio 1.10, so the sum of the first five terms is

$$s_5 = \frac{1000(1 - 1.10^5)}{1 - 1.10} = 6105.10$$

We found that the total payment to the football player for 18 games under Plan A is $1,710,000. The payments under Plan B form a geometric sequence with first term 2 and common ratio 2, so the sum of the payments under Plan B is

$$s_{18} = \frac{2(1 - 2^{18})}{1 - 2} = 524{,}286$$

Thus, total earnings for the 18 games for Plan A are more than three times those for Plan B. However, if the season has 20 games, the sum of the payments for Plan A is

$$s_{20} = \frac{20(10{,}000 + 200{,}000)}{2} = 2{,}100{,}000 \text{ dollars}$$

and the sum of the payments for Plan B is

$$s_{20} = \frac{2(1 - 2^{20})}{1 - 2} = 2{,}097{,}150 \text{ dollars}$$

In this case, the plans are nearly equal.

Infinite Geometric Series

If we add the terms of an infinite geometric sequence, we have an **infinite geometric series**. It is perhaps surprising that we can add an infinite number of positive numbers and get a finite sum. To see that this is possible, consider a football team that has only 5 yards remaining to score a touchdown. How many penalties on the opposing team will it take for the sum of the penalties to be 5 yards (giving them a touchdown), if each penalty is "half the distance to the goal"? The answer is that the team can never reach the goal line because every penalty is half the distance to the goal, which always leaves some distance remaining to travel. Thus, the sum of any number of penalties is less than 5 yards. To see this, let's find the sum of the first 10 such penalties. The first few penalties are $\frac{5}{2}, \frac{5}{4}$, and $\frac{5}{8}$, so the sequence of penalties has first term $\frac{5}{2}$ and common ratio $\frac{1}{2}$ and the sum of the first n penalties is

$$s_n = \frac{a_1(1 - r^n)}{1 - r} = \frac{\left(\frac{5}{2}\right)\left[1 - \left(\frac{1}{2}\right)^n\right]}{1 - \frac{1}{2}}$$

The sum of the first 10 penalties is found by letting $n = 10$, getting

$$\frac{\frac{5}{2}\left[1 - \left(\frac{1}{2}\right)^{10}\right]}{1 - \frac{1}{2}} \approx 4.9951$$

We can see that this sum is less than 5 yards, but close to 5 yards. Notice that as n gets larger, $\left(\frac{1}{2}\right)^n$ gets smaller, approaching 0, and the sum approaches $\frac{5/2(1 - 0)}{1 - (1/2)} = 5$. Thus, the sum of an *infinite number* of penalties is the finite number 5 yards.

In general, if $|r| < 1$, it can be shown that r^n approaches 0 as n gets large without bound. Thus, $1 - r^n$ approaches 1, and $\frac{a_1(1 - r^n)}{1 - r}$ approaches $\frac{a_1}{1 - r}$.

Sum of an Infinite Geometric Series

The sum of an infinite geometric series with common ratio r, $|r| < 1$, is

$$S = \frac{a_1}{1 - r}.$$

where a_1 is the first term.
If $|r| \geq 1$, the sum does not exist (is not a finite number).

We can use this formula to compute the sum of an infinite number of "half the distance to the goal" penalties from the 5-yard line.

$$S = \frac{\frac{5}{2}}{1 - \frac{1}{2}} = 5$$

EXAMPLE 5 ▶ **Infinite Geometric Series**

Find the sum represented by the following series, if possible.

a. $81 + 9 + 1 + \cdots$

b. $\sum_{i=1}^{\infty} \left(\frac{2}{3}\right)^i$

SOLUTION

a. This is an infinite geometric series with first term $a_1 = 81$ and common ratio $r = \dfrac{1}{9} < 1$, so the sum is

$$S = \frac{81}{1 - \dfrac{1}{9}} = \frac{729}{8}$$

b. This is the sigma notation for an infinite series. Each new term results in one higher power of $\dfrac{2}{3}$ (because i increases by one), so the common ratio is $\dfrac{2}{3} < 1$. The first term is $\dfrac{2}{3}$ (the value when $i = 1$), so the sum is

$$S = \frac{\dfrac{2}{3}}{1 - \dfrac{2}{3}} = 2$$

SKILLS CHECK 8.4

1. Find the sum of the first 6 terms of the geometric sequence 9, 3, 1, ...

2. Find the sum of the first 7 terms of the arithmetic sequence 1, 3, 5, ...

3. Find the sum of the first 10 terms of the arithmetic sequence 7, 10, 13, ...

4. Find the sum of the first 20 terms of the arithmetic sequence 5, 12, 19, ...

5. Find the sum of the first 15 terms of the arithmetic sequence with first term −4 and common difference 2.

6. Find the sum of the first 10 terms of the arithmetic sequence with first term 50 and common difference 3.

7. Find the sum of the first 15 terms of the geometric sequence with first term 3 and common ratio 2.

8. Find the sum of the first 12 terms of the geometric sequence with first term 48 and common ratio $\dfrac{1}{2}$.

9. Find the sum of the first 10 terms of the geometric sequence 5, 10, 20, ...

10. Find the sum of the first 15 terms of the geometric sequence 100, 50, 25, ...

11. Find the sum represented by the series $1024 + 256 + 64 + 16 + \cdots$, if possible.

12. Find the sum represented by the series $64 + 32 + 16 + \cdots$, if possible.

In Exercises 13–16, evaluate the sums.

13. $\displaystyle\sum_{i=1}^{6} 2^i$

14. $\displaystyle\sum_{i=1}^{5} 4^i$

15. $\displaystyle\sum_{i=1}^{4} \frac{1+i}{i}$

16. $\displaystyle\sum_{i=1}^{5} \left(\frac{1}{2}\right)^i$

17. Find the sum of the series $\displaystyle\sum_{i=1}^{\infty} \left(\frac{3}{4}\right)^i$, if possible.

18. Find the sum $\displaystyle\sum_{i=1}^{\infty} \left(\frac{5}{6}\right)^i$, if possible.

19. Find the sum $\displaystyle\sum_{i=1}^{\infty} \left(\frac{4}{3}\right)^i$, if possible.

20. Find the following sum, if possible.

$$800 + 600 + 450 + \cdots$$

EXERCISES 8.4

21. Profit A new firm loses $2000 in its first month, but its profit increases by $400 in each succeeding month for the rest of the year. What is the sum of its profit for the first year?

22. Salaries Suppose you are offered a job with a relatively low starting salary but with a bonus that is $1500 for the first year and increases by $1500 each year after that. What is the total amount of your bonuses over the next 7 years?

23. Bee Ancestry A female bee hatches from a fertilized egg whereas a male bee hatches from an unfertilized egg. Thus, a female bee has a male parent and a female parent whereas a male bee has only a female parent. Therefore, the numbers of ancestors of a male bee follow the Fibonacci sequence

$$1, 2, 3, 5, 8, 13, \ldots$$

Use this sequence to determine the total number of ancestors of a male bee back four generations.

24. Salaries Suppose you are offered two identical jobs, one paying a starting salary of $40,000 with yearly raises of $2000 and one paying a starting salary of $36,000 with yearly raises of $3000.

  a. Which job will pay more over the first 3 years? How much more?

  b. Which job will pay more over the first 12 years? How much more?

  c. What factor is important in deciding which job to take?

25. Clocks A grandfather clock strikes a chime indicating each hour of the day, so it chimes 3 times at 3:00, 10 times at 10:00, and so on.

  a. How many times will it chime in a 12-hour period?

  b. How many times will it chime in a 24-hour day?

26. Depreciation Under the sum-of-the-year's-digits depreciation method, annual depreciation for 4 years is found by multiplying the value of the property by a fraction whose denominator is the sum of the 4 years' digits and whose numerator is 4 for the first year's depreciation, 3 for the second year's, 2 for the third year's, and so on.

  a. Write a sequence that gives the depreciation for each year on a truck with value $36,000.

  b. Show that the sum of the depreciations for the 4 years adds to 100% of the original value.

27. Profit Suppose a new business makes a profit of $2000 in its first month, and its profit increases by 10% in each of the next 11 months. How much profit did it earn in its first year?

28. Profit If changing market conditions cause a company earning a profit of $8,000,000 this year to project decreases of 2% of its profit in each of the next 5 years, what is the sum of the yearly profits that it projects for the next 5 years?

29. Pumps A pump removes $\frac{1}{3}$ of the water in a container with every stroke. After 4 strokes, what amount of water is still in a container that originally had 81 cm^3 of water?

30. Salaries If you accept a job paying $32,000 for the first year with a guaranteed 8% raise each year, how much will you earn in your first 5 years?

31. Email Chain Letters Suppose you receive an email chain letter with 5 names on it, and to keep the chain unbroken, you mail a dollar to the person whose name is at the top of the list, cross out the top name, add your name to the bottom, and mail it to 5 friends.

  a. How many friends will receive your letter?

  b. If your friends cross the second person from the original list off the list and mail out 5 letters each, how many people will receive letters from your friends?

  c. How many "descendants" of your letters will receive email letters in the next two levels if no one breaks the chain?

  d. Describe the type of sequence that gives the number of people receiving the letter at each level.

32. Email Chain Letters Suppose the email chain letter in Exercise 31 were to go through 12 unbroken levels.

  a. How many people would receive a letter asking for money in the 12th mailing?

  b. What is the significance of this number?

33. Credit Card Debt Kirsten has a $10,000 credit card debt. She makes a minimum payment of 10% of the balance at the end of each month. If she makes no new purchases, how much will she owe at the end of 1 year?

34. Credit Card Debt Kirsten has a $10,000 credit card debt. She makes a minimum payment of 10% of the balance at the end of each month. If she makes no new purchases, in how many months will she have half of her debt repaid?

35. Bouncing Ball A ball is dropped from 128 feet. It rebounds $\frac{1}{4}$ of the height from which it falls every time it hits the ground. How far will it have traveled up and down when it hits the ground for the fifth time?

36. Bouncing Ball A ball is dropped from 64 feet. If it rebounds $\frac{3}{4}$ of the height from which it falls every time it hits the ground, how far will it have traveled up and down when it hits the ground for the fourth time?

37. Depreciation A new car costing $35,000 depreciates by 16% of its current value each year.

   a. Write an expression that gives the sum of the depreciations for the first n years.

   b. What is the value of the car at the end of the nth year?

38. Loans An interest-free loan of $15,000 requires monthly payments of 8% of the unpaid balance. What is the unpaid balance after

   a. 1 year?

   b. 2 years?

39. Annuities Find the 8-year future value of an annuity with a contribution of $100 at the end of each month into an account that pays 12% per year, compounded monthly.

40. Annuities Find the 10-year future value of an ordinary annuity with a contribution of $300 per quarter into an account that pays 8% per year, compounded quarterly.

41. Present Value of Annuities The present value of an annuity with n payments of R dollars, at interest rate i per period, is

$$A = R(1 + i)^{-1} + R(1 + i)^{-2}$$
$$+ R(1 + i)^{-3} + \cdots + R(1 + i)^{-(n-1)}$$
$$+ R(1 + i)^{-n}$$

Treat this as the sum of a geometric series with n terms, $a_1 = R(1 + i)^{-n}$, and $r = (1 + i)$, and show that the sum is

$$A = R\left[\frac{1 - (1 + i)^{-n}}{i}\right]$$

8.5 Preparing for Calculus

KEY OBJECTIVE

- Apply the algebra skills from Chapters 1–6 to the solution of calculus problems

SECTION PREVIEW Calculus

In this section, we show how algebra is used in the context of a calculus course; that is, we show how algebra skills that we have already acquired are applied in the solution of calculus problems. Three fundamental concepts in calculus are the limit, derivative, and integral. Although we do not define or use these concepts here, we do illustrate that algebra skills are necessary for computing, simplifying, and applying these concepts. ■

EXAMPLE 1 ▶ **Simplifying Expressions**

The following expressions represent the derivatives of functions. Simplify these expressions by multiplying to remove parentheses and combining like terms.

   a. $(x^3 - 3x)(2x) + (x^2 - 1)(3x^2 - 3)$

   b. $\dfrac{x^3(20x^4) - (4x^5 - 2)(3x^2)}{(x^3)^2}$

SOLUTION

a. $(x^3 - 3x)(2x) + (x^2 - 1)(3x^2 - 3) = 2x^4 - 6x^2 + 3x^4 - 3x^2 - 3x^2 + 3$

$$= 5x^4 - 12x^2 + 3$$

b. $\dfrac{x^3(20x^4) - (4x^5 - 2)(3x^2)}{(x^3)^2} = \dfrac{20x^7 - (12x^7 - 6x^2)}{x^6}$

$$= \dfrac{8x^7 + 6x^2}{x^6} = \dfrac{8x^5 + 6}{x^4}$$

EXAMPLE 2 ▶ **Evaluating Functions**

In calculus, the derivative of $f(x)$ is denoted by $f'(x)$, and the value of the derivative at $x = a$ is denoted $f'(a)$. If $f'(x) = 3x^2 - 4x + 2$, find $f'(-1)$ and $f'(3)$.

SOLUTION

$$f'(-1) = 3(-1)^2 - 4(-1) + 2 = 9$$

$$f'(3) = 3(3)^2 - 4(3) + 2 = 17$$

EXAMPLE 3 ▶ **Slope of a Tangent Line**

Calculus can be used to find the slope of the tangent to a curve. The slope of the tangent to the curve $y = f(x)$ at $x = a$ is $f'(a)$. If $f'(x) = 4x^3 - 3x^2 + 1$, find the slope of the tangent at $x = 2$.

SOLUTION

The slope of the tangent to the curve at $x = 2$ is $f'(2) = 4(2)^3 - 3(2)^2 + 1 = 21$.

EXAMPLE 4 ▶ **Equation of a Tangent Line**

Calculus can be used to find the slope of the tangent to a curve, and then we can write the equation of the tangent line. If the slope of the line tangent to the graph of $f(x) = x^2$ at $(1, 1)$ is 2, write the equation of this line (Figure 8.24).

Figure 8.24

SOLUTION

The point-slope form of the equation of a line is

$$y - y_1 = m(x - x_1)$$

Using the given point and the slope of the tangent line gives the equation

$$y - 1 = 2(x - 1), \quad \text{or} \quad y = 2x - 1$$

EXAMPLE 5 ▶ **Difference Quotient**

a. An important step in finding the derivative of a function involves reducing fractions like

$$\frac{(2(x + h) + 1) - (2x + 1)}{h}$$

which is called the difference quotient. Reduce this fraction, if $h \neq 0$.

b. An important calculation in calculus is the evaluation of the difference quotient $\frac{f(x + h) - f(x)}{h}$, where $h \neq 0$. Find $\frac{f(x + h) - f(x)}{h}$ if $f(x) = 3x + 2$.

SOLUTION

a. $\dfrac{(2(x + h) + 1) - (2x + 1)}{h} = \dfrac{2x + 2h + 1 - 2x - 1}{h} = \dfrac{2h}{h} = 2$

b. $\dfrac{f(x + h) - f(x)}{h} = \dfrac{(3(x + h) + 2) - (3x + 2)}{h}$

$$= \frac{3x + 3h + 2 - 3x - 2}{h} = \frac{3h}{h} = 3$$

EXAMPLE 6 ▶ **Solving Equations**

Finding where the derivative is zero is an important skill in calculus.

a. Find where $f'(x) = 0$ if $f'(x) = 3x - 6$.

b. Find where $f'(x) = 0$ if $f'(x) = 8x - 4$.

SOLUTION

a. $f'(x) = 0$ where $0 = 3x - 6$, or where $6 = 3x$, at $x = 2$.

b. $f'(x) = 0$ where $0 = 8x - 4$, or where $4 = 8x$, at $x = \dfrac{1}{2}$.

EXAMPLE 7 ▶ **Rewriting Radicals**

To perform calculus operations on functions, it is frequently desirable or necessary to write a radical function in the form $y = ax^b$.

a. Convert the function $y = 3\sqrt[3]{x^4}$ to this form.

b. Convert the function $y = 2x\sqrt{x}$ to this form.

SOLUTION

a. $3\sqrt[3]{x^4} = 3x^{4/3}$, so the form is $y = 3x^{4/3}$.

b. $2x\sqrt{x} = 2x \cdot x^{1/2} = 2x^{1+1/2} = 2x^{3/2}$, so the form is $y = 2x^{3/2}$.

EXAMPLE 8 ▶ Rewriting Fractional Exponents

Sometimes a derivative contains fractional exponents, and simplification requires that all fractional exponents be converted to radicals. Write the expression $f'(x) = -3x^{1/3} + 2x^{1/2} + 1$ without fractional exponents.

SOLUTION

$$f'(x) = -3x^{1/3} + 2x^{1/2} + 1 = -3\sqrt[3]{x} + 2\sqrt{x} + 1$$

EXAMPLE 9 ▶ Evaluating the Difference Quotient

An important calculation in calculus is the evaluation of the difference quotient $\dfrac{f(x + h) - f(x)}{h}$, where h is close to, but not equal to, 0. Find each of the following, if $f(x) = 3x^2 - 2x$, and observe the resulting function as h gets "very close" to 0.

a. $f(x + h)$ **b.** $f(x + h) - f(x)$ **c.** $\dfrac{f(x + h) - f(x)}{h}$

d. The function that the result in (c) approaches as h approaches 0

SOLUTION

a. $f(x + h) = 3(x + h)^2 - 2(x + h) = 3(x^2 + 2xh + h^2) - 2(x + h)$
$$= 3x^2 + 6xh + 3h^2 - 2x - 2h$$

b. $f(x + h) - f(x) = (3x^2 + 6xh + 3h^2 - 2x - 2h) - (3x^2 - 2x)$
$$= 6xh + 3h^2 - 2h$$

c. $\dfrac{f(x + h) - f(x)}{h} = \dfrac{6xh + 3h^2 - 2h}{h} = 6x + 3h - 2$

d. As h approaches 0, $6x + 3h - 2$ approaches $6x + 3(0) - 2 = 6x - 2$.

EXAMPLE 10 ▶ Solving Equations

Finding where the derivative is zero is an important skill in calculus.

a. Find where $f'(x) = 0$ if $f'(x) = 3x^2 + x - 2$.

b. Find where $f'(x) = 0$ if $f'(x) = 3x^2 + 5x - 2$.

SOLUTION

a. To find where $f'(x) = 0$, we solve $3x^2 + x - 2 = 0$.
$$3x^2 + x - 2 = 0$$
$$(3x - 2)(x + 1) = 0$$
$$3x - 2 = 0 \quad \text{or} \quad x + 1 = 0$$
$$x = \frac{2}{3} \quad \text{or} \quad x = -1$$

b. To find where $f'(x) = 0$, we solve $3x^2 + 5x - 2 = 0$.

$$3x^2 + 5x - 2 = 0$$
$$(3x - 1)(x + 2) = 0$$
$$3x - 1 = 0 \quad \text{or} \quad x + 2 = 0$$
$$x = \frac{1}{3} \quad \text{or} \quad x = -2$$

EXAMPLE 11 ▶ Simplifying Expressions by Factoring

After a derivative is found, the expression frequently must be simplified. Simplify the following expression by factoring:

$$x[2(6x - 1)6] + (6x - 1)^2$$

SOLUTION

Factoring $(6x - 1)$ from both terms gives

$$x[2(6x - 1)6] + (6x - 1)^2 = (6x - 1)[x \cdot 2 \cdot 6 + (6x - 1)]$$
$$= (6x - 1)(18x - 1)$$

EXAMPLE 12 ▶ Simplifying Expressions by Factoring

Simplify by factoring:

$$x^4[6(3x^2 - 2x)^5(6x - 2)] + (3x^2 - 2x)^6(4x^3)$$

SOLUTION

Each term contains the factor $(3x^2 - 2x)$ to some power, and it would be difficult to raise this factor to the fifth and sixth power, so we factor $(3x^2 - 2x)^5$ from each term.

$$x^4[6(3x^2 - 2x)^5(6x - 2)] + (3x^2 - 2x)^6(4x^3)$$
$$= (3x^2 - 2x)^5\{x^4[6(6x - 2)] + (3x^2 - 2x)(4x^3)\}$$
$$= (3x^2 - 2x)^5(36x^5 - 12x^4 + 12x^5 - 8x^4) = (3x^2 - 2x)^5(48x^5 - 20x^4)$$

Because $4x^4$ is a factor of $(48x^5 - 20x^4)$, we can write the expression as $4x^4(12x - 5)$ $(3x^2 - 2x)^5 = 4x^4(12x - 5) \cdot [x(3x - 2)]^5 = 4x^9(12x - 5)(3x - 2)^5$.

In calculus, it is sometimes useful to find two functions whose composition gives a function. For example, if $y = (3x - 2)^2$, we can write this function as $y = u^2$, where $u = 3x - 2$. Going backwards from a function to two functions whose composition gives the function is called finding the *decomposition* of the function. Another example of decomposition is writing $y = \dfrac{1}{x^3 - 1}$ as $y = \dfrac{1}{u}$, where $u = x^3 - 1$. It is possible to write several decompositions for a given function. For example, $y = \dfrac{1}{x^3 - 1}$ could also be written as $y = \dfrac{1}{u - 1}$, where $u = x^3$.

EXAMPLE 13 ▶ Composition of Functions

An equation of the form $y = (ax^k + b)^n$ can be thought of as the composition of functions and can be written as $y = u^n$, where $u = ax^k + b$. Use this idea to write two functions whose composition gives $y = (x^4 + 5)^6$.

SOLUTION

We can let $y = u^6$, where $u = x^4 + 5$. Then $y = (x^4 + 5)^6$.

EXAMPLE 14 ▶   **Function Decomposition**

Find two functions, f and g, such that their composition is $(f \circ g)(x) = \sqrt{x^2 + 1}$.

SOLUTION

One decomposition is found by letting $f(u) = \sqrt{u}$ and $g(x) = x^2 + 1$; then $(f \circ g)(x) = f(g(x)) = \sqrt{x^2 + 1}$.

EXAMPLE 15 ▶   **Rewriting Expressions Using Negative Exponents**

Many functions must be written in a form with no variables in the denominator in order to take the derivative. Use negative exponents where necessary to write the following function with all variables in the numerator of a term. That is, write each term in the form cx^n.

$$f(x) = \frac{4}{x^2} + \frac{3}{x} - 2x$$

SOLUTION

Using the definition $a^{-n} = \frac{1}{a^n}, a \neq 0$, we get

$$f(x) = 4\frac{1}{x^2} + 3\frac{1}{x} - 2x = 4x^{-2} + 3x^{-1} - 2x$$

EXAMPLE 16 ▶   **Rewriting Expressions Using Positive Exponents**

Sometimes a derivative contains negative exponents, and simplification requires that all exponents be positive. Write the expression $f'(x) = -3x^{-4} + 2x^{-3} + 1$ without negative exponents.

SOLUTION

$$f'(x) = -3x^{-4} + 2x^{-3} + 1$$

$$= -3\frac{1}{x^4} + 2\frac{1}{x^3} + 1 = -\frac{3}{x^4} + \frac{2}{x^3} + 1$$

EXAMPLE 17 ▶   **Using Logarithmic Properties**

Taking derivatives involving logarithms frequently requires simplifying by converting the expressions using logarithmic properties. Rewrite the following expressions as the sum, difference, or product of logarithms without exponents.

a. $\log\left[x^2(2x + 3)^2\right]$

b. $\ln\dfrac{\sqrt{2x - 4}}{3x + 1}$

SOLUTION

a. $\log\left[x^2(2x+3)^2\right] = \log x^2 + \log(2x+3)^2$

$$= 2\log x + 2\log(2x+3)$$

b. $\ln\dfrac{\sqrt{2x-4}}{3x+1} = \ln\sqrt{2x-4} - \ln(3x+1)$

$$= \ln(2x-4)^{1/2} - \ln(3x+1)$$

$$= \frac{1}{2}\ln(2x-4) - \ln(3x+1)$$

EXAMPLE 18 ▶ **Reducing Fractions**

Evaluating limits is a calculus skill that frequently requires reducing fractions, as in the following cases. Reducing these fractions may require factoring.

a. Simplify $\dfrac{x^2-4}{x-2}$, if $x \neq 2$. **b.** Simplify $\dfrac{x^2-7x+12}{x-4}$, if $x \neq 4$.

c. Simplify $\dfrac{x^3-4x}{2x^2-x^3}$, if $x \neq 0, x \neq 2$.

SOLUTION

a. $\dfrac{x^2-4}{x-2} = \dfrac{(x-2)(x+2)}{x-2} = x+2$, if $x \neq 2$

b. $\dfrac{x^2-7x+12}{x-4} = \dfrac{(x-3)(x-4)}{x-4} = x-3$, if $x \neq 4$

c. $\dfrac{x^3-4x}{2x^2-x^3} = \dfrac{x(x^2-4)}{x^2(2-x)} = \dfrac{x(x-2)(x+2)}{x^2(2-x)} = \dfrac{x(x-2)(x+2)}{-x^2(x-2)} = -\dfrac{x+2}{x}$,

if $x \neq 0, x \neq 2$

EXAMPLE 19 ▶ **Simplifying Expressions**

After a derivative is taken, the expression frequently must be simplified. Simplify the following expression by multiplying and combining like terms:

$$\frac{(4x-3)(2x-3) - (x^2-3x)(4)}{(4x-3)^2}$$

SOLUTION

$$\frac{(4x-3)(2x-3) - (x^2-3x)(4)}{(4x-3)^2} = \frac{8x^2 - 12x - 6x + 9 - 4x^2 + 12x}{(4x-3)^2}$$

$$= \frac{4x^2 - 6x + 9}{(4x-3)^2}$$

EXAMPLE 20 ▶ **Solving Equations**

Finding where the derivative is zero is an important skill in calculus.

a. Find where $f'(x) = 0$ if $f'(x) = 2x^3 - 2x$.

b. Find where $f'(x) = 0$ if $f'(x) = 4x^3 - 10x^2 - 6x$.

c. Find where $f'(x) = 0$ if $f'(x) = (x^2 - 1)3(x - 5)^2 + (x - 5)^3(2x)$.

d. Find where $f'(x) = 0$ if $f'(x) = \dfrac{x[3(x - 2)^2] - (x - 2)^3}{x^2}$.

SOLUTION

a. Solving $0 = 2x^3 - 2x$ gives

$$0 = 2x^3 - 2x$$
$$0 = 2x(x^2 - 1)$$
$$0 = 2x(x - 1)(x + 1)$$
$$x = 0, x = 1, \text{ and } x = -1 \text{ are the solutions.}$$

Thus, $f'(x) = 0$ at $x = 0$, $x = 1$, and $x = -1$.

b. Solving $0 = 4x^3 - 10x^2 - 6x$ gives

$$0 = 4x^3 - 10x^2 - 6x$$
$$0 = 2x(2x^2 - 5x - 3)$$
$$0 = 2x(2x + 1)(x - 3)$$
$$x = 0, x = -\frac{1}{2}, \text{ and } x = 3 \text{ are the solutions.}$$

Thus, $f'(x) = 0$ at $x = 0$, $x = -\dfrac{1}{2}$, and $x = 3$.

c. Each term contains the factor $(x - 5)$ to some power. Rather than raise this factor to the second and third power, we use factoring to simplify this expression. Factoring $(x - 5)^2$ from each term gives the following:

$$0 = (x^2 - 1)3(x - 5)^2 + (x - 5)^3(2x)$$
$$0 = (x - 5)^2[1(3)(x^2 - 1) + (x - 5)(2x)]$$
$$0 = (x - 5)^2(3x^2 - 3 + 2x^2 - 10x)$$
$$0 = (x - 5)^2(5x^2 - 10x - 3)$$

One of the solutions is $x = 5$. The remaining solutions can be found with the quadratic formula.

$$x = \frac{10 \pm \sqrt{100 - 4(5)(-3)}}{2(5)} = \frac{10 \pm 4\sqrt{10}}{10} = \frac{5 \pm 2\sqrt{10}}{5}$$

Thus, $f'(x) = 0$ at $x = 5$, $x = \dfrac{5 + 2\sqrt{10}}{5}$, and $x = \dfrac{5 - 2\sqrt{10}}{5}$.

d. To find where the fraction equals 0, we set the numerator equal to 0 and solve.

$$0 = x[3(x - 2)^2] - (x - 2)^3$$
$$0 = (x - 2)^2[3x - (x - 2)]$$
$$0 = (x - 2)^2(2x + 2)$$
$$0 = (x - 2)^2 \quad \text{or} \quad 0 = 2x + 2$$
$$0 = x - 2 \qquad\qquad -2x = 2$$
$$x = 2 \qquad\quad \text{or} \quad x = -1$$

Checking these values in the rational function $f'(x) = \dfrac{x[3(x - 2)^2] - (x - 2)^3}{x^2}$

verifies that $f'(x) = 0$ at $x = 2$ and at $x = -1$.

SKILLS CHECK 8.5

The expressions in Exercises 1–4 represent the derivatives of functions. Simplify the expressions.

1. $f'(x) = (2x^3 + 3x + 1)(2x) + (x^2 + 4)(6x^2 + 3)$

2. $f'(x) = (x^2 - 3x)(3x^2) + (x^3 - 4)(2x - 3)$

3. $f'(x) = \dfrac{x^2(3x^2) - (x^3 - 3)(2x)}{(x^2)^2}$

4. Simplify the numerator of

$$f'(x) = \frac{(x^2 - 4)(3x^2 - 6x) - (x^3 - 3x^2 + 2)(2x)}{(x^2 - 4)^2}$$

In calculus, the derivative of $f(x)$ is denoted by $f'(x)$, and the value of the derivative at $x = a$ is $f'(a)$. Evaluate the derivatives in Exercises 5–7.

5. Find $f'(2)$ if $f'(x) = 4x^3 - 3x^2 + 4x - 2$.

6. Find $f'(-1)$ if
$$f'(x) = (x^2 - 1)(6x) + (3x^2 + 1)(2x)$$

7. Find $f'(3)$ if $f'(x) = \dfrac{(x^2 - 1)(2) + (2x)(2x)}{(x^2 - 1)^2}$.

8. Calculus can be used to find the slope of the line tangent to a curve. The slope of the tangent to the curve $y = f(x)$ at $x = a$ is $f'(a)$. Find the slope of the tangent to $y = f(x)$ at $x = 2$ if $f'(x) = 8x^3 + 6x^2 - 5x - 10$.

9. The derivative of a function can be used to find the slope of the line tangent to its graph. If the slope of the line tangent to the graph of $f(x) = x^3$ at $(2, 8)$ (see the following figure) is 12, write the equation of this tangent line.

An important calculation in calculus is the simplification of the difference quotient $\dfrac{f(x + h) - f(x)}{h}$, where $h \neq 0$. In Exercises 10 and 11, find each of the following.

a. $f(x + h)$ **b.** $f(x + h) - f(x)$

c. $\dfrac{f(x + h) - f(x)}{h}$

10. $f(x) = 3x - 2$ **11.** $f(x) = 4x + 5$

12. Consider the function $f(x) = 12 - 2x^2$.

a. Find $\dfrac{f(x + h) - f(x)}{h}$.

b. What does $\dfrac{f(x + h) - f(x)}{h}$ approach as h becomes very close to 0?

Writing the equation defining the function in the form $y = f(x)$ usually makes it easier to find the derivative. Write each of the equations in Exercises 13–16 in this form by solving for y.

13. $8x^2 + 4y = 12$ **14.** $3x^2 - 2y = 6$

15. $9x^3 + 5y = 18$ **16.** $12x^3 = 6y - 24$

The vertex of a parabola occurs where the derivative of its equation is 0, so an equation associated with the equation of a parabola can be solved to find the x-coordinate of the vertex. For each parabola whose equation and "associated equation" are given in Exercises 17–20:

a. Find the solution of the associated equation.

b. Compare the solution with the x-coordinate of the vertex found by evaluating $x = -\dfrac{b}{2a}$ for the parabola with equation $y = ax^2 + bx + c$.

c. Find the vertex of the parabola.

Equation of Parabola	Associated Equation
17. $y = x^2 - 2x + 5$	$0 = 2x - 2$
18. $y = 5 + x - x^2$	$0 = 1 - 2x$
19. $y = 6x^2 - 24x + 15$	$0 = 12x - 24$
20. $y = -(4 + 3x + x^2)$	$0 = -3 - 2x$

21. Between $x = -3$ and $x = 10$, the largest value of the expression $y = x^3 - 9x^2 - 48x + 15$ occurs where $3x^2 - 18x - 48 = 0$.

a. Find the values of x that satisfy
$$3x^2 - 18x - 48 = 0$$

b. Find the value of $y = x^3 - 9x^2 - 48x + 15$ at each of the solutions found in part (a).

c. Which of the solutions from part (a) gives the maximum value of $y = x^3 - 9x^2 - 48x + 15$ on $[-3, 10]$? Test the expression at other values of x to confirm your conclusion.

22. Between $x = -2$ and $x = 4$, the largest value of the expression $y = x^3 - 3x^2 - 9x + 5$ occurs where $3x^2 - 6x - 9 = 0$.

a. Find the values of x that satisfy

$$3x^2 - 6x - 9 = 0$$

b. Find the value of $y = x^3 - 3x^2 - 9x + 5$ at each of the solutions found in part (a).

c. Which of the solutions from part (a) gives the maximum value of $y = x^3 - 3x^2 - 9x + 5$ on $[-2, 4]$? Test the expression at other values of x to confirm your conclusion.

To perform calculus operations on functions, it is frequently desirable or necessary to write the function in the form $y = cx^b$, where b is a constant. Write each of the functions in Exercises 23–26 in this form.

23. $y = 3\sqrt{x}$ **24.** $y = 6\sqrt[3]{x}$

25. $y = 2\sqrt[3]{x^2}$ **26.** $y = 4\sqrt{x^3}$

Write each of the functions in Exercises 27–32 without radicals by using fractional exponents.

27. $y = \sqrt[3]{x^2 + 1}$ **28.** $y = \sqrt[4]{x^3 - 2}$

29. $y = \sqrt[3]{(x^3 - 2)^2}$ **30.** $y = \sqrt{(5x - 3)^3}$

31. $y = \sqrt{x} + \sqrt[3]{2x}$ **32.** $y = \sqrt[3]{(2x)^2} + \sqrt{(4x)^3}$

Equations containing y^2 occasionally must be solved for y. Sometimes this can be accomplished with the root method, and sometimes a method such as the quadratic formula is necessary. Solve the equations in Exercises 33–36 for y.

33. $y^2 + 4x - 3 = 0$ **34.** $5x - y^2 = 12$

35. $y^2 - 6x + y = 0$ **36.** $2x - 2y^2 + y + 4 = 0$

An equation of the form $y = (ax^k + b)^n$ can be thought of as the composition of the functions $f(u) = u^n$ and $u(x) = ax^k + b$ and can be written as $y = u^n$, where $u = ax^k + b$. This form is very useful in calculus. In Exercises 37–39, use this idea to fill in the blanks.

37. $y = (4x^3 + 5)^2$ can be written as $y = u^-$, where $u(x) = $ _____.

38. $y = $ _____ can be written as $y = u^7$, where $u(x) = 5x^2 - 4x$.

39. $y = \sqrt{}$ can be written as $y = u^-$, where $u(x) = x^3 + x$.

40. If $f(u) = u^6$ and $u = x^4 + 5$, then express $6u^5 \cdot 4x^3$ in terms of x alone.

41. If $f(u) = u^{3/2}$ and $u = x^2 - 1$, then express $\frac{3}{2}u^{1/2} \cdot 2x$ in terms of x alone.

In calculus it is often necessary to write expressions in the form cx^n, where n is a constant. Use negative exponents where necessary to write the expressions in Exercises 42–44 with all variables in the numerator of a term. That is, write each term in the form cx^n if the term contains a variable.

42. $y = 3x + \dfrac{1}{x} - \dfrac{2}{x^2}$ **43.** $y = \dfrac{3}{x} - \dfrac{4}{x^2} - 6$

44. $y = 5 + \dfrac{1}{x^3} - \dfrac{2}{\sqrt{x}} + \dfrac{3}{\sqrt[3]{x^2}}$

45. Write $y = \dfrac{1}{(x^2 - 3)^3}$ with no denominator.

Sometimes a derivative contains negative exponents, and simplification requires that all exponents be positive. Write each of the expressions in Exercises 46–49 without negative exponents.

46. $f'(x) = 8x^{-4} + 5x^{-2} + x$

47. $f'(x) = -6x^{-4} + 4x^{-2} + x^{-1}$

48. $f'(x) = -3(3x - 2)^{-3}$

49. $f'(x) = (4x^2 - 3)^{-1/2}(8x)$

Rewrite each of the expressions in Exercises 50–52 as the sum, difference, or multiple of logarithms.

50. $\ln\dfrac{3x - 2}{x + 1}$ **51.** $\log x^3(3x - 4)^5$

52. $\ln\dfrac{\sqrt[4]{4x + 1}}{4x^2}$

Find where the derivative is zero in Exercises 53–56.

53. Find where $f'(x) = 0$ if $f'(x) = 3x^2 - 3x$.

54. Find where $f'(x) = 0$ if $f'(x) = 3x^3 - 14x^2 + 8x$.

55. Find where $f'(x) = 0$ if

$$f'(x) = (x^2 - 4)[3(x - 3)^2] + (x - 3)^3(2x)$$

56. Find where $f'(x) = 0$ if

$$f'(x) = \dfrac{(x + 1)[3(2x - 3)^2 2] - (2x - 3)^3}{(x + 1)^2}$$

The Binomial Theorem

SECTION PREVIEW   **Binomial Expansions**

We introduced some binomial expansions in the Chapter 6 Toolbox. We restate them here, along with some additional expansions.

$$(x + y)^0 = 1$$
$$(x + y)^1 = x + y$$
$$(x + y)^2 = x^2 + 2xy + y^2$$
$$(x + y)^3 = x^3 + 3x^2y + 3xy^2 + y^3$$
$$(x + y)^4 = x^4 + 4x^3y + 6x^2y^2 + 4xy^3 + y^4$$
$$(x + y)^5 = x^5 + 5x^4y + 10x^3y^2 + 10x^2y^3 + 5xy^4 + y^5$$

In this section, we will show how the expansions of powers of binomials can be generalized, with the Binomial Theorem giving the formula for these expansions. The Binomial Theorem uses factorial notation, which is also introduced in this section. ∎

Binomial Coefficients

By looking at the expansions above, we can observe several facts for all but the special case $(x + y)^0 = 1$.

1. The exponents of the first and last terms are the same as the power on the binomial, and they are also equal to the sum of the exponents on every term.

2. The exponent on x decreases by one on each term, and the exponent on y increases by one on each term.

3. In general, the variables in the expansion $(x + y)^n$ are

$$x^n, x^{n-1}y, x^{n-2}y^2, x^{n-3}y^3, \ldots, x^{n-k}y^k, \ldots, xy^{n-1}, y^n$$

We can also observe a special pattern for the coefficients of the terms. Writing just the coefficients gives the following pattern, which is known as **Pascal's triangle**.

$$
\begin{array}{ccccccccccc}
 & & & & & 1 & & & & & \\
 & & & & 1 & & 1 & & & & \\
 & & & 1 & & 2 & & 1 & & & \\
 & & 1 & & 3 & & 3 & & 1 & & \\
 & 1 & & 4 & & 6 & & 4 & & 1 & \\
1 & & 5 & & 10 & & 10 & & 5 & & 1
\end{array}
$$

Observe that except for the 1's at the ends of the rows, each number is the sum of the numbers to the left and right in the row above it. For example, the 6 in the fifth row is equal to the sum of the two 3's to the left and right in the row above it, and each 4 in the fifth row is the sum of 3 and 1.

Using the observations above, we can write the expansion of $(x + y)^6$ as

$$(x + y)^6 = x^6 + 6x^5y + 15x^4y^2 + 20x^3y^3 + 15x^2y^4 + 6xy^5 + y^6$$

We can use Pascal's triangle to find the coefficients of $(x + y)^n$ for larger values of n, but we have an alternative method that is more direct, especially for large values of n. This method uses **factorial notation**, which is defined as follows.

Factorial Notation

For any positive integer n, we define n factorial as

$$n! = n(n - 1)(n - 2)(n - 3) \cdots (3)(2)(1)$$

and we define $0! = 1$

EXAMPLE 1 ▶ **Evaluating Factorials**

Evaluate the following factorials.

a. $3!$ **b.** $6!$ **c.** $8!$ **d.** $0!$ **e.** $(2!)(3!)$ **f.** $\dfrac{6!}{4!\,2!}$

SOLUTION

a. $3! = 3 \cdot 2 \cdot 1 = 6$

b. $6! = 6 \cdot 5 \cdot 4 \cdot 3 \cdot 2 \cdot 1 = 720$

c. $8! = 8 \cdot 7 \cdot 6 \cdot 5 \cdot 4 \cdot 3 \cdot 2 \cdot 1 = 40{,}320$

d. $0! = 1$

e. $(2!)(3!) = (2 \cdot 1)(3 \cdot 2 \cdot 1) = 2 \cdot 6 = 12$

f. $\dfrac{6!}{4!\,2!} = \dfrac{6 \cdot 5 \cdot 4 \cdot 3 \cdot 2 \cdot 1}{4 \cdot 3 \cdot 2 \cdot 1 \cdot 2 \cdot 1} = \dfrac{6 \cdot 5}{2 \cdot 1} = 15$

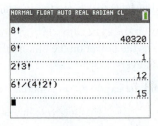

Figure 8.25

We can use a graphing calculator to evaluate factorials. We type the number and then press **MATH**, move to PROB, and choose 4:!. See Figure 8.25.

Notice that $\dfrac{6!}{4!\,2!} = 15$, which is the coefficient of $x^2 y^4$ in

$$(x + y)^6 = x^6 + 6x^5 y + 15x^4 y^2 + 20x^3 y^3 + 15x^2 y^4 + 6xy^5 + y^6$$

In fact, the coefficient of $x^{n-k} y^k$ in the expansion of $(x + y)^n$ is equal to $\dfrac{n!}{k!\,(n - k)!}$.

EXAMPLE 2 ▶ **Finding Binomial Coefficients**

a. Find the coefficient of $x^2 y^5$ in the expansion of $(x + y)^7$.

b. Find the coefficient of $x^6 y^3$ in the expansion of $(x + y)^9$.

SOLUTION

a. The coefficient of $x^2 y^5$ is $\dfrac{7!}{5!\,2!} = \dfrac{7 \cdot 6 \cdot \cancel{5} \cdot \cancel{4} \cdot \cancel{3} \cdot \cancel{2} \cdot \cancel{1}}{\cancel{5} \cdot \cancel{4} \cdot \cancel{3} \cdot \cancel{2} \cdot \cancel{1} \cdot 2 \cdot 1} = \dfrac{7 \cdot 6}{2 \cdot 1} = 21.$

b. The coefficient of $x^6 y^3$ is $\dfrac{9!}{3!\,6!} = \dfrac{9 \cdot 8 \cdot 7 \cdot \cancel{6} \cdot \cancel{5} \cdot \cancel{4} \cdot \cancel{3} \cdot \cancel{2} \cdot \cancel{1}}{3 \cdot 2 \cdot 1 \cdot \cancel{6} \cdot \cancel{5} \cdot \cancel{4} \cdot \cancel{3} \cdot \cancel{2} \cdot \cancel{1}} = \dfrac{9 \cdot 8 \cdot 7}{3 \cdot 2 \cdot 1} = 84.$

The number $\dfrac{n!}{k!\,(n - k)!}$ is called a **binomial coefficient** and can be written as $\dbinom{n}{k}$ or as $_nC_k$. Thus $\dbinom{6}{4} = \dfrac{6!}{4!\,2!} = \dfrac{6 \cdot 5 \cdot \cancel{4} \cdot \cancel{3} \cdot \cancel{2} \cdot \cancel{1}}{\cancel{4} \cdot \cancel{3} \cdot \cancel{2} \cdot \cancel{1} \cdot 2 \cdot 1} = \dfrac{6 \cdot 5}{2 \cdot 1} = 15$ and $_7C_2 = \dfrac{7!}{2!\,5!} = \dfrac{7 \cdot 6}{2 \cdot 1} = 21.$ Note that $\dfrac{n!}{k!\,(n - k)!} = \dfrac{n!}{(n - k)!\,k!}$, so $\dbinom{n}{k} = \dbinom{n}{n - k}$ and

$_nC_k = {}_nC_{n-k}.$

Putting these ideas together, we see that the term of the expansion of $(x + y)^n$ containing $x^{n-k}y^k$ is

$$\binom{n}{k}x^{n-k}y^k = \frac{n!}{k!\,(n-k)!}x^{n-k}y^k$$

and we can write the expansion with the following equation, which is called the **Binomial Theorem**.

Binomial Theorem

For any positive integer n,

$$(x + y)^n = x^n + \binom{n}{1}x^{n-1}y^1 + \binom{n}{2}x^{n-2}y^2 + \binom{n}{3}x^{n-3}y^3 + \cdots +$$

$$\binom{n}{k}x^{n-k}y^k + \cdots + \binom{n}{n-1}x^1y^{n-1} + y^n$$

Because the exponent of y increases from 0 in the first term of the expansion of $(x + y)^n$ to n in the last term, we can write the expansion in sigma notation as

$$(x + y)^n = \sum_{k=0}^{n}\binom{n}{k}x^{n-k}y^k = \sum_{k=0}^{n}\frac{n!}{k!\,(n-k)!}x^{n-k}y^k$$

Fortunately, graphing calculators can be used to evaluate $\binom{n}{k} = {_nC_k}$ for given values of n and k. See Example 3.

EXAMPLE 3 ▶ **Using the Binomial Theorem**

Expand $(x + y)^8$.

SOLUTION

We will first calculate the coefficients for the expansion with a graphing calculator by entering the power n, pressing the **MATH** key, moving to PROB, selecting nCr, and entering the exponent of y. The values for sample terms are shown in Figures 8.26 (a) and (b). In the figure, we can see the *symmetry* of the coefficients, which means we need only find half of the coefficients to know what all of them are. For example, we see that $_8C_2 = {_8C_{8-2}} = {_8C_6}$ and we know that $_8C_8 = 1$ because $_8C_0 = 1$.

Using these coefficients in the Binomial Theorem gives

$$(x + y)^8 = x^8 + 8x^7y + 28x^6y^2 + 56x^5y^3 + 70x^4y^4 + 56x^3y^5 + 28x^2y^6 + 8xy^7 + y^8$$

(a)   (b)

Figure 8.26

EXAMPLE 4 ▶ Finding a Particular Term in a Binomial Expansion

Find the sixth term of $(2x - y)^9$.

SOLUTION

This binomial has the form $(a + b)^9$ with $a = 2x$ and $b = -y$, so we will first find the sixth term of $(a + b)^9$. The sixth term is the term $\binom{n}{k}a^{n-k}b^k$, in which $k = 5$, because the second term has the power $k = 1$. The coefficient of this term, found with factorials or with a graphing calculator, is

$$\binom{9}{5} = {}_9C_5 = 126$$

Thus the sixth term of $(a + b)^9$ is $126a^{9-5}b^5 = 126a^4b^5$. To find the sixth term of $(2x - y)^9$, we substitute $2x$ for a and $-y$ for b.

$$126(2x)^4(-y)^5 = 126(16x^4)(-y^5) = -2016x^4y^5$$

SKILLS CHECK 8.6

Evaluate the following.

1. $\dfrac{8!}{4!\,4!}$

2. $\dfrac{7!}{4!\,3!}$

3. $\dfrac{9!}{0!\,9!}$

4. $\dfrac{7!}{7!\,0!}$

5. $\dfrac{9!}{3!\,6!}$

6. $\dfrac{11!}{7!\,4!}$

7. $\binom{12}{2}$

8. $\binom{13}{10}$

9. $\binom{10}{3}$

10. $\binom{9}{7}$

11. ${}_{11}C_8$

12. ${}_{14}C_{12}$

Expand the following binomial powers.

13. $(a + b)^4$

14. $(m + n)^6$

15. $(x - y)^4$

16. $(x - y)^5$

17. $(x + 2y)^6$

18. $(2x + y)^4$

19. $(3x + 4y)^5$

20. $(5x + 6y)^4$

21. $(3a + b)^4$

22. $(2r - 3s)^5$

23. $(x + y)^7$

24. $(x - y)^7$

Write the indicated term of each binomial expansion.

25. The fifth term of $(m + n)^6$

26. The fourth term of $(a + b)^6$

27. The eighth term of $(x + y)^{10}$

28. The ninth term of $(6x + y)^{11}$

29. The third term of $(3m - n)^5$

30. The fourth term of $(r - 3s)^6$

31. The fourteenth term of $(m - n)^{16}$

32. The twelfth term of $(2r + s)^{14}$

33. The middle term of $(x + y)^{10}$

34. The middle term of $(x + y)^{12}$

SECTION PREVIEW   Circles and Parabolas

In earlier sections, we have discussed the algebraic properties of **circles** and **parabolas**. We will study the geometric properties of these two **conic sections** in this section and of two others, the **ellipse** and the **hyperbola**, in the next section. These special curves are called conic sections because they can be formed by the intersection of a plane and a (double-napped) right circular cone. The intersection curves are illustrated in Figure 8.27. Note that if the plane passes through the vertex of the cone, the intersection is a point or lines, and conic sections are not formed. The resulting point or lines formed by the intersection are called *degenerate conics*.

Figure 8.27 Conic Sections

The geometric properties of these conic sections, which we will discuss further in this chapter, were discovered by the Greeks around 200 BC. New applications of conic sections were developed in the 17th century. Using data from Tyco Brahe's 16th-century observations of planetary movement, Johannes Kepler determined that planets move in elliptical orbits about the sun. We now know that many comets have parabolic paths, and other applications of parabolas include parabolic searchlights, bridge cables, and bridge arches.

Conic sections can be defined as intersections of planes and cones as the Greeks defined them, as the graphs of the **second-degree equation**

$$Ax^2 + Bxy + Cy^2 + Dx + Ey + F = 0$$

or as sets of all points satisfying specific geometric properties. We will use this last definition in our discussion of conic sections in this section and the next. ■

Distance and Midpoint Formulas

Our definitions of the conic sections use the distance between two points, so we will develop the formula used to find this distance. The notation (x_a, y_b) is used to denote a fixed but arbitrary point in the coordinate plane, so the points $P_1(x_1, y_1)$ and $P_2(x_2, y_1)$ have the same y-coordinates but different x-coordinates, while the points $P_2(x_2, y_1)$ and $P_3(x_2, y_2)$ have the same x-coordinates but different y-coordinates. (See Figure 8.28.)

From our definition of distance on a number line, the distance between P_1 and P_2 on the horizontal line is $|x_2 - x_1|$, and the distance between P_2 and P_3 on the vertical line is $|y_2 - y_1|$. By the Pythagorean Theorem, $c^2 = a^2 + b^2$, the distance between P_1 and P_3 is

$$d = \sqrt{|x_2 - x_1|^2 + |y_2 - y_1|^2}$$

Figure 8.28

But since $|a|^2 = a^2$ for every real number a, we have the following formula for the distance between two points.

Distance Formula

The distance between the points $P_1(x_1, y_1)$ and $P_2(x_2, y_2)$ in a plane is

$$d = \sqrt{(x_2 - x_1)^2 + (y_2 - y_1)^2}$$

EXAMPLE 1 ▶   Using the Distance Formula

a. Find the distance from $(-3, 2)$ to $(1, -4)$.

b. Use the distance formula to show that the points $A(-3, 2)$, $B(1, -4)$, and $C(3, -7)$ lie on a line.

SOLUTION

a. $d = \sqrt{[1 - (-3)]^2 + (-4 - 2)^2} = \sqrt{16 + 36} = \sqrt{52} = 2\sqrt{13}$

b. The three points lie on the same line if the sum of the distances between two pairs of points equals the distance between the third pair of points. The distances are

$$\overline{AB} = \sqrt{(1 + 3)^2 + (-4 - 2)^2} = \sqrt{52} = 2\sqrt{13}$$

$$\overline{BC} = \sqrt{(3 - 1)^2 + (-7 + 4)^2} = \sqrt{13}$$

$$\overline{AC} = \sqrt{(3 + 3)^2 + (-7 - 2)^2} = \sqrt{117} = 3\sqrt{13}$$

$\overline{AB} + \overline{BC} = \overline{AC}$, so the points lie on a line.

The midpoint formula is also useful in the study of conic sections. For instance, the center of a circle can be found by the midpoint formula if the endpoints of one of its diameters are known.

Midpoint Formula

The midpoint M of the line segment joining $P_1(x_1, y_1)$ and $P_2(x_2, y_2)$ is

$$M\left(\frac{x_1 + x_2}{2}, \frac{y_1 + y_2}{2}\right)$$

For example, if the endpoints of a diameter of a circle are $(3, 6)$ and $(5, -2)$, the center is at the midpoint of the diameter, $M\left(\dfrac{3 + 5}{2}, \dfrac{6 + (-2)}{2}\right) = M(4, 2)$.

Circles

Each of the conic sections is defined as the set of all points in a plane satisfying a given geometric definition. We define the **circle** as the set of all points in a plane that are equidistant from a fixed point called the center. Recall the distance formula, which states that the distance between the point $P_1(x_1, y_1)$ and the point $P_1(x_2, y_2)$ in a plane is

$$d = \sqrt{(x_2 - x_1)^2 + (y_2 - y_1)^2}$$

Thus any point (x, y) on a circle of radius r with center at (h, k) satisfies the equation

$$r = \sqrt{(x - h)^2 + (y - k)^2}$$

Squaring both sides of this equation gives the equation of the circle of radius r with center at (h, k).

$$r^2 = (x - h)^2 + (y - k)^2$$

Standard Form of the Equation of a Circle

The equation

$$(x - h)^2 + (y - k)^2 = r^2$$

is the standard form of an equation of a circle with center at (h, k) and radius r.

Figures 8.29 and 8.30 show the graphs of $x^2 + y^2 = 3^2$ and $(x + 2)^2 + (y - 3)^2 = 4^2$, respectively. Note that the center of the first circle is at the origin, and the center of the second is at $(-2, 3)$.

Figure 8.29   **Figure 8.30**

EXAMPLE 2 ▶   ## Finding the Equation of a Circle

Write the equation of the circle with center at $(-3, 4)$ and radius 5.

SOLUTION

The standard form of the equation is

$$[x - (-3)]^2 + (y - 4)^2 = 5^2 \quad \text{or} \quad (x + 3)^2 + (y - 4)^2 = 25$$

EXAMPLE 3 ▶   **Finding the Equation of a Circle**

Write the equation of the circle centered at $(3, -1)$ and passing through $(4, 6)$.

SOLUTION

The radius is the distance between $(3, -1)$ and $(4, 6)$:

$$r = \sqrt{(4 - 3)^2 + [6 - (-1)]^2} = \sqrt{1 + 49} = \sqrt{50}$$

Thus, substituting in $(x - h)^2 + (y - k)^2 = r^2$ gives the equation

$$(x - 3)^2 + (y + 1)^2 = 50$$

The *general form* of the equation of a circle is

$$Ax^2 + Ay^2 + Dx + Ey + F = 0$$

By completing the square on this equation, we can convert it to the standard form

$$(x - h)^2 + (y - k)^2 = r^2$$

EXAMPLE 4 ▶   **Converting the Equation of a Circle to Standard Form**

Write $x^2 + y^2 + 4x - 6y - 3 = 0$ in standard form and locate its center and radius.

SOLUTION

The equation is equivalent to

$$(x^2 + 4x) + (y^2 - 6y) = 3$$

We add 4 to complete the square on the x portion and 9 to complete the square on the y portion, and we add 13 on the right side of the equation to preserve the equality:

$$(x^2 + 4x + 4) + (y^2 - 6y + 9) = 3 + 4 + 9$$

Rewriting the trinomials as binomials squared gives

$$(x + 2)^2 + (y - 3)^2 = 16$$

The center is at $(-2, 3)$ and the radius is 4. (See Figure 8.31.)

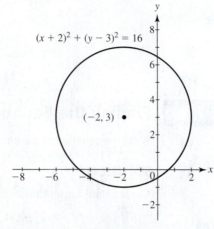

Figure 8.31

Sometimes we encounter functions of the form

$$y = \sqrt{4 - x^2}$$

If we square both sides of this equation, we get the equation

$$y^2 = 4 - x^2 \quad \text{or} \quad x^2 + y^2 = 4$$

whose graph is a circle with center at the origin and radius 2. Because $y = \sqrt{4 - x^2}$ gives only nonnegative values for y, its graph is the upper half of the circle. This graph is the solid curve shown in Figure 8.32(a).

(a)   **(b)**

Figure 8.32

We can graph the equation $x^2 + y^2 = 4$ with technology by graphing the two functions $y = \sqrt{4 - x^2}$ and $y = -\sqrt{4 - x^2}$ on the same axes, as shown Figure 8.32(b).

Note that if the conversion of a quadratic equation from general form to standard form results in an equation of the form $(x - h)^2 + (y - k)^2 = 0$, its graph is a circle of radius 0 (the point (h, k)).

Parabolas

We have seen that a function of the form $y = ax^2 + bx + c \ (a \neq 0)$ has as its graph a parabola that opens upward if $a > 0$, and opens downward if $a < 0$. The following geometric definition applies to parabolas opening in any direction.

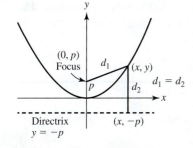

Figure 8.33

> A **parabola** is the set of all points in a plane that are equidistant from a fixed point called the **focus** and a fixed line called the **directrix**. The midpoint between the focus and the directrix is the **vertex**, and the line passing through the vertex and the focus is the **axis** of the parabola. See Figure 8.33.

We can use this definition to derive the standard equation of a parabola with the vertex at the origin. See Exercises 63 and 64.

> ### Standard Equation of a Parabola with Vertex at Origin
>
> The parabola with vertex at $(0, 0)$, focus at $(0, p)$, and directrix $y = -p$ has equation
>
> $$x^2 = 4py \quad \text{(Its axis is vertical; see Figure 8.33.)}$$
>
> The parabola with vertex at $(0, 0)$, focus at $(p, 0)$, and directrix $x = -p$ has equation
>
> $$y^2 = 4px \quad \text{Its axis is horizontal; see Figure 8.34.}$$

Figure 8.34

Note that the equation for a parabola with a vertical axis and vertex $(0, 0)$ can be solved for y, yielding the function

$$y = \frac{1}{4p}x^2$$

Figure 8.35

and the parabola can be graphed using technology as well as by hand. It will open up if $p > 0$ and down if $p < 0$. The graph of the equation $y^2 = 4px$ is a parabola with a horizontal axis, vertex at $(0, 0)$, and focus at $(p, 0)$. It opens to the right if $p > 0$ and to the left if $p < 0$. The equation $y^2 = 4x$ has the form $y^2 = 4px$ with $p = 1$. Thus the graph is a horizontal parabola that opens to the right. The focus is located at $(1, 0)$, and the directrix is $x = -1$. See Figure 8.34.

The equation $y^2 = 4px$ can also be written in the form

$$y = \sqrt{4px} \quad \text{and} \quad y = -\sqrt{4px}$$

The first function gives the upper half of a parabola opening to the right or left, and the second function gives the bottom half. Graphing both functions with technology or by hand on the same axes gives the parabola that opens right or left. See the graphs of $y = \sqrt{4x}$ and $y = -\sqrt{4x}$ in Figure 8.35. Graphing these two functions on the same axes gives the graph of $y^2 = 4x$.

The graphs of equations of the form $x^2 = 4py$ or $y^2 = 4px$ are parabolas with vertices at $(0, 0)$. If the graph of a parabola is the same as the graph of one of these equations except that its vertex is shifted from $(0, 0)$ to (h, k), then its equation is the same except that x is replaced by $x - h$, and y is replaced by $y - k$.

EXAMPLE 5 ▶ **Finding the Equation of a Parabola**

Write the equation of the parabola with vertex at $(-1, 3)$ and focus at $(-3, 3)$.

SOLUTION

The axis contains the vertex at $(-1, 3)$ and focus at $(-3, 3)$, so it is horizontal. The distance from the vertex to the focus is $-3 - (-1) = -2$, so $p = -2$. Thus the standard form of the equation of the parabola is

$$(y - 3)^2 = 4(-2)[x - (-1)] \quad \text{or} \quad (y - 3)^2 = -8(x + 1)$$

EXAMPLE 6 ▶ **Graphing the Equation of a Parabola**

Graph the equation $y^2 + 4x - 6y + 5 = 0$.

SOLUTION

There is a y^2-term and no x^2-term, so the graph is a parabola that opens to the right or left, with standard form $(y - k)^2 = 4p(x - h)$. We write the equation in this form by completing the square as follows.

$$y^2 - 6y = -4x - 5$$
$$y^2 - 6y + 9 = -5 - 4x + 9$$
$$(y - 3)^2 = -4x + 4$$
$$(y - 3)^2 = -4(x - 1)$$

Hence the vertex point is $(1, 3)$. If we select y-values near the vertex, calculate the x values, and plot the points, we obtain the graph shown in Figure 8.36.

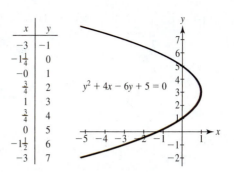

x	y
-3	-1
$-1\frac{1}{4}$	0
-0	1
$\frac{3}{4}$	2
1	3
$\frac{3}{4}$	4
0	5
$-1\frac{1}{2}$	6
-3	7

$y^2 + 4x - 6y + 5 = 0$

Figure 8.36

Figure 8.37

The definition of a parabola is useful in many applications of the parabolic shape. Figure 8.37 shows the parabolic cable on a bridge. Parabolic dishes have the property that entering rays are reflected to the focus and that rays initiated from the focus are reflected outward in parallel lines. These parabolic dishes have applications to satellite transmission, sideline microphones for sporting events, and spotlights.

EXAMPLE 7 ▶ | **Design of a Satellite Dish**

A satellite dish is a paraboloid formed by rotating a parabola about its axis. Its design results in incoming signals parallel to the axis being reflected to the receiver at the focus of the parabola. If the dish is 32 inches across and 4 inches deep, where should the receiver be placed?

SOLUTION

We can represent a cross section of the dish with a parabola that has its vertex at $(0, 0)$ and contains the points $(16, 4)$ and $(-16, 4)$. This parabola has its axis along the y-axis, and we can use the equation $x^2 = 4py$ and one of the points to find p.

$$16^2 = 4p(4)$$
$$256 = 16p$$
$$p = 16$$

Thus the focus is at $(0, 16)$, and the receiver should be 16 inches from the vertex of the dish.

SKILLS CHECK 8.7

1. a. Write the equation of the circle with center at $(-1, 2)$ and radius 3.

   b. Graph the circle.

2. Write the equation of the circle with center at $(2, 5)$ and radius $\sqrt{5}$, and graph the circle.

In Exercises 3–8, find the radius and center of a circle with the given equation.

3. $x^2 + y^2 = 36$

4. $x^2 + y^2 = 49$

5. $(x - 2)^2 + (y + 1)^2 = 14$

6. $(x + 1)^2 + (y - 1)^2 = 1$

7. $(x - 0.5)^2 + (y + 3.2)^2 = 7$

8. $(x + 1.3)^2 + \left(y - \dfrac{3}{2}\right)^2 = \dfrac{9}{16}$

In Exercises 9–26, graph the equation.

9. $x^2 + y^2 - 49 = 0$ 10. $x^2 + y^2 - 18 = 0$

11. $x^2 + y^2 - 6 = 8$ 12. $x^2 + (y - 2)^2 = 20$

13. $(x + 1)^2 + (y + 3)^2 = 1$

14. $(x - 4)^2 + (y - 1)^2 = \dfrac{1}{4}$

15. $\left(x + \dfrac{1}{2}\right)^2 + (y - 2)^2 = 6$

16. $(x + 2)^2 + y^2 = 1$ 17. $x^2 + (y + 4)^2 = 16$

18. $(x - 6)^2 + (y - 6)^2 = 12$

19. $x^2 + y^2 + 4x - 2y - 4 = 0$

20. $x^2 + y^2 - 6x + 10y - 2 = 0$

21. $x^2 + y^2 - x - 4y + 4 = 0$

22. $4x^2 + 4y^2 + 12x - 8y = 3$

23. $x^2 + y^2 - y = 0$ 24. $x^2 - x + y^2 - 2y = 0$

25. $y = \sqrt{9 - x^2}$ 26. $y = \sqrt{16 - x^2}$

In Exercises 27–32, write the equation of the circle with the given center and radius.

27. $(4, 0); 4$ 28. $(-2, 3); 5$ 29. $(-3, -1); 2$

30. $(0, -3); 13$ 31. $(5, -2); 1/2$ 32. $(1, -1); \sqrt{3}$

In Exercises 33–36, write the equation of the shape described.

33. The upper semicircle centered at $(0, 0)$ with radius 3

34. The lower semicircle centered at $(0, 0)$ with radius 5

35. The left semicircle centered at $(0, -1)$ with radius 1

36. The right semicircle centered at $(0, 0)$ with radius 6

37. Write the equation of the circle that has the ends of a diameter at $(1, 6)$ and $(4, -1)$.

38. Write the equation of the circle centered at $(2, -3)$ and passing through the point $(1, 1)$.

39. Write the equation of the circle that is tangent to the x-axis at the point $(-4, 0)$ and passes through the point $(0, 2)$.

40. Write the equation of the circle of radius 5 that is tangent to both axes and passes through the point $(-8, 1)$.

In Exercises 41–60, graph each equation.

41. $y^2 - 12x = 0$ 42. $y^2 - 8x = 0$

43. $y^2 + 4x = 0$ 44. $y^2 + 8x = 0$

45. $x^2 - 4y = 0$ 46. $x^2 - 8y = 0$

47. $x^2 + 12y = 0$ 48. $x^2 + 4y = 0$

49. $y = \sqrt{x - 3}$ 50. $y = \sqrt{2x - 2}$

51. $y = \sqrt{3 - x}$ 52. $y = \sqrt{2 - 3x}$

53. $y^2 - 6y + 4x + 5 = 0$

54. $x^2 + 10x + 8y + 49 = 0$

55. $4x^2 + 12x + 32y + 25 = 0$

56. $4y^2 - 4y + 24x + 37 = 0$

57. $y = x^2 + 10x + 25$ 58. $y = x^2 + x$

59. $y = 3 + x - 2x^2$ 60. $y = 9 - x^2$

EXERCISES 8.7

61. *Bridges* A cable of a suspension bridge is suspended between two towers that are 300 feet apart and 80 feet above the road. The cable is in the shape of a parabola whose low point is 20 feet above the road. Find the equation of this cable if the point $(0, 0)$ is the point on the road directly below the low point of the cable.

62. *Satellite Dishes* A satellite dish is 20 feet across and 5 feet deep. How far from the vertex of the dish should the receiver be placed to receive radio signals coming in parallel to the axis of the dish?

63. *Parabolas* The points on a parabola satisfy the equation $\sqrt{(x - 0)^2 + (y - p)^2} = y + p$ because the distance from the focus $(0, p)$ to any point on the curve, (x, y), must equal the distance from (x, y) to the directrix $y = -p$. Show that this equation can be simplified to obtain the equation $x^2 = 4py$.

64. *Parabolas* Use the definition of a parabola to show that the equation of a parabola with vertex at $(0, 0)$, focus at $(p, 0)$, and directrix $x = -p$ can be written in the form $y^2 = 4px$.

SECTION PREVIEW Ellipses and Hyperbolas

As we mentioned in the previous section, Johannes Kepler determined that planets moved in elliptical orbits about the sun. We now know that objects in space can move in elliptical or hyperbolic paths, as well as parabolic paths. In particular, we know there are more than 1000 comets, many of which have elliptical paths, and many which have hyperbolic paths. Other applications include elliptical whispering chambers and hyperbolic antennas for radar searches. ∎

Ellipses

An **ellipse** is defined as the set of all points in a plane such that the sum of their distances from two fixed points (called **foci**) is a constant (see Figure 8.38). We can see how this definition results in an ellipse by placing two thumbtacks on a corkboard, fastening the ends of a fixed length of string to the tacks, and moving a pencil against the string to sketch the ellipse.

Figure 8.38

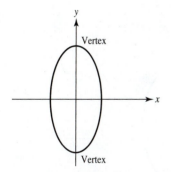

Figure 8.39

The line segment through the foci that ends at its intersection with the ellipse is the **major axis**. For the ellipse in Figure 8.38, the points of intersection are the **vertices** $(a, 0)$ and $(-a, 0)$, and the length of the major axis is $2a$. The line segment through the center perpendicular to the major axis and ending at its intersection with the ellipse is the **minor axis**. The length of the minor axis is $2b$, with $a > b > 0$. The major and minor axes are axes of symmetry for the ellipse. An ellipse with a horizontal major axis is a **horizontal ellipse** (see Figure 8.38), and an ellipse with a vertical major axis is a **vertical ellipse** (see Figure 8.39).

If the foci are on the x-axis (or on the y-axis) equidistant from the origin, we can use the distance formula and the definition of an ellipse to find the following. (See Exercise 53.)

> ### Standard Forms of Equations of Ellipses
>
> The ellipse centered at the origin with vertices at $(a, 0)$ and $(-a, 0)$, foci at $(c, 0)$ and $(-c, 0)$, and endpoints of the minor axis at $(0, b)$ and $(0, -b)$ has the equation
>
> $$\frac{x^2}{a^2} + \frac{y^2}{b^2} = 1 \qquad \text{(See Figure 8.38.)}$$
>
> The ellipse centered at the origin with vertices at $(0, a)$ and $(0, -a)$, foci at $(0, c)$ and $(0, -c)$, and endpoints of the minor axis at $(b, 0)$ and $(-b, 0)$ has the equation
>
> $$\frac{y^2}{a^2} + \frac{x^2}{b^2} = 1 \qquad \text{(See Figure 8.39.)}$$

EXAMPLE 1 ▶ **Graphing the Equation of an Ellipse**

Graph the equation $9x^2 + 25y^2 = 225$.

SOLUTION

This equation can be written in the form

$$\frac{x^2}{25} + \frac{y^2}{9} = 1$$

so we see that it is an ellipse centered at the origin. Since $a^2 = 25$ and $b^2 = 9$, the ellipse intersects the x-axis at 5 and -5, and it intersects the y-axis at 3 and -3. The sketch is shown in Figure 8.40.

Figure 8.40

The graph of

$$\frac{x^2}{4} + \frac{y^2}{16} = 1$$

is a vertical ellipse with center at the origin and with the ends of the longer (major) axis 4 units up and 4 units down from the center. The shorter (minor) axis ends 2 units to the left and 2 units to the right of the center. The graph is shown in Figure 8.41.

The graph of

$$\frac{(x-1)^2}{4} + \frac{(y+2)^2}{16} = 1$$

is the graph in Figure 8.41 shifted to the right 1 unit and down 2 units. The center is shifted from $(0, 0)$ to $(1, -2)$; the *lengths* of the axes are unchanged (see Figure 8.42).

Figure 8.41   **Figure 8.42**

In general, we have the following.

Horizontal and Vertical Shifts of Ellipses

The graph of the equation

$$\frac{(x-h)^2}{a^2} + \frac{(y-k)^2}{b^2} = 1$$

is the graph of the equation $\dfrac{x^2}{a^2} + \dfrac{y^2}{b^2} = 1$ with the center shifted h units horizontally and k units vertically.

The *general form* of the equation of an ellipse with its major axis parallel to the *x*-axis or the *y*-axis is

$$Ax^2 + Cy^2 + Dx + Ey + F = 0 \ (A \neq C, AC > 0)$$

As with the circle, we can complete the square to write the equation in its *standard form*.

EXAMPLE 2 ▶ **Converting the Equation of an Ellipse to Standard Form**

Write the standard form and sketch the graph of $4x^2 + y^2 + 16x - 2y + 1 = 0$.

SOLUTION

From the coefficients of x^2 and y^2 we see that this equation represents an ellipse. We obtain the standard form as follows.

$$4x^2 + y^2 + 16x - 2y + 1 = 0$$

$$4x^2 + 16x + y^2 - 2y = -1 \qquad \text{Grouping}$$

$$4(x^2 + 4x + 4) + (y^2 - 2y + 1) = -1 + 16 + 1 \qquad \text{Factoring and completing squares}$$

(Note that adding 4 inside the first parentheses actually adds $4 \cdot 4$ to the left side, so 16 must also be added to the right side.)

$$4(x + 2)^2 + (y - 1)^2 = 16 \qquad \text{Factoring}$$

$$\frac{(x + 2)^2}{4} + \frac{(y - 1)^2}{16} = 1 \qquad \text{Dividing both sides by 16}$$

Thus the center is $(-2, 1)$; the major axis ends 4 units above and 4 units below the center. The minor axis ends 2 units to the left and 2 units to the right of the center (see Figure 8.43).

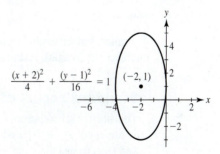

Figure 8.43

As with semicircles, equations of the top half of ellipses with the form

$$y = \sqrt{b^2 - \frac{b^2}{a^2}x^2}$$

can often be graphed with technology or by squaring both sides of the equation. In particular, squaring both sides of

$$y = \sqrt{9 - \frac{9}{25}x^2}$$

Figure 8.44

gives

$$y^2 = 9 - \frac{9}{25}x^2$$

$$\frac{9}{25}x^2 + y^2 = 9 \quad \text{or} \quad \frac{x^2}{25} + \frac{y^2}{9} = 1$$

Thus the graph of

$$y = \sqrt{9 - \frac{9}{25}x^2}$$

is the upper half of this ellipse, because only nonnegative values of y are possible (see Figure 8.44).

Hyperbolas

A hyperbola is defined as the set of all points in a plane such that the difference of their distances from two fixed points (called foci) is a constant. If the foci are on the x-axis equidistant from the origin, the distance formula can be used to develop the equation of this **horizontal hyperbola**, which intercepts the x-axis at its vertices a and $-a$ (see Figure 8.45).

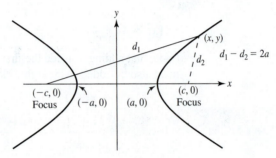

Figure 8.45

Figure 8.45 shows the graph of the horizontal hyperbola with equation

$$\frac{x^2}{a^2} - \frac{y^2}{b^2} = 1 \text{ with } b^2 = c^2 - a^2$$

A similar, but different, equation results in a **vertical hyperbola** (in which the curve intersects the y-axis). A comparison of horizontal and vertical hyperbolas follows.

Standard Forms of Equations of Hyperbolas

The horizontal hyperbola centered at the origin with x-intercepts at $(a, 0)$ and $(-a, 0)$ and foci at $(c, 0)$ and $(-c, 0)$ has the equation

$$\frac{x^2}{a^2} - \frac{y^2}{b^2} = 1$$

with $b^2 = c^2 - a^2$.

The vertical hyperbola centered at the origin with x-intercepts at $(0, a)$ and $(0, -a,)$ and foci at $(0, c)$ and $(0, -c)$ has the equation

$$\frac{y^2}{a^2} - \frac{x^2}{b^2} = 1$$

with $b^2 = c^2 - a^2$.

Figure 8.46 also shows the graph of $\dfrac{x^2}{a^2} - \dfrac{y^2}{b^2} = 1$. Note that the graph does not intersect the y-axis and that the curve approaches, but does not touch, the *asymptotes* (the dashed lines in Figure 8.46). The equations for the asymptotes are found by solving for y, which yields

$$y = \pm \frac{b}{a}\sqrt{x^2 - a^2}$$

and observing that x gets much larger than a as $|x| \to \infty$, so $\sqrt{x^2 - a^2} \to \sqrt{x^2}$. Hence y approaches $\pm \left(\dfrac{b}{a}\right)x$ as $|x| \to \infty$. Thus, if the hyperbola is centered at the origin, the asymptotes are lines through the origin with slopes b/a and $-b/a$, respectively.

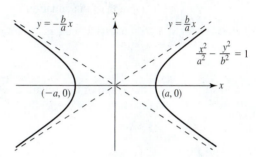

Figure 8.46

| EXAMPLE 3 ▶ | Graphing the Equation of a Hyperbola |

Graph the equation $9x^2 - 4y^2 = 36$.

SOLUTION

Dividing both sides of the equation by 36 gives the equation in the standard form:

$$\frac{x^2}{4} - \frac{y^2}{9} = 1$$

We see that its graph is a horizontal hyperbola with

Center: $(0, 0)$

Vertices: 2 units to left and 2 units to the right of center: $(-2, 0)$, $(2, 0)$

Asymptotes: Through $(0, 0)$ with slopes $3/2$ and $-3/2$

Graph: See Figure 8.47.

Figure 8.47

The standard forms of the equations of hyperbolas that are shifted horizontally and vertically follow.

> ### Horizontal and Vertical Shifts of Hyperbolas
>
> The graph of the equation
>
> $$\frac{(x-h)^2}{a^2} - \frac{(y-k)^2}{b^2} = 1$$
>
> is the graph of the equation $\frac{x^2}{a^2} - \frac{y^2}{b^2} = 1$ with the center shifted h units horizontally and k units vertically.
>
> The graph of the equation
>
> $$\frac{(y-k)^2}{a^2} - \frac{(x-h)^2}{b^2} = 1$$
>
> is the graph of the equation $\frac{y^2}{a^2} - \frac{x^2}{b^2} = 1$ with the center shifted h units horizontally and k units vertically.

The graph of $\dfrac{(x-1)^2}{4} - \dfrac{(y+2)^2}{9} = 1$ is the graph of $\dfrac{x^2}{4} - \dfrac{y^2}{9} = 1$ shifted to the right 1 unit and down 2 units. The center is shifted from $(0, 0)$ to $(1, -2)$; the vertices are still 2 units to the left and to the right of the center, and the asymptotes still pass through the center with slopes $3/2$ and $-3/2$. This graph is shown in Figure 8.48.

Figure 8.48

For graphing purposes, we can convert the general form of a hyperbola,

$$Ax^2 + Cy^2 + Dx + Ey + F = 0 \quad (AC < 0)$$

to its standard form by completing the square.

EXAMPLE 4 ▶ **Converting the Equation of a Hyperbola to Standard Form**

Write the standard form and sketch the graph of

$$x^2 - 4y^2 - 2x - 8y + 1 = 0$$

SOLUTION

Note from the coefficients of x^2 and y^2 that this equation represents a hyperbola. We obtain the standard form as follows.

$$(x^2 - 2x) - 4(y^2 + 2y) = -1 \qquad \text{Grouping}$$
$$(x^2 - 2x + 1) - 4(y^2 + 2y + 1) = -1 + 1 - 4 \qquad \text{Completing squares}$$
$$(x - 1)^2 - 4(y + 1)^2 = -4$$

$$\frac{(x - 1)^2}{-4} + \frac{(y + 1)^2}{1} = 1$$

$$\frac{(y + 1)^2}{1} - \frac{(x - 1)^2}{4} - = 1 \qquad \text{Standard form}$$

This is the standard form for the equation of a vertical hyperbola with

Center: $(1, -1)$

Vertices: 1 unit above and 1 unit below center: $(1, 0)$, $(1, -2)$

Asymptotes: Through center with slopes $\pm 1/2$

Graph: See Figure 8.49.

$$\frac{(y + 1)^2}{1} - \frac{(x - 1)^2}{4} = 1$$

$m = \frac{1}{2}$ $m = -\frac{1}{2}$

Figure 8.49

Rectangular Hyperbolas

Although we have not discussed quadratic equations containing xy terms, there is one other important equation form that has the hyperbola as its graph. This equation has the form $xy = c$, and its graph is called a **rectangular hyperbola**. If $c > 0$, the graph has the form shown in Figure 8.50(a), and if $c < 0$, the graph has the form shown in Figure 8.50(b).

$xy = c$, or $y = \frac{c}{x}$
$(c > 0)$

$xy = c$, or $y = \frac{c}{x}$
$(c < 0)$

(a)   (b)

Figure 8.50

SKILLS CHECK 8.8

In Exercises 1–16, sketch the graph of each equation. In Exercises 1–12, give the center and the endpoints of the major axis.

1. $\dfrac{x^2}{9} + \dfrac{y^2}{4} = 1$

2. $\dfrac{x^2}{9} + \dfrac{y^2}{5} = 1$

3. $4x^2 + y^2 = 16$

4. $6x^2 + 49y^2 - 294 = 0$

5. $x^2 + 4y^2 = 1$

6. $2x^2 + 3y^2 = 6$

7. $\dfrac{(x + 2)^2}{4} + \dfrac{(y - 1)^2}{16} = 1$

8. $\dfrac{(x - 1)^2}{9} + y^2 = 1$

9. $x^2 + 4y^2 + 4x - 8y + 4 = 0$

10. $2x^2 + y^2 - 4x + 6y + 7 = 0$

11. $9x^2 + y^2 + 6y = 0$ **12.** $4x^2 + 9y^2 - 24x = 0$

13. $y = \dfrac{\sqrt{1 - 4x^2}}{3}$

14. $y = -\sqrt{4 - 9x^2}$

15. $x = \dfrac{-2\sqrt{225 - y^2}}{3}$ **16.** $x = 2\sqrt{1 - y^2}$

In Exercises 17–20, determine whether the graph of each equation is a circle or an ellipse.

17. $\dfrac{y^2}{16} + \dfrac{x^2}{25} = 1$ **18.** $\dfrac{x^2}{50} + \dfrac{y^2}{50} = 1$

19. $9x^2 + 9y^2 + 18x - 36y - 180 = 0$

20. $4x^2 + 2y^2 + 16x - 8y - 82 = 0$

21. Sketch the graph of the ellipse with center at $(0, 0)$, vertical major axis 10 units long, and minor axis 2 units long.

22. Sketch the graph of the ellipse that has a minor axis half the length of its major axis and that has major axis endpoints at $(3, 16)$ and $(3, -12)$.

In Exercises 23–46, sketch the graph of each equation. Give the center and the slopes of the asymptotes in Exercises 23–34.

23. $\dfrac{x^2}{16} - \dfrac{y^2}{25} = 1$ **24.** $\dfrac{x^2}{25} - \dfrac{y^2}{9} = 1$

25. $y^2 - x^2 = 1$ **26.** $5y^2 - x^2 - 20 = 0$

27. $y^2 - 4x^2 = 4$ **28.** $y^2 - 4x^2 - 6 = 16$

29. $\dfrac{(x+1)^2}{36} - \dfrac{(y+4)^2}{36} = 1$

30. $\dfrac{(y+3)^2}{4} - \dfrac{(x-0.5)^2}{9} = 1$

31. $x^2 - 4y^2 - 24y - 40 = 0$

32. $4x^2 - 4y^2 + 36x + 77 = 0$

33. $5x^2 - 4y^2 + 20x - 16y + 24 = 0$

34. $x^2 - 4y^2 - 8x + 56y - 164 = 0$

35. $y = -\sqrt{1 + x^2}$ **36.** $y = \sqrt{4 + 9x^2}$

37. $x = \dfrac{\sqrt{9 + y^2}}{3}$ **38.** $x = \dfrac{-2\sqrt{25 + y^2}}{5}$

39. $xy = 16$ **40.** $xy = 2$

41. $xy = -\dfrac{1}{2}$ **42.** $xy = -9$

43. $(x - 3)y = 4$ **44.** $(x + 1)(y - 1) = 9$

45. The left half of the hyperbola $x^2 - y^2 = 1$

46. The upper half of the hyperbola $y^2 - 9x^2 = 36$

EXERCISES 8.8

47. *Bridges* An arch under a bridge is the upper half of an ellipse with a width of 100 feet and a height of 16 feet at its highest point. What is the height of the arch 19 feet from its center?

48. *Bridges* An arch of a bridge across a stream is in the shape of half an ellipse. If the span is 50 feet wide and 12 feet high, write the equation of an ellipse whose upper half represents this arch.

49. *Halley's Comet* The orbit of Halley's Comet is an ellipse 36.18 AU (astronomical units) long and 9.12 AU wide. Write the equation of an ellipse that represents this path.

50. *Roman Colosseum* The outside of the Roman Colosseum is an ellipse with a major axis of length 620 feet and a minor axis of length 513 feet. Write the equation of this ellipse.

51. *Ohm's Law* Ohm's law states that $E = IR$, where E is the voltage in a circuit, I is the current, and R is the resistance. In a circuit where voltage is a constant 120 volts, sketch the graph of current as a function of resistance.

52. *Boyle's Law* Boyle's law states that for a perfect gas at a constant temperature, the product of its pressure P and its volume V is a constant. Sketch the graph of pressure as a function of volume if their product is 24.

53. *Ellipses* The ellipse in Figure 8.38 is the set of all points such that the sum of the distances from the foci, $(-c, 0)$ and $(c, 0)$, to any point (x, y) on the ellipse is a constant, $2a$. Thus any point (x, y) on this ellipse satisfies the equation

$$\sqrt{(x + c)^2 + y^2} + \sqrt{(x - c)^2 + y^2} = 2a$$

Show that if we let $b^2 = a^2 - c^2$, this equation can be written in the form

$$\frac{x^2}{a^2} + \frac{y^2}{b^2} = 1$$

54. *Hyperbolas* The hyperbola in Figure 8.45 is the set of all points such that the difference of the distances from the foci, $(-c, 0)$ and $(c, 0)$, to any point (x, y) on the hyperbola is a constant, $\pm 2a$. Thus any point (x, y) on this hyperbola satisfies the equation

$$\sqrt{(x + c)^2 + y^2} - \sqrt{(x - c)^2 + y^2} = 2a$$

Show that if we let $b^2 = c^2 - a^2$, this equation can be written in the form

$$\frac{x^2}{a^2} - \frac{y^2}{b^2} = 1$$

chapter 8 SUMMARY

In this chapter, we graphed linear inequalities in two variables and systems of linear inequalities in two variables as regions in the two-dimensional plane. We discussed an important application of linear inequalities, linear programming, where we seek a point or points in a region determined by constraints that give an optimal value for an objective function.

We also discussed sequences, which are special discrete functions whose domains are sets of positive integers. Sequences can be used to solve numerous problems involving discrete inputs, including investment problems. We also discussed series, which are the sums of terms of a sequence. Series are useful in finding the future value of annuities if the interest is compounded over discrete time periods.

We showed how important algebra is in the development and application of calculus. Finally, we discussed the Binomial Theorem and conic sections (circles, parabolas, ellipses, and hyperbolas).

Key Concepts and Formulas

8.1 Systems of Inequalities

Linear inequalities in two variables	The solution to a linear inequality in two variables is the set of ordered pairs that satisfy the inequality.
Systems of linear inequalities	The solution to a system of linear inequalities in two or more variables consists of the values of the variables that satisfy all the inequalities.
Borders	The lines that bound the solution region are called the borders of the system of linear inequalities.
Solution region	The region that includes the half-plane containing the ordered pairs that satisfy an inequality is called the solution region. The border line is included in the solution region if the inequality includes an equal sign.
Corner	A corner of the graph of a solution to a system of inequalities occurs at a point where boundary lines intersect.
Systems of nonlinear inequalities	As with systems of linear inequalities in two variables, the solutions of systems of nonlinear inequalities in two variables are regions in a plane.

8.2 Linear Programming: Graphical Methods

Linear programming	Linear programming is a technique that can be used to solve problems when the constraints on the variables can be expressed as linear inequalities and a linear objective function is to be maximized or minimized.
Constraints	The inequalities that limit the values of the variables in a linear programming application are called the constraints.
Feasible region	The constraints of a linear programming application form a feasible region in which the solution lies.
Feasible solution	Any point in the region determined by the constraints is called a feasible solution.
Optimal value	A maximum or minimum value of an objective function is called an optimal value.

Solving a linear programming problem	1. If a linear programming problem has a solution, then the optimal value (maximum or minimum) of an objective function occurs at a corner of the feasible region determined by the constraints.
	2. When a feasible region for a linear programming problem is closed and bounded, the objective function has a maximum and a minimum value.
	3. When the feasible region is not closed and bounded, the objective function may have a maximum only, a minimum only, or neither.

8.3 Sequences and Discrete Functions

Sequence	A sequence is a function whose outputs result uniquely from positive integer inputs. Because the domains of sequences are positive integers, sequences are discrete functions.
Terms of the sequence	The ordered outputs corresponding to the integer inputs of a sequence are called the terms of the sequence.
Infinite sequence	If the domain of a sequence is the set of all positive integers, the outputs form an infinite sequence.
Finite sequence	If the domain of a sequence is a set of positive integers from 1 to n, the outputs form a finite sequence.
Arithmetic sequence	A sequence is called an arithmetic sequence if there exists a number d, called the common difference, such that $a_n = a_{n-1} + d$ for $n > 1$.
nth term of an arithmetic sequence	The nth term of an arithmetic sequence is given by $a_n = a_1 + (n - 1)d$, where a_1 is the first term of the sequence, n is the number of the term, and d is the common difference between the terms.
Geometric sequence	A sequence is called a geometric sequence if there exists a number r, the common ratio, such that $a_n = ra_{n-1}$ for $n > 1$.
nth term of a geometric sequence	The nth term of a geometric sequence is given by $a_n = a_1 r^{n-1}$, where a_1 is the first term of the sequence, n is the number of the term, and r is the common ratio of consecutive terms.

8.4 Series

Series	A series is the sum of the terms of a sequence.
Finite series	A finite series is defined by $a_1 + a_2 + a_3 + \cdots + a_n$, where $a_1, a_2, \ldots a_n$ are the terms of a sequence.
Infinite series	An infinite series is defined by $a_1 + a_2 + a_3 + \cdots + a_n + \cdots$.
Sum of n terms of an arithmetic sequence	The sum of the first n terms of an arithmetic sequence is given by the formula $s_n = \dfrac{n(a_1 + a_n)}{2}$, where a_1 is the first term of the sequence and a_n is the nth term.
Sum of n terms of a geometric sequence	The sum of the first n terms of a geometric sequence is $s_n = \dfrac{a_1(1 - r^n)}{1 - r}$, where a_1 is the first term of the sequence and r is the common ratio, where $r \neq 1$.
Infinite geometric series	An infinite geometric series sums to $S = \dfrac{a_1}{1 - r}$, where a_1 is the first term and r is the common ratio, if $|r| < 1$. If $|r| \geq 1$, the sum does not exist (is not a finite number).

8.5 Preparing for Calculus

Algebra review	Algebra topics from Chapters 1–6 are presented in calculus contexts.

8.6 The Binomial Theorem

Pascal's triangle	Pascal's triangle can be used to find the coefficients of $(x + y)^n$ for larger values of n.
Factorial notation	For any positive integer n, we define n factorial as $$n! = n(n - 1)(n - 2)(n - 3) \cdots (3)(2)(1)$$ and we define $0! = 1$.
Binomial coefficients	The coefficient of $x^{n-k}y^k$ in the expansion of $(x + y)^n$ is equal to $\dfrac{n!}{k!(n - k)!}$.
Binomial theorem	For any positive integer n, $$(x + y)^n = x^n + \binom{n}{1}x^{n-1}y^1 + \binom{n}{2}x^{n-2}y^2 + \binom{n}{3}x^{n-3}y^3 + \cdots$$ $$+ \binom{n}{k}x^{n-k}y^k + \cdots + \binom{n}{n-1}x^1 y^{n-1} + y^n$$

8.7 Conic Sections: Circles and Parabolas

Conic sections	Conic sections can be defined as intersections of planes and cones as the Greeks defined them, as the graphs of the second-degree equation $Ax^2 + Bxy + Cy^2 + Dx + Ey + F = 0$, or as sets of all points satisfying specific geometric properties.
The distance formula	The distance between the points $P_1(x_1, y_1)$ and $P_2(x_2, y_2)$ in a plane is $d = \sqrt{(x_2 - x_1)^2 + (y_2 - y_1)^2}$.
The midpoint formula	The midpoint M of the line segment joining $P_1(x_1, y_1)$ and $P_2(x_2, y_2)$ is $M\left(\dfrac{x_1 + x_2}{2}, \dfrac{y_1 + y_2}{2}\right)$.
Equation of a circle	The equation $(x - h)^2 + (y - k)^2 = r^2$ is the standard form of an equation of a circle with center at (h, k) and radius r.
Parabolas	A parabola is the set of all points in a plane that are equidistant from a fixed point called the focus and a fixed line called the directrix. The midpoint between the focus and the directrix is the vertex, and the line passing through the vertex and the focus is the axis of the parabola.
Equation of a parabola	The parabola with vertex at $(0, p)$ and directrix $y = -p$ has equation $x^2 = 4py$, and its axis is vertical.
	The parabola with vertex at $(p, 0)$ and directrix $x = -p$ has equation $y^2 = 4px$, and its axis is horizontal.
Shifted graphs of parabolas	The graphs of equations of the form $x^2 = 4py$ or $y^2 = 4px$ are parabolas with vertices at $(0, 0)$. If a parabola is shifted, its graph is the same as the graph of one of these equations except that its vertex is shifted from $(0, 0)$ to (h, k), and its equation is the same except that x is replaced by $x - h$, and y is replaced by $y - k$.

8.8 Conic Sections: Ellipses and Hyperbolas

Ellipses	An ellipse is defined as the set of all points in a plane such that the sum of their distances from two fixed points (called foci) is a constant.
Horizontal ellipse	The ellipse centered at the origin with vertices at $(a, 0)$ and $(-a, 0)$, foci at $(c, 0)$ and $(-c, 0)$, and endpoints of the minor axis at $(0, b)$ and $(0, -b)$ has the equation $$\frac{x^2}{a^2} + \frac{y^2}{b^2} = 1.$$
Vertical ellipse	The ellipse centered at the origin with vertices at $(0, a)$ and $(0, -a)$, foci at $(0, c)$ and $(0, -c)$, and endpoints of the minor axis at $(b, 0)$ and $(-b, 0)$ has the equation $$\frac{y^2}{a^2} + \frac{x^2}{b^2} = 1.$$
Shifted graphs of ellipses	The graph of the equation $$\frac{(x - h)^2}{a^2} + \frac{(y - k)^2}{b^2} = 1$$ is the graph of the equation $\frac{x^2}{a^2} + \frac{y^2}{b^2} = 1$ with the center shifted h units horizontally and k units vertically.
Hyperbolas	A hyperbola is defined as the set of all points in a plane such that the difference of their distances from two fixed points (called foci) is a constant.
Shifted graphs of hyperbolas	The graph of the equation $\frac{(x - h)^2}{a^2} - \frac{(y - k)^2}{b^2} = 1$ is the graph of the equation $\frac{x^2}{a^2} - \frac{y^2}{b^2} = 1$ with the center shifted h units horizontally and k units vertically. The graph of the equation $\frac{(y - k)^2}{a^2} - \frac{(x - h)^2}{b^2} = 1$ is the graph of the equation $\frac{y^2}{a^2} - \frac{x^2}{b^2} = 1$ with the center shifted h units horizontally and k units vertically.

chapter 8   SKILLS CHECK

Graph the inequalities in Exercises 1 and 2.

1. $5x + 2y \leq 10$

2. $5x - 4y > 12$

Graph the solutions to the systems of inequalities in Exercises 3–6. Identify the corners of the solution regions.

3. $\begin{cases} 2x + y \leq 3 \\ x + y \leq 2 \\ x \geq 0, y \geq 0 \end{cases}$

4. $\begin{cases} 3x + 2y \leq 6 \\ 3x + 6y \leq 12 \\ x \geq 0, y \geq 0 \end{cases}$

5. $\begin{cases} 2x + y \leq 30 \\ x + y \leq 19 \\ x + 2y \leq 30 \\ x \geq 0, y \geq 0 \end{cases}$

6. $\begin{cases} 2x + y \leq 10 \\ x + 2y \leq 11 \\ x \geq 0, y \geq 0 \end{cases}$

Graph the solutions to the systems of nonlinear inequalities in Exercises 7 and 8.

7. $\begin{cases} 15x - x^2 - y \geq 0 \\ y - \dfrac{44x + 60}{x} \geq 0 \\ x \geq 0, y \geq 0 \end{cases}$

8. $\begin{cases} x^3 - 26x + 100 - y \geq 0 \\ 19x^2 - 20 - y \leq 0 \\ x \geq 0, y \geq 0 \end{cases}$

9. Minimize $g = 3x + 4y$ subject to the constraints

$$\begin{cases} 3x + y \geq 8 \\ 2x + 5y \geq 14 \\ x \geq 0, y \geq 0 \end{cases}$$

10. Maximize the function $f = 7x + 12y$ subject to the constraints

$$\begin{cases} 7x + 3y \le 105 \\ 2x + 5y \le 59 \\ x + 7y \le 70 \\ x \ge 0, y \ge 0 \end{cases}$$

11. Find the maximum value of $f = 3x + 5y$ and the values of x and y that give that value, subject to the constraints

$$\begin{cases} 2x + y \le 30 \\ x + y \le 19 \\ x + 2y \le 30 \\ x \ge 0, y \ge 0 \end{cases}$$

Determine whether the sequences in Exercises 12–14 are arithmetic or geometric. Then state the value of either r or d.

12. $\dfrac{1}{9}, \dfrac{2}{3}, 4, 24, \ldots$

13. $4, 16, 28, \ldots$

14. $16, -12, 9, \dfrac{-27}{4}$

15. Find the fifth term of the geometric sequence with first term 64 and eighth term $\dfrac{1}{2}$.

16. Find the sixth term of the geometric sequence $\dfrac{1}{9}, \dfrac{1}{3}, 1, \ldots$.

17. Find the sum of the first 10 terms of the geometric sequence with first term 5 and common ratio -2.

18. Find the sum of the first 12 terms of the sequence $3, 6, 9, \ldots$.

19. Find $\displaystyle\sum_{I=1}^{\infty} \left(\dfrac{4}{5}\right)^i$, if possible.

20. Simplify the expression

$$(x^3 + 2x + 3)(2x) + (x^2 - 5)(3x^2 + 2)$$

21. Between $x = 0$ and $x = 10$, the smallest value of the expression $y = 3x^4 - 32x^3 - 54x^2$ occurs where $x^3 - 8x^2 - 9x = 0$. Find the values of x that satisfy $x^3 - 8x^2 - 9x = 0$.

22. Write $y = x + \dfrac{1}{x^2} - \dfrac{2}{\sqrt{x^3}} + \dfrac{3}{\sqrt[3]{x}}$ with each term in the form cx^n.

23. Write $f'(x) = 8x^{-3} + 5x^{-1} + 4x$ with no negative exponents.

24. Find where $f'(x) = 0$ if $f'(x) = 2x^3 + x^2 - x$.

25. Evaluate $\dfrac{6!}{5!3!}$.

26. Evaluate $\dbinom{9}{4}$.

27. Evaluate $_{10}C_7$.

28. Expand the binomial power: $(x + 2y)^4$

29. Expand the binomial power: $(2a - 3b)^6$

30. Write the fourth term of $(m - n)^7$.

31. Write the sixth term of $(2x + 3y)^9$.

32. Write the equation of a circle with center at $(2, -1)$ and radius 3. Graph the circle.

33. Find the radius and center of the circle $x^2 - 6x + y^2 - 2y = 15$.

34. Write the equation of the lower semicircle centered at $(0, 0)$ with radius 6.

35. Graph the equation $y^2 + 6x = 0$.

36. Graph the equation $y = x^2 - 8x + 16$.

37. Graph the equation $4y^2 - 4y + 12x + 37 = 0$.

38. Graph the equation $\dfrac{x^2}{4} + \dfrac{y^2}{9} = 1$, and give the center and the endpoints of the major axis.

39. Graph the equation $\dfrac{(x-2)^2}{16} + \dfrac{(y+1)^2}{4} = 1$, and give the center and the endpoints of the major axis.

40. Sketch the graph of $y = -\sqrt{9 - 4x^2}$.

41. Sketch the graph of the equation $\dfrac{x^2}{9} - \dfrac{y^2}{25} = 1$.

42. Sketch the graph of the equation $16y^2 - 9x^2 = 64$.

43. Sketch the graph of the equation $xy = -12$.

44. **Manufacturing** Ace Manufacturing produces two types of DVD players: the Deluxe model and the Superior model, which also plays videotapes. Each Deluxe model requires 3 hours to assemble and $40 for assembly parts, and each Superior model requires 2 hours to assemble and $60 for assembly parts. The company is limited to 1800 assembly hours and $36,000 for assembly parts each month. If the profit on the Deluxe model is $30 and the profit on the Superior model is $40, how many of each model should be produced to maximize profit? What is the maximum profit?

45. **Manufacturing** A company manufactures two grades of steel, cold steel and stainless steel, at the Pottstown and Ethica factories. The table gives the daily cost of operation and the daily production capabilities of each of the factories, as well as the number of units of each type of steel that is required to fill an order.

    a. How many days should each factory operate to fill the order at minimum cost?

    b. What is the minimum cost?

	Units/ Day at Pottstown	Units/ Day at Ethica	Required for Order
Units of Cold Steel	20	40	At least 1600
Units of Stainless Steel	60	40	At least 2400
Cost/Day for Operation ($)	20,000	24,000	

46. **Nutrition** A laboratory wishes to purchase two different feeds, A and B, for its animals. Feed A has 2 units of carbohydrates per pound and 4 units of protein per pound, and feed B has 8 units of carbohydrates per pound and 2 units of protein per pound. Feed A costs $1.40 per pound, and feed B costs $1.60 per pound. If at least 80 units of carbohydrates and 132 units of protein are required, how many pounds of each feed are required to minimize the cost? What is the minimum cost?

47. **Manufacturing** A company manufactures leaf blowers and weed wackers. One line can produce 260 leaf blowers per day, and another line can produce 240 weed wackers per day. The combined number of leaf blowers and weed wackers that shipping can handle is 460 per day. How many of each should be produced daily to maximize company profit if the profit is $5 on the leaf blowers and $10 on the weed wackers?

48. **Profits** A woman has a building with 60 two-bedroom and 40 three-bedroom apartments available to rent to students. She has set the rent at $800 per month for the two-bedroom units and $1150 for the three-bedroom units. She must rent to one student per bedroom, and zoning laws limit her to at most 180 students in this building. How many of each type of apartment should she rent to maximize her revenue?

49. **Loans** A finance company has at most $30 million available for auto and home equity loans. The auto loans have an annual return rate of 8% to the company, and the home equity loans have an annual rate of return of 7% to the company. If there must be at least twice as many auto loans as home equity loans, how much should be put into each type of loan to maximize the profit to the finance company?

50. **Salaries** Suppose you are offered two identical jobs, job 1 paying a starting salary of $20,000 with yearly raises of $1000 at the end of each year and job 2 paying a starting salary of $18,000 with yearly increases of $1600 at the end of each year.

    a. Which job will pay more in the fifth year on the job?

    b. Which job will pay more over the first 5 years?

51. **Drug in the Bloodstream** Suppose a 400-milligram dose of heart medicine is taken daily. During each 24-hour period, the body eliminates 40% of the drug (so that 60% remains in the body). Thus, the amount of drug in the body just after the 21 doses, over 21 days, is given by

$$400 + 400(0.6) + 400(0.6)^2 + 400(0.6)^3 + \cdots$$
$$+ 400(0.6)^{19} + 400(0.6)^{20}$$

Find the level of the drug in the bloodstream at this time.

52. **Chess Legend** Legend has it that when the king of Persia offered to reward the inventor of chess with any prize he wanted, the inventor asked for one grain of wheat on the first square of the chessboard, with the number of grains doubled on each square thereafter for the remaining 63 squares.

    a. How many grains of wheat would there be on the 64th square?

    b. What is the total number of grains of wheat for all 64 squares? (This is enough wheat to cover Alaska more than 3 inches deep in wheat.)

53. **Compound Interest** Find the future value of $20,000 invested for 5 years at 6%, compounded annually.

54. **Annuities** Find the 5-year future value of an ordinary annuity with a contribution of $300 per month into an account that pays 12% per year, compounded monthly.

55. **Profit** The weekly total cost function for a product is $C(x) = 3x^2 + 1228$ dollars and the weekly revenue is $R(x) = 177.50x$, where x is the number of hundreds of units. Use graphical methods to find the number of units that gives profit for the product.

56. **Bridges** The towers of a suspension bridge are 800 feet apart and rise 160 feet above the ground. The cable between the towers has the shape of a parabola, and the cable just touches the sides of the road midway between the towers. What is the height of the cable 100 feet from a tower?

57. **Archway** An archway is the upper half of an ellipse and has a height of 20 feet and a width of 50 feet. Can a truck 14 feet high and 10 feet wide drive under the archway without going into the other lane?

58. **Construction** Two houses are designed so that they are shaped and positioned like the branches of the hyperbola whose equation is $625y^2 - 900x^2 = 562,500$, where x and y are in yards. How far apart are the houses at their closest point?

Group Activity
▶ EXTENDED APPLICATIONS

Salaries

In the 1970s, a local teachers' organization in Pennsylvania asked for a $1000 raise at the end of each year for the next 3 years. When the school board said it could not afford to do this, the teachers' organization then said that the teachers would accept a $300 raise at the end of each 6-month period. The school board initially responded favorably because the members thought it would save them $400 per year per teacher.

As an employee, would you rather be given a raise of $1000 at the end of each year (Plan I) or be given a $300 raise at the end of each 6-month period (Plan II)? To answer the question, complete the table for an employee whose base salary is $40,000 (or $20,000 for each 6-month period), and then answer the questions below.

1. Find the total additional money earned for the first 3 years from the raises of Plan I.
2. Find the total additional money earned for the first 3 years from the raises of Plan II.
3. Which plan gives more money from the raises, and how much more? Which plan is better for the employee?
4. Find the total additional money earned (per employee) from raises for each of Plan I and Plan II if they are extended to 4 years. Which plan gives more money in raises, and how much more? Which plan is better for the employee?
5. Did the school board make a mistake in thinking that the $300 raise every 6 months instead of $1000 every year would save money? If so, what was the mistake?
6. If there were 200 teachers in the school district, how much extra would Plan II cost over Plan I for the 4 years?

Year	Period (in months)	Salary Received per 6-Month Period ($) Plan I	Salary Received per 6-Month Period ($) Plan II
1	0–6	20,000	20,000
	6–12	20,000	20,300
2	12–18	20,500	20,600
	18–24	20,500	20,900
3	24–30		
	30–36		
Total for 3 years			

Appendix A

Basic Graphing Calculator Guide

Operating the TI-84 Plus and TI-84 Plus C Silver Edition

Turning the Calculator On and Off

ON	Turns the calculator on
2ND **ON**	Turns the calculator off

Adjusting the Display Contrast

2ND ⌃	Increases the contrast (darkens the screen)
2ND ⌄	Decreases the contrast (lightens the screen)

Note: If the display begins to dim (especially during calculations), then batteries are low and you should replace the batteries or charge the calculator.

The TI-84 Plus keyboards are divided into four zones: graphing keys, editing keys, advanced function keys, and scientific calculator keys (Figure 1).

Homescreen

Graphing keys

Editing keys (Allow you to edit expressions and variables)

Advanced function keys (Display menus that access advanced functions)

Scientific calculator keys

Figure 1

Keystrokes on the TI-84 Plus, and TI-84 Plus C Silver Edition

ENTER	Executes commands or performs a calculation
2ND	Pressing the 2ND key *before* another key accesses the character located above the key and printed in blue
ALPHA	Pressing the ALPHA key *before* another key accesses the character located above the key and printed in green
2ND ALPHA	Locks in the ALPHA keyboard
CLEAR	Pressing CLEAR once while typing on a line clears the line
	Pressing CLEAR after pressing ENTER clears the screen
2ND MODE	Returns to the homescreen
DEL	Deletes the character at the cursor
2ND DEL	Inserts characters at the underline cursor
X,T,θ,n	Enters an X in Function mode, a T in Parametric mode, a θ in Polar mode, or an n in Sequence mode
STO►	Stores a value to a variable
^	Raises to an exponent
2ND ^	The number π
(−)	Negative symbol
MATH 1	Converts a rational number to a fraction
MATH ► 1: abs(	Computes the absolute value of the number or expression in parentheses
2ND ENTER	Recalls the last entry
ALPHA ·	Used to enter more than one expression on a line
2ND (−)	Recalls the most recent answer to a calculation
x^2	Squares a number or an expression
x^{-1}	Inverse; can be used with a real number or a matrix
2ND x^2	Computes the square root of a number or an expression in parentheses
2ND LN	Returns the constant e raised to a power
ALPHA 0	Space
2ND ◁	Moves the cursor to the beginning of an expression
2ND ▷	Moves the cursor to the end of an expression

Creating Scatter Plots

Clearing Lists

Press **STAT** and under EDIT press 4:ClrList (Figure 2). Press **2ND** **1** (L1) **ENTER** to clear List 1. Repeat the line by pressing **2ND** **ENTER** and moving the cursor to L1. Press **2ND** **2** (L2) and **ENTER** to clear List 2 (Figure 3).

Another way to clear a list in the STAT menu is, after accessing the list, to press the up arrow until the name of the list is highlighted and then press **CLEAR** and **ENTER**. *Caution:* Do not press the **DEL** key. This will not actually delete the list, but hide it from view. To view the list, press **2ND** **DEL** (INS) and type the list name.

Entering Data into Lists

Press **STAT** and under EDIT press 1:Edit. This will bring you to the screen where you enter data into lists.

Enter the *x*-values (input) in the column headed L1 and the corresponding *y*-values (output) in the column headed L2 (Figure 4).

Create a Scatter Plot of Data Points

Go to the Y= menu and turn off or clear any functions entered there. To turn off a function, move the cursor over the = sign and press **ENTER**.

Press **2ND** **Y=**, 1:Plot1, highlight On, and then highlight the first graph type (Scatter Plot). Enter Xlist:L1, Ylist:L2, and pick the point plot mark and color you want (Figure 5).

Choose an appropriate window for the graph and press **GRAPH** or **ZOOM**, 9:ZoomStat to plot the data points (Figure 6).

Tracing Along the Plot

Press **TRACE** and the right arrow to move from point to point. The *x*- and *y*-coordinates will be displayed at the bottom of the screen.

In the upper left-hand corner of the screen, Plot1:L1,L2 will be displayed. This tells you that you are tracing along the scatter plot (Figure 7).

Turning Off Stat Plot

After creating a scatter plot, turn Stat Plot1 off by pressing **Y=**, moving up to Plot1 with the cursor, and pressing **ENTER**. Stat Plot1 can be turned on by pressing **ENTER** with the cursor on Plot1. It will be highlighted when it is on.

Graphing Equations

Setting Windows

The window defines the highest and lowest values of *x* and *y* on the graph of the function that will be shown on the screen. To set the window manually, press the **WINDOW** key and enter the values that you want (Figure 8).

The values that define the viewing window can also be set by using **ZOOM** keys (Figure 9). Frequently the standard window (**ZOOM** 6) is appropriate. The standard window sets values Xmin = −10, Xmax = 10, Ymin = −10, Ymax = 10. Often a decimal or integer viewing window (**ZOOM** 4 or **ZOOM** 8) gives a better representation of the graph.

Figure 2

Figure 3

Figure 4

Figure 5

Figure 6

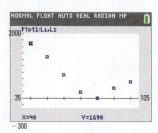

Figure 7

The window should be set so that the important parts of the graph are shown and the unseen parts are suggested. Such a graph is called complete.

Graphing Equations

To graph an equation in the variables x and y, first solve the equation for y in terms of x. If the equation has variables other than x and y, solve for the dependent variable and replace the independent variable with x.

Press the ⟨ Y= ⟩ key to access the function entry screen. You can input up to ten functions in the ⟨ Y= ⟩ menu (Figure 10). Enter the equation, using the **X,T,θ,n** key to input x. Use the subtraction key ⟨ − ⟩ between terms. Use the negation key ⟨ (-) ⟩ when the first term is negative. Use parentheses as needed so that what is entered agrees with the order of operations (Figure 10).

To change the color of the graph, use the left arrow ⟨ ◁ ⟩ to move to the left of Y_1, press **ENTER**, and choose the color and/or graph style you want. Colors are changed by using the right arrow ⟨ ▷ ⟩ and **ENTER** (Figure 11).

The = sign is highlighted to show that Y1 is *on* and ready to be graphed. If the = sign is not highlighted, the equation will remain but its graph is "turned off" and will not appear when ⟨ GRAPH ⟩ is pressed. (The graph is "turned on" by repeating the process.)

To erase an equation, press **CLEAR**. To return to the homescreen, press **2ND** **MODE** (QUIT).

Determine an appropriate viewing window.

Pressing ⟨ GRAPH ⟩ or a ⟨ ZOOM ⟩ key will activate the graph (Figure 12).

Finding Function Values

Using TRACE, VALUE Directly on the Graph

Enter the function to be evaluated in Y1. Choose a window that contains the x-value whose y-value you seek (Figure 13).

Turn all Stat Plots off.

Press ⟨ TRACE ⟩ and then enter the selected x-value followed by **ENTER**. The cursor will move to the selected value and give the resulting y-value if the selected x-value is in the window.

If the selected x-value is not in the window, ERROR:INVALID will occur.

If the x-value is in the window, the y-value will be given even if it is not in the window (Figures 14 and 15).

Figure 8

Figure 9

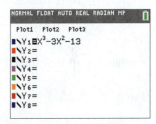

Figure 10 The function $y = x^3 - 3x^2 - 13$ is entered in Y1.

Figure 11

Figure 12 The graph of the function using the window $[-10, 10]$ by $[-25, 10]$

Figure 13 To evaluate $y = -x^2 + 8x + 9$ when $x = 3$ and when $x = -5$, graph the function using the window $[-10, 10]$ by $[-10, 30]$.

Figure 14

Figure 15

Using the TABLE ASK Feature

Enter the function with the ⟨Y=⟩ key. (Note: The = sign must be highlighted.) Press **2ND** ⟨WINDOW⟩ (TBLSET), move the cursor to Ask opposite Indpnt:, and press **ENTER** (Figure 16). This allows you to input specific values for x. Pressing **DEL** will clear entries in the table.

Then press **2ND** ⟨GRAPH⟩ (TABLE) and enter the specific values (Figure 17).

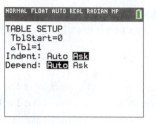

Figure 16

Figure 17

Making a Table of Values with Uniform Inputs

Press **2ND** ⟨WINDOW⟩ (TBLSET), enter an initial x-value in the table (TblStart), and enter the desired change in the x-value in the table (ΔTbl). Select Auto for Indpnt and Depend (Figure 18).

Enter **2ND** ⟨GRAPH⟩ (TABLE) to get the list of x-values and the corresponding y-values. The value of the function at the given value of x can be read from the table (Figure 19).

Use the up or down arrow to find the x-values where the function is to be evaluated (Figure 20).

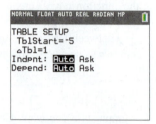

Figure 18

Using Y-VARS

Use the ⟨Y=⟩ key to store Y1 = $f(x)$. Press **2ND** (QUIT).

Press **VARS**, Y-VARS 1,1 to display Y1 on the home-screen. Then press ⟨ (⟩, the x-value, ⟨) ⟩, and **ENTER** (Figure 21).

Alternatively, you can enter the x-values needed as follows: Y1 ({value 1, value 2, etc.}) **ENTER** (Figure 22). Values of the function will be displayed.

Figure 19

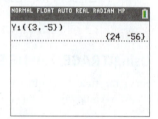

Figure 20

Figure 21

Figure 22

Finding Intercepts of Graphs

Solve the equation for y. Enter the equation with the ⟨Y=⟩ key.

Finding the y-Intercept

Press ⟨TRACE⟩ and enter the value 0. The resulting y-value is the y-intercept of the graph (Figure 23).

Finding the x-Intercept(s)

Set the window so that the intercepts to be located can be seen. The graph of a linear equation will cross the x-axis at most one time; the graph of a quadratic equation will cross the x-axis at most two times; etc.

To find the point(s) where the graph crosses the x-axis, press **2ND** ⟨TRACE⟩ to access the CALC menu. Select 2:zero (Figure 24).

Figure 23 The y-intercept of the graph of the function $y = -x^2 + 8x + 9$ is 9.

Figure 24

Answer the question "Left Bound?" with **ENTER** after moving the cursor close to and to the left of an *x*-intercept (Figure 25).

Answer the question "Right Bound?" with **ENTER** after moving the cursor close to and to the right of this *x*-intercept (Figure 26).

To answer the question "Guess?" press **ENTER**.

Note that vertical dotted lines show where the left bound and the right bound have been marked (Figure 27).

The coordinates of the *x*-intercept will be displayed (Figure 28). Repeat to get all *x*-intercepts (Figure 29). (Some of these values may be approximate.)

Finding Maxima and Minima

To locate a local maximum on the graph of an equation, graph the function using a window that shows all the turning points (Figure 30).

Press **2ND** **TRACE** to access the CALC menu. Select 4:maximum. The calculator will ask the question "Left Bound?" (Figure 31). Move the cursor close to and to the left of the maximum and press **ENTER**. Next the calculator will ask the question "Right Bound?" (Figure 32). Move the cursor close to and to the right of the maximum and press **ENTER**. Note that vertical dotted lines show where the left bound and the right bound have been marked. When the calculator asks "Guess?" move the cursor close to the maximum and press **ENTER** (Figure 33). The coordinates of the maximum will be displayed (Figure 34).

Note: When selecting a left bound or a right bound, you may also type in a value of *x* if you wish.

To locate a local minimum on the graph of an equation, press **2ND** **TRACE** to access the CALC menu. Select 3:minimum, and follow the same steps as above (Figure 35).

Figure 25

Figure 26

Figure 27

Figure 28

Figure 29

Figure 30 The graph of the function $y = x^3 - 27x$ on the window $[-10, 10]$ by $[-70, 70]$

Figure 31

Figure 32

Figure 33

Figure 34

Figure 35

Solving Equations Graphically

The *x*-Intercept Method

To solve an equation graphically using the *x*-intercept method, first rewrite the equation with 0 on one side of the equation. Using the Y= key, enter the nonzero side of the equation equal to Y1. Then graph the equation with a window that shows all points where the graph crosses the *x*-axis.

For example, to solve the equation

$$\frac{2x - 3}{4} = \frac{x}{3} + 1$$

enter

$$Y1 = \frac{2x - 3}{4} - \frac{x}{3} - 1$$

and graph using the window $[-10, 15]$ by $[-5, 5]$ (Figures 36 and 37).

The graph will intersect the *x*-axis where $y = 0$ — that is, when *x* is a solution to the original equation. To find the point(s) where the graph crosses the *x*-axis and the equation has solutions, press **2ND** TRACE to access the CALC menu (Figure 38). Select 2:zero.

Select "Left Bound?" (Figure 39), "Right Bound?" (Figure 40), and "Guess?" (Figure 41). The coordinates of the *x*-intercept will be displayed.

The *x*-intercept is the solution to the original equation (Figure 42).

Repeat to get all *x*-intercepts (and solutions).

The Intersection Method

To solve an equation graphically using the intersection method, under the Y= menu, assign the left side of the equation to Y1 and the right side of the equation to Y2. Then graph the equations using a window that contains the point(s) of intersection of the graphs.

To solve the equation

$$\frac{2x - 3}{4} = \frac{x}{3} + 1$$

enter the left and right sides as shown in Figure 43 and graph using the window $[-10, 20]$ by $[-5, 10]$ (Figure 44).

Press **2ND** TRACE to access the CALC menu (Figure 45). Select 5:intersect.

Answer the question "First curve?" by pressing **ENTER** (Figure 46) and "Second curve?" by pressing **ENTER** (Figure 47). (Or press the down arrow to move to one of the two curves.)

Figure 36

Figure 37

Figure 38

Figure 39

Figure 40

Figure 41

Figure 42

Figure 43

Figure 44

Figure 45

Figure 46

Figure 47

To the question "Guess?" move the cursor close to the desired point of intersection and press **ENTER** (Figure 48). The coordinates of the point of intersection will be displayed. Repeat to get all points of intersection.

The solution(s) to the equation will be the values of x from the points of intersection (Figure 49).

Figure 48

Figure 49 The solution to the equation is $x = 10.5$.

Solving Systems of Equations

To find the solution of a system of equations graphically, first solve each equation for y and use the ⟨ **Y=** ⟩ key with Y1 and Y2 to enter the equations. Graph the equation with an appropriate window.

For example, to solve the system

$$\begin{cases} 4x + 3y = 11 \\ 2x - 5y = -1 \end{cases}$$

graphically, we solve for y:

$$y_1 = \frac{11}{3} - \left(\frac{4}{3}\right)x$$

$$y_2 = \left(\frac{2}{5}\right)x + \frac{1}{5}$$

We graph using ⟨ **ZOOM** ⟩ 4:ZDecimal and then Intersect under the CALC menu to find the point of intersection. If the two lines intersect in one point, the coordinates give the x- and y-values of the solution (Figure 50). The solution of the system above is $x = 2$, $y = 1$. If the two lines are parallel, there is no solution; the system of equations is inconsistent.

Figure 50

For example, to solve the system

$$\begin{cases} 4x + 3y = 4 \\ 8x + 6y = 25 \end{cases}$$

we solve for y (Figure 51) and obtain the graph in Figure 52. This system has no solution. Note that if the lines are parallel, then when we solve for y the equations will show that the lines have the same slope and different y-intercepts.

Figure 51

Figure 52

If the two graphs of the equations give only one line, every point on the line gives a solution to the system and the system is dependent. For example, to solve the system

$$\begin{cases} 2x + 3y = -6 \\ 4x + 6y = -12 \end{cases}$$

we solve for y (Figure 53) and obtain the graph in Figure 54.

This system has many solutions and is dependent. Note that the two graphs will be the same graph if, when we solve for y to use the graphing calculator, the equations are equivalent.

Figure 53

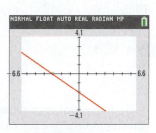

Figure 54

Solving Inequalities Graphically

To solve a linear inequality graphically, first rewrite the inequality with 0 on the right side and simplify. Then, under the Y= menu, assign the left side of the inequality to Y1, so that $Y1 = f(x)$. For example, to solve $3x > 6 + 5x$, rewrite the inequality as $3x - 5x - 6 > 0$ or $-2x - 6 > 0$. Set $Y1 = -2x - 6$.

Figure 55

Graph this equation so that the point where the graph crosses the x-axis is visible. Note that the graph will cross the axis in at most one point because the function is of degree 1. (Using ⟨ZOOM⟩ OUT can help you find this point.)

Using ⟨ZOOM⟩ 4, the graph of $Y1 = -2x - 6$ is shown in Figure 55.

Use the ZERO command under the CALC menu to find the x-value where the graph crosses the x-axis (the x-intercept). This value can also be found by finding the solution to $0 = f(x)$ algebraically.

The x-intercept of the graph of $Y1 = -2x - 6$ is $x = -3$ (Figure 56).

Figure 56

Observe the inequality as written with 0 on the right side. If the inequality is $<$, the solution to the original inequality is the interval (bounded by the x-intercept) where the graph is below the x-axis. If the inequality is $>$, the solution to the original inequality is the interval (bounded by the x-intercept) where the graph is above the x-axis.

The solution to $-2x - 6 > 0$, and thus to the original inequality $3x > 6 + 5x$, is $x < -3$.

The region above the x-axis and under the graph can be shaded by pressing **2ND** **PRGM** (DRAW), 7:Shade(, and then entering (0, Y1) to shade on the homescreen (Figure 57).

Figure 57

The x-interval where the shading occurs is the solution.

Piecewise-Defined Functions

A piecewise-defined function is defined differently over two or more intervals.

To graph a piecewise-defined function

$$y = \begin{cases} f(x) & \text{if } x \le a \\ g(x) & \text{if } x > a \end{cases}$$

press ⟨Y=⟩ and enter

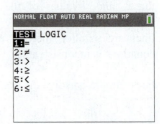

Figure 58

$$Y1 = (f(x))/(x \le a)$$

and

$$Y2 = (g(x))/(x > a)$$

Note that the inequality symbols are found by pressing **2ND** **MATH** to access the TEST menu (Figure 58).

To illustrate, we will enter the function

Figure 59

$$f(x) = \begin{cases} x + 7 & \text{if } x \le -5 \\ -x + 2 & \text{if } x > -5 \end{cases}$$

(Figure 59) and graph using an appropriate window (Figure 60).

Evaluating a piecewise-defined function at a given value of x requires that the correct equation ("piece") be selected. For example, $f(-6)$ and $f(3)$ are shown in Figures 61 and 62, respectively.

Figure 60

Modeling—Regression Equations

Finding an Equation That Models a Set of Data Points

To illustrate, we will find the equation of the line that is the best fit for the average daily numbers of inmates in the Springfield Detention Center for the years 2010–2015, given by the data points (2010, 96), (2011, 109), (2012, 119), (2013, 116), (2014, 137), (2015, 143).

Press **STAT** and under EDIT press 1:Edit. Enter the x-values (inputs) in the column headed L1 and the corresponding y-values (outputs) in the column headed L2 (Figure 63).

Press **2ND** **Y=** 1:Plot 1, highlight On, and then highlight the first graph type. Enter Xlist:L1, Ylist:L2, and pick the point plot mark you want (Figure 64).

Press **GRAPH** with an appropriate window or **ZOOM**, 9:ZoomStat to plot the data points (Figure 65).

Observe the point plots to determine what type of function would best model the data.

The graph looks like a line, so use the linear model, with LinReg.

Press **STAT**, move to CALC, and select the function type to be used to model the data (Figure 66). Enter L1 for Xlist, L2 for Ylist, and for Store RegEQ press the **VARS** key, go over to Y-VARS, and select 1:Function and 1:Y1 (Figure 67). Press **ENTER** twice. The coefficients of the equation will appear on the screen (Figure 68), and the linear regression equation will appear as Y1 on the Y= screen (see Figure 69).

To see how well the equation models the data, press **GRAPH**. If the graph does not fit the points well, another type function may be used to model the data.

For some models, r is a diagnostic value that gives the **correlation coefficient**, which determines the strength of the relationship between the independent and dependent variables. To display this value, select DiagnosticOn after pressing **2ND** **0** (CATALOG) (Figure 68).

Report the equation in a way that makes sense in the context of the problem, with the appropriate units and the variables identified.

The equation that models the average daily number of inmates in the Springfield Detention Center is $y = 9.0286x - 18{,}050$ (rounded to four decimal places), where x is the year and y is the number of inmates (Figures 69 and 70).

Figure 61

Figure 62

Figure 63

Figure 64

Figure 65

Figure 66

Figure 67

Figure 68

Figure 69

Figure 70

Using a Model to Find an Output

To use the model to find output values inside the data range (*interpolation*) or outside the data range (*extrapolation*), evaluate the function at the desired input value. This may be done using TABLE (Figure 71), Y-VARS (Figure 72), or (TRACE) (Figure 73). (Note that you may have to change the window to see the *x*-value to which you are tracing.)

To predict the average daily inmate population in the Beaufort County jail in 2020, we compute Y1 (2020). According to the model, approximately 188 inmates were, on average, in Springfield jail in 2020.

When TRACE is used, the window must be changed so that $x = 2020$ is visible (Figure 73).

Using a Model to Find an Input

To use the model to estimate an input for a given output, set the model function equal to Y1 and the desired output equal to Y2 and solve the resulting equations for the input variable. An approximate solution can be found using TABLE or **2ND** (TRACE) (CALC), 5:Intersect.

To use the model to determine in what year the population will be 260, we must solve the equations Y1 = $f(x)$, Y2 = 260. TABLE may be used to find the *x*-value when *y* is approximately 260 (Figure 74).

We can also solve the equation Y1 = Y2 using the intersect method (Figures 75 and 76).

According to the model, the population of the jail will reach 260 in the year 2028.

Complex Numbers

Calculations that involve complex numbers of the form $a + bi$ will be displayed as complex numbers only if MODE is set to a + bi (Figure 77).

Complex numbers can be added (Figure 78), subtracted (Figure 79), multiplied (Figure 80), divided (Figure 81), and raised to powers (Figure 82). The number *i* is accessed by pressing **2ND** (·).

Operations with Functions

Combinations of Functions

To find the graphs of combinations of two functions $f(x)$ and $g(x)$, enter $f(x)$ as Y1 and $g(x)$ as Y2 under the Y= menu. Consider the example $f(x) = 4x - 8$ and $g(x) = x^2$. To graph $(f + g)(x)$, enter Y1 + Y2 as Y3

Figure 71 Using TABLE

Figure 72 Using Y-VARS

Figure 73 Using TRACE

Figure 74

Figure 75

Figure 76

Figure 77

Figure 78

Figure 79

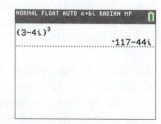

Figure 80

Figure 81

Figure 82

under the Y= menu (Figure 83). Place the cursor on the = sign beside Y1 and press **ENTER** to turn off the graph of Y1. Repeat with Y2. Press **GRAPH** with an appropriate window (Figure 84).

To graph $(f - g)(x)$, enter Y1 − Y2 as Y3 under the Y= menu (Figure 85). Place the cursor on the = sign beside Y1 and press **ENTER** to turn off the graph of Y1. Repeat with Y2. Press **GRAPH** (Figure 86).

To graph $(f * g)(x)$, enter Y1*Y2 as Y3 under the Y = menu (Figure 87). Place the cursor on the = sign beside Y1 and press **ENTER** to turn off the graph of Y1. Repeat with Y2. Press **GRAPH** (Figure 88).

To graph $(f/g)(x)$, enter Y1/Y2 as Y3 under the Y = menu (Figure 89). Place the cursor on the = sign beside Y1 and press **ENTER** to turn off the graph of Y1. Repeat with Y2. Press **GRAPH** with an appropriate window (Figure 90).

To evaluate $f + g, f - g, f * g$, or f/g at a specified value of x, enter Y1, the value of x enclosed in parentheses, the operation to be performed, Y2, and the value of x enclosed in parentheses, and press **ENTER** (Figure 91). Or, if the combination of functions is entered as Y3, enter Y3 and the value of x enclosed in parentheses (Figure 92), or use TABLE.

The value of $(f + g)(3)$ is shown in Figures 91 and 92.

Note: Entering $(Y1 + Y2)(3)$ does not produce the correct result (Figure 93).

Composition of Functions

To graph the composition of two functions $f(x)$ and $g(x)$, enter $f(x)$ as Y1 and $g(x)$ as Y2 under the Y= menu. Consider the example $f(x) = 4x - 8$ and $g(x) = x^2$.

To graph $(f \circ g)(x) = f(g(x))$, enter Y1(Y2) as Y3 under the Y= menu (Figure 94). Place the cursor on the = sign beside Y1 and press **ENTER** to turn off the graph of Y1. Repeat with Y2. Press **GRAPH** with an appropriate window (Figure 95).

Figure 83

Figure 84

Figure 85

Figure 86

Figure 87

Figure 88

Figure 89

Figure 90

Figure 91

Figure 92

Figure 93

Figure 94

Figure 95

To graph $(g \circ f)(x)$, enter Y2(Y1) as Y4 under the Y= menu (Figure 96). Turn off the graphs of Y1, Y2, and Y3. Press GRAPH with an appropriate window (Figure 97).

To evaluate $(f \circ g)(x)$ at a specified value of x, enter Y1(Y2(, the value of x, and two closing parentheses (Figure 98), and press ENTER. Or, if the combination of functions is entered as Y3, enter Y3 and the value of x enclosed in parentheses (Figure 99). The value of $(f \circ g)(-5) = f(g(-5))$ is shown in Figures 98 and 99.

Figure 96

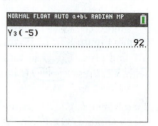

Figure 97

Inverse Functions

To show that the graphs of two inverse functions are symmetrical about the line $y = x$, graph the functions on a square window. Enter $f(x)$ as Y1 and the inverse function $g(x)$ as Y2, and press ENTER. Press GRAPH with an appropriate window and different colored curves. Enter Y3 = x under the Y= menu and graph using ZOOM, 5:Zsquare. For example, the graphs of $f(x) = x^3 - 3$ and its inverse $g(x) = \sqrt[3]{x + 3}$ are symmetric about the line $y = x$, as shown in Figures 100 and 101.

To graph a function $f(x)$ and its inverse, under the Y= menu, enter $f(x)$ as Y1 (Figure 102). Choose a square window. Press 2ND PRGM (DRAW) (Figure 103), 8:Draw-Inv (Figure 104), press VARS, move to Y-VARS, and press ENTER three times. The graph of $f(x)$ and its inverse will be displayed.

To clear the graph of the inverse, press 2ND PRGM (DRAW), 1:ClrDraw. The graphs of $f(x) = 2x - 5$ and its inverse are shown on a square window in Figure 105.

Figure 98

Figure 99

Figure 100

Figure 101

Figure 102

Figure 103

Finance

Future Value of an Investment

The TVM Solver displays the time-value-of-money variables. Given four TVM variables, the TVM Solver solves for the fifth variable.

To find the future value of a present amount, press the APPS key, select 1:Finance and next select 1:TVM Solver (Figures 106 and 107). Then set the variables as follows:

Set N = the total number of compounding periods.

Set I% = the annual percentage rate.

Set PV = lump-sum investment (with a negative sign because the sum is leaving the investor and going into the investment).

Set PMT = 0 because no additional money is being invested.

Set both P/Y (payments/year) and C/Y (compounding periods per year) equal to the number of compounding periods per year.

Figure 104

Figure 105

Figure 106

Figure 107

Highlight END, move the cursor to FV, and press **ALPHA** **ENTER** (SOLVE) to get the future value.

For example, to find the future value of $15,000 invested at 4% compounded quarterly for 10 years, use the settings in Figure 108. We use N = 10 years × 4 quarters per year, I% = 4, PV = −15000 to indicate the amount that is being invested, PMT = 0 because no additional money is being invested, and P/Y and C/Y = 4 compounding periods per year.

The future value, FV in Figure 109, is $22,332.96.

Figure 108

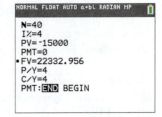

Figure 109

Present Value of an Investment

Use the same steps as above, except set FV = the desired future value, and solve for PV.

To find the present value of $20,000 at 8% compounded semiannually for 6 years, use the settings in Figure 110.

Move the cursor to PV, and press **ALPHA** **ENTER** (SOLVE) to get the present value.

The present value, PV in Figure 111, is $12,491.94. This value is negative because it is the investment leaving the investor.

Figure 110

Figure 111

Future Value of an Ordinary Annuity

To find the 5-year future value of an ordinary annuity with a contribution of $500 per quarter into an account that pays 8% per year compounded quarterly, press the **APPS** key, select **1:Finance**, and next select **1:TVM Solver**.

Set N = the total number of compounding periods = 5*4 = 20.

Set I% = the annual percentage rate = 8.

Set PV = lump-sum investment = 0.

Set PMT = quarterly payment = −500 (negative because it is leaving the investor).

Set both P/Y (payments/year) and C/Y (compounding periods per year) equal to the number of compounding periods per year = 4 Highlight END.

See the settings in Figure 112.

Move the cursor to FV, and press **ALPHA** **ENTER** (SOLVE) to get the future value.

The future value, FV in Figure 113 on the next page, is $12,148.68.

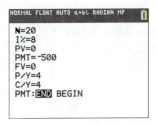

Figure 112

Present Value of an Ordinary Annuity

We can find the present value of an annuity by using the same steps used to find future value, with the inputs for PV and FV changed.

To find the amount to put in an annuity that will provide $2000 at the end of each month for 20 years at 6% compounded monthly, press the **APPS** key, select **1:Finance**, and next select **1:TVM Solver.**

Set N = the total number of compounding periods = 240.

Set I% = the annual percentage rate = 6.

Set FV = value after all payments are made = 0.

Set PMT = quarterly payment = 2000.

Set both P/Y (payments/year) and C/Y (compounding periods per year) equal to the number of compounding periods per year = 12. Highlight END.

See the settings in Figure 114.

Move the cursor to PV, and press **ALPHA** **ENTER** (SOLVE) to get the present value.

The amount needed to fund the annuity is $279,161.54, found in Figure 115.

Figure 113

Figure 114

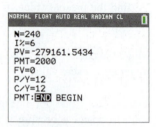

Figure 115

Matrices

Entering Data into Matrices

To enter data into matrices, press **2ND** **x⁻¹** to get the MATRIX menu (Figure 116). Move the cursor to EDIT. Select the number of the matrix into which the data are to be entered. Enter the dimensions of the matrix, and enter the value for each entry of the matrix (Figure 117). Press **ENTER** after each entry.

For example, to enter as [A] the matrix

$$\begin{bmatrix} 1 & 2 & 3 \\ 2 & -2 & 1 \\ 3 & 1 & -2 \end{bmatrix}$$

enter 3's to set the dimension and enter the elements of the matrix (Figure 118).

To perform operations with the matrix or leave the editor, first press **2ND** **MODE** (QUIT).

To view the matrix, access MATRIX, enter the name of the matrix (Figure 119), and press **ENTER** (Figure 120).

Figure 116

Figure 117

Figure 118

Figure 119

Figure 120

The Identity Matrix

To display an identity matrix of order n (an $n \times n$ matrix consisting of 1's on the main diagonal and 0's elsewhere), access MATRIX, move to MATH, select 5:identity(, and enter the order of the identity matrix desired (Figure 121). The identity matrix of order 2 is shown in Figure 122.

An identity matrix can also be created by entering the numbers directly with MATRIX and EDIT.

Figure 121

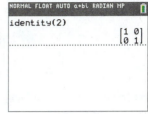

Figure 122

Adding and Subtracting Matrices

To find the sum of two matrices [A] and [D], enter the values of the elements of [A] using MATRIX and EDIT (Figure 123). Press **2ND** **MODE** (QUIT). Enter the values of the elements of [D] using MATRIX and EDIT (Figure 124). Press **2ND** **MODE** (QUIT).

Use MATRIX and NAMES to enter [A] + [D], and press **ENTER**. If the matrices have the same dimensions, they can be added (or subtracted). If they do not have the same dimensions, an error message will occur.

For example, Figure 125 shows the sum

$$\begin{bmatrix} 1 & 2 & 3 \\ 2 & -2 & 1 \\ 3 & 1 & -2 \end{bmatrix} + \begin{bmatrix} 7 & -3 & 2 \\ 4 & -5 & 3 \\ 0 & 2 & 1 \end{bmatrix}$$

To find the difference of the matrices [A] and [D], enter [A] – [D] and press **ENTER** (Figure 126).

Figure 123

Figure 124

Figure 125

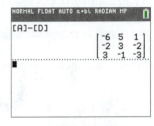

Figure 126

Multiplying a Matrix by a Real Number

We can multiply a matrix [D] by a real number (scalar) k by entering k[D] (or $k*$[D]). In Figures 127 and 128, we multiply the matrix [D] by 5.

Figure 127

Figure 128

Multiplying Two Matrices

To find the product of matrices [C][A], access MATRIX, move to EDIT, select 1: [A], enter the dimensions of [A] (Figure 129), and enter the elements of [A] (Figure 130). Press **2ND** **MODE** (QUIT). Do the same for matrix [C] (Figure 131) and access **2ND** **MODE** (QUIT). Access MATRIX [C], $*$, MATRIX [A], and **ENTER** (Figure 132). Or press MATRIX [C], MATRIX [A], and **ENTER** (Figure 133).

Figure 129

Figure 130

Figure 131

Figure 132

Figure 133

The product

$$\begin{bmatrix} 1 & 2 & 4 \\ -3 & 2 & -1 \end{bmatrix} \begin{bmatrix} 1 & 2 & 3 \\ 2 & -2 & 1 \\ 3 & 1 & -2 \end{bmatrix}$$

is shown in Figures 132 and 133.

Note that [A][C] does not always equal [C][A]. The product [A][C] may be the same as [C][A], may be different from [C][A], or may not exist.

In Figures 134 and 135, [A][C] cannot be computed because the matrices have dimensions such that they cannot be multiplied.

Finding the Inverse of a Matrix

To find the inverse of a matrix, enter the elements of the matrix using MATRIX and EDIT (Figure 136). Press **2ND** **MODE** (QUIT). Access MATRIX, enter the name of the matrix, and then press the **x⁻¹** key and **ENTER** (Figure 137).

Figure 138 shows the inverse of

$$E = \begin{bmatrix} 2 & 0 & 2 \\ -1 & 0 & 1 \\ 4 & 2 & 0 \end{bmatrix}$$

To see the entries as fractions, press **MATH**, 1:Frac, and **ENTER** (Figures 139 and 140).

Not all matrices have inverses. Matrices that do not have inverses are called singular matrices (Figures 141–143).

Solving Systems—Reduced Echelon Form

To solve a system of linear equations by using rref under the MATRIX MATH menu, create an augmented matrix [A] with the coefficient matrix augmented by the constants (Figure 144). Use the MATRIX menu to produce a reduced row echelon form of [A], as follows:

1. Access MATRIX and move to the right to MATH (Figure 145).

2. Scroll down to B:rref(and press **ENTER**, or press **ALPHA** B (Figure 146). Access MATRIX, 1: [A] to get rref([A]). Press **ENTER** to get the reduced echelon form.

Figure 134

Figure 135

Figure 136

Figure 137

Figure 138

Figure 139

Figure 140

Figure 141

Figure 142

Figure 143

Figure 144

Figure 145

Figure 146

The system

$$\begin{cases} 2x - y + z = 6 \\ x + 2y - 3z = 9 \\ 3x - 3z = 15 \end{cases}$$

is solved in Figure 147.

Figure 147

If each row in the coefficient matrix (the first three columns) contains a 1 with the other elements 0's, the solution is unique and the number in column 4 of a row is the value of the variable corresponding to a 1 in that row. The solution to the system above is unique: $x = 4$, $y = 1$, and $z = -1$.

If the bottom row contains all zeros, the system has many solutions. The values for the first two variables are found as functions of the third.

If there is a nonzero element in the augment of row 3 and zeros elsewhere in row 3, there is no solution to the system.

Figure 148

Linear Programming

To solve a linear programming problem involving two constraints graphically, write the inequalities as equations, solved for y. Graph the equations. The inequalities $x \geq 0$, $y \geq 0$ limit the graph to Quadrant I, so choose a window with Xmin = 0 and Ymin = 0. Use (TRACE) or intersect (accessed by pressing **2ND** (TRACE)) to find the corners of the region, where the borders intersect.

For example, to find the region defined by the inequalities

$$5x + 2y \leq 54$$
$$2x + 4y \leq 60$$
$$x \geq 0, \ y \geq 0$$

write $y = 27 - 5x/2$ and $y = 15 - x/2$, graph, and find the intersection points as shown in Figures 148 to 150.

Figure 149

Figure 150

The corners of the region determined by the inequalities are $(0, 15)$, $(6, 12)$, and $(10.8, 0)$.

Change the icons to the left of Y1= and Y2= to show regions below the lines (Figure 151). Then graph these inequalities. The cross-hatched region is the solution region (Figure 152).

Evaluating the objective function $P = 5x + 11y$ at the coordinates of each of the corners determines where the objective function is maximized or minimized.

$$\text{At } (0, 15), f = 165$$
$$\text{At } (6, 12), f = 162$$
$$\text{At } (10.8, 0), f = 54$$

The maximum value of f is 165 at $x = 0$, $y = 15$.

Figure 151

Figure 152

Sequences and Series

Evaluating a Sequence

To evaluate a sequence for different values of n, press **MODE** and highlight Seq (Figure 153). Press **ENTER** and **2ND** **MODE** (QUIT). Store the formula for the sequence (in quotes) in u, using **STO▶** u. (Press the **X,T,Θ,n** key to enter n for the formula, and **2ND** 7 to get u.) Press **ENTER**.

For example, to evaluate the sequence with nth term $n^2 + 1$ at $n = 1, 3, 5,$ and 9, we store the formula as shown in Figure 154.

Enter u({$a, b, c, \ldots$}) to evaluate the sequence at $a, b, c, \ldots$, and press **ENTER** (Figure 155).

To generate a sequence after the formula is defined, enter u($n$start, $n$stop, step) and press **ENTER**. For the sequence formula above, evaluate every third term beginning with the second term and ending with the eleventh term (Figure 156).

Figure 153

Figure 154

Figure 155

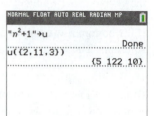

Figure 156

Finding the nth Term of an Arithmetic Sequence

To find the nth term of an arithmetic sequence with first term a and common difference d, press **MODE** and highlight Seq (Figure 157). Press **ENTER** and press **2ND** **MODE** (QUIT).

Press **Y=**. At u(n) =, enter the formula for the nth term of an arithmetic sequence, using the **X,T,Θ,n** key to enter n. The formula is $a + (n - 1)d$, where a is the first term and d is the common difference.

For example, to find the 12th term of the arithmetic sequence with first term 10 and common difference 5, substitute 10 for a and 5 for d, as shown in Figure 158, to get u(n) = $10 + (n - 1)5$.

Press **2ND** **MODE** (QUIT). To find the nth term of the sequence, press **2ND** 7, followed by the value of n in parentheses, to get u(n); then press **ENTER** (Figure 159).

Additional terms can be found in the same manner (Figure 160).

Figure 157

Figure 158

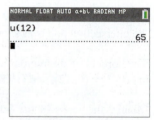

Figure 159   The 12th term

Finding the Sum of an Arithmetic Sequence

To find the sum of the first n terms of an arithmetic sequence, press **MODE** and highlight Seq (Figure 161). Press **ENTER** and press **2ND** **MODE** (QUIT).

Press **Y=**. At v(n) =, enter the formula for the sum of the first n terms of an arithmetic sequence, using the

Figure 160   The 8th term

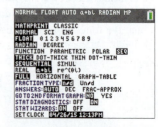

Figure 161

X,T,Θ,n key to enter n. The formula is $(n/2)(a + (a + (n-1)d))$, where a is the first term and d is the common difference (Figure 162).

Press 2ND MODE (QUIT). To find the sum of the first n terms of the sequence, press 2ND 8, followed by the value of n in parentheses, to get v(n); then press ENTER. For example, the sum of the first 12 terms of the arithmetic sequence with first term 10 and common difference 5 is shown in Figure 163.

Other sums can be found in the same manner (Figure 164).

Finding the nth Term and Sum of a Geometric Sequence

To find the nth term of a geometric sequence with first term a and common ratio r, press MODE and highlight Seq (Figure 165). Press ENTER and 2ND MODE (QUIT).

Press Y=. At u$(n) =$, enter the formula for the nth term of a geometric sequence, using the X,T,Θ,n key to enter n. The formula is ar^{n-1}, where a is the first term and r is the common ratio.

For example, to find the geometric sequence with first term 40 and common ratio $1/2$, we substitute 40 for a and $(1/2)$ for r (Figure 166).

Press 2ND MODE (QUIT). To find the nth term of the sequence, press 2ND 7, followed by the value of n in parentheses, to get u(n); then press ENTER. The 8th and 12th terms of the geometric sequence above are shown in Figures 167 and 168, respectively. To get a fractional answer, press MATH, 1:Frac. Additional terms can be found in the same manner.

To find the sum of the first n terms of a geometric sequence, press Y=. At v$(n) =$, enter the formula for the sum of the first n terms of a geometric sequence, using the X,T,Θ,n key to enter n. The formula is $a(1 - r^n)/(1 - r)$, where a is the first term and r is the common ratio. Press 2ND MODE (QUIT). To find the sum of the first n terms of the sequence, press 2ND 8, followed by the value of n in parentheses, to get v(n); then press ENTER.

For example, to find the sum of the first 12 terms of the geometric sequence with first term 40 and common ratio $1/2$, substitute 40 for a and $(1/2)$ for r (Figure 169).

To get a fractional answer, press MATH, 1:Frac (Figure 170).

Other sums can be found in the same manner (Figure 171).

Figure 162

Figure 163 The sum of the first 12 terms

Figure 164 The sum of the first 8 terms

Figure 165

Figure 166

Figure 167

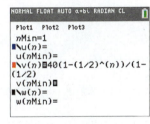

Figure 168

Figure 169

Figure 170 The sum of the first 12 terms

Figure 171 The sum of the first 8 terms

Appendix

B

Basic Guide to Excel 2010 and Excel 2013

This appendix describes the basic features and operation of Excel. More details of specific features can be found in the *Graphing Calculator and Excel® Manual*.

The Excel steps shown in this guide are for Excel 2010 and Excel 2013.

To find additional information about how to use Excel 2010 menus and buttons, use 2010 Excel Help as follows:

1. Select File and then Help; then select Microsoft Office Help.

2. Select Getting Started with Excel 2010, and click on WHERE TO GO based on your situation (previous knowledge of Excel).

For Excel 2013, click the ⑦ icon in the upper right corner of the screen for Excel Help.

Excel Worksheet

When you start up Excel by using the instructions for your software and computer, the screen in Figure 1 will appear. The components of the **spreadsheet** are shown, and the grid shown is called a **worksheet**. By clicking on the tabs at the bottom, you can move to other worksheets.

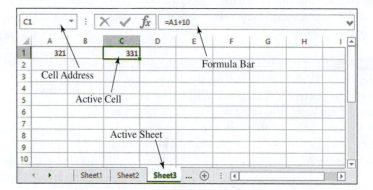

Figure 1

Addresses and Operations

Notice the letters at the top of the columns and the numbers identifying the rows. The cell addresses are given by the column and row; for example, the first cell has address A1. You can move from one cell to another with arrow keys, or you can select a cell with a mouse click. After you enter an entry in a cell, press ENTER to accept the entry. To edit the contents of a cell, make the edits in the formula bar at the top. To delete the contents, press the delete key.

File operations such as "open a new file," "saving a file," and "printing a file" are similar to those in Word. For example, <CTRL>S saves a file. You can also format a cell entry by selecting it and using menus similar to those in Word.

Working with Cells

Cell entries, rows containing entries, and columns containing entries can be copied and pasted with the same commands as in Word. For example, a highlighted cell can be copied with <CTRL>C. Sometimes an entry exceeds the width of the cell containing it, especially if it is text. To widen the cells in a column, move the mouse to the right side of the column heading until you see the symbol ↔; then hold down the left button and move the mouse to the right (moving to the left makes the cells narrower). If entering a number results in #####, the number is too long for the cell, and the cell should be widened.

You can work with a cell or with a range of cells. To select a range of cells,

1. Click on the beginning of the range of cells, hold down the left mouse button, and drag to the end of the desired range.

2. Release the mouse button. The range of cells will be highlighted. If the range of cells were from A2 through B6, the range would be indicated by A2:B6.

Creating Tables of Numbers in Excel

Entering Independent Values

Type the column heading, such as *x*, in cell A1, and enter each data value in a cell of the column (Figure 2).

Figure 2

Fill-Down Method for Values That Change by a Constant Increment

1. Type in the first two numbers (in C2 and C3, for example).

2. Select the cells C2 and C3 (C2:C3) (Figure 3).

Figure 3

3. Move the mouse to the lower right corner until the mouse becomes a thin + sign.

4. Drag the mouse down to the last cell where data are required (Figure 4). Press the right arrow to remove the highlight.

Figure 4

Evaluating a function

1. Put headings on the two columns (x and $f(x)$, for example).
2. Begin a formula (in cell B2, for example) by entering =; then enter the function performing operations with data in cells by entering the operation side of the formula with appropriate cell address(es) representing the variable(s) (Figure 5). Use * to indicate multiplication and ^ to indicate power.
3. Enter the value(s) of the variable(s) for which the function is to be evaluated and press ENTER (Figure 6).

| Figure 5 | Figure 6 |

4. Changing the input values will give different function values (Figure 7). Press ENTER to complete the process (Figure 8).

| Figure 7 | Figure 8 |

Creating a Table for a Function Using Fill Down

1. Put headings on the two columns (x and $f(x)$, for example).
2. Fill the inputs (x-values) as described on page 685.
3. Enter the formula for the function as described above.
4. Select the cell containing the formula for the function (B2, for example) (Figure 9).
5. Move the mouse to the lower right corner until there is a thin + sign.
6. Drag the mouse down to the last cell where the formula is required, and the values will be displayed (Figure 10). Using an arrow to move to a cell to the right will remove the highlight from the outputs.

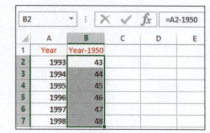

| Figure 9 | Figure 10 |

The worksheet in Figure 11 has headings typed in, numbers entered with Fill Down in column A, inputs entered in column B, the numbers in column D created with Fill Down using the formula A5 − 1950, and column B copied in column E.

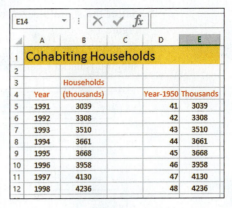

Figure 11

Graphing a Function of a Single Variable ($y = f(x)$, for example)

1. Use the function to create a table containing values for x and $f(x)$.
2. Highlight the two columns containing the values of x and $f(x)$ (Figure 12).

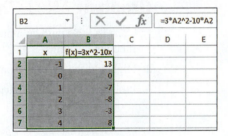

Figure 12

3. Select the Insert tab.
4. Select the scatter plot icon in the Charts group to get the Scatter menu (Figure 13).

Figure 13

5. Select the smooth curve option under Scatter, and the graph will be as shown in Figure 14.

6. To add or change titles or legends, double click the item to be changed or use the subgroups under Chart Tools.

Figure 14

Graphing More Than One Function

1. Create a table with an input column and one column for each function, with multiple function headings.

2. Highlight the columns containing the variable and function values (Figure 15).

3. Proceed with the same steps as are used to graph a single function (Figure 16).

Figure 15

Figure 16

Changing Graphing Windows

1. To change the *x*-scale,

   a. Double click on the *x*-axis, and choose Axis Options. In 2010, it opens as a dialog box with Axis Options as the first option. In 2013, it appears as a pane on the right-hand side of the screen. Once there, choose the bar chart icon and open the Axis Options dropdown menu.

   b. Change the minimum and maximum values to the desired values. (In 2010, you will have to uncheck the Auto boxes first.)

2. To change the *y*-scale, double click on the *y*-axis and proceed as for the *x*-axis.

3. For shading options, double click on the plot area to open Format Plot Area options. In 2010 this opens as a dialog box, and in 2013 it opens as a pane on the right-hand side of the screen.

4. To change the Chart Title, double click on it.

Graphing Discontinuous Functions

An Excel graph will connect all points corresponding to values in the table, so if the function you are graphing is discontinuous for some x-value a, enter x-values near this value and leave (or make) the corresponding $f(a)$ cell blank. Then graph the function using the steps described on page 687 (Figure 17).

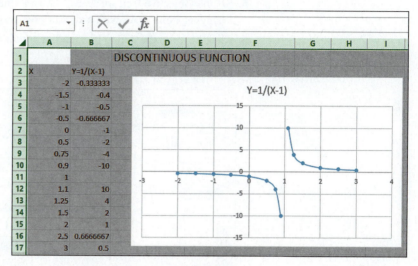

Figure 17

Finding a Minimum or Maximum of a Function with Solver

Finding the Minimum of a Function (if it has a minimum)

1. Enter the inputs (x-values) in column A and the functional values in column B, and use them to graph the function over an interval that shows a minimum (Figure 18). (If a minimum exists but does not show, add more points.)

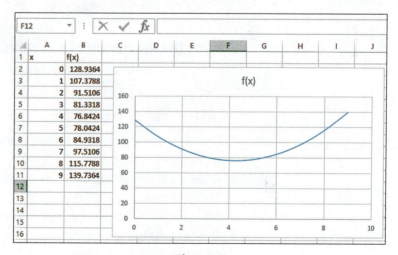

Figure 18

2. Use any value for x in A2 and the function formula in B2 (Figure 19).

Figure 19

3. Select the Data tab, and select ?Solver from the Analysis group. [Note: Solver is an Add-In. To install it, select the File tab and choose Options. Choose Add-Ins, and in the Manage Box, select Excel Add-Ins and click Go. In the Add-Ins Available box, select Solver Add-In and click OK. If Solver Add-In isn't available, choose Browse to find it.]

4. Click on Min in the Solver dialog box.

5. Set the Objective Cell to B2 by choosing the box and clicking on B2. Then set the Changing Cells to A2 with a similar process (Figure 20).

6. Click Solve in the dialog box, and the minimum will show in B2. Click OK to save the solution (Figure 21).

	A	B
1	x	f(x)
2	4.289081	76.60467

Figure 20   **Figure 21**

Finding the Maximum of a Function (if it has a maximum)

Use the process above, but in step 4 click on Max in the Solver dialog box.

Scatter Plots of Data

1. Enter the inputs (*x*-values) in column A and the outputs (*y*-values) in column B.

2. Highlight the two columns.

3. Select the Insert tab and select the scatter plot icon in the Charts group to get the Scatter menu.

4. Select the points option under Scatter. The graph will appear (Figure 22).

5. To add or change titles or legends, double click the item to be changed or use the subgroups under Chart Tools.

Figure 22

Finding the Equation of the Line or Curve That Best Fits a Given Set of Data Points

1. Create the scatter plot of the data in the worksheet.
2. Right click on one of the data points and select Add Trendline from the menu that appears (Figure 23).

Figure 23

3. Select the desired regression type and check Display Equation on chart (Figure 24). [Note: If Polynomial is selected, choose the appropriate Order (degree).]

4. Close the box, and the function and its graph will appear (Figure 25).

Figure 24

Figure 25

Figure 26

Figure 27

Figure 28

Solving a Linear Equation of the Form $f(x) = 0$ Using the x-Intercept Method

The solution will also be the x-intercept of the graph and the zero of the function.

1. Enter an x-value in A2 and the function formula in B2, as described on page 689 (Figure 26).

2. Highlight B2.

3. Select the Data tab.

4. Select What-if Analysis in the Data Tools group.

5. Select Goal Seek under What-if Analysis.

6. In the Goal Seek dialog box,

    a. Click the Set cell box and click on the B2 cell.

    b. Enter 0 in the To value box.

    c. Click the By changing box and click on the A2 cell (Figure 27).

7. Click OK to see the Goal Status box.

8. Click OK again. The solution will be in A2 and 0 will be in B2. The solution may be approximate.

Solving a Quadratic Equation of the Form $f(x) = 0$ Using the x-Intercept Method

The solutions will also be the x-intercepts of the graph and the zeros of the function. Because the graph of a quadratic function $f(x) = ax^2 + bx + c$ can have two intercepts, it is wise to graph the function using an x-interval with the x-coordinate $\left(h = \dfrac{-b}{2a} \right)$ near the center.

1. Enter x-values centered around h in column A and use the function formula to find the values of $f(x)$ in column B, as described on page 686 (Figure 29).

2. Graph the function as described on page 687, and observe where the function values are at or near 0.

3. Select the Data tab and choose What-If Analysis in Data Tools. Select Goal Seek, enter the address of a cell with a function value in column B at or near 0, enter 0 in To value, and enter the corresponding cell in column A in By changing cell (Figure 30). Click OK to find the x-value of the solution in column A. The solution may be approximate.

Figure 29　　　　　　　　　　　　　　　　　　　　Figure 30

4. One solution is $x = 0.5$ (Figure 31).*

5. After finding the first solution, repeat the process using a second function value at or near 0.

Figure 31

---

* **Notes:** (1) The solution $x = 0.50001$, giving $f(x) = -7.287\text{E}-05$, is an approximation of the exact solution $x = 0.5$, which gives $f(x) = 0$. (2) Solutions of other equations (logarithmic, etc.) can be found with similar solution methods.

Solving an Equation of the Form $f(x) = g(x)$ Using the Intersection Method

If One Solution Is Sought

1. Enter a value for the input variable (x, for example) in cell A2 and the formula for each of the two sides of the equation in cells B2 and C2, respectively.

2. Enter "=B2−C2" in cell D2 (Figure 32).

3. Select the Data tab and choose What-If Analysis in Data Tools. Select Goal Seek to find x when B2−C2=0.

4. In the dialog box,

   a. Click the Set cell box and click on the D2 cell.

   b. Enter 0 in the To value box.

   c. Click the By changing cell box and click on the A2 cell (Figure 33).

Figure 32	Figure 33

5. Click OK to find the x-value of the solution in cell A2 (Figure 34). The solution may be approximate.

Figure 34

If Multiple Solutions Are Sought (as in the case of a quadratic equation)

1. Proceed as in steps 1 and 2 above.

2. Enter values for $f(x)$ and for $g(x)$, and use them to graph the function over an interval that shows points of intersection. Input values near those of these points will give column D values that are at or near 0. Entering x-values with $h = \dfrac{-b}{2a}$ near the center is useful when solving quadratic equations.

3. Select the Data tab and choose What-If Analysis in Data Tools. Use Goal Seek to enter in D2 the address of a cell with a function value at or near 0, and complete the process as above.

4. After finding the first solution, repeat the process using a second function value at or near 0.

Investments

Finding the Future Value of a Lump-Sum Investment

To find the future value of a lump-sum investment:

1. Type the headings Principal, Number of Periods, Payment, Annual Rate, Periods per year, and Periodic Rate into row 1, cells A1 through F1, and enter their values (with interest rates as decimals) in row 2. Enter the investment under Principal as a negative number, and divide the annual rate by the periods per year to get the periodic rate.

2. Type the heading Future Value in cell A4.

3. In cell B4, type the formula =fv(F2,B2,C2,A2, 0) and press ENTER to compute the future value.

The future value of $10,000 invested at 8% per year for 10 years, compounded daily, is shown in Cell B4 of Figure 35.

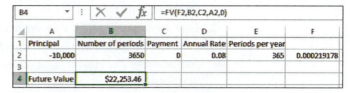

Figure 35

Finding the Present Value of an Investment

To find the present value (investment) that gives the given future value:

1. Type the headings Future Value, Number of Periods, Payment, Annual Rate, Periods per year, and Periodic Rate into row 1, cells A1 through F1, and enter their values (with interest rates as decimals) in row 2. Enter the investment as a negative number, and divide the annual rate by the periods per year to get the periodic rate

2. Type the heading Present Value in cell A4.

3. In cell B4, type the formula =pv(F2,B2,C2,A2, 0) and press ENTER to compute the present value.

The lump sum that must be invested at 10% per year, compounded semiannually, to get $15,000 in 7 years is shown as the present value in cell B4 of Figure 36. This present value will be negative because it leaves the investor.

Figure 36

Annuities

Finding the Future Value of an Ordinary Annuity

To find the future value of an ordinary annuity:

1. Type the headings Principal, Number of Periods, Payment, Annual Rate, Periods per year, and Periodic Rate into row 1, cells A1 through F1, and enter their values (with interest rates as decimals) in row 2. Enter the periodic investment under payment as a negative number, and divide the annual rate by the periods per year to get the periodic rate.

2. Type the heading Future Value in cell A4.

3. In cell B4, type the formula =fv(F2,B2,C2,A2, 0) and press ENTER to compute the future value.

The future value of an ordinary annuity with a contribution of $500 per quarter into an account that pays 8% per year, compounded quarterly, is shown in Cell B4 of Figure 37.

Figure 37

Finding the Present Value of an Ordinary Annuity

To find the present value of an ordinary annuity:

1. Type the headings Principal, Number of Periods, Payment, Annual Rate, Periods per year, and Periodic Rate into row 1, cells A1 through F1, and enter their values (with interest rates as decimals) in row 2. Divide the annual rate by the periods per year to get the periodic rate.

2. Type the heading Present Value in cell A4.

3. In cell B4, type the formula =pv(F2,B2,C2,A2, 0) and press ENTER to compute the present value. The present value is negative because the money leaves the investor to start the annuity.

The lump sum necessary to establish an ordinary annuity that provides $2000 at the end of each month for 20 years, if it earns 6%, compounded monthly, is shown in cell B4 of Figure 38.

Figure 38

Solving a System of Two Linear Equations in Two Variables

1. Write the two equations as linear functions in the form $y = mx + b$.

2. Enter a value for the input variable (x) in cell A2 and the formula for each of the two equations in cells B2 and C2, respectively.

3. Enter "=B2−C2" in cell D2 (Figure 39).

4. Select the Data tab and choose What-If Analysis in Data Tools. Select Goal Seek to find x when B2−C2=0, as on page 694.

5. In the dialog box,

   a. Click the Set cell box and click on the D2 cell.

   b. Enter 0 in the To value box.

   c. Click the By changing cell box and click on the A2 cell.

6. Click OK in the Goal Seek dialog box, getting the solution.

7. The x-value of the solution will be in cell A2, and the y-value will be in both B2 and C2 (Figure 40).

Figure 39   **Figure 40**

Matrices

Excel is useful to add, subtract, multiply, and find inverses of matrices. To perform operations with a matrix, enter each of its elements in a cell of a worksheet. Addition and subtraction of matrices are intuitive with Excel, but finding products and inverses requires special commands.

Adding and Subtracting Matrices (steps for two 3 × 3 matrices)

1. Type a name A in A1 to identify the first matrix.
2. Enter the elements of matrix A in the cells B1:D3.
3. Type a name B in A5 to identify the second matrix.
4. Enter the elements of matrix B in the cells B5:D7.
5. Type a name A+B in A9 to indicate the matrix sum.
6. Type the formula "=B1+B5" in B9 and press ENTER (Figure 41).
7. Use Fill Across to copy this formula across the row to C9 and D9.
8. Use Fill Down to copy the row B9:D9 to B11:D11, which gives the sum (Figure 42).

Figure 41

Figure 42

9. To subtract the matrices, change the formula in B9 to "=B1−B5" (Figure 43) and proceed as with addition (Figure 44).

Figure 43

Figure 44

Finding the Product of Two Matrices
(steps for two 3 × 3 matrices)

1. Enter the names and elements of the matrices as described on page 697.

2. Enter the name A×B in A9 to indicate the matrix product (Figure 45).

3. Select a range of cells that is the correct size to contain the product (B9:D11 in this case).

4. Type "=mmult(" in the formula bar, and then select the cells containing the elements of matrix A (B1:D3 in this case) (Figure 46).

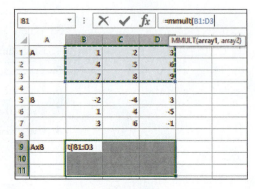

| **Figure 45** | **Figure 46** |

5. Staying in the formula bar, type a comma, select the matrix B elements (B5:D7 in this case), and close the parentheses.

6. Hold the CTRL and SHIFT keys down and press ENTER, which will give the product (Figure 47).

Figure 47

Finding the Inverse of a Matrix
(steps for a 3 × 3 matrix)

1. Enter the name A in A1 and the elements of the matrix in B1:D3, as described on page 697.

2. Enter the name "Inverse(A)" in A5 and select a range of cells that is the correct size to contain the inverse (B5:D7 in this case).

3. In the formula bar, enter "=minverse(", select matrix A (B1:D3), and close the parentheses (Figure 48).

4. Hold the CTRL and SHIFT keys down and press ENTER, which will give the inverse (Figure 49).

Figure 48   **Figure 49**

Solving Systems of Linear Equations with Matrix Inverses

A system of linear equations can be solved by multiplying the matrix containing the augment by the inverse of the coefficient matrix. Following are the steps used to solve

$$\begin{cases} 2x + y + z = 8 \\ x + 2y \quad\;\; = 6 \\ 2x \quad\;\; + z = 5 \end{cases}$$

1. Enter the coefficient matrix A in B1:D3.

2. Compute the inverse of A in B5:D7, as described above (Figure 49).

3. Enter B in cell A9 and enter the matrix containing the augment in B9:B11.

4. Enter X in A13, Y in A14, and Z in A15 and select the cells B13:B15.

5. In the formula bar, type "=mmult(", select matrix inverse(A) in B5:D7, type a comma, select matrix B in B9:B11, and close the parentheses (Figure 50).

6. HOLD the CTRL and SHIFT keys down and press ENTER, which will give the solution.

7. Matrix X gives the solution $x = 0$, $y = 3$, $z = 5$ (Figure 51).

Figure 50   **Figure 51**

Credits

TEXT CREDITS

Chapter 1   p. 25, Exercise 55: Projected Prevalence of Alzheimer's Disease Cases in the U.S. from 2000 through 2050 from A National Alzheimer's Strategic Plan: The Report of the Alzheimer's Study Group. Copyright © 2009 by Alzheimer's Association. Used by permission of Alzheimer's Association.

Chapter 2   p. 137, Exercise 46: Age at First Marriage and First Birth. Copyright © 2013 by the National Marriage Project. Used by permission of National Marriage Project.

Chapter 5   p. 373, Exercise 31: Vehicle Temperature from Department of Geosciences. Copyright by Jan Null. Used by permission of Jan Null.

PHOTO CREDITS

Endpaper   Pearson Education, Inc.

Chapter 1   p. 1, Richard Levine/Alamy

Chapter 2   p. 84, Stephen Coburn/Shutterstock **p. 104,** Christopher Marin/Shutterstock   **p. 136,** CandyBox Images/Shutterstock

Chapter 3   p. 159, Jon Barlow/Pearson Education, Inc. **p. 166,** Andrew Park/Shutterstock   **p. 181,** Georgerudy/123RF   **p. 230,** Rafael Ramirez Lee/ Shutterstock

Chapter 4   p. 239, Blend Images/Shutterstock **p. 300,** Nejron/123RF

Chapter 5   p. 308, Natursports/Shutterstock **p. 342,** Kyodo News/AP Images   **p. 343,** Artwork by Liz Bradford   **p. 406,** Monkey Business Images/Shutterstock

Chapter 6   p. 418, Itsallgood/Fotolia

Chapter 7   p. 507, Chungking/Fotolia

Chapter 8   p. 587, Fuse/Getty Images   **p. 618,** Eric Isselee/Shutterstock

Appendix A   p. 664, Texas Instruments, Inc.

Screenshots from Texas Instruments. Courtesy of Texas Instruments.

Screenshots from Microsoft® Excel®. Used by permission of Microsoft Corporation.

Answers to Selected Exercises

Chapter 1 Functions, Graphs, and Models; Linear Functions

Toolbox Exercises

1. $\{1, 2, 3, 4, 5, 6, 7, 8\}$; $\{x \mid x \in N, x < 9\}$ **2.** Yes **3.** No
4. No **5.** Yes **6.** Yes **7.** Integers, rational **8.** Rational
9. Irrational **10.** $x > -3$ **11.** $-3 \le x \le 3$ **12.** $x \le 3$
13. $(-\infty, 7]$ **14.** $(3, 7]$ **15.** $(-\infty, 4)$

16.

17.

18.

19.

20. **21.**

22. **23.**

24. Yes; 4 **25.** No **26.** No **27.** Yes; 6
28. $-3x^2$, -3; $-4x$, -4; constant 8
29. $5x^4$, 5; $7x^3$, 7; constant -3
30. $3z^4 + 4z^3 - 27z^2 + 20z - 11$
31. $7y^4 - 2x^3y^4 - 3y^2 - 2x - 9$
32. $4p + 4d$ **33.** $-6x + 14y$ **34.** $-ab - 8ac$
35. $x - 6y$ **36.** $6x + 9xy - 5y$ **37.** $3xyz - 5x$
38. $-12a^2x^5y^3$ **39.** $4x^3y^4 + 8x^2y^3z - 6xy^3z^2$
40. $2x^2 - 11x - 21$ **41.** $k^2 - 6k + 9$
42. $16x^2 - 49y^2$ **43.** $12 - 5x$ **44.** $4x + 1$

1.1 Skills Check

1. a. Input **b.** Output **c.** $D = \{-9, -5, -7, 6, 12, 17, 20\}$;
$R = \{5, 6, 7, 4, 9, 10\}$ **d.** Each input gives exactly one output.
3. 5; 9 **5.** No; input 9 for y gives two outputs for x.
7. a. $f(2) = -1$ **b.** $f(2) = -2$ **c.** $f(2) = -3$

9.

x	y
0	2
-2	-4

11. a. -7 **b.** 3 **c.** 18

13. Yes; $D = \{-1, 0, 1, 2, 3\}$; $R = \{5, 7, 2, -1, -8\}$ **15.** No
17. Yes **19.** No; one value of x, 3, gives two values of y. **21.** b
23. b **25.** $D = \{-3, -2, -1, 1, 3, 4\}$; $R = \{-8, -4, 2, 4, 6\}$
27. $D = [-10, 8]$; $R = [-12, 2]$
29. $D = (-\infty, \infty)$; $R = [-4, \infty)$ **31.** 15; 22 **33.** No

35. $F = \dfrac{9}{5}C + 32$ **37. a.** Subtract 32. **b.** Multiply by 5.

c. Divide by 9.

1.1 Exercises

39. a. Yes; a, p; there is one premium p for each age a. **b.** No
41. Yes; y, p; there is one percent p for each year y.
43. a. Yes; there is one price for each barcode. **b.** No; numerous
items with the same price have different barcodes. **45.** There is one
output for each input. **47. a.** Yes **b.** The days 1–14 of May
c. $\{171, 172, 173, 174, 175, 176, 177, 178\}$ **d.** May 1, May 3
e. May 14 **f.** 3 days **49. a.** 1096.78; if the car is financed over
3 years, the payment is \$1096.78 **b.** \$42,580.80; $C(5) = 42,580.80$
c. 4 **d.** \$3096.72 **51. a.** 22 million **b.** 11; approximately
11 million women were in the workforce in 1930. **c.** $\{$1930, 1940,
1950, 1960, 1970, 1980, 1990, 2000, 2005, 2010, 2015$\}$ **d.** For
each new indicated year, the number of women increased.
53. a. 26.1, 29.2 **b.** 21.5, 26.6 **c.** 1980; the median age at first
marriage was 24.7 in 1980. **d.** Increase **55. a.** 5.7 million
b. 7.7; in 2030, 7.7 million U.S. citizens age 65 and older are
expected to have Alzheimer's disease. **c.** 2040 **d.** Increasing.
57. a. 56.0; the birth rate in 1995 for U.S. girls ages 15–19 was 56.0
per 1000. **b.** 2005 **c.** 1990 **d.** The birth rate increased until 2007,
then decreased. **59. a.** 6400; the revenue from the sale of 200 hats is
\$6400. **b.** \$80,000; $R(2500) = 80,000$ **61. a.** 876.35; the charge
for 1000 kWh is \$876.35. **b.** \$1304.85 **63. a.** 1200; the daily profit
from the sale of 100 bicycles is \$1200. **b.** \$1560 **65. a.** All real
numbers except $-\dfrac{3}{7}$ **b.** Positive values of n **67. a.** $0 \le p < 100$

b. 355,500; 2,133,000 **69. a.** 8640; 11,664 **b.** $0 < x < 27$, so volume is positive and box exists **c.** Testing values in the table shows a
maximum volume of 11,664 cubic inches when $x = 18$ inches.

x	y
10	6,800
15	10,800
20	11,200
21	10,584
19	11,552
18	11,664
17	11,560

The dimensions of the box are 18 in. by
18 in. by 36 in.

1.2 Skills Check

1. a.

x	-3	-2	-1	0	1	2	3
$y = x^3$	-27	-8	-1	0	1	8	27

b. 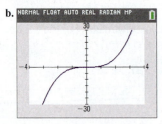   **c.** Graphs are the same.

3. Answers vary.

x	$f(x)$
-2	-7
-1	-4
0	-1
1	2
2	5

5. Answers vary.

x	$f(x)$
-4	8
-2	2
0	0
2	2
4	8

7. Answers vary.

x	$f(x)$
-2	$-1/4$
-1	$-1/3$
0	$-1/2$
1	-1
2	undefined
3	1
4	$1/2$

9. ; yes

11. ; yes

13. ; yes

15. a.   **b.** View (b) is better.

17. a.   **b.** View (b) is better.

19. Letting y vary from -5 to 100 gives one view.

Turning point coordinates: $(0, 50)$

21. Setting y from -200 to 300 gives one view.

Turning point coordinates: $(-5, 175)$, $(3, -81)$

23.

25.

t	S
12	51.9
16	72.7
28	135.1
43	213.1

27.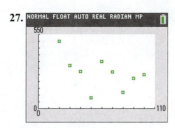

41. a. 5;11 **b.** 2015; 79.29 **c.** 0;15

43. a.

b. 10,984,200

45. a.

b. 2398.8 **c.** \$3183 billion

29. a. **b.**

c. Yes; yes **31. a.** 300 **b.** 2020

47.

$$P(x)$$

$$P(x) = 1500x - 8000 - 0.01x^2$$

1.2 Exercises

33.

49. **a.** 0 to 30 **b.** 23.4; 43.35

35. a.

$$S = 100 + 64t - 16t^2$$

b. 148 ft and 148 ft; ball rising and falling
c. 164 ft, in 2 seconds

51. a. 299.9 million, or 299,900,000

b.

Years after 2000	Population (millions)
0	275.3
10	299.9
20	324.9
30	351.1
40	377.4
50	403.7
60	432.0

c.

37.

$$C(x) = 15,000 + 100x + 0.1x^2$$

53. a. **b.**

39. a. 76.16; the model predicts that there will be 76.16 million women in the workforce in 2020. **b.** 80; 84,060,000

55. a. 2.26 billion barrels

b. **c.**

1.3 Skills Check

1. b   **3.** $-\dfrac{1}{2}$ **5.** 2

7. a. x: 3; y: -5
b.

9. a. x: $\dfrac{3}{2}$; y: 3
b.

11. 0; undefined   **13. a.** Positive   **b.** Undefined
15. a. Slope 4; y-intercept 8
b.

17. a. Slope 0; y-intercept $\dfrac{2}{5}$
b.

19. a. Slope 4; y-intercept 5
b. Rising
c.

21. a. Slope -100; y-intercept 50,000
b. Falling
c.

23. 4   **25.** -15 **27.** -2 **29. a.** ii   **b.** i   **31. a.** 0   **b.** 0

1.3 Exercises

33. Yes; it is written in the form $y = mx + b$; x, the number of years after 1990   **35. a.** It is written in the form $y = mx + b$.
b. 0.077; the life expectancy is projected to increase by approximately 0.1 year per year.   **37. a.** 72.5   **b.** 9.57; in 1980 the marriage rate for unmarried women was 9.57 per 1000.
c.

39. x: 50; R: 3500

41. a. Constant function   **b.** $y = 11.81$ **c.** 0   **d.** 0   **43. a.** 0.328
b. From 2010 through 2040, the disposable income is projected to increase by $328 million per year.   **45. a.** 20.86; approximately 21 million women were in the workforce in 1950.   **b.** 0.79
c. The rate of change of the number of women in the workforce is 790 thousand per year.   **47. a.** 0.057   **b.** From 1990 to 2050, the percent of the U.S. population that is black increased by 0.057 percentage point per year.   **49. a.** Positive   **b.** As the temperature increases, the number of chirps increases.   **51. a.** $0.975 billion per year   **b.** $6.3 billion   **c.** No; the amount is negative.
53. a. $-61,000$ **b.** $-61,000$ per year   **55.** $58 per unit
57. a. 0.56   **b.** $0.56 per ball   **c.** The cost will increase by $0.56 for each additional ball produced in a month.   **59. a.** 1.60   **b.** $1.60 per ball   **c.** The revenue will increase by $1.60 for each additional ball sold in a month.   **61.** $19 per unit

1.4 Skills Check

1. $y = 4x + \dfrac{1}{2}$ **3.** $y = \dfrac{1}{3}x + 3$ **5.** $y = \dfrac{-3}{4}x - 3$ **7.** $x = 9$

9. $y = x + 3$ **11.** $y = 2$ **13.** $y = \dfrac{4}{5}x + 4$ **15.** $y = -3x + 6$

17. $y = \dfrac{3}{2}x + \dfrac{23}{2}$ **19.** $y = 3x + 1$ **21.** $y = -5$

23. $y = -15x + 12$ **25.** $y = \dfrac{2}{3}x + 7$ **27.** 1   **29.** -3 **31.** -7

33. -15 **35.** $4x + 2h$ **37. a.** Yes; a scatter plot shows that data fit along a line.   **b.** $y = 3x + 555$

1.4 Exercises

39. $y = 12.00 + 0.1034x$ dollars   **41.** $y = -3600t + 36,000$
43. a. $H = -0.65t + 143$ **b.** 117   **45. a.** $25,000   **b.** $5000
c. $s = 26,000 - 5000t$ **47.** $y = x$; $x =$ deputies; $y =$ cars
49. $P(x) = 58x - 12,750$; $x =$ number of units
51. $V = -61,000x + 1,920,000$ **53. a.** $y = 160x - 8560$ for $66 \le x \le 70$ **b.** 2640   **55. a.** 0.045 percentage point per drink
b. $y = 0.045x$ **57. a.** $g(t) = 0.565t - 1057.95$ **b.** Yes
c. They are the same.   **59. a.** $p = 0.67x - 1317.9$ **b.** 35.5%
61. a. 0.79   **b.** 0.79 million per year   **c.** $y = 0.79x + 18.4$
63. a. 26,105   **b.** 26,105   **c.** $y = 26,105x - 50,952,863$ **d.** No
65. a. No   **b.** Yes, they appear to be on a line.   **c.** -0.085
d. $y = -0.085x + 174.75$ **67. a.** 0.326 million per year
b. $y = 0.326x - 498.9$ **c.** 159.62 million   **69.** 32 ft/sec
71. $64 - 16h$

Chapter 1 Skills Check

1. Each value of x is assigned exactly one value of y.
2. $D = \{-3, -1, 1, 3, 5, 7, 9, 11, 13\}$;
$R = \{9, 6, 3, 0, -3, -6, -9, -12, -15\}$

3. $f\{3\} = 0$ **4.** Yes; $y = -\dfrac{3}{2}x + \dfrac{9}{2}$ **5. a.** -2 **b.** 8   **c.** 14

6. a. 1   **b.** -10

7.

$f(x) = -2x^3 + 5x$

8.

$y = 3x^2$

9. Standard window:

$[0, 40]$ by $[0, 5000]$ gives a better graph:

10.

11. a. $x \geq 4$ **b.** All real numbers except 6

12. Perpendicular   **13.** $-\dfrac{7}{6}, \dfrac{6}{7}$ **14.** $-\dfrac{11}{6}$

15. a. $(0, -4)$ and $(6, 0)$ **b.**

$2x - 3y = 12$

16. $\dfrac{2}{3}$ **17.** Slope -6, y-intercept 3   **18.** -6 **19.** $y = \dfrac{1}{3}x + 3$

20. $y = -\dfrac{3}{4}x - 3$ **21.** $y = x + 4$ **22.** 3   **23. a.** $5 - 4x - 4h$

b. $-4h$ **c.** -4 **24. a.** $10x + 10h - 50$ **b.** $10h$ **c.** 10

Chapter 1 Review

25. a. Yes   **b.** $f(1992) = 82$; in 1992, 82% of black voters supported the Democratic candidate for president.   **c.** 1964; in 1964, 94% of black voters supported the Democratic candidate for president.
26. a. $\{1960, 1964, 1968, 1972, 1976, 1980, 1984, 1992, 1996, 2000, 2004, 2008, 2012\}$ **b.** No; 1982 was not a presidential election year.
27.

28. a. 0.15   **b.** 0.15 percentage point per year   **c.** No

29. a. As each input value changes by 5000, the output value changes by 89.62.   **b.** 448.11; the monthly payment to borrow \$25,000 is \$448.11.   **c.** $A = 20,000$ **30. a.** $D = \{10,000, 15,000, 20,000, 25,000, 30,000\}$; $R = \{179.25, 268.87, 358.49, 448.11, 537.73\}$
b. No   **31. a.** $f(28,000) = 501.882$; if \$28,000 is borrowed, the payment is \$501.88.   **b.** Yes   **32. a.** $f(1960) = 15.9$; a 65-year-old woman in 1960 was expected to live 15.9 more years, or 80.9 years.
b. $65 + 19.4 = 84.4$ years   **c.** 1990   **33. a.** 16.9; a 65-year-old man in 2020 is expected to live 16.9 more years, or 81.9 years.
b. 77.8 years   **c.** $g(1990) = 15$ **34.** No; it cannot be written in the form $y = mx + b$.   **35. a.** $h = -0.8t + 176$ **b.** Yes   **c.** Yes
36. a. 0.016 billion barrels per year   **b.** $y = 0.016x - 30.22$
37. $f(x) = 4500$ **38. a.** $f(x) = 33.8$ **b.** It is a constant function.
39. a. \$67,680   **b.** \$47,680   **c.** MR: 564; MC: 64   **d.** $m = 64$
e.

$R(x) = 564x$

$C(x) = 40,000 + 64x$

40. a. $P(x) = 500x - 40,000$ **b.** \$20,000   **c.** 80   **d.** 500
e. Marginal revenue minus marginal cost   **41. a.** $y = 300,000$; the initial value of the property is \$300,000.   **b.** $x = 100$; the value of the property after 100 years is zero dollars.   **42. a.** \$4.40 per unit
b. Slope 4.4   **c.** $P(x) = 4.4x - 205$ **d.** \$4.40 per unit
e. Approximately 47

Chapter 2 Linear Models, Equations, and Inequalities

Toolbox Exercises

1. Division Property; $x = 2$ **2.** Addition Property; $x = 18$
3. Subtraction Property; $x = 5$ **4.** Addition Property; $x = 3$
5. Multiplication Property; $x = 18$ **6.** Division Property; $x = -2$
7. Subtraction Property and Division Property; $x = -10$
8. Addition Property and Multiplication Property; $x = 32$ **9.** $x = 3$
10. $x = \dfrac{3}{5}$ **11.** $x = 16$ **12.** $x = -4$ **13.** $x = 2$ **14.** $x = 1$
15. $x = \dfrac{1}{2}$ **16.** $x = 1$ **17.** $x = 4$ **18.** $x = 5$ **19.** $x = 3$
20. $x = 5$ **21.** Conditional   **22.** Contradiction   **23.** Identity
24. Conditional   **25.** $x > -\dfrac{6}{5}$ **26.** $x \leq -2$ **27.** $x > -12$
28. $x > -12$ **29.** $x < \dfrac{-8}{19}$ **30.** $x < \dfrac{6}{11}$ **31.** $x > \dfrac{19}{3}$
32. $x > -16$ **33.** 6   **34.** 4   **35.** 2   **36.** 5   **37.** 5   **38.** 2
39. a. $f(x) = 0.008x^2 - 632.578x + 480.650$
b. $f(x) = 0.00754x^2 - 633x + 481$
c. $f(x) = 0.007543x^2 - 632.6x + 480.7$

2.1 Skills Check

1. $x = -\dfrac{37}{2}$ **3.** $x = 10$ **5.** $x = -\dfrac{13}{24}$ **7.** $x = -63$ **9.** $t = -1$
11. $x = \dfrac{25}{36}$ **13.** $x = -\dfrac{8}{3}$ **15.** $x = \dfrac{16}{3}$ **17.** $x = \dfrac{190}{113}$
19. a. -20 **b.** -20 **c.** -20 **21. a.** 4   **b.** 4   **c.** 4
23. a. 2   **b.** -34 **c.** 2   **25. a.** 40   **b.** 40
27. $2x - 5 = 3(x - 2)$; $x = 1$ **29. a.** 25   **b.** 25   **c.** 25
31. a. -8.25 **b.** -8.25 **c.** -8.25 **33.** $x = 3$ **35.** $s = -5$
37. $t = -4$ **39.** $t = \dfrac{17}{4}$ **41.** $r = \dfrac{A - P}{Pt}$ **43.** $F = \dfrac{9}{5}C + 32$
45. $n = \dfrac{5m}{2} - \dfrac{P}{4} - \dfrac{A}{2}$

47. $y = \dfrac{5x - 5}{3}$

49. $y = \dfrac{6 - x^2}{2}$

2.1 Exercises

51. 60 months, or 5 years   **53.** $6000   **55.** $54,620   **57.** 2020
59. 2026   **61.** 2020   **63.** $x = 55$, so 2015   **65. a.** 20   **b.** 60
67. 1990   **69.** 2028   **71.** 97   **73.** $78.723 billion   **75.** 25%

77. $1698   **79.** $t = \dfrac{A - P}{Pr}$ **81.** 10%

83. $613.33   **85.** Yes; 2π **87. a.** 4/9   **b.** 72 kg

2.2 Skills Check

1. No; data points do not lie close to a line.   **3.** Approximately; all the
data points do not lie on the line, but are close to the line.
5.

7. Exactly; first differences of the outputs are equal.   **9.** $y = 1.5x + 2.5$
11.

13. $y = 2.419x - 5.571$

15.

17. $y = 1.577x + 1.892$

19. $y = -1.5x + 8$ **21. a.** Exactly   **b.** Nonlinear   **c.** Approximately

2.2 Exercises

23. a. Discrete   **b.** Continuous   **c.** No; a line will not fit the points well.
25. a. No; increases from one year to the next are not constant.
b. Yes; increases from one year to the next are constant.
c. $S = 80x - 4280$ **27. a.** Yes; the first differences are constant for

uniform inputs.   **b.** Two   **c.** $D = \dfrac{5w}{11}$ **d.** 95 mg

29. a. No   **b.** $y = 0.328x + 6.316$ **c.** $13.850 billion

31. a.

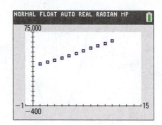

b. $y = 2375x + 39,630$ **c.** $m = 2375$; the average annual wage is
expected to increase $2375 per year.
33. a. $y = 0.465x + 16.696$ **b.** 24.6%   **c.** 2041

35. a.

37. a.

b. $y = 1.890x + 247.994$

b. $y = 579x + 9410$

c. excellent fit

c. is not an excellent fit

d. 308.476   **e.** 2057

39. a.

c. Good fit

b. $y = 0.077x + 13.827$

d. 19.4 years   **e.** 2043

41. a. $y = 0.297x + 2.043$
b. excellent fit

43. a.

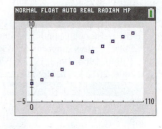

c. 0.3 percentage point per year   **b.** $y = 0.0715x + 2.42$

c. yes

d. 6.5 billion   **e.** 2056

45. a. $y = 138.217x + 97.867$ **b.** \$3967.947 billion
c. 2023

2.3 Skills Check

1. a. No   **b.** Yes   **3.** $(1, 1)$ **5.** $x = -14, y = -54$

7. $x = \dfrac{-4}{7}, y = \dfrac{4}{7}$ **9.** Many solutions; the two equations have the

same graph.   **11.** $x = 2, y = -2$ **13.** $x = 1, y = -1$

15. $x = 2, y = 1$ **17.** $x = \dfrac{4}{7}, y = \dfrac{12}{7}$ **19.** $x = 4, y = 3$

21. Dependent; many solutions   **23.** No solution

25. $x = 2, y = 4$ **27.** $x = \dfrac{32}{11}, y = \dfrac{-14}{11}$ **29.** $x = 3, y = -2$

31. $x = 2, y = 1$ **33.** No solution

2.3 Exercises

35. 60 units   **37.** 682 units

39. a. \$260   **b.** 30

41. a. $y = 0.69x + 2.4$ **b.** $y = -0.073x + 7.7$ **c.** 7

43. 2072   **45.** $x = 19.8$, so 2010

47. \$669.8 million in 2013, \$805.3 million in 2016

49. a. $x + y = 2400$ **b.** $30x$ **c.** $45y$ **d.** $30x + 45y = 84,000$
e. 1600 \$30 tickets, 800 \$45 tickets

51. a. \$75,000 at 8%, \$25,000 at 12%   **b.** 12% account is probably
more risky; investor might lose money.   **53.** \$200,000 at 6.6%,
\$50,000 at 8.6%   **55.** 60 cc of 10% solution, 40 cc of 5% solution
57. 3 glasses of milk, 2 servings of meat   **59.** 2031; 11%

61. a. $p = -\dfrac{1}{2}q + 155$ **b.** $p = \dfrac{1}{4}q + 50$ **c.** \$85

63. 78   **65. a.** $300x + 200y = 100,000$ **b.** 250 in the first group,
125 in the second group   **67.** 700 units at \$30; when the price is \$30,
the amount demanded equals the amount supplied equals 700.

2.4 Skills Check

1. $x \le 3$; $(-\infty, 3]$

3. $x \le \dfrac{-1}{7}$; $\left(-\infty, -\dfrac{1}{7}\right]$

5. $x < \dfrac{20}{23}$; $\left(-\infty, \dfrac{20}{23}\right)$

7. $x < \dfrac{61}{5}$; $\left(-\infty, \dfrac{61}{5}\right)$

9. $x \ge \dfrac{-30}{11}$; $\left[-\dfrac{30}{11}, \infty\right)$

11. $x \ge 2.8\overline{6}$; $[2.8\overline{6}, \infty)$

13. $(-\infty, -2)$ **15.** $[-8, \infty)$ **17. a.** $x = -1$ **b.** $(-\infty, -1)$

19. $\dfrac{22}{3} \le x < 12$ **21.** $\dfrac{5}{2} \le x \le 10$ **23.** No solution

25. $x \ge \dfrac{32}{11}$ or $x \le \dfrac{3}{4}$ **27.** $72 \le x \le 146$

2.4 Exercises

29. a. $V = 12,000 - 2000t$ **b.** $12,000 - 2000t < 8000$
c. $12,000 - 2000t \ge 6000$ **31.** $C \le 0$ **33.** More than \$22,000

35. 80 through 100   **37.** After 2018   **39.** After age $82\dfrac{1}{2}$

41. $1350 \le x \le 1650$ **43.** In 2022 and after
45. a. $y = 20,000x + 190,000$ **b.** $11 \le x \le 14$ **c.** No
47. More than 2000 units   **49.** At least 1500 feet
51. $364x - 345,000 > 0$ **53.** 2015 and after   **55.** After 2023

Chapter 2 Skills Check

1. $x = \dfrac{34}{5}$ **2.** $x = -14.5$ **3.** $x = -138$ **4.** $x = \dfrac{5}{28}$ **5.** $x = \dfrac{52}{51}$

6. $x = 5$ **7. a.** 15   **b.** 15   **c.** 15   **8.** $y = \dfrac{3 + m - 3Pa}{-3P}$

9. $y = \dfrac{4x - 6}{3}$

10.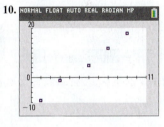

11. $y = 2.895x - 11.211$

12.

13. No   **14.** $x = 2, y = -3$ **15.** $x = \dfrac{-3}{13}, y = \dfrac{-15}{13}$

16. Many solutions   **17.** No solution   **18.** $x = -3, y = 5$

19. $x = \dfrac{1}{2}, y = -4$ **20.** $x < -\dfrac{4}{5}$ **21.** $x \le \dfrac{25}{28}$ **22.** $6 \le x < 18$

Chapter 2 Review

23. a. Yes   **b.** Yes   **c.** $P = f(A) = 0.018A + 0.010$
24. a. 504.01; the predicted monthly payment on a car loan of \$28,000
is \$504.01.   **b.** Yes   **c.** \$27,895   **25.** 2030   **26.** $f(x) = 4500$
27. a. $P(x) = 4.4x - 205$ **b.** At least 47 units   **28. a.** \$22,000
b. More than \$22,000 per month   **29.** \$25,440 each
30. \$300,000 in the safe account, \$120,000 in the risky account

31. a. $y = 0.3x - 14$ **b.**

c. Yes   **d.** 9.4%   **e.** age 88

32. More than 120 units
33. a. $P(x) = 500x - 40,000$ **b.** $x > 80$ **c.** More than 80 units
34. a. $x > 10$ **b.** 10 years after its purchase
35. a. $p = -1.873t + 60.643$ **b.** 23.2%; extrapolation
36. a. $y = 0.0638x + 15.702$ **37. a.** $y = 0.0655x + 12.324$

b.

b.

c. 22; the projected life span of a 65-year-old woman is 87 in 2049.
d. 2002 and after

c. 20.8; the projected life span of a 65-year-old man is 86 in 2080.
d. 2144   **e.** Before 2007

38. a. $y = 80x - 4280$
for $66 \le x \le 70$

b.

c. exact

39. a. $y = 72.25x - 3751.94$
for $62 \le x \le 70$

b.

c. approximate

40. a.

; linear equation is reasonable.

b. $y = 0.209x - 0.195$

c.

; the line seems to fit the data points very well.

41. 2005 and before; 2025 and after
42. a. no   **b.** $y = -0.652x + 142.881$ **c.** yes; $y = -0.65x + 143$
43. Between 17 and 31 months   **44.** $140,000 at 8%, $100,000 at 12%
45. 25 units   **46.** Each dosage of medication A contains 2.8 mg, and each dosage of medication B contains 4.2 mg.
47. $p = \$100$; $q = 80$ pairs   **48.** $p = \$79$; $q = 710$ units

49. a. $x + y = 2600$ **b.** $40x$ **c.** $60y$ **d.** $40x + 60y = 120,000$
e. 1800 $40 tickets, 800 $60 tickets   **50. a.** $x + y = 500,000$
b. $0.12x$ **c.** $0.15y$ **d.** $0.12x + 0.15y = 64,500$ **e.** $350,000 in the 12% property, $150,000 in the 15% property

Chapter 3 Quadratic, Piecewise-Defined, and Power Functions

Toolbox Exercises

1. $\dfrac{9}{4}$ **2.** $\dfrac{8}{27}$ **3.** $\dfrac{1}{100}$ **4.** $\dfrac{1}{64}$ **5.** $\dfrac{1}{8}$ **6.** $\dfrac{1}{256}$
7. $\dfrac{x^8}{16y^4}$ **8.** $\dfrac{x^{10}}{y^{15}}$ **9.** $\dfrac{1}{1024x^{10}}$ **10.** 1   **11.** $\dfrac{x^6z^3}{y^3}$ **12.** $\dfrac{a^{18}c^{12}}{b^6}$
13. 6   **14.** 4   **15. a.** $x^{3/2}$ **b.** $x^{3/4}$ **c.** $x^{3/5}$ **d.** $3^{1/2}y^{3/2}$ **e.** $27y^{3/2}$
16. a. $\sqrt[4]{a^3}$ **b.** $-15\sqrt[8]{x^5}$ **c.** $\sqrt[8]{(-15x)^5}$ **17.** $36x^2 - 60x + 25$
18. $49s^2 - 4t^2$ **19.** $64w^2 + 32w + 4$ **20.** $36x^2 - 12xy + y^2$
21. $4y^2 + 20yz + 25z^2$ **22.** $9x^4 - 25y^2$ **23.** $3x(x - 4)$
24. $12x^3(x^2 - 2)$ **25.** $(3x - 5m)(3x + 5m)$
26. $(x - 3)(x - 5)$ **27.** $(x - 7)(x + 5)$
28. $(x - 2)(3x + 1)$ **29.** $(2x - 5)(4x - 1)$
30. $3(2n + 1)(n + 6)$ **31.** $3b(3a - 4ab + 6b)$
32. $8x(xy - 20 + 6x)$ **33.** $4y^2(3yz + x^2 - 2z^3)$
34. $(x^2 + 2)(x^2 + 4)$ **35.** $(x + 4)^2$ **36.** $(x - 2)^2$
37. $(2x + 5)^2$ **38.** $(x^2 - 3y)^2$
39. $5x(x^4 - 16) = 5x(x^2 + 4)(x + 2)(x - 2)$
40. $3(y - 2)(y + 2)(y^2 + 3)$ **41.** $(3p + 2)(6p - 1)$
42. $(5x - 3)(x - 2y)$ **43. a.** Imaginary   **b.** Pure imaginary
c. Real   **d.** Real   **44. a.** Imaginary   **b.** Real   **c.** Pure imaginary
d. Imaginary   **45.** $a = 4, b = 0$ **46.** $a = 15, b = -3$
47. $a = 2, b = 4$

3.1 Skills Check

1. a. Quadratic   **b.** Up   **c.** Minimum
3. Not quadratic   **5. a.** Quadratic   **b.** Down   **c.** Maximum
7. a.

b. Yes

9. a.

b. Yes

11. a.

b. Yes

13. a.

$y = 0.01x^2 - 8x$

b. No; the complete graph will be a parabola.

15. $y = (x - 2)^2 - 4$ **17.** y_1

19. $y = -5x^2 + 10x + 8$

21. a. $(1, 3)$ **b.**

$y = (x - 1)^2 + 3$

23. a. $(-8, 8)$ **b.**

$y = (x + 8)^2 + 8$

25. a. $(4, -6)$ **b.**

$f(x) = 2(x - 4)^2 - 6$

27. a. $(2, 12)$ **b.**

$y = 12x - 3x^2$

29. a. $(-3, -30)$ **b.**

$y = 3x^2 + 18x - 3$

31. a. $x = 10$ **b.**

$y = 2x^2 - 40x + 10$

c. $(10, -190)$

33. a. $x = -80$ **b.**

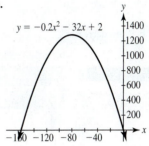

$y = -0.2x^2 - 32x + 2$

c. $(-80, 1282)$

35.

$y = x^2 + 24x + 144$

37.

$y = -x^2 - 100x + 1600$

39.

$y = 2x^2 + 10x - 600$

41. $x = 1, x = 3$ **43.** $x = -10, x = 11$ **45.** $x = -2, x = 0.8$

3.1 Exercises

47. a.

b. It increases.   **c.** Profit decreases.

49. a.   **b.**

c. Shipments will decrease.

51. a.

b. \$796.87 billion
c. Extrapolation

53. a. $t = 2, S = 69.2$ **b.** Two seconds after the ball is thrown, it reaches its maximum height of 69.2 meters.   **c.** The function is increasing until $t = 2$ seconds. The ball rises for two seconds, at which time it reaches its maximum height. After two seconds, the ball falls toward the ground.   **55. a.** Vertex: $(18.5, 2.291)$; the maximum number of barrels of crude oil projected to be produced during this period is 2.291 billion barrels during 2029.   **b.** 2.271 billion barrels

c.

57. a. 2000   **b.** \$37,000   **59. a.** 37,500   **b.** \$28,125,000
61. a. Yes   **b.** $A = 100x - x^2$; maximum area of 2500 sq ft when x is 50 ft

63. a.

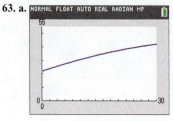

b. $(48.55, 45.671)$; the percent of Americans who are obese will reach a maximum of 45.671% in 2049.   **c.** The percent of Americans who are obese will continue to increase until 2049 and then begin to decrease.

65. a. Minimum   **b.** $(10.184, 33.621)$; in 2001, the number of people in the United States who lived below the poverty level was at a minimum, 33,621,000.

c.

67. a.

b. Decreasing
c. 25   **d.** When wind speed is 0 mph, amount of pollution is 25 oz per cubic yard.

69. The t-intercepts are 3.5 and -3.75. This means that the ball will strike the pool in 3.5 seconds; -3.75 is meaningless.
71. a. $y = -16t^2 + 32t + 3$ **b.** 19 feet
73. a.

Rent ($)	Number of Apartments	Revenue ($)
1200	100	120,000
1240	98	121,520
1280	96	122,880
1320	94	124,080

b. Yes
c. \$1600

75. a. $x = 53.5, y = 6853.3$ **b.** World population will be maximized at 6,853,300,000 in 2044.   **c.** Until 2044

3.2 Skills Check

1. $x = 5, x = -2$ **3.** $x = 8, x = 3$ **5.** $x = -3, x = 2$

7. $t = \dfrac{3}{2}, t = 4$ **9.** $x = -2, x = \dfrac{1}{3}$ **11.** $-\dfrac{1}{3}, 2$ **13.** $\dfrac{3}{2}, -\dfrac{3}{2}$

15. $-2, \dfrac{2}{3}$ **17.** $x = -2, x = -5$ **19.** $x = \dfrac{2}{3}, x = 2$

21. $x = \dfrac{1}{2}, x = -4$ **23.** $w = -\dfrac{1}{2}, w = 3$ **25.** $x = 8, x = 32$

27. $s = 50, s = -15$ **29.** $x = \pm\dfrac{3}{2}$ **31.** $x = \pm 4\sqrt{2}$

33. $x = 2, 8$ **35.** $x = 2 \pm \sqrt{13}$ **37.** $x = 1, x = 2$

39. $x = \dfrac{5 \pm \sqrt{17}}{2}$ **41.** $x = 1, x = -\dfrac{8}{3}$ **43.** $x = -3, x = 2$

45. $x = \dfrac{2}{3} \approx 0.667, x = -\dfrac{3}{2} = -1.5$

47. $x = \dfrac{-1}{2} = -0.5, x = \dfrac{2}{3} \approx 0.667$

49. $x = \pm 5i$ **51.** $x = 1 \pm 2i$ **53.** $x = -2 \pm 2i$
55. a. Positive   **b.** 2   **c.** $x = 3, x = -2$ **57. a.** 0   **b.** 1 with multiplicity 2   **c.** $x = 3$ **59.** 49, two   **61.** 0; one with multiplicity 2

3.2 Exercises

63. $t = 2$ sec and 4 sec   **65.** 20 or 90
67. a. $P(x) = 520x - 10,000 - x^2$ **b.** $-\$964$ (loss of \$964)
c. \$5616   **d.** 20 or 500   **69. a.** $s = 50, s = -50$ **b.** There is no
particulate pollution.   **c.** $s = 50$; speed is positive.
71. a. 0 mL or 100 mL   **b.** When x is zero, there is no amount of
drug in a person's system, and therefore no sensitivity to the drug.
When x is 100 mL, the amount of drug in a person's system is so
high that the person may be overdosed on the drug and therefore
have no sensitivity to the drug.   **73.** \$100 is the price that gives
demand $=$ supply $= 97$ trees   **75.** 2019
77. a. In 1990, energy consumption in the United States was 87.567
quadrillion BTUs.
b.   **c.** Yes; energy consumption in
the United States will again be
87.567 when $x = 78.5$, during
2049.

79. In 2014 and in 2025   **81. a.** \$62.85 billion   **b.** 2014
c. \$264.998 billion; extrapolation; that the model remains valid
83. a.   **b.** \$4246.02
c. 2016

$G(x) = 11.532x^2 - 259.978x + 1666.555$

85. a.   **b.** $x = 5.7187$ and
$x = 16.07$, so in 2006
and 2017   **c.** No;
the model indicates a
maximum in 2011.

$S(x) = -1.751x^2 + 38.167x + 388.997$

87. 2023

3.3 Skills Check

1.   **3.**

5.   **7.**

9.
$$y = \begin{cases} -1 \text{ if } x < 0 \\ 1 \text{ if } x \geq 0 \end{cases}$$

11. a.   **b.** Piecewise, step function

13. a.   **b.** 5, 16
c. All real numbers

15. a. 0, 3, -3, 5
b.

17. a.   **b.** 2, 5
c. All real numbers

$f(x) = |x|$

19. a. 5   **b.** 6   **21. a.** 0   **b.** 29   **23. a.** Increasing   **b.** Increasing
25. Concave down   **27.** Concave up
29.

31.

33. They are the same.   **35.** $x = \dfrac{7}{2}, x = -\dfrac{5}{2}$ **37.** $x = \dfrac{1}{7}$
39. 16   **41.** 80

3.3 Exercises

43. a. $f(x) = \begin{cases} 7.10 + 0.06747x & \text{if } 0 \le x \le 1200 \\ 88.06 + 0.05788(x - 1200) & \text{if } x > 1200 \end{cases}$

b. \$71.87   **c.** \$110.05   **45. a.** $P(x) = \begin{cases} 98 & \text{if } 0 < x \le 1 \\ 119 & \text{if } 1 < x \le 2 \\ 140 & \text{if } 2 < x \le 3 \\ 161 & \text{if } 3 < x \le 4 \end{cases}$

b. 119; the postage on a 1.2-oz first-class letter in a large envelope is 1.19.
c. $0 < x \le 4$ **d.** 119, 140   **e.** \$1.19, \$1.40
47. a. 18,266; 14,832; 5430
b.

$f(x) = \begin{cases} 84.3x + 12,365 & \text{if } x < 80 \\ -376.1x + 48,681 & \text{if } x \ge 80 \end{cases}$

49. a. Power   **b.** 155.196; in 1995, there were 155,196 female
physicians.   **c.** 358,244
51. a. 61.925%, 67.763%
b.   **c.** Yes, but far in the future

$f(t) = 61.925t^{0.041}$

53. a. Increasing   **b.** Concave down; the function is increasing at a
decreasing rate   **c.** 2032   **55. a.** Increase   **b.** 23.0%   **c.** 2012
57. a.   **b.** \$0.367, or 37 cents

$y = 34.394x^{-1.1088}$

59. 115.2 pounds   **61.** 6%   **63.** \$13,854

3.4 Skills Check

1. $y = 2x^2 - 3x + 1$ **3.** $y = x^2 - 0.5x - 3$
5. $y = \dfrac{x^2}{2} - \dfrac{x}{3} + 6$ **7.** $y = -16x^2 + 32x + 48$
9. $y = 2x^2 - 3x + 1$ **11.** $f(x) = 3x^2 - 2x$
13. The x-values are not equally spaced.
15. $y = 99.9x^2 + 0.64x - 0.75$
17. a. $y = 3.545x^{1.323}$ **b.** $y = 8.114x - 8.067$ **c.** Power function
19. a. $y = 1.292x^{1.178}$ **b.** $y = 2.065x - 1.565$ **c.** They are both
good fits.   **21.** $y = 2.98x^{0.614}$

3.4 Exercises

23. a. $y = 0.221x^2 - 0.381x + 4.180$
b.   **c.** \$15.3 billion

25. a. $y = 9.286x^2 - 147.619x + 605.595$
b.   ; excellent fit   **c.** \$957 billion

27. a. $y = 0.0052x^2 - 0.62x + 15.0$ **b.** 50 mph   **c.** No; the model
estimates the wind chill temperature is a minimum near 59.4 mph, so
the model predicts warmer temperatures for winds greater than 60 mph.
29. a. $y = -0.00834x^2 + 0.690x - 6.81$
b.   **c.** at age 42 ($x = 41.38$)

31. a.

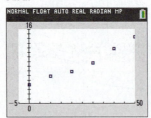

b. $y = 0.002107x^2 + 0.09779x + 4.207$

c. ; good fit   **d.** 16.3 million   **e.** 2035

47. a.

33. a. $Q(t) = -0.032t^2 + 0.682t + 5.41$ **b.** 9.047 billion
c. 8.994 billion; model outputs are decreasing in 2022.   **d.** $x = 10.7$, in 2021   **e.** No; the number of subscribers should continue to increase.   **35. a.** $y = -0.224x^2 + 22.005x + 370.705$
b. 911 in 2020.   **37. a.** $y = 0.068x^2 + 1.746x + 96.183$
b. ; excellent fit   **c.** 126.98   **d.** 2041

39. Equation b
41. a. $y = 0.00785x^2 - 1.22x + 66.792$
b.   **c.** 2078

43. a.

b. Yes; $y = 0.052x^2 - 3.234x + 66.656$ **c.** $y = 0.11x^{2.038}$
d. Quadratic model   **45. a.** $y = 0.117x^2 - 3.792x + 45.330$
b. $y = 0.711x^{1.342}$
c. Quadratic   power

The quadratic model is the better fit.

b. $y = 1.428x^{1.408}$ **c.** $y = -0.001x^2 + 0.058x + 1.473$
d. 2.4   **e.** 2.1   **f.** Quadratic model

Chapter 3 **Skills Check**

1. a. $(5, 3)$ **b.**

2. a. $(-7, -2)$ **b.**

3. a. $(1, -27)$ **b.**

4. a. $(-2, -18)$ **b.**

5. a. $(15, 80)$ **b.**

6. a. $(30, -400)$ **b.**

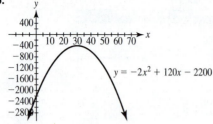

$y = -2x^2 + 120x - 2200$

7. a. $(0.05, -60.0005)$ **b.**

$y = x^2 - 0.1x - 59.998$

8. a. $(-0.2, -100)$ **b.**

$y = x^2 + 0.4x - 99.96$

9. $x = 4, x = 1$ **10.** $x = \frac{1}{2}, x = -\frac{2}{3}$ **11.** $x = -\frac{4}{5}, x = 1$

12. $x = -2, x = \frac{2}{3}$ **13.** $x = 3, x = 1$ **14.** $x = \frac{1}{2}, x = -\frac{3}{2}$

15. a. $4, -2$ **b.** $4, -2$ **16. a.** $-5, 1$ **b.** $-5, 1$

17. $2, -2$ **18.** $9, -1$ **19.** $z = 2 \pm i\sqrt{2}$ **20.** $w = 2 \pm i$

21. $x = \dfrac{5 \pm i\sqrt{23}}{8}$ **22.** $x = \dfrac{-1 \pm i\sqrt{3}}{4}$

23.

$f(x) = \begin{cases} 3x - 2 & \text{if } x < -1 \\ 4 - x^2 & \text{if } x \geq -1 \end{cases}$

24.

$f(x) = \begin{cases} 4 - x & \text{if } x \leq 3 \\ x^2 - 5 & \text{if } x > 3 \end{cases}$

25.

$f(x) = 2x^3$

26.

$f(x) = x^{3/2}$

27.

$f(x) = \sqrt{x - 4}$

28.

$f(x) = \dfrac{1}{x} - 2$

29.

$y = x^{4/5}$

30.

$y = \sqrt[3]{x} + 2$

31. a. Increasing **b.** Decreasing **32. a.** Concave up
b. Concave down **33.** $x = 10, x = -6$ **34.** $x = 5, x = -8$
35. $y = 2x^2 - 3x - 2$ **36.** $y = -2x^2 + 4x + 7$
37. $y = 3.545x^{1.323}$ **38.** $y = 1.043x^2 - 0.513x + 0.977$
39. 128 **40.** $-26, -4, -5$

Chapter 3 Review

41. a. 3100 **b.** \$84,100 **42. a.** 4050 **b.** \$312,050
43. a. 2 sec **b.** 256 ft **44. a.** 1.5 sec **b.** 82.05 m
45. a. 2004 **b.** 137 thousand **c.** 2009 **46.** 30 or 120
47. 5 sec **48.** 400 or 3700
49. a.

$y = \begin{cases} 0.08x^2 - 2.64x + 22.35 & \text{if } 15 \leq x \leq 45 \\ -0.525x + 89.82 & \text{if } 45 < x \leq 110 \end{cases}$

b. 42.57%   **c.** Yes; the model gives 32.07%.
50. a. $y = -0.00482x^2 + 0.754x + 8.512$ **b.** 22.8%
c. $x = 38.0$, so in 2038
51. a.

$f(x) = \begin{cases} 3.607x^2 - 16.607x + 254 & \text{if } 9 \leq x \leq 15 \\ -43.25x^2 + 1361.65x - 9881.75 & \text{if } 15 < x \leq 19 \end{cases}$

b. 648 **c.** 376 **d.** 2010

52. 2506 ha

53. a. $y = 0.065x^2 + 1.561x + 6.689$

b.   **c.** Yes   **d.** 2018

54. a. $y = 2.475x^2 + 299.256x + 1594.282$ **b.** 2003   **c.** 2033

55. a. $y = 15.8425x^{0.5176}$

b.

 ; good fit   **c.** 111 million

56. a. $y = 0.05143x^2 + 3.18286x + 65.40000$ **b.** 16 yr

57. a.

b. Quadratic is most appropriate; $y = 7.897x^2 - 50.650x + 2267.916$

c.

d. 91.5 billion dollars per year; 217.8 billion dollars per year
e. 5.2%; 6.1%; yes   **58. a.** $y = 749.95x^{0.168}$ **b.** Concave down
c. 2043   **59. a.** $y = 2.095x + 6.265$, good fit
b. $y = 0.037x^2 + 1.026x + 12.694$ **c.** Quadratic   **d.** Linear; $69.1 billion; quadratic: $77.0 billion
60. a.

b. $y = -0.99x^2 + 24.27x + 114.11$
c. $y = 0.93x^2 - 11.55x + 256.47$

d. $y = \begin{cases} -0.99x^2 + 24.27x + 114.11, \text{ if } 6 \le x \le 13 \\ 0.93x^2 - 11.55x + 256.47, \text{ if } 13 < x \le 22 \end{cases}$

e. $257.5 billion: $454.1 billion

Chapter 4 Additional Topics with Functions

Toolbox Exercises

1. $(-\infty, 0) \cup (0, \infty)$; $(-\infty, 0) \cup (0, \infty)$ **2.** $(-\infty, \infty)$; $\{k\}$
3. $(-\infty, 0) \cup (0, \infty)$ **4.** $(0, \infty)$; $(-\infty, 0)$ **5.** $[0, \infty)$
6. $(-\infty, \infty)$ **7.** Increasing   **8.** Increasing   **9.** Decreasing
10. Increasing   **11.** Decreasing   **12.** Increasing
13. Cubing   **14.** Square root

15. Power function, with power $1/6$ **16.** Power function, with power -3

17. Not a function   **18.** Not a function   **19.** Function
20. Function   **21.** Function   **22.** Function

4.1 Skills Check

1. a.   **b.** Vertical shift 5 units up

3. a.   **b.** Horizontal shift 4 units right

5. a.

b. Horizontal shift 2 units left, vertical shift 1 unit down

15. a.

b. Vertical stretch using a factor of 3

17. Shifted 2 units right and 3 units up   **19.** $y = (x + 4)^{3/2}$
21. $y = 3x^{3/2} + 5$

23. $f(x) = \sqrt{x + 6} + 3$ **25.** $f(x) = \dfrac{4}{x} - 3$

7. a.

b. Horizontal shift 2 units right, vertical shift 1 unit up

27. $f(x) = -\sqrt{x} + 2$

9. a.

b. Reflection across x-axis, vertical shift 5 units up

29. $g(x) = -x^2 + 2$ **31.** $g(x) = |x + 3| - 2$ **33.** y-axis; even   **35.** x-axis; not a function   **37.** Yes   **39.** y-axis
41. Origin   **43.** Origin   **45.** x-axis, y-axis, origin   **47.** Even
49. Even   **51.** Odd   **53.** Even
55.

; symmetric about the line $y = -1$

11. a.

b. Vertical shift 3 units down

4.1 Exercises

57. a. $y = x^2$ **b.** $M(3) = 25.199$; in 2003, 25,199,000 people 12 and older in the United States used marijuana.

c.

13. a.

b. Vertical compression using a factor of 1/3

59. a. Linear   **b.** Reciprocal; vertical stretch using a factor of 30,000 and shift down 20 units

61. a. Decrease   **b.** Stretch vertically by a factor of 30,000, then vertical shift 150 units up
63. a. Shift 10 units left, shift 1 unit down, reflect about the *x*-axis, and vertical stretch using a factor of 1000.
b.

$$P = -1000\left(\frac{1}{t+10} - 1\right)$$

65. $C(x) = 306.472(x + 5)^{0.464}$
67. $T(x) = -13.898(x + 10)^2 + 255.467(x + 10) + 5425.618$
69. a. 715.994 million   **b.** $C(t) = 0.00061(t + 5)^{3.9}$
c. $t = 31$; yes   **71. a.** \$12,342   **b.** $S(x) = 2029\sqrt{x + 5}, x \geq -5$
c. $x = 32$; yes

4.2 Skills Check

1. a. $2x - 1$ **b.** $4x - 9$ **c.** $-3x^2 + 17x - 20$ **d.** $\dfrac{3x - 5}{4 - x}$
e. All real numbers except 4   **3. a.** $x^2 - x + 1$
b. $x^2 - 3x - 1$ **c.** $x^3 - x^2 - 2x$ **d.** $\dfrac{x^2 - 2x}{1 + x}$
e. All real numbers except -1
5. a. $\dfrac{x^2 + x + 5}{5x}$ **b.** $\dfrac{-x^2 - x + 5}{5x}$ **c.** $\dfrac{x + 1}{5x}$ **d.** $\dfrac{5}{x(x + 1)}$
e. All real numbers except 0 and -1 **7. a.** $\sqrt{x} + 1 - x^2$
b. $\sqrt{x} - 1 + x^2$ **c.** $\sqrt{x}(1 - x^2)$ **d.** $\dfrac{\sqrt{x}}{1 - x^2}$
e. All real numbers $x \geq 0$ except 1   **9. a.** -8 **b.** 1   **c.** 196   **d.** 3.5
11. a. $6x - 8$ **b.** $6x - 19$ **13. a.** $\dfrac{1}{x^2}$ **b.** $\dfrac{1}{x^2}$
15. a. $\sqrt{2x - 8}$ **b.** $2\sqrt{x - 1} - 7$ **17. a.** $|4x - 3|$ **b.** $4|x - 3|$
19. a. x **b.** x **21. a.** 2   **b.** 1   **23. a.** -2 **b.** -1 **c.** -3
d. -3 **e.** 2

4.2 Exercises

25. a. $P = 66x - 3420$ **b.** \$6480   **c.** $\overline{P}(x) = \dfrac{66x - 3420}{x}$
d. \$43.20   **27. a.** Cost is quadratic; revenue is linear.
b. $P = 1020x - x^2 - 10,000$ **c.** Quadratic
29. a. $P = 520x - x^2 - 10,000$ **b.** 260   **c.** \$57,600
31. a. $\overline{C}(x)$ is $C(x) = 50,000 + 105x$ divided by $f(x) = x$; that is,
$\overline{C}(x) = \dfrac{C(x)}{x}$.   **b.** \$121.67   **33. a.** $\overline{C}(x) = \dfrac{3000 + 72x}{x}$
b. \$102 per printer   **35. a.** $S(p) + N(p) = 0.5p^2 + 78p + 12,900$
b. $0 \leq p \leq 100$ **c.** 23,970   **37. a.** $B(8) = 162$
b. $P(8) = \$7.54$ **c.** $(B \cdot P)(8) = \$1221.48$
d. $W(x) = (B \cdot P)(x) = 6(x + 1)^{3/2}(8.5 - 0.12x)$
39. a. $P(x) = 160x - 32,000$ **b.** \$64,000   **c.** \$160 per unit
41. a. No. Adding the percentages is not valid since the percentages are based on different populations of people. Adding the number of males and the number of females completing college and then dividing by the total is a legitimate approach.   **b.** 64.78%; 68.79%

c. 65.25%; 66.3%; the percentages are relatively close but not the same.   **43. a.** Meat put in container   **b.** Meat ground   **c.** Meat ground and ground again   **d.** Meat ground and put in container
e. Meat put in container, then both ground   **f.** d
45. $p(s(x)) = x - 18.5$ **47.** 6.00 Russian rubles
49. $\dfrac{g(x)}{f(x)}$ **51.** $f(x) + g(x)$ **53.** 40%

4.3 Skills Check

1. Yes;   Domain   Range   **3.** No inverse

5. a. x **b.** Yes   **7.** Yes
9.

x	$f(x)$		x	$f^{-1}(x)$
-1	-7		-7	-1
0	-4		-4	0
1	-1		-1	1
2	2		2	2
3	5		5	3

11. No   **13.** Yes   **15.** Yes   **17.** No   **19. a.** $f^{-1}(x) = \dfrac{x + 4}{3}$
b. Yes   **21.** -2 **23.** $g^{-1}(x) = \dfrac{x - 1}{4}$ **25.** $g^{-1}(x) = \sqrt{x + 3}$
27.   **29.** Yes; yes

31.

4.3 Exercises

33. a. $t^{-1}(x) = x - 34.5$ **b.** $8\dfrac{1}{2}$ **35. a.** $f^{-1}(x) = \dfrac{x}{1.0136}$;
dividing Canadian dollars by 1.0136 gives U.S. dollars   **b.** \$500

37. a. $f^{-1}(x) = \dfrac{2.97 - x}{0.085}$; the years after 2000 when the percentage taking antidepressants is $x\%$ **b.** 2008   **39. a.** $W^{-1}(x) = \sqrt[3]{500x}$
b. Given the weight, the inverse function calculates the length.
c. 10 in.   **d.** Domain: $x > 0$; range: $W^{-1}(x) > 0$ **41.** Make my day   **43.** No; two checks could be written for the same amount.

45. a. Yes   **b.** $f^{-1}(x) = \sqrt[3]{\dfrac{3x}{4\pi}}$ **c.** Domain: $x > 0$; range:
$f^{-1}(x) > 0$ **d.** To calculate the radius of a sphere given the volume
e. 25 in.   **47. a.** $f^{-1}(x) = (0.01877x)^{1/0.170}$ **b.** 18
49. a. No   **b.** $x > 0$; it is a distance, $x \neq 0$.   **c.** Yes
d. $I^{-1}(x) = \sqrt{\dfrac{300{,}000}{x}}$; 2 ft   **51. a.** $f^{-1}(x) = 0.6154x$; converts
U.S. dollars to U.K. pounds   **b.** \$1000

4.4 Skills Check

1. $x = 1$ **3.** $x = -7$ **5.** $x = 6$ **7.** $x = -2, x = 6$
9. $x = 9$ **11.** $x = 23, x = -31$ **13.** $2^{2/5}$ or $\sqrt[5]{4}$
15. $2, -2, 1, -1$ **17.** $\dfrac{1}{6}, -\dfrac{1}{5}$ **19.** $\dfrac{27}{8}, -64$ **21.** $-4 < x < 0$
23. $-3 \le x \le 3$ **25.** $4 < x < 5$ **27.** $x \le -2$ or $x \ge 6$
29. $-1 < x < 7$ **31.** $x \le 0.314$ or $x \ge 3.186$
33. $x \le -0.914$ or $x \ge 1.314$ **35.** $x < 0.587$ **37.** $x < 5$
39. $-1 < x < 2$ **41.** $x \le 4$ or $x \ge 8$ **43. a.** $x \le -2$ or $x \ge 3$
b. $-2 < x < 3$ **45. a.** No solution   **b.** All real numbers
47. $-3 \le x \le 2.5$

4.4 Exercises

49. $100 < x < 4000$ units   **51.** $100 < x < 8000$ units
53. $2 < t < 8.83$ sec   **55.** 1983 to 1995
57. 2020 and after   **59.** 50 to 69.23 mph
61. 1995 to 2005   **63.** 1990 to 2007
65. 2004 to 2018   **67.** 1985 to 2012

Chapter 4 Skills Check

1. Shift 8 units right and 7 units up
2. Shift 1 unit left, vertical stretch using a factor of 2 units, reflect about the x-axis
3. a.

b. The graph of $g(x)$ can be obtained from the graph of $f(x)$ by shifting 2 units left and 3 units down.

4. $[-2, \infty)$ **5.** $y = (x - 6)^{1/3} + 4$ **6.** $y = 3x^{1/3} - 5$
7. f   **8.** c   **9.** e   **10.** Origin   **11.** y-axis   **12.** Odd
13. $3x^2 + x - 4$ **14.** $9 - x^3 - 6x$ **15.** $18x^3 - 42x^2 + 20x$
16. $\dfrac{5 - x^3}{6x - 4}$, $x \neq \dfrac{2}{3}$ **17.** 38   **18.** $108x^2 - 174x + 68$
19. $18x^2 - 30x - 4$ **20.** 188   **21. a.** x; x
b. They are inverse functions.   **22.** $f^{-1}(x) = \dfrac{x + 2}{3}$
23. $g^{-1}(x) = x^3 + 1$ **24.**

25. Yes   **26.** No   **27.** $x = -\dfrac{3}{8}$ **28.** $x = -2$ **29.** $\dfrac{1}{16}, 1$ **30.** 4, 11

31. $[-2, 9]$ **32.** $x \le -3$ or $x \ge \dfrac{1}{2}$ **33.** $-2 \le x \le 6$
34. $x \ge 4.5$ or $x \le -3$ **35.** $x < 20$
36. $x \ge 16\sqrt{2} - 2$ or $x \le -16\sqrt{2} - 2$

Chapter 4 Review

37. a.

b. 64 in.

38. a. 4 sec   **b.** 380 ft   **c.** Stretched vertically using a factor of 16, and reflected across the x-axis, then shifted 4 units right and 380 units up.
39. a. 2007   **b.** 2004 to 2009   **40.** 2007 to 2010   **41.** 1990 to 2010
42. a. $\overline{C}(x) = \dfrac{30x + 3150}{x}$ **b.**

43. a.

b. The function, and therefore the average cost, is decreasing.
c. Vertical stretch using a factor of 50,000 and shift 120 units up

44. a. Demand function
b.

45. a. Demand function
b.

46. a. Reflect across the x-axis, compress vertically using a factor of 0.2, shift right 10.3 units, and shift up 48.968 units
b.

c. After 2015

47. a. $R(E(t)) = 0.002805t^2 + 0.35706t + 1.104065$; the function calculates the revenue for Southwest Airlines given the number of years after 1990. **b.** $R(E(3)) \approx 2.2$; in 1993, Southwest Airlines had revenue of \$2.2 billion. **c.** $E(7) = 24.042$; in 1997, Southwest Airlines had 24,042 employees. **d.** \$3.7 billion

48. a. $f^{-1}(x) = \dfrac{x + 2.886}{0.554}$ **b.** The mean sentence length can be found as a function of the mean time in prison.

49. a. $f^{-1}(x) = (0.01877x)^{1/0.170}$ **b.** 15

50. a. Reciprocal function **b.** Function P, which gives \$0.449

51. More than 200 and less than 6000 units **52.** 2021 through 2025

Chapter 5 Exponential and Logarithmic Functions

Toolbox Exercises

1. a. x^7 **b.** x^5 **c.** $256a^4y^4$ **d.** $\dfrac{81}{z^4}$ **e.** $2^5 = 32$ **f.** x^8 **2. a.** y^6

b. w^6 **c.** $216b^3x^3$ **d.** $\dfrac{125z^3}{8}$ **e.** 3^5, or 243 **f.** $16y^{12}$ **3.** 10 **4.** 256

5. $\dfrac{1}{x^7}$ **6.** $\dfrac{1}{y^8}$ **7.** $\dfrac{1}{c^{18}}$ **8.** $\dfrac{1}{x^8}$ **9.** a **10.** b^2 **11.** $x^{1/6}$ **12.** $y^{1/15}$

13. $\dfrac{6}{ab^2}$ **14.** $\dfrac{-8a^2}{b^2}$ **15.** $\dfrac{x^{10}}{4}$ **16.** $\dfrac{8y^{18}}{27}$ **17.** $\dfrac{-7}{a^2b}$ **18.** $\dfrac{-6y^2}{x}$

19. 4.6×10^7 **20.** 8.62×10^{11} **21.** 9.4×10^{-5} **22.** 2.78×10^{-6}

23. 437,200 **24.** 7,910,000 **25.** 0.00056294 **26.** 0.0063478

27. 3.708125×10^6 **28.** 6.460833515×10^1 **29.** $x^{4/3}$ **30.** $y^{13/20}$

31. $c^{5/3}$ **32.** $x^{9/8}$ **33.** $x^{1/4}$ **34.** $y^{1/8}$ **35.** $\dfrac{b}{27c^2}$ **36.** $\dfrac{27x}{64y^{5/4}}$

37. $\dfrac{x^{2/3}y}{z^{1/2}}$ **38.** $\dfrac{16x^{12}y^{2/3}}{81z^2}$

5.1 Skills Check

1. c, d, e

3. a.

b. 2.718; 0.368; 54.598
c. x-axis
d. 1

5.

7.

9.

11.

13. B **15.** A **17.** E

19. Vertical shift, 2 units up

21. Reflection across the y-axis

23. Vertical stretch using a factor of 3

25. The graph of $y = 3 \cdot 4^{(x-2)} - 3$ is the graph of $y = 3(4^x)$ shifted to the right 2 units and shifted down 3 units.

27. Horizontal shift, 3 units right

29. Vertical compression with factor $\frac{1}{2}$

31. Reflection across the y-axis

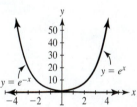

33. Horizontal compression; outputs are cubed

35. a.

b. 1.624; 88.669
c. Decay

5.1 Exercises

37. a. $12,000 **b.** $8603.73 **c.** No

39. a.   **b.** $130,311.57

41. a.   **b.** In about 11.45 yr

c.

t (year)	S ($)
10	17,804.33
20	39,624.26
22	46,499.50

43. a. 376.84 g **b.**
c. About 24.5 yr

45. a.   **b.** $1000

b. $1000 **c.** Sales declined drastically after the end of the ad campaign. (Answers may vary.) **47. a.** $14,339.44 **b.** Because of future inflation, they should plan to save money. (Answers may vary.)
49. a. $122,140.28 **b.** In about 14 yr **51. a.** Increasing **b.** 57,128
c. 61,577 **d.** Approximately 858 people per year
53. a. About 88.6 g **b.** About 19,034 yr
55. a.   **b.** 50

5.2 Skills Check

1. $3^y = x$ **3.** $e^y = 2x$ **5.** $\log_4 x = y$ **7.** $\log_2 32 = 5$
9. a. 0.845 **b.** 4.454 **c.** 4.806
11. a. 5 **b.** 2 **c.** 3 **d.** 3 **e.** 4

13. a. -3 **b.** 0 **c.** 1 **d.** -4
15.   **17.**

19. a. $y = \log_4 x$ **b.** Graphs are symmetric about the line $y = x$.

21. $a^x = a; x = 1$ **23.** 14 **25.** 12 **27.** 2.2146 **29.** 2.322
31. $\ln(3x - 2) - \ln(x + 1)$

33. $\dfrac{1}{3}\log_3(4x + 1) - \log_3(4) - 2\log_3(x)$

35. $\log_2 x^3 y$ **37.** $\ln \dfrac{(2a)^4}{b}$

5.2 Exercises

39. a. 57; 78 **b.** Improved health care and better diet. (Answers may vary.) **41. a.** 77.271 million **b.** Increasing
c.

43. $75.50 **45. a.** 47.1%; 47.5% **b.** Increasing **47.** ≈ 6.9 yr

49. 35 quarters, or $8\dfrac{3}{4}$ yr **51.** 9 yr **53.** 4.4

55. $10^{6.4}I_0 \approx 2,511,886 I_0$ **57.** $10^{7.1}I_0 \approx 12,589,254 I_0$
59. About 14.1 times as intense **61.** 158.5 times as intense
63. Decibel reading of higher intensity sound is 20 more than the other. **65.** $10^{14} I_0$ **67.** Intensity of painful sound is 100 times the intensity of other. **69.** $[H^-] = 10^{-7.79} \approx 0.0000000162$
71. 251.2 times as acidic

5.3 Skills Check

1. $x \approx 3.204$ **3.** $x = 7.824$ **5.** $x \approx 1.819$ **7.** $x \approx 0.374$

9. $x \approx 1.204$ **11.** 1.6131 **13.** 0.1667 **15.** $x = \dfrac{10}{3}$ **17.** $x = 2$

19. $x \approx 3.5$ **21.** $x \approx 0.769$ **23.** $x = 8$ **25.** $x = e^{1.5} \approx 4.482$

27. $x = \dfrac{e^6}{2} \approx 201.71$ **29.** $x = 50$ **31.** $x = \dfrac{1}{3}$ **33.** $x = 40$

35. $x = 5$ **37.** $x < 5$ **39.** $x \geq 9$

5.3 Exercises

41. 5   **43. a.** $\ln \dfrac{S}{25,000} = -0.072x$ **b.** 6 weeks   **45. a.** \$3200

b. 9 days   **47. a.** \$1,457,837   **b.** over-estimate; it estimates the cost for 2015 at \$6.2 million   **49.** 13.86 yr   **51.** 23.11 yr   **53. a.** 500 g

b. ≈ 24.51 yr   **55.** 3 hours   **57.** 2029   **59.** 7 units   **61.** $t = \dfrac{\ln 2}{\ln (1.07)}$

63. 13 yr   **65.** 5 yr   **67. a.** 2008   **b.** 2008; yes   **69.** 2026
71. 2019
73. a.

Annual Interest Rate	Rule of 72 Years	Exact Years
2%	36	34.66
3%	24	23.10
4%	18	17.33
5%	14.4	13.86
6%	12	11.55
7%	10.29	9.90
8%	9	8.66
9%	8	7.70
10%	7.2	6.93
11%	6.55	6.30

b. As interest rate increases, the estimate gets closer to actual value.
75. a. ≈ 11.9 **b.** 6 yr
77. 16 yr   **79.** 46 months
81. ≈ 2075 yr
83. a. 16 weeks   **b.**

5.4 Skills Check

1. $y = 2(3^x)$ **3.** Not exponential   **5.** $f(x) = 4^x$
7. a.   **b.** Linear model

9. Exponential function   **11.** $y = 0.876(2.494^x)$
13. a.   **b.** Logarithmic model

15. a.   **b.** $y = 3.671x^{0.505}$

c. $y = -0.125x^2 + 1.886x + 1.960$ **d.** $y = 3.468 + 2.917 \ln x$

5.4 Exercises

17. a. $y = 30,000(1.04^t)$ **b.** \$54,028   **19. a.** $y = 20,000(0.98^x)$
b. \$18,078   **21. a.** $y = 492.439(1.070^x)$ **b.** \$20,100.8 billion
c. 2014
23. a. $y = 11.682(1.748^x)$
b.   ; good fit

c. 5438 million cubic feet
d. 2022
25. a. $y = 13.410(1.040^x)$ **b.**

c. 137.2 thousand dollars   **d.** 2053 $(x = 72.75)$
27. a. $y = 11.027 + 14.304 \ln x$
b. $y = -0.0018x^2 + 0.4880x + 46.2495$ **c.** Logarithmic model

d. Logarithmic: 79.0; quadratic: 78.6
29. a. $y = 3.885 - 0.733 \ln x$ **b.** $y = 2.572(0.982^x)$
c.

31. a.

b. $y = 68.371 + 13.424 \ln x$; the input cannot be 0 for a logarithmic function

c.

d. 119.5°F   **33. a.** $y = -681.976 + 251.829 \ln x$
b. 31.5%   **c.** $y = 0.627x^2 - 7.400x - 26.675$
d. Quadratic function   **35. a.** Exponential decay
b. $y = 16,278.587(0.979^x)$ **c.** $y = 20,022.816x^{-1}$ **d.** Power function   **e.** 2100 gal

5.5 Skills Check

1. 49,801.75   **3.** 17,230.47   **5.** 26,445.08   **7.** 12,311.80
9. 1,723,331.03   **11.** 1123.60; 1191.00; 1191.00

13. $P = S\left(1 + \dfrac{r}{k}\right)^{-kn}$

5.5 Exercises

15. a. $16,288.19   **b.** $88,551.38
17. a.   **b.** 8 yr

19. $32,620.38   **21. a.** $33,194.62   **b.** More; compounded more frequently   **23.** $49,958.02   **25. a.** $20,544.33   **b.** $29,446.80
27. a. $28,543.39   **b.** Continuous compounding yields $903.41 more.
29. a. ≈ 7.27 yr   **b.** ≈ 6.93 yr   **31. a.** $2954.91   **b.** $4813.24
33. $6152.25
35. a.

Years	0	7	14	21	28
Future Value ($)	1000	2000	4000	8000	16,000

b. $y = 1000(1.104)^x$ **c.** $1640.01; $2826.02   **37.** $29,303.36
39. $5583.95   **41.** $4996.09   **43.** $24,215.65
45. $t \approx 17.39$; 17 yr, 5 mo   **47.** 12 yr and after   **49.** 8%   **51.** 5%

5.6 Skills Check

1. $P = \dfrac{S}{(1 + i)^n}$ **3.** $A = R\left(\dfrac{1 - (1 + i)^{-n}}{i}\right)$

5. $R = A\left(\dfrac{i}{1 - (1 + i)^{-n}}\right)$

5.6 Exercises

7. $52,723.18   **9.** $21,824.53   **11.** $486,043.02   **13.** $9549.11
15. $7023.58   **17.** $530,179.96   **19.** $372,845.60
21. a. $702,600.91   **b.** $121,548.38   **c.** $100,000 plus the annuity has a larger present value than $800,000.   **23. a.** $198,850.99
b. $576,000   **c.** $377,149.01   **25. a.** 2%   **b.** 16   **c.** $736.50
27. a. $1498.88   **b.** $639,596.80   **c.** $289,596.80

5.7 Skills Check

1. 73.83   **3.** 995.51; 999.82
5. a.   **7. a.**

b. 25; 99.99   **c.** Increasing
d. 100

b. 5
c. 100

5.7 Exercises

9. a.   **b.** 5
                                          c. 5000

11. a.   **b.** 29.56%
                                          c. 86.93%
                                          d. 89.786%

13. a. Approximately 218   **b.** 1970   **c.** Seventh day

15. a. $y = \dfrac{89.786}{1 + 4.6531e^{-0.8256x}}$ **b.** Yes

c. $y = 14.137x + 17.624$ **d.** Logistic model

17. a. $y = \dfrac{10.641}{1 + 53.050e^{-2.201x}}$ **b.**

Good fit

c. 10.641 million; This number is close to, but less than, the projections in the table.

19. a. $y = \dfrac{82.488}{1 + 0.816e^{-0.024x}}$ **b.** 67.9; 77.7   **c.** 82.5

21. a. 4000   **b.** Approximately 9985

c. 10,000

23. a. 21,012   **b.**
c. 40,000

25. a. 10   **b.** 100   **c.** 1000   **d.** 6 yr after the company was formed
27. 10 days   **29.** 11 yr

Chapter 5 Skills Check

1. a. $f(x) = 4e^{-0.3x}$ **b.** ≈ 80.342; ≈ 0.19915

2. Decreasing
3.

4.

5. Graph of $y = 3^{(x-1)} + 4$ is graph of $f(x) = 3^x$ shifted right
1 unit and up 4 units.   **6.** Increasing   **7. a.** $y = 500$ **b.** $x = 20$
8. $y = \log_6 x$ **9.** $3x = \log_7 y$ **10.** $x = 4^y$ **11.** $x = 10^y$
12. $x = e^y$ **13.** $y = \log_4 x$ **14.** 1.3424   **15.** 4.0254   **16.** 1
17. 4   **18.** 4   **19.** -3 **20.** 3.6309   **21.** 1.9358

22.

23.

24. 5.8289   **25.** 0.2012   **26.** 2.8813   **27.** 2
28. $\ln(2x - 5)^3 - \ln(x - 3) = 3\ln(2x - 5) - \ln(x - 3)$

29. $\log_4 \dfrac{x^6}{y^2}$ **30.** Exponential function; $y = 0.810(2.470^x)$

31. 4926.80   **32.** 32,373.02

33. a.

b. 222.22; 1973.76
c. 2000

34. a.

b. 50
c. 500

Chapter 5 Review

35. a. Exponential growth   **b.**

36. $1515.72   **37.** 10 weeks   **38. a.** 3 on Richter scale
b. $3,162,278 I_0$ **39.** Approximately 1259 times as intense

40. 10 yr   **41. a.** $x = 7 \log_2 \dfrac{S}{1000}$ **b.** 30 yr   **42.** 2023

43. One million times as loud
44. a. $P(x) = 10(1.26^x) - 2x - 50$
b. 10; $P(x)$

45. 7 weeks   **46. a.** $y = 54.62$ g   **b.** 8445 years ago   **47.** 14 yr
48. $\approx \$4451.08$ **49.** 15 yr   **50. a.** $y = 2.7(1.057^x)$
b. $4.7 trillion   **51. a.** $y = 172(1.04^x)$ **b.** 376.9 million   **c.** 2022
52. a. $y = 81.041(0.981^x)$; decreasing
b.

53. a. $y = 133.240 + 9.373 \ln x$
b.

c. 36% of GDP in 1980

c. 164.8 million

54. a. $y = 4.337 + 40.890 \ln x$

55. a. $y = \dfrac{120}{1 + 5.26e^{-0.0639x}}$ **b.** 63.9 million   **c.** 120 million

d. 2037   **56. a.** $y = \dfrac{21,240}{1 + 0.913e^{-0.2882x}}$ **b.** 20,986
57. a. $y = -823.535 + 558.720 \ln x$ **b.** 1194   **c.** in 2025
58. $30,072.61   **59.** $34,426.47   **60.** $274,419.05   **61.** $209,281.18
62. $6899.32   **63.** $66.43   **64.** $773.16   **65. a.** 26.989%; 31.699%
b. 44.742%   **66. a.** Approximately 1184   **b.** 16 days   **67. a.** 1298
b. 3997   **c.** 4000   **68. a.** Approximately 17,993   **b.** 18,000
69. 20 years   **70.** 5%   **71.** 4%

Chapter 6 Higher-Degree Polynomial and Rational Functions

Toolbox Exercises

1. a. Fourth   **b.** 3   **2. a.** Third   **b.** 5   **3. a.** Fifth
b. -14 **4. a.** Sixth   **b.** -8 **5.** $4x(x-7)(x+5)$
6. $-x^2(2x+1)(x-4)$ **7.** $(x-3)(x+3)(x-2)(x+2)$
8. $(x-4)(x+4)(x^2-5)$ **9.** $2(x^2-2)^2$
10. $3x(x-2)^2(x+2)^2$ **11.** $\dfrac{1}{3}$ **12.** $\dfrac{x-3}{4}$

13. $2y+2$ **14.** $\dfrac{4x^2-3}{x-1}$ **15.** $\dfrac{x-2}{x+4}$ **16.** $\dfrac{3x+2}{x-1}$ **17.** $\dfrac{20x}{y}$

18. $\dfrac{1}{x^2}$ **19.** $2x^2-7x+6$ **20.** $-x(x+2)$ **21.** $\dfrac{x-2}{2x}$ **22.** $\dfrac{1}{2y}$

23. $\dfrac{-(x+1)^2}{2x(x-3)}$ **24.** $\dfrac{3}{y}$ **25.** $\dfrac{6}{x-1}$ **26.** $\dfrac{3x^3+x-2}{x^3}$

27. $\dfrac{4x^2+2x+4}{x^3}$ **28.** $\dfrac{4a-4}{a^3-2a^2}$ **29.** $\dfrac{8x^4-16x^3+32x^2-59x}{x^4-16}$

30. $\dfrac{x-2}{x}$ **31.** $\dfrac{x^3-x^2-2x+4}{x^2(x-2)}$ **32.** $\dfrac{2x^2-11x-3}{(x-1)(x-4)(x-5)}$

33. $\dfrac{2x^2+9x-7}{4x^2-2x}$ **34.** $\dfrac{y+x}{y-x}$ **35.** $\dfrac{66}{3y+4}$ **36.** $\dfrac{x+1}{x^2}$

37. $\dfrac{x+1}{(x-1)(x-2)}$ **38.** $x^4-x^3+2x^2-2x+2, \text{R}(-3)$

39. a^3+a^2 **40.** $3x^3-x^2+6x-2, \text{R}(17x-5)$ **41.** $x^2+1, \text{R}2$

6.1 Skills Check

1. a.

b. Window (b)

3. a.

b. Window (b)

5. a. $-2, 1, 2$ **b.** Positive   **c.** Cubic   **7. a.** $-1, 1, 5$ **b.** Negative
c. Cubic   **9. a.** $-1.5, 1.5$ **b.** Positive   **c.** Quartic   **11.** C
13. E   **15.** F   **17. a.** Degree 3; leading coefficient 2
b. Opening up to right and down to left
c.

19. a. Degree 3; leading coefficient -2 **b.** Opening down to right and up to left   **c.**

21. a.

b. Yes; three x-intercepts show, along with the y-intercept and 2 turning points.

23. a.

b.

25. a.

b.

c. Window (b)

27. a.

b. 3   **c.** No

29. Answers will vary. One such graph is the graph of $f(x) = -4x^3 + 4$, shown below.

31. Answers will vary. One such graph is the graph of $f(x) = x^4 - 3x^2 - 4$, shown below.

33. a.

b. $(-2.67, 14.48)$ **c.** $(0, 5)$

35. Maximum: $(1, 1)$; minima: $(0, 0)$, $(2, 0)$

6.1 Exercises

37. a. 2 turning points

c.

b. Both x and R need to be nonnegative.

d. \$10,000

39. a.

c.

b. 60 units; \$28,800

d. Part (a)   **e.** $0 < x < 60$

41. a.

Rate, r	Future Value, $S(\$)$
0.00	2000.00
0.05	2315.25
0.10	2662.00
0.15	3041.75
0.20	3456.00

b.

c. \$794
d. 10% is much more likely than 20%.

43. a.

b. 25,506   **c.** No

45. a.

b. 9.064 billion   **c.** 2024

47. a.

b. 9.4   **c.** $83 \,(x = 83.7)$ **d.** $34 \,(x = 34.6)$
49. a. 2000   **b.** \$90,000

6.2 Skills Check

1. $y = x^3 - 2x^2$
3. $y = x^4 - 4x^2$
5. a.

b. Cubic

7. a.

b.

c. Cubic
9. a. $y = 35x^3 - 333.667x^2 + 920.762x - 677.714$
b. $y = 12.515x^4 - 165.242x^3 + 748x^2 - 1324.814x + 738.286$
11. $y = x^4 - 4x^2 - 3x + 1$ **13.** Not exactly
15. $y = 0.565x^3 + 2.425x^2 - 4.251x + 0.556$

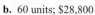

6.2 Exercises

17. a. $y = -0.00556x^3 - 0.145x^2 + 4.725x + 50.740$
b. 61.8%   **c.** Good fit

19. a.

b. $y = -0.00000486x^3 + 0.000825x^2 + 0.0336x + 2.11$
c.

d. 7.737   **e.** 2026
f. It is not a good predictor, because it predicts that the population will decrease.

21. a. $y = 0.00000537x^3 + 0.000369x^2 - 0.0844x + 26.0$
b. Age 30 (30.5)

23. a. $y = -0.000487x^3 - 0.162x^2 + 19.8x + 388$
b.

c. 913 million   **d.** 2020; 914 million

25. a. $y = 0.000078x^3 - 0.01069x^2 - 0.1818x + 64.6848$
b. 2000

27. a.

b. $y = -0.000233x^3 + 0.0186x^2 + 2.33x + 235$

c. Excellent fit   **d.**

The model does not seem appropriate for the increased time period, because it predicts that the population of the United States will begin to decrease, and this does not seem likely.

29. a.

b. $y = 2.481x^3 - 52.251x^2 + 528.682x + 5192.580$
c.

d. \$19,894 per person   **e.** 2021

31. a. $y = -0.00007x^3 + 0.00567x^2 + 0.863x + 16.009$ **b.** 2046

33. a.

b. $y = 0.117x^2 - 3.79x + 45.3$
c. $y = 0.00188x^3 - 0.0954x^2 + 2.79x - 2.70$
d.   Quadratic   Cubic

Cubic is a better fit.

35. a.

b. $y = -0.02189x^3 + 1.511x^2 - 13.67x + 347.4$
c.

d. 628.4 million metric tons   **e.** 2035

37. a. $y = -0.169x^2 + 0381x + 99.9$
b. $y = 0.0576x^3 - 0.774x^2 + 1.96x + 99.2$

c.   Quadratic   Cubic

The cubic is a better fit.

d. The quadratic model indicates that after 2017, the number of households using pay TV will continue to decrease, and the cubic model indicates that the number will begin to increase again. Given probable advances in technology, the quadratic model that predicts a continued decrease is more appropriate.

6.3 Skills Check

1. $\dfrac{3}{2}, -1, 6$ **3.** $-1, 4, \dfrac{5}{2}$ **5.** $0, 4, -4$ **7.** $0, 2$
9. $0, 1, -1$ **11.** $4, 3, -3$ **13.** $2, -2, \dfrac{4}{3}$ **15.** 2
17. $2, -2$ **19.** $0, \sqrt{2}, -\sqrt{2}$
21. $0, 5, -5$ **23.** $\sqrt{3}, -\sqrt{3}$
25. a. $-3, 1, 4$ **b.** $(x + 3)(x - 1)(x - 4)$
27. a. $-1, 2, 3$ **b.** $(x + 1)^2(x - 2)(x - 3)$
29. a. $-1, 1, 5$ **b.** $(x + 1)(x - 1)(x - 5)$
31. $-2, 5, \dfrac{3}{4}$

6.3 Exercises

33. a. $0, 20$ **35. a.** $0, 1000$
b. Yes   **b.** Yes

37. a.

Rate	Future Value
4%	$2249.73
5%	$2315.25
7.25%	$2467.30
10.5%	$2698.47

b.

c. 10%   **d.** 20%
39. a. x in.   **b.** $18 - 2x$ in. by $18 - 2x$ in.
c. $V = (18 - 2x)(18 - 2x)x = 324x - 72x^2 + 4x^3$
d. $x = 0, x = 9$ **e.** Neither
41. $2, 20$
43. a.

t	0	0.1	0.2	0.3
s (cm/sec)	810	240	30	0

b. 0.3 sec; yes

45. a. 160.6 million   **b.** 2017   **47.** 2014   **49.** 2042

6.4 Skills Check

1. $x^3 - x^2 - 3x - 6 - \dfrac{8}{x - 3}$ **3.** $2x^3 - x^2 - x - \dfrac{7}{x - 1}$
5. No   **7.** No   **9.** 1 (a double solution)   **11.** $-2, 4$
13. $4, -2, -5$ **15.** $-2, 1, \dfrac{1}{3}$ **17.** $\pm 1, \pm 2, \pm 3, \pm 4, \pm 6, \pm 12$
19. $\pm 1, \pm 2, \pm 4, \pm\dfrac{1}{3}, \pm\dfrac{2}{3}, \pm\dfrac{4}{3}, \pm\dfrac{1}{9}, \pm\dfrac{2}{9}, \pm\dfrac{4}{9}$ **21.** $-1, 3, 4$
23. $\dfrac{1}{3}, -1, -\dfrac{4}{3}$ **25.** $x = 0, x = \dfrac{-7 \pm \sqrt{89}}{2}$
27. $w = 1, w = 2 \pm \sqrt{2}$ **29.** $z = 2, z = -1 \pm i\sqrt{3}$

6.4 Exercises

31. a. $-0.2x^2 + 56x + 1200$ **b.** 300 units
33. a.   **b.** 8

c. $-0.1x^2 + 49.9x + 50$ **d.** $x = 8, x = 500, x = -1$
e. 8 units, 500 units
35. 10 units
37. a. $0.4566x^3 - 14.3085x^2 + 117.2978x - 136.1544 = 0$
b. $x = 12$ **c.** $0.4566x^2 - 8.8293x + 11.3462$
                        d. $x = 17.953, x = 1.384$
                        e. 1982, 1992, 1998

39. a. In 2010, the U.S. per capita out-of-pocket cost for healthcare was $946.   **b.** No   **41.** 2009

6.5 Skills Check

1. a. $x = 5$ **b.** $y = 0$ **3. a.** $x = \dfrac{5}{2}$ **b.** $y = -\dfrac{1}{2}$
5. a. $x = 1, x = -1$ **b.** None
7. c; no value of x makes the denominator 0.
9. E   **11.** F   **13.** C
15. a. $y = 1$ **b.** $x = 2$ **c.**

17. a. $y = 0$ **b.** $x = -1, x = 1$ **19.** A hole in the graph
c.

21. Vertical: $x = -1$; horizontal: none; slant: $y = x - 1$

41. a.

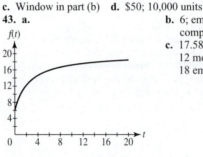

b.

c. Window in part (b)   **d.** $50; 10,000 units

43. a.   **b.** 6; employees when the
          company starts
       c. 17.586; employees after
          12 months; approximately
          18 employees

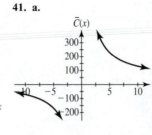

23. Vertical: $x = 1$; horizontal: none; slant: $y = x$

45. a. $p = 100$ **b.** It is impossible to remove 100% of the impurities.
47. a. Yes; $p = -2$

25. $(0, 0)$ and $(2, 4)$ **27.** $(-2, -4)$ and $(2, 4)$

b.

Price/ Unit ($)	5	20	50	100	200	500
Weekly Sales Volume	13,061	1322	237	62	16	3

c. Nonnegative values of p; no
d. $V = 0$; weekly sales approach 0 as price increases.

29. a.   **31. a.**   **49. a.**   **b.** $p = 0$
                c. As the price drops, the
                quantity demanded increases.

b. $y = 0, y = -8$ **b.** $y = -3, y = -1$ **51. a.** $S = \dfrac{x^2 + 40x + 160}{4x}$ **b.** 4 hours or 40 hours
c. $x = 4, x = 3.5$ **c.** $x = -1$
d. $x = 4, x = 3.5$ **d.** $x = -1$

33. $x = \dfrac{1}{2}$ **35.** 80

6.5 Exercises

37. a. $55.80   **b.** $57.27   **c.** $55   **d.** No; for example, at 600 units
the average cost is $56.67 per unit.   **39. a.** $80,000   **b.** -20; no

53. a.

b. $H = \dfrac{3}{2}$ **c.** 17 days

57. a. 48.0%; yes   **b.** 57%   **c.** 2022

55. a.

b. 0.927   **c.** Does not

6. a.

b. No

c.

d. Maximum: $(-8, 360)$; minimum: $(0.667, 34.5)$
7. $0, -4, 4$ **8.** $0, 2, -2$ **9.** $0, 5, -4$ **10.** $0, 7, 8$
11. $-1, 1, 5$ **12.** $\dfrac{3}{4}, 2, -2$ **13.** $-1, 2, 3$ **14.** $12, -2.5, \dfrac{1}{3}$
15. 6   **16.** 5, 1   **17.** $4x^3 + 5x^2 + 10x + 22$ R36
18. $1, -4, \dfrac{1}{2}$ **19.** $2, \dfrac{1}{3}, -2$ **20.** $1, -3, -\dfrac{1}{2}$

21. a. y-intercept: $\left(0, \dfrac{1}{2}\right)$; x-intercepts: $(-1, 0), (1, 0)$
b. Vertical asymptote: $x = -2$; no horizontal asymptote
c. $y = 2 - x$
d.

22. a. x-intercept $\left(\dfrac{2}{3}, 0\right)$; y-intercept $\left(0, \dfrac{2}{3}\right)$
b. Horizontal asymptote: $y = 3$; vertical asymptote: $x = 3$
c.

6.6 Skills Check

1. $-4 \le x \le 4$ **3.** $x < 0$ or $x > 2$ **5.** $-1 \le x \le 1$ or $x \ge 3$
7. $0 < x < 1$ **9.** $x \le -3$ or $-1 < x \le 2$ **11.** $x > 4$
13. $x < 5$ **15.** $x = -5$ or $x \ge 0$ **17. a.** $x < -3$ or $0 < x < 2$
b. $-3 \le x \le 0$ or $x \ge 2$ **19.** $\dfrac{1}{2} \le x \le 3$

6.6 Exercises

21. $0 < x < 20$; more than 0 and less than 20 units
23. a. $0 < x < 18$ or $x > 18$ **b.** More than 0 and less than 18 cm
25. $0 \le x \le 2$ or $x \ge 10$; between 0 and 2 units or at least 10 units
27. $x \ge 20$; at least \$20,000   **29.** $0.10 \le r \le 0.20$; between 10%
and 20%   **31.** $10 \le x \le 194$; between 10 and 194 units
33. $p > 80$; above \$80 per unit   **35. a.** $0 \le t \le 15$ **b.** First 15
months

Chapter 6 Skills Check

1. 4   **2.** Quartic   **3.** Yes

4. a.

b.

b. $(0, -4)$ **c.** $(2, -8)$

5. a.

23.

Maximum: $(0, 0)$;
minimum: $(8, 16)$

24.

Maximum: $(3, -6)$;
minimum: $(-1, 2)$

25. a.

b. $y = 1; y = 3.8$
c. $x = 2, x = -0.875$
d. $2, -\dfrac{7}{8}$

c.

d. 8.578 million

26. $-3, 3, -2, 2$ **27.** $1, -1 \pm i\sqrt{3}$

28. $-2, -\dfrac{1}{4} \pm \dfrac{\sqrt{3}}{4}i$ **29.** $x = 0, x \geq 5$ **30.** $x \geq 4$ or
$-1 \leq x \leq 2$ **31.** $x > 3$ or $x < 0$ **32.** $-2 \leq x < -1$

Chapter 6 **Review**

33. a.

b.

c. \$23,750
34. 84,051 units
35. a.

Rate, r	Future Value S (\$)
0.01	5307.60
0.05	6700.48
0.10	8857.81
0.15	11,565.30

b.

c. \$6072.11   **36.** Intensity 4
37. a.

b. \$12,513   **c.** 2023

38. a.

b. $y = -0.0000596x^3 + 0.00933x^2 - 0.308x + 5.63$

39. a.

b. $y = -0.00000110x^4 + 0.000278x^3 - 0.0203x^2 + 0.318x + 13.4$
c. Very good fit   **d.** 1966

40. 0.05
41. a. $0, 9$ **b.** No sides if $x = 0$; no box material left if $x = 9$.
c. $0 < x < 9$
42. a.

b. $-0.2x^2 + 19.7x + 30$ **c.** $4, -1.5, 100$
d. 400 units, 10,000 units
43. a.

b. $y = 10.3x^3 - 400x^2 + 5590x - 8690$
c.

d. 23,574

44. a. $C = 0$. **b.** As t increases without bound, concentration approaches 0. **c.** 0.15 when $t = 1$
45. a. $\overline{C}(0)$ does not exist. If no units are produced, an average cost per unit cannot be calculated. **b.** $\overline{C}(x) = 50$; as number of units increases without bound, average cost approaches \$50. **c.** Decreases

46. a.

b. 20 units; \$1200

47.

48. a. $C = \dfrac{200x + 20x^2 + 180}{x}$ **b.** 5 **c.** 3

49. a.

b. 0.69 (69%)

c. $f(10) = 1.015$; this indicates that there are more smartphone users than cell phone users, which does not make sense because a smartphone is a special type of cell phone. Thus, the model is not valid after 2009.
50. 50 or more hours
51. a.

b. $y = 0.00001828x^4 - 0.003925x^3 + 0.3031x^2 - 9.907x + 118.2$
c.

d. From 29 to 43, and from 65 to 77
52. At least 50 units but no more than 605 units
53. Between 2000 and 5000 units, inclusive
54. \$4134 or more

Chapter 7 Systems of Equations and Matrices

Toolbox Exercises

1. 12 **2.** $\dfrac{1}{10}$ **3.** No **4.** Yes **5.** Yes **6.** Yes

7. 99 **8.** 78 **9.** No **10.** Yes **11.** No **12.** No
13. No **14.** Yes **15.** Yes **16.** No **17.** Parallel planes
18. Same plane **19.** Neither **20.** Same plane

7.1 Skills Check

1. $x = 2, y = 2, z = 3$ **3.** $x = -14, y = 9, z = -2$
5. $x = 24, y = -16, z = 10$ **7.** $x = -1, y = 1, z = 1$
9. $x = \frac{1}{2}, y = \frac{1}{2}, z = \frac{3}{4}$ **11.** $x = 2, y = 4, z = 14$
13. $x = 1, y = 3, z = 1$ **15.** $x = 2, y = 2, z = 2$
17. *Infinitely many solutions of the form*
$x = 5z - 14, y = 2z - 4, z = z$
19. *Infinitely many solutions of the form*
$x = 20 - 10z, y = 16 - 8z, z = z$
21. *Infinitely many solutions of the form*
$x = -\dfrac{7}{4}z - \dfrac{5}{4}, y = \dfrac{1}{4}z - \dfrac{1}{4}, z = z$ **23.** Inconsistent system,
no solution **25.** *Infinitely many solutions of the form*
$x = 10z + 11, y = -2z - 2, z = z$

7.1 Exercises

27. 30 compact, 20 midsize, 10 luxury **29. a.** $x + y = 2600$
b. $40x$ **c.** $60y$ **d.** $40x + 60y = 120,000$ **e.** 1800 \$40 tickets, 800
\$60 tickets **31. a.** $x + y + z = 400,000$
b. $0.075x + 0.08y + 0.09z = 33,700$ **c.** $z = x + y$ **d.** \$60,000 at
7.5%, \$140,000 at 8%, \$200,000 at 9%
33. A: 100 units; B: 120 units; C: 80 units
35. a. First equation: sum of investments is \$500,000; 2nd equation: sum of interest earned is \$49,000
b. $x = 50,000 + 2z, y = 450,000 - 3z$ **c.** $0 \le z \le 150,000$
37. Inconsistent system; can't use all hours
39. Food I $= 132 - (4.4)$ food III, food II $= (3.2)$ food III $- 64$,
where $20 \le$ food III ≤ 30 g

7.2 Skills Check

1. $\begin{bmatrix} 1 & 1 & -1 & | & 4 \\ 1 & -2 & -1 & | & -2 \\ 2 & 2 & 1 & | & 11 \end{bmatrix}$ **3.** $\begin{bmatrix} 5 & -3 & 2 & | & 12 \\ 3 & 6 & -9 & | & 4 \\ 2 & 3 & -4 & | & 9 \end{bmatrix}$

5. $x = -1, y = 4, z = -2$ **7.** $x = 3, y = 2, z = 1$
9. $x = 2, y = 1, z = 3$ **11.** Inconsistent system, no solution
13. *Infinitely many solutions of the form* $x = 2 - 3z$,
$y = 5z + 5, z = z$
15. $x = 3, y = -2, z = 1$ **17.** $x = -\dfrac{25}{3}, y = -\dfrac{25}{3}, z = \dfrac{14}{3}$
19. $x = 0, y = -1, z = 2$ **21.** $x = 40, y = 22, z = -19, w = 1$
23. $x = 0, y = 3, z = 2$
25. *Infinitely many solutions of the form*
$x = \dfrac{32}{19} - \dfrac{22z}{19}, y = \dfrac{10}{19} - \dfrac{14z}{19}, z = z$
27. *Infinitely many solutions of the form*
$x = \dfrac{11}{8} - \dfrac{31z}{8}, y = \dfrac{1}{8} + \dfrac{11z}{8}, z = z$
29. *Infinitely many solutions of the form*
$x = \dfrac{23}{14} + \dfrac{z}{2}, y = -\dfrac{4}{7} + z, z = z$
31. *Infinitely many solutions of the form*
$x = -2 + 3z, y = 4 - 4z, z = z, w = 0$

7.2 Exercises

33. 2400 \$40 seats, 800 \$70 seats, 400 \$100 seats

35. a. $\begin{cases} 15x + 10y + 5z = 100 \\ y = 2x \\ z = 3x \end{cases}$

b. 2 points for true-false, 4 points for multiple choice, 6 points for essay
37. 3 units of plan I, 2 units of plan II, 1 unit of plan III
39. 5 g of food I, 6 g of food II, 8 g of food III
41. 200 Type A clients, 100 Type B clients, 200 Type C clients

43. a. $\begin{cases} 2x + 8y + 6z = 140, \\ 3x + 5y + 4z = 110 \end{cases}$

b. Not possible

45. a. $\begin{cases} x + y + z = 4 \\ 40{,}000x + 30{,}000y + 20{,}000z = 100{,}000 \end{cases}$

b. No   **c.** $x = z - 2$, $y = 6 - 2z$, $2 \le z \le 3$ **d.** 2 at \$30,000, 2 at \$20,000, and 0 at \$40,000 or 1 at \$40,000, 0 at \$30,000, and 3 at \$20,000

47. a. $\begin{cases} x + y + z = 500{,}000 \\ .04x + .08y + .10z = 35{,}000 \end{cases}$

b. $x = 0.5z + 125{,}000$ and $y = 375{,}000 - 1.5z$ with nonnegative values of x, y, and z **c.** \$225,000 at 8%, \$100,000 at 10%, and \$175,000 at 4%   **49.** Traffic from intersection A to intersection B is 650 less than the traffic from intersection D to intersection A. Traffic from intersection B to intersection C is 100 less than the traffic from intersection D to intersection A. Traffic from intersection C to intersection D is 1200 plus the traffic from intersection D to intersection A.
51. a. At A, $400{,}000 = x_1 + x_2$; at B, $x_1 = x_4 + 100{,}000$; at D, $x_3 + x_4 = 100{,}000$ **b.** $x_1 = 100{,}000 + x_4$, $x_2 = 300{,}000 - x_4$, $x_3 = 100{,}000 - x_4$, where x_4 is the number of gallons flowing from B to D; $x_4 \le 100{,}000$

7.3 Skills Check

1. A and D, A and E, D and E, B and F

3. $\begin{bmatrix} 3 & 6 & -1 \\ 6 & 5 & 3 \\ -3 & 8 & 7 \end{bmatrix}$ **5.** $\begin{bmatrix} 3 & 9 & -6 \\ 9 & 3 & 12 \\ -15 & 9 & 18 \end{bmatrix}$

7. $\begin{bmatrix} 0 & -6 & 10 \\ -6 & 4 & -18 \\ 24 & -2 & -22 \end{bmatrix}$

9. a. $AD = \begin{bmatrix} 7 & 5 & -4 \\ 17 & 33 & 6 \\ 11 & 27 & -2 \end{bmatrix}$, $DA = \begin{bmatrix} 6 & 12 & 14 \\ 20 & 10 & 4 \\ 12 & 14 & 22 \end{bmatrix}$

b. No   **c.** Yes

11. a. $DE = ED = \begin{bmatrix} 10 & 0 & 0 \\ 0 & 10 & 0 \\ 0 & 0 & 10 \end{bmatrix}$ **b.** 3×3 identity matrix

13. $\begin{bmatrix} 1 + 2a & 5 + 3b \\ 3 - c & 2 - 2d \end{bmatrix}$ **15.** $\begin{bmatrix} 3a - 2 & 3b - 4 \\ 3c - 6 & 3d - 8 \end{bmatrix}$

17. $m \times k$ **19. a.** BA **b.** 4×3

21. $AB = \begin{bmatrix} a + 3b + 5c & 2a + 4b + 6c \\ d + 3e + 5f & 2d + 4e + 6f \end{bmatrix}$;

$BA = \begin{bmatrix} a + 2d & b + 2e & c + 2f \\ 3a + 4d & 3b + 4e & 3c + 4f \\ 5a + 6d & 5b + 6e & 5c + 6f \end{bmatrix}$

23. $AB = \begin{bmatrix} -3 & -7 \\ 4 & 5 \end{bmatrix}$; $BA = \begin{bmatrix} 11 & 16 \\ -7 & -9 \end{bmatrix}$

25. $AB = \begin{bmatrix} -2 & 0 \\ 5 & 19 \end{bmatrix}$; $BA = \begin{bmatrix} 6 & 1 & 10 \\ 10 & 11 & 14 \\ 1 & 6 & 0 \end{bmatrix}$

7.3 Exercises

27. a. $B = \begin{bmatrix} 1341.3 & 148.7 & 93.3 \\ 73.8 & 69.4 & 69.2 \\ 1.56 & 2.16 & 3.05 \\ 19.6 & 41.8 & 20.9 \end{bmatrix}$

$A = \begin{bmatrix} 1211.5 & 192.4 & 165.5 \\ 80.8 & 79.0 & 78.4 \\ 1.88 & 1.68 & 1.95 \\ 8.3 & 12.5 & 10.2 \end{bmatrix}$

b. $C = \begin{bmatrix} -129.8 & 43.7 & 72.2 \\ 7 & 9.6 & 9.2 \\ 0.32 & -0.48 & -1.10 \\ -11.3 & -29.3 & -10.7 \end{bmatrix}$

c. The negative entries in row 4 indicate that the infant mortality rates of all three countries are projected to decline, which is a positive change for the countries.
29.

	Under Age 5	5–13	14–17	18–64	Over Age 64
2015	475	796	388	-686	-5613
2020	492	669	374	41	-6029
2025	500	909	379	842	-6644
2030	502	932	425	1671	-7357

The matrix shows the projected differences between male and female populations for different ages and years.
b. For ages 18–64 in 2015, and for those over age 64 in all years 2015–2030.
31. a.

Year	Euros	Yen
2000	0.89	102.51
2015	0.86	113.75

b. Yen

33. $BA = \begin{bmatrix} 1210 \\ 980 \\ 594 \end{bmatrix}$ $\begin{matrix} \text{Cost} \\ \text{Singles} \\ \text{Males 35–55} \\ \text{Females 65+} \end{matrix}$

35. a.

	DeTuris	Marriott
Dept. A	54,000	49,600
Dept. B	42,000	42,400

b. Dept. A should purchase from Marriott; dept. B should purchase from DeTuris   **37.** 55% Republican; 45% Democrat
39. a.

	16–19	20–24	25–34	45–54	55–64	65 and older
Men	449	706	917	955	1003	768
Women	436	663	737	722	742	598

b. The 2×2 matrix is $\begin{bmatrix} 1.1 & 0 \\ 0 & 1.25 \end{bmatrix}$

	16–19	20–24	25–34	45–54	55–64	65 and older
Men	493.90	776.60	1008.70	1050.50	1103.30	844.80
Women	545	828.75	921.25	902.50	927.50	747.50

7.4 Skills Check

1. a. $AB = BA = \begin{bmatrix} 1 & 0 \\ 0 & 1 \end{bmatrix}$ **b.** They are inverses.

3. $AB = BA = \begin{bmatrix} 1 & 0 & 0 \\ 0 & 1 & 0 \\ 0 & 0 & 1 \end{bmatrix}$ **5.** $A^{-1} = \begin{bmatrix} 7 & -3 \\ -2 & 1 \end{bmatrix}$

7. $A^{-1} = \begin{bmatrix} -1/6 & -1/3 & 1 \\ -1/3 & 1/3 & 0 \\ 1/3 & 2/3 & -1 \end{bmatrix}$ **9.** $A^{-1} = \begin{bmatrix} -\frac{1}{3} & -1 & \frac{1}{3} \\ 1 & 1 & 0 \\ -\frac{1}{3} & 0 & \frac{1}{3} \end{bmatrix}$

11. $A^{-1} = \begin{bmatrix} 0.9 & 0.2 & -0.7 \\ -0.5 & 0 & 0.5 \\ 0.7 & -0.4 & -0.1 \end{bmatrix}$

13. $C^{-1} = \begin{bmatrix} 1 & 0 & 1 & 1 \\ 0 & 1 & 1 & 1 \\ 0 & 0 & 1 & 0 \\ 0 & 0 & 0 & 1 \end{bmatrix}$ **15.** $X = \begin{bmatrix} 10 \\ 20 \end{bmatrix}$

17. $x = 24, y = 27, z = 14$ **19.** $x = 1, y = 1, z = 1$
21. $x = 2, y = 1, z = -1$
23. $x_1 = 34, x_2 = 16, x_3 = -40, x_4 = 96$

7.4 Exercises

25. Company X: 225,000; company Y: 45,000
27. 50% Republican, 50% Democrat

29. a. $\begin{cases} x + y + z = 400{,}000 \\ x + y - z = -100{,}000 \\ x - \dfrac{1}{2}y = 0 \end{cases}$ **b.** \$50,000; \$100,000; \$250,000

31. \$50,000 at 6%, \$100,000 at 8%, \$250,000 at 10%
33. 47% from business loans, 27% from auto loans, 32% from home loans
35. a. j u s t _ d o _ i t
     10 21 19 20 27 4 15 27 9 20
b. 124, 52; 156, 59; 124, 35; 168, 69; 116, 49
37. 80, 27, 50; 156, 63, 106; 260, 138, 168; 156, 96, 108; 212, 111, 146   **39.** Vote early   **41.** Mind your manners   **43.** Monday night

7.5 Skills Check

1. -5 **3.** 0 **5.** -9 **7.** -1 **9.** Yes **11.** No **13.** $(2, 1)$
15. $(1.8, -0.4)$ **17.** Inconsistent **19.** Infinitely many solutions
21. $(1, 1, 1)$ **23.** $(1, 3, 2)$ **25.** 4 **27.** 21

7.5 Exercises

29. a. A $= 5.5$ mg and B $= 8.8$ mg for patient I
b. A $= 10$ mg and B $= 16$ mg for patient II
31. 2 Deluxe, 8 Premium, and 32 Ultimate
33. 60 compact, 40 midsize, and 20 large
35. Species III $=$ any amount (between 1800 and 2300)
     Species I $= 6900 - 3$(Species III)
     Species II $= \frac{1}{2}$ (Species III) $- 900$
37. \$400,000 at 9%, \$280,000 at 8%, \$120,000 at 7.5%

7.6 Skills Check

1. $(0, 0), (-3, 9)$ **3.** $(2, 0), (-4, 4)$ **5.** $(4, 8), (-4, -8)$
7. $(2, 6), (6, 2)$ **9.** $(2, 8), (-5, -6)$ **11.** $(5, 33), (-8, 26.5)$

13. $(4, 3)$ **15.** $\left(1, \dfrac{9}{4}\right), \left(-10, -\dfrac{1}{2}\right)$
17. $(2, 0), (5, 0), (-10, -1800)$ **19.** $(2, 3)$
21. a. $x^3 - x^2 - 4 = 0$ **b.** Synthetic division by $x - 2$ gives $x^2 + x + 2$, which has no real zeros; thus, 2 is the only real solution to the system. **23.** Possible answers: $(3, 2), (2, 3.25), (-0.44, 4.20), (-4.56, -0.95)$

7.6 Exercises

25. \$290; 1200 units **27.** \$4700; 6500 units **29.** 18,000 units and 10,000 units **31.** 30 units and 5425 units **33.** 15 in. by 12 in.
35. 10 cm by 10 cm by 20 cm **37.** 10 yr

Chapter 7 Skills Check

1. $x = 2, y = 2, z = 4$ **2.** $x = 3, y = -2, z = 1$
3. $x = 1, y = 1, z = -1$ **4.** $x = 2, y = -1, z = 3$
5. $x = 3, y = 1, z = 2$ **6.** $x = 1, y = 2, z = -1$
7. Infinitely many solutions of the form $x = 41z - 160$, $y = -18z + 70$, $z = z$ **8.** Infinitely many solutions of the form $x = 31z - 110, y = 50 - 14z, z = z$ **9.** Inconsistent system, no solution **10.** Inconsistent system, no solution
11. $x = 1, y = 3, z = -2, w = 1$

12. $x = 2, y = 1, z = -2, w = -1$ **13.** $\begin{bmatrix} -1 & 5 & 2 \\ -1 & 1 & 5 \end{bmatrix}$

14. $\begin{bmatrix} -3 & 1 & 0 \\ -5 & 3 & -1 \end{bmatrix}$ **15.** $\begin{bmatrix} 10 & 15 \\ -5 & 10 \\ 15 & -10 \end{bmatrix}$ **16.** Not possible

17. $\begin{bmatrix} 4 & 14 & 1 \\ -3 & 11 & -1 \end{bmatrix}$ **18.** $\begin{bmatrix} -13 & 12 & 8 \\ -4 & 1 & 3 \\ 0 & 5 & -1 \end{bmatrix}$

19. $\begin{bmatrix} -4 & -2 \\ -2 & -9 \end{bmatrix}$ **20.** $\begin{bmatrix} 10 & 6 & -6 \\ 9 & 25 & 0 \\ 3 & 15 & 10 \end{bmatrix}$

21. $\begin{bmatrix} \frac{1}{8} & \frac{3}{8} & -\frac{3}{8} \\ \frac{1}{8} & \frac{1}{40} & \frac{7}{40} \\ -\frac{1}{4} & \frac{3}{20} & \frac{1}{20} \end{bmatrix}$ **22.** $\begin{bmatrix} \frac{4}{3} & \frac{1}{3} & -1 \\ -1 & 0 & 1 \\ -\frac{2}{3} & -\frac{2}{3} & 1 \end{bmatrix}$

23. $\begin{bmatrix} 1 & 0 & -1 \\ -1 & 1 & 1 \\ 2 & -2 & -1 \end{bmatrix}$ **24.** $\begin{bmatrix} -0.4 & 0.2 & 0.2 & 0 \\ -0.8 & 0.4 & -0.6 & 1 \\ 0.6 & 0.2 & 0.2 & -1 \\ 0 & 0 & 0 & 1 \end{bmatrix}$

25. $x = 2, y = 3, z = -1$ **26.** $x = -16, y = 20, z = -8$
27. $x_1 = 2, x_2 = 4, x_3 = -4, x_4 = 12$
28. $x_1 = -2, x_2 = 0, x_3 = -2, x_4 = 8$ **29.** -5 **30.** -19

31. $(-19, -15)$ $\left(\dfrac{89}{3}, \dfrac{41}{3}, \dfrac{11}{3}\right)$ **33.** $x = 1, y = 0; x = 4, y = 12$

34. Possible answers: $x = 10, y = 20; x = -10, y = 20$; $x \approx 6.32, y = 50; x \approx -6.32, y = 50$

Chapter 7 Review

35. 2800 at \$40, 800 at \$60, 400 at \$100 **36.** 2.8 mg of medication A, 4.2 mg of medication B, 3.5 mg of medication C **37.** \$350,000 at 12%, \$150,000 at 15%, \$250,000 at 10% **38.** \$80,000 at 12%, \$40,000 at 16%, \$240,000 at 8% **39.** \$110,000; \$185,000; \$885,000

40. 5 units of portfolio I, 10 units of portfolio II, 12 units of portfolio III **41.** \$150,000; \$100,000; \$125,000 **42.** 20 g of food I, 30 g of food II, 40 g of food III **43.** 3 passenger planes, 4 transport planes, 4 jumbo planes

44. If x = tech, y = balanced, z = utility,

$$x = \frac{2800 - z}{3}, \ y = 200 - \left(\frac{2}{7}\right)z, \ 0 \le z \le 700$$

45. If x = number of units of product A,
y = number of units of product B,
z = number of units of product C, $x = 252 - 1.2y$,
$0 \le y \le 210, z = 74$

46. $x = 2z, y = 2000 - 4z$ for $0 \le z \le 500; A$ = twice number of $C, B = 2000 - 4$ times number of $C, 0 \le C \le 500$

47. a $A = \begin{bmatrix} 240{,}326 & 312{,}032 \\ 226{,}079 & 301{,}610 \\ 215{,}907 & 292{,}651 \\ 198{,}289 & 281{,}292 \\ 163{,}665 & 249{,}257 \\ 128{,}892 & 204{,}658 \end{bmatrix}$ **b.** $B = \begin{bmatrix} 294{,}158 & 346{,}063 \\ 280{,}529 & 332{,}553 \\ 277{,}594 & 324{,}264 \\ 262{,}874 & 315{,}325 \\ 229{,}986 & 277{,}637 \\ 176{,}654 & 226{,}248 \end{bmatrix}$

c. $C = \begin{bmatrix} -53{,}832 & -34{,}031 \\ -54{,}450 & -30{,}943 \\ -61{,}687 & -31{,}613 \\ -64{,}585 & -34{,}033 \\ -66{,}321 & -28{,}380 \\ -47{,}762 & -21{,}590 \end{bmatrix}$ **d.** No **e.** Mexico

48. a. $\begin{cases} x + y + z = 800{,}000 \\ .06x + .07y + .10z = 71{,}600 \end{cases}$

b. $x = 3z - 1{,}560{,}000$ and $y = 2{,}360{,}000 - 4z$ with $x, y,$ and z nonnegative; $z \ge 520{,}000$ and $z \le 590{,}000$
c. \$210,000 at 6%, \$0 at 7%, and \$590,000 at 10%
d. \$280,000 at 7%, \$0 at 6%, and \$520,000 at 10%
49. a. $y = 0.586x + 450.086, x$ in thousands of \$
b. $y = 0.578x + 464.216, x$ in thousands of \$
c. For each thousand-dollar increase in family income, the critical reading SAT score increases by 0.586 point and the math SAT score increase by 0.578 point. **50. a.** $x + y = 10$
b. $0.20x + 0.05y = 0.155(10)$ **c.** 7 cc of 20%, 3 cc of 5%
51. \$592; 14 units **52.** 20 units and approximately 1252 units

Chapter 8 Special Topics in Algebra

8.1 Skills Check

1. $y \le 5x - 4$

3. $6x - 3y \ge 12$

5. $4x + 5y \le 20$

7. C **9.** D
11. Corners: $(0, 0), (0, 5), (4, 0), (3, 2)$
13. Corners: $(0, 5), (1, 2), (2, 0)$
15. Corners: $(0, 5), (1, 2), (3, 1), (6, 0)$

17. $\begin{cases} y \le 8 - 3x \\ y \le 2x + 3 \\ y > 3 \end{cases}$ **19.** $\begin{cases} 2x + y < 5 \\ 2x - y > -1 \\ x \ge 0; y \ge 0 \end{cases}$

Corners: $(0, 3), (1, 5), (5/3, 3)$

Corners: $(0, 0), (0, 1), (2.5, 0),$ $(1, 3)$

21. $\begin{cases} x + 2y \ge 4 \\ x + y \le 5 \\ 2x + y \le 8 \\ x \ge 0, y \ge 0 \end{cases}$ **23.** $\begin{cases} x^2 - 3y < 4 \\ 2x + 3y < 4 \end{cases}$

Corners: $(3, 2), (4, 0), (0, 2),$ $(0, 5)$

Corners: $(-4, 4), (2, 0)$

25. $\begin{cases} x^2 - y - 8x \le -6 \\ y + 9x \le 18 \end{cases}$

Corners: $(-4, 54), (3, -9)$

8.1 Exercises

27. a. $x + y \ge 780$
b.

29. a. $240x + 150y \le 36{,}000$
b.

31. a. $\begin{cases} 0.12x + 0.009y \geq 7.56 \\ x + y \geq 100 \\ x \geq 0, y \geq 0 \end{cases}$

b. Corners: $(0, 840)$, $(60, 40)$, $(100, 0)$

33. $\begin{cases} x + y \geq 780 \\ 78x + 117y \leq 76,050 \\ x \geq 0, y \geq 0 \end{cases}$

Corners: $(780, 0)$, $(975, 0)$, $(390, 390)$

9. Maximum: 140 at $(4, 2)$ **11.** Maximum: 1200 at all points on the line segment from $(2, 6.5)$ to $(0, 7.5)$ **13.** Minimum: 140 at $(2, 2)$ **15.** Minimum: 161 at $(3, 1)$ **17.** Minimum: 70 at $(0, 7)$

8.2 Exercises

19. Maximum monthly profit is \$31,200 with 975 Turbo and 0 Tornado models.   **21.** Maximum daily profit is \$396 with 9 Safecut and 6 Safecut Deluxe chainsaws.   **23. a.** 100 minutes on TV and 80 minutes on radio   **b.** \$94,000   **25.** Assembly line 1 for 50 days and assembly line 2 for 0 days; \$1,000,000   **27.** 15 Van Buren and 0 Jefferson models, 5 Van Buren and 8 Jefferson models, or 10 Van Buren and 4 Jefferson models; \$900,000   **29.** Operating facility A for 8 weeks and facility B for 2 weeks minimizes the cost at \$160,000.   **31.** 4 servings of diet A and 10 servings of diet B minimizes the substance at 0.71 oz.

35. a. $\begin{cases} 80x + 40y \geq 3200 \\ 20x + 20y \geq 1000 \\ 100x + 40y \geq 3400 \\ x \geq 0, y \geq 0 \end{cases}$

b. Corners: $(30, 20)$, $(0, 85)$, $(50, 0)$, $(10, 60)$

37. $\begin{cases} 4x + 10y \leq 1600 \\ 6x + 7y \leq 1600 \\ x \geq 0, y \geq 0 \end{cases}$

Corners: $(0, 0)$, $(0, 160)$, $(150, 100)$, $(266\ 2/3, 0)$

8.3 Skills Check

1. 5, 7, 9, 11, 13, 15   **3.** 10, 5, $\dfrac{10}{3}$, $\dfrac{5}{2}$, 2   **5.** 9, 11, 13   **7.** 25   **9.** 24, 48, 96, 192   **11.** 2430   **13.** 5, 3, 1, -1 **15.** 2, 7, 17, 37

8.3 Exercises

17. \$10,500   **19. a.** $300 + 60n$ **b.** 360, 420, 480, 540, 600, 660   **21. a.** First by \$2000   **b.** They are the same.   **c.** \$800   **23.** \$1050, \$1102.50, \$1157.63, \$1215.51   **25. a.** \$20,000   **b.** $50,000 - 10,000n$ **c.** \$40,000, \$30,000, \$20,000, \$10,000, \$0   **27.** 320,000   **29. a.** geometric   **b.** $2^{10} = 1024$ **c.** 9   **31.** \$7,231,366.37   **33.** 15 years   **35.** \$22,253.46   **37.** 13, 21, 34, 55   **39.** \$1100, \$990, \$891, \$801.90

39. a. $\begin{cases} 4x + 6y \leq 480 \\ 2x + 6y \leq 300 \\ x \geq 0, y \geq 0 \end{cases}$

41. a. $\begin{cases} x + y \leq 1400 \\ x \geq 500 \\ y \geq 750 \\ x \geq 0, y \geq 0 \end{cases}$

8.4 Skills Check

1. $\dfrac{364}{27}$ **3.** 205   **5.** 150   **7.** 98,301   **9.** 5115   **11.** $\dfrac{4096}{3}$ **13.** 126   **15.** $\dfrac{73}{12}$ **17.** 3   **19.** Not possible, infinite

b.

Corners: $(90, 20)$, $(0, 50)$, $(120, 0)$, $(0, 0)$

b.

Corners: $(500, 750)$, $(500, 900)$, $(650, 750)$

8.4 Exercises

21. \$2400   **23.** 11   **25. a.** 78   **b.** 156   **27.** \$42,768.57   **29.** 16 cm^3 **31. a.** 5   **b.** 25   **c.** 125; 625   **d.** Geometric, with $r = 5$ **33.** \$2824.30   **35.** 213 feet   **37. a.** $35,000(1 - 0.84^n)$ **b.** $35,000(0.84)^n$ **39.** \$15,992.73   **41.** kth term: $R(1 + i)^{-n}(1 + i)^{k-1}$

$$s_n = \frac{R(1 + i)^{-n}(1 - (1 + i)^n)}{1 - (1 + i)}$$

$$= R\left[\frac{(1 + i)^{-n} - (1 + i)^0}{-i}\right]$$

$$= R\left[\frac{(1 + i)^{-n} - 1}{-i}\right]$$

$$= R\left[\frac{1 - (1 + i)^{-n}}{i}\right]$$

8.2 Skills Check

1. Maximum: 382 at $(10, 38)$; minimum: 0 at $(0, 0)$
3. Maximum: 22 at $(4, 3)$; minimum: 0 at $(0, 0)$
5. a.

Corners: $(0, 0)$, $(0, 18)$, $(22, 0)$, $(12, 10)$
b. Maximum: 90 at $(0, 18)$

7. a.

Corners: $(0, 10)$, $(5, 5)$, $(15, 0)$
b. Minimum: 20 at $(0, 10)$

8.5 Skills Check

1. $10x^4 + 33x^2 + 2x + 12$ **3.** $\dfrac{x^3 + 6}{x^3}$ **5.** 26   **7.** $0.8125 = \dfrac{13}{16}$ **9.** $y = 12x - 16$ **11. a.** $4(x + h) + 5$ **b.** $4h$ **c.** 4   **13.** $y = 3 - 2x^2$ **15.** $y = \dfrac{18 - 9x^3}{5}$ **17. a.** $x = 1$ **b.** Same   **c.** $(1, 4)$ **19. a.** $x = 2$ **b.** Same   **c.** $(2, -9)$

21. a. $x = -2, x = 8$ **b.** $y = 67$ at $x = -2$; $y = -433$ at $x = 8$
c. $x = -2$ **23.** $y = 3x^{1/2}$ **25.** $y = 2x^{2/3}$ **27.** $y = (x^2 + 1)^{1/3}$
29. $y = (x^3 - 2)^{2/3}$ **31.** $y = x^{1/2} + (2x)^{1/3}$

33. $y = \pm\sqrt{3 - 4x}$ **35.** $y = \dfrac{-1 \pm \sqrt{1 + 24x}}{2}$

37. $2; 4x^3 + 5$ **39.** $x^3 + x; \dfrac{1}{2}$ **41.** $3x\sqrt{x^2 - 1}$

43. $y = 3x^{-1} - 4x^{-2} - 6$ **45.** $y = (x^2 - 3)^{-3}$

47. $f'(x) = -\dfrac{6}{x^4} + \dfrac{4}{x^2} + \dfrac{1}{x}$ **49.** $f'(x) = \dfrac{8x}{(4x^2 - 3)^{1/2}}$

51. $3\log x + 5\log(3x - 4)$ **53.** $x = 0, x = 1$

55. $x = 3, x = \dfrac{3 + \sqrt{69}}{5}, x = \dfrac{3 - \sqrt{69}}{5}$

8.6 Skills Check

1. 70   **3.** 1   **5.** 84   **7.** 66   **9.** 120   **11.** 165
13. $a^4 + 4a^3b + 6a^2b^2 + 4ab^3 + b^4$
15. $x^4 - 4x^3y + 6x^2y^2 - 4xy^3 + y^4$
17. $x^6 + 12x^5y + 60x^4y^2 + 160x^3y^3 + 240x^2y^4 + 192xy^5 + 64y^6$
19. $243x^5 + 1620x^4y + 4320x^3y^2 + 5760x^2y^3 + 3840xy^4 + 1024y^5$
21. $81a^4 + 108a^3b + 54a^2b^2 + 12ab^3 + b^4$
23. $x^7 + 7x^6y + 21x^5y^2 + 35x^4y^3 + 35x^3y^4 + 21x^2y^5 + 7xy^6 + y^7$
25. $15m^2n^4$ **27.** $120x^3y^7$ **29.** $270m^3n^2$
31. $-560m^3n^{13}$ **33.** $252x^5y^5$

8.7 Skills Check

1. a. $(x + 1)^2 + (y - 2)^2 = 9$
b.

$(x + 1)^2 + (y - 2)^2 = 9$

3. $C(0, 0); r = 6$ **5.** $C(2, -1); r = \sqrt{14}$
7. $C(0.5, -3.2); r = \sqrt{7}$

9.

$x^2 + y^2 - 49 = 0$

11.

$x^2 + y^2 - 6 = 8$

$r = \sqrt{14}$

13.

$(x + 1)^2 + (y + 3)^2 = 1$

15.

$r = \sqrt{6}$

$\left(x + \dfrac{1}{2}\right)^2 + (y - 2)^2 = 6$

17.

$x^2 + (y + 4)^2 = 16$

Center

19.

$(-2, 1)$ $r = 3$

$x^2 + y^2 + 4x - 2y - 4 = 0$

21.

Center $\left(\dfrac{1}{2}, 2\right)$

$r = \dfrac{1}{2}$

$x^2 + y^2 - x - 4y + 4 = 0$

23.

Center $\left(0, \dfrac{1}{2}\right)$

$r = \dfrac{1}{2}$

$x^2 + y^2 - y = 0$

25.

$y = \sqrt{9 - x^2}$

27. $x^2 - 8x + y^2 = 0$ **29.** $x^2 + 6x + y^2 + 2y + 6 = 0$
31. $4x^2 - 40x + 4y^2 + 16y + 115 = 0$ **33.** $y = \sqrt{9 - x^2}$
35. $x = -\sqrt{-y^2 - 2y}$ **37.** $\left(x - \dfrac{5}{2}\right)^2 + \left(y - \dfrac{5}{2}\right)^2 = \dfrac{58}{4}$
39. $(x + 4)^2 + (y - 5)^2 = 25$

41.

$y^2 = 12x$

43.

$y^2 + 4x = 0$

45.

$x^2 - 4y = 0$

47.

$x^2 + 12y = 0$

49.

$y = \sqrt{x - 3}$

51.

$y = \sqrt{3 - x}$

53.

$y^2 - 6y + 4x + 5 = 0$

Vertex $(1, 3)$

55.

Vertex $\left(-\frac{3}{2}, -\frac{1}{2}\right)$

$y = -\frac{4}{32}x^2 - \frac{12}{32}x - \frac{25}{32}$

13.

$y = \dfrac{\sqrt{1 - 4x^2}}{3}$

15.

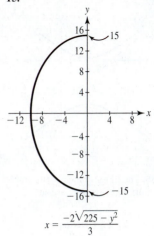

$x = \dfrac{-2\sqrt{225 - y^2}}{3}$

57.

$y = x^2 + 10x + 25$

Vertex $(-5, 0)$

59.

Vertex $\left(\frac{1}{4}, 3\frac{1}{8}\right)$

$y = 3 + x - 2x^2$

8.7 Exercises

61. $x^2 = 375(y - 20)$

8.8 Skills Check

1. $(0,0)$; $(3,0)$, $(-3,0)$

3. $(0,0)$; $(0,-4)$, $(0,4)$

17. Ellipse   **19.** Circle

21.

$x^2 + \dfrac{y^2}{25} = 1$

23.

$m = -\frac{5}{4}$ $m = \frac{5}{4}$

$\dfrac{x^2}{16} - \dfrac{y^2}{25} = 1$

$\dfrac{x^2}{9} + \dfrac{y^2}{4} = 1$

$4x^2 + y^2 = 16$

5. $(0,0)$; $(1,0)$, $(-1,0)$

7. $(-2,1)$; $(-2,5)$, $(-2,-3)$

25.

$m = -1$ $m = 1$

$y^2 - x^2 = 1$

27.

$m = -2$ $m = 2$

$y^2 - 4x^2 = 4$

$x^2 + 4y^2 = 1$

$(-2, 1)$

$\dfrac{(x + 2)^2}{4} + \dfrac{(y - 1)^2}{16} = 1$

9. $(-2,1)$; $(0,1)$, $(-4,1)$

11. $(0,-3)$; $(0,0)$, $(0,-6)$

29.

$m = -1$ $m = 1$

$(-1, -4)$

$\dfrac{(x + 1)^2}{36} - \dfrac{(y + 4)^2}{36} = 1$

31.

$m = \frac{1}{2}$ $m = -\frac{1}{2}$

$\dfrac{x^2}{4} - (y + 3)^2 = 1$

$(-2, 1)$

$\dfrac{(x + 2)^2}{4} + (y - 1)^2 = 1$

$x^2 + \dfrac{(y + 3)^2}{9} = 1$

$(0, -3)$

33.

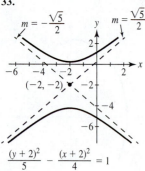

$$\frac{(y+2)^2}{5} - \frac{(x+2)^2}{4} = 1$$

35. The lower branch of the graph of the equation in Exercise 25.

37.

39.

41.

43.

45.

8.8 Exercises

47. 14.8 feet   **49.** $\dfrac{x^2}{327.25} + \dfrac{y^2}{20.79} = 1$

51.

Chapter 8 Skills Check

1.

2.

3. $\begin{cases} 2x + y \le 3 \\ x + y \le 2 \\ x \ge 0, y \ge 0 \end{cases}$

4. $\begin{cases} 3x + 2y \le 6 \\ 3x + 6y \le 12 \\ x \ge 0, y \ge 0 \end{cases}$

Corners: $(0,0), (0,2), (1,1),$ $(1.5, 0)$

Corners: $(0,0), (0,2), (1, 1.5),$ $(2, 0)$

5. $\begin{cases} 2x + y \le 30 \\ x + y \le 19 \\ x + 2y \le 30 \\ x \ge 0, y \ge 0 \end{cases}$

6. $\begin{cases} 2x + y \le 10 \\ x + 2y \le 11 \\ x \ge 0, y \ge 0 \end{cases}$

Corners: $(0,0), (0,15),$ $(15,0), (11,8), (8,11)$

Corners: $(0,0), (0,5.5), (3,4),$ $(5,0)$

7. $\begin{cases} 15x - x^2 - y \ge 0 \\ y - \dfrac{44x + 60}{x} \ge 0 \\ x \ge 0, y \ge 0 \end{cases}$

8. $\begin{cases} x^3 - 26x + 100 - y \ge 0 \\ 19x^2 - 20 - y \le 0 \\ x \ge 0, y \ge 0 \end{cases}$

9. Minimum: 14 at $(2,2)$ **10.** Maximum: 168 at $(12,7)$
11. Maximum: 79 at $(8, 11)$ **12.** Geometric; $r = 6$

13. Arithmetic; $d = 12$ **14.** Geometric; $r = -\dfrac{3}{4}$

15. 4   **16.** 27   **17.** -1705 **18.** 234   **19.** 4
20. $5x^4 - 9x^2 + 6x - 10$ **21.** $x = 0, x = 9, x = -1$

22. $y = x + x^{-2} - 2x^{-3/2} + 3x^{-1/3}$

23. $f'(x) = \dfrac{8}{x^3} + \dfrac{5}{x} + 4x$ **24.** $x = 0, x = -1, x = \dfrac{1}{2}$

25. 1   **26.** 126   **27.** 120   **28.** $x^4 + 8x^3y + 24x^2y^2 + 32xy^3 + 16y^4$

29. $64a^6 - 576a^5b + 2160a^4b^2 - 4320a^3b^3 + 4860a^2b^4 - 2916ab^5 + 729b^6$

30. $-35m^4n^3$ **31.** $489{,}888x^4y^5$

32. $(x - 2)^2 + (y + 1)^2 = 9$

33. $r = 5$, center at $(3, 1)$ **34.** $y = -\sqrt{36 - x^2}$

35.

36.

37.

$4y^2 - 4y + 12x + 37 = 0$

38. Center at $(0, 0)$, end points of major axis $(0, 4)$ and $(0, -4)$

$\dfrac{x^2}{4} + \dfrac{y^2}{9} = 1$

39. Center at $(2, -1)$, end points of major axis $(-2, -1)$ and $(6, -1)$

$\dfrac{(x - 2)^2}{16} + \dfrac{(y + 1)^2}{4} = 1$

40.

$y = -\sqrt{9 - 4x^2}$

41.

$\dfrac{x^2}{9} - \dfrac{y^2}{25} = 1$

42.

$16y^2 - 9x^2 = 64$

43.

$xy = -12$

Chapter 8 Review

44. \$25,200 profit at 360 units of each   **45. a.** Pottstown, 20 days; Ethica, 30 days   **b.** \$1,120,000   **46.** Minimum cost is \$48 with 32 units of feed A and 2 units of feed B   **47.** 220 leaf blowers and 240 weed wackers   **48.** 60 two-bedroom and 20 three-bedroom apartments   **49.** \$30 million in auto loans and no home equity loans   **50. a.** Job 2   **b.** Job 1   **51.** Approximately 999.98 mg   **52. a.** 9.22337×10^{18} grains   **b.** 1.84467×10^{19} grains   **53.** \$26,764.51   **54.** \$24,500.90   **55.** Between 800 and 5200 units   **56.** 90 feet   **57.** Yes   **58.** 60 yards

Index

Index of Applications